Global Conference on Polymer and Composite Materials 2016 (PCM 2016)

IOP Conference Series: Materials Science and Engineering
Volume 137

Hangzhou, China
20 - 23 May 2016

ISBN: 978-1-5108-2731-8
ISSN: 1757-8981

Printed from e-media with permission by:

Curran Associates, Inc.
57 Morehouse Lane
Red Hook, NY 12571

Some format issues inherent in the e-media version may also appear in this print version.

Printed by Curran Associates, Inc. (2016)

For permission requests, please contact the Institute of Physics
at the address below.

Institute of Physics
Dirac House, Temple Back
Bristol BS1 6BE UK

Phone: 44 1 17 929 7481
Fax: 44 1 17 920 0979

techtracking@iop.org

Additional copies of this publication are available from:

Curran Associates, Inc.
57 Morehouse Lane
Red Hook, NY 12571 USA
Phone: 845-758-0400
Fax: 845-758-2633
Email: curran@proceedings.com
Web: www.proceedings.com

Table of contents

Volume 137

2016 Global Conference on Polymer and Composite Materials (PCM 2016)
20–23 May 2016, Hangzhou, China

Accepted papers received: 21 June 2016
Published online: 27 July 2016

Preface

Papers

IOP Conf. Series: Materials Science and Engineering **137** (2016) 011001 doi:10.1088/1757-899X/137/1/011001

Preface

The *2016 Global Conference on Polymer and Composite Materials (PCM 2016)*, sponsored by Wuhan Advanced Materials Society, and Ningbo Adhesives and Products Industry Association, was held from May 20 to May 23 2016 in Hangzhou, China. The technical program consisted of 8 international keynote speakers, 22 invited speakers, 57 oral presentations, and a poster session with 47 works.

The present issue of IOP Conference Series: Materials Science and Engineering (MSE) records the proceedings of PCM 2016 and contains 71 specially selected manuscripts submitted to the PCM2016 conference. The electronic submission and handling of manuscripts via the conference website, including the selection of reviewers and evaluation of manuscripts, were identical to the procedures applied to manuscripts submitted as regular contributions for publication.

The organization of this conference and the preparation of proceedings volume would have been impossible without the tremendous efforts and dedication of many individuals, especially from Ms. Zha Yuanyuan, who oversaw the organization of the conference and the program, and a large team of reviewers with their timely submission of quality reports. We express our sincere thanks to all authors and presenters for their contributions. We also thank very much our sponsors for their generous support.

The 2017 Global Conference on Polymer and Composite Materials (PCM2017) will be held during May 29-31, 2017, at the University of Macau, China. Macau is a special administrative region on the southern coast of China situated across the Pearl River Delta, at about 65 kilometers southeast of Hong Kong. Being a Portuguese territory until 1999, it reflects a mix of Portuguese and Chinese influences, with casinos and extravagant malls that have earned it the nickname of *"Las Vegas of Asia."* All participants are welcome to the renewed and vibrant PCM2017! More information will be available at the conference website: http://www.cpcmconf.org.

Prof. Dr. Esteban Broitman
General Chairman of PCM2016
Linköping, Sweden, June 2016

IOP Conf. Series: Materials Science and Engineering **137** (2016) 011002 doi:10.1088/1757-899X/137/1/011002

Peer review statement

All papers published in this volume of *IOP Conference Series: Materials Science and Engineering* have been peer reviewed through processes administered by the proceedings Editors. Reviews were conducted by expert referees to the professional and scientific standards expected of a proceedings journal published by IOP Publishing.

DSC and curing kinetics study of epoxy grouting diluted with furfural -acetone slurry

H Yin[1, 2, 3], D W Sun[1, 2], B Li[1, 2], Y T Liu[1, 2], Q P Ran[1, 2] and J P Liu[1, 2]

[1]Jiangsu Sobute New Materials Co. Ltd., Nanjing 211103, Jiangsu, China
[2]State Key Laboratory of High Performance Civil Engineering Materials, Jiangsu Research Institute of Building Science, Nanjing, 210008, Jiangsu, China

E-mail: yinhao@cnjsjk.cn

Abstract. The use of furfural-acetone slurry as active diluents of Bisphenol-A epoxy resin (DGEBA) groutings has been studied by dynamic and non-isothermal DSC for the first time. Curing kinetics study was investigated by non-isothermal differential scanning calorimetries at different heating rates. Activation enery (E_a) was calculated based on Kissinger and Ozawa Methods, and the results showed that E_a increased from 58.87 to 71.13KJ/mol after the diluents were added. The furfural-acetone epoxy matrix could cure completely at the theoretical curing temperature of 365.8K and the curing time of 139mins, which were determined by the kinetic model parameters.

1. Introduction

Crack is one of the most common deterioration problems of concrete dams, which can be caused by several factors, such as large temperature gradients, shrinkage, foundation deformation, and hydrostatic overload, etc. [1]. Some cracks emerge in construction period and some are formed during the operating period [2]. Cracks can not only affect the building appearance, but also can bring about a series of serious diseases, like leakage dissolution, environment water erosion, freeze-thaw damage and steel corrosion. Interacting with cracks, these problems cause a vicious cycle doing great harm to the durability of dam structures. Therefore, a scientific, reasonable and effective treatment of cracks is urgent for dam foundation improvement in order to solve these problems.

Because of their high mechanical strength, good adhesive properties, extremely low shrinkage and superior chemical resistance under corrosive conditions, at present, epoxy resin chemical grouting materials are the most widely used for reinforcement and anti-seepage projects of concrete structures and foundations in construction.

Although the addition of reactive diluents can reduce the viscosity of epoxy resin, such as butyl glycidyl ether [3], allyl glycidyl ether [4], and cyclohexanediol diglycidyl ether [5], they can't improve the low permeability of epoxy slurry, due to the high surface tension of the epoxy slurry itself. Acetone and furfural having low viscosity and epoxy resin can be well dissolved in them to form a homogeneous solution system. Under the action of alkali curing agent, acetone and furfural can react to generate furan resin, forming an interpenetrating network structure with epoxy resin.

Compared with general epoxy grouts, the furfural–acetone diluted epoxy grouting has a lower viscosity and small contact angle indicating a low surface tension properties, a privileged wetting and infiltration capacity, a long time for operation, and a high consolidation strength. Therefore this more effective acetone-furfural method has been widely applied into construction industry in China [6-7]. However, this promising material has not been drawing enough attentions, particularly with regard to the curing kinetics study of the epoxy grouting diluted by furfural -acetone slurry.

2. Material and methods

2.1. *Materials*

Bisphenol-A epoxy resin (DGEBA) was bought from the NanYa Resin Company with the epoxy equivalent of 187 g/mol and room temperature (25 °C) viscosity of about 12–15 Pa·s. The used curing agent was an Ancamine 2758 supplied by Air Products and Chemicals, Inc, which is a Mannich-Base polyamine. The furfural and acetone were bought from Sinopharm Chemical Reagent Company Ltd. and all materials were used as received.

2.2. *Preparation of epoxy grouting blends*

Two epoxy grouting systems were established. Sample A was prepared by mixing DGEBA and Ancamine 2758 at a mass rate of 100:43.9 at 25°C under vigorous mechanical stirring for about 5mins. Sample B was obtained by blending DGEBA and Ancamine 2758 with furfural-acetone slurry under the same conditions. The mass rate of the four raw materials was 100:54.8:48:29. The curing samples were well mixed until a homogeneous system was obtained.

2.3. *DSC characterization*

Differential scanning calorimeters (DSC) was used to monitor the curing process of the epoxy resin and curing agent. The epoxy resin and curing agent should be mixed homogeneously by a vigorous stirring. About 9-11 mg of the above mixture was put into the DSC pan and put into the DSC sample chamber immediately. An empty DSC pan was selected as a reference when DSC was working. The epoxy resin and curing agent were non-isothermally cured in DSC at different heating rate of 5, 7.5, 10 or 15°C/min, respectively. The temperature and heat was calibrated with high-pure indium standard beforehand [8-10].

3. Result and discussion

3.1. *Dynamic DSC analysis*

Curing kinetics parameters of epoxy matrix play an important role in understanding the curing reaction such as the activation energy and reaction order and so on. The apparent activation energy is a parameter to determine whether the curing process can take place or not. The curing behavior can be carried out only when the involved molecules obtain greater energy than the apparent activation energy does. By calculating the reaction order, the curing mechanism of the epoxy matrix can be roughly estimated.

DSC is widely used in the kinetics analyzing thermosetting reaction [11-16]. The curing kinetics parameters can be analyzed by the following Eq. (1):

$$d\alpha/dt = \beta d\alpha/dT = k(T)f(\alpha) \qquad (1)$$

where $d\alpha/dt$ is the rate of conversion, β is heat flow, T is the absolute temperature, $f(\alpha)$ is the function of the kinetic model, and k(T) is the function of the temperature which can be described by the Arrhenius Equation as Eq. (2):

$$k(T)=A\exp(-E_a/RT) \qquad (2)$$

where A is the pre-exponential factor, E_a is the activation energy, R is the gas constant.

Various kinetic models can be used to describe the curing kinetics. The frequently used autocatalytic Kamal model [17-19] was applied in this work to analyze the cuing kinetics mechanism (see Eq. (3)).

$$f(\alpha)= (1-\alpha)^n \qquad (3)$$

where n is the reaction order.

So the curing kinetic model equation can be easily changed into Eq. (4) based Eq. (1) and Eq. (2) and Eq. (3):

$$d\alpha/dt = A\exp(-E_a/RT)\,(1-\alpha)^n \qquad (4)$$

As shown in Eq. (4) as mentioned above, calculation of the activation energy, reaction order and pre-exponential factor are prerequisite to determine the specific kinetic model. The two most common theories of activation energy are both Kissinger method [20-22] and Ozawa method [23, 24]. Both of them can be applied to give the activation energy E_1 and E_2, respectively, by the Eq. (5) and Eq. (6):

$$\ln(\beta/T_p^2) = \ln(AR/E_1)-E_1/RT_p \qquad (5)$$

$$\ln\beta = -1.052E_2/RT_p + C \qquad (6)$$

where C is a constant.

Fig. 1 and 2 show the dynamic DSC analysis curves measured for sample A and B, respectively, with four heating rates (β) 5, 7.5, 10, 15 K min-1. Compared with the two curves, it can be seen that the peak temperature (T_p) increased by the addition of furfural and acetone. This phenomenon is accordance with the coordinated reaction of the aldehyde or carbonyl groups and amino groups, which slow the curing rate and result into more heating energy if necessary.

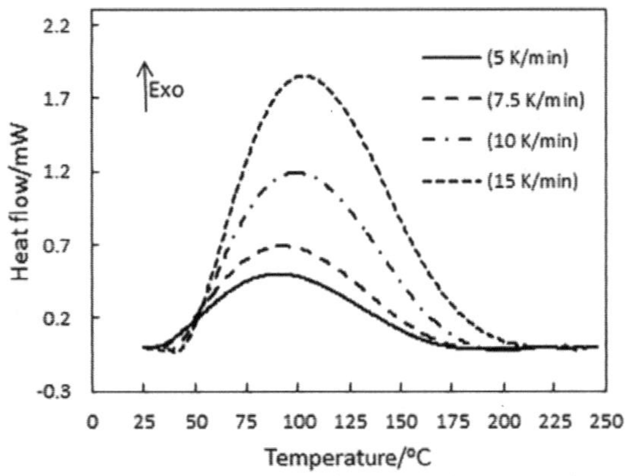

Figure 1. DSC curve of sample A.

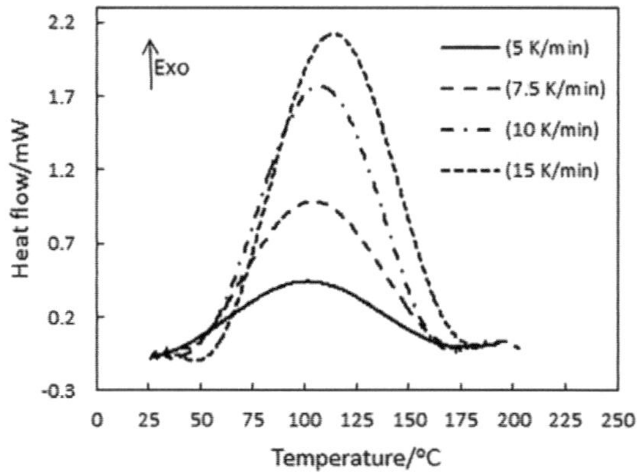

Figure 2. DSC curve of sample B.

Table 1. DSC analysis of curing process at different fates.

Curing system	β(K /min)	T_i(°C)	T_p(°C)	T_f(°C)	LN(β/T_p^2)	1000/T_p(K^{-1})
A	5	37.2	85.9	173.9	-10.16	2.79
	7.5	39.9	94.2	179.5	-9.80	2.72
	10	44.1	99.2	188.4	-9.54	2.68
	15	50.0	105.3	205.7	-9.16	2.64
B	5	39.8	100.6	176.5	-10.24	2.68
	7.5	41.6	105.2	182.1	-9.86	2.64
	10	45.5	111.0	191.2	-9.60	2.60
	15	51.1	117.6	199.6	-9.23	2.56

The main information about the dynamic DSC analysis curves is summarized as Table 1. Obviously, the heating rate has a great influence on the shape of the exothermic curves. With the increase of heating rate of the curing reaction, the thermal effect was enhanced and reacting time was shortened, causing a thermal lag. Therefore, shifting the initial curing temperature(T_i), peak temperature(T_p) and finishing temperature(T_f) to the high temperature zone were observed.

3.2. Average activation energy

According to Kissinger method, with Eq. (5), the peak temperature of different heating rates, E_1 can be obtained from the slope of the plot of ln(β/T_P^2) versus 1/T_p, whereas sample A and sample B were 57.38 and 69.83kJ mol^{-1}, respectively. Fig. 3 shows the results. The E_2 can be estimated form the slope of the plot of lnβ versus 1/Tp according to Ozawa method with Eq. (6), whereas sample A and sample B were 60.36 and 72.42 kJ/mol, respectively. Fig. 4 shows the results.

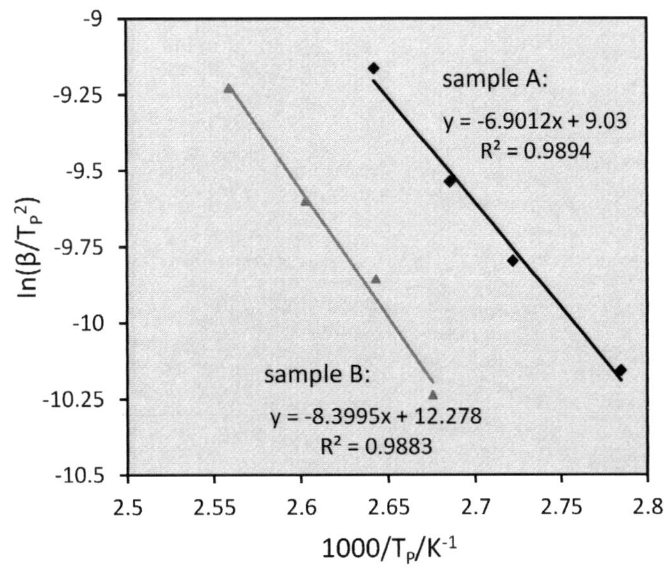

Figure 3. Kissinger's plot of ln(β/T_p) versus 1/T_p for sample A and B.

In this study, the final activation energy E_a is an average result of E_1 and E_2. Hence, the activation energy (E_a) of sample A and sample B were 58.87 and 71.13 kJ/mol respectively. The addition of the furfural acetone slurry made the activation energy increase a lot. This is the very reason that the set time of furfural acetone system is much longer, which does make an important contribution to the permeability of the grouting materials because of the long time for viscosity increase.

3.3. The kinetics equation of the curing reaction

Generally speaking, Crane kinetic model is used to determine the reaction order, which can be described by Eq. (7):

$$d(\ln\beta)/d(1/T_p) = -(E_a/nR + 2T_p) \quad (7)$$

The Eq. (6) can be changed into Eq. (8) by mathematical integration because of the ignored item $2T_p$ where $E_a/nR \gg 2T_p$.

$$\ln\beta = -E_a/nRT_p + C \quad (8)$$

where C is a constant. Hence, the reaction order n can be obtained from the slope of the plot of lnβ versus $1/T_p$.

Fig. 4 shows the plot of ln(β) versus $1/T_p$ for sample A and B, while the reaction orders of sample A and B curing systems are 0.9270 and 0.9336, respectively, in accordance with the average activation energy calculated above. Both of them are underwent complicated reactions due to the non-integer reaction orders.

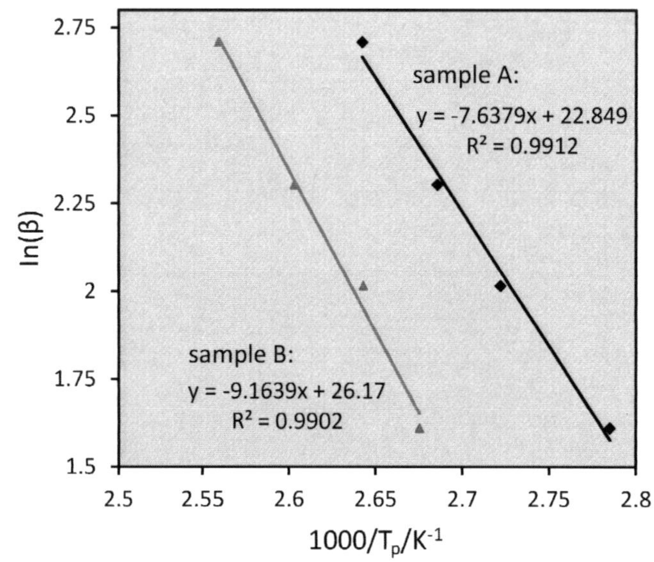

Figure 4. Ozawa's plot of ln(β) versus $1/T_p$ for sample A and B.

The pre-exponential factor(A) can be calculated approximately by Kissinger method from the intercept of the plot of $\ln(\beta/T_P^2\ p)$ versus 1/Tp as Eq. (5), while sample A and sample B are 5.76×107 and 1.56×109, respectively. Curing reaction kinetic equation of sample A and sample B can be described as Eq. (9) and Eq. (10):

$$d\alpha/dt = 5.76\times10^7\exp(-7.08/T)\ (1-\alpha)^{0.9270} \quad (9)$$

$$d\alpha/dt = 1.56\times10^9\exp(-8.56/T)\ (1-\alpha)^{0.9336} \quad (10)$$

3.4. The study of the curing process

Finally, T-β Extrapolation method was used to determine the curing process parameters including gelling temperature, curing temperature and post-curing temperature, which can be obtained from the intercepts of the plot of K versus β. According to the data given in Table 1, T-β fitting curve is shown in Fig. 5. The gelling temperature, curing temperature and post-curing temperature are 306.72K, 365.75K and 438.45K, respectively.

The curing time is also an important condition to determine the finally application of the products. Similarly, the theoretical curing time can be obtained from the kinetic model parameters by the way of Eq. (11):

$$\alpha(t) = 1\text{-}(A\exp(-E_a/RT)(n-1)t+1)^{1/1-n} \quad (11)$$

At the end of the curing reaction, the conversion rate α(t) was set as 100%. After the calculated parameters were put into Eq. (11), theoretical cure time was 139 mins at the curing temperature of about 365.8K.

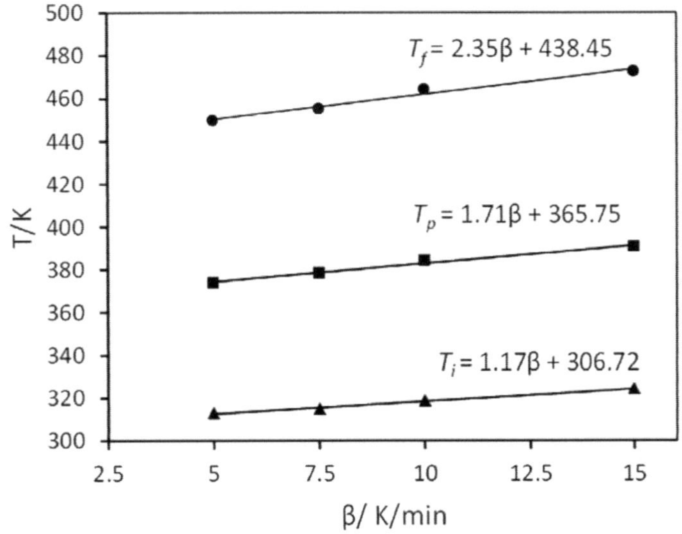

Figure 5. Relationship between curing temperature and heating rate in T-β Extrapolation method.

4. Conclusions

In this work, the differential scanning calorimetries have been applied to determine the kinetic parameters of furfural-acetone epoxy grouting for the first time. The results show the E_a of the DGEBA, Ancamine 2758 with and without furfural -acetone slurry are 71.13 and 58.87 KJ/mol, the reaction order are 0.9336 and 0.9270, the pre-exponential factor(A) are 5.76×10^9 and 1.56×10^7, respectively. The furfural-acetone slurry can improve the activation energy, exhibiting exactly the opposite point of the well-known active diluents based on glycerol ether, both curing reactions of them are complicated models, and then curing reaction kinetic equations are also described. The furfural-acetone epoxy matrix could cure completely with the theoretical curing temperature being 65.8K and the theoretical curing time being 139 mins, which are all determined by all the above kinetic model parameters.

Acknowledgement

This work was financially supported by the Natural Science Foundation of Jiangsu Province, China (BK 20140110 and BK 20130097) and the National Natural Science Foundation of China (51208236).

References

[1] Gallagher S and Leger P 2014 Mechanical Resistance of Cracked Mass Concrete Repaired by Grouting: Experimental Study *ACI Struct. J.* **111**(4) 799-807

[2] Wang Z J, Liu S H, Vallejo L and Wang L J 2014 Numerical analysis of the causes of face slab cracks in Gongboxia rockfill dam *Eng. Geol.* **181** 224-32

[3] Chen Z K, Yang G, Yang J P, Fu S Y, Ye L and Huang Y G 2009 Simultaneously increasing cryogenic strength, ductility and impact resistance of epoxy resins modified by n-butyl glycidyl ether *Polymer* **50**(5) 1316-23

[4] Oral I, Guzel H and Ahmetli G 2013 Determining the mechanical properties of epoxy resin

(DGEBA) composites by ultrasonic velocity measurement *J. Appl. Polym. Sci.* **127**(3) 1667-75

[5] Li C, Liu M H, Liu Z Y, Qing M L and Wang G 2014 DSC and curing kinetics of epoxy resin using cyclohexanediol diglycidyl ether as active diluents *J. Therm. Anal. Calorim.* **116**(1) 411-6

[6] Xing H G, Yang X G, Dang Y H, Yao X and Zhou J W 2014 Experimental Study of Epoxy Resin Repairing of Cracks in Fractured Rocks *Polym. & Polym. Compos.* **22**(5) 459-66

[7] Wang Z H, Yin G D, Qin J, Gao M, Cao L P and Wu A X 2008 An Efficient Method for the Selective Iodination of alpha,beta-Unsaturated Ketones *Synthesis-Stuttgart.* **2008**(22) 3675-81

[8] Roşu D, Mustaţă F and Caşcaval C N 2001 Investigation of the curing reactions of some multifunctional epoxy resins using differential scanning calorimetry *Thermochim. Acta.* **370**(1-2) 105-10

[9] Roşu D, Caşcaval C N, Mustaţă F and Ciobanu C 2002 Cure kinetics of epoxy resins studied by non-isothermal DSC data *Thermochim. Acta.* **383**(1-2) 119-27

[10] Xiong X, Ren R, Liu S, Lu S and Chen P 2014 The curing kinetics and thermal properties of epoxy resins cured by aromatic diamine with hetero-cyclic side chain structure *Thermochim. Acta.* **595**(0) 22-7

[11] Jankovic B 2008 Kinetic analysis of the nonisothermal decomposition of potassium metabisulfite using the model-fitting and isoconversional (model-free) methods *Chem. Eng. J.* **139**(1) 128-35

[12] Kandelbauer A, Wuzella G, Mahendran A, Taudes I and Widsten P 2009 Model-free kinetic analysis of melamine-formaldehyde resin cure *Chem. Eng. J.* **152**(2-3) 556-65

[13] Wan J T, Bu Z Y, Xu C J, Fan H and Li B G 2011 Model-fitting and model-free nonisothermal curing kinetics of epoxy resin with a low-volatile five-armed starlike aliphatic polyamine *Thermochim. Acta.* **525**(1-2) 31-9

[14] Vyazovkin S, Burnham A K, Criado J M and Perez-Maqueda L A 2011 Popescu C, Sbirrazzuoli N. ICTAC Kinetics Committee recommendations for performing kinetic computations on thermal analysis data *Thermochim. Acta.* **520**(1-2) 1-19

[15] Vyazovkin S, Chrissafis K, Di Lorenzo M L, Koga N, Pijolat M, Roduit B, Sbirrazzuoli N and Sunol J J 2014 Kinetics Committee recommendations for collecting experimental thermal analysis data for kinetic computations *Thermochim. Acta.* **590**(20) 1-23

[16] Cai H, Li P, Sui G, Yu Y, Li G, Yang X and Ryu S 2008 Curing kinetics study of epoxy resin/flexible amine toughness systems by dynamic and isothermal DSC *Thermochim. Acta.* **473**(1–2) 101-5

[17] Jubsilp C, Punson K, Takeichi T and Rimdusit S 2010 Curing kinetics of Benzoxazine-epoxy copolymer investigated by non-isothermal differential scanning calorimetry *Polym Degrad and Stabil.* **95**(6) 918-24

[18] Sun H, Liu Y Y, Wang Y S and Tan H F 2014 Curing Behavior of Epoxy Resins in Two-Stage Curing Process by Non-Isothermal Differential Scanning Calorimetry Kinetics Method *J. Appl. Polym. Sci.* **131**(17) 40711-16

[19] Cao J, Hu J J, Fan H, Wan J T and Li B G 2014 Novel silicone-phenyl contained amine curing agent for epoxy resin: 1. Non-isothermal cure and thermal decomposition *Thermochim. Acta.* **593**(10) 30-6

[20] Kissinger H E 1956 Variation of Peak Temperature with Heating Rate in Differential Thermal Analysis *J. Res. Natl. Bur. Stand.* **57**(4) 217-21

[21] Kissinger H E 1957 Reaction Kinetics in Differential Therml Analysis *Anal. Chem.* **29**(11) 1702-6

[22] Hardis R, Jessop L P, Peters F E and Kessler M R 2013 Cure kinetics characterization and monitoring of an epoxy resin using DSC, Raman spectroscopy, and DEA *Compos. Part a-Appl. Sci. & Manufac.* **49**(6) 100-8

[23] Nam J D and Seferis J C 1993 Application of the Kinetic Composite Methodology to Autocatalytic-Type Thermoset Prepreg Cures *J. Appl. Polym. Sci.* **50**(9) 1555-64

[24] Omrani A, Rostami A A, Khostavan S and Vazifeshenas Y 2012 Preparation, characterization and application of advanced isoconversional kinetics to epoxy/1,4-Bis(3-aminopropoxy) butane/MWCNT nanocomposite *Compos. Part a-Appl. Sci. & Manufac.* **43**(3) 381-7

A novel catalyst of warm-cloud seeding to enhance precipitation

H X Zhu, X D Li[1] and R J Yang

School of Materials Science & Engineering, Beijing Institute of technology, Beijing 100081, China

E-mail: bitlxd@bit.edu.cn

Abstract. Water is necessary for sustaining human life. In many regions of the world, traditional water sources cannot meet increasing water demands. Warm-cloud seeding is an efficient way to augment water supplies. In this paper, we explore two new hygroscopicity catalysts: Poly acrylamide (PAM) and Sodium polyacrylate (PAAS). We designed a series of experiments to investigate the effects of these catalysts together with common catalyst salt powder (NaCl). The experiment was held in a cloud chamber built in our laboratory. The results show that: 1) Catalysed by NaCl, a dose of 0.91g/m^3 can obtain the best precipitation efficiency and enhancement rate at 70.8% and 142%, respectively; 2) A 1.36g/m^3 dose catalysed by PAM and PAAS exhibits optimal performance at 76.7% and 70.4% precipitation efficiency, respectively; 3) Under the same conditions, PAM shows better catalytic effects than NaCl does.

1. Introduction

Water is one of the most basic necessities for human life. In many regions of the world, traditional water cannot meet increasing water demands [1]. Warm-cloud seeding can efficiently augment water supplies. Cotton and Pielke [2] reasoned that the Collision-Coalescence Process (CCP) is the most effective warm cloud precipitation process. This concept has been previously used in programs in the United States and other countries [2-4], and now is still widely used in India and Southeast Asian countries [5].

Hygroscopic salt like NaCl and $CaCl_2$ was used in most of these experiments. Many researchers have performed experimental and theoretical studies on cloud microstructures modified by NaCl and $CaCl_2$. Drofa [6] studied experimental and theoretical NaCl cloud microstructure modifications in a large cloud chamber. Drofa found that seeding at low salt powder consumption rates could initiate

IOP Conf. Series: Materials Science and Engineering **137** (2016) 012002 doi:10.1088/1757-899X/137/1/012002

precipitation from otherwise non-precipitating warm convective clouds, mainly by adding large cloud drops to the distribution tail.

The disadvantages of this approach are that it requires large salt quantities, and salt dispersion can corrode soil and spreading tools. Poly acrylamide (PAM) and Sodium polyacrylate (PAAS) are super absorbent resins, which contain strong hydrophilic groups and a three-dimensional network structures. They can absorb the water as hundreds of times of the weight as their own, there is no wonder that they are significantly more hygroscopic than NaCl. Moreover, they are not corrosive to soil. Therefore, PAM and PAAS may be preferable to NaCl for warm-cloud precipitation.

2. Experimental Procedure

2.1. Processing of catalyst

NaCl, PAM, and PAAS were separately milled in a planetary ball mill, then sieved by a 360 mesh sieve tray. After sieving, we put aside the powder that can penetrate the sieve tray, and dried the powders in a vacuum drying oven at 60 °C for 6 hours. A Malvern laser particle size analyzer Mastersizer 2000 was used to measure the catalysts' average diameters. Table 1 gives the purity, manufacturer and particle size of major reagents.

Table 1. Major reagents used in the experiment

Reagent	Purity	Manufacturer	volume average diameter
PAM	AR	Tianjin Fuchen Chemical Reagents Factory	27.868 μm
PAAS	AR	Tianjin Kemiou Chemical Reagent Co. Ltd.	42.856 μm
NaCl	AR	Beijing Tongguang Fine Chemicals Company	27.214 μm

2.2. The cloud chamber and its usage for measurements

Experimental studies on the catalyst cloud seeding effects were performed in a cloud chamber, which was designed for modelling clouds and fogs similar to those found in a real atmosphere. The chamber is an airtight cylinder 14 meters tall and with a 1-meter diameter; it is made of polymethyl methacrylate (PMMA). The chamber's total volume is 10.99 m^3. Figure 1 shows the sketch and physical map of the cloud chamber. It is equipped with an ultrasonic fog making machine, a series of wireless temperature and humidity sensors, light sensors, lasers, and a data acquisition unit.

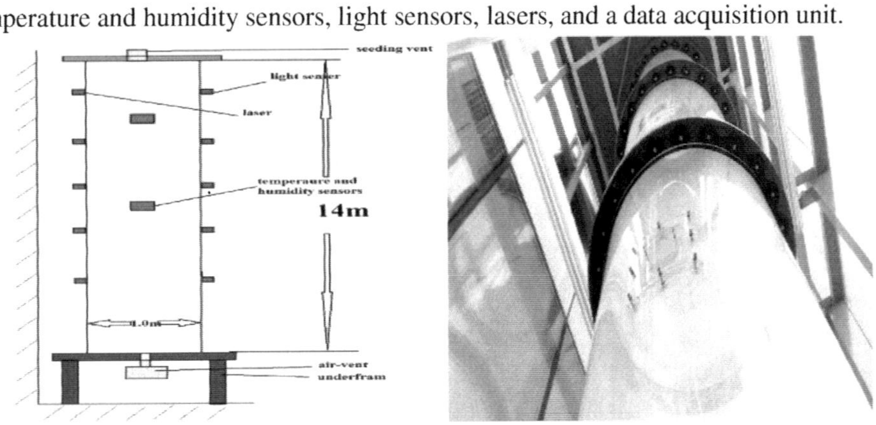

Figure 1. Sketch map and physical map of the cloud chamber

The light sensors monitored laser transmittance in the chamber, and the laser transmittance was proportional to the cloud's water vapor content in the chamber. The higher the water vapor content is, the lower the laser transmittance is. Therefore, we can use laser transmittance to measure the catalytic effects of catalyst powders.

To measure rain enhancement, an experiment was designed. After seeding the catalyst powder into the chamber, a 10cm diameter culture dish was placed at the bottom of the chamber. We weighed the dish every two minutes. Higher water collection resulted in higher weights, which indicated that the experimental catalyst had a better effect on rain enhancement.

2.3. Experimental procedure

First, we opened the sensors and the ultrasonic fog making machine to create a cloud in the chamber. The machine was closed after 1 hour when the cloud reached its saturation condition. Then, a certain mass of catalyst powder (the control group was not seeded) was seeded from the top of the chamber. Meanwhile, a culture dish was placed at the bottom of the chamber and weighed every 2 or 5 minutes. A computer recorded the temperature, humidity, and illumination data every 5 seconds. We stopped the experiment after 50 minutes, opened the chamber's ventilation, and cleaned the chamber.

3. Results and Discussion

3.1. Catalyzed by NaCl

This section discusses the effects of catalyst mass on precipitation. Precipitation efficiency is the ratio of water that reaches the ground against the total condensation in the cloud [7]. We derived the temperature and relative humidity (RH, units: %) in the chamber from the temperature and humidity sensor; therefore, the absolute humidity (AH, units: g/cm^3) can be determined from the humidity table. The water content can then be easily obtained using the below formula [8]:

$$M = AH \times V \tag{1}$$

Figure 2(a) shows the precipitation of different masses of NaCl seeded into the chamber. There was not a significant increase when 0.18 g/m^3 NaCl was seeded, but the precipitation experienced a sharp rise in the first 2 minutes when the amount of NaCl reached 0.45 g/m^3. Seeding 0.91 g/m^3 and 1.36 g/m^3 NaCl produced similar results, but the results of the former were slightly higher. Figure 2(b) shows this tendency. Higher amounts of NaCl seeded into the chamber resulted in faster cloud dispersal, and more water dropping to the ground. When 1.36 g/m^3 NaCl was seeded, the cloud dispersal was faster than it was when 0.91 g/m^3 was seeded, however, the precipitation had the opposite effect. The reason is that when cloud dispersal is very rapid, a significant amount of water droplets falls, so the number of droplets in the chamber is reduced, and the chances of collision and coalescence are reduced.

Table 2 summarizes the precipitation of different NaCl masses seeded within 50 minutes. From the table, we can see that the precipitation enhancement and efficiency increased with the increase of NaCl mass. They reached their peak value when the NaCl mass was 0.91 g/m^3, while they fell slightly when NaCl mass reached 1.36 g/m^3. This change indicates that 1.36 g/m^3 of NaCl is excessive for catalyzing.

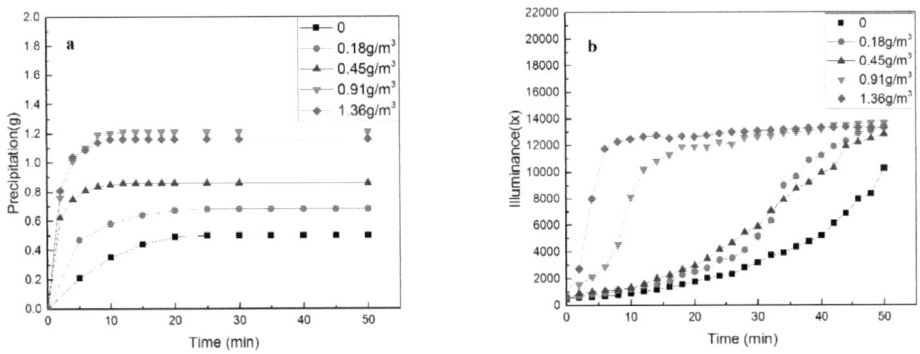

Figure 2. (a) Precipitation of different mass of NaCl seeded (b) Illuminance of different mass of NaCl seeded

Table 2. Catalytic effect of different mass of NaCl (T:20℃, RH:90%)

Mass of NaCl	Precipitation	Rate of Precipitation Enhancement	Precipitation Efficiency
0 g/m^3	0.50g	0	29.3%
0.18 g/m^3	0.68g	36%	39.8%
0.45 g/m^3	0.86g	72%	50.4%
0.91 g/m^3	1.21g	142%	70.8%
1.36 g/m^3	1.16g	132%	67.9%

3.2. Catalyzed by PAM

Figure 3(a) shows the precipitation of different PAM masses seeded into the chamber. Unlike clouds catalyzed by NaCl, there was a significant increase when 0.18 g/m^3 NaCl was seeded. Also, there was a sharp rise in the first 2 minutes when PAM reached 0.91 g/m^3 and 1.36 g/m^3 respectively. Figure 3(b) also shows the same tendency. Seeding more PAM into the chamber resulted in faster cloud dispersal.

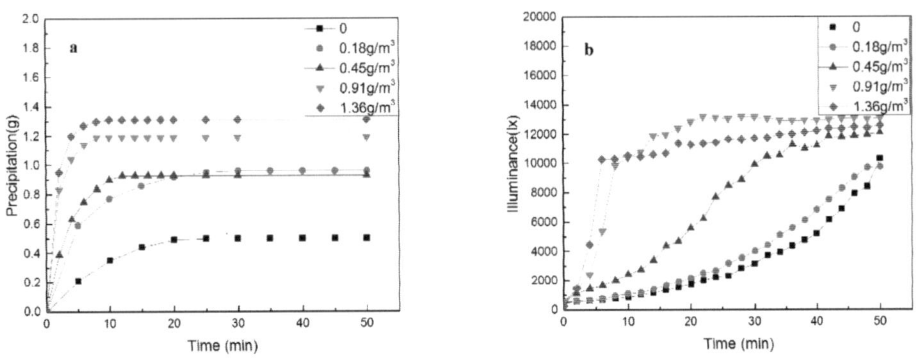

Figure 3. (a) Precipitation of different mass of PAM seeded (b) Illuminance of different mass of PAM seeded

Table 3 summarizes the precipitation of different PAM masses seeded within 50 minutes. The precipitation enhancement and efficiency increased as PAM mass increased. The precipitation enhancement and efficiency reached their peak values when PAM mass was 1.36 g/m^3. Compared with NaCl with the same conditions, PAM created higher precipitation enhancement and efficiency values, except for 0.91g/m^3. PAM therefore had better precipitation enhancement.

Table 3. Catalytic effect of different mass of PAM (T:20℃, RH:90%)

Mass of NaCl	Precipitation	Rate of Precipitation Enhancement	Precipitation Efficiency
0 g/m^3	0.50g	0	29.3%
0.18 g/m^3	0.96g	92%	56.2%
0.45 g/m^3	0.93g	84%	54.4%
0.91 g/m^3	1.19g	138%	69.7%
1.36 g/m^3	1.31g	162%	76.7%

3.3. Catalyzed by PAAS

Figure 4(a) shows the precipitation of different PAAS masses seeded into the chamber. As can be seen, precipitation increased as more PAAS was seeded into the chamber. In Figure 4(b), the 5 curves roughly conform. This indicates that PAAS seeding does not significantly affect cloud dispersal.

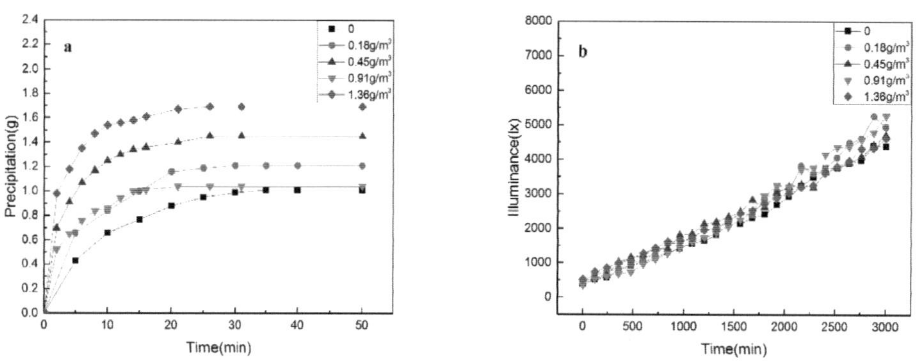

Figure 4. (a) Precipitation of different mass of PAAS seeded (b) Illuminance of different mass of PAAS seeded

Table 4. Catalytic effect of different mass of PAAS (T:25℃, RH:95%)

Mass of PAAS	Precipitation	Rate of Precipitation Enhancement	Precipitation Efficiency
0 g/m^3	1.01g	0	42.1%
0.18 g/m^3	1.21g	19.8%	50.4%
0.45 g/m^3	1.45g	43.6%	60.4%
0.91 g/m^3	1.04g	3.0%	43.3%
1.36 g/m^3	1.69g	67.3%	70.4%

Table 4 summarizes the precipitation of different PAAS masses seeded within 50 minutes. PAAS seeding had much lower precipitation enhancement rates compared with those of NaCl and PAM seeding. This indicates that PAAS is not as effective as NaCl and PAM as a catalyst.

4. Conclusion

The conclusions drawn from the experiments carried out in the cloud chamber are as follows:

- NaCl and PAM seeding can enhance warm cloud precipitation. NaCl's optimal seeding mass was 0.91 g/m^3, reaching a 70.8% precipitation efficiency rate and a 142% precipitation enhancement rate respectively. PAM seeding was more effective than NaCl seeding under the same condition. Its optimal seeding mass was 1.36 g/m^3, resulting in 76.7% precipitation efficiency and 162% precipitation enhancement rate.
- PAAS seeding cannot disperse the cloud, and its ability in precipitation enhancement was much lower than that of NaCl and PAM. It is not an effective warm cloud catalyst.

In summary, the experiment showed that PAM was effective at precipitation enhancement. It can be used as a novel catalyst in warm cloud seeding. However, the experiment was carried out in the cloud chamber; more experiments in the field are needed for further researches.

References

[1] Roelof T. 1999 A Review of Cloud Seeding Experiments to Enhance Precipitation and Some New Prospects. *Bull. Amer. Mete. Soc.* **80** 805-820.

[2] Cotton W R. 1982 Modification of precipitation from warm clouds- a review *Bull. Amer. Mete. Soc.* **63** 146-160.

[3] Biswas K. R. and Dennis A. S. 1971 Formation of rain shower by salt seeding *J. Appl. Meteor.* **10** 780–4.

[4] Bowen E. G. 1952 A new method of stimulating convective clouds to produce rain and hail *Quart. J. Roy. Meteor. Soc.* **78** 37–45.

[5] Murty A. S. P. 1989 An overview of warm cloud modification research in India *Proc. Fifth WMO Conf. on 'Weather Modification and Applied Cloud Physics', Beijing, China, World Meteorological Organization* 521–4.

[6] Drofa A. S., Ivanov V. N., Rosenfeld D. and Shilin A. G. 2010 Studying an effect of salt powder seeding used for precipitation enhancement from convective clouds *Atmos. Chem. Phys.* **10** 8011–23.

[7] Sui C. H., Li X. F. and Yang M. J. 2007 On the definition of precipitation efficiency *J. Atmos. Sci.* **64** 4506-13.

[8] Nguyen J. L., Schwartz J. and Dockery D. W. 2014 The relationship between indoor and outdoor temperature, apparent temperature, relative humidity, and absolute humidity, *Indoor Air* **24** 103-112.

Preparation and degradation mechanisms of biodegradable polymer: a review

S H Zeng[1], P P Duan[1], M X Shen[1,2,3], Y J Xue[1] and Z Y Wang[1,2,3]

[1] College of Mechanics & Materials, Hohai University, Nanjing 211100, P R China

[2] Nantong Ocean Research and Offshore Engineering Institute of Hohai University, Nantong 226300, P R China

E-mail: mxshen@hhu.edu.cn; tigerwang8810@126.com

Abstract. Polymers are difficult to degrade completely in Nature, and their catabolites may pollute the environment. In recent years, biodegradable polymers have become the hot topic in people's daily life with increasing interest, and a controllable polymer biodegradation is one of the most important directions for future polymer science. This article presents the main preparation methods for biodegradable polymers and discusses their degradation mechanisms, the biodegradable factors, recent researches and their applications. The future researches of biodegradable polymers are also put forward.

1. Introduction

Synthesized polymeric materials with low density, high specific strength, good wear and corrosion resistance, etc., which are being widely used in industry, agriculture, national defense, transportation, are difficult to be broken down naturally and completely in the natural environment [1]. If not recycled properly, they would do harm to the environment. The conventional treatment methods for waste polymeric materials include landfilling, incineration, blending with new materials after granulation, chemical degradation, recycling, etc. [2]. However, some of these methods are not cost-efficient, others can cause serious environmental pollution. Therefore, in recent years, the research and development of new biodegradable polymeric materials have become a hot topic in science and industry.

Biodegradable polymeric materials can be resistant to degradation during usage period and have a biodegradable property at the end of their useful life [3]. The biggest advantages of biodegradable polymeric materials are preserving fossil resources on one hand, and on the other hand reducing the environmental pollution in a sustainable development context [4]. Biodegradable polymers can be degraded by microorganisms under either aerobic or anaerobic conditions and ultimately converted to the bio-decomposited products including CH_4, H_2O and some inorganic compounds[5]. The processes can be promoted by abiotic chemical reactions such as hydrolysis, photodegradation, oxidation, etc. [6].

The field of biodegradable polymers has developed rapidly since 1980s. According to the statistics from European Plastic, the worldwide production capacity for bioplastics will increase from around 1.4 million tons in 2011 to approximately 6.0 million tons in 2017. Even though the speed-up rate is obvious, the output of bioplastics only account for about 1.7% of the whole polymer industry, which indicates that biodegradable polymers have a great and promising future. There are three kinds of biodegradable polymers: natural polymers, microbiologically synthesized polymers, and chemically synthesized biodegradable polymers [7]. According to their origins, biodegradable polymeric materials can be further classified into two species, the petroleum resources and biological resources respectively (renewable resources). The biodegradable polymers have been mainly applied now into agriculture, medicine, tissue engineering, industry, etc. [8].

This review mainly focuses on recent progress in biodegradable polymeric materials including preparation methods, biodegradation mechanisms, and influencing factors. It provides a general overview of the theme in the last five years. Finally, the applications and the future trends of biodegradable polymers are also briefly summarized.

2. Preparation of biodegradable polymeric materials

2.1. Traditional methods

2.1.1. Modified natural polymers. The polysaccharide polymers that are readily biodegradable in Nature such as starch, cellulose, chitosan and chitin can be transformed into new biodegradable polymers by co-blending modification. The biodegradable films have demonstrated an improved mechanical properties, low water vapor permeability and solubility [9]. For example, biodegradable blend films were developed by casting rice starch-chitosan solution on levelled trays [10]. However, the performances of thermal and mechanical properties of these products are quite poor. Many efforts have been made to obtain the desired functions through various modification methods [11].

2.1.2. Chemically synthesized polymers. These polymers are chemically synthesized with the chemical structures similar to natural polymers. These polymeric chains containing ester, amide and peptide bonds can be readily biodegradable. Luo et al. [12] have reported using L-lactic acid and glycerol as starting materials to synthesize a biodegradable material (poly(lactic acid-co-glycerol)) under the optimal synthetic conditions. However, the required synthesis condition was harsh and bad and led to many by-products, therefore this chemical synthesis method is not cost-efficient for biodegradable polymers [13].

2.1.3. Microbiologically synthesized polymers. Microorganism can fabricate a series of complex polymeric materials using some organic matters (like glucose or starch) as food sources. These polymers contain the various types of silk, polysaccharides, polyesters, etc. The separation of these products is difficult due to their similar chemical properties. Bacterially synthesized polyhydroxyalkanoates (PHAs) as highly biocompatible and biodegradable plastics have attracted wide attention and can be prepared through many renewable resources [14].

2.2. Enzymatic synthesis

Enzymatic synthesis is a new method for the preparation of biodegradable polymeric materials. Some enzymes show different properties and can catalyze some special polymerization reactions. This kind of polymerization process may not produce any by-products due to the high specificity of the enzymes, which makes the product separation easy to perform. Furthermore, the enzyme can be recycled, and the mild catalytic reaction conditions (generally at room temperature and atmospheric pressure) reduce the process cost significantly [15]. Using the stereospecificities of the enzymes, some special products, which cannot be produced with traditional methods, can also be made [16].

One example is the aliphatic polyesters synthesized by lipase-catalyzed polymerization, which has become an important synthetic method for polymeric materials [17]. By enzymatic synthesis, fully biodegradable polymers such as polyester, polysaccharide, polyamide, etc., can also be produced [18].

2.3. Chemo-enzymatic synthesis

Enzymatic synthesis can not only endow the polymer with both high specificity and stereoselectivity, but also adjust the polymer molecular weight [19]. The chemo-enzymatic synthetic method combines the conventional polymerization with the highly efficient and regioselective enzymatic approach, and is therefore a very attractive strategy for the preparation of high molecular weight biodegradable polymers [20]. Cai et al. [21] have used an efficient chemo-enzymatic route to fabricate some optically active polymeric prodrugs for non-steroidal anti-inflammatory drugs with high molecular weight.

3. Biodegradation mechanism

3.1. Abiotic involvement

The phenomenon of "biodegradation" can be explained by biological activity as the predominant factor. However, the organic matter is decomposed due to the synergism of biotic and abiotic factors in Nature [22]. Biodegradable polymers exposed to outdoor conditions (e.g., weather, aging or burying) can undergo more or less transformations such as in mechanical, chemical, light, thermal forms, etc. [23]. This exposure reduces the performance of the biodegradable polymers. In most cases, abiotic involvement contributes to weakening the macromolecular structure of biodegradable polymers, and in this way favor the degradation. Moreover, these abiotic parameters may be useful to initiate the biodegradation process [24]. Consequently, the involvement of the abiotic conditions should not be neglected.

3.2. Biotic involvement

So far, several biodegradable mechanisms of biodegradable polymers have been reported, three main biodegradable mechanisms are, (1) physical biodegradation: polymers are eroded by microorganisms and then destroyed due to cell growth; (2) Chemical biodegradation: with the direct action of biological enzymes, microorganisms erode the polymer chain which results in the depolymerization. Usually, biochemical degradation can be divided into two steps. The first step, extra-cellular enzymes (exoenzymes) acting on the surface of materials break the polymer chains and generate small molecule compounds (acids, esters, etc.) with the molecular weight of less than 500 daltons. The second step, the small molecule compounds are assimilated by the microorganisms through metabolic routes, and then transformed into biomass and bioenergy, until final products of CO_2 or CH_4 and water can be obtained [4]. In general, the second step is often concomitant with the first one; (3) The interaction between microorganisms and polymers form new macromolecules. Polymers are usually not degraded by a single mechanism, but by a complex combination of biophysical, biochemical and physicochemical mechanisms [25].

Since microorganisms play an important role in the biodegradable process, proper humidity is always necessary for the polymer to achieve a biodegradable property [26]. Polymers can be further biodegraded under either aerobic or anaerobic conditions or the combination of both [27]. In the presence of oxygen, aerobic microorganisms contribute to the degradation of polymeric materials, with CO_2, water and microbial biomass as the final products. Equation (1) shows the mass balance equation for aerobic biodegradation. In contrast, anaerobic microorganisms can degrade the biodegradable polymers under the anaerobic environment. The final products will be CO_2, water, CH_4 or other hydrocarbons, or microbial biomass. Equation (2) shows the total carbon equation for anaerobic biodegradation. When the primary metabolites are some inorganic species, e.g., N_2, CO_2, H_2O, salts, etc., the degradation is called mineralization [28].

$$C_T + O_2 \rightarrow CO_2 + C_R + C_B \quad (1)$$

$$C_T \rightarrow CH_4 + CO_2 + C_R + C_B \quad (2)$$

Where, C_T is the total carbon content of biodegradable polymers, C_R is any residual products of polymers left or any by-products produced during the degradation process, gaseous CO_2 is a measurable product, and C_B is the microbial biomass from the reproduction and growth of microorganisms.

In addition, , the effect of enzymes occupies a key position. When the active enzymes from microorganisms are secreted and permeated into the active sites of polymer chains, hydrolysis reactions occur on the macromolecular chains, which may break into short chains or smaller molecules, even disappear, and polymers undergo complete degradation [29]. Results have been found that microbial enzymes tend to decompose the polymer structures, such as amide, enamine, ester, urea, urethane linkages, etc.[30].

4. Effect factors of biodegradation

Based on in-depth researches, some scientists have found that the polymer biodegradation depends not only on the polymers characteristics including chemical structure, morphology, molecular weight but also on some environmental conditions, e.g., temperature, humidity, pH value, radiation, etc.

4.1. Polymeric configuration

4.1.1. Chemical structure. Biodegradability is ultimately controlled by the chemical structures which directly affects the ability of degradation. Functional groups' biodegradation ability may be descendant in such order: Aliphatic ester, Peptide bond > Carbamate > Aliphatic ether > Methylene.

The polymers with hydrophilic groups (amino, hydroxyl, carboxyl and amide groups) are easily biodegradable under certain humidity, and those amphiphilic diblock copolymers are more biodegradable than polymers are with only one hydrophobic block [31]. Linear polymers are more easily biodegradable than branched and cross-linked polymers [32].

4.1.2. Aggregation structure. Polymers with flexible chains are easily biodegradable, and the amorphous areas are easier to degrade than crystallization areas. Unsaturated compounds are hard to biodegrade [33], while aliphatic polyester are easier to biodegrade [34]. Moreover, rigid aromatic polyesters such as polyethylene terephthalate (PET) are bioinert [35]. Blend-modification usually can help to achieve biodegradation [36]. Recently, some special plasticizers and additives which can play important roles in polymer degradation have attracted wide attention in this field [37].

4.1.3. Polymerization degree. Typically, the molecular weight of polymers has an important effect on its biodegradation. When polymers have a higher molecular weight, the rate of biodegradation may be slower [38]. If the molecular weight reaches an upper limit or goes beyond the scope that microbial cells can take, the biodegradation will not occur in polymers [39]. The synthetic plastics are usually too large to allow the cells to enter, but natural polymers do not have these problems. In most cases, alkane-based polymers with the molecular weight of exceeding 400–500 daltons must be degraded into smaller molecules by biological or chemical means or photodegradation before biodegradation [40].

4.2. Environmental conditions

Water is necessary for the growth of microorganisms. Polymeric materials can be biodegraded only under certain humidity.

Temperature has a dual effect on biodegradation. In a certain range, high temperature speeds up microbial metabolic processes and causes the vigorous growth, which is helpful to degradation on one hand. On the other hand the temperature has a greater effect on the biological activities of proteins and enzymes. Therefore, each kind of microorganisms has its own optimal temperature for growth and reproduction [41].

The pH value also has a great influence on the growth of microorganisms [42]. Optimal pH can accelerate microbial metabolism, eventually speed up the degradation rate. For example, the biodegradation of poly(lactic-co-glycolic acid) (PLGA) and polyglycolic acid (PGA) mainly depends on bacteria, actinomyces, fungi and other microbes under pH 5–9 [43].

Moreover, biodegradable materials produce free radicals or ions after γ-ray radiation, which can also speed up the degradation [44].

5. Applications

5.1. Industry
With high heat resistance, waterproof, chemical erosion resistance, dirty resistance and strong coloring power, biodegradable polymers can be widely used in making leather, fiber, and packaging film, especially the biodegradable packaging films attracting increasing attention [45]. Compared with other biodegradable polyesters, poly-ε-caprolactones (PCL) enjoys easy availability and lower price, which can be widely applied as an environment-friendly material. Some polysaccharide-based biopolymers, e.g., pullulan, starch, chitosan, etc., have been investigated and used as packaging films [46]. Another biodegradable polymer, poly-3-hydroxybutyrate (PHB) can be used to fabricate the packing materials or disposable products [14].

5.2. Agriculture
Biodegradable polymers can be used as agricultural mulches, the planting support, gradual releasing agents of the agricultural pesticide, ropes and fishing nets in fish culture [47]. Under appropriate conditions, biodegradable materials especially chitin and chitosan materials can degrade into compost or organic muck, which is beneficial to plant growth and the soil environment [48]. Some natural polymers, e.g., starch, lignin, cellulose, chitin, etc., are being applied in the releasing agent system, and the PCL is being used as a support material for agricultural planting [49].

5.3. Medicine and pharmacy
As a biomaterial, biodegradable polymers have three important characteristics: mechanical resistance, biological absorbability, and biocompatibility [50]. Biodegradable polymers are mainly being used as the releasing agents of drugs, absorbable sutures, and also in orthopedic surgery and tissue engineering [51].

In drug releasing delivery systems, biodegradable polymers, such as proteins, gelatin, chitin and their derivatives, polylactic acid (PLA), PHB, etc., have been extensively used as drug carriers due to their biological absorbability. PGA, PLA, and their copolymers are used to prepare absorbable sutures, which can automatically degrade and be absorbed by organisms in wound healing. Polybutylene succinate can be used as a promising substance to repair bone and cartilage [52]. Biodegradable fixing materials can also avoid osteoporosis and secondary operation caused by using stainless steel [53]. Moreover, antibiotic drugs, bone growth factors, and regulatory proteins can be implanted into the biodegradable materials to prevent infection and promote bone healing during the treatment period [54]. Poly(propylene fumarate) (PPF) is the best biodegradable scaffold for the bone tissue regeneration so far [55]. PLGA with a good cell adhesion has been applied in the bone tissue engineering [56].

5.4. Other fields
Biodegradable polymers can be used in other specific fields such as electronics[57], automotive sectors, sports and leisure [58], construction sectors [59], etc., even with some unusual applications [60]. A huge and promising market in the future will be expected.

6. Conclusions
The development of biodegradable polymers is still in the early stage yet with a promising prospect in the future. The researches of biodegradable polymer will help to alleviate the environmental pollution and also have a critical significance in relieving the shortage of oil resources. Compared with traditional plastics, biodegradable polymers are still expensive and have relatively lower mechanical properties. Therefore, at present, only limited biodegradable polymer materials are being used.With the development of smart design, more and more new preparation methods will be explored, and better biodegradable materials will be developed.

Future researches of biodegradable polymeric materials should be anticipated in the following aspects, (1) To control the degradation rates of biodegradable polymers and balance the properties between the material performance and the shelf life; (2) To explore new methods to make existing polymers biodegradable and synthesize biodegradable polymers with novel structures; (3) To further improve biodegradability of biodegradable polymers and reduce the cost to broaden their applications; (4) To develop biodegradable composites based on natural polymers.

Acknowledgments
We are grateful for the support of the program of Jiangsu Province in six main fields (No. XNY48-038), and the guide program of Nantong (No. 2013400103).

References

[1] Kumbar S, Laurencin C and Deng M 2014 *Natural and Synthetic Biomedical Polymers* (Oxford: Elsevier) pp 1-31
[2] Hamad K, Kaseem M and Deri F 2013 *Polym. Degrad. Stabil.* **98** 2801-12
[3] Yates M R and Barlow C Y 2013 *Resour. Conserv. Recy.* **78** 54-66
[4] Luckachan G E and Pillai C 2011 *J. Polym. Environ.* **19** 637-76
[5] Rizzarelli P and Carroccio S 2014 *Anal. Chim. Acta* **808** 18-43
[6] Cho H, Moon H, Kim M, Nam K and Kim J 2011 *Waste Manage.* **31** 475-80
[7] Ebnesajjad S 2012 *Handbook of biopolymers and biodegradable plastics: properties, processing and applications* (New York: William Andrew)
[8] Doppalapudi S, Jain A, Khan W and Domb A J 2014 *Polym. Advan. Technol.* **25** 427-35
[9] Andrade-Mahecha M M and Menegalli F C 2012 *Carbohyd. Polym.* **88** 449-58
[10] Bourtoom T and Chinnan M S 2008 *LWT-Food Sci. Technol.* **41** 1633-41
[11] Simi C K and Abraham T E 2010 *Colloid Polym. Sci.* **288** 297-306
[12] Luo S H, Wang Z Y, Mao C X and Huo J P 2011 *J. Polym. Res.* **18** 2093-102
[13] Gunatillake P, Mayadunne R and Adhikari R 2006 *Biotechnology annual review* **12** 301-47
[14] Arrieta M P, López J, Hernández A and Rayón E 2014 *Eur. Polym. J.* **50** 255-70
[15] Gross R A, Ganesh M and Lu W 2010 *Trends Biotechnol.* **28** 435-43
[16] Uyama H and Kobayashi S 2006 *Adv. Polym. Sci.* **194** 133-58
[17] Zhang J X, Shi H, Wu D, Zhang A J, Yang Y and Li Q S 2014 *Process. Biochem.* **49** 797-806
[18] Kobayashi S and Makino A 2009 *Chem. Rev.* **109** 5288-353
[19] Kobayashi S 2013 *Adv. Polym. Sci.* **262** 141-66
[20] Yang Y, Zhang J X, Wu D, Zhou Y L, Shi W and Li Q S 2014 *Biotechnol. Adv.* **32** 642-51
[21] Cai X Q, Wang N and Lin X F 2006 *Polymer* **47** 6491-95
[22] Leja K and Lewandowicz G 2010 *Pol. J. Environ. Stud.* **19** 255-66
[23] Smitthipong W, Chollakup R and Nardin M 2014 *Bio-Based Composites for High-Performance Materials: From Strategy to Industrial Application* (Boca Raton: CRC Press) p 205

[24] Kijchavengkul T, Auras R and Fernandez R T 2010 *Polym. Degrad. Stabil.* **95** 2641-47

[25] Revie R W 2011 *The Uhlig's Corrosion Handbook, 3rd Edition* (New York: Wiley) pp 421-38

[26] Gautam R, Bassi A and Yanful E 2007 *Appl. Biochem. Biotech* **141** 85-108

[27] Leja K and Lewandowicz G 2010 *Pol. J. Environ. Stud.* **19** 255-66

[28] Mohan K 2011 *Journal of Biochemical Technology* **2** 210-15

[29] Reis R L and Román J S 2005 *Biodegradable systems in tissue engineering and regenerative medicine* (Boca Raton: CRC Press) pp 177-201

[30] Araceli L T, Gerardo G S, Raúl R H and Aguilar C N 2012 *J. Polym. Environ.* **20** 258-65

[31] Nottelet B, Di Tommaso C and Möller M 2010 *J. Polym. Sci. Polym. Chem.* **48** 3244-54

[32] Wong A D, DeWit M A and Gillies E R 2012 *Adv. Drug Deliver. Rev* **64** 1031-45

[33] Ohlmann D M, Tschauder N, Dierker M and Gooßen L J 2012 *J. Am. Chem. Soc.* **134** 13716-29

[34] Cameron D J and Shaver M P 2011 *Chem. Soc. Rev.* **40** 1761-76

[35] Nair L S and Laurencin C T 2007 *Prog. Polym. Sci.* **32** 762-98

[36] Kijchavengkul T, Auras R, Rubino M, Alvarado E, Camacho Montero J R and Rosales J M 2010 *Polym. Degrad. Stabil.* **95** 99-107

[37] Gomes L B, Klein J M and Grisa A M C 2014 *Mater. Res-Ibero-Am. J.* **17** 121-26

[38] Gleadall A, Pan J, Kruft M A and Kellomäki M 2014 *Acta Biomater.* **10** 2233-40

[39] Santerre J, Woodhouse K, Laroche G and Labow R 2005 *Biomaterials.* **26** 7457-70

[40] Lucas N, Bienaime C, Belloy C and Nava-Saucedo J E 2008 *Chemosphere* **73** 429-42

[41] Nakayama M, Okano T, Sakai K and Yokoyama M 2006 *J. Control Release.* **115** 46-56

[42] Zhang X, Qiu W H and Chen H Z 2012 *Bioresource. Technol.* **123** 30-35

[43] Hild N, Tawakoli P N, Buchalla W, Stark W J and Mohn D 2013 *Acta Biomater.* **9** 9118-25

[44] Akter N, Khan R A, Salmieri S, and Lacroix M 2012 *Radiat. Phys. Chem.* **81** 995-98

[45] Thakar R and Dhawale A 2014 *Popular Plastics & Packaging* **59** 27-31

[46] Vieira M G A, da Silva M A, dos Santos L O and Beppu M M 2011 *Eur. Polym. J.* **47** 254-63

[47] Kosaric N and Sukan F V 2014 *Biosurfactants: Production and Utilization-Processes, Technologies, and Economics* (Boca Raton: CRC Press) pp 313-326

[48] Lopez O, Garcia M, Villar M and Albertengo L 2014 *LWT-Food Sci. Technol.* **57** 106-15

[49] Roy A, Singh S K, Bajpai J and Bajpai A K 2014 *Cent. Eur. J. Chem.* **12** 453-69

[50] Tian H Y, Tang Z H, Zhuang X L, Chen X S and Jing X B 2012 *Prog. Polym. Sci.* **37** 237-80

[51] Domb A J 2014 *Polym. Advan. Technol.* **25** 425-26

[52] Puppi D, Chiellini F, Piras A and Chiellini E 2010 *Prog. Polym. Sci.* **35** 403-40

[53] Kluin O S, van der Mei H C, Busscher H J and Neut D 2013 *Expert Opin. Drug Del.* **10** 341-51

[54] Dickson G, Buchanan F, Marsh D and McCaigue M 2007 *Technol. Health Care* **15** 57-67

[55] Cai Z Y, Zhang T, Di L Z, Xu D H and Yang D A 2011 *Biomed. Microdevices* **13** 623-31

[56] Zhao L, Wu C, Lin K and Chang J 2012 *Bio-med. Mater. Eng.* **22** 289-300

[57] Irimia-Vladu M, Sariciftci N S and Bauer S 2011 *J. Mater. Chem.* **21** 1350-61

[58] Ammala A 2013 *Int. J. Cosmetic Sci.* **35** 113-24

[59] Yazdi M F A, Zakaria R, Mustaffar M and Yahya K 2013 *Desalin. Water Treat.* **52** 1-6

[60] Sun M, Hong C Y and Pan C Y 2012 *J. Am. Chem. Soc.* **134** 20581-84

Extraction and characterization of the auricularia auricular polysaccharide

Q T Zhang[1]

Shanxi Institute of International Trade & commerce, Shanxi, 712046, China

E-mail: zqt0526@163.com

Abstract. To study a new protein drugs carrier, the Auricularia auricular polysaccharide (AAP) was extracted and purified from Auricularia auricular, and then characterized by the micrOTOF-Q mass spectrometer, UV/Vis spectrophotometer, moisture analyzer and SEM. The results showed that the AAP sample was water- soluble and white flocculence, its molecular weight were 20506.9 Da~63923.7 Da, and the yield, moisture, and total sugar contents of the AAP were 4.5%, 6.2% and 90.12%(w/w), respectively. The results of the SEM revealed that the AAP dried by vacuum were spherical particles with a smooth surface, and the AAP freeze-dried had continuous porous sheet shape with the loose structure.

1. Introduction

To date, over 160 therapeutic protein drugs have been licensed, and increasing protein drugs that will be approved by the regulatory agencies are forecasted to increase significantly in the next few years [1, 2]. The physicochemical and biological properties of protein drugs are unlike those of conventional ones, such as molecular weight, biological half-life, conformational stability, physicochemical stability, solubility, oral bioavailability, dose requirement, and administration [3]. In addition, the disadvantage of the oral protein drugs is that its low bioavailability. The reasons for this are that the poor protein transmembrane transport of gastrointestinal mucosa, easy destruction by the rigor pH environment and the abundant enzyme system of gastrointestinal [4, 5]. Therefore, the oral protein drugs delivery systems have caused the extensive concern.

In order to get over the above hurdles and increasing the gastrointestinal absorption of the oral protein drugs, we have been trying to make some of the nanoparticles (NPs) from natural biodegradable polymers, which have been exploited and measured for protein drugs[6, 7]. However, biodegradable polymer nanoparticles made from polyglycolic acid and polylactide and their copolymers, are usually obtained by using organic solvents, high temperatures, forces and are inactivated by physical and chemical denaturation. Furthermore, after the formulation has been administered, changes in the microenvironment within the nanoparticles due to polymer degradation can dramatically affect the tertiary structure of the protein [8, 9].

A method is to encapsulate protein drugs within Polyelectrolyte polysaccharide nanoparticles, so that the degradation of protein drugs by the gastrointestinal pH and Enzyme System is avoided, they can be transported to the sensitive target for release, enhance their tissue and cell penetrating across the gastrointestinal mucosa , reduced reticuloendothelial system phagocytic [10-12]. Nature polysaccharides possess plenty of promising properties, including excellent biodegradability, high biocompatibility, poor toxicity, safety, abundant availability and low production cost [13]. In particular, most of the natural polysaccharides have hydrophilic groups such as –OH, –COOH, $-CONH_2$ and $-SO_3H$, they are apt to form non-covalent bonds with the biological tissues, that can prolong residence time in the site of absorption and promote the absorption of the protein drugs [14, 15]. Auricularia auricular polysaccharide (AAP) is found rich in Auricularia auricular, as significant bioactive substances, broad physiological activity, favorable medicinal value, and a research highlight in the field of medicine [16]. However, AAP as a novel nature polysaccharide has been rarely investigated and reported in the application of the drug carriers.

In the present work, the AAP was extracted and purified from Auricularia auricular and characterized. The results of this work should provide valuable information for selecting protein drugs carrier.

2. Materials and methods

2.1. Materials

Bovine serum albumin (BSA) and bovine hemoglobin (BHb) were purchased from Sigma–Aldrich (St. Louis, MO, USA). Coomassie brilliant blue G-250 was obtained from Regent Chemicals Co., Ltd (Tianjin, China). Auricularia auricular was bought from Carrefour supermarket (Shenyang, China), which was cultivated in Liaoning Province, China. Low molecular weight chitosan was purchased from Golden-Shell Pharmaceutical Co., Ltd (Zhejiang, China), and the degree of deacetylation was 90%. All other reagents and chemicals were of analytical grade.

2.2. Extraction and purification of AAP

The AAP was extracted and purified by modified water extraction and alcohol precipitation method as follows [17]: The Auricularia auricular was defatted by reflux, and then dried. The resulting powder was weighed and extracted, followed by precipitating and washing with ethanol, and then crude gray polysaccharide was obtained by vacuum-dried. 2% crude polysaccharide solution was prepared by removing protein via the Savage method (chloroform: n-butanol=4:1). With pH adjusted to 8.0 as well as ammonia, color materials were removed by hydrogen peroxide. Afterward, the decolourization of polysaccharide solution was dialyzed against distilled water. The purified AAP was obtained by concentrating or freeze-drying dialysates eventually.

2.3. Characterization of AAP

2.3.1. Polysaccharide content. The AAP contents were measured by a phenol-sulfuric acid method using D-glucose as a standard [18]. The percentage of AAP extraction yield(%) was calculated with the formula as follows:

$$Yield(\%) = \frac{W_2}{W_1} \times 100$$

Where W_2 was the polysaccharides content of extraction, and W_1 represented dried sample weight.

The protein of the polysaccharide was detected by UV-4802 Double Beam UV/Vis Spectrophotometer (Unico, Shanghai, China). The moisture content of the AAP was obtained by a moisture analyzer (Shuangquan, Shanghai, China).

2.3.2. Molecular weight (Mw) The molecular weight of AAP was determined by time-of-flight mass spectrometer (Bruker micrOTOF-Q, Ettlingen, Germany). LC-Q-TOF/MS instrument parameters: electrospray ionization (ESI) detection mode, capillary voltage: 4.5 kV, capillary export voltage: -500V, collision cell RF voltage: 800Vpp, capillary temperature: 180℃, N_2 as the fog and auxiliary gas and flow rate of 4mL/min. Argon as the collision gas for CID experiments, the collision energy was 20eV. The analytical data were collected in full scan mode from 50 m/z to 900 m/z.

2.3.3. SEM. The morphology and structure characteristics of the AAP were observed by SEM (SSX-550, Shimadzu, Japan). The sample was thinly sprinkled onto a metal stub and vacuum coated with a thin layer of gold in an argon atmosphere. The coated samples were examined at an acceleration voltage of 15kV.

2.3.4. DSC. Thermal properties of the samples were surveyed by DSC (DSC-60, Shimadzu, Japan). The samples were scanned at a heating speed of 5℃/min over a temperature range of 20-300℃ with a nitrogen purge of 40ml/min in an aluminum pan and sealed hermetically, using aluminum oxide as the reference. Melting point and crystallization point corresponded to the maximum and minimum of the DSC curves respectively.

3. Results and discussion

3.1. Physicochemical properties of AAP

Purified AAP were obtained by hot-water extraction, ethanol precipitation, removed via Sevag method, dialyzed with distilled water and lyophilized by freeze-drying. The yield, moisture, and total sugar contents in the AAP were 4.5%, 6.2% and 90.12% (w/w), respectively. The AAP sample was water-soluble and white flocculence. AAP was not detected with a significant absorption peak in the wavelength of 280nm by the UV/Vis Spectrophotometer, which indicated that protein impurities have been reduced to the very low amount [19]. This result may be due to the process of vacuum freeze-drying, because of the water precipitated from polysaccharide solution under the condition of the solution freezing, which changed the solution system and made protein molecules denaturation and finally led to its lower solubility [20].

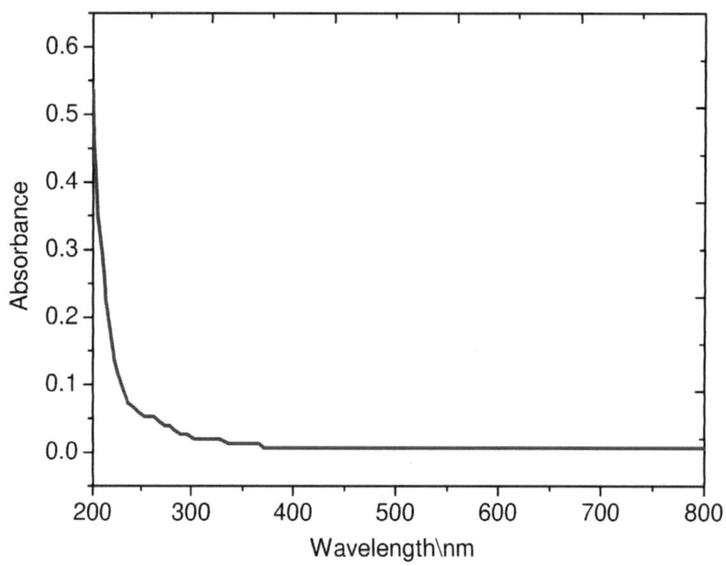

Figure 1. UV Spectrum of AAP in water

3.2. Molecular weight and SEM

The mass spectrometer was applied to determine the molecular weight of AAP. Figure 2, 3 exhibited the mass spectrometer of AAP, and Mw of AAP ranged from 20,506.9Da to 63,923.7Da. The surface topography and structure of a polysaccharide would be affected by different methods of extraction, purification and preparation [21]. Figure 4 revealed the surface morphology of AAP dried by vacuum and lyophilization. The freeze-dried AAP had continuous porous sheet shape with loose structure, whereas vacuum one had a smooth surface and spherical shape with a tight structure. Comparing the two AAP products, we could speculate that the freeze-dried AAP dissolved much easier and faster than the other one did [22].

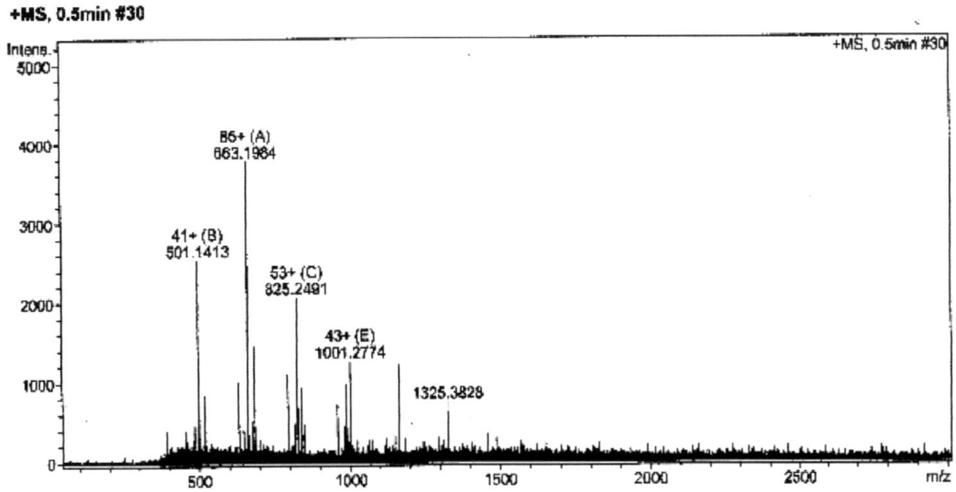

Figure 2. Mass spectrometer of AAP under positive model with ESI source

Component	Molecular Mass	Molecule	Absolute Abundance	Relative Abundance
A	56281.1089	56281.1089 Mr	5927	100.00
B	20506.9015	20506.9015 Mr	3247	54.78
C	43685.0381	43685.0381 Mr	2449	41.33
D	25999.9693	25999.9693 Mr	2315	39.06
E	43015.1381	43015.1381 Mr	2569	43.34
F	63923.7047	63923.7047 Mr	1981	33.42
G	24537.8055	24537.8055 Mr	1236	20.85
H	60160.0246	60160.0246 Mr	1659	27.99
I	23471.2823	23471.2823 Mr	981	16.55
J	25585.7662	25585.7662 Mr	1803	30.42

Figure 3. Mw range of AAP with Flight mass spectrometer

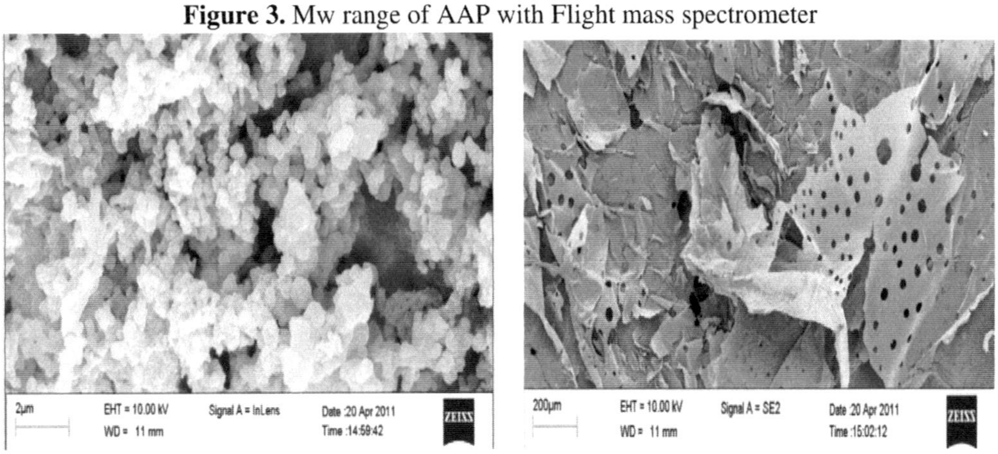

Figure 4. SEM images of AAP dried by vacuum (left) and lyophilization (right).

4. Conclusion

As an innovative idea of the application of AAP serving as the drug carriers, this study demonstrated the polysaccharide extracted and purified from Auricularia auricular. The results revealed that the AAP sample was water-soluble and white flocculence, its molecular weight was 20506.9 Da～63923.7 Da, and the yield, moisture, and total sugar contents of the AAP were 4.5%, 6.2% and 90.12%(w/w), respectively. The AAP dried by vacuum were spherical particles with a smooth surface, and the AAP dried by freeze-drying had continuous porous sheet shape with a loose structure.

References

[1] Han C, Davis C B and Wang B 2010 *Evaluation of drug candidates for preclinical development: pharmacokinetics, metabolism, pharmaceutics, and toxicology* (Pulisher location: John Wiley & Sons, Manhattan, New York)
[2] Research and Markets Global Protein Therapeutics Market Analysis -2011, press@researchandmarkets.com.
[3] Wang T, Xu Q, Wu Y, Zeng A, Li M and Gao H 2009 *Carbohydr. Res.* **344** 908-914
[4] Harush O, Rozentur E, Benita S and Altschuler Y 2008 *Biomacromolecules* **9** 435-443
[5] Sandri G, Bonferoni M, Rossi S, Ferrari F, Gibin S, Zambito Y, Di Colo G and Caramella C 2007 *Eur. J. Pharm. Biopharm.* **65** 68-77
[6] Papadimitriou S, Achilias D and Bikiaris D 2012 *Int. J. Pharm.* **430** 318-327
[7] Balasse E, Odot J, Gatouillat G, Andry M and Madoulet C 2008 *Int. J. Pharm.* **353** 131-138
[8] Ravi K, Bakowsky U and Lehr C 2004 *Biomaterials* **25** 1771-1777
[9] Salmaso S and Caliceti P 2013 *Int. J. Pharm.* **440** 111-123
[10] Jadhav S and Singhal R 2014 *Carbohydr. Polym.* **105** 49-56
[11] Li Q, Xia B, Branham M, Ha W, Wu H, Peng S, Ding L, Li B and Zhang S 2011 *Carbohydr. Polym.* **86** 120-126
[12] Mo R, Jiang T, Di J, W. Tai W and Gu Z 2014 *Chem. Soc. Rev.* **43** 3595-3629
[13] Luo Y and Wang Q 2014 *Int. J. Biol. Macromol.* **64** 353-367
[14] Liu Z, Y. Jiao Y, Wang Y, Zhou C and Zhang Z 2008 *Adv. Drug Delivery Rev.* **60** 1650-1662
[15] Song Y, Zhou J, Li Q, Guo Y and Zhang L 2009 *Macromol. Biosci.* **9** 857-863
[16] Zeng W, Zhang Z, Gao H, Jia L and Chen W 2012 *Carbohydr. Polym.* **89** 694-700
[17] Zeng F, Zhao C, Pang J, Lin Z, Huang Y and Liu B 2013 *J. Food Sci.* **78** 1470-1475
[18] Dubois M, Gilles K, Hamilton J, Rebers P and Smith F 1956 *Anal. Chem.* **28** 350-356
[19] Hu H, H. Liang H and Wu Y 2015 *Carbohydr. Polym.* **127** 94-100
[20] Nam J and Park Y 2001 *J. Appl. Polym. Sci.* **81** 3008-3021
[21] Elijah.I.Nep, Barbara.R.Conway 2010 *J. Excipients Food Chem.* 2010, **1**(1) 30-40
[22] Kong L, Yu L, Feng T, Yin X, Liu T and Dong L 2015 *Carbohydr. Polym.* **125** 1-8

The effect of spin-orbit interaction on the electron spin transport in inorganic semiconductors and the comparison of organic semiconductors

H Li[1], S Y Wu, X R Liang and M Y Zhou

Naval Aeronautical and Astronautical University Department of Basic Sciences, Yantai, Shandong 264001, China

E-mail: 346521161@qq.com

Abstract. In theory, the influence of spin-orbit interaction on the electron spin transport in inorganic semiconductors was discussed, which was compared with that of organic semiconductors. Results reveal that the internal mechanism of spin precession in inorganic semiconductor is different from that of the organic semiconductor, while the influence of the spin-orbit interaction on the carrier spin transport is same.

1. Introduction

There are some spin-dependent interactions in the spin electronics. The first one is the classical interaction between a pair of electron magnetic dipoles usually in the intensity of 1K, which is very weak; The second one is the interaction between the spin and the external magnetic field, which is called the Zeeman effect[1-3]. Generally speaking, Only the magnetic field is in the order of tens of Tesla order can obvious spin level splitting appear; Third, spin-orbit interaction[4-7]; The fourth kind of interaction is the one between Heisenberg and Pauli; The fifth kind of interaction is the hyperfine interaction between the electron and the nuclear spins. Any one of these five interactions can cause the spin to be reversed. In accordance with different demands, the main task of the spin electronics is to control spin flip

Spin-orbit interaction is due to the asymmetric effects of electrons in motion. The spin moment will be affected effectively by the intrinsic magnetic field, which is spin-orbit interaction[8].

$$H_{SO} = -\vec{\mu}_s \cdot \vec{B} = \vec{\mu}_s \cdot (\vec{v}/c^2) \times \nabla V(r) = \frac{e\hbar}{4m_e c^2} \vec{\sigma} \cdot (\vec{v} \times \nabla V(r)) \tag{1}$$

In this formula, e, m_e, c are the electronic power, the electron mass, and the speed of light, respectively. $\vec{\sigma}$ is Pauli matrix, and $\vec{v}$ is motion velocity of electrons. I analzsed the effect of the spin-orbit interaction on the spin transport of organic semiconductors, it found that the spin cycles would occur when the organic semiconductor carrier polaron moved along the polymer chain due to the spin-orbit interaction[9]. What is the difference between the spin-orbit interaction and the electron spin transport when we use the same model to describe the inorganic semiconductor? In this paper, we will discuss this problem.

2. Models and Formulas in the Organic Semiconductor

In the one-dimensional organic semiconductor with non-degenerate ground state, the Hamiltonian includes two parts, electronic partion and lattice partion[10-11]:

$$H = H_e + H_{lattice} \tag{2}$$

The electronic part is represented by the extended SSH model:

$$H_e = -\sum_{n,s}\left[t_0 - \alpha(u_{n+1} - u_n) - (-1)^n t_e\right]\left(e^{-i\gamma A} C^+_{n+1,s} C_{n,s} + e^{i\gamma A} C^+_{n,s} C_{n+1,s}\right) \tag{3}$$

in which t_0 is the transition integral of electrons in adjacent grid points n and $n+1$, α the electron-phonon coupling constant, u_n the position deviation of the n lattice point(CH group) , t_e degenerate parameter, and $\gamma = \frac{ea}{\hbar c}$. e,a,c the electronic power, the lattice constant and the speed of light respectively. Driving electric field $E_x(t) = -\partial_t A(t)/c$, and A is vector-potential, which is a scalar in one dimensional system.

The crystal lattice is represented in the classical manner:

$$H_{lattice} = \frac{K}{2}\sum_n (u_{n+1} - u_n)^2 + \frac{M}{2}\sum_n \dot{u}_n^2 \tag{4}$$

K is the elastic coefficient of the lattice, M represents the effective quality of a lattice point.

The spin-orbit interaction is introduced by the gate voltage perpendicular to the one-dimensional organic semiconductors. We assume that the one-dimensional organic semiconductor is along the x direction, the gate voltage points to the z direction, and the corresponding Hamiltonian is expressed as:

$$H_{SO} = i\beta\left(\sigma_x \frac{\partial}{\partial y} - \sigma_y \frac{\partial}{\partial x}\right) \tag{5}$$

in which β is the spin-orbit interaction strength, which is controlled by the gate voltage, and σ is the Pauli matrix. For the one-dimensional organic semiconductor along the x direction, only the second of the bracket is retained in equation (5). The one-dimensional tight binding model of spin-orbit interaction is obtained after the two quantization [12]:

$$H_{SO} = -t_{SO}\sum_n\left[C^+_{n+1,\uparrow}C_{n,\downarrow} - C^+_{n+1,\downarrow}C_{n,\uparrow} + C^+_{n,\downarrow}C_{n+1,\uparrow} - C^+_{n,\uparrow}C_{n+1,\downarrow}\right] \tag{6}$$

in which t_{SO} describes quantities of the Rashba spin-orbit coupling. $t_{SO} = \beta/2a$, a is a lattice constant, $\beta = |e|\hbar^2 E_{SO}/4m_e^2c^2$[13], E_{SO} denotes Rashba electric field, m_e is electron mass, and $C^+_{n,s}(C_{n,s})$ is the generation (disappearance) operator of an electron at a lattice point n.

From the equation (6), we can see that this type of spin-orbit interaction is closely related to the movement of carriers. Due to the existence of driving electric field, the spin-orbit interaction Hamiltonian is correspondingly changed into the following form:

$$H_{SO} = -t_{SO}\sum_n\left[e^{-i\gamma A}C^+_{n+1,\uparrow}C_{n,\downarrow} - e^{-i\gamma A}C^+_{n+1,\downarrow}C_{n,\uparrow} + e^{i\gamma A}C^+_{n,\downarrow}C_{n+1,\uparrow} - e^{i\gamma A}C^+_{n,\uparrow}C_{n+1,\downarrow}\right] \tag{7}$$

3. Effect of Spin-orbit Interaction on Electron Spin Transport in Inorganic Semiconductors

Contrary to the inorganic semiconductor, the organic one have a larger disorder[14-19], because of the strong electron-lattice interaction of organic materials, the structure is "soft". Any changes in the density of charge may cause restructure of system structure[20-22]. The injected electrons (or holes) are not in the form of an extended state like inorganic semiconductor, but form localized states such as soliton, polaron, etc. and at the same time, the local level is formed in the band gap, so the electron can not easily jump to the continuous conduction band. We know that the spin cycles will occur due to the spin-orbit interaction in the organic semiconductor. Based on the analysis of the charge occupation in the eigen energy of polaron, the phenomenon of spin precession in organic semiconductors occurs only in polaron energy, which is also called internal level precession[23]. The internal mechanism of spin precession in inorganic semiconductor is different from that of the organic semiconductor, the electronic spin precession of inorganic semiconductor happened in different energy levels of electrons.

In order to compare the effect of spin-orbit interaction in organic semiconductors, we use a similar model to describe the inorganic semiconductors: one dimensional inorganic chain along the x direction, and electrons can only move in this direction, Rashba electric field point to the z direction, which is used to control the strength of the spin-orbit interaction(t_{SO}). Driving electric field is along the -x direction, which is introduced through the index terms of the transition integral in the electronic Hamiltonian. In the model, the biggest difference between the organic and inorganic semiconductors lies in the "soft" structure. The lattice points of organic semiconductors are movable, while the lattice points of inorganic semiconductors are fixed. In inorganic semiconductors, thus, the Vibronic coupling constant $\alpha = 0$, and the whole chain is rigid. The Hamiltonian of the inorganic semiconductor only has an electronic part:

$$H_e = -\sum_{n,s} t_0 \left[e^{-i\gamma A} C^+_{n+1,s} C_{n,s} + e^{i\gamma A} C^+_{n,s} C_{n+1,s} \right] \tag{8}$$

The Hamiltonian form of the spin-orbit interaction has not changed.

We get the electron's energy eigen value by Fourier transformation:

$$E(k) = E_0(k) \pm 2t_{SO} \sin(ka - \gamma A) \tag{9}$$

Rashba spin-orbit interaction causes the energy band of the spin in the inorganic semiconductor to split in the $\vec{y}$ direction and the $-\vec{y}$ direction. The split is different from that caused by the Zeeman effect, because it does not eliminate spin merger.

Due to the splitting of the energy level, the electrons of different wave vectors (k_1 and k_2) have the same energy, i.e. $E^+(k_1) = E^-(k_2)$. k_1 corresponds to spin intrinsic state $\begin{pmatrix} 1 \\ i \end{pmatrix}$ in the $\vec{y}$ direction. k_2 corresponds to spin intrinsic state $\begin{pmatrix} 1 \\ -i \end{pmatrix}$ in the $-\vec{y}$ direction. An electron pointing to the $\vec{z}$ direction (Spin intrinsic state is $\begin{pmatrix} 1 \\ 0 \end{pmatrix}$) gets into to one dimensional inorganic semiconductor, and the electron will move in the $\vec{x}$ direction due to the driving electric field. Due to the Rashba spin orbit interaction, an effective magnetic field is generated in the $\vec{y}$ direction, and the electronic spin will rotate around the $\vec{y}$ axis.

The electron wave function is expressed as

$$\psi = \begin{pmatrix} 1 \\ i \end{pmatrix} e^{ik_1 L} + \begin{pmatrix} 1 \\ -i \end{pmatrix} e^{ik_2 L} \tag{10}$$

in which, $\vec{k}_1, \vec{k}_2$ are in the $\vec{x}$ direction, L is the distance of the electronic has gone through.

The projection of electron spin in the $\vec{z}$ direction:

$$S_z = \left| \langle (1,0) | \psi \rangle \right|^2 - \left| \langle (0,1 | \psi \rangle) \right|^2 = 4\cos(k_2 - k_1)L \tag{11}$$

Owing to $E(k) = E_0(k) \pm 2t_{SO} \sin(ka - \gamma A)$, $E_0(k) = -2t_0(\cos ka - \gamma A)$

When $E^+(k_1) = E^-(k_2)$, we get: $k_2 - k_1 = \frac{2}{a} \arctan \frac{t_{SO}}{t_0}$,

So $S_z = 4\cos(k_2 - k_1)L = 4\cos\frac{2L}{a}\arctan\frac{t_{SO}}{t_0}$.

We see that in one dimensional inorganic semiconductors, electrons move along the $\vec{x}$ direction, Affected by the Rashba spin-orbit interaction, the moving track of the projection of the electron spin in the $\vec{z}$ direction satisfies the cosine law.

After the spin precesses in a cycle, i.e. $S_z = 1$, the distance that electrons have moved in this cycle:

$$L = \frac{C}{\arctan\frac{t_{SO}}{t_0}} \quad \text{(C is constant)} \tag{12}$$

The bigger t_{SO} is, the smaller L will be, i.e. the faster the electron spin precesses, the smaller spin precession cycle will be, and vice versa. We can conclude that the electron spin precession cycle is inversely proportional to the strength of the Rashba spin orbit interaction.

4. Conclusion

After studying one dimensional organic semiconductor and one dimensional inorganic semiconductor, we find that the internal mechanism of spin precession in inorganic semiconductor is different from that of the organic semiconductor, while the effect of Rashba spin-orbit interaction on the spin evolution in the charge motion is same,i.e. the moving track of the projection of the electron spin in the $\vec{z}$ direction meets the cosine law, the spin precession period is inversely proportional to the spin orbit interaction strength.

References

[1] Hichri A and Jaziri S 2004 *Physica.* **350** 305
[2] Liu W W 2015 *Physical Experiment of College*. **6** 15
[3] Peng Y H and Huang T T 2011 *Journal of XiangNan University*. **2** 64
[4] Bychkov Y A and Rashba E I 1984 *JETP Lett.* **39** 78
[5] Dresselhaus G 1955 *Phys. Rev.* **100** 580
[6] Yang J,Dai B F and Li X 2011 *College physics*. **8** 65
[7] Chen G P 2015 *Acta Physica Sinica.* **3** 203
[8] Sun Q F, Guo H and Wang J 2004 *Phys. Rev. B.* **69** 054409
[9] Li H,Qu L S, Zhou M Y and Liang X R 2012 One dimensional spin precession of Polaron in organic polymer *College Physics*. **31** 5
[10] Brazovskii S A and Kirova N N 1981 *Sov. Phys. JETP.* **33** 4
[11] Su W P, Schrieffer J R and Heeger A J 1979 *Phys. Rev. Lett.* **42** 1698
[12] Sun Q F 2005 *Phys. Rev. B.* **71** 165310
[13] Hyun C L 2005 *Phys. Rev. B.* **72** 245338
[14] Mastour N, Mejatty H and Bouchriha H 2015 *Superlattice and Microstructure*.**82** 461
[15] Saidi H, Ridene S and Bouchriha H 2015 *Int.J. Mod. Phys. B.* **29** 1550054
[16] Ridene R, Mastour N, Gammra D and Bouchriha H 2015 *Int.J. Mod. Phys. B.***29** 1550211
[17] Bai W F,Wu Y Q,Xin W and Eerdun C L 2014 *Acta Physica Sinica.***17** 329
[18] Mamat J, Dulat S and Hekim M 2015 *Journal of Xinjiang University*.**2** 68
[19] Gong S J and Duan C G 2015 *Acta Physica Sinica*.**18** 107
[20] Sun X, Fu R L, Yonemitsu K and Nasu K 2000 *Phys. Rev. Lett.* **84** 2830
[21] Zhao H X, Zhao H and Chen Y G 2015 *Journal of Nanchang Hangkong University*. **4** 31
[22] Zhang X M, Shigemi M, Takahide K,Ma Q L, Mikihiko O, Hiroshi N, Yasuo A,Terunobu M and Shakin V A 2013 *Nature communications*.**50** 4306

[23] Lei J 2009 Dynamics of spin injection and transport in organic polymer *JiNan: Shandong University press*. 46-48

Effect of silanecoupling agent on properties of biocomposites based on poly(lactic acid)and durian rind cellulose

P Penjumras[1,5,6], R AbdulRahman[1,2,3,6], R A Talib[1] and K Abdan[4]

[1]Department of Process and Food Engineering, Faculty of Engineering,Universiti Putra Malaysia, 43400 UPM Serdang, Selangor, Malaysia
[2]Department of Food Technology, Faculty of Food Science and Technology, Universiti Putra Malaysia, 43400 UPM Serdang, Selangor, Malaysia
[3] Halal Products Research Institute, Universiti Putra Malaysia UPM Serdang, Selangor, Malaysia;
[4]Department of Biological and Agricultural Engineering, Faculty of Engineering, Universiti Putra Malaysia, 43400 UPM Serdang, Selangor, Malaysia
[5]Department of Food Science and Technology, Maejo University-Phrae Campus, 54140 Phrae, Thailand

E-mail: russly@upm.edu.my;p_atp@hotmail.com

Abstract. Durian rind cellulose reinforced poly(lactic acid) (PLA) biocomposites were prepared using Brabender internal mixer followed by hot compression molding technique. Cellulose was previously treated by 3-aminopropyltriethoxysilane for improving the compatibility with PLA matrix. The silane-grafting of cellulose was confirmed via Fourier transform infrared spectroscopy (FTIR) with the presence of Si-O-Si, Si-C, and Si-O-C bonds. The silane-treated cellulose was subsequently introduced into PLA matrix, and the effects of cellulose surface modification on mechanical, thermal and morphological properties, and water absorption of biocomposites were studied. It was found that silane-treated cellulose reinforced biocompositeshave superior mechanical properties compared with untreated cellulose reinforced biocomposites. The lowest crystallization temperature of silane-treated biocomposites was confirmed via Differential scanning calorimetry (DSC). Scanning electron microscopy (SEM) investigation also showed that adhesion of cellulose and PLA matrix was improved by modification of cellulosesurfaceusing3-aminopropyltriethoxysilanewhich can result in less water absorption into biocomposites.

1. Introduction

Biocomposite material is defined as the combination of natural filler and amatrix. A matrix is obtained from petroleum-based polymer, such as polyethylene, polypropylene, and biopolymer, such as polylactic acid (PLA) and polyhydroxyalkanoate (PHA) [1, 2]. However, biocomposite materials made from petroleum-based polymers cannot break down completely; only the natural filler will decompose whereas the matrix remains and affects the environment negatively. Therefore, the development of a fully biodegradable polymer using natural fiber as a reinforcement or filler in biopolymer has been attracting extra attention home and abroad in recent years [3]. Compared with conventional reinforcement materials produced from inorganic fiber, natural fiber has various advantages, including low density, durability and comparable strength, and most importantly, renewability and biodegradability [4]. However, high costs due to low volume production affect the widespread usage of these biocomposite materials for the manufacturing scale. Thus, they are suitable for the packaging industry [5, 6] and minor strength requirement applications including short-life-cycle products and indoor used products for example, device using in household, food packaging and disposable utensils [2, 7].

In the current study, the matrix phase was poly (lactic acid), or PLA, and the filler phase was cellulose from durian rinds. The combination of the two components is aimed at developing a composite suitable for light and short-term packaging applications. PLA is considered as a fully biodegradable polymer for commercial use. It is made from the fermentation of renewable agricultural raw materials to produce lactic acid. It has high performance in mechanical properties and good processability compared with other biodegradable materials [8,9]. Moreover, its packaging performance characteristics are similar to polyethylene terephthalate (PET) [10, 11]. Therefore, it has become one of the most promising biodegradable polymers to replace conventional synthetic polymers for industrial plastic applications [12].

Durian fruit called as "king of fruits" has been well-known and most consumed in Southeast Asia region. . However, two-thirds of the fruit; the seeds and peels are not edible making large-scale of biological wastes [13, 14]. In common practice, the wastes are incinerated or dumped in landfills without awareness of their negative impact [15]. In long-term, the conversion of these wastes into added value product should be an alternative way. Durian rinds have been found to be a good source of cellulose [16]. Meanwhile, the study of using cellulose from durian rinds as the reinforcing material in PLA has not been explored deeply.

Although natural fibers provide varying advantages for biocomposites, the using of natural fiber as a reinforced material is also restricted. Amiscibility gap due to incompatibility between surface of the fiber and the matrix becomes a main bottleneck [8, 17]. The fiber performs high hydrophilic behavior owing to its rich in polar of hydroxyl and also other polar groups while the matrix is, generally, non-polar and demonstrates hydrophobicity [18]. Thus, the surface adhesion of two components is generally somewhat poor [17, 19, 20]. This weakness affects the reduction of stress transfer ability from the matrix phase to filler phase and resulting decrease of strength [21]. Therefore, an enhancement of the compatibility of natural fillers and biodegradable polymers is needed.

Various developments of the adherence between the cellulose and PLA surfaces have been done. The surface modification of cellulose via esterification or graft copolymerization onto cellulose substrates, and the use of compatibilizers, such as maleated polylactide and isocyanate, has been studied [8, 22].

To enhance the property of fiber reinforced composites, a commercially coupling agent named silane is widely used for fiber treating [17]. Generally speaking, coupling agent represents two functional groups. The available hydroxyl (-OH) groups on the surface of natural fiber react with the former one by hydrogen bonding whereas the latter improves mechanical properties by covalent bonding between the fiber phase and the matrix phase [14, 17]. Hydrolization of ethoxy group from 3-aminopropyltriethoxysilane (APS) in water and solvent produces a silanol. Then, the stable covalent bonding is formed by reaction between silanol and -OH group on the surface of fiber, as shown in figure 1 [22, 23]. On the contrary, COO-sites on the hydrolyzed PLA structure form hydrogen bonding with amino groups were given by APS [22, 23, 24]. The aim of the present research is to investigate the effects of surface modification of cellulose derived from durian rinds on the mechanical and thermal properties, morphological feature, and water absorption behavior of biocomposites.

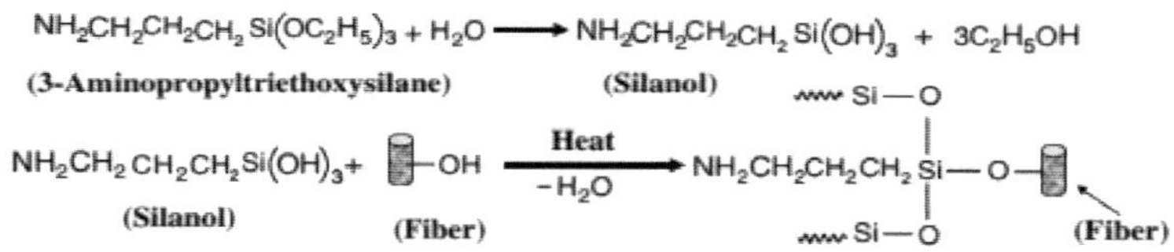

Figure 1. The chemical bonding of the silanol and the natural fiber [10, 22].

2. Materials and methods

2.1. Materials and chemicals

Durian rinds (cultivar D159 Monthong) were collected from Phatthalung Province, Thailand. Poly(lactic acid) (PLA) resin (Ingeo biopolymer 2003Dfood packaging-grade with >98% purity; specific gravity of 1.24 and 6 g/10 min melting flow index at 210 °C) was purchased from Natureworks (USA). Reagent-grade of glacial acetic acid (CH_3COOH) and sodium hydroxide (NaOH), and technical-grade of 80% purity sodium chlorite ($NaClO_2$) and 99% purity 3-aminopropyltriethoxysilane (APS) were supplied by Fisher Chemicals Sdn Bhd. (Malaysia).

2.2. Extraction of cellulose from durian rinds

The durian rinds were first sun-dried and then ground. Two-step process according to [14, 25, 26] with slight modifications was used to produce cellulose. First, bleaching was used to remove lignin. Twenty grams of the sample was rinsed with tap water for preliminary cleaning. The sample was subsequently soaked in 640 ml of hot distilled water contained in a 1,000-ml beaker. The sample was then immersed in a water bat hat 70 °C. Then, 8 g of $NaClO_2$ and 4 ml of CH_3COOH were added into the sample. At an hour interval, these two chemicals were continuously added to the beaker for 5 h until the color of sample became white which indicated that lignin was completely removed from the sample. The bleached sample is known as holo cellulose. Finally, holo cellulose was then filtered, and tap water was used to remove a residue chlorine dioxide and yellow color. Mercerization is the second step in which cellulose was derived from converting of holo cellulose using alkali solution. Alkali solution, 80 ml of 17.5 % w/v NaOH was poured into holo cellulose. A glass rod was used to stir the mixture and was then left for 5 min at ambient temperature. Every 5 min, another 40 ml of 17.5% w/v NaOH was added for three times and left for another 30 min. Next, the sample was mixed with 240 ml of distilled and left 1h before filtering. Then, 800 ml of 8.3% w/v NaOH was added and left for 5 min. The sample was then rinsed with water. The adding 120 ml of 10% v/v acetic acid for 5 min purposed to neutralize sample. Next, the sample was filtered and residue acetic acid was removed using distilled water. The sample was then dried to remove moisture in oven for overnight at 80°C. Finally, the cellulose was ground using a grinder and then passed through a sieve. The ground cellulose was collected with size between sieves 250μm and 125 μm and kept in an airtight box at room temperature.

2.3. Preparation of silane- treated cellulose

The cellulose was treated with silane coupling agent according to [22] with slight modification. First, the solution was prepared. Distilled water was mixed with ethanol with a ratio of 40:60 w/w for 600 ml. Next, 5 wt.% of APS (compared with the weight of cellulose) was hydrolysed in water-ethanol solution. The solution was adjusted pH to between 3.5 and 4 using an acetic acid and was then stirred 1 h. The APS solution was poured onto10 g of dried cellulose and left for 3 h. Then, the treated cellulose was filtered and dried in ambient condition for 3 days. Finally, residue moisture was then expelled using an oven at 80°C for 12 h.

2.4. Preparation of biocomposites

An internal mixer (Brabender Plastograph EC, Germany) at 165 °C was used to prepare a mixture of PLA-cellulose with ratio of 65:35 w/w. The blending was processed for 15 min using screw speed of 50 rpm. Before further step, the biocomposites were oven dried to avoid pores' formation at 80 °C for 16 h. Then, the molded sheets were then compressed at 160 °C using a hot press machine as shown in figure 2. For tensile testing specimens with 1-mm thickness, the process was conducted at 5 min of preheating, 5 min of compressing, and 2 min of cooling.

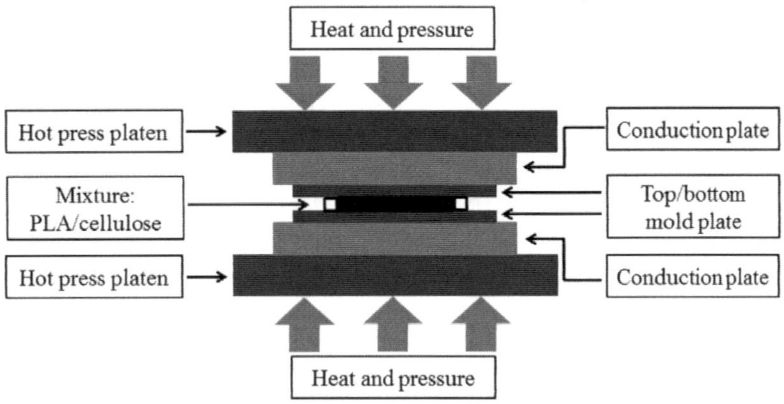

Figure 2. Schematic representation of the hot pressing machine.

2.5. Characterization

Fourier transform infrared spectroscopy (FTIR). The functional groups of untreated and APS-treated cellulose were identified via the FTIR Perkin Elmer, Spectrum One FT-IR Spectrometer (USA).The test was carried out in the wavenumber range of 4000 to 650 cm^{-1} with the transmittance mode at resolution of 4 cm^{-1}. All spectra were recorded.

Tensile measurement. Tensile properties were measured according to ASTM 1882L using an instron universal testing machine Model 5566 (USA).Three important properties; tensile strength, modulus of Young and elongation at break were recorded. The rectangular specimen with63 mm length, 12.7 mm width, and 1 mm thickness was prepared. The test was run at a cross speed of 5 mm/min. At least five replications of each treatment were tested.

*Differential scanning calorimetry (DSC).*Thermal properties were analyzed via the DSC model Perkin Elmer-DSC7 (USA).Ten mg of specimen was packed in aluminum cup and sealed. The sealed sample was then placed into heating chamber of the DSC. The DSC was heated with 5 °C/min from 30 to 200 °C. Nitrogen gas was flushed with flow rate 10 ml/min.

Morphological characterization. Characterization of the cracked surfaces of biocomposites was carried out using the SEM model S-3400N, Hitachi (Japan).All surfaces were then scanned at 15 kV after the first coated with a thin layer of gold in a vacuum.

Water absorption behaviour. Five specimens of each treatment with 30 mm length, 12.7 mm width, and 3 mm thickness were prepared. The rectangular specimens were then oven dried over night at 80°Cand were then cooled in a desiccator. The weight and the thickness of each specimen were recorded. The specimens were soaked in distilled water at ambient temperature. At 24 h intervals, each specimen was then removed from water. The tissue paper was used to wipe off the surface water and subsequently weighted by electronic balance to a precision of 0.0001 g. The measurement was terminated at 50 days. The percentage of water uptake (Mt) was computed as formulation (1) [27].

$$Mt\ (\%) = \frac{(Wt-Wo)}{Wo} \times 100\% \tag{1}$$

Where W_t is the weight of specimen at the given test time and W_o is the weight of the specimen after oven dried. All data from five replications were averaged [28].

The diffusion of water was performed theoretically by Fick's law as formulation (2) [28].

$$\log\left(\frac{Mt}{M\alpha}\right) = \log(k) + nlog(t) \tag{2}$$

Where M_t is the percentage of water uptake at given test time and M_α is the percentage of water uptake at the equilibrium point. The coefficient k and n were computed from slope of curve and intercept of the log plot of (M_t/M_α) against time (h) [29].

Typically, at early stages $M_t/M_\alpha \leq 0.5$, the diffusion phenomenon was demonstrated as formulation (3)[29].

$$\frac{Mt}{M\alpha} = 4\left(\frac{Dt}{\pi L^2}\right)^{1/2} \tag{3}$$

Where L is the specimen thickness and D is the diffusion coefficient.

3. Results and Discussion

3.1. Chemical modification of durian rind-derived cellulose

The untreated and treated cellulose show a considerable difference in IR spectra as seen in figure 3.Both samples presented the hydrogen bonded -OH stretching at 4000-3200cm^{-1}, the -CH stretching at 3000-2850 cm^{-1}, and the -OH bending at 1638 cm^{-1} attributed to adsorbed water [19, 30]. These bands related to cellulose characteristic [19, 30]. However, the result was found that the peak related to –OH group of APS-treated cellulose was lower in intensity since available-OH groups from cellulose bonded with APS. The comparison of obtained spectra gave a clear confirmation of the effectiveness of surfaces modification using APS. The amine groups were found in the APS-treated cellulose at the peak of 1562cm^{-1} due to hydroxyl groups of both silanol and cellulose being bonded with these amine groups from APS [12, 19, 31]. After cellulose was modified with APS, the highest peak at 1030 cm^{-1} was observed which indicated the Si-O-Si or Si-O-cellulose was investigated [32].These linkages are generally present in the 1200- 1000cm^{-1} region [6, 12, 19, 33]. The peak showed more apparent because of the Si-O-Si bond lapped with the C-O stretching from cellulose structure [17]. The additional peaks at 852 cm^{-1} and 802 cm^{-1} were observed after the modification of cellulose using APS. These peaks are indicated to the Si-O-C bond [6] and Si-C stretching [8] respectively. In a comparison of intensity of peak at 1638cm^{-1}, which is related to absorbed water [19], it was found that the intensity decreased after APS treatment. This result concluded that APS grafting onto the surface of cellulose is advantageous as APS is completely combined with the hydroxyl group of cellulose and resulting in reducing water uptake into cellulose.

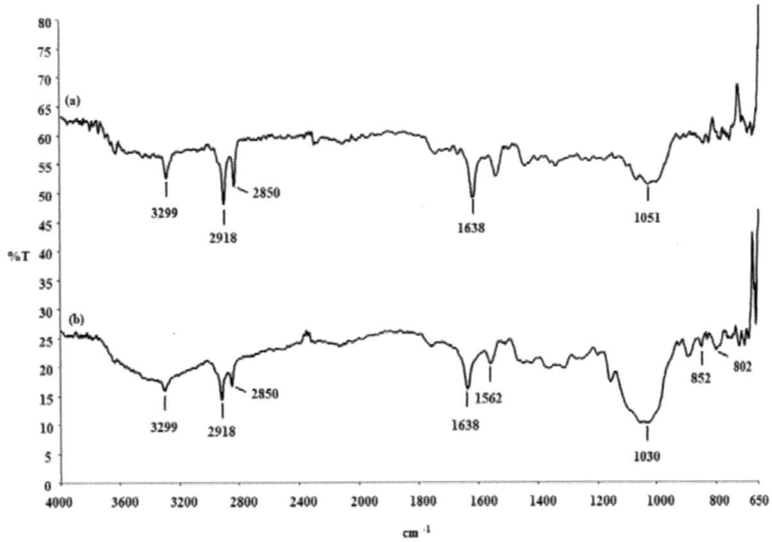

Figure 3.FTIR spectra of (a) untreated cellulose and (b) APS-treated cellulose.

3.2. Mechanical properties of biocomposites

In our previous research, we found that the optimal condition considering the mechanical properties was found to be at cellulose loading, mixing temperature, and mixing time of 35 wt%, 165 °C, and 15 mins with a desirability of 94.16%. At this optimal condition, the tensile strength was found to be 46.207 MPa, which was lower than PLA [34] because of the difference of the surface between cellulose and PLA matrix [17, 19, 20]. To improve mechanical properties of biocomposites, this research used 3-aminopropyltriethoxysilane (APS) to modify the surface of cellulose. The results are presented in table 1.

Table 1. Tensile properties of PLA and biocomposites.

Sample	Tensile strength (MPa)	Tensile strength changes (%)	Modulus of Young (GPa)	Modulus of Young change (%)	Elongation at break (%)	Elongation at break changes (%)
PLA	52.42±1.05		3.61±0.05		3.65±0.37	
PLA-UC	47.04±2.02	-10.26	4.84±0.43	+25.41	2.58±0.40	-29.32
PLA-TC	55.37±0.94	+5.63	6.40±0.22	+77.29	2.44±0.19	-33.15

*UC means untreated cellulose and TC means APS-treated cellulose.
Changes of properties were compared with neat PLA.

The neat PLA had tensile strength of 52.42 MPa. With the addition of 35 wt% untreated cellulose, it was shown that tensile strength decreased to 47.04 MPa. The biocomposites reinforced with 35 wt.% treated cellulose exhibited an increase of tensile strength to 55.37 MPa, or 5.63%, compared with neat PLA, which indicated that the treatment led to a better load transfer and homogenization of biocomposites microstructure, providing for better continuity and a stronger interface between the matrix and filler. The similar results in PLA/flax composites were found by Le Moigne et al. [33].

Modulus of Young for PLA is 3.61 GPa. The Modulus of Young of biocomposites reinforced increased to 4.84 GPa and 6.40 GPa, respectively. Way et al. [35] have suggested that the modulus of Young of biocomposites increased due to the modulus of fiber being much higher than PLA. The use of 3-aminopropyltriethoxysilane (APS) provides the hydrophobicity of the cellulose fiber surface by the presences of a high polar amine group. Its polar amine group toward the fiber surface forms strong hydrogen bonds, and thus, the configuration leaves the hydrophobic methylene group exposed at the surface [17]. Therefore, adhesion between the hydrophobic matrix surface and hydrophilic fiber surface can be improved via APS treatment. The improved adhesion reduces the deformation of elastic zone of matrix, then resulting in increase of the modulus of Young [31, 35].

Comparison with neat PLA, the elongation at break reduced with continuous adding cellulose. However, the results present that elongation at break remained unaffected by silane treatment. Some researchers have obtained the similar results; composites based on PLA and flax [36], PLA and jute fiber [37], and PLA reinforced with lysine-diisocyanate treated by bamboo fiber [38]. Therefore, two main factors affecting the elongation at break are homogenization of fiber phase and matrix phase [37] and fiber dispersion ability [38]. Basically, the elongation at break of material related to the strain property. Therefore, the decrease of elongation at break by adding of cellulose indicated that the strain of biocomposites was lower than neat PLA [35]. Generally, the incorporation of higher amount of fiber makes fiber agglomeration and then restricts the mobility of macromolecules resulting in lower resistance to breaking [35, 39]. Therefore, the elongation at break of fiber reinforced PLA biocomposites decreases.

3.3. Thermal Properties

The DSC curves of PLA and both biocomposites (PLA-untreated cellulose and PLA-treated cellulose) are shown in figure 4. The detailed data; the glass transition temperature (T_g), crystallization temperature (T_c), melting temperature (T_m), enthalpy of fusion (ΔH_m), enthalpy of crystallization (ΔH_c) and degree of crystallinity (X_c) are summarized in table 2. T_g was reported as the temperature at the midpoint while T_c and T_m were reported as the maximum peak.

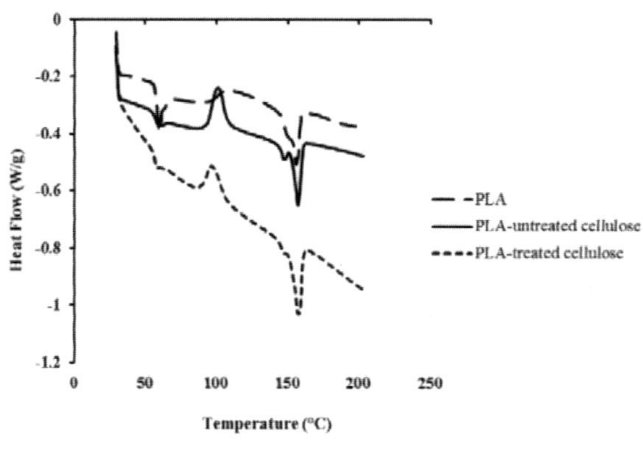

Figure 4. DSC curves of PLA and biocomposites.

The T_g and T_m of neat PLA were 56.96° C and 156.26 °C, respectively. The addition of 35 wt.% untreated and treated cellulose affecting the T_g decreased slightly to 56.20 °C and 55.97 °C, respectively, whereas the T_m increased slightly to 157.59 °C and 158.14 °C, respectively. The similar results were demonstrated by González et al. [24] and Lee et al. [40]. They found that adding kenaf fiber reduced the T_g and increased the T_m with comparison with neat PLA.

Table 2.DSC results of PLA and biocomposites.

Specimens	T_g (°C)	T_c (°C)	T_m (°C)	ΔH_m (J/g)	ΔH_c (J/g)	X_c (%)
PLA	56.96	109.74	156.26	18.01	14.63	15.73
PLA-UC	56.20	101.24	157.59	20.31	20.86	22.43
PLA-TC	55.97	96.75	158.14	23.34	15.94	17.14

*UC means untreated cellulose and TC means APS-treated cellulose.

The PLA had a T_c of 109.74 °C. The T_c of untreated and treated cellulose reinforce PLA biocomposites decreased to 101.24 °C and 96.75 °C, respectively. Based on infinite size, PLA crystal has melting enthalpy of 93 J/g, therefore $\Delta H_c/93$ was used to calculate X_c[41].The both biocomposites showed higher X_c than neat PLA as seen in table 2. The X_c of PLA was 15.73% whereas the X_c of untreated and treated cellulose reinforced PLA biocomposites increased to 22.43% and 17.14%, respectively. The decrease in T_c and increase in crystallinity confirmed the previous observations related to the role of cellulose as nucleating agents [2, 40, 42]. In terms of the crystallization behavior, the results showed significant effects of APS treatment on cellulose as a reinforcing material. The DSC curves demonstrated that both biocomposites showed dual melting peaks. This phenomenon may be attributed to lamellae rearrangement, resulting in imperfect crystal formation on melting endotherms of PLA [35]. Generally, the polymer is always melted to a slightly above T_g. Therefore, the drop in T_g could be an advantage for packaging industry that involves at lower temperature for sheet thermoforming to avoid thermal degradation [6].

3.4. Fracture morphologies of biocomposites

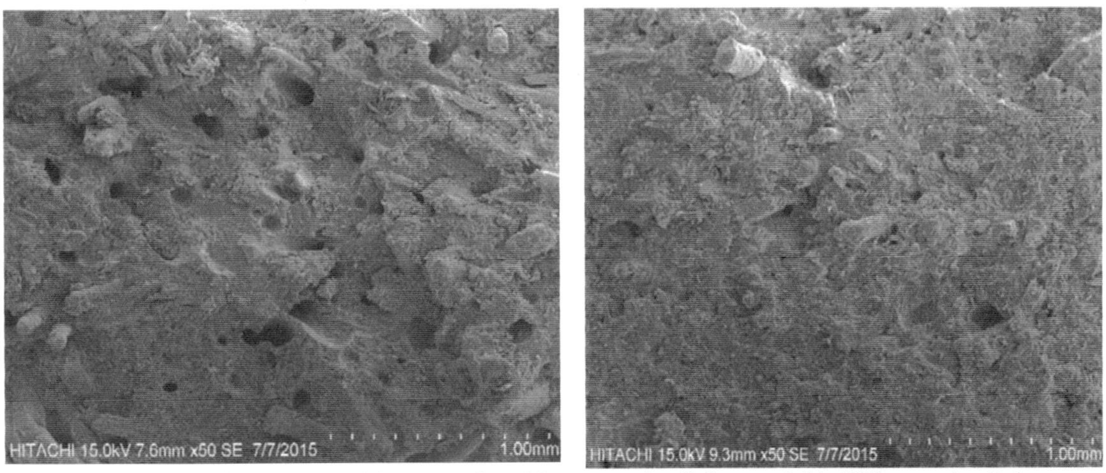

Figure 5. SEM of cellulose- PLA biocomposites (a) untreated cellulose and (b) APS-treated cellulose.

The coupling agent is needed to improve the surface of cellulosic filler and enhance chemical reaction with the polymer matrix. Moreover, it was found that coupling agent provides well-disperse of the natural fiber in the matrix phase[40]. Figure 5 shows the scanned fractured surfaces of biocomposites by SEM. Many holes were observed after the cellulose was pulled out from the PLA matrix as shown in figure 5(a). Meanwhile, APS-treated cellulose-PLA biocomposites showing better adhesion exist at interphase as

APS-treated cellulose still embedded in the matrix as seen in figure 5(b). Besides, fewer cavities present on the surface and the pull out of filler from PLA matrix decreased. Similar results were obtained by other authors; silane-modified bamboo fiber and natural rubber composites [39], maleic anhydride (MA) grafted PLA and wood composited [43].

3.5. Water absorption behavior

The major disadvantage of the natural fiber reinforced materials is their hygroscopic behavior which is extremely responsive to water, affecting their mechanical properties which in return makes them perform poorly [17, 26]. Figure 6 represents the percentage of water uptake againstsoaking time.The result demonstrated that the water uptake by PLA was less than 1% with the maximum of 0.52%. Both biocomposites absorbed water rapidly in the first 15 days of soaking. After that, the curves showed a gradual increase and finally reached an equilibrium point, where the weight of biocomposites remained constant. However, treatment of the 35 wt.%. biocomposites with APS displayed reduced water absorption compared with the untreated cellulose biocomposites. The maximum water absorptions of untreated and APS-treated cellulose-PLA biocomposites were 5.91% and 4.05%, respectively.

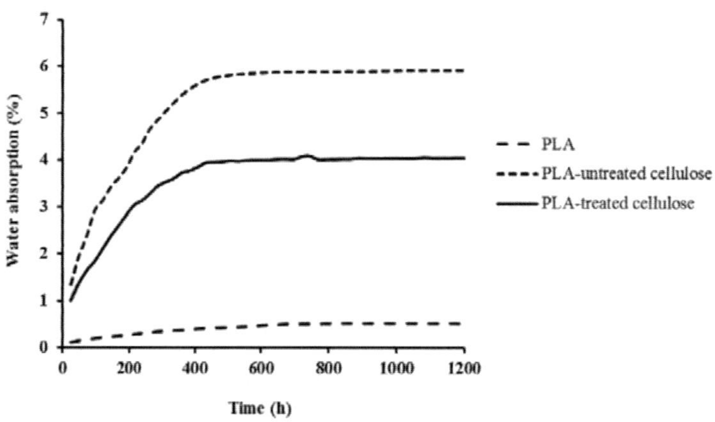

Figure 6. Water absorption curves of PLA and biocomposites.

Similar results have been observed by other authors who found that water was absorbed by PLA up to 1% [26, 38, 44]. Meanwhile, water absorbed by biocomposites increased with the addition of cellulose and their values were very high compared with pure PLA, due to the number of hydroxyl group in the biocomposites increasing[25, 26]. Our previous paper found that after mercerization with an alkali solution, durian cellulose demonstrated more –OH groups than untreated durian rind [14]. The alkali treatment increased the accessibility of free hydroxyl groups, resulting in their better hydrophilic characteristics [26, 45]. It then became the cause for water uptake in biocomposites affecting their physical, mechanical and thermal properties. Therefore, moisture uptake due to their hygroscopic property is the important factor when using natural fiber as filler in composite system. It is important that this problem be addressed so that natural fiber may be considered as a viable reinforcement in composite materials [46].

Silane is known as a commercial coupling agent available in large-scale within a large number of functional groups, such as the amine group forming hydrogen bonding to COO-site on PLA structure [22, 24]. Additionally, alkosilane groups react to hydroxyl groups in cellulosic material. Thus, adding treated cellulose with a coupling agent can reduce water uptake of biocomposites. Such results include the lysine-based diisocyanate modified bamboo fiber as filler in composite based on poly (lactic acid) (PLA) and poly (butylene succinate) (PBS)[38], 3-glycidoxypropyl trimethoxy silane in biocomposites of kenaf fibers in polylactide [40], and silane treated oil palm – derived fiber reinforced PLA and polycaprolactone (PCL) composites [47].

Figure 7 presents the diffusion curves fitting of PLA and biocomposites. The coefficient k and n were computed as formulation (2) and were summarized in table 3.

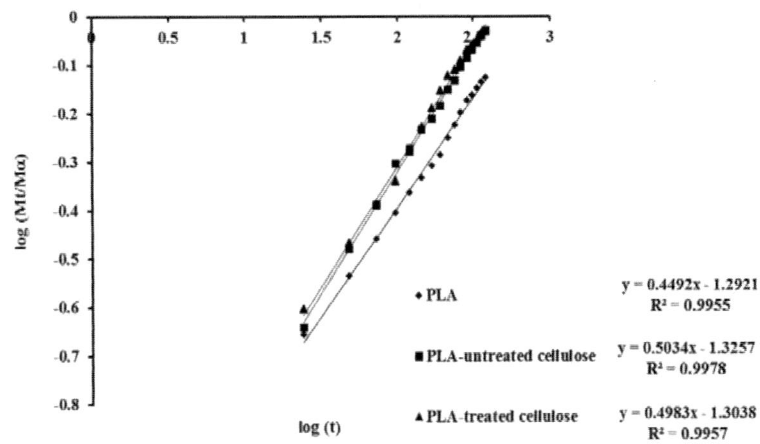

Figure 7. Diffusion curves fitting of PLA and biocomposites.

In general, three known mechanisms of diffusion behavior are Fickian diffusion with n=0.5, relaxation controlled with n > 0.5, and non-Fickian or anomalouswith0.5 < n < 1[29, 48, 49]. All formulations had n-values less than 0.5. Thus, the diffusion behaviors of all formulations corresponded to the Fickian model. The diffusion coefficient is the most significant parameter of the Fickian model. This value indicates the penetration efficacy of water into the composite structures [28].

Table 3. Water absorption behavior parameters of PLA and biocomposites.

Sample	n	k (h^2)	Maximum water absorption (%)	Water diffusion coefficient; D ($\times10^{-13}$ m^2s^{-1})
PLA	0.4492	0.0510	0.52	7.0269
PLA-UC	0.5034	0.0472	5.91	12.6607
PLA-TC	0.4983	0.0497	4.05	10.6790

*UC means untreated cellulose and TC means APS-treated cellulose.

Although many advantages of natural fiber have been reported; low density, innocuousness, high stiffness, renewable and biodegradable [4], however, both biocomposites displayed disadvantages of higher water diffusion coefficient than neat PLA. The immiscibility due to its poor interaction of the natural fiber and the matrix phase is a major limitation [17]. This limitation makes water penetrate into the gaps at the interfaces [50, 51]. However, the results also demonstrated that the biocomposites prepared from APS-treated cellulose had a lower water diffusion coefficient compared with untreated cellulose biocomposites. The similar result was also found in modified empty fruit bund oil palm fiber reinforced polypropylene (PP) biocomposites [53] Thusly, it can be concluded that the gaps due to its poor interaction between cellulose and PLA were successfully narrowed using the silane coupling agent.

4. Conclusion

Cellulose from durian rinds was successfully modified with3-aminopropyltriethoxysilane (APS). The surface modification was confirmed via FTIR. From the above experimental results and analysis, it could be realized that the tensile strength and the modulus of Young of cellulose reinforced PLA biocomposites were improved. The improvement was attributed to better interfacial between filler and matrix as given morphological feature leading to lower water uptake into biocomposites. In addition, it could be summarized that cellulose showed the role of nucleating agent in PLA to generate crystallization with lower temperature as investigated in DSC curves.

Acknowledgement

The authors are sincerely grateful for financial aid from the German Academic Exchange Service (DAAD) and Southeast Asian Regional Center for Graduate Study and Research in Agriculture (SEARCA). In addition, we acknowledge Department of Process and Food Engineering, Faculty of Engineering and Institute of Tropical Forestry and Forest Products, Universiti Putra Malaysia for laboratory equipment and all technical staff for their support.

References

[1] John M Jand Thomas S 2008 A review *Carbohyd. Polym.* **71** 343
[2] Chun K S, Husseinsyah S and Osman H 2012 *J. Polym. Res.* **19-9859** 1
[3] Ave´rous L and Le Digabel F 2006 *Carbohyd. Polym.* **66** 480
[4] Santiagoo R, Ismail H and Hussin K 2011 *Biores.* 63714
[5] Mohantya A K, Misraa M and Hinrichsen G 2000 An overview *Macromol. Mater. Eng.* **276-277** 1
[6] Tee Y B, Talib R A, Abdan K, Chin, N L, Basha R K and Md Yunos K F 2013 *Biores.* **8** 4468
[7] Graupner N 2008 *J Mater. Sci.* **43** 5222
[8] Lu T, Liu S, Jiang M, Xu X, Wang Y, Wang Z, Gou J, Hui D and Zhou Z 2014 *Compos. Part B Eng.* **62** 191
[9] Suryanegara L, Nakagaito A N and Yano H 2009 *Compos. Sci. Technol.* **69** 1187
[10] Johansson C, Bras J, Mondragon J, Nechita P, Plackett D, Šimon P, Svetec D G, Virtanem S, Baschetti M C, Breen C, Clegg F and Aucejo S 2012 *Biores.* 72506
[11] Ahmed J, Varshney S Kand Auras R 2010 *J. Food Sci.* **75** N17
[12] Frone AN, Berlioz S, Chailon J-F, Panaitescu DM and Donescu D 2011 *Polym. Compos.* **32** 976
[13] Amid B T and Mirhosseini H 2012 *Food. Chem.* **132** 1258
[14] Penjumras P, Abdul Rahman R, Talib R A and Abdan K 2014 *Agric. Agric. Sci. Procedia* **2** 237
[15] Foo K Y and HameedB H2011*Biomass Bioenerg.* **35** 2470
[16] Rachtanapun P, Luangkamin S, Tanprasert K and SuriyatemR2012 *Food Sci. Technol.* **48** 52
[17] Abdelmouleh M, Boufi S, Belgacem M N and Dufresne A 2007 *Compos. Sci. Technol.* **67** 1627
[18] Suarez J C M, Coutinho F M B and Sydenstricker T H 2003 *Polym. Test.* **22** 819
[19] Lu J, Askeland P and Drzal L T2008 *Polymer* **49** 1285
[20] Xie Y, Hill C A S, Xiaoa Z, Militz H, and MaiC2010 *Compos Part A Appl. S.* **41** 806
[21] Kim H-S, Lee B-H, Lee S, Kim H-J and Dorgan J R 2010 *J. Therm. Anal. Calorim.* **104** 331
[22] Huda M S, Drzal L T, Mohanty, A K and Misra M 2008 *Compos. Sci. Technol.* **68** 424

[23] Huda M S, Drzal L T, Mohanty, A K and Misra M 2008 *Compos. Interface.* **15** 169
[24] González D, Santos V and Parajo´ J C 2011 *J. Thermoplas. Compos.* **25** 1005
[25] Tawakkal I S M A, Talib R A, Abdan K and Chin N L 2010 *Asian J. Chem.* **22** 6652
[26] Tawakkal I S M A, Talib R A, Abdan K and Chin N L 2012 *Biores.* **7** 1643
[27] Alomayria T, Assaedia H, Shaikhc F U A and Lowa I M 2014 *J. Asian Ceram. Soc.* **2** 223
[28] Kord B, Danesh M A, Veysi R and Shams M 2011 *Biores.* **6** 4190
[29] Alamri H and Low I M 2012 *Polym. Test.* **31** 620
[30] Zhao Y, Qiu J, Feng H and Zhang M 2012 *J. Appl. Polym. Sci.***125** 3211
[31] Ifuku S and Yano H 2015 *Int. J. Biol. Macromol.* **74** 428
[32] Qu P, Gao Y, Wu G-F and Zhang L-P 2010 *Biores.* **5** 1811
[33] Le Moigne N, Longerey M, Taulemesse J M, Bénézet J C and Bergeret A 2014 *Ind. Crop. Prod.* **52** 481
[34] Penjumras P, Abdul Rahman R, Talib R Aand Abdan K 2015 *Sci. World J.* **2015-293609**1
[35] Way C, Wu D-Y, Cram D, Dean Kand PalomboE 2013 *J.Polym. Environ.* **21** 54
[36] Oksman K, Skrifvars Mand Selin J F 2003 *Compos. Sci. Technol.***63**1317
[37] Plackett D, Andersen T L, Pedersen W B and Nielsen L 2003 *Compos. Sci. Technol.* **63** 1287
[38] Lee S-Hand Wang S 2006 *Compos. Part A Appl. S.* **37** 80
[39] Ismail H, Shuhelmy Sand Edynm M R 2002 *Eur. Polym. J.* **38** 39
[40] Lee B-H, Kim H-S, Lee S, Kim H J and Dorgan J R 2009 *Compos. Sci. Technol.* **69** 2573
[41] Cao X, Mohamed A, Gordon S H, Willett J Land Sessa D J2003*Thermochim.Acta* **406** 115
[42] FroneA N, Berlioz S,Chailan J Fand Panaitescu D M 2013 *Carbohyd. Polym.* **91** 377
[43] Csikós Á, Faludi G, Domján A, Renner K, Móczó J and Pukánszky B 2015*Eur. Polym. J.* **68** 592
[44] Finkenstadt V L, Liu C-K, Evangelista R, Liu L S, Cermak S C, Hojilla-Evangelista M and Willet J L 2007 *Ind. Crop. Prod.* **26** 36
[45] Vilaseca F, Mendez J A, Pèlach A, Llop M, Cañigueral N, Gironès J, Turon Xand Mutjé P 2007 *Proc. Biochem.* **42** 329
[46] Herrera-Franco PJ and Valadez-Gonza´lez A 2005 *Compos. Part B Eng.* **36** 597
[47] Eng C C, Ibrahim N A, Zainuddin N, Ariffin A and Yunus W M Z W 2014 *Sci. World J.* **2014- 21318** 1
[48] Ghasemi I and Kord B 2009 *Iran. Polym. J.* **18** 683
[49] Kord B 2011 *J. Thermoplast. Compos.* **26** 296
[50] Vilakati G D, Mishra A K, Mishra S B, Mamba B B and Thwala J M 2012 *Polym. Eng. Sci.* **52** 760
[51] Dhakal H N, Zhang Z Y and Richardson M O W 2007 *Compos. Sci. Technol.* **67** 1674
[52] Kittikorn T, Strömberg E, Ek M and Karlsson S 2013 *Biores.* **8** 2998

3D Numerical study on the hollow profile polymer extrusion forming based on the gas-assisted technique

Z Ren[1,2,3], X Y Huang[2] and H S Liu[2]

[1]Key Laboratory of Optic-Electronic and Communication, Jiangxi Science and Technology Normal University, Nanchang, 330038, China
[2]School of Mechanical and Electrical Engineering, Nanchang University, Nanchang, 330031, China

E-mail: renzhong0921@163.com

Abstract. In this study, gas-assisted extrusion method was introduced into the extrusion of the hollow profiles. To validate the feasibility of the new extrusion method, 3D numerical simulation of the hollow profiles based on gas-assisted technique was carried out by using the finite element method. The Phan-Thien-Tanner (PTT) mode was selected as the construction equation. In the simulations, the physical field distributions of four different extrusion modes were obtained and analyzed. Results showed that the extrudate effect of traditional no gas-assisted mode was poor because the extrudate swell phenomenon is obvious and the physical field values are larger. For the gas-assisted of the inner wall, the extrudate swell of the melt was more obvious than that of the traditional no gas-assisted mode on account of the no-slip boundary condition on the outer wall. For the gas-assisted of the outer wall, the dimple effect of the inner wall is more obvious owing to the no-slip boundary condition on the inner wall. However, the extrusion effect of the double walls gas-assisted mode is very good because of the full-slip effect on the both walls.

1. Introduction

During the process of using plastic products, the hollow profiles [1, 2] are usually used in many fields, the engineering fields in particular, for example, architectural decoration, doorframe, etc. In general, the hollow profiles are made from the process of polymer melt extrusion. Due to the high viscosity of the melt, the high shear rate, shear stress and tensile stress in the die channel, the hollow profiles are easily distorted, such as the extrudate swell [3], the great deformation [4], and the melt fracture [5]. Up to now, the extrusion forming of the hollow profiles have been studied by some researchers, for example, Mu et al. [6, 7] used the finite element method and the differential multimode Phan-Thien and Tanner (mPTT) constitutive model to study the extrudate swell behavior of the hollow profiles. Zhang et al.[8] employed the finite element method to reversely predict and design the geometric structure of the die for the hollow profiles. Alfaro et al. [9] made use of the natural element method to study the extrusion behavior of the hollow profiles. However, the studies on the L-typed hollow profiles are very few. In this paper, to ascertain the mechanism of the extrusion deformation for the hollow profiles, the extrudate swells of the L-typed hollow profiles were studied through the finite element method. Moreover, to overcome the extrudate deformation of the L-typed hollow profiles, the gas-assisted extrusion technique was introduced and applied to the L-typed hollow profiles. To validate the feasibility of the gas-assisted extrusion technique, the numerical studies of four different extrusion modes were analyzed and compared with each others from the physical field distributions, such as velocity, pressure, shear rate, and normal stress.

2. Numerical model and method

2.1. Geometric model

The geometric model and the finite element mesh of the L-typed hollow profiles are shown in Figure 1(a) and (b), respectively. In Figure 1(a), the face of $ABCDEF - A'B'C'D'E'F'$ is the inlet face of the die, the face of $HIJKLM - H'I'J'K'L'M'$ is the outlet of the die, the face of $NRSTUV - N'R'S'T'U'V'$ is the exit of the melt. The length of the inside die is about 30 mm, the length of the outside die is also about 30 mm, the high of AF is 30 mm, the wide of EF is 20 mm, the thickness of the hollow profile is 3 mm, and the width of the hollow part is 4 mm.

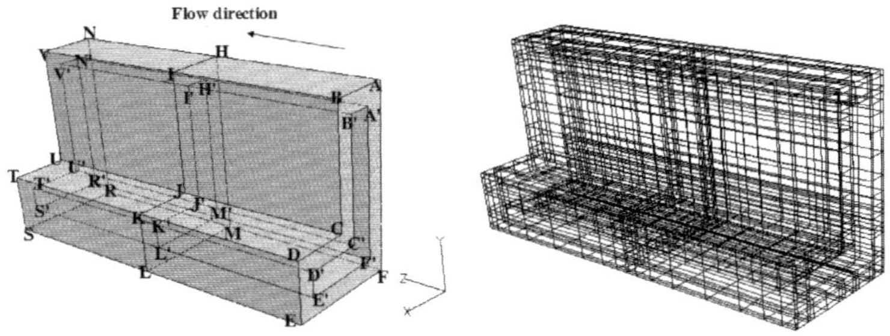

(a) geometric model (b) finite element mesh

Figure 1. Geometric model and mesh of the L-typed hollow profiles

2.2. Governing equations

In the numerical simulations, the polymer melt is regarded as the non-compressible, iso-thermal, steady, laminar and viscoelastic Non-Newtonian fluid. The gas is considered as the compressible, iso-thermal, steady, laminar and Newtonian fluid. Both the inertia, the gravity force, and the viscous dissipation effects and the diffuse effect of the gas enters into the melt all could be neglected.
The mass and the momentum equations are as follows:

$$\nabla \cdot v = 0 \tag{1}$$

$$-\nabla p + \nabla \cdot \tau = 0 \tag{2}$$

Where ∇ is Hamilton operator, v is the velocity of melt, p is the pressure drop, τ is the extra stress tension.

In this paper, PTT constitutive model [10] was used because it not only described the characteristic of the shear-thinning, but also represented the elastic properties of the viscoelastic fluid. PTT constitutive equation is as follows:

$$\exp\left[\frac{\varepsilon\lambda}{(1-\eta_r)\eta} tr(\tau_1)\right]\tau_1 + \lambda\left[\left(1-\frac{\xi}{2}\right)\overset{\nabla}{\tau_1} + \frac{\xi}{2}\overset{\Delta}{\tau_1}\right] = 2(1-\eta_r)\eta D \tag{3}$$

where $\eta_r = \eta_2 / \eta$ is the viscosity ratio, η is the total viscosity of the melt, η_2 is the Newtonian viscosity component of the melt, λ is the relaxation time, ε and ξ are the parameters of the melt correlated with the material tensile and the shear characteristics, respectively. $\overset{\nabla}{\tau_1}$ is the upper convected derivative of the extra stress tensor τ_1, η_1 is the Non-Newtonian component viscosity of the melt, and D is the strain-rate of the tensor.

In the simulations, for the PTT construction equation, its parameters can be seen from Table 1.

Table 1. The parameters of PTT construction equation

Parameters	η(Pa.s)	λ(s)	ε	ξ	η_r
Values	2700	0.2	0.18	0.23	0.12

2.3. Boundary conditions

- For the inlet face of the melt, the flow of melt is full-developed, steady and laminar flow, which is in accordance with the following relationship

$$\partial v_z / \partial z = 0, v_x = v_y = 0 \tag{4}$$

Where v_x, v_y and v_z are the flow velocities of melt at the direction of x, y and z direction, respectively.

- For the inner and outer wall of the mlet, no-slip wall boundary condition is suitable for the traditional no gas-assisted extrusion mode, which meets the following relationship,

$$v_n = v_s = 0 \tag{5}$$

But for the gas-assisted extrusion, the full-slip boundary condition is used meeting the following relationship,

$$v_s = 0, f_n = 0 \tag{6}$$

In this paper, the compressible air was used as the assisted gas in the gas-assisted polymer extrusion forming, and the temperature of gas should be heated to reach the temperature of the melt.

- For the free surface of the melt, zero normal stress and entangle velocity are imposed, the following equation should be expressed as following, i.e.,

$$f_n = 0, f_s = 0 \text{ and } v_n = 0 \tag{7}$$

- For the exit of melt, zero trance force and entangle velocity are imposed, i.e.,

$$f_n = 0, \ v_s = 0 \tag{8}$$

2.4. Numerical method

In this paper, the numerical simulations were carried out by using the finite element CFD software package POLYFLOW [11]. Because of the highly non-linear form of the differential viscoelastic constitutive equation used to describe material behavior, the position of the boundaries and free surface are unknown, and the position of the free surface becomes part of the solution, the nonlinear algebraic equation must be used to describe this class of problems, which result in the difficulties in solving the equations like the N-S equation. In order to solve these nonlinear equations, the coupled computing method of velocities-pressure was adopted. Since the total extra-stress tensor of differential constitutive equation was decomposed into the viscoelastic components and purely-viscous component, the elastic viscous split stress algorithm incorporating with the streamline upwind (EVSS-SU) algorithm [12] was used into the computing of the differential viscoelastic constitutive equations, and the momentum equations. And the implicit Euler discretization method [13] was incorporated into the integration of partial differential equations. In the interpolation scheme of finite element computing, the quadratic velocities, and linear pressure were used. For the free boundary and their moving-mesh problems, the coupled moving surface computing method and the Optimesh 3D remeshing method [14] were used. Moreover, to easily get the convergence solutions of the nonlinear equations, and the evolution schemes [11] of the material parameters, we also made use of boundary conditions and moving boundaries for instance, relaxation time of polymer melt, and the free boundary. In the convergence solving the nonlinear governing equations, the classical Newton-Raphson iterative algorithm [15] was used, the convergence condition is 1.0×10^{-4}.

3. Numerical results and analysis

3.1. X velocity distributions

To verify the feasibility of the gas-assisted extrusion method, the physical field distributions of four different extrusion modes are obtained, the X velocity distributions are shown in Figure 2.

From Figure 2(a), it is seen that the X velocity directions of two outer walls at the outlet of die are negative. The extrudate effect of traditional no gas-assisted extrusion method is poor, due to the extrudate swell effect, and the inner wall was adhered after the melt was extruded out from the die. From Figure 2(b), it is seen that the X positive velocities of two outer walls are very large, which result in the extrudate swell phenomenon being large. From Figure 2(c), it can be seen that the adhesion phenomenon is more obvious than that of the traditional no gas-assisted extrusion. However, as for the double gas-assisted extrusion mode, the extrudate effect is good, no extrudate swell and distortion phenomena occur on the extrudate because the velocity distribution of the melt at the outlet of the die is zero.

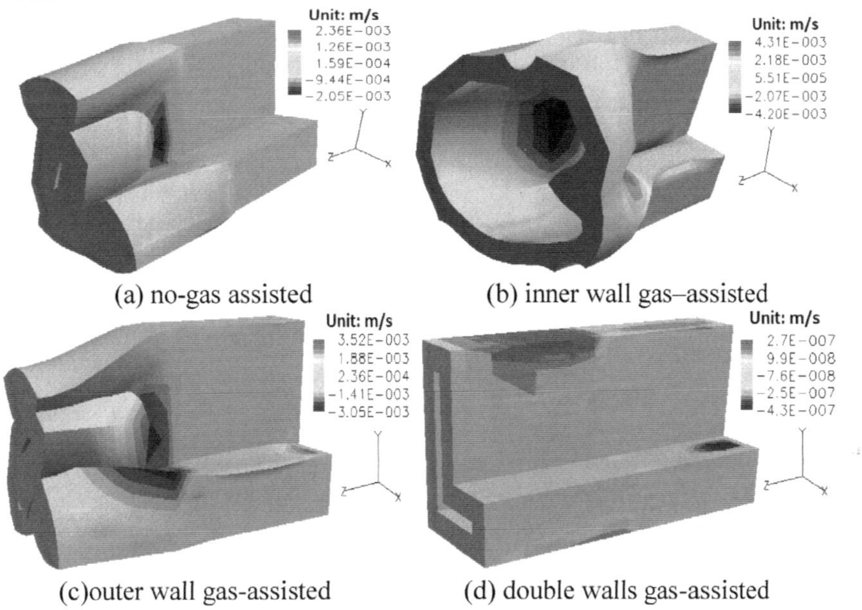

Figure 2. X velocity distributions of four different extrusion modes

3.2. Pressure distributions

From Figure 3, it is seen that the pressure gradient of traditional no gas-assisted extrusion mode is largest, which is the result of the no-slip wall boundary condition and the large flow resistance. But for the gas-assisted extrusion modes, it can be seen that the pressure gradients are decreased. In addition, for the double walls gas-assisted extrusion mode, the pressure gradient is very little, which leads to little extrusion deformation of hollow profiles.

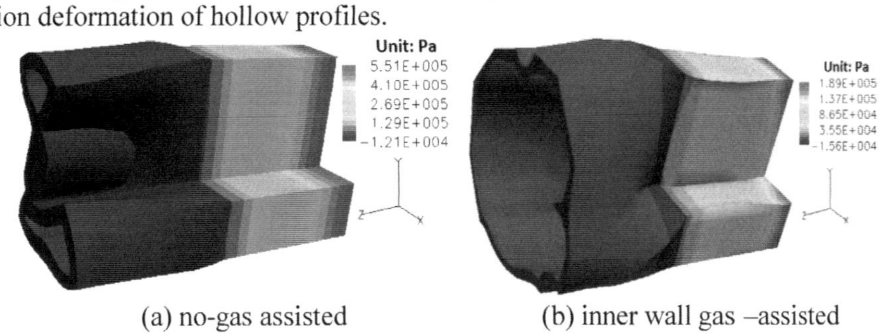

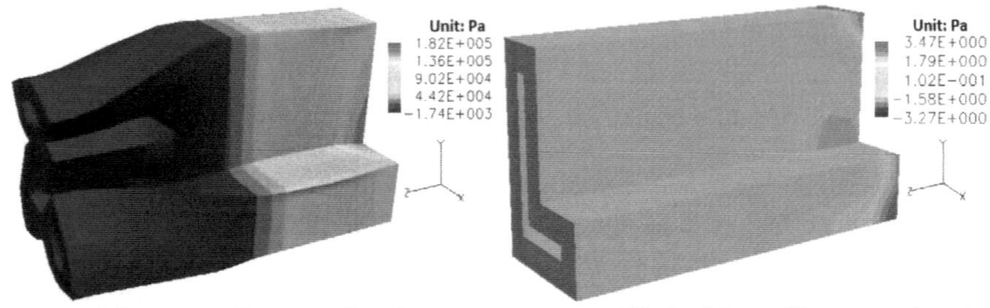

(c)outer wall gas-assisted (d) double walls gas-assisted

Figure 3. Pressure distributions of four different extrusion modes

3.3. Shear rate distributions

From Figure 4, it is seen that compared with others, the shear rate of no gas-assisted mode is the largest. As we all know, the larger both shear rate and velocity gradient are, the larger flow resistance is. While larger shear rate will result in larger shear stress on the wall of the die. But for the gas-assisted mode, the shear rates of melt are decreased, especially for the double walls gas-assisted mode.

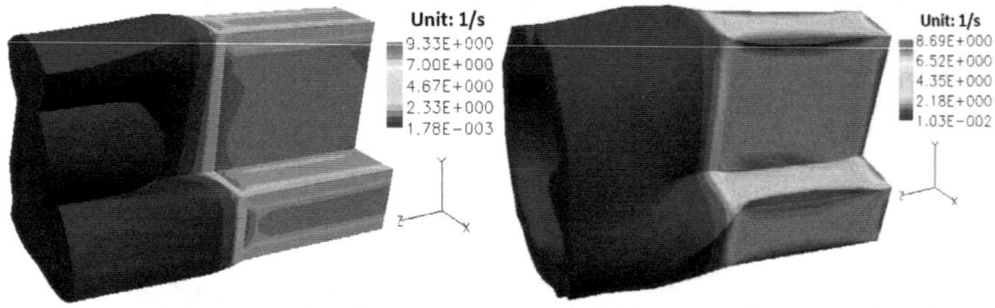

(a) no-gas assisted (b) inner wall gas –assisted

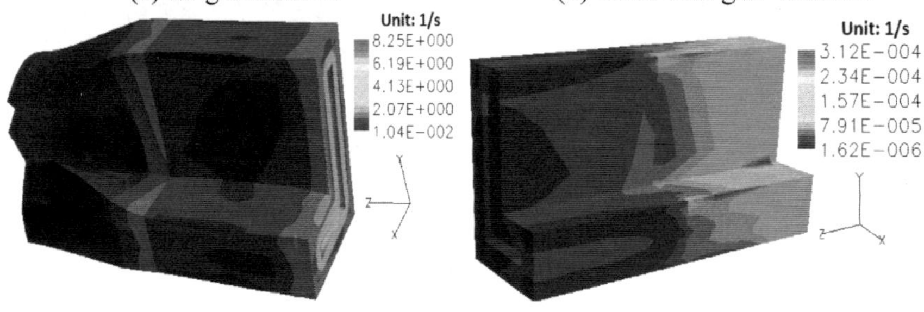

(c)outer wall gas-assisted (d) double walls gas-assisted

Figure 4. Shear rate distributions of four different extrusion modes

3.4. Normal stress distributions

As we know, the extrudate swell of the melt is related to the normal stress. To study the mechanism of overcoming the extrudate swell for gas-assisted extrusion mode, the first normal stress distributions of four different extrusion modes were obtained, which are shown in Figure 5 from which, it is seen that the first normal stress is large at the outlet of the die. However, for three kinds of gas-assisted extrusion mode, the first normal stress decreased. And for the double gas-assisted mode, the first normal stress is very little, which is so little that it can be neglected, which in return results in the elimination of the extrudate swell phenomenon for the hollow profiles.

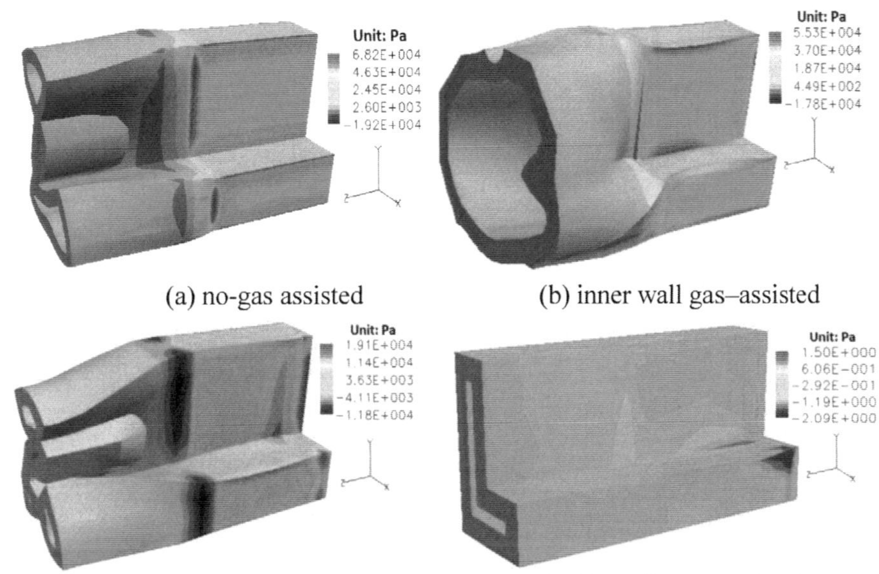

(a) no-gas assisted (b) inner wall gas–assisted

(c)outer wall gas-assisted (d)double walls gas-assisted

Figure 5. Normal stress distributions of four different extrusion modes

4. Conclusions

Extrudate swell is a common phenomenon in the processing of polymer melt, which also occurs in the hollow profiles extrusion. To overcome the extrudate swell of hollow profiles, the gas-assisted mode was introduced and analyzed by using the numerical simulation method in the study. And to validate the feasibility of gas-assisted mode, the physical field distributions of four different gas-assisted modes were obtained and compared with others. Research results indicated that the extrusion effect of the traditional no gas-assisted mode is poor, the extrudate swell is obvious, and the inner wall is easily adhered. For the gas-assisted in single wall, for example, for the inner wall gas-assisted mode, the extrudate swell phenomenon of the melt was more obvious than that of the traditional no-gas assisted mode. For the outer wall gas-assisted mode, the adhesion effect was more obvious. However, as for the double walls gas-assisted mode, the extrudate swell and the deformation are all greatly eliminated. Therefore, the double wall gas-assisted technique has a potential value in the process of the hollow profiles.

Acknowledgements

This work is supported by the Natural Science Foundation Project of Jiangxi Province (No. 20151BAB202011), the Science and Technology Pillar Program Project of Jiangxi Province (No. 20132BBG70103), Top-notch Talent Foundation Project of JXSTNU (No. 2014QNBJRC004), and 2011 Cooperative Innovation Center Project of Jiangxi Province (No. 2014XTCX005).

References

[1] Reggiani B, Segatori A, Donati L, et al. 2014 Comparison of bulge test vs. conical expansion test for hollow extruded profile characterization. *Key Eng. Mater.* **585** 111-119

[2] Sun Y D, Chen Q R, Sun W J 2015 Numerical simulation of extrusion process and die structure optimization for a complex magnesium doorframe. *Int. J. Adv. Manuf. Tech.* **80** 495-506

[3] Konaganti V K, Ansari M, Mitsoulis E, et al. 2015 Extrudate swell of a high-density polyethylene melt: II. Modeling using integral and differential constitutive equations. *J.Non-Newton. Fluid Mech.* **225** 94-105

[4] Shibakov V G, Pankratov D L, Andreev A P, et al. 2015 Prediction of samples failure during severe plastic deformation by multiple extrusion. *IOP Conf. Ser.: Mater. Sci. Eng.* **86** 12032-12036

[5] Kwon Y 2015 Melt fracture modeled as 2D elastic flow instability. *Rheol. Acta* **54** 445-453

[6] Mu Y, Zhao G, Wu X, Zhai J 2013 Finite-element simulation of polymer flow and extrudate swell through hollow profile extrusion die with the multimode differential viscoelastic model. *Adv. Polym.Tech.***32** E1-E19

[7] Mu Y, Zhao G Q, Zhang C R 2010 Numerical investigation of viscoelastic flow and swell behaviors of polymer melts in the hollow profile extrusion process. *Adv. Mater.s Res.* **97-101** 209-213

[8] Zhang M, Huang C Z, Jia Y X, Liu J L 2014The inverse prediction for profile extrusion die based on the finite element method. *Adv. Mater.s Res.* **941-944** 2332-35

[9] Alfaro I, Gagliardi F, Olivera J, et al. 2009 Simulation of the extrusion of hollow profiles by natural element methods. *Int. J.Mater. Form.* **2** 597-600

[10] Thien N P, Tanner R I 1977 A new constitutive equation derived from network theory. *J.Non-Newton. Fluid Mech.* **2** 353-365

[11] ANYSYS Inc.2012 ANYSYS POLYFLOW Version 14.5 User's Guide. Canonsburg, PA, USA

[12] Sun J, Phan-Thien N P, Tanner R I 1996 An adaptive viscoelastic stress splitting scheme and its applications: AVSS/SI and AVSS/SUPG. *J.Non-Newton. Fluid Mech.* **65** 75-91

[13] He Y N 2013 Euler implicit/explicit iterative scheme for the stationary Navier–Stokes equations. *Numer. Math.* **123** 67-96

[14] Su Y Y, Rwei S P, Wu L Y, Lin Y T, An T C, Lin W P, Chien S P, Lin L Y 2011 Shaping conjugated hollow fibers using a four-segmented arc spinneret. *Polym. Eng. Sci.* **51** 704–711

[15] Caddemi S, Martin J B 1991 Convergence of the Newton-Raphson algorithm in elastic-plastic incremental analysis. *Int. J. Numer. Meth. Eng.* 66 71-76

IOP Conf. Series: Materials Science and Engineering **137** (2016) 012008 doi:10.1088/1757-899X/137/1/012008

Adsorption of mercury in coal-fired power plants gypsum slurry on TiO_2/chitosan composite material

P Gao, B B Gao, J Q Gao, K Zhang, Y J Chen,Y P Yang[1] and H W Chen

School of Energy, Power and Mechanical Engineering, North China Electric Power University, Baoding 071000, China

E-mail: yyp@ncepu.edu.cn

Abstract. In this study, a simple method was used to prepare a chitosan adsorbent to mix with KI and TiO_2. Gravimetric analysis (TG), scanning electron microscopy (SEM) and Fourier transform infrared spectroscopy (FTIR) were used to characterize the samples before and after adsorption of Hg^{2+}. A mercury adsorption experiment was also conducted in the gypsum slurry. The results show that using hydrobromic acid as a solvent of adsorbent resulted in a better adsorption effect than using acetic acid alone. Also, the sample (CS-KI/TiO_2-HBr) had a maximum mercury adsorption capacity when the pH=5 and the t=50℃. The characterization experiments showed that the thermal stability of composite materials declined and the TiO_2 uniformly dispersed in the surface of the samples with a lamellar structure, generating a lot of cracks and recesses that increased the reactive sites. Furthermore, when the TiO_2 reacted with CS, it resulted in Ti-C, Ti-O and Ti-N bonds. The Br- can prevent the growth of TiO_2 crystal grains and strengthen the ability of I^- to remove mercury. The adsorption isotherm and kinetic results indicated that the adsorption behaviour of CS-KI/TiO_2-HBr as it removes Hg^{2+} is an inhomogeneous multilayer adsorption process. The surface adsorption and intraparticle diffusion effects are both important in the Hg^{2+} adsorption process.

1. Introduction

Mercury is a toxic heavy metal element that usually exists in the lithosphere, hydrosphere and atmosphere. It has been a long-term pollution problem because of its bio-accumulation and migration. Coal-fired power plants are the main source of anthropogenic mercury emissions. In the coal-fired flue gas particle state, elemental mercury (Hg^0) and oxidized mercury (Hg^{2+}) are the main forms of mercury. Wet desulfurization (WFGD), a technology that uses limestone or lime as a sorbent, has been found to be effective to simultaneously remove SO_2 and Hg^{2+} from the flue gas, which reduces mercury emission. Especially, the installation of denitration equipment can enhance the oxidation of Hg^0 to Hg^{2+}, but results in mercury enrichment in the gypsum slurry. However, this method also has many disadvantages. The enriched Hg^{2+} is converted to Hg^0 again by the reducing substances in gypsum slurry (such as sulfite ions), and this converted mercury is then re-released into the atmosphere [1]. The reuse of desulfurizing gypsum will release the mercury again. Any additional mercury that is in the wastewater from the desulfurization process is discharged into the environment, which will cause water pollution [2].

Studies have shown that the mercury in the desulfurization slurry mainly exists in the form of mercury sulfide (HgS) and mercury halide [3]. The desulfurization slurry undergoes a rise in temperature and pH, which promotes the transformation of Hg^{2+} to Hg^0 [4, 5]. The research of Raquel [1] shows that Fenton reagent can make sulfite ions into sulfate ions, and the addition of sodium sulfide (Na_2S) and 2,4,6 trimercaptotriazinetrisodium (TMT) can prevent the conversion of Hg^{2+}. Sun et al. [5, 6] compared the inhibitory effect of Na_2S, TMT and sodium dithiocarbamate (DTCR) for the reduction of Hg^{2+}, and found that DTCR has the best inhibition effect.

Polymer chitosan contains a lot of $-NH_2$ and -OH groups, and is a good adsorbent of heavy metal ions. Its amino and hydroxyl groups can act as chelating sites to effectively adsorb Hg^{2+}. However, chitosan has a low adsorption capacity and poor selectivity for heavy metal before it is modified. In this paper, a composite chitosan adsorbent was prepared by mixing KI and TiO_2 to realize adsorption of Hg^{2+}. Its adsorption for mercury in the gypsum liquid was tested. The objective of this work is to synthesize a novel sorbent that is suitable for mercury removal in flue gas after FGD in power plant. The determination of sorption isotherms and kinetics of Hg^{2+} sorption on modified CS sorbent is important to find the controlling steps in uptake kinetics and achieve a complete adsorption mechanism.

2. Materials and methods

2.1. Materials
Chitosan (CS) (degreeof deacetylation>85%), hydrobromic acid (HBr), 36% acetic acid (HAc), titanium dioxide (TiO_2), potassium iodide (KI), sodium hydroxide (NaOH), deionized water, gypsum.

2.2. Preparation of the adsorbents
20 ml 2% HBr (or HAc) was taken. 1 g CS was added. The mixture was stirred until it was solved; then, 10 ml deionized water (including 0.2g TiO_2 and 0.2g KI) was added. The mixture was magnetically stirred for 10 min in a 50℃ water bath pot to form sol solution. Next, the sol was added to a 10%wt NaOH solution drop wise. The particles were filtered and washed until neutral, then allowed to dry at 50 ℃. Then, the CS-KI/TiO_2-HBr (or CS-KI/TiO_2-HAc) gel particles were acquired.

2.3. Characterization of the adsorbents
The adsorption capacity of mercury was measured by a solid mercury analyzer (Lumex RA915+, Russia); Thermal stability analysis used a thermogravimetric analyzer (SDTQ-600, TA company in the United States); Microscopic morphology was observed by scanning electron microscopy (Dutch Phenom-Word company PhenomProX); Adsorbent structure was measured by a Fourier transform infrared spectrometer (American PerkinElmer Spectrum 100).

2.4. The experiment of mercury removal in the gypsum slurry
5g dry gypsum and 45 ml deionized water or 45 ml 1ppm mercury standard solution were used. They were configured to 10% gypsum slurry A and B and stirred for 10 min. It was determined that the pH of the solution A was 5. Then, 10% NaOH was added dropwise to the B solution, adjusting it until the pH was identical to solution A. 15 mg samples were placed into the A and B solution, and stirring was continued for 60 min in a 30℃ water bath pot for the adsorption experiment. The samples were dried at 50 ℃. Finally, the adsorption capacity of a single sample was tested by the mercury analyzer.

2.5. The experiments for mercury removal by CS-KI/TiO_2-HBr
The studies show that the adsorption isotherms of chitosan are mainly in line with the Langmuir model. The increase of temperature was helpful for the adsorption of Hg^{2+} within a certain temperature range, and the ability to adsorb Hg^{2+} was influenced by the Ph [7]. From the former study, CS-KI/TiO_2-HBr was found to have a better adsorption capacity for Hg^{2+} in the gypsum slurry. So, CS-KI/TiO_2-HBr was used as the adsorbent to study the sorption behavior of Hg^{2+}. The experimental conditions were: total solution 50 ml, Hg^{2+} concentration 1ppm, water bath temperature 30℃, adsorption time 1 hour and continuous stirring. The adsorption kinetics experiment was carried out at 303 K and pH 5.0 for 6h.

IOP Conf. Series: Materials Science and Engineering **137** (2016) 012008 doi:10.1088/1757-899X/137/1/012008

Adsorption isotherms could describe adsorbate distribution when the adsorption process reaches equilibrium. This is therefore effective to clarify the adsorption mechanism. Thus, the experiment was carried out at 303 K and pH 5.0, and the concentrations of Hg^{2+} were 0.5,1,3,5,10,20,40,60, and 80ppm, at a volume of 50ml.

3. Results and discussion

3.1. Mercury removal in the gypsum slurry

The test results are shown in Figure 1. As the figure shows, when the sample is less than 3 mg, especially in the low concentration mercury solution (A), the adsorption capacity is clearly influenced by the sample quality. Although the quality is lower, the adsorption capacity is enhanced. When the quality of the single sample is greater than 3 mg, there is not a significant change to the adsorption capacity. Also, the adsorption capacity of CS-KI/TiO_2-HBr is better than CS-KI/TiO_2-HAc, proving that the presence of Br^-can enhance the Hg^{2+} adsorption of CS.

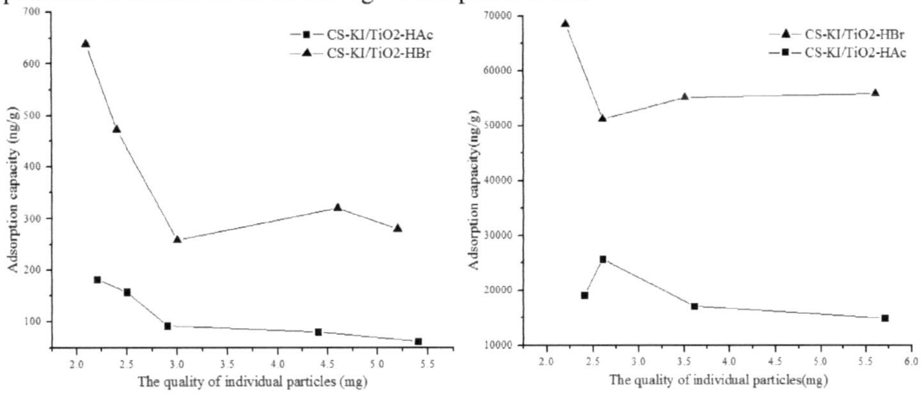

Figure 1. (a) The original gypsum slurry (b) Add 1ppm Hg^{2+} to gypsum slurry.

3.2. Effects of solution pH and reaction temperature

Figure 2 shows the experimental results under different pH conditions. As can be seen from the figure, CS-KI/TiO_2-HBr had the largest adsorption capacity when the pH was between 5 and 7. At the same time, when the pH was more than 7 or less than 5, it greatly affected the adsorption capacity. The reason maybe that H^+ makes the -NH_2 of the CS protonate, and then it loses reactive activity. Also, it is easy for the OH^- in the alkaline solution to react with Hg^{2+} to produce $Hg(OH)_2$, which then decomposes into HgO precipitation. Thus, the concentration of Hg^{2+} in the solution will decrease and affect the Hg^{2+} adsorption on Chitosan. The reaction equations are as follows:

$$Hg^{2+}+2OH^-=HgO\downarrow+H_2O \quad (1)$$

Figure 3 shows that the adsorption capacity of CS-KI/TiO_2-HBr increased with an increase in temperature when it was under 50℃; however, when the temperature was over 50℃, the adsorption capacity reduced sharply. This is consistent with the findings in other literature [7]. In contrast, it is clear from Figures 2 and 3 that the absorption capacity of CS-KI/TiO_2-HBr in the gypsum slurry was lower at 1/3 than in water. It may be that other metal ions or nonmetallic ions in the gypsum slurry occupy the active site of the adsorbent, thus reducing the adsorption of Hg^{2+}. Additionally, the insoluble particles in the gypsum slurry occupy the micropores of the adsorbent, which can reduce the adsorption capacity of the adsorbent.

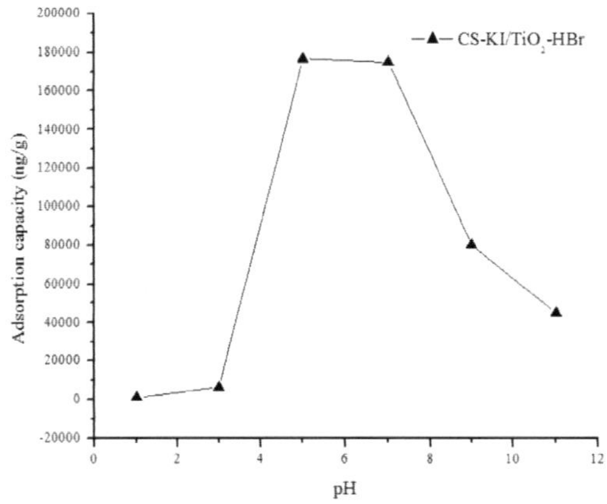

Figure 2. The mercury adsorption under different pH conditions.

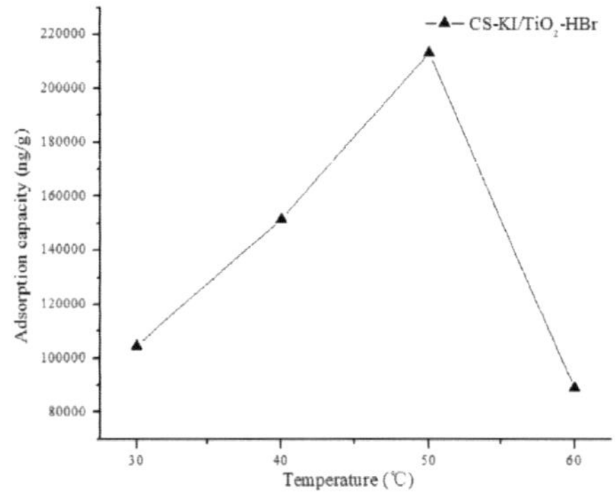

Figure 3. The mercury adsorption under different temperatures (PH=5).

3.3. *Heat stability experiment*

A 12 mg sample was placed into the thermogravimetric analyzer, and the temperature was increased from 30 ℃ to 800 ℃at a rate of 20 ℃/min, used N_2 as shielding gas. As shown in Figure 4, the thermal stability of the two samples decreased, and there was no obvious weightlessness peak of the physical adsorption of water. The decomposition temperature of CS-KI/TiO_2-HAc is between 248℃ and 331℃. The maximum weight loss rate was at 279℃. The thermal stability of CS-KI/TiO_2-HBr is better than that of CS-KI/TiO_2-HAc. Its decomposition temperature is from 235℃ to 355℃, and at 288℃, its decomposition rate was the fastest.

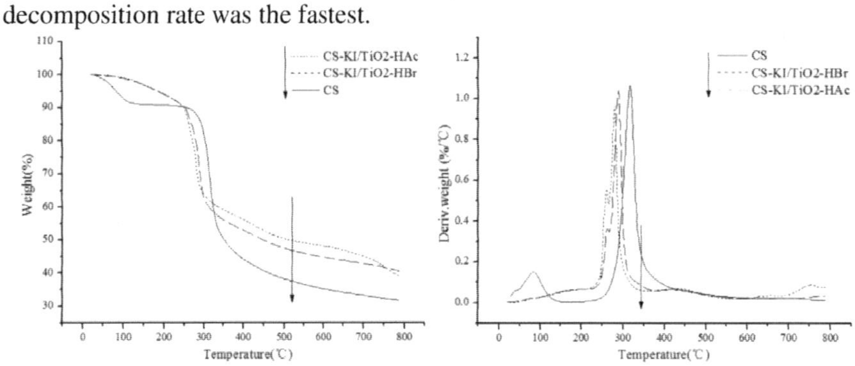

Figure 4. (a) TG of the adsorbents (b) DTG of the adsorbents.

The decrease in thermal stability was likely due to the CS taking a hydrolysis reaction by acid catalysed in the process of preparation. With the molecular chain of CS degradation, fracture, and shortening, the crystallinity and the thermal stability will decrease [8]. As compared to CS, the DTG diagrams of the composite adsorbents showed two adjacent weight loss peaks. The first weightless peak maybe formed by the pyrolysis of iodine. The presence of Br- reduced the pyrolysis of the iodine.

3.4. Scanning electron microscopy (SEM)

Figure 5 shows the microscopic morphology of the CS and the modified CS surface magnified about 20000 times by scanning electron microscopy. As Figure 5 (a) demonstrates, the fold and reticular formation are obvious on the surface of CS. As can be seen from Figures 5 (b) and (c), TiO_2 distributed well in the surface of the sample by the sheet structure, and the sample's surface formed a large number of cracks and grooves, which provide more contact area to absorb the mercury. Also, many KI, which are much smaller than TiO_2, joined together to provide a new reactive site of absorbing the mercury. A comparison of Figures 5 (b) and (c) demonstrates that the grain size of TiO_2 layer of CS-KI/TiO_2-HBr is smaller than the CS-KI/TiO_2-HAc. It can increase the adsorbent's surface area and promote adsorption of mercury, indicating that the presence of HBr can inhibit the growth of TiO_2 grains and that the use of HBr is better than HAc.

Figure 5. SEM of original (a) CS, (b) CS-KI/TiO_2-HAc, (c) CS-KI/TiO_2-HBr.

Figure 6 shows the scanning electron micrographs of Hg^{2+} loaded adsorbents. Agglomerate and spherical particles were loaded on the adsorbents' surface to reduce the porous structure. This was especially true of the CS-HAc (as Figure 6(d) shows); it has no co-ions, but has amultihole surface. The white grain of $HgCl_2$ is full of the porous surface of CS-HAc. Figures 6(e) and 6(f) are the SEM of CS sorbents which are modified by different acids and supplemented by KI and TiO_2. Though the existence of TiO_2 and KI influences observation of mercury adsorption, it can be clearly seen that many tiny white grains are well distributed on the surface of the sorbents. The cracks and grooves of the sorbents are full of white tiny spherical particles. This can be explained by inhomogeneous multilayer adsorption on the surface of the sorbent.

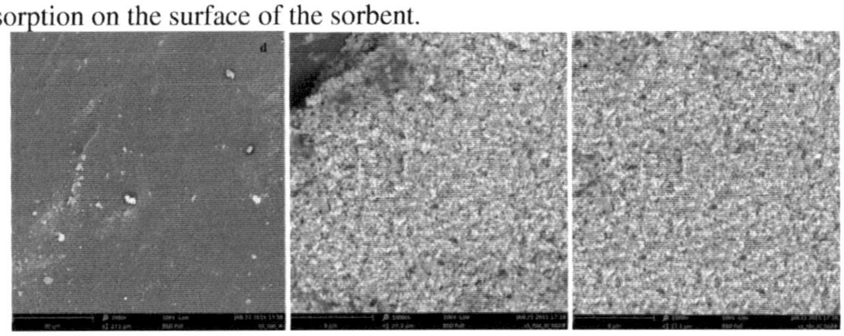

Figure 6. SEM of Hg^{2+} loaded adsorbents (d) CS-HAc, (e) CS- KI/TiO_2-HAc, (f) CS-KI/TiO_2-HBr.

3.5. Fourier transform infrared spectroscopy(FTIR)

The infrared spectrum can be used to analyze molecular structure and chemical bonding. The infrared spectrum can reflect the change of the samples after modification, and can be used to analyze the effective functional groups of adsorbents. As shown in Figure 7, after modification, the O-H stretching vibration peak of the CS was at 3373cm^{-1}, the C-O stretching vibration peak of the CS was at 1084cm^{-}

IOP Conf. Series: Materials Science and Engineering **137** (2016) 012008 doi:10.1088/1757-899X/137/1/012008

^{1}, and the C-H stretching vibration peak of CS was at 2918 cm^{-1} [9, 10]. The O-H, C-O and C-H stretching vibration peak of CS all have undergone place migration. The peaks became sharp, perhaps because the addition of TiO_2 produced Ti-O bonds or Ti-C bonds. Also, the amide Ⅰ band at 1661 cm-1 migrated to low wave about 5cm^{-1}, which perhaps formed the Ti-N bonds. Besides a new peak appeared at 1450 cm^{-1} in the modified samples. It is possible that TiO_2 and the CS formed the Ti-O bonds. At the same time, the C-O vibration peak at 1084 cm^{-1} of C_3 moved to 1071 cm^{-1} and the peak intensity greatly increased, which is a typical TiO_2 characteristic peak. The modified samples also produced double I^- peaks, which replaced H+ at 900 cm^{-1}. All the samples formed C-Br absorption peaks near 667cm^{-1} because the KBr had a reaction with the sample during the process of making the tablet.

The sorbent structure underwent significant changes, as shown by the contrast between Figures 7 and 8. The large band around 3410 cm^{-1} is commonly attributed to the combination of stretching vibrations of OH and NH_2 groups that shift to higher wave numbers after adsorption of Hg^{2+}, especially CS-KI/TiO_2-HBr. The peaks at around 1661$cm^{-1}$and 1076 cm^{-1} are assigned to –NH_2 and –C–OH [11, 12], which all shift to lower wave numbers after adsorbing Hg^{2+}. Because of the chelating effects, new chemical bonds are formed between CS and Hg^{2+} [13-16]. The bands at 2922 and 2818cm^{-1} are attributed to C-Ti disappearing in CS-KI/TiO_2-HBr. This indicates that C-Ti is covered by Hg^{2+}. The band at 1071 cm^{-1}, which is a typical TiO_2 characteristic peak, is weak in CS-KI/TiO_2-HBr. Combined with Figure 6(f), it leads to the conclusion that the TiO_2 surface maybe covered with a large number of mercury compounds like HgI_2 or $HgBr_2$. The chemical reaction mechanism is as follows:

$$Hg^{2+} + 2Br^- \rightarrow HgBr_2 \tag{2}$$

$$Hg^{2+} + 2I^- \rightarrow HgI_2 \tag{3}$$

The Hg^{2+} adsorption on the composite sorbent is inhomogeneous multilayer adsorption. The surface adsorption and intraparticle diffusion effects in the Hg^{2+} adsorption process are both important.

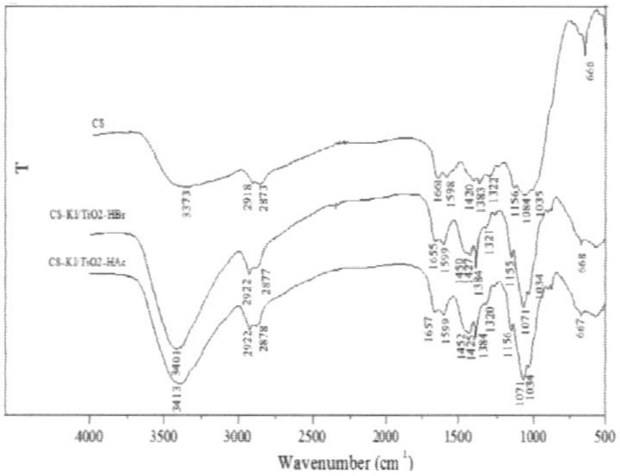

Figure 7. FTIR of the adsorbents.

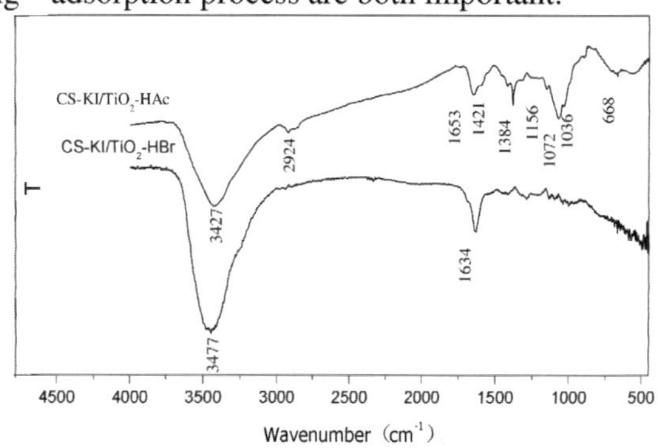

Figure 8. FTIR of the adsorbents after adsorption Hg^{2+}.

3.6. Adsorption equilibrium study

3.6.1. Adsorption kinetics. Adsorption kinetics is a fundamental and significant tool to evaluate adsorption behaviour [9, 10, 17, 18]. Typical kinetic adsorption results are shown in Figure 9.

From above, the metal ions are chelated onto the adsorbents during the adsorption process. On the basis of the pseudo-first order, pseudo-second order and intraparticle diffusion kinetic models, the kinetic data simulation was taken to further study the adsorption mechanism. The equations are expressed in Table 1, where k1 is the pseudo-first order rate constant(min^{-1}) of sorption, q_e and q(t) (mg/g) are the amounts of Hg^{2+} sorbed at equilibrium and time t, k_2 is the pseudo-second order rate constant (g $mg^{-1}min^{-1}$), and Ki is the intraparticle diffusion rate (mg $g^{-1}min^{-0.5}$).

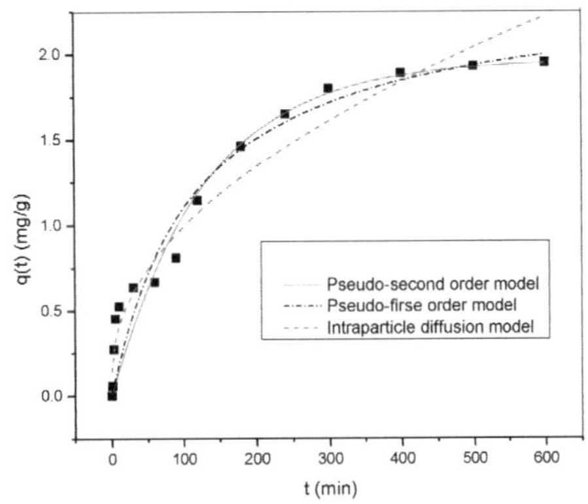

Figure 9. Adsorption kinetics of CS-KI/TiO_2-HBr for removal of Hg^{2+} (PH 5; T: 303K).

After fitting by the kinetic models, the parameters and the correlation coefficients (R^2) are listed in Table 2. The R^2 value of the intraparticle diffusion model is closer to 1.0 than the other two, indicating that is the most suitable model to describe the adsorption of CS-KI/TiO_2-HBr for Hg^{2+}. However, all three models had very similar R^2 values. The pseudo-second order equation was slightly better than the pseudo-first order model. In addition, the intraparticle diffusion is very important because the sorbent is a composite material. The addition of TiO_2, KI and HBr made the adsorbents own a special surface morphology that had large number of cracks and grooves. That will make the adsorption of Hg^{2+} easier because of the surface adsorption and intraparticle diffusion effects.

Table 1. Kinetics model equations.

Type	Non-Linear form	Linear form	Plot
Pseudo-first order model	$q_t=q_e(1-e^{-K_1t})$	$\log(q_e-q_t)=\log q_e-(\frac{K_1}{2.303})t$	$\log(q_e-q_t)$ vs. t
Pseudo-second order model	$q_t=\frac{K_2t}{1+K_2q_et}$	$\frac{t}{q_t}=\frac{1}{K_2q_e^2}+(\frac{1}{q_e})t$	$(\frac{t}{q_t})$ vs. t
Intraparticle diffusion model	-	$q_t=K_it^{0.5}+C$	q_t vs. $t^{0.5}$

Table 2. Kinetic parameters for Hg^{2+} sorption.

Adsorbent	q_{max} (mg/g)	Pseudo-first order model			Pseudo-second order model			Intraparticle diffusion model		
		K_1 (min^{-1})	q_e (mg/g)	R^2	K_2 (min^{-1})	q_e (mg/g)	R^2	K_i (mg/g·$min^{-0.5}$)	C (mg/g)	R^2
CS-KI/TiO_2-HBr	1.95	0.0077	1.97	0.934	0.0035	2.39	0.939	0.085	0.14	0.952

3.6.2. Adsorption equilibrium analysis. The results of the adsorption isotherms are shown in Figure 10. From Figure 10, the equilibrium Hg uptake of CS-KI/TiO_2-HBr is approximately 1.98mg/g.

The Langmuir and Freundlich models are also used to further study the adsorption process. The former is based on the assumption of monolayer adsorption on a homogeneous surface. The latterised based on the assumption that adsorption energy decays exponentially with coverage rise. This condition is usually present in inhomogeneous multilayer adsorption on the surface [9 10]. Their equations are expressed in Table 3.

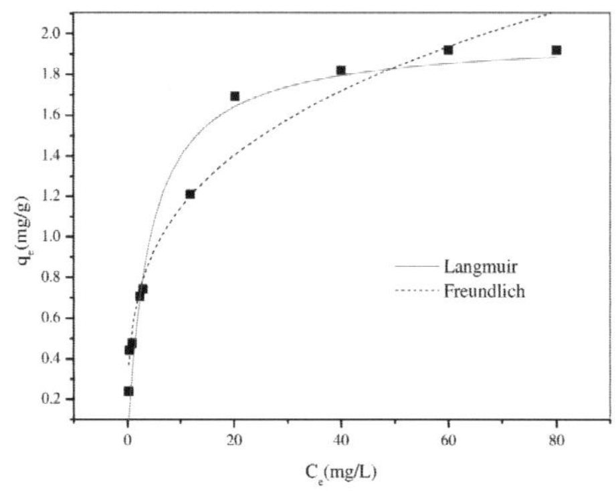

Figure 10. Adsorption isotherms of CS-KI/TiO_2-HBr for removal of Hg^{2+} (PH 5; T: 303K).

Table 3. Adsorption isotherm equations.

Type	Non-Linear form	Linear form	Plot
Langmuir	$q_{eq}=\frac{q_{m,L}K_LC_{eq}}{1+K_LC_{eq}}$	$\frac{C_{eq}}{q_{eq}}=\frac{C_{eq}}{q_{m,L}}+\frac{1}{K_Lq_{m,L}}$	$\frac{C_{eq}}{q_{eq}}$ vs. C_{eq}
Freundlich	$q_{eq}=K_FC_{eq}^{1/n}$	$\ln q_{eq}=\ln K_f+\frac{1}{n}\ln C_{eq}$	$\ln q_{eq}$ vs. $\ln C_{eq}$

The fitting results are listed in Table 4. Evidently, the correlation coefficient (R2) of the Freundlich model is closer to 1.0, which is much higher than the Langmuir model. That is to say that the mercury ions first interacted with the adsorbent surface at sites with strong binding forces, only then moving to sites where the binding force is weak. As the adsorption process progresses, the binding force between the adsorbent and the molecule becomes increasingly weak. The Hg^{2+} adsorption on the composite sorbent is inhomogeneous multilayer adsorption, as Figure 6 demonstrates.

Table 4. Parameters of the adsorption isotherm.

Sorbent	q_{max} (mg/g)	Langmuir model			Freundlich model		
		$q_{m,L}$ (mg/g)	K_L (L/mg)	R^2	n	K_F	R^2
CS-KI/TiO_2-HBr	1.95	1.98	0.24	0.950	3.39	0.58	0.963

4. Conclusion

A novel Chitosan adsorbent compound for mixing with KI and TiO_2 was developed for use in mercury adsorption experiments on gypsum slurry. The results show that the composite adsorbents successfully adsorbed both low and the high concentration Hg^{2+} in the gypsum slurry. HBr as a solvent in composite adsorbent for gypsum slurry Hg^{2+} adsorption performed better than HAc in preparing the adsorbent. The presence of Br^- can prevent the growth of TiO_2 crystal grain and strengthen the ability of I^- to remove mercury and enhance the adsorption ability of chitosan for Hg^{2+}. Also, TiO_2 may form Ti-C, Ti-O and Ti-N bonds with CS. The lamellar structure of TiO_2 promotes the adsorption of Hg^{2+} by the chitosan. The adsorption isotherms and kinetics simulation indicated that the adsorption behaviour of CS-KI/TiO_2-HBr to remove the Hg^{2+} was an inhomogeneous multilayer adsorption process. The surface adsorption and intraparticle diffusion effects both are important to the Hg^{2+} adsorption process. However, the inhibition effect of Hg^{2+} into Hg0 was not verified for the composite materials in this work. Therefore, further research should focus on this inhibition effect.

Acknowledgments

This research was financially supported by the Fundamental Research Founds for the Central Universities, China (No. 2014MS108) and the National High Technology and Development Program of China (863 Program) (No. 2013AA065404).

References

[1] González R O, Somoano M D and Tarazona M R M 2013 *Fuel* **105** 112
[2] Sun M Y, Cheng G H, Lu R J, Tang T M, Baig S A and Xu X H 2014 *Fuel Process Technol.* **118** 28
[3] Rallo M, Lopez M A and Perry R M 2010 *Fuel* **89** 2157
[4] Schuetze J, Kunth D, Weissbach S and Koeser H 2012 *Environ. Sci. Technol.* **46** 3008
[5] Sun M Y, Lou Z, Cheng G H, Baig S A, Fang L, Zhou X X, ShenY X and Xu X H 2015 *Fuel* **140** 136
[6] Sun M Y, Hou J A, Tang T M, Lu R J, Ch L H and Xu X H 2012 *Fuel Process Technol.* **104** 160
[7] Hu H Y and Zhu H 2012 *PROG CHEM* **24** 2212
[8] Jiang T D 2006 *Chitosan* (Beijing: Chemical Industry Press) p28
[9] Elwakeel K Z and Guibal E 2015 *CHEM ENG J* **281** 345
[10] Li K, Wang Y W, Huang M, Yan H, Yang H, Xiao S J and Li A M 2015 *J COLLOID INTERF SCI* **455** 261
[11] Qi L F, Xu Z R, Jiang X, Hu C H and Zou X F 2004 *Carbohydro. Res.* **339** 2693
[12] Paulino A T, Pereira A G B, Fajardo A R, Erickson K, Kipper M J, Muniz E C, Belfiore L A and Tambourgi E B 2012 *Carbohydr. Polym.* **90** 1216
[13] Zhang X, Wu T X, Song X P, Zhang Y X and Wang G Z 2015 *J. Chin. Ceram. Soc.* **43** 1143
[14] Gao P, Xiang J and Zhang A C 2010 *China Environmental Science* **30** 733
[15] Zhang A C, Sun L S, Xiang J, Gao P, Hu S, Xu C F and Fu P 2010 *Journal of Engineering Thermophysics* **3** 11607
[16] Gao P, Xiang J, Mao J B, Liu X F, Wang X, Chen W, Zhao Q S and Xu C F 2006 *Proceeding of the CSEE* **26** 88
[17] Raji F and Pakizeh M 2014 *Appl. Surf. Sci.* **301** 568
[18] Dou B, Dupont V, Pan W and Chen B 2011 *Chem. Eng. J.* **166** 631

Study on deposition technique and properties of Pd/Ag alloy film sensor supported on ceramic substrate

Z T Geng[1, 3], Q He[1] and C G Jin[2]

[1]School of Energy Power and Mechanical Engineering, North China Electric Power University, Beijing100089, China
[2]KAIENTE Technology CO., Ltd. Beijing 102606, China

E-mail: qhgzt@126.com

Abstract. Developing high-quality hydrogen sensitive material is the core part of hydrogen sensor, whose performance is determined by the sensitive response, reproducibility and recovery of hydrogen material etc. In order to overcome the defects of hydrogen embrittlement in previous hydrogen sensor which were based on the pure palladium, sliver as the second component added to the palladium was studied. Using photochemical etching technology to produce a bent metal mask, the mask is put on the ceramic substrate. Firstly, the thin film of Ta_2O_5 as a transition layer grew on the ceramic substrate. Then, a series of Pd/Ag alloy film sensors were prepared, and each performance characterization of Pd/Ag alloy film was studied. Testing results indicated that the thin film had a good linear output performance at 0~30% hydrogen concentration range, and demonstrates a high responsiveness and good repeatability. With temperature increasing, the strength of the responsive signal of the Pd/Ag alloy film decreases and its responsive time was also shortened.

1. Introduction

Developing high-quality hydrogen sensitive material is the core part of hydrogen sensor, whose performance is determined by the sensitive response, reproducibility and recovery of hydrogen material [1]. In order to overcome the defects of hydrogen embrittlement in previous hydrogen sensor which were based on the pure palladium, sliver as the second component added to the palladium was studied [2]. Pd/Ag alloy has good selectivity and permeability on hydrogen at room temperature. Therefore, the Pd/Ag alloy has become an important source of hydrogen sensitive material for hydrogen sensor. BaBaK had done a lot of research on the hydrogen sensitivity properties of palladium alloy system, such as the change of the electrical properties of Pd/Ni, Pd/Au, Pd/Ag and other alloys with hydrogen concentration [3]. Encountered hydrogen containing environments, the Pd/Ag alloy materials will form the palladium hydrogen structure which results in the electrical and optical properties changes of hydrogen sensitive materials. So, when the hydrogen concentration changes, the detection signal of Pd/Ag alloy is also changed.

In this paper, a thin film hydrogen sensor was designed and prepared by using Pd/Ag alloy as hydrogen sensitive material, and its sensing properties were tested.

2. Experimental

2.1. The working principle of hydrogen sensor

Gas sensitive material is the key to gas sensor. There are different sensing properties and detection principles in gas sensitive materials. Hydrogen sensor detection principle is based on physical or chemical property change of gas sensitive material after the adsorption of hydrogen. Gas sensitive material of traditional hydrogen sensor is palladium or palladium alloy. When palladium was exposed to hydrogen environment, palladium hydrogen compound (Pd-H) was formed after palladium film had contacted with hydrogen. When palladium was from hydrogen, hydrogen molecules could strip from the metal palladium, the conductivity and many other features of palladium were back to the original form. Therefore, reaction of palladium with hydrogen is a reversible process. Working principle diagram of palladium alloy hydrogen sensor is shown in Fig.1 for the resistance measurement [4-6].

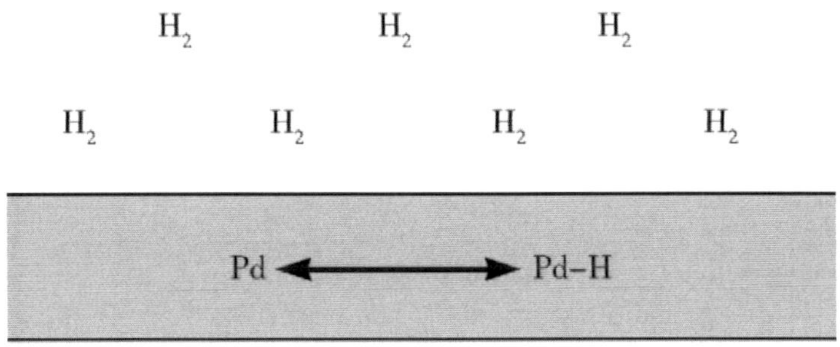

Figure.1 Palladium resistive hydrogen sensor operating principle

Due to limited range of pure palladium hydrogen storage, when the hydrogen concentration is high, the pure palladium will produce hydrogen embrittlement which leads off the film, and then the pure palladium hydrogen sensor fails. Adding silver to palladium will stabilize hydrogen palladium alloy, it can prevent the hydrogen compound phase transition from α to β phase. Pd/Ag alloy has good reversibility and rapid response capability at room temperature [7, 8]. Thus, at present, Pd/Ag alloy becomes an important source of the hydrogen gas sensor sensitive material.

2.2. Experiment methods

Pd/Ag alloy film preparation of Hydrogen sensor is prepared by the production of JGP-450B type magnetron sputtering deposition system. The Pd/Ag alloy coating process steps are as follows.

First, the alumina ceramic substrates were cleaned by absolute ethanol ultrasonic in 5min, washed with distilled water, and then dried.

Second, the metal masks were put on alumina ceramic substrates with the magnets below, and then the substrates were put into sputtering equipment. Pumping the vacuum to 5×10^{-4}Pa, the argon work pressure was adjusted to 0.5Pa. Next, the Ta_2O_5 transition layer was sputtered onto the substrate by selecting RF sputtering for 20s.

Last, pre sputtered Pd/Ag alloy for about 10min by using DC magnetron deposition, then Pd/Ag alloy film was sputtered on alumina ceramic substrate for 35s. After the Pd/Ag alloy sputtering well, the samples were annealed in argon gas atmosphere for one hour, and the annealing temperature was kept at 450 C.

3. Results and discussion

3.1. Relationship between film thickness and the resistance value

Five Pd/Ag alloy thin film sensors were obtained by magnetron sputtering at different sputtering time, as shown in Fig.2. Then the thickness of the Pd/Ag alloy film on alumina ceramic substrate was measured with the American XP-1 AMBIOS surface probe scanning step instrument. The film resistance was measured by the universal meter. The results of the measurement were shown in table1.

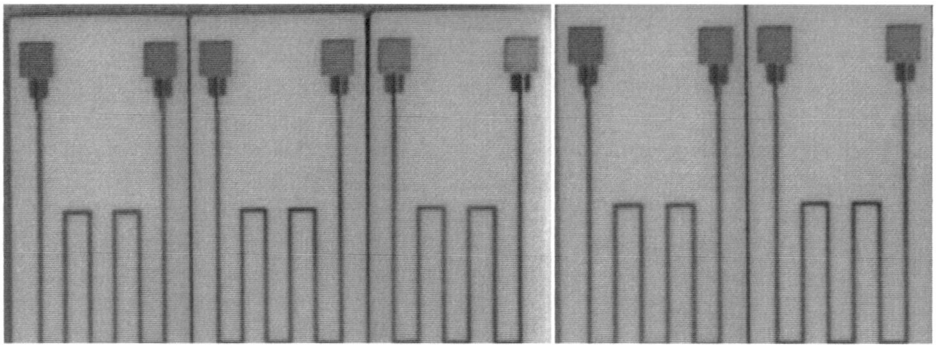

Figure2. Pd/Ag alloy thin film hydrogen sensors

Table1. Pd/Ag alloy membrane thickness and the resistance value

Sputtering time(s)	Resistance value(KΩ)	Film thickness(nm)
20	28.4	50
30	24.2	100
45	22.0	200
60	12.8	250
90	9.0	300

Pd/Ag alloy film resistance can be reduced with the increase of sputtering time, while Pd/Ag alloy film thickness increases. The Pd/Ag alloy film thickness is proportional to the sputtering time, and the resistance is inversely proportional to the film thickness.

3.2. Test of Pd/Ag alloy membrane response characteristic

Using three groups of different thickness of Pd/Ag alloy film as the hydrogen sensor sensitive material, hydrogen sensor amplitude and response time impacted by alloy membrane thickness were tested. Under the normal temperature and atmospheric pressure, 30% standard hydrogen gas was bubbled into three groups Pd/Ag alloy samples for 10 minutes, and then high purity nitrogen as the dehydrogenation was put into the container. In this process, the amplitude and response time of the sensors with different film thickness were recorded, the best film thickness range for Pd/Ag alloy hydrogen sensor was verified by comparing the results. Fig.3 shows the experimental results of hydrogen sensor output recorded in real-time by testing software.

- For the Pd/Ag alloy sensor with thin film thickness (less than 50nm), its response time for hydrogen is relatively fast, but its response amplitude is low, and the absorption of hydrogen is easy to reach saturation, as shown in Fig.3 (a)
- For the Pd/Ag alloy sensor with larger film thickness (more than 300nm), its response time of the hydrogen is relatively slow, and its response amplitude is low. It is shown in Fig.3 (c).
- For the hydrogen sensor with (100 ~ 300 nm) film thickness, as shown in Fig.3 (b), sensor performance is relatively stable and reproducible for 30% H_2. It can be seen from the figure that the response time of sensor is about 32s, and the hydrogen desorption time of the sensor is

100s. Its response time is fast, and it is very suitable for reaction film as a hydrogen sensor.

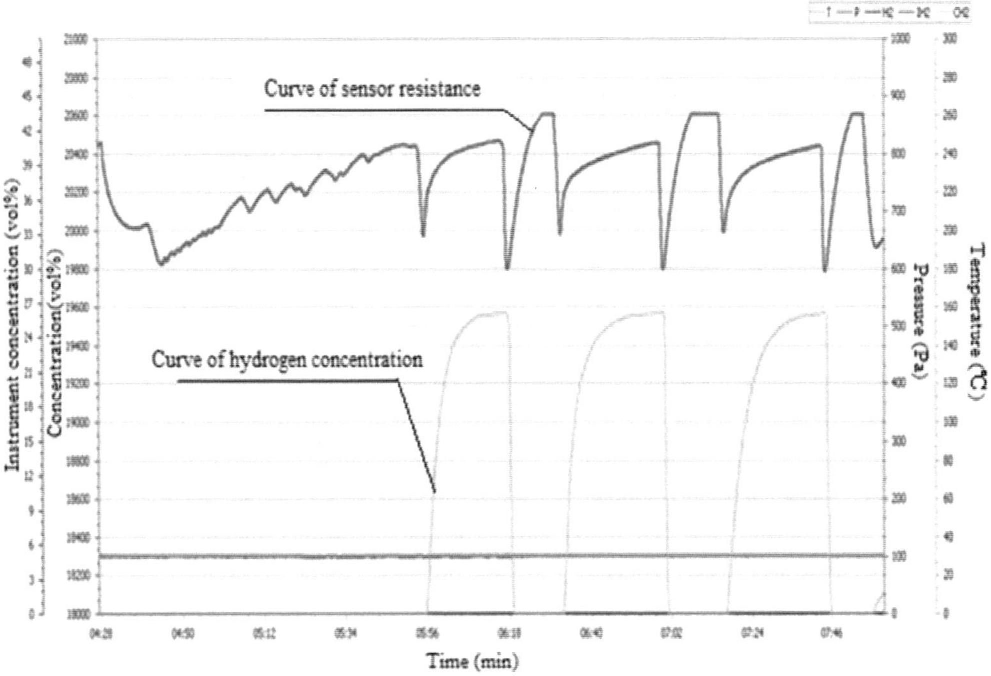

(a) 50nm

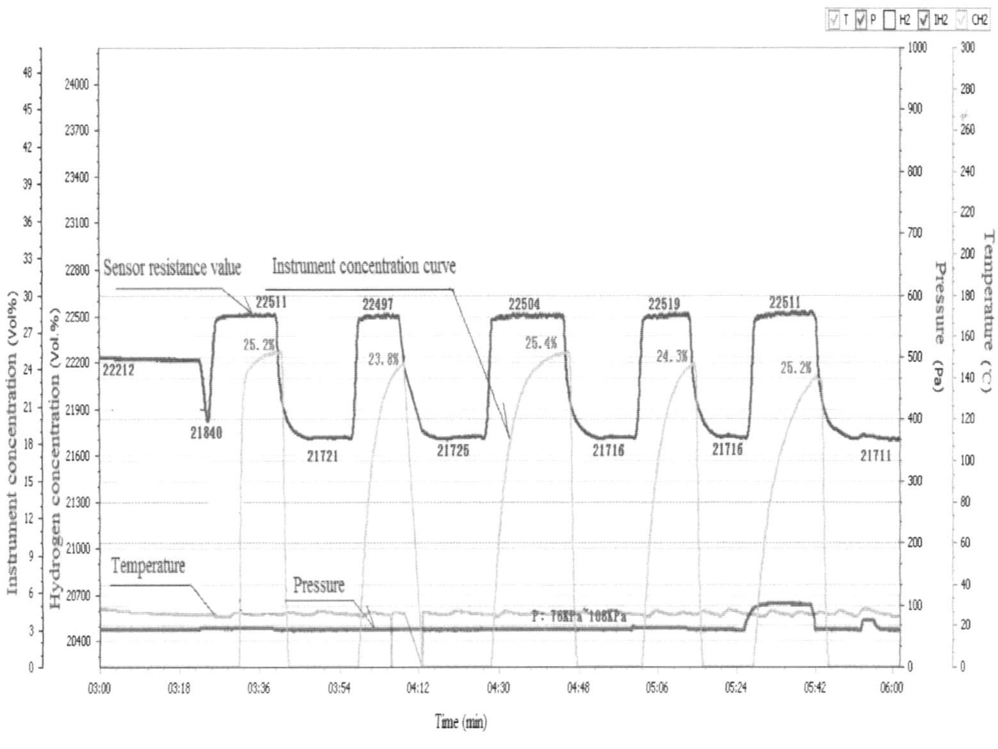

(b) 200nm

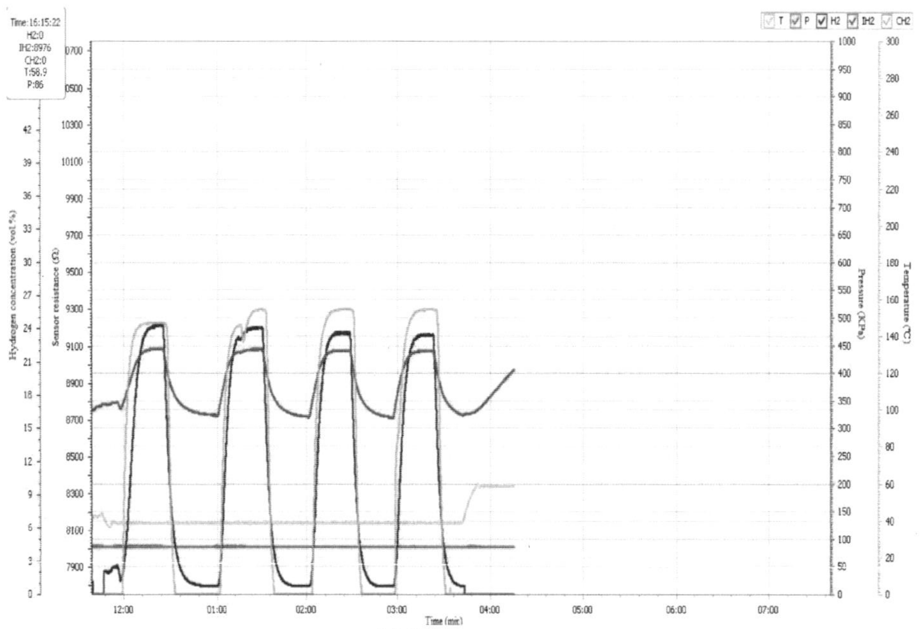

(c) 300nm

Figure3. Response curve of different Pd/Ag alloy film thickness

3.3. Basal temperature effects on response performance

Pd/Ag alloy samples with different substrate temperature (200 ℃, 300℃, 450℃) were passed into 30%H_2/N_2 standard gas in 10 mins, with high purity nitrogen as the dehydrogenation gas. Pd/Ag alloy films with different substrate temperature were compared with the hydrogen concentration response amplitude and response time, and then Pd/Ag alloy hydrogen sensor optimum substrate temperature was verified. Fig.4 shows the results of the hydrogen sensor output at different substrate temperature.

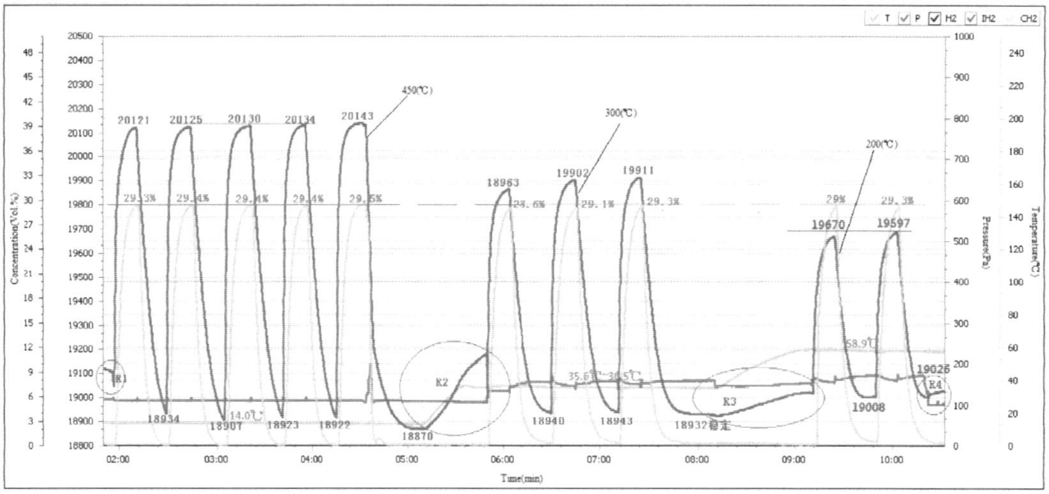

Figure4. Results of the hydrogen sensor output at different substrate temperature

The experimental results indicate that there are not many differences among the response time of the three kinds of samples, while the response amplitude is different, the higher the substrate temperature is, the higher the response amplitude is, the better the response is.

4. Conclusion

Pd/Ag alloy has good selectivity and permeability on hydrogen. Pd/Ag alloy system was selected as hydrogen sensitive material, and the film hydrogen sensors were prepared by using magnetron sputtering technology.

- The test results show that Pd/Ag alloy film resistance value is inversely proportional to the thin film thickness.
- The result of experiments shows that the sensor output signal has a good liner relationship in 0~30% hydrogen concentration. The sensor has high sensitivity, and its response time is about 32s, and the recovery time is about 100s.
- The experimental results indicate that there are not many differences among the response time of the three kinds of samples, while the response amplitude is different, the higher the substrate temperature is, the higher the response amplitude is, and the better the stability is.

References

[1] Feng Y 2006 *Research on tiny Pd-Ag alloy hydrogen sensor preparation and application.* (Hangzhou: Zhejiang University)

[2] Zhang ZD 2013 *Hydrogen sensor and detection technology.* (Harbin: HarbinUniversity of Technology)

[3] Tao C Y, Tang J J and Du J et.al 2005 Hydrogen sensitive materials and the research progress of hydrogen sensor*Material Review* **02** 11.

[4] Nakano S, Yanmara S, Chinashi S U, et.al. 2005 Effect of hydrogen on the electrical resistance of melt-spun $Mg_{90}Pd_{10}$ amorphous alloy *Sensor Actuat B* **104** 75.

[5] Kumar P, Malhotra L K 2004 Palladium capped samarium thin films as potential hydrogen sensors*Mater. Chem. Phys.* **88** 106.

[6] Kumar P, Malhotra L K 2004 Electrochemistry induced switching in SmHx thin films*Appl. Surf. Sci.***23** 461.

[7] Sayago I, Terrado E, Lafuente E, et al2005 Hydrogen sensors based on carbon nanotubes thin films *Synth. Met.* **01** 15.

[8] Shu J, Grandjean B P A, Neste A, et al. 1991Catalytic palladium-based membrane reactors *The Can. J. Chem. Eng.* **05** 1036.

Synthesis and characterization of PEG-P(MAA-SS-VCL) nanoparticles

L L Yu[1,3], K Yang[1], R H Mu[2], N Zhang[1] and L Su[1]

[1]Department of Pharmacy, Xi'an Medical University, Xi'an, shaanxi, 710021, China
[2]College of Environmental and Chemical Engineering, Xian Polytechnic University, Xian 710048, China

E-mail: yulili0218@163.com

Abstract. The PEG-P(MAA-SS-VCL) nanoparticles were obtained using disulfide containing dimethacrylate (SS) as cross-linking agent, using polyethylene glycol methyl acrylate (PEGMA), N-Vinyl-ε-caprolactam (VCL), and methacrylic acid (MAA) as monomers via homogeneous polymerization in aqueous. The PEG-P(MAA-SS-VCL) nanoparticles were characterized by FT-IR and TGA. The particle size and morphology variation in different environments were detected by dynamic light scattering (DLS) and scanning electron microscopy (SEM). It is the very method that PEG-P(MAA-SS-VCL) nanoparticles can be obtained in this study.

1. Introduction

Nanoparticles which can release drug at the targeted site have attracted much attention in recent years.[1] Nanoparticles which can respond to certain triggers including pH[2], redox[3], temperature[4], specific enzymes[5], and so on, have become a research hotspot. The literature reports show that disulfide bond is redox sensitivity, while poly (methacrylic acid) (PMAA) has good pH sensitivity. [2] Herein, we report on the design and fabrication of a redox and pH dual responsive nanoparticle by incorporating disulfide bond containing network into pH-responsive PMAA particle matrix. The preparation and the property of the nanoparticles were investigated by FT-IR, TGA, DLS, and SEM.

2. Materials and methods

2.1. Materials

Vinyl caprolactam (VCL), Methoxypolyethylene glycols 350 ($PEG_{350}OH$) were purchased from Alfa Aesar. 2, 2'-dithiodiethanol was provided by Alarich. Methacryloyl chloride and methyl acrylic acid (MAA) were purchased from Energy Chemical. Glutathione (GSH) was obtained from Shanghai source biological technology co., LTD. Tetrahydrofuran (THF), triethylamine (TEA), and dichloromethane (DCM) were distilled over CaH_2. Poly(ethylene glycol) methyl ether methacrylate (PEGMA) and the cross-linking agent (2, 2'-dithiodiethanol dimethacrylate, SS) were obtained according to the literatures [6, 7].

2.2. *Preparation of PEG-P(MAA-SS-VCL) nanoparticles*
0.97 g VCL, 0.03g MAA, 20 mg SDS, 25 mg $NaHCO_3$ and a certain amount SS (20 mg, 40 mg, 80 mg, 120 mg, and 160 mg respectively) were dissolved in 100 mL water in round bottom flask. The mixture was stirred for 30 mins at 70°C in N_2. Then, Potassium persulfate (KPS) solution (25 mg KPS in 2 mL distilled water) and 10 mg PEGMA were added to the mixture under N_2. Keeping the original temperature for another 6 hs, the solution was transferred into a dialysis bag, and dialyzed against 500 mL water for 24 hs, the dialysis medium was exchanged every 2 hs. The dried nanoparticles were obtained by freeze drying the dialyzed solution at -70 °C.

2.3. *Measurement*
The compositions of PEG-P(MAA-SS-VCL) nanoparticles with different amount of cross-linking agent were determined by FT-IR spectra (Bruker, Tensor 27). TGA data were recorded on a Thermogravimetry analyzer (TA, Q500) under a flowing nitrogen atmosphere at a heating rate of 20 °C/min from 100 to 650 °C. Particle size and Zeta potential analysis were also determined by DLS (Malvern, ZEN3600). The surface morphology of PEG-P(MAA-SS-VCL) nanoparticles was characterized by SEM using Nanoscope IIIa.

3. Results and discussion

3.1. *Synthesis of PEG-P(MAA-SS-VCL) nanoparticles*
Homogeneous free radical polymerization is such an effective method for preparation of nanoparticles, so that this method has been employed to the preparation of PEG-P(MAA-SS-VCL) nanoparticles using PEGMA, VCL, MAA as monomers, and SS as cross-linking agent. The chemical formation of PEG-P(MAA-SS-VCL) nanoparticles with different cross-linking degree were confirmed by FT-IR spectra as illustrated in Fig. 1. The FT-IR data displayed that strong peaks for C=O stretching vibration at 1728 cm^{-1} and 1631 cm^{-1} were attributed to PMAA and PVCL. Specific peaks for C-N (1440 cm^{-1}) belonged to PVCL, and PEG (1144 cm^{-1}) was observed in all nanoparticles. Remarkably, the intensity peak ratio (1728 cm^{-1} to 1631cm^{-1}) for nanoparticles with 20 and 40 mg cross-linking agent was much higher than samples with more cross-linking agent, suggesting that the reaction rate of MAA and VCL was relating to the dosage of the cross-linking agent. As in Fig. 1, a high cross-linking agent dosage was favorable for the reaction rate of MAA in the nanoparticles. Disulfide in both nanoparticles could not be confirmed by FT-IR spectra, for the signal for disulfide bond is rather weak. Thus, the content of disulfide for nanoparticles was determined according to Ellman's titration, which was 51, 87, 115, 156, and 205μM/g, respectively, while for nanoparticles with different cross-linker concentration (20 mg, 40 mg, 80 mg, 120 mg, and 160 mg), things were different and that was why the existence of disulfide in nanoparticles was validated.

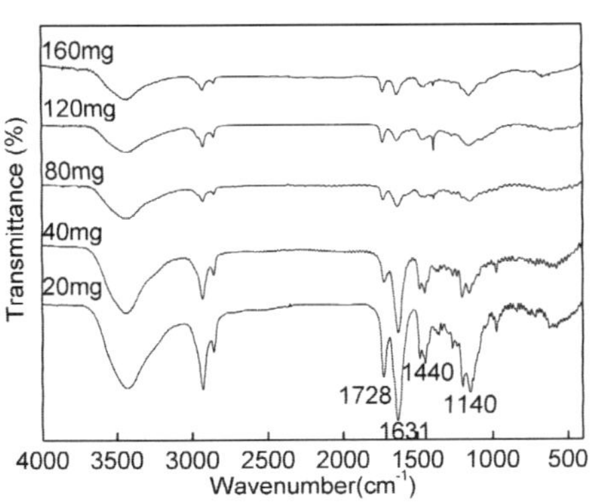

Figure 1. FT-IR spectra of PEG-P (MAA-SS -VCL) with different crosslinking degree.

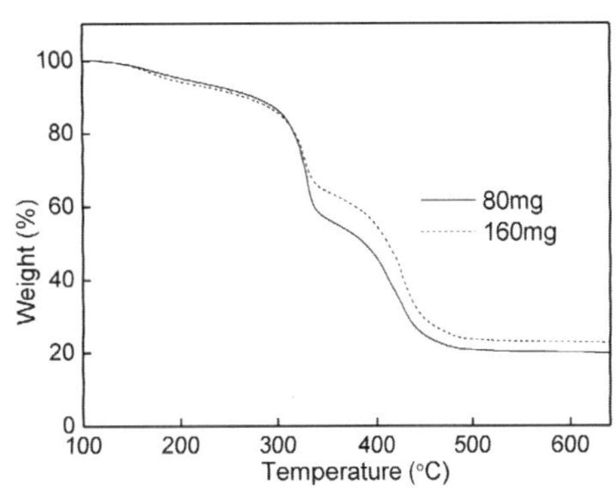

Figure 2. TGA thermograms of PEG-P(MAA-SS-VCL)-80 mg and PEG-P(MAA-SS-VCL)-60 mg.

Furthermore, the structural differences for PEG-P(MAA-SS-VCL) nanoparticles with different cross-linker concentration (80 mg and 160 mg) were also confirmed by TGA data twice. As shown in Fig. 2, pyrolysis temperature of PEG-P(MAA-SS-VCL) nanoparticles was 302 °C, and the thermal decomposition processes of nanoparticles were composed of two stages. Meanwhile, pyrolysis temperature of PEG-P (MAA-SS-VCL) with 80 mg crosslink agent was lower than PEG-P(MAA-SS-VCL) with 160 mg (Fig. 2), which indicated that the cross-linked structure was conducive to the thermal stability of nanoparticles.

3.2. GSH and pH triggered variation of PEG-P(MAA-SS-VCL) nanoparticles

As shown in Table 1, particle size variation is not pH dependent, while GSH concentration has a certain effect on the particle size. At low GSH level, the effect is not significant, but the particle size has a trend to be larger at the high GSH level. Because of disulfide bond breakage at the high GSH level, the contact between the polymer chain segments can be weakened with the nanoparticles' structureswelling. In addition, with the increase of the amount of cross-linking agent the change trend can be weakened too.

Table 1. Particle size variation in PEG-P(MAA-SS-VCL) nanoparticles in different condition.

	SS=20 mg		SS=40 mg		SS=80 mg		SS=120 mg		SS=160 mg	
	Size (nm)	PDI	Size (nm)	PDI	Size (nm)	PDI	Size (nm)	PDI	Size (nm)	PDI
pH=2.2	165.0	0.10	189.9	0.04	166.3	0.16	127.8	0.04	169.3	0.08
pH=5.0	233.9	0.25	233.3	0.20	161.7	0.13	130.5	0.08	176.0	0.12
pH=7.4	178.4	0.09	490.7	0.02	160.7	0.15	155.6	0.26	181.2	0.07
GSH 0.002 mmol/L	169.1	0.10	199.8	0.05	167.8	0.18	140.3	0.11	183.5	0.06
GSH 2 mmol/L	173.4	0.14	294	0.17	254.7	0.24	138	0.11	207.3	0.18
GSH 10 mmol/L	459.5	0.35	933.4	0.53	242.4	0.31	144.4	0.04	192.5	0.07

Table 2. Zeta potential variation in PEG-P(MAA-SS-VCL) nanoparticles in different condition.

	pH=2.2	pH=5.0	pH=7.4	GSH 0.002 (mmol/L)	GSH 2 (mmol/L)	GSH 10 (mmol/L)
SS=20 mg	-20.8	-24.4	-28.5	-35.2	-11.90	-8.20
SS=40 mg	-13.4	-45.0	-33.3	-36.2	-10.10	-7.16
SS=80 mg	-16.2	-28.9	-37.7	-23.7	-6.88	-5.85
SS=120 mg	-15.5	-15.3	-36.2	-32.8	-14.20	-10.40
SS=160 mg	-23.2	-36.8	-40.8	-34.3	-10.30	-13.50

As shown in Table 2, when the pH value decreased, the Zeta potential of nanoparticles decreased obviously, this was attributed to the dissociation degree for carboxylic acid groups in nanoparticles declining in the acidic environment, and negative charge on the surface of nanoparticles declining. In addition, when GSH concentration increased, Zeta potential of nanoparticles increased, this showed that the fracture of crosslinking agent was not conducive to the stability of system.

In Fig.3a, PEG-P(MAA-SS-VCL) nanoparticles maintained good spherical morphology, and the particle size distribution was relatively uniform. The particle diameter was around 200 nm in diameter as shown in SEM image. After 24 hs treatment by 10 mmol/L GSH solution, the spherical shapes crashed. In addition, nanoparticles were bonding in the medium with pH 5.4.

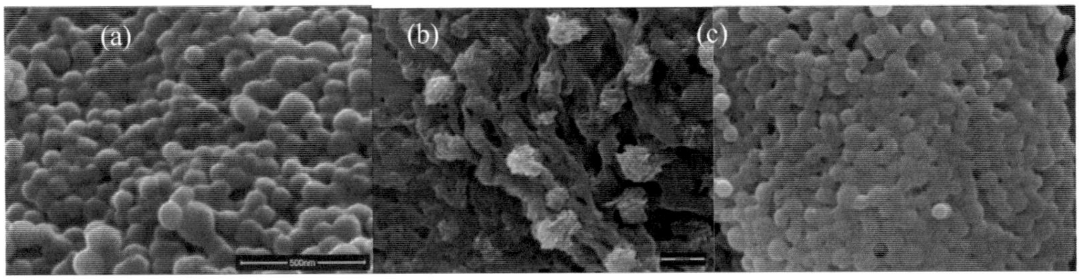

Figure 3. SEM images of nanoparticles (a), nanoparticles in 10 mmol/L GSH (b), and nanoparticles in pH 5.4.

4. Conclusion

In summary, the novel redox responsive biodegradable PVCL based on nanoparticles (PEG-P(MAA-SS-VCL)) were developed successfully. The structures and properties of nanoparticles with redox sensitive network were characterized by FT-IR, DLS, SEM, and TGA. The redox sensibility of nanoparticles was confirmed by SEM and DLS.

Acknowledgments

This work was financially supported by the national natural science foundation of China (81302706), the student's innovation and entrepreneurship training programs of Shaanxi province (No. 1757 and 1649), discipline construction funds of Xi'an Medical University, and students' scientific research training foundation of Xi'an Medical University (No. 2015DXS1-12 and 2015DXS2-19). The authors of the manuscript extend our sincere gratitude to them.

References

[1] Klinger D and Landfester K 2011 Dual Stimuli-Responsive Poly (2-hydroxyethyl methacrylate-co-methacrylic acid) Microgels Based on Photo-Cleavable Cross-Linkers: pH-Dependent Swelling and Light-Induced Degradation. *Macromolecules* **44** 9758-72.

[2] Basak D and Ghosh S 2013 pH-Regulated Controlled Swelling and Sustained Release from the Core Functionalized Amphiphilic Block Copolymer Micelle. *ACS Macro Letters* **2** 799-804.

[3] Guerry A, Cottaz S, Fleury E, Bernard J, and Halila S 2014 Redox-stimuli responsive micelles from DOX-encapsulating polycaprolactone-g-chitosan oligosaccharide. *Carbohydrate Polymers* **112** 746-52.

[4] Bhargava P, Tu Y, Zheng J X, Xiong H, Quirk R P, and Cheng S Z 2007 Temperature-induced reversible morphological changes of polystyrene-block-poly (ethylene oxide) micelles in solution. *J Am Chem Soc* **129** 1113-21.

[5] Rao J and Khan 2013 A Enzyme Sensitive Synthetic Polymer Micelles Based on the Azobenzene Motif. *Journal of the American Chemical Society* **135** 14056-9.

[6] Lili Y, Ruihua M, Li L, Fei L, Lin Y, and Li S 2016 Intracellular Doxorubicin Delivery of a Core Cross-linked, Redox-responsive Polymeric Micelles. *International Journal of Pharmaceutics* **498** 195-204.

[7] Yu L, Mu R, You J, Liang F, and Yao L 2015 Swelling Kinetics of pH and GSH Dual-Responsive P(mPEGMA-DEAM-SS) Hydrogel. *Polymer Materials Science & Engineering* **31** 19-24.

IOP Conf. Series: Materials Science and Engineering **137** (2016) 012011 doi:10.1088/1757-899X/137/1/012011

3D printed auxetic forms on knitted fabrics for adjustable permeability and mechanical properties

N Grimmelsmann, H Meissner and A Ehrmann[1]
Bielefeld University of Applied Sciences, Faculty of Mathematics and Engineering Sciences, Bielefeld, Germany

E-mail: andrea.ehrmann@fh-bielefeld.de

Abstract. The 3D printing technology can be applied into manufacturing primary shaping diverse products, from models dealing as examples for future products that will be produced with another technique, to useful objects. Since 3D printing is nowadays significantly slower than other possibilities to manufacture items, such as die casting, it is often used for small parts that are produced in small numbers or for products that cannot be created in another way. Combinations of 3D printing with other objects, adding novel functionalities to them, are thus favourable to a complete primary shaping process. Textile fabrics belong to the objects whose mechanical and other properties can notably be modified by adding 3D printed forms. This article mainly reports on a new possibility to change the permeability of textile fabrics by 3D printing auxetic forms, e.g. for utilising them in textile filters. In addition, auxetic forms 3D printed on knitted fabrics can bring about mechanical properties that are conducive to tensile constructions.

1. Introduction

The 3D printing technology has been developing rapidly ever since last year. Besides the well-known methods of stereolithography or selective laser sintering, especially the fused deposition modelling (FDM) technique is of great importance for many companies as well as individuals due to its relatively low costs [1]. While often being considered as the base for a new industrial revolution [2,3], these primary shaping technologies are still much slower than conventional production methods, such as die casting.

In FDM printing, a filament is transported into a heated extruder nozzle where it is molten; the liquefied material is deposited on the printer bed line by line to cool down and harden. By lowering the printing plate, the second layer can be printed on top of the first one, etc. [4].

Most FDM printers can work with ABS (Acrylonitrile butadiene styrene) and PLA (polylactic acid); often other materials such as polyamide, polycarbonate or wax can be handled [5].

To reduce the printing time, combinations of FDM printing with other objects, such as textiles, can be used [6-8]. While earlier experiments concentrated on inserting textile fibres into FDM printed objects [7] or printing on woven fabrics [8] to create new multi-material systems with increased mechanical properties, this article explores the possibilities to generate filters as well as garments and other objects with adjustable permeability.

In common materials, longitudinally stretching leads to contracting laterally. Auxetic materials, however, have a negative Poisson's ratio and behave vice versa – they expand laterally when stretched longitudinally [9]. This unexpected behaviour does not only bring about several useful mechanical and sound-proofing properties [10] but also allows for using auxetic structures to open and close pores in a membrane uniformly in both in-plane directions. As one application, membranes with auxetic behaviour can be used as filter materials, since the permeability of circular pores is superior to the value over elliptical ones [11]. This idea is of special interest for filtering particles which are not circular themselves and should either pass a pore or be retained by the filter, independent of the orientation in which they reach the pore.

Soft and responding as auxetic knitted fabrics [12,13] are, they indeed have some defects, making them problematic for technical applications, while a combination of knitted fabrics or flexible membranes with stiffer auxetic patterns can be used to solve this problem. Additionally, mechanical properties, such as bending stiffness of the textile, are significantly altered by imprinted auxetic forms.

As a model system, investigations of knitted fabrics with imprinted auxetic patterns are depicted in this article, using different material and structure combinations.

2. Experimental

The auxetic forms used for printing were designed with Autocad Inventor®. After exporting them as STL files, they were imported in the repetier host and sliced into a layer model. During the slicing process, all printing parameters, such as printing velocity, layer thickness, flow rate, etc. could be defined. The sliced result was sent to the 3D printer.

The objects shown in this article were created using an FDM printer Orcabot XXL by Prodim (The Netherlands). As filament, Taulman 645 Nylon (FilamentWorld) with a filament diameter of 1.75 mm was used which is more flexible than ABS or PLA yet not as stretchable as most "soft" filaments. The extruder temperature was set to 250 °C and the printing bed temperature to 60 °C. The layer height was 0.2 mm. The structures were filled.

The knitted fabrics used as bases were created on a single bed hand knitting machine Silver Reed SK-280 (gauge E5.5). Yarns from natural fibres (cotton 75 %, linen 25 %) as well as man-made fibre yarns (acrylic 50 %, polyester 50 %) were used to create single jersey (single-face) fabrics. In addition the man-made fibre yarns were also used on a double bed hand knitting machine Orion to produce double-face fabrics.

3. Results

First, tests were performed to investigate whether the combination of knitted fabrics and 3D printed auxetic forms allowing for stretching the pores of the textile isotropically or not. Fig. 1 shows a sample with five layers of nylon printed in a typical auxetic shape on a single-face knitted fabric. It was recognized that the nylon layers could not be detached from the fabric by hand; both materials build a stable connection.

Figure 1. Auxetic shape printed on a single-face fabric.

The differences between relaxed and stretched form are depicted in Fig. 2. While the left panel shows the relaxed state, stretching the textile in horizontal (left-right) direction results in the stitches being stretched vertically (top-bottom), too. In some parts, this effect is even disproportionately high.

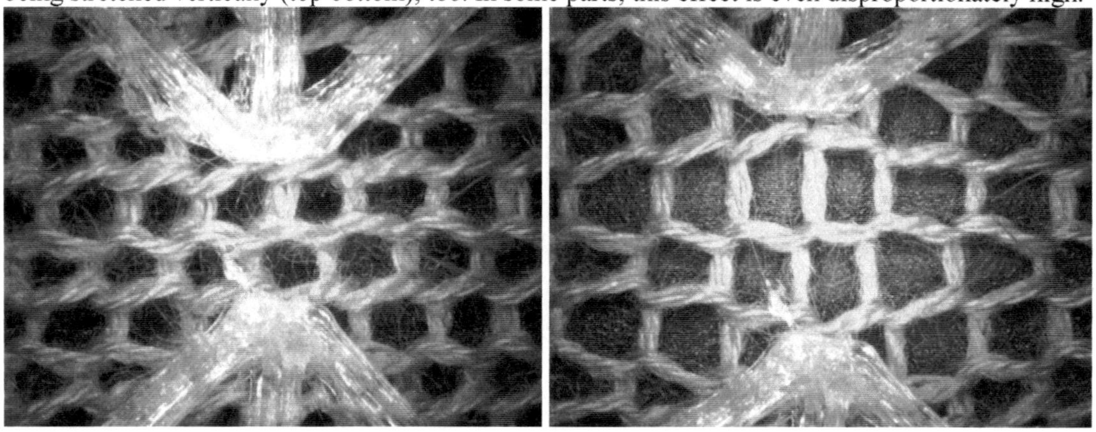

Figure 2. Microscopic pictures of the auxetic shape printed on a single-face fabric, depicted in Fig. 1, in relaxed (left panel) and horizontally stretched state (right panel).

Opposite to this behaviour, the pure knitted fabric – without auxetic printed forms on top – shows the typical performance of "usual" materials, as depicted in Fig. 3: While stretched longitudinally, it shrinks vertically; the shape of the pores clearly differs from the nearly square form shown in Fig. 2.

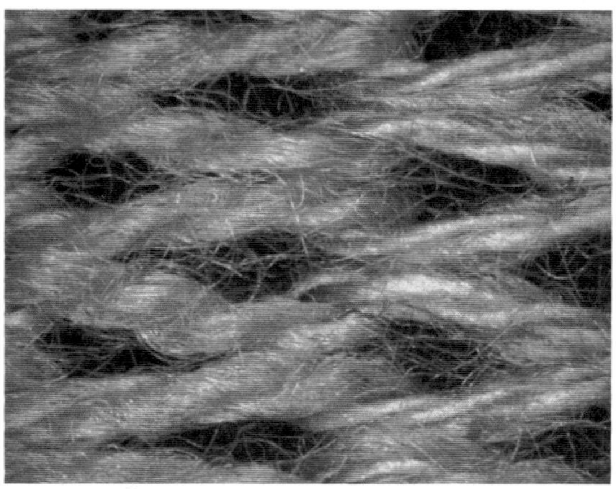

Figure 3. Horizontally stretched single-face fabric without auxetic 3D print on top.

While the principle functionality is apparently reached with this material combination, Fig. 2 nevertheless depicts that the desired proportionality of stretching in both directions is not obtained in this example. Thus, further experiments were conducted with varying parameters of both the knitted fabric and the auxetic 3D print to evaluate their influence on the shape of the pores in the different areas of the auxetic form.

First, different yarns were examined. Fig. 4 shows a single-face fabric knitted from cotton/linen. Apparently, the structure is quite "dense", with only small holes between the yarns. The auxetic structure is again printed with five layers of nylon.

Stretching the fabric (Fig. 4, right panel) results in small changes only. Significant differences in the pore sizes are invisible. For such a dense knitted fabric – with correspondingly low elasticity – the auxetic potential of the 3D print is apparently insufficient.

In comparison, Fig. 5 depicts the same auxetic structure, printed with a height of five layers on a single-face fabric produced from acrylic/polyester yarn. This yarn has a significantly smaller diameter and thus a lower bending stiffness, additionally resulting in a fabric with larger open areas. Stretching the fabric with 3D print laterally is possible without much force, which in return can make the pores open relatively homogenously, as already examined in Fig. 2 for a similar yarn.

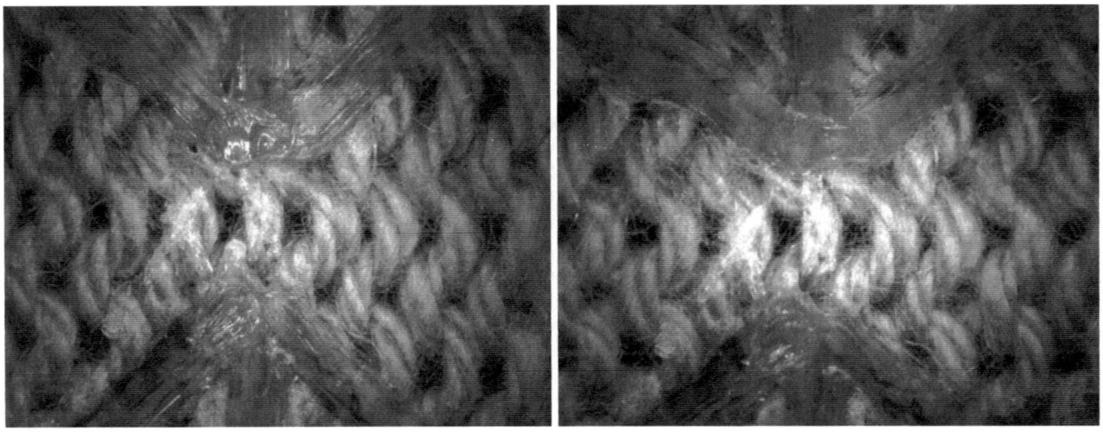

Figure 4. Microscopic pictures of the auxetic shape printed on a single-face fabric from cotton/linen, relaxed (left panel) and stretched (right panel).

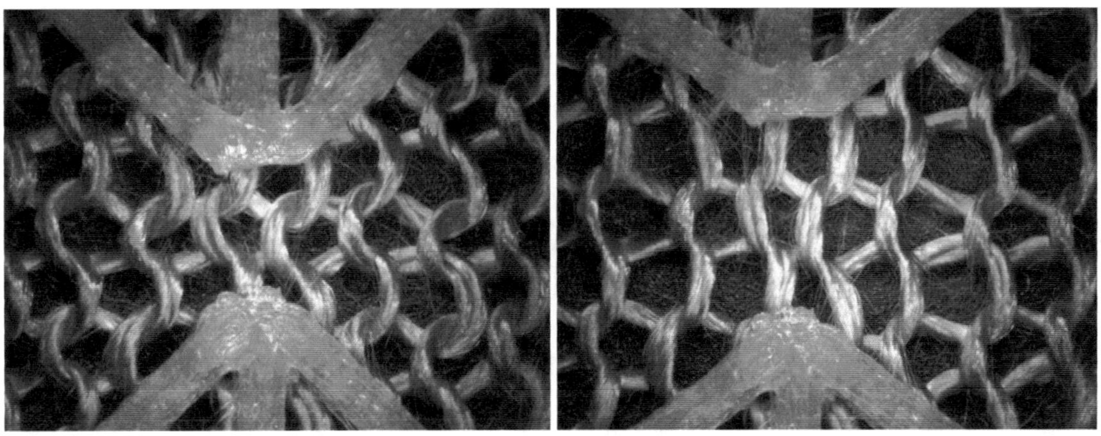

Figure 5. Microscopic pictures of the auxetic shape printed on a single-face fabric from acrylic/polyester, relaxed (left panel) and stretched (right panel).

Apparently it is necessary to choose relatively elastic structures to allow for the auxetic structure on top to significantly influence the textile fabric.

Next, different knitted structures were compared. As fundamental examples, single-face (single jersey) and double-face were chosen. While Figures 1-5 show single jersey fabrics, Fig. 6 depicts a double-face knitted fabric in relaxed and stretched state. Double-face fabrics are much more stretchable than single-face fabrics. Concerning the above-described influence of the in-plane elasticity of the textile on the auxetic performance of the whole system, a double-face fabric should show an even stronger auxetic behaviour.

Fig. 6, however, shows clearly that another factor has to be taken into account: Since the double-face fabric has a zigzag structure, only each other stitch (i.e. only the stitches on the front side of the knitted fabric, not those on the back) gets in contact with the imprinted auxetic structure. This results in a higher irregularity of the elongated textile. In the stretched state (Fig. 6, right panel), the zigzag structure partly unfolds, while the areas where neighbouring front stitches are connected by the printed nylon cannot expand. This behaviour results in large open areas alternating with nearly completely relaxed stitches. Such performance may be attractive for distinct applications; however, it is not useful for the uniformly tailorable permeability.

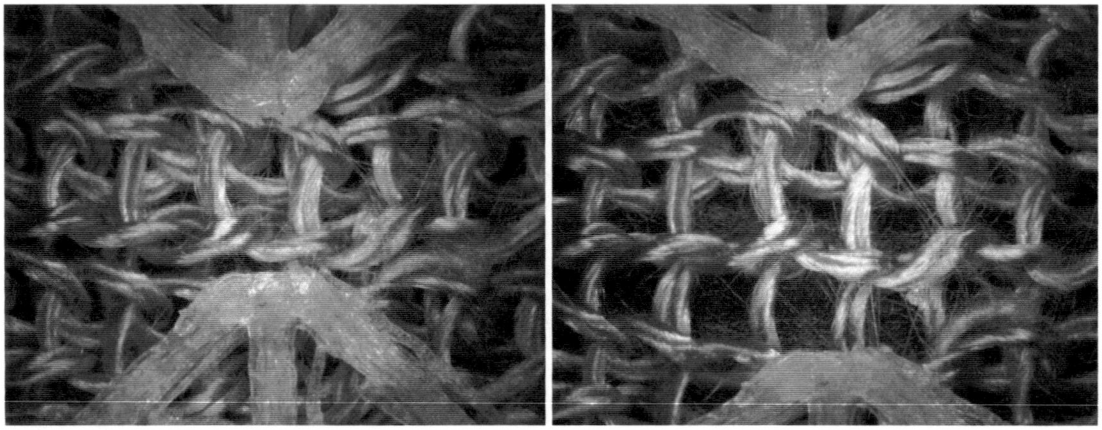

Figure 6. Microscopic pictures of the auxetic shape printed on a double-face fabric from acrylic/polyester, relaxed (left panel) and stretched (right panel).

The auxetic performance of the multi-material system is not only based on the mechanical properties of the knitted fabric but also on those of the 3D print. It is of importance to use a 3D printed shape that is on the one hand not too stiff in the connection points so that the angles between neighbouring strips can be varied by an external force without breaking the material. On the other hand, it must be rigid enough to influence the knitted fabric in the desired way.

Fig. 7 shows the same knitted fabric as in Fig. 5 with a 3D print of two layers thickness, unlike Fig. 5 where the 3D print has a thickness of five layers. Stretching the system now leads to a significantly lower elongation since the mechanical properties of the knitted fabric have a much higher influence on the whole system. Apparently, the 3D printed structure should not be too weak to dominate the multi-material system's characteristics.

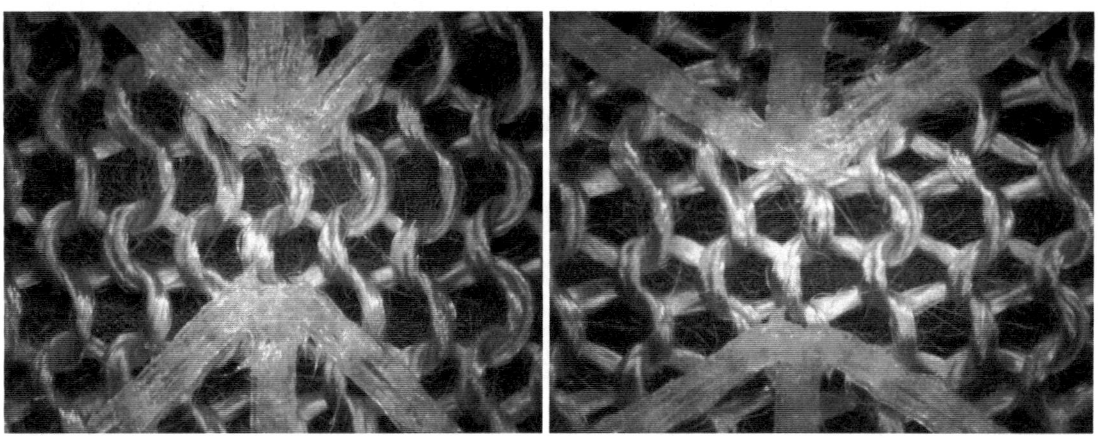

Figure 7. Microscopic pictures of the auxetic shape printed with a thickness of two layers on a single-face fabric from acrylic/polyester, relaxed (left panel) and stretched (right panel).

Finally, the influence of the angles in the chosen auxetic structure is investigated. Fig. 8 shows the same material combination as Fig. 5, with the x-axis scaling reduced to 75 % (left panel) or the length in y-direction decreased to 75 % (right panel), respectively. Both pictures are taken in a stretched state. Comparing them with Fig. 5 (right panel), apparently scaling down in y-direction supports the homogeneity of the pores while scaling down in x-direction slightly increases the irregularities of the single holes in the fabric. Therefore, the ideal structure can have other angles than the original 45° orientations of the connection lines.

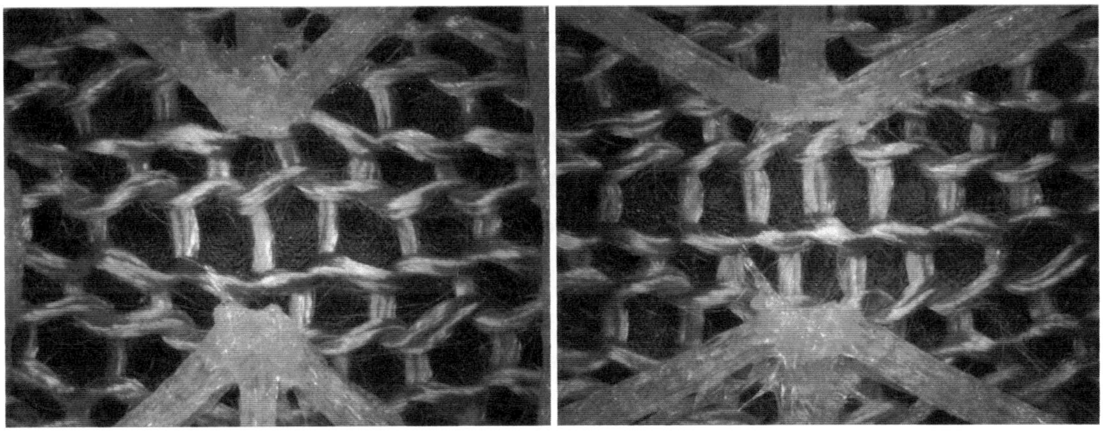

Figure 8. Microscopic pictures of the auxetic shape printed on a single-face fabric from acrylic/polyester (stretched states), scaled down by 25 % in the x-direction (left panel) or in the y-direction (right panel), respectively.

Table 1 gives an overview of the ellipticity of the pores, defined as the aspect ratio of the pores with the idea to tailor the permeability by tailoring the pore size. For each sample, the average ellipticity of 50 pores was measured, using images taken on three different positions of the samples.

Table 1. Aspect ratios of different fabrics with auxetic forms, printed on single-face knitted fabrics, in relaxed and stretched states.

Aspect ratio	**relaxed**	**stretched**
Pure knitted fabric	1.30	2.77
Auxetic form on knitted fabric, 2 layers	1.28	1.49
Auxetic form on knitted fabric, 5 layers	1.27	1.31
Auxetic form scaled in x-direction on knitted fabric, 5 layers	1.32	1.25
Auxetic form scaled in y-direction on knitted fabric, 5 layers	1.28	1.19

As Table 1 shows, the aspect ratio of the pores is nearly unaltered for all samples with auxetic forms printed on them, which means that elongating this sample does not significantly change the pores' shape. The auxetic form printed with two layers only shows a slightly enhanced aspect ratio in the stretched state. The smallest aspect ratio can be found for the stretched sample with the auxetic form scaled in y-direction, as already assumed due to the images in Fig. 8. Transferring this finding to other – not knitted – base materials, however, necessitates additional tests in order to find the ideal auxetic structures and angles.

Moreover, 3D printing auxetic (and other geometric) structures on knitted fabrics helping to tailor their permeability for the possible application in filter materials etc. should be mentioned here. The 3D printed structure can also significantly modify the mechanical properties of the textile fabric under investigation, such as bending stiffness, elasticity, elongation, etc. Fig. 9 shows the enhanced bending rigidity of the samples used in this study due to the 3D printed structure.

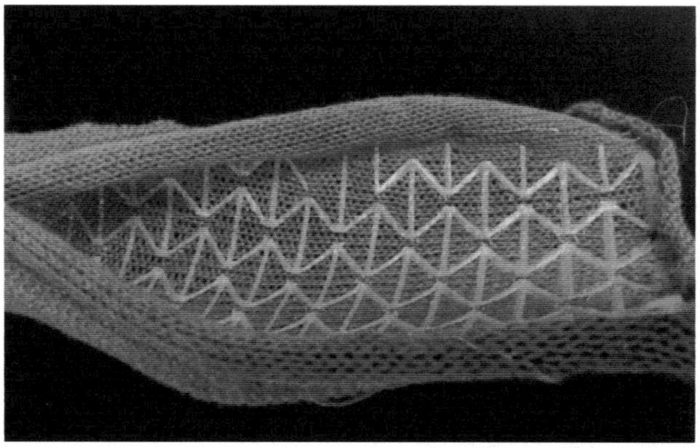

Figure 9. Single-face fabric with increased stiffness due to auxetic 3D print.

The mechanical properties of the knitted fabrics with additional 3D printed auxetic forms can be measured, e.g., by stress-strain experiments. Fig. 10 shows measured forces vs. relative elongations for the samples mentioned in Table 1 as well as the pure auxetic 3D print (without knitted fabric). While the fabric alone needs only a very small force to be elongated by 20 %, the pure 3D print shows higher forces for the same elongation. All combinations of both materials, however, generate much higher forces, combined with large hysteresis loops between increasing and decreasing elongation. This shows that the mechanical properties of the new multi-material system differ from those of both partners alone.

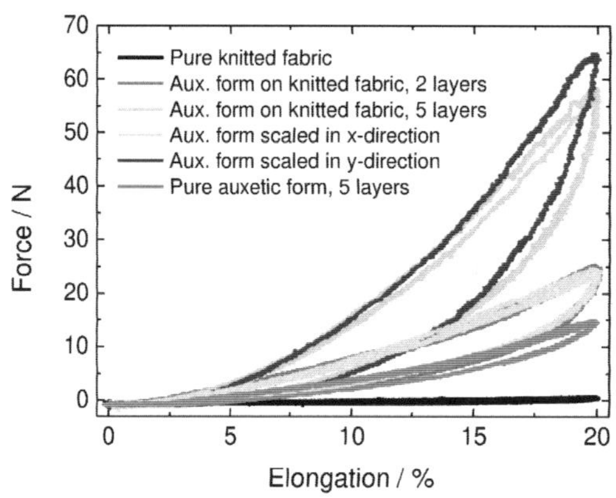

Figure 10. Comparison of the stress-strain curves for different combinations of 3D printed auxetic forms and knitted fabrics as well as for both materials alone.

Thus, a forthcoming project will focus on the examination of the influence of 3D printing diverse geometrical forms on the mechanical properties of different textile fabrics to create lightweight forms with tailorable bending stiffness and elongation, e.g. for tensile constructions and other architectural textiles.

4. Conclusion

Combining 3D printing with traditional technologies like knitting can bring multi-material systems with novel characteristics, such as homogenously variable permeability or enhanced mechanical properties. In a recent project, we have examined the possibilities to influence the permeability of knitted fabrics and other elastic, porous materials such as membranes by 3D printing auxetic structures on top. The interplay between both material partners has shown to determine the properties of the final composite, allowing for modifying the pore sizes and shapes. Thus, it is necessary to adjust carefully the mechanical properties of both partners in such multi-material systems to tailor the character of the final product.

In future projects, other auxetic forms will be tested about their influence on the knitted fabrics' pore sizes and shapes. Additionally, the mechanical properties of such composites will be investigated and explored further to develop new stiff, lightweight structures for textile buildings and other applications.

References

[1] Chua C K, Leong K F and Lim C S 2003 *Rapid Prototyping: Principles and Applications, 2nd Edition* (Singapore: World Scientific Publishing Co. Pte. Ltd.)

[2] Bermann B 2012 *3-D printing: The new industrial revolution*, Business Horizons **55** 155-162

[3] Rifkin J 2012 *The Third Industrial Revolution: how the internet, green electricity, and 3-D printing are ushering in a sustainable era of distributed capitalism*, World Financial Review **1**

[4] Novakova-Marcincinova L 2012 Application of Fused Deposition Modeling Technology in 3D Printing Rapid Prototyping Area, *Manuf. and Ind. Eng.* **11** 35-37

[5] Noorani R 2005 *Rapid Prototyping: Principles and Applications*, John Wiley & Sons, New Jersey

[6] Melnikova R, Ehrmann A and Finsterbusch K 2014 3D printing of textile-based structures by Fused Deposition Modelling (FDM) with different polymer materials, *IOP Conf. Ser.: Mater. Sci. Eng.* **62** 012018

[7] Richter C, Schmülling S, Ehrmann A and Finsterbusch K 2015 FDM printing of 3D forms with embedded fibrous materials, *Design, Manufacturing and Mechatronics* 961-969

[8] Sabantina L, Kinzel F, Ehrmann A and Finsterbusch K 2015 Combining 3D printed forms with textile structures – mechanical and geometrical properties of multi-material systems, *IOP Conf. Ser.: Mater. Sci. Eng.* **87** 012005

[9] Evans K E, Nkansah M A, Hutchinson I J and Rogers S C 1991 Molecular Network Design, *Nature* **353** 124

[10] Grima J N, Attard D, Gatt R and Cassar R N 2009 A Novel Process for the Manufacture of Auxetic Foams and for Their re-Conversion to Conventional Form, *Adv. Eng. Mater.* **11** 533-535

[11] Scholz C, Wirner F, Götz J, Rüde U, Schröder-Turk G E, Mecke K, and Bechinger C 2012 Permeability of porous materials determined from the Euler characteristic, *Phys. Rev. Lett.* **109**, 264504

[12] Rant D, Rijavec T and Pavko-Cuden A 2013 Auxetic Textiles, *Acta Chim. Slov.* **60** 715-723

[13] Rant D, Ciobanu R, Blaga M and Pavko-Cuden A 2014 Compression of foldable links-links knitted structures, *Tekstil ve Konfeksiyon* **24** 349-355

IOP Conf. Series: Materials Science and Engineering **137** (2016) 012012 doi:10.1088/1757-899X/137/1/012012

Effects of epoxy/hardener stoichiometry on structures and properties of a diethanolamine-cured epoxy encapsulant

Y F Li[1], M Z Xiao, Z Wu, K Peng, C M Han, W Xiang and J Y Dai

Institute of Electrical Engineering, Chinese Academy of Engineering Physics, Mianyang 621900, P R China

Email: yuefangli_caep@163.com

Abstract. For the epoxy encapsulant cured by diethanolamine, optimal epoxy/hardener stoichiometry could hardly be predicted due to the complex curing mechanisms. In this paper, the influences of stoichiometry were investigated by FTIR, DMA and tensile testing. The results showed that stoichiometry has a dominating effect on both T_g and tensile properties of the cured epoxy. The largest T_g, highest crosslink density as well as excellent ductility appeared in epoxy encapsulant cured with 14 wt% diethanolmine. When the content of diethanolamine was lower than 14 wt%, epoxy encapsulants showed smaller glycidyl conversion even with long-duration post-cure. Larger tensile strength and modulus were also observed in the glycidyl-rich epoxies, which could be explained by anti-plasticization effect. The amine-rich epoxy, however, had extremely high glycidyl conversion and presented brittle tensile behavior. A diethanolamine content of 12-14 wt% for the epoxy encapsulant is suggested to obtain optimal thermal and tensile properties.

1. Introduction

Epoxy resins have broad applications in the impregnations, potting and encapsulations of high-voltage electronics, primarily due to their facile preparation, excellent thermal/mechanical properties and remarkable electrical insulation [1]. The resin/hardener ratio, also named stoichiometry, is the basis of an epoxy encapsulant formulation, which plays a dominating role in determining the properties of the cured resin [2, 3]. Particularly, stoichiometry in the interfacial area tends to deviate from bulk epoxy [4], affecting distinctly the interfacial performance of the epoxy encapsulants.

For typical epoxy hardeners, such as amines and anhydrides, a stoichiometric formulation could be readily predicted from the amount of reactive hydrogen of the hardener, which usually presents maximum T_g [5,6]. However, stoichiometric formulations do not necessarily guarantee optimal mechanical properties. Meyer et al associated improved fracture toughness and higher modulus of epoxy with a stoichiometric offset with excess glycidyl, taking 4,4'-diaminodiphenyl sulfone as the curing agent [5]. More ductile tensile behavior and enhanced impact strength have been reported in the amine-rich epoxy resins cured by triethylene tetramine [7].

Diethanolamine (DEA) has been widely used as epoxy hardener, which has the merits of long induction period, low curing exotherm and small curing stress [8]. The multiple hydroxyls in DEA also enhance the adhesion of epoxy to other materials. Unlike typical amine-based hardeners, DEA cures epoxy with complex mechanisms that mainly rely on self-catalyzed condensations and homo-polymerizations, making direct prediction of stoichiometry impossible. Thus there is a demand to experimentally acquire the stoichiometric formulation of DEA-cured epoxy encapsulants and establish the stoichiometry-structure-property correlations.

In this paper, stoichiometry effect upon DEA-cured epoxy encapsulant was studied through analyzing structural, thermal and tensile properties. The epoxy resin that yielded the highest T_g and crosslink density was considered to be the stoichiometric formulation. The abundance of glycidyl content was found to be essential in obtaining epoxy resins with optimized thermal and tensile properties. A proper range of DEA content for epoxy encapsulant was suggested according to the stoichiometry effect.

2. Experimental

2.1. Materials and sample processing

A diglycidyl ether bisphenol-A (DGEBA) based epoxy precursor (WSR 618) was purchased from Nantong Xingchen Synthetic Material. DEA was purchased from Aladdin. The chemical structures of DGEBA and DEA are shown in figure 1. To prepare epoxy encapsulants, DEA was mixed with DGEBA at fixed ratios shown in table 1, and the mixture was then degassed, casted into aluminum molds and cured.

Table 1. DEA/DGEBA formulations and cure schedules.

Sample	DEA content (wt%) [a]	Molar ratio of DEA/DGEBA	Cure schedule
D08	8	0.15:1	65 °C (10 h) + 90 °C (32 h)
D10	10	0.18:1	65 °C (10 h) + 90 °C (16h)
D12	12	0.22:1	65 °C (10 h) + 90 °C (16h)
D14	14	0.26:1	65 °C (10 h) + 90 °C (16h)
D16	16	0.30:1	65 °C (10 h) + 90 °C (16h)
D18	18	0.33:1	65 °C (10 h) + 90 °C (16h)
D20	20	0.37:1	65 °C (10 h) + 90 °C (16h)

[a] Based on the weight of DGEBA.

2.2. Instruments and characterizations

Attenuated total reflectance Fourier transform infrared (ATR-FTIR) analysis was performed on a FTIR spectrometer (Nicolet 6700) equipped with a SMART ITR accessory.

Dynamic mechanical analysis (DMA) was performed on a DMA Q800 (TA Instrument) using three point bending clamp. Measurements were conducted over a temperature ranging from -100 °C to 150 °C at a heating rate of 3°C min^{-1} and a frequency of 1.0 Hz.

Bulk density was measured at 25°C using a Mettler-Toledo XS204 balance adapted with a density determination kit.

Tensile tests were performed using dumbbell-shaped samples according to ASTM D638-96 on a MTS 800 instrument equipped with a 10 kN load cell. A crosshead speed of 5 mm·min^{-1} was used for all measurements.

3. Results and discussion

3.1. Reaction mechanism and ATR-FTIR analysis

It is known that both secondary amines and hydroxyls in DEA are reactive to glycidyls, while the reactivity of amine is much higher than that of hydroxyl [8]. As a result, reactions between DGEBA and DEA virtually occurred in a two-step manner. As shown in figure 1, Amine-glycidyl condensations takes place soon after mixing DEA with DGEBA, wherein adducts are generated containing tertiary amines with glycidyl ends (namely DEA-endcapped DGEBA). Crosslinked networks could only be formed through etherifications, including tertiary-amine-catalyzed hydroxide-glycidyl condensations and epoxy homo-polymerizations.

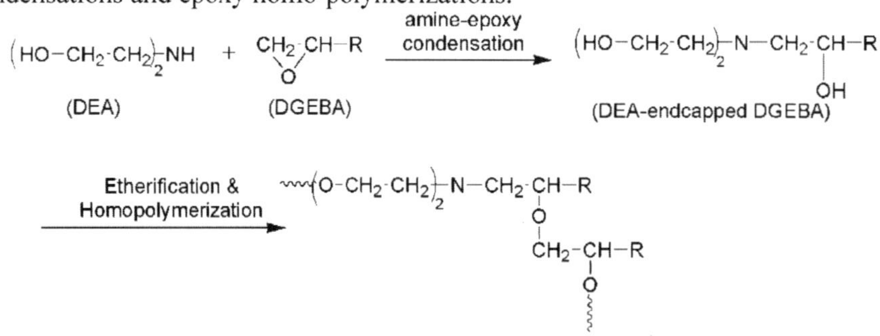

Figure 1. Schematics of DEA/DGEBA curing reactions.

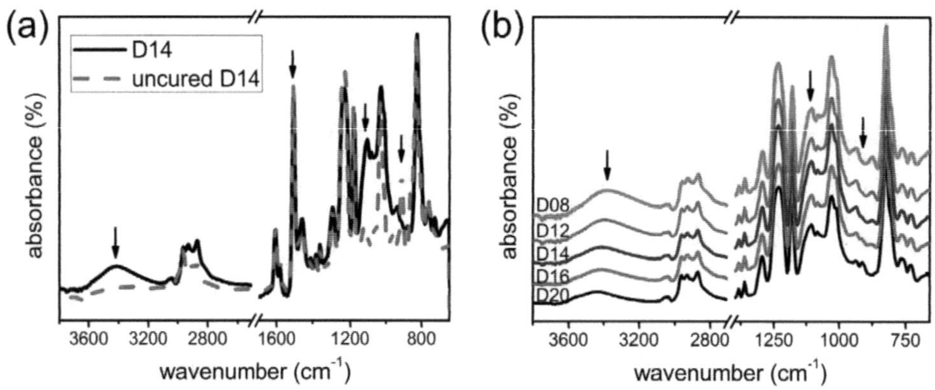

Figure 2. ATR-FTIR spectra of DEA/DGEBA before/after cure (a) and with varying stoichiometry (b).

Table 2. Normalized absorptions at 3400 cm^{-1} (A_{3400}) and 1109 cm^{-1} (A_{1109}) in ATR-FTIR.

Sample	A_{3400}/A_{1509}	A_{1109}/A_{1509}
D08	1.27	0.39
D10	1.36	0.40
D12	1.63	0.42
D14	1.67	0.47
D16	2.24	0.43
D18	2.47	0.41
D20	2.65	0.42

ATR-FTIR spectra taken before and after cure in figure 2(a) showed three major absorption changes at 3400 cm^{-1}, 1109 cm^{-1} and 913 cm^{-1}, which denoted the accumulation of hydroxyls in hydrogen-bonding states, generation of aliphatic ethers, and depletion of glycidyls. These absorption signals also changed when adjusting DEA content as shown in figure 2(b). All the characteristic absorption peak areas were normalized to the peak area of the aromatic ring at 1509 cm^{-1} (A_{1509}) to make quantitative comparisons. As listed in table 2, the normalized absorption around 3400 cm^{-1}

enhanced with increasing DEA content, indicating an increment of hydroxyl pendants with DEA/DGEBA stoichiometry. The normalized absorption at 1109 cm^{-1} reached maximum in D14, implying a most abundance of ether linkage in the formulation. Glycidyl conversion derived from the absorption peak at 913 cm^{-1} was taken as a measure for the cure degree of each epoxy formulation (table 3). Both D08 and D10 had a relatively-low glycidyl conversion around 90%. It should be noted that D08 epoxy underwent a long-duration post-cure treatment, otherwise the resin would be extremely brittle and could hardly be detached from the casting mold. Apparently, the deficient amount of DEA in D08 hindered the efficient construction of the epoxy network. Under the same curing schedule, increasing DEA content led to much higher glycidyl conversion in the cured epoxy resin. In the case of D18 and D20, no glycidyl signals could be detected from ATR-FTIR, indicating a complete cure.

Table 3. Major physical properties of epoxies fabricated with varying stoichiometry.

Sample	T_g(°C)	E'_{T_g+50} (MPa)	M_c ($g \cdot mol^{-1}$)	n ($mol \cdot m^{-3}$)	A_{beta}(min)[a]	Glycidyl conversion (%)	Density ($g \cdot cm^{-3}$)
D08	80	17.4	693	1730	0.1996	89.2	1.195
D10	79	16.3	738	1625	0.3026	90.6	1.195
D12	91	22.8	544	2207	0.4803	97.2	1.195
D14	93	24.8	502	2389	0.5614	98.6	1.195
D16	90	21.2	583	2057	0.6773	99.8	1.196
D18	77	19.1	627	1913	0.7553	100	1.197
D20	72	13.8	857	1400	0.6711	100	1.198

[a] Area of beta-relaxation peak obtained from tanδ curve of DMA.

3.2. DMA study

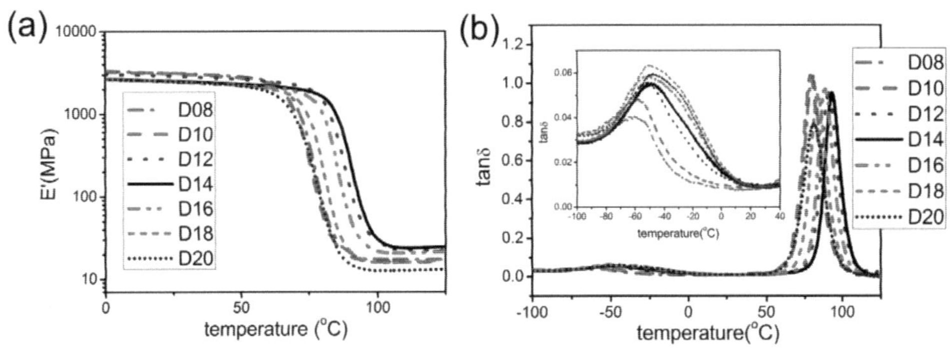

Figure 3. Storage modulus (a) and tanδ (b) curves for stoichiometric and off-stoichiometric DEA/DGEBA epoxies.

For the stoichiometric and off-stoichiometric epoxies, evolutions of storage modulus and *tanδ* as a function of temperature were studied by DMA, as shown in figure 3. It appears that T_g was significantly affected by stoichiometry, as shown in table 3. The maximum T_g appeared in D14, and the T_gs of D12 and D16 were only slightly lower than that of D14. However, T_g dropped sharply when the DEA content decreased to below 12 wt% or increased to above 16 wt%. The smaller T_g in D08 and D10 should be due to the lower glycidyl conversion and subsequent inefficient construction of crosslink network. The smaller T_g in D18 and D20, which had extremely high glycidyl conversion, might be related to complex reaction mechanisms as discussed later.

In figure 3(a), no matter under T_g or in the rubbery state, the storage modulus varied with DEA/DGEBA stoichiometry. According to previous reports [2, 9], the apparent crosslink densities (n) and the molecular weight between crosslinks (M_c) of epoxies could be estimated from the rubbery modulus at T_g+50 °C (E'_{T_g+50}) using the following equation:

$$E'_{T_g+50} = 3nRT = 3q\rho RT / M_c \quad (1)$$

Where q is front factor and is usually taken as 1 when crosslinks of the networks are mobile, ρ is density, and R is gas constant. It should be emphasized that equation (1) is an empirical equation and only provide an approximate trend for comparative evaluation [5].

The estimation results of M_c and n are also listed in table 3. As the DEA content increased from 8 wt% to 20 wt%, M_c changed in a reverse manner to T_g. The highest crosslink density as well as the lowest M_c also occurred in D14 epoxy. Therefore, D14 should represent the stoichiometric formulation for DEA/DGEBA encapsulant.

Broad beta-relaxations were observed for all formulations (insert in figure 3(b)), typically in the temperature range between -80 °C and 20 °C. Particularly, both width and height of the beta-relaxation peak increased monotonously with the DEA content. The origins of beta-relaxation in epoxy are complex, potentially involving motions of hydroxylether, flips of diphenolpropane ring and even motions of crosslinks [10].

3.3. Interpretation of crosslink structures

Based on the above analysis, the network structure evolution of epoxies with varying DEA/DGEBA stoichiometry was depicted in figure 4.

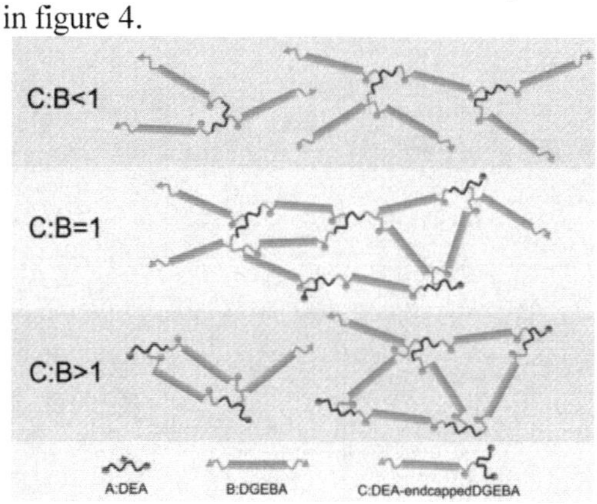

Figure 4. Schematic illustration of network structures in stoichiometric and off-stoichiometric states.

At low DEA content, the formulation has less amount of DEA-endcapped DGEBA adducts and more remnant glycidyl pendants. According to literatures [11], both tertiary amines and secondary hydroxyls formed in DEA-endcapped DGEBA have acceleration effects on epoxy etherifications. Etherification reactions are thus less favorable at low DEA content, wherein an epoxy network with smaller glycidyl conversion and lower crosslink density is generated, as has been proved by DMA and ATR-FTIR.

As the DEA content increases, increment of crosslink density is expected due to the abundant tertiary amines and hydroxyls. Maximum T_g appeared in D14, the stoichiometric DEA/DGEBA formulation, where the molar ratio between DEA-endcapped DGEBA and intact DGEBA was about 1. Under this condition, the number of hydroxyls (including the primary hydroxyls from DEA and the secondary hydroxyls in endcapped DGEBA) approximates the number of residual glycidyls (including intact DGEBA and endcapped DGEBA).

At high DEA content, the amine-glycidyl condensation consumes more glycidyls to form DEA-endcapped DGEBA adducts. The crosslinking reaction rate is increased due to the further increased concentrations of tertiary amines and hydroxyls. The crosslink density is decreased, however, primarily due to the lack of abundant glycidyls for chain growth. Another possible reason is that undesirable chain transfers and chain terminations in epoxy homo-polymerization become more favorable in the presence of high amount of tertiary amines [12], which also hinders chain propagations.

3.4. *Analysis of tensile properties*

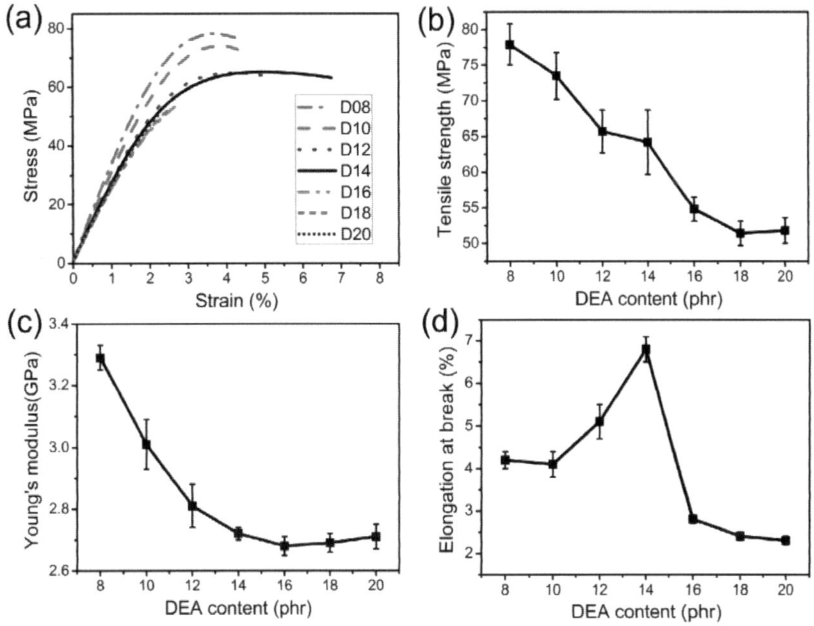

Figure 5. Stress-strain curves (a), tensile strength (b), Young's modulus (c) and strain at break (d) of DEA/DGEBA as a function of DEA content.

The stress-strain curve shapes of DEA/DGEBA epoxies also changed with stoichiometry, as shown in figure 5(a). The glycidyl-excess formulations exhibited yielding under large tension, resulting in higher elongation at break. Both maximum tensile strength and maximum Young's modulus appeared in the DEA-deficient D08, as shown in figure 5. The high modulus in epoxies with lower T_g/crosslink density has been explained by the concept of free volume [5]. The bulk densities of DEA/DGEBA epoxies were herein measured as a coarse measure of room temperature free volume [4], which only showed slightly variations among different formulations (table 3). It seemed that the tiny density variations in epoxies were too small to take charge of such a distinct difference in mechanical behavior. The so-called anti-plasticization effect reported by Won et al in amine-cured epoxy [13,14] provides a reasonable explanation. Anti-plasticization effect appears as a simultaneous decrease of modulus and increase of T_g, and had been associated with the local mobility of crosslinked chains [13]. In the case of D08 and D10, epoxy conversions were only around 90%. The pendant glycidyl chain ends could have played the role of internal plasticizer by reducing crosslink density. Similar to the results of Won [13], smaller beta-relaxation intensity occurred in the glycidyl-excess epoxy, as shown in figure 3(b). The lower intensity of high temperature part in beta-relaxation peak in D08 and D10 suggested a suppressed short-range molecular motion by the antiplasticizer. We therefore ascribe the ductile

IOP Conf. Series: Materials Science and Engineering **137** (2016) 012012 doi:10.1088/1757-899X/137/1/012012

tensile behavior and higher Young's modulus of the glycidyl-rich resins to the anti-plasticization effect.

Epoxies with DEA-excess formulations, on the contrary, fractured in a brittle manner under tension. As a result, lower tensile strength and ultra-small elongation at break were obtained for all the DEA-excess epoxies (Figure 5). Previous studies on aromatic amine-cured epoxies have suggested that extremely-high glycidyl conversion in the fully-cured epoxy leads to decrease of epoxy toughness [15]. In addition, faster crosslinking reaction rate and undesirable chain transfer/terminations in amine-rich formulations were reported to induce local stress buildup [5], inhomogeneity [12] and micro-defects, which could also cause reduction of the ductility and toughness. As a matter of fact, notable defects in DEA-excess epoxies were observed when applying epoxy resin of large quantity to potting applications. Thus both the high glycidyl conversion, as well as network inhomogeneity in DEA-rich epoxies could contribute to the decreasing tensile strength and brittle tensile fractures.

4. Conclusions

This paper demonstrates that DEA/DGEBA stoichiometry has profound influence on both the structures and the thermal/mechanical properties of epoxy encapsulants. The epoxy possessed highest T_g and largest crosslink density when DEA content reached 14 wt%, while either increasing or decreasing DEA content resulted in reducing of T_g. All the DEA-deficient formulations showed smaller glycidyl conversion in FTIR, weaker beta-relaxation in DMA, and larger strength and modulus in tensile testing, which was in support of an anti-plasticization effect of glycidyl pendants in epoxy network. Brittle tensile behavior was observed in the DEA-rich formulations, which was ascribed to the ultra-high glycidyl conversion and network inhomogeneity caused by the complex curing mechanism. The present results suggested an optimal DEA content of 12~14 wt% for higher T_g and better tensile properties of the epoxy encapsulant.

References

[1] Mohan P 2013 A Critical Review: The Modification, Properties, and Applications of Epoxy Resins *Polym.-Plast. Technol.* **52** 107

[2] Pramanik M, Fowler E W and Rawlins J W 2014 Another look at epoxy thermosets correlating structure with mechanical properties *Polym. Eng. Sci.* **54** 1990

[3] Bignotti F, Pandini S, Baldi F and De Santis R 2011 Effect of the resin/hardener ratio on curing, structure and glass transition temperature of nanofilled epoxies *Polym. Composite* **32** 1034

[4] Vanlandingham M R, Eduljee R F and Gillespie J 1999 Relationships between stoichiometry, microstructure, and properties for amine-cured epoxies *J. Appl. Polym. Sci.* **71** 699

[5] Meyer F, Sanz G, Ecerza A, Mondragon I and Mijovic J 1995 The effect of stoichiometry and thermal history during cure on structure and properties of epoxy networks *Polymer* **36** 1407

[6] Garcia F G, Soares B G, Pita V, Sanchez R and Rieumont J 2007 Mechanical properties of epoxy networks based on DGEBA and aliphatic amines *J. Appl. Polym. Sci.* **106** 2047

[7] d'Almeida J R and Monteiro S N. 1998 The influence of the amount of hardener on the tensile mechanical behavior of an epoxy system *Polym. Advan. Technol.* **9** 216

[8] Adolf D and Chambers R 1997 Verification of the capability for quantitative stress prediction during epoxy cure *Polymer* **38** 5481

[9] Crawford E and Lesser A J 1998 The effect of network architecture on the thermal and mechanical behavior of epoxy resins *J. Polym. Sci. Pol. Phys.* **36** 1371

[10] Heux L, Halary J L, Laupretre F and Monnerie L 1997 Dynamic mechanical and 13C n.m.r. investigations of molecular motions involved in the beta relaxation of epoxy networks based on DGEBA and aliphatic amines *Polymer* **38** 1767

[11] Xu L and Schlup J R 1998 Etherification versus amine addition during epoxy resin/amine cure: an in situ study using near-infrared spectroscopy *J. Appl. Polym. Sci.* **67** 895

[12] Galante M J, Vazquez A and Williams R J 1991 Macro- and microgelation in the homopolymerization of diepoxides initiated by tertiary amines *Polym. Bull.* **27** 9

[13] Won Y, Galy J, Gerard J, Pascault J, Bellenger V and Verdu J 1990 Internal antiplasticization in copolymer and terpolyrner networks based on diepoxides, diamines and monoamines *Polymer* **31** 1787

[14] Vassileva E and Friedrich K 2003 Epoxy/Alumina nanoparticle composites. I. dynamic mechanical behavior *J. Appl. Polym. Sci.* **89** 3774

[15] Weeks T S, Adolf D and McCoy J D 1999 Cohesive failure in partially cured epoxies *Macromolecules* **32** 1918

Investigation of laser heating effect of metallic nanoparticles on cancer treatment

G S Shan[1,4], X M Liu[1,4], H J Chen[2], J S Yu[1], X D Chen[3], Y Yao[1], L M Qi[1] and Z J Chen[1]

[1]School of Electronic Engineering, Beijing University of Posts and Telecommunications, Beijing 100876, China
[2] SYSU-CMU Shunde International Joint Research Institute, Shunde 510006, China
[3]School of Electronic Engineering and Computer Science, Queen Mary University of London, London E14NS, United Kingdom

E-mail: guangshuai_shan@qq.com; xiaoming_liu@bupt.edu.cn

Abstract. Metallic nanoparticles can be applied for hyperthermia therapy of cancer treatment to enhance the efficacy because of their high absorption rate. The absorption of laser energy by metallic nanoparticles is strongly dependent on the concentration, shape, material of nanoparticles and the wavelength of the laser. However, there is no systematic investigation on the heating effect involving different material, concentration and laser wavelength. In this paper, gold nanoparticles (AuNPs), sliver nanoparticles (AgNPs) and sliver nanowires (AgNWs) with different concentrations are heated by 450nm and 532nm wavelength laser to investigate the heating effect. The result shows that the temperature distribution of heated metallic nanoparticles is non-uniform.

1. Introduction

Nanomaterials have been widely investigated for biomedical applications, such as drug delivery, bioimaging, photothermal cancer treatment[1-5]. Metallic nanoparticles, due to their nature of good biocompatibility, being easy of preparation, and strong absorption and scattering effect of light laser, are ideal candidates for photothermal cancer treatment.[6-7] A laser is an important tool for metallic nanoparticle heating.[8] There are many factors affecting g the heating efficiency of metallic nanoparticles, such as the wave length of the laser, the shape, concentration, and material of metallic nanoparticles.[9-10] However, there is no comprehensive study regarding different materials, concentration, and morphology. To narrow this knowledge gap, a preliminary experimental scheme for investigation of the heating effects of 450nm, 532nm laser of different concentration AuNPs, AgNPs and AgNWs from 25μmol/L to 150μmol/L in the interval of 25μmol/L has been designed.

2. Experimental Procedure

The experimental setup for laser heating is illustrated as figure 1. A cuvette was placed in a chamber with two 50mm by 50mm square holes. The laser was aligned to metallic nanoparticle dispersion through the top hole, and the lateral was left for thermal imager detection. A fiber optical temperature meter monitored the temperature change of dispersion. The optical fiber was attached to the inner wall of the cuvette to avoid the direct illumination with an accuracy of ± 0.1℃.

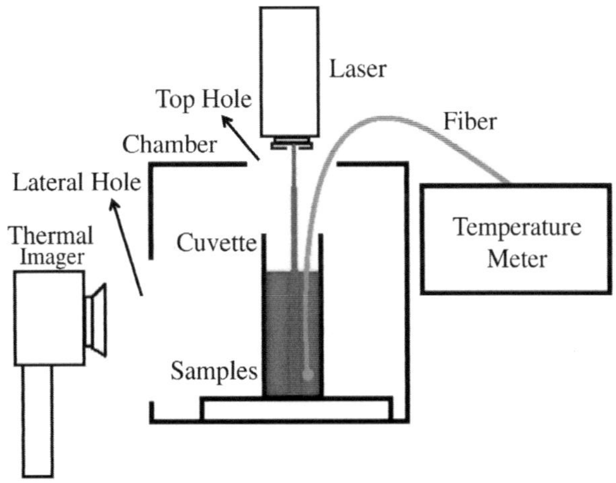

Figure 1. Experimental set up for laser heating.

The experimental procedure is as follows:

- Drop 4ml sample into the cuvette with a dropper;
- Put the cuvette in the chamber for 10 mins to achieve thermal equilibrium;
- Record the temperature of the original sample by using the fiber temperature meter and thermal imager respectively;
- Illuminate the sample for 10 mins with the 450nm/532nm laser;
- Record rise in temperature of the sample by using the fiber temperature meter and thermal imager respectively.

3. Results and Discussion

The original samples are shown in figure 2 (a), (b) and (c). The temperature profiles of dispersion recorded by thermal imager are presented in figure 2 (d). As for AgNPs and AuNPs, it clearly shows that non-uniform temperature distribution can be observed for high concentration samples, while relative uniform heating can be found for low concentration samples. However, in the AgNWs samples, the temperature distribution is always uniform whatever the concentration is. Taking the AgNPs samples heated by 532nm laser as an example, the temperature distributions of samples with the concentration of 25, 50, and 75μmol/L are uniform, as can be seen from the temperature profile. For higher concentrations, 100, 125, and 150μmol/L, however, layered temperature distribution can be observed. In light of this phenomenon, the temperature measurement was conducted with the sensor placed at the top, middle and bottom layers, respectively. The temperature rises of dispersion with different concentration were plotted in figure 3.

For AgNPs samples, the temperature increases with the concentration whatever the concentration is. However, the bottom layers are always less heated than their upper layers. In comparison, for AuNPs, the temperature rises in the upper area increase with the concentration. However, the middle layer and bottom layer show a turning point depending on the wave length of the laser. The rise in temperature of AgNWs is less than that of eithor the AgNPs or AuNPs.

It is seen that the AuNP dispersions are more heated than AgNP samples with the same concentration. For instance, samples at the concentration of 100 μ mol/L, heated by 532nm laser, the temperature rise of the upper, middle and bottom layer of AuNPs are 9.3, 7.3 and 4.9℃, which are stronger than the rise of AgNPs, 5.6, 5.2 and 4.3℃ respectively. The AgNP samples are more heated than AgNW dispersions with the same concentration. For instance, the maximum temperature rise of AgNWs samples heated by 450nm laser is 1.9℃, which is even weaker than the minimum, 2.1℃, of

IOP Conf. Series: Materials Science and Engineering **137** (2016) 012013 doi:10.1088/1757-899X/137/1/012013

AgNPs under the same condition. It seems AuNPs are better for photothermal therapy. However, from the layered temperature, we can make the conclusion that AuNPs with worse penetration depth. are more absorptive than AgNPs.

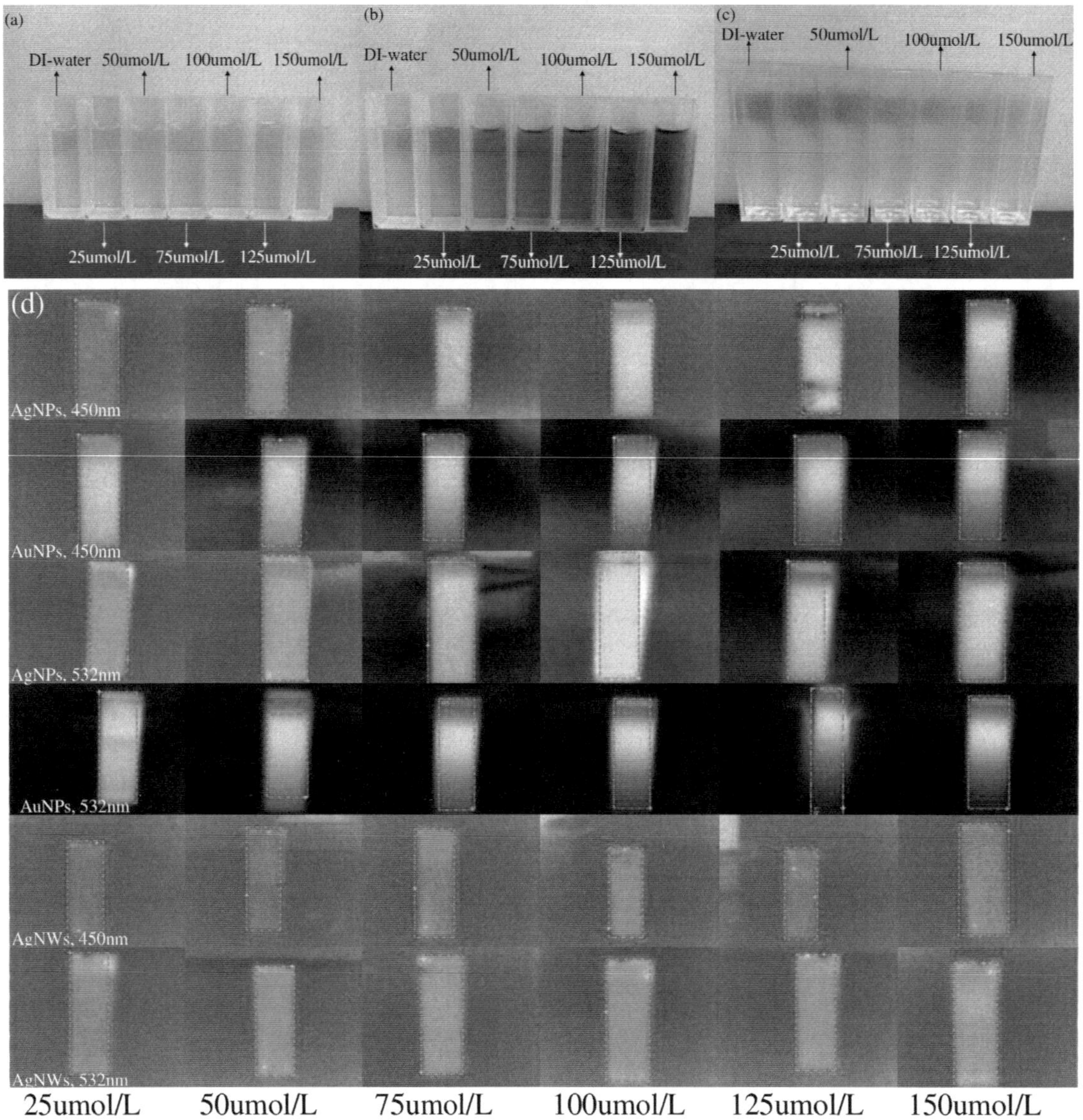

Figure 2.(a) AgNPs original samples (b) AuNPs original samples (c) AgNWs original samples (d) Temperature profiles recorded by the thermal imager.

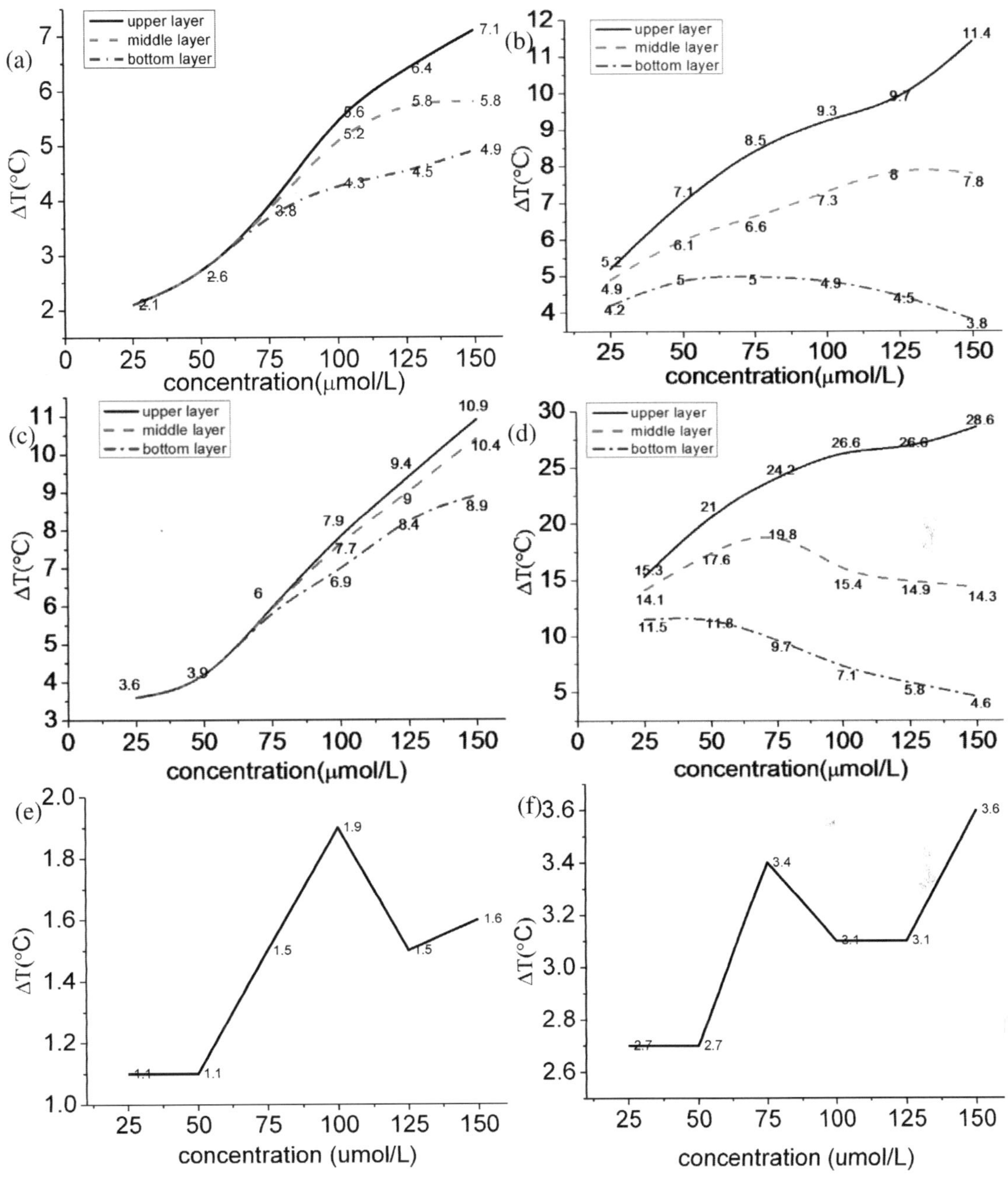

Figure 3. The temperature rises with different concentrations (a) AgNPs, 450nm (b) AuNPs, 450nm (c) AgNPs, 532nm (d)AuNPs, 532nm (e) AgNWs, 450nm (f) AgNWs, 532nm

4. Conclusion

The laser heating effect of metallic nanoparticles with different concentrations has been investigated in this paper. The conclusion is that the rise temperature of AgNPs and AuNPs with high concentrations are layered. The AuNP dispersions are more heated than AgNP and AgNW samples with the same concentration. The AuNPs with worse penetration depth are more absorptive than AgNPs..

5. Acknowledgments

This work is supported by the National Nature Science Foundation of China under the contract number of 61401031, and the fund from Shunde Government, Guangdong Province, China, under the contract number of 20140401.

References

[1] Cho K, Wang X, Nie S, Chen Z and Shin D-M 2008 Therapeutic nanoparticles for drug delivery in cancer *Clin. Cancer Res.* **14** 1310

[2] Tiwari P M, Vig K, Dennis V A an d Singh S R 2011 Functionalized gold nanoparticles and their biomedical application *Nanomaterials* **1** 31

[3] Lim Z J, Li J J, Ng C, Yung L L and Bay B 2011 Gold nanoparticles in cancer therapy *Acta Pharmacol. Sinica* **32** 983

[4] Muddineti O S, Ghosh B and Biswas S 2015 Current trends in using polymer coated gold nanoparticles for cancer therapy *Inter. J. Pharm* **484** 252

[5] Homberger M and Simon U 2010 On the application potential of GNP in nanoelectronics and biomedicine *Phil.Trans.R.Soc.A* **368** 1405

[6] Cai W, Gao T, Hong H and Sun J 2008 Applications of gold nanoparticles in cancer *Nanotech. Sci App.* **1** 17

[7] Jain P K, Lee K S, EI-Sayed I H and EI-Sayed M A 2006 Calculated Absorption and Scattering Properties of Gold Nanoparticles of Different Size, Shape, and Composition: Applications in Biological Imaging and Biomedicine *J.Phys.Chem.B* **110** 7238

[8] Fasla B, Benmouna R and Benmouna M 2014 On the Hyper Thermal Therapy of Tumor Tissues by Direct Laser Heating and Gold Nano Particles *J.Biomater.Nanobiotech.* **5** 44

[9] Qin Z P and Bischof J C 2014 Thermophysical and biological responses of gold nanoparticle laser heating *Chem.Soc.Rev.* **41** 1191

[10] Pustovalov V K 2005 Theoretical study of heating of spherical nanoparticle in media by short laser pulses *Chem. Phys.* **308** 103

IOP Conf. Series: Materials Science and Engineering **137** (2016) 012014 doi:10.1088/1757-899X/137/1/012014

Effects of process parameters on the molding quality of the micro-needle array

Z J Qiu[1], Z Ma and S Gao

State Key Laboratory of Precision Measuring Technology & Instruments, Tianjin University, Tianjin, CHINA

E-mail: qiuzhongjun@tju.edu.cn

Abstract. Micro-needle array, which is used in medical applications, is a kind of typical injection molded products with microstructures. Due to its tiny micro-features size and high aspect ratios, it is more likely to produce short shots defects, leading to poor molding quality. The injection molding process of the micro-needle array was studied in this paper to find the effects of the process parameters on the molding quality of the micro-needle array and to provide theoretical guidance for practical production of high-quality products. With the shrinkage ratio and warpage of micro needles as the evaluation indices of the molding quality, the orthogonal experiment was conducted and the analysis of variance was carried out. According to the results, the contribution rates were calculated to determine the influence of various process parameters on molding quality. The single parameter method was used to analyse the main process parameter. It was found that the contribution rate of the holding pressure on shrinkage ratio and warpage reached 83.55% and 94.71% respectively, far higher than that of the other parameters. The study revealed that the holding pressure is the main factor which affects the molding quality of micro-needle array so that it should be focused on in order to obtain plastic parts with high quality in the practical production.

1. Introduction

At present, the rapid development of microsystem technology sets higher requirements for the manufacturing of micro parts. An efficient manufacturing processes of micro parts with low-cost, reliability and stability is expected [1]. As a key technique for microsystems, micro-injection molding process is available for manufacturing micro components. It offers the capability to massively produce micro components with relatively low costs and short-cycle times, and the potential for full automation and accurate replication [2]. Therefore, micro-injection molding process attracts more and more attentions home and abroad. There is no wonder that its products which are called micro-molded parts are widely applied in information and communication technology, medical and biotechnology, micro-sensor technology and daily life [3].

The concept of micro-needle was put forward in the 1970s. Due to the limitation of process conditions, micro-needle was not applied in transdermal drug delivery and blood extraction fields until 1998. In recent years it started to get wide attentions. Micro-needles can be characterized into two groups [4]: (1) in-plane design where the micro-needle shank is in the same plane as the substrate and (2) out-of-plane design where micro-needles are made perpendicular to the substrate. With the development of various technologies, micro-needles or micro-needle arrays with different types and different sizes have been researched and manufactured. Sammoura et al. used plastic injection molding to fabricate an in-plane, open-channel micro-needle [4]. Mold insert temperature control is identified as the major reason for the discrepancy between experiments and simulation results. Sha et al. investigated the effects of three process (barrel temperature, mold temperature and injection speed) and one geometric factors on the surface quality of micro-pins in three different polymer materials [5]. The experimental results revealed that the effects of these factors on the process replication capabilities are not consistent for different polymer materials. Zhang and Lu analyzed the morphology and performance of polypropylene (PP) microstructures manufactured by micro injection molding and fabricated micro-needles with ϕ130μm×depth 250μm dimensions [3].

Micro-needle arrays belong to micro components with high surface-to-volume ratios and high aspect ratios. Compared with conventional injection molding, micro-injection molding is confronted with large challenges and more difficulties, which leads to micro parts with poor replication quality. In addition other micro-molded parts face the same problem. Many studies that used different experimental conditions and test parts including micro-needles were carried out [6]. The results showed that individual process parameter has a significant effect on the molding quality of the produced components. Sha et al. studied the effects of five process (barrel temperature, mold temperature, injection speed, holding pressure, the existence of air evacuation) and one size factors on the achievable aspect ratios, and the roles they play in producing micro components in different polymer materials [7]. The results revealed that the barrel temperature and the injection speed are key factors affecting the aspect ratios of micro features replicated in PP and Acrylonitrile-Butadiene-Styrene (ABS). Lin and Young observed that compared with increasing holding pressure, increasing mold temperature is more effective to improve the polymer melt to fill into the microstructure with high aspect ratio [8]. Zhao et al. found that metering size and holding time are the process parameters that have the most significant effects on part quality [1].

In this paper, the injection molding process of the micro-needle array was studied and the orthogonal experiment and the contribution analysis were carried out. The shrinkage ratios and warpages of micro needles under different process conditions were compared to find the effects of the process parameters on the molding quality of the micro-needle array. The single parameter method [9], namely one parameter changed while the rest of the parameters kept constant, was used to analyze the significant process parameter to provide theoretical guidance for practical production and to realize the goals of high quality, low costs and low rejection ratio.

2. Molding defects

Due to the molding material characteristics, the cavity complexity, the machine performance and the process parameters, molding defects appear inevitably in the plastic parts, such as short shots, bubbles, weld lines, ring grooves, warpage, shrinkage and stress marks. The molding quality, particularly the shrinkage and the deformation of the micro needles, will have a great influence on the part demolding behavior. Because of the shrinkage and deformation of the molded parts, the contact pressure between the part and the mold may increase [10]. Moreover, the part-mold friction force becomes lager, leading to a difficult demolding behavior. The needles may be damaged and seriously fractured. Therefore, deformation and shrinkage are the defects that cannot be ignored in the process of injection molding, which need to be studied deeply.

2.1. Shrinkage

The shrinkage refers to the reduction of the shape size caused by the characteristic of heat-expansion and cold-contraction, the elastic recovery of the plastic and demolding deformation. According to the shrinkage ratios, the plastic parts can match well with the original size to ensure product sizes within the tolerances of the cavities [11].

2.2. Warpage

The phenomenon that the shapes of the plastic parts deviate from the mold cavity boundaries is called warpage deformation [11]. It is one of the common defects which appear in plastic parts. With the development of the plastics industry, the appearance quality and functional performance of plastic parts meet with higher and higher requirements. As one of the most important indices to evaluate the quality of the products, warpage deformation attracts more and more attention.

3. Experimental setup

3.1. Test part design and material

Figure 1 presents a micro-needle array whose diameter and height is ϕ100μm and 1mm respectively, perpendicular to a substrate whose diameter and height are 5mm and 1mm respectively. The micro needles are regarded as the microstructures. The number of micro-needle array is set 3×3 for convenient analysis and low computational complexity.

The dimension of the micro needles is in the micrometer range while that of the substrate is in the millimeter. Due to the great gap in size between micro-needle and substrate, the difficulty for the injection molding of the micro needles becomes much larger. Short shots and deformation defects are easier to arise, leading to bad needles with poor quality. As a result, the effect mechanism of process parameters on the molding quality must be clear in order to obtain micro-needle arrays with high quality.

A commonly used material in injection molding, PP is selected to conduct the planned experiments. Due to its good flowability and stability, PP is widely used for spring elements, clips, gear wheels, and small machinery products.

Figure 1. The micro-needle array model obtained through the SolidWorks software

3.2. Evaluation indices

To investigate the effects of the processing parameters on part replication, this experimental research was focused on the shrinkage ratio and the warpage of the micro needles.

From the bottom to the top of a micro needle, the shrinkage ratio and the warpage are both increasing gradually (shown in figure 2. (a) and (b)). Because the shrinkage ratios and warpages of all the needles are different, we can't use a needle's values to represent the entire value of the array. In order to better describe the shrinkage ratios and warpages of the entire array and reduce the data contingency, we take the averages of top values as the evaluation indices as shown in figure 2. (c) and (d). All the values were obtained under the same process condition that the injection speed, mold temperature, melt temperature, holding pressure and holding time is 40mm/s, 40° C, 220° C, 40MPa and 4s respectively.

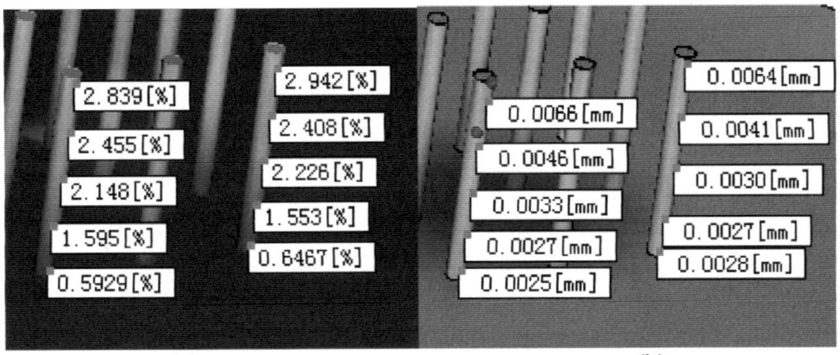

(a) (b)

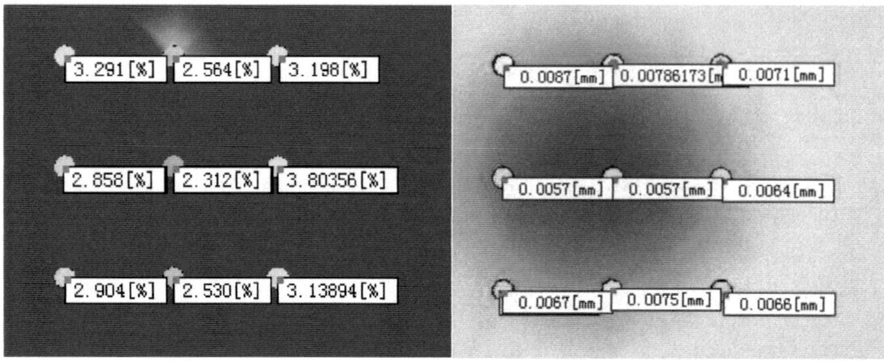

(c) (d)

Figure 2. The shrinkage ratios and warpages of micro needles

3.3. Design of experiments

As the significant parameters, injection speed (S), mold temperature (Tmol), melt temperature (Tm), holding pressure (P) and holding time (T) were investigated in this study. Several pre-experiments were conducted before the orthogonal experiment to get the processing windows and promise complete filling.

According to the levels of each factor (table 1), a L16 orthogonal experiment (table 2) was conducted to reduce the trial numbers without repetition and to ensure that the experimental results were representative of the considered processing windows for the selected material [7].

Table 1. The experimental factors and levels

Factors / levels	S(mm/s)	Tmol(°C)	Tm(°C)	P(MPa)	T(S)
1	40	40	220	40	4
2	60	55	235	60	6
3	80	70	250	80	8
4	100	85	265	100	10

3.4. Results and analysis

From table 2, it is obviously observed that as the parameters change, the shrinkage and warpage of the micro-needle array both have different changes. To find the significant process parameters affecting the molding quality, a contribution analysis method was used. First, we finished analyzing variance through the orthogonal assistant software. After getting sum of square of deviations (Q) of a factor, we calculated its contribution rate. The contribution rate refers to the proportion of the factor's Q to the sum of each factor' Q. The contribution rates of each parameter on shrinkage ratio and warpage were calculated according to the following equations and listed in table 3.

$$Qs = Q_1 + Q_2 + Q_3 + Q_4 + Q_5$$

$$R_n = \frac{Q_n}{Qs} \times 100\%, n = 1,2,3,4,5$$

Q is sum of square of deviations of a factor, R*n* is the contribution rate of the factor.

Table 2.The results of the L16 orthogonal experiment

trials	S	Tmol	Tm	P	T	Shrinkage ratio (%)	Warpage (μm)
1	1	1	1	1	1	2.956	6.92
2	1	2	2	2	2	2.714	7.10

3	1	3	3	3	3	2.236	4.96
4	1	4	4	4	4	2.220	5.86
5	2	1	2	3	4	0.574	2.24
6	2	2	1	4	3	7.134	11.49
7	2	3	4	1	2	2.492	6.70
8	2	4	3	2	1	3.346	8.39
9	3	1	3	4	2	2.829	6.86
10	3	2	4	3	1	2.293	5.50
11	3	3	1	2	4	3.902	9.81
12	3	4	2	1	3	2.889	8.64
13	4	1	4	2	3	3.207	7.82
14	4	2	3	1	4	3.221	7.60
15	4	3	2	4	1	3.040	6.30
16	4	4	1	3	2	2.643	5.93

Table 3. The contribution rates of each parameter on shrinkage ratio/warpage.

Factors	sum of square of deviations(Q)	degree of freedom	Standard Deviation	Contribution rate (%)
S	1.050/1.830	3	0.350/0.610	8.61/2.67
Tmol	0.293/0.635	3	0.064/0.212	2.41/0.93
Tm	0.211/0.145	3	0.070/0.048	1.73/0.21
P	10.184/64.778	3	3.395/21.59	83.55/94.71
T	0.451/1.010	3	0.150/0.337	3.70/1.48
Sum(Qs)	12.189/68.398	15		

From table 3, we can find that each parameter has an effect on the shrinkage and warpage of the micro-needle array. But, the degrees of effect are different. The contribution rate of holding pressure on the shrinkage ratio and warpage are 83.55% and 94.71% respectively, much higher than the other parameters. This indicates that holding pressure is the most significant factor affecting the molding quality of the micro-needle array. In practical production, holding pressure should be focused on in order to obtain plastic parts with high quality.

4. Effects of holding pressure on molding quality

In the previous section, combined with the orthogonal experiment and the contribution analysis method, the holding pressure was found to be the most significant process parameter affecting the molding quality of the micro-needle array. In order to know how the holding pressure affects the molding quality clearly, a single parameter experiment on holding pressure, namely the holding pressure changed while the rest of the process parameters kept constant, was designed and conducted, expecting to provide feasible theoretical guidance for practical production and reduce rejection rate.

The holding pressure refers to the injection molding pressure on the melt in the holding phase. It can help to reduce the shrinkage, improve the density of plastic parts and overcome the defects on the surface of the plastic [11]. At the same time, it can also prevent the reverse flow of melt in the cavity.

In the experiment, other process parameters were kept constant and only holding pressure was changed. The process condition was: Injection speed 60mm/s, Mold temperature 55°C, Melt temperature 235°C and Holding time 4s. Figure 6 presents the change of shrinkage ratio and warpage obtained through the same method in the section 3.2.

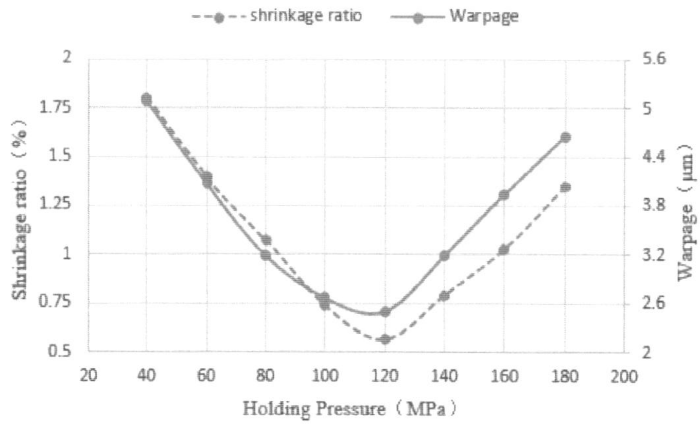

Figure 3. The variation trends of the Shrinkage ratio and warpage on holding pressure

In figure 3, the shrinkage ratio and warpage value of the micro needle array both present a parabolic trend with continuous increase of the holding pressure. Due to the shrinkage of the melt in the cavity, gap appears between the melt and the mold. The purpose of the holding stage is to drive a small amount of melt into the cavity to compensate the gap and ensure the cavity wholly filled [12]. In the experiment, the peak value of the injection pressure was observed to be a constant value when the mold temperature, melt temperature and injection speed were all kept constant, no matter how the holding pressure and holding time were changed. Because of the characteristic of heat-expansion and cold-contraction, the melt in cavity will shrink quickly with the heat loss. And the gate cools and solidifies in a short time. Therefore, if the holding pressure is low at this point, the cavity will not get sufficient melt before the gate solidifies absolutely. As a result, high shrinkage ratios and warpages will arise in the micro-needle array. With the continuous increase of the holding pressure, the pressure transfers effectively in the cavity to make more melt flow into the cavity to reduce shrinkage and warpage. The pressure can drive a large quantity of melt into the cavity when it exceeds a certain value, leading to part expansion, which also causes obvious warpage deformation. Combined with the variation tendency of the shrinkage ratio and the warpage versus the holing pressure, the conclusion can be obtained: increasing the holding pressure can effectively decrease the shrinkage ratio and the warpage of the micro-needle array. However it should be noted that too much pressure will not only lead to part expansion but also incur higher requirements for the equipment. Of course, in practical production, the holding pressure must be given at an appropriate value, neither too high nor too low, to reduce shrinkage and warpage as much as possible.

5. Conclusions

Micro-needle array is a kind of typical injection products with microstructures. Owing to its tiny micro-features size and high aspect ratios, it is more likely to produce short shots and deformation defects, leading to poor molding quality during micro injection molding. The injection molding process of micro-needle array was studied in this paper to find the effects of the process parameters on the shrinkage ratio and warpage of the micro-needle array and to provide theoretical guidance for practical production of high-quality products, combined with the orthogonal experiment and the contribution analysis method. The study showed that the shrinkage ratio and the warpage are both increasing from the bottom to the top of the micro needles gradually. The holding pressure was proved to be the most significant factor affecting the molding quality, whose effectiveness is much larger than that of other factors. The shrinkage ratio and warpage value of the micro-needle array both displayed a parabolic trend with the continuous increase of the holding pressure, reaching a minimum value at one point. When the holding pressure is too low, the cavity can't get enough melt to be completely filled, leading to obvious shrinkage and warpage deformation. Then the phenomenon gets improved as the holding pressure rises. When the pressure exceeds a certain value, the melt is compressed excessively. This can cause part expansion and obvious warpage deformation. Therefore, the holding pressure should be focused on in order to reduce the shrinkage and warpage and obtain plastic parts with high quality in the practical production.

Acknowledgements
The authors would like to thank the Natural Science Foundation of China (Nos. 51375336 &50905125) and Key Program of Tianjin Municipal National Natural Science of China (No. 12JCZDJC27900) for the financial support.

References

[1] Zhao J, Mayes R H, Chen G, Xie H and Chan P S 2003 *J. Polym. Eng. Sci.***43** 1542-54
[2] AttiaU M, Marson S and Alcock J R 2009 *J. Microfluid. Nanofluid.* 71-28
[3] Zhang K F and Lu Z 2008 *J. Microsyst. Technol.***14** 209-14
[4] Sammoura F, Kang J J, Heo Y M, Jung T S and Lin L W 2007 *J. Microsyst. Technol* **13** 517-22
[5] Sha B, Dimov S, Griffiths C and Packianather M S 2007 *J. J. Mater. Process. Technol.* **183** 284-96
[6] Bellantone V, Surace R, Trotta G and Fassi I 2013 *J. Int. J. Adv. Manuf. Technol.* **67** 1407-21
[7] Sha B, Dimov S, Griffiths C and Packianather M S 2007 *J. Int. J. Adv. Manuf. Technol.* **33** 147-56
[8] Lin H Y and Young W B 2009 *J. Appl. Math. Model.* **33** 3746-55
[9] Lin Y H, Huang C F and Cheng H 2013 *J. Polym. Eng. Sci.* **53** 212-18
[10] Griffiths C A, Dimov S S and Scholz S G 2014 *J. J. Manuf. Sci. Eng.* **136** 213-26
[11] Chen Y X， Chen R X and Wu S J 2012 *Moldflow 2012 completely learning handbook* (Beijing: Publishing House of Electronics Industry)
[12] Wang L D, Liu C and Xu Z 2012 Polymer Micro-nanofabrication Technology (Beijing: National Defend Industry Press)

Research on the damping properties of $Fe_{12}O_{19}Sr$/the polyurethane elastomer composite

Y Li, Yan Qin[1], P C Sun and Z X Huang

Key Laboratory of Advanced Technology for Specially Functional Materials Ministry of Education, Institute of Materials Science and Engineering, Wuhan University of Technology, Wuhan Hubei, 430070, China

E-mail: qinrock@sina.com

Abstract. Magnetic elastomer composite is a promising damping material.In this paper, both strontium ferrite ($Fe_{12}O_{19}Sr$) powders and polyurethane elastomer which were mixed by mechanical blending method were used as the magnetic filler and as the matrix respectively, , the properties of the magnetic damping composite materials were studied. The results show that the magnetic properties of the magnetic elastomers composite are enhanced with the ferrite loading. The mechanical properties and Shore hardness are highly influenced by mass fraction of ferrite particles. The damping properties of magnetic elastomer composite reach best when the strontium ferrite loading is 15phr, and the damping properties deteriorate when the loading continue increasing. The damping properties of the composites with the X direction of magnetization are better than that with Y direction of magnetization.

1. Introduction

Damping materials are functional ones, which can absorb mechanical vibration energy and change vibration energy into heat[1]. Among the damping materials, magnetic elastomers material as a high-performance damping material is highly valued [2]. The magnetic damping mechanism is magnetic hysteresis; magnetic domains play irreversible movement under alternating stress, so the mechanical vibration energy change into heat energy and then dissipate [3]. It can play the role of elastic damping of ordinary elastomers, while interconversion of elastic magnetic energy can also generate magnetic-mechanical coupling damping, so the magnetic elastomers composite has effective damping function[4]. Chinese researchers have also started a study on magnetic damping elastomers, but the products have a certain gap compared with similar foreign products in magnetic properties and damping performances and other aspects[5].

Therefore, this article intends to adopt polyurethane elastomers as the matrix, strontium ferrite powders as a functional body, and to study a new high-performance magnetic damping composites by adding different strontium ferrite loading ratios.

2. Experimental

2.1 Materials

ALL materials were used as received. Polyurethane elastomers were purchased from Ausbond (China) Co., Ltd. Strontium ferrite powders ($Fe_{12}O_{19}Sr$) are from Wuhan Xinda Magnetic Materials CO., LTD.

2.2 Sample preparation

The strontium ferrite loading is 5phr, 15phr, 25phr, 30phr respectively. The strontium ferrite and polyurethane elastomers were blended by pre-mixer at the temperature of 25℃ for 5 mins, then transferred to a kneader, kneaded 15mins at 25℃, removed and placed in the mold then the mixture was cured in the oven at the temperature of 70℃ for 3 hours.

2.3 Characterization

The mechanical properties and Shore hardness were measured by RGM 4100 type universal testing machine and TH200 Shore A hardness tester. The dynamic mechanical performances were obtained in shear mode by using the STDA 861e dynamic mechanical analyzer (DMA) at the frequency of 10Hz, at the temperature ranging from -80℃ to +120℃, as well as the heating rate being 4℃/min. The magnetic properties were tested under the condition of 1TGS for 20mins at the temperature of 25℃ by using JDAW-2000B-type vibrating sample magnetometer (VSM), and the VSM samples were the same as the DMA samples, both were small and round whose diameter was 6.60mm and thickness was 5.00mm. The set up used for these measurements can be seen in the Figure. 1.

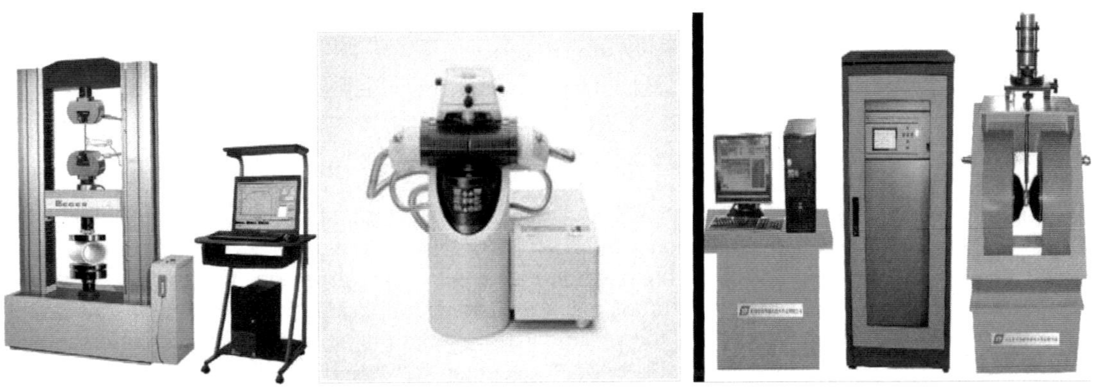

(a)universal testing machine (b) dynamic mechanical analyzer(c) vibrating sample magnetometer

Figure. 1. Set up used for these measurements

3. Results and Discussion

3.1 Magnetic properties of magnetic elastomer composite

Figure.2 shows the hysteresis loops for magnetic elastomer composites with a function of ferrite content. With the addition of strontium ferrite, the coercivity of magnetic elastomers composite remain unchanged, and remanence enhances accordingly. Strontium ferrite is hard magnetic powder, whose remanence, saturation magnetization and coercivity are higher than that of other ferrites. As can be seen from Figure.2, the hysteresis loop of strontium ferrite is broad, and performs strong anti-demagnetization.

The variation of coercive force H_c and saturation magnetization B_r of magnetic elastomers composite with a function of ferrite content are shown in Figure.3. When the magnetic powder loading is small, the interaction between the magnetic powder particles is not active, so coercive force H_c of the composite is basically unchanged. Saturation magnetization B_r is determined by the ferrite mass fraction, thus, saturation magnetization B_r enhances with the increasing of ferrite mass fraction.

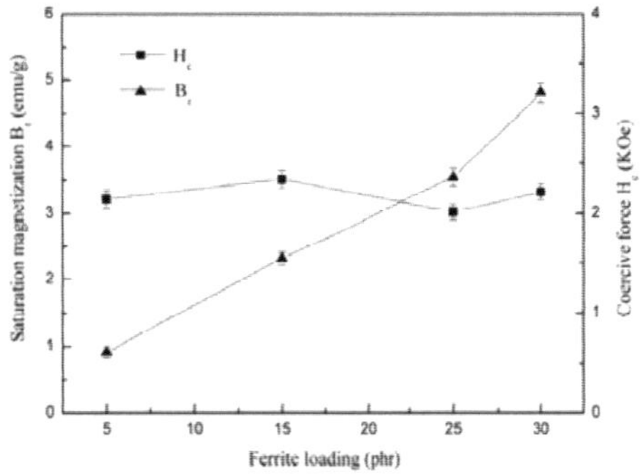

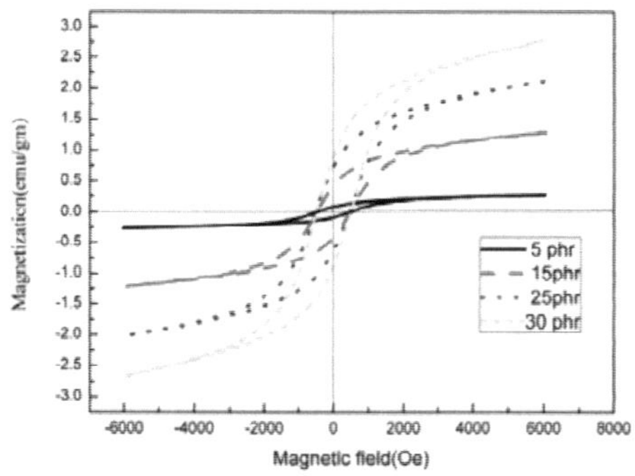

Figure.2. Hysteresis loops for magnetic elastomers composite with a function of ferrite content

Figure. 3. Coercive force H_c and saturation magnetization B_r with a function of ferrite content

3.2 The mechanical properties of magnetic elastomer composite

The mechanical properties of magnetic elastomers composites are strongly influenced by magnetic filler and the polymer matrix properties, such as the type, size, shape, dispersion and loading of magnetic filler, as well as the interfacial bonding between the ferrite and the matrix [6].

The variations of elongation at break and tensile strength of magnetic elastomers composite as the function of ferrite loading are shown in Figure.4. The tensile strength of the magnetic elastomers composite deteriorates with the addition of ferrite. The tensile strength of pure polyurethane is 3.18MPa, when the ferrite loading is 30phr, while the tensile strength of the composite is down to 1.3MPa. This is because that the interface of magnetic powder – elastomer increase with increasing ferrite loading, the interactions between the strontium ferrite and polyurethane matrix is inferior, and the gap between filler and matrix also leads to a decline in tensile strength. With the increase of ferrite loading, the elongation at break of composite decrease. The elongation at break of pure polyurethane is 920.53%, when the ferrite loading is 30phr, whereas the elongation at break is only 446.32%. The increasing of ferrite loading leads to the decreasing of the relative density of polymer matrix in composite, which makes the elongation at break decrease.

The variation of Shore A hardness of magnetic elastomers composite as the function of ferrite loading is shown in Figure.5. We can find that the Shore A hardness of pure polyurethane is 30, and the Shore A hardness of composite gradually increase with the increase of ferrite loading. This is because the hardness of strontium ferrite is higher than that of the polyurethane, ferrite particles effectively fill the hole of the polymer matrix on one hand, on the other hand, rigid particles are uniformly dispersed in a resin matrix, the material can lead to crazing when subjected to damage, thus consuming a large amount of impact energy and better withstanding the external stress [7].

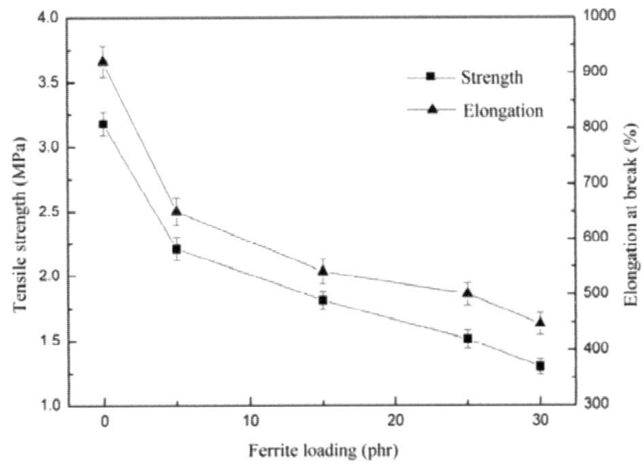

Figure. 4. Elongation at break and tensile strength with a function of ferrite content

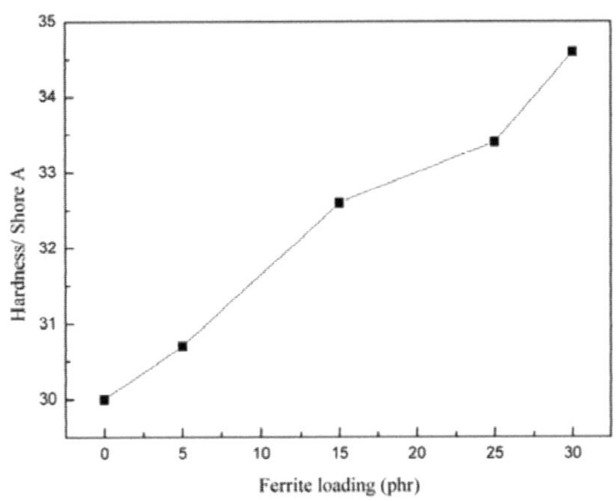

Figure. 5. Shore A hardness with a function of ferrite content

3.3 The damping properties of magnetic elastomer composite

The relationship between temperature and tan δ of magnetic elastomers composites with different content of strontium ferrite powder loading is presented in Figure.6. When temperature increases, the damping factor of the magnetic elastomers gradually enhances. Glass transition temperature of magnetic elastomers is -45°C, while the damping factor increases dramatically with rising temperature. The damping factor reaches a maximum at -15°C and then fall. When temperature exceeds 85°C, the damping factor is reduced to about 0.08 and substantially unchanged, and we can see the characteristic temperature of the polyurethane elastomer is not affected by the loading of strontium ferrite powders. The operating temperature of magnetic elastomers is determined by the matrix.

The maximum damping factor of the magnetic elastomers is shown in Table.1. Both Viscoelastic damping and magnetic damping contribute to the damping property of magnetic elastomers. When the ferrite loading rises, magnetic damping is greater than that under the condition of increasing amount of viscoelastic damping, therefore the whole damping properties of the composite are improved, and reach their best at the loading of 15phr. When the content of magnetic powders continues increasing, it could, on one hand, not only enhance magnetic damping property of the composite, but also increases the proportion of the magnetic powder in the matrix, on the other hand, decreases the volume ratios of polyurethane matrix occupying in the same volume of composites, which results in deterioration of the viscoelastic damping property of magnetic elastomer composites. In addition, the improvement of magnetic damping is insufficient to compensate for the decline of viscoelastic damping, which in return, leads to the decrease of damping properties of the composite.

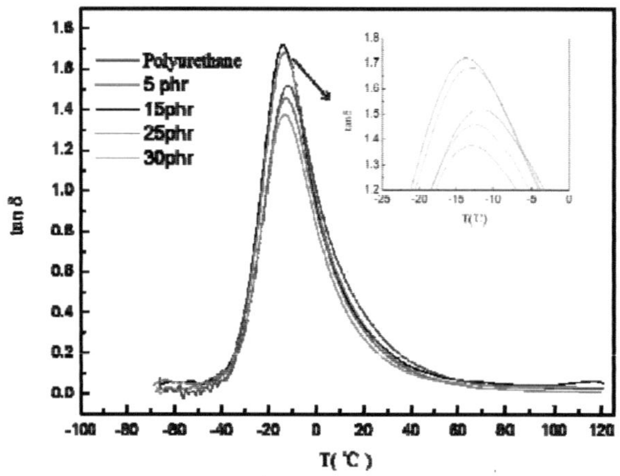

Figure. 6. tanδ-T curves of magnetic elastomers with a function of ferrite content

Table.1. $(\tan\delta)_{max}$ of magnetic elastomers with a function of ferrite content

Ingredient	polyurethane	5phr	15phr	25phr	30phr
$(\tan\delta)_{max}$	1.51	1.68	1.72	1.45	1.37

3.4 The damping properties analysis of composites with different directions of magnetization

The samples of composites with different ferrite loading were characterized by DMA, and the magnetic field of 1 TGS was exerted in different directions. It can be seen in the Figure.7.

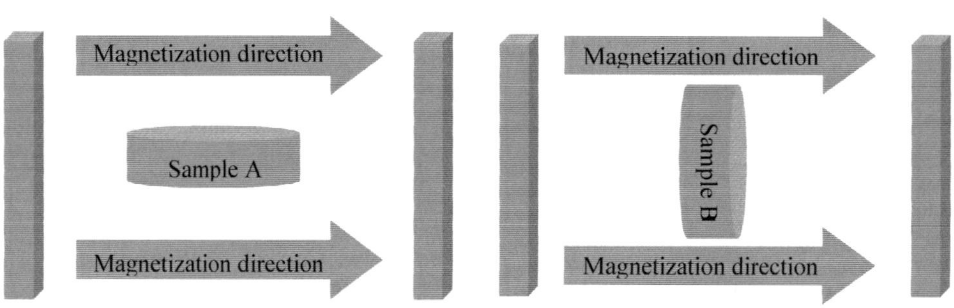

(a) X direction (b) Y direction

Figure. 7. different directions of magnetization

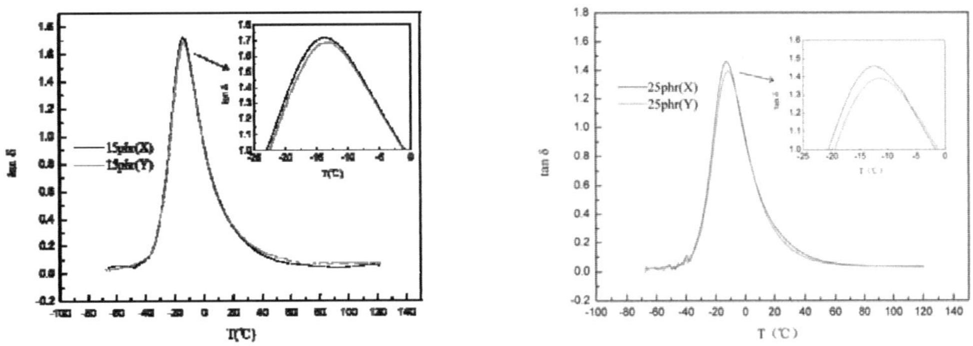

(a) 15%phr (b) 25%phr

Figure. 8. tan δ -T curves of magnetic elastomers with X and Y magnetization direction

In Figure.8 (a) we can see that the max tanδ value of the sample which is given the magnetization of X direction is higher than the sample given the magnetization of Y direction; it also can be seen in Figure.8 (b). When the loss factor of the sample is tested in the shear mode, the direction of the force goes along the X direction, so it means the external magnetic field in different directions affects different stress. The changes in the macro magnetic field, the vortex formation, the reversible and irreversible movement of the domain walls, all these are different from each other leading to the little changes of the loss factor in the X and Y magnetization direction, which may play a role in practical applications.

4. Conclusions

- With the addition of strontium ferrite, the coercivity of magnetic elastomers composite remain unchanged, while remanence enhances accordingly. Thus the magnetic properties of the composite improve.
- The elongation at break and tensile strength of the magnetic elastomers composite decrease with the increasing ferrite loading, but the Shore A hardness increase oppositely..
- With the increase of the strontium ferrite, the damping properties enhance accordingly, when the ferrite loading is 15phr, and the damping properties of the composite have the best performance, while when the ferrite loading continues increasing, the damping properties deteriorate.
- Given the different direction of magnetization, the damping properties of the composites with the X direction of magnetization are better than that with Y direction of magnetization.

References

[1] Deng J G, Wang J H and He C L 2001 The Research Progress of Polyurethane Damping Materials *Poly Ind* **16** 1-4
[2] Wang R Y and Zheng Y S 2003 The Research Progress of Rubber Damping Materials *China Rubber Ind* **50** 88-93
[3] Liu X D and Wu B R 1993 Dynamic Observation of Magnetic Domain in Ferromagnetic Damping Alloys *J Iron Steel Res* **5** 45-47
[4] Wang Y Y, Chen X R and Huang G S 2004 The Lastest Research Progress of Multiplex Damping Materials *Mater Rev* **18** 54-56

[5] Li Y N 2005The Mixing Process and Properties of Magnetic Rubber *World Rubber Ind* **32** 26

[6] Xiao J and Otaigbe J U 2000 Polymer bonded magnet. II. Effect of liquid crystal polymer and surface modification on magneto-mechanical properties *Polym. Comps* **21** 332

[7] Gao X R 2012 Research on Synthesis of Polyurethane Matrix Constrained Damping Composites and Their Properties. *MA Dissertation,* Department of Materials Science and Chemical Engineering, Harbin Engineering University

A method of quantitative characterization for the component of C/C composites based on the PLM video

Y X Li[1,3], L H Qi[1], Y S Song[1] and H J Li[2]

[1]School of Mechatronic Engineering, Northwestern Polytechnical University, Xi'an, Shaanxi, China
[2]Carbon/Carbon Composites Research Center, Northwestern Polytechnical University, Xi'an, Shaanxi, China

E-mail: crystal1507@163.com

Abstract. PLM video is used for studying the microstructure of C/C composites, because it contains the structure and motion information at the same time. It means that PLM video could provide more comprehensive microstructure features of C/C composites, and then the microstructure could be quantitatively characterized by image processing. However, several unavoidable displacements still exist in the PLM video, which could occur during the process of image acquisition. Therefore, an image registration method was put forward to correct the displacements by the phase correlation, and further to achieve the quantitative characterization of component combined with image fusion and threshold segmentation based on the PLM video of C/C composites. Specifically, PLM video was decomposed to a frame sequence firstly. Then a series of processes was carried out on this basis, including selecting the frame as equal interval, segmenting the static and dynamic regions and correcting the relative displacements between the adjacent frames. Meanwhile, the result of image registration was verified through image fusion, and it indicates that the proposed method could eliminate the displacements effectively. Finally, some operations of image processing were used to segment the components and calculate their fractions, thus the quantitative calculation was achieved successfully.

1. Introduction

Carbon/carbon (C/C) composites, as a kind of new high-performance materials, have been widely used in aerospace and military fields [1-3]. They have excellent mechanical and thermodynamic properties, especially under high-temperature environment [4-7]. However, complex technology parameters and high cost have made them hardly be used in the civilian fields. Therefore, it is necessary to conduct studies on the relationship between the technology, microstructure and properties for process optimization and performance design. In particular, the most important issue is how to characterize the microstructure of C/C composites quantitatively [8-11].

Qualitative description of microscopic images could not illustrate one problem or phenomenon completely in common research. So image processing is used, as a powerful tool, for characterizing the microstructure quantitatively through image feature describing and analyzing, by more and more researchers. In C/C composites, carbon fibers (CF) are reinforced by the carbon matrix [12]. It means that the fiber and matrix phases are all formed by carbon elements. So it is difficult to distinguish the fiber and matrix on the images of Metallurgical microscopy or Computed tomography(CT). Polarized light microscope (PLM) image is the most intuitive and convenient tool, which could observe the fiber, matrix and pores easily by eyes, during the microstructural research.

An image segmentation method developed by Miaoling Li was used for segmenting the PLM image based on a sequence of PLM images in different angle. However, the condition for image acquisition was very strict [13]. In fact, the extinction crosses of C/C composites have obvious moving feature by rotating the analyser of microscope. PLM video could store the structure and movement information of C/C composites at the same time, so it is important to analyze and characterize the microstructure quantitatively. Unluckily, some unavoidable displacements between the adjacent frames of video still exist, during the analyzer rotating. That would lead to some serious errors and even a wrong result in the subsequent image processing. Therefore, to eliminate the displacement between the adjacent frames, a displacement correction method based on the phase correlation algorithm was put forward in this paper. And then the component fraction was calculated successfully by image fusion, threshold segmentation and morphological processing, etc.

2. Materials and method

2.1. Materials

C/C composite samples were prepared by isothermal chemical vapor infiltration (ICVI). The temperature was 1150°C and the total pressure was 2.5kPa. The prepared sample was mounted with resin and grounded coarsely by water proof abrasive papers firstly, and then polished by diamond paste for 30 minutes and the water for 45 minutes successively.

During the image acquisition process, the magnification times of the objective was kept at 50× for the Leica DMLP. To acquire the moving process of extinction cross, the PLM video was collected when the analyzer was slowly rotated from 0° to 360°and stored as a format of AVI, the resolution was 1280×960 and the frame rate was 5fps.

2.2. Phase correlation algorithm

Since the primary displacement is caused by the translational motion for the PLM video of C/C composites, a phase correlation algorithm, which is mainly used to calculate the displacement between the adjacent frames, was introduced to solve the image registration problem in this paper.

Suppose that $f_1(x,y)$ and $f_2(x,y)$ are two image matrices with the same size, and a translational motion relationship is between them. The mathematical expression is shown as follow:

$$f_2(x,y) = f_1(x - x_0, y - y_0) \tag{1}$$

The expression (1) can be transformed to (2) by fourier transformation. Where $f_1(x,y)$ and $f_2(x,y)$ are forms of and after the successive fourier transformation.

$$F_2(u,v) = F_1(u,v)\mathrm{e}^{-j(ux_0+vy_0)} \tag{2}$$

The cross power spectrum of them is shown in (3), where $F_2^*(u,v)$ is the complex conjugate of $F_2(u,v)$.

$$\frac{F_1(u,v)F_2^*(u,v)}{\left|F_1(u,v)F_2^*(u,v)\right|} = \mathrm{e}^{-j(ux_0+vy_0)} \tag{3}$$

A two dimensional pulse function $W(x-x_0, y-y_0)$ can be obtained by inverse fourier transformation from $\mathrm{e}^{-j(ux_0+vy_0)}$. The translational parameters of x_0 and y_0 can be determined by the peak position of pulse function $W(x-x_0, y-y_0)$, which is the key of phase correlation algorithm[14].

3. Results and discussion

3.1. Pre-processing of PLM video image

PLM video is decomposed into frames to extract the information of the PLM image by the home-made soft and storied as the format of JPG firstly. Considering the mode of manual sampling cannot keep the analyzer rotation with a constant speed and the data of frames are too massive, the frames are need to be selected at regular intervals, and as in this paper the interval is 10°. The average gray value of each frame is calculated, and then the front and end static parts are removed according to their different gray values. In particular, the difference of gray value is close to 0for the static parts, and the remainder parts usually have a certain value. The moving part of average gray value is fitted by cosine curve, because in fact, the variation of average gray value is consistent with the cosine curve. To further reduce the error during the process of fitting, the data of gray value is fitted piecewise in this paper.

Although the removing speed of image acquisition is not consistent, the extreme points still correspond to the angles of 90°, 180° and 270°. We could divide the intervals to (0~90°), (90~180°), (180~270°) and (270~360°) based on the positions of extreme points and fit the average gray values by the cosine function successively. Figure1 shows the relationship during the process of sampling. The x-data (frame number) of fitting curve is further divided by the 10° sampling intervals, as the data of the fitting curve represents the situation that the angle of analyzer is uniform motion. The corresponding y-data (average gray value) could be obtained by the fitting curve, which is just the theoretical average gray value uniform motion. So the actual frame number could be selected from the regional average gray value data by the corresponding y-data.

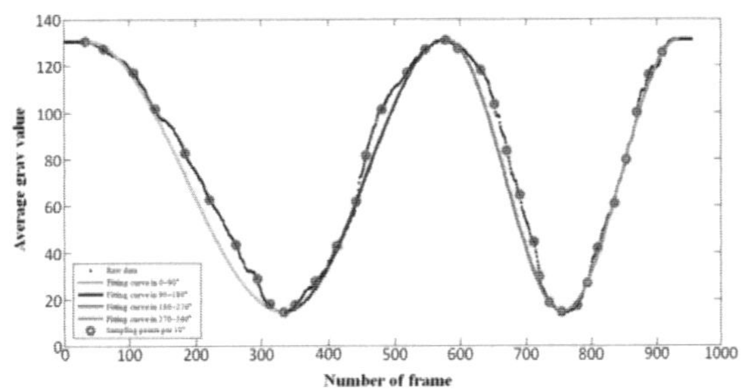

Figure 1. The average gray value of PLM video

3.2. Segmentation of the static and dynamic regions

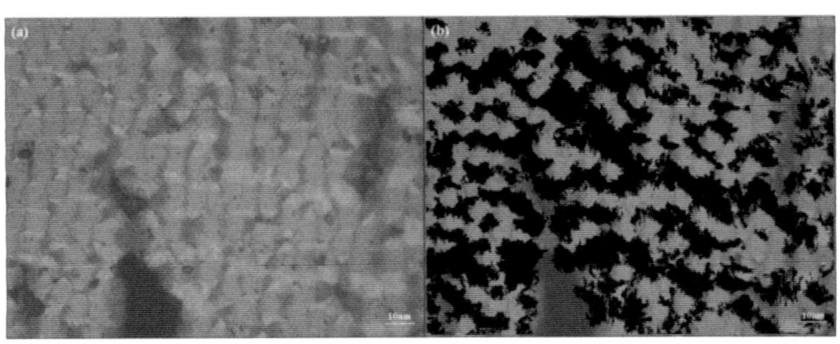

Figure 2. Segmentation of the static and moving regions: (a) raw image. (b) segmented image.

Considering the movement of extinction crosses would change the content of image, they have a larger area on the image, a processing of image segmentation is needed for reducing error between the dynamic and static regions, in which the dynamic region is belong to part of pyrocarbon, and the static region contains CF and pore. The methods of threshold segment and morphological processing are used to segment them, and a new frame sequence that only contains the information of static region is introduced to the next step processing, as the figure 2(b) shown.

3.3. Correction of the relative displacement by phase correlation method

For each new frame, the projections of gray value along the directions of X and Y are calculated respectively, and the cross-correlated analysis is carried out. Figure3 shows the result of cross-correlated function, which indicates the matching degree of two data sequence on the different relative positions. The data of cross-correlated function on the X direction illustrate correlation between the two groups of gray value projective data when the one group data shift a certain value relative to the other group, in which the point of maximum correlation is just the position of maximum matching degree for the two frames on the X direction. The relative displacement could be easily obtained by calculating the distance between the matching position and the origin, as the figure 3(a) shown. Meanwhile, the relative displacement on the Y direction could be obtained by the same method, as the figure 3(b) shown.

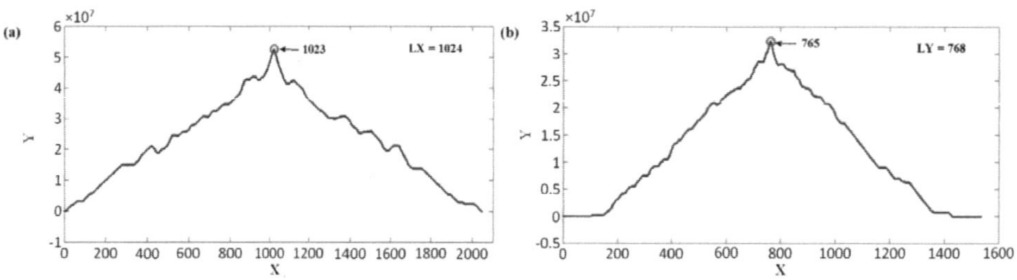

Figure 3. Calculation of the matching positions on the X and Y directions: (a) matching position on the X direction. (b) matching position on the Y direction.

Table1. The relative displacement of frame sequence

Number	1	2	3	4	5	6	7	8	9
sx	-1	1	-2	3	-2	-2	-2	1	-2
sy	-3	1	2	0	2	1	7	0	-1
Number	10	11	12	13	14	15	16	17	
sx	-3	1	-3	-1	-3	3	-1	-4	
sy	5	0	1	0	0	-3	1	0	

Table.1 shows the results of the relative displacement for all of the adjacent frames in the interval(0~180°), in which the *sx* and *sy* refer to the displacements on the X and Y directions successively. Thus, through correcting the frame sequence according to the data in Table.1, the registration of the frame sequence could be achieved by the phase correlation method. In addition, a processing of image fusion is carried out for verifying the results of registration, because the relative displacement between the adjacent frames is tiny and difficult to distinguish by eyes. The figure4 shows the comparative results of image fusion, in which the fusion image of registered frames is more clear than the unregistered frames. It means that the relative displacements between the adjacent frames exist really for the unregistered frame sequence, and they have been eliminated well by the phase correlation method.

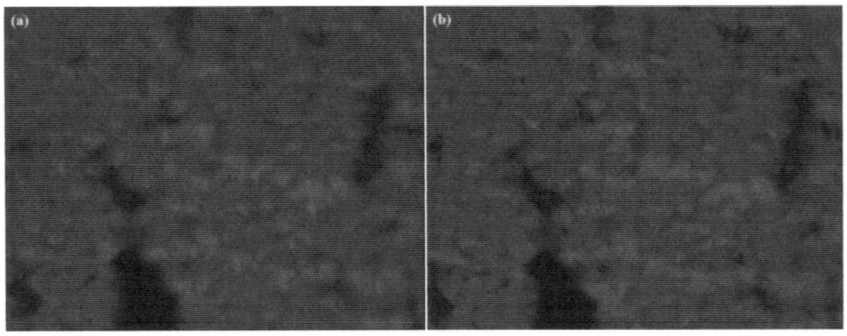

Figure 4. The contrast of the registered and unregistered PLM video: (a) fusion result of the unregistered PLM video. (b) fusion result of the registered PLM video.

3.4. Calculation of the component fraction

The regions of CF, pore and matrix show different gray value in the fusion image because of the different microstructure, especially for the matrix with different texture. So each component could be segmented by the methods of threshold segmentation and morphological processing according to their different gray value. As the figure 5 shown, the regions of CF and pore are segmented from the PLM image, and fractions of pore, CF and pyrocarbon are in order of 5.36%, 48.60% and 46.04%. Thus, the quantitative characterization of component fraction is achieved by a series of image processing methods based on PLM video.

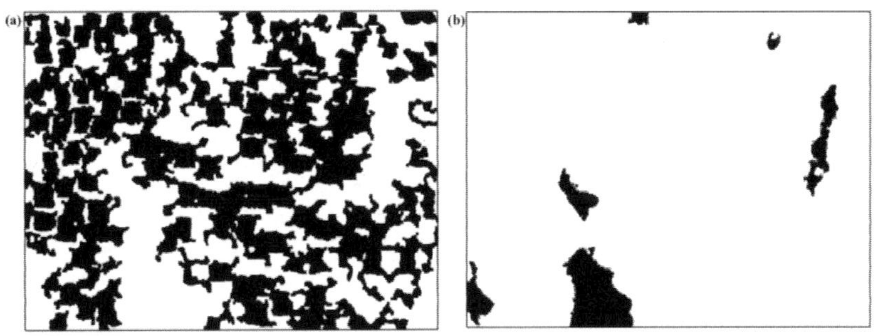

(a) (b)

Figure 5. Segmentation of components: (a) CF region of C/C composites. (b) pore region of C/C composites.

4. Conclusion

A quantitative characterization method for the component of C/C composites based on PLM video was researched by a series of imaging processing methods and the phase correlation method in this paper, the main results are as follows:

- The phase correlation method was introduced to calculate the relative displacement between the two adjacent frames, and the frame sequence was registered on this basis.
- The component fraction was characterized quantitatively by a series of imaging processing methods.

References

[1] Oberlin A 2002 Pyrocarbons *J.Carbon* **40**7-14.
[2] Li H J 2001 Carbon/Carbon Composites *New Carbon Mater.* **16** 79-80.
[3] Lu X F and Xiao P 2014 Short time oxidation behaviour and residual mechanical properties of C/C composites modified by in situ grown carbon nanofibers *Ceram. Int.* **40** 10705-9.

IOP Conf. Series: Materials Science and Engineering **137** (2016) 012016 doi:10.1088/1757-899X/137/1/012016

[4] Yu S, Zhang FQ, Xiong X, Li Y P, Tang N and Koizhumi Y et al. 2013 Tribological properties of carbon/carbon composites with various pyrolytic carbon microstructures *Wear* **304** 103-8.

[5] Xu G Z, Li H J,Bai R C,Wei J, and Zhai Y Q 2008 Influence of the matrix texture on the fracture behavior of 2D carbon/carbon composites. *Mater. Sci. Eng. A* **478** 319–323.

[6] Ren J J, Li K Z, Zhang S Y, Yao X Y, and Li H J 2016 Preparation of high texture three-dimensional braided carbon/carbon composites by pyrolysis of ethanol and methane. *Ceram. Int.* **42** 2887–91.

[7] Feng L, Li K Z, Zhao Z G, Li H J, Zhang L L and Lu J H et al.2016 Three-dimensional carbon/carbon composites with vertically alignedcarbon nanotubes: Providing direct and indirect reinforcements to thepyrocarbon matrix. *Mater. Des.* **92** 120-8.

[8] Song K N, ZhangY Y, Meng J S, Green E C Tajaddod N and Li H et al. 2013Structural polymer-based carbon nanotube composite fibers: understanding the processing–structure–performance relationship *Mater.* **6** 2543-2577.

[9] Zhang Y Y, Song K N, Meng J S, and Minus M L 2013 Tailoring polyacrylonitrile interfacial morphological structure by crystallization in the presence of single-wall carbon nanotubes *ACS Appl. Mater.Interfaces* **5** 807-814.

[10] Yin X W, Gotman I, Gutmanas E Y, and Weiss R 2015 Microstructure and mechanical properties of carbon–carbon compositesmodified by reactive diffusion treatment in niobium powder. *Surf. Coat. Technol.* **281** 76-81.

[11] Fang H C, Xiao P, Xiong X, and Yu G J 2016 Microstructures, mechanical and oxidation behaviors ofC/C composites modified by NiAl alloy. *Trans. Nonferrous Met. Soc. China* **26** 196-202.

[12] Zhang Y, Tajaddod N, Song K, and Minus M L 2015 Low temperature graphitization of interphase polyacrylonitrile (PAN). *Carbon* **91** 479-93.

[13] Li M L,Q L H, and Li H J 2012 An Imaging Technique Using Rotational Polarization Microscopy for the Microstructure Analysis of Carbon/Carbon Composites. *Microsc. Res. Tech.* **75**65-73.

[14] Zheng Z B and Ye Z F 2006 Image Registration Algorithm Based on Phase-Correlation. *Journal of Data Acquisition and Processing* **21** 444-49.

IOP Conf. Series: Materials Science and Engineering **137** (2016) 012017 doi:10.1088/1757-899X/137/1/012017

3D printing of weft knitted textile based structures by selective laser sintering of nylon powder

M Beecroft[1]

Dept. of Design, Manchester School of Art, Manchester Metropolitan University, Benzie Building, Cavendish Street, Manchester, M15 6BR England

E-mail: m.beecroft@mmu.ac.uk

Abstract. 3D printing is a form of additive manufacturing whereby the building up of layers of material creates objects. The selective laser sintering process (SLS) uses a laser beam to sinter powdered material to create objects. This paper builds upon previous research into 3D printed textile based material exploring the use of SLS using nylon powder to create flexible weft knitted structures. The results show the potential to print flexible textile based structures that exhibit the properties of traditional knitted textile structures along with the mechanical properties of the material used, whilst describing the challenges regarding fineness of printing resolution. The conclusion highlights the potential future development and application of such pieces.

1. Introduction

There has been an explosion of hype and interest around 3D printing for textiles and fashion over the last few years. We've seen fully printed dresses on the catwalk [1] alongside the recent fully printed fashion collection printed on Desktop printers [2]. However, the majority of these works act as a form of body sculpture rather than functional flexible textiles. There have been more subtle examples combining 3D printed structures with traditional textiles [3]. This paper builds on previous research into the possibility of printing knitted textile based structures by Selective Laser Sintering that exhibit the flexibility, stretch and strength of traditional textiles [4]. Current published research has found SLS limited in terms of flexibility [5]. This paper challenges these results by demonstrating the potential to print fully flexible weft knitted structures using SLS of nylon powder.

In order to print 3D structures, a 3D Computer Aided Design (CAD) drawing is required. This 3D CAD drawing needs to consider the wall thickness and distance between objects, particularly interlocking structures i.e. chainmail, knitting, weaving. The 3D CAD drawing is then sliced into layers by the machines host software then printed out layer by layer. In the SLS process a C02 laser is used to fuse fine powder into solid material. The laser is directed by a computer guided mirror and builds objects in layers of 0.1mm, the building platform lowers down by this measurement each time allowing the next layer of powder to be rolled onto the surface. The non-sintered powder acts as a support for the build [6]. The excess powder is then removed by high pressure after printing. This paper explores the results of SLS using nylon powder to create both single-faced and double-faced weft knit structures with various thickness, documenting the effect on flexibility, stretch and strength.

2. Experimental Method

As previously mentioned despite the hype, the printing of flexible textile structures is still in its infancy. Previous research papers have explored the potential of printing weft knitted structures but have found SLS of Nylon too rigid to create fully flexible structures. This paper builds on earlier research in which weft knitted structures were combined with machine knitted textiles [7].

In order to print 3D structures, a 3D CAD drawing is required. Initial investigations utilised the free software Autodesk 123D design to generate a 3D model of the weft knitted structure. This model was created by using the spline tool to draw a single knitted loop. A polygon circle was then used to extrude along the spline. This single loop was then copied and pasted to create the horizontal rows (courses) of loops. The row of loops was then copied, angled and then pasted. This method was repeated to create the desired vertical rows (wales) of loops. This method resulted in the initial structures demonstrating alternating rows (courses) of plain knit and purl knit.

Due to the limitations of this software, McNeel Rhinoceros software has been used to generate the 3D CAD models discussed in this paper. Once drawn, these designs were run through the Magic 3D printing software to make sure all neighboring and intersecting parts have a distance of at least 0.4mm.

The printed structures discussed in this paper have been printed by London based bureau Digits 2 Widgets using an EOS Formiga P1 machine. The machines are calibrated up to ±0.15% on the X and Y-axis and built with 0.1mm layers on the Z-axis [8].The material used is nylon powder (Nylon PA12)mixed to a ratio of 1:1 of new to used powder. The mechanical properties of Nylon PA12 include good dimensional stability and superior flexibility [9].

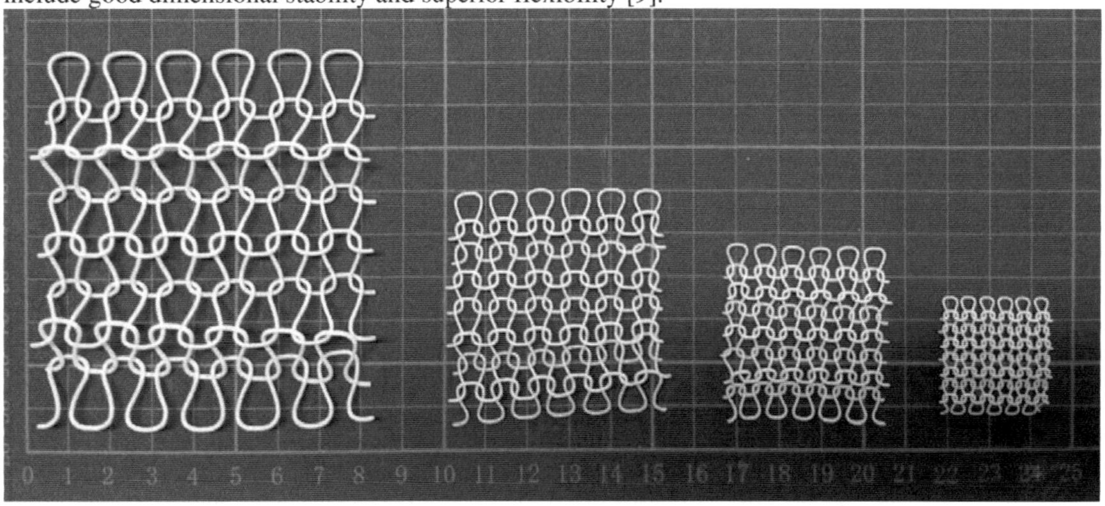

Figure 1. Initial results of weft knitted structures printed form Autodesk 123D CAD model.

3. Results

The first experiments aimed to reproduce plain weft knit single-faced structures at various thickness to test the flexibility, stretch and strength (Fig. 1), later experiments focused on developing the 3D structure to produce an interlock knit structure (double-faced weft knitted structure).

3.1. 3D Printing single-faced weft knitted structure

To print the weft knitted structure a 3D CAD model was created in Rhinoceros. A single loop unit was initially drawn as curves then piped with thickness (Figs. 2 and 3). Once created the cell unit can be repeated to create desired width (courses) and length (wales).

IOP Conf. Series: Materials Science and Engineering **137** (2016) 012017 doi:10.1088/1757-899X/137/1/012017

Figure 2. CAD Line drawing of repeat loop unit.

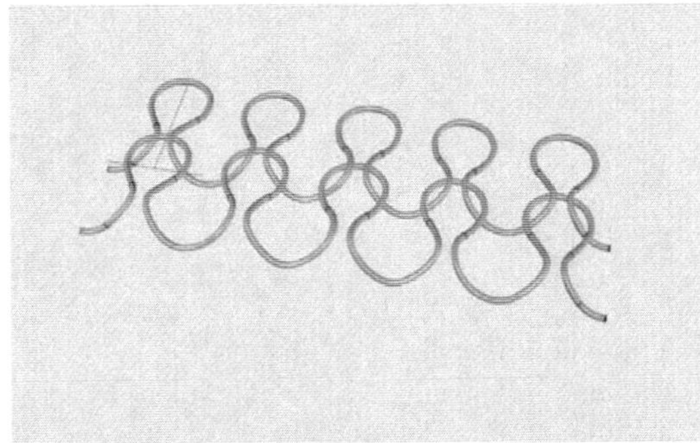

Figure 3. Piped version of 3D CAD model.

Once drawn the 3D CAD model was run through Magic software to ensure all neighboring and intersecting parts have a minimum distance of 0.4mm. Initially the model was piped with a thickness of 0.4mm thickness. The printed structure was too fine so the piece wasn't strong enough to retain its structure. Following this the piece was reprinted with a pipe thickness of 0.5mm. This enabled the printed piece to be strong enough to hold the knitted structure and flexible enough to articulate like a fabric. Figs. 4 and 5 show the sample contracted and expanded. The piece also exhibits stretch horizontally across the courses. Figs. 6 and 7 demonstrate the flexibility of the printed piece. Due to the mechanical properties of the Nylon PA12 the piece bounces back to its original form.

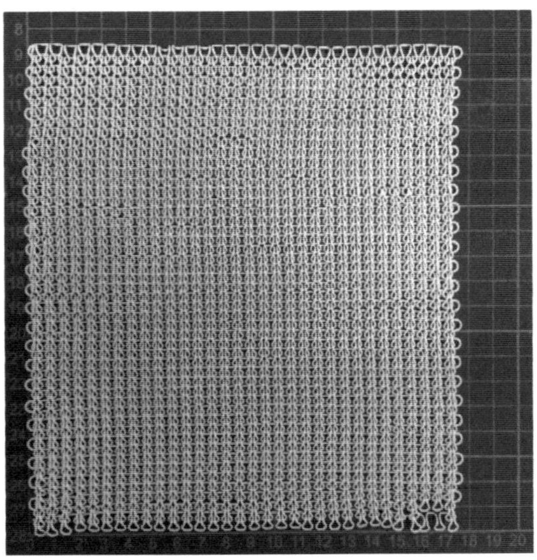

Figure 4. Weft knitted structure contracted.

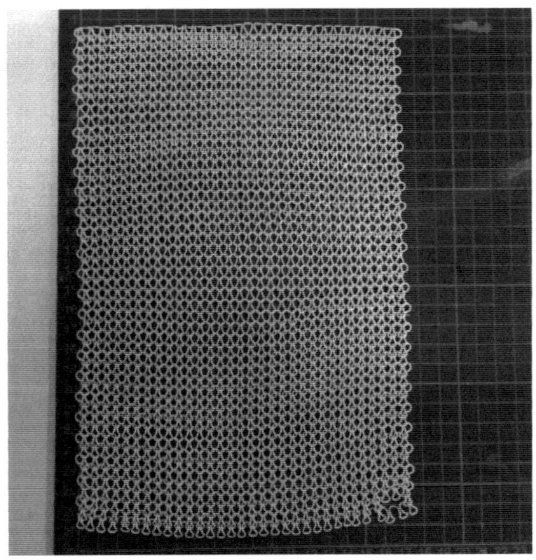

Figure 5. Weft knitted structure expanded.

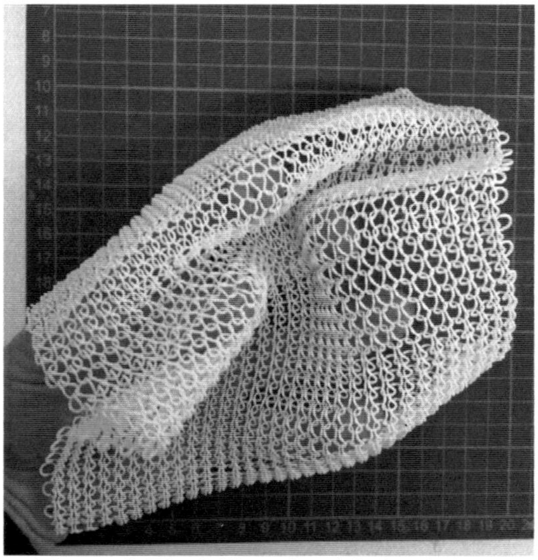

Figure 6. Flexibility of weft knitted piece.

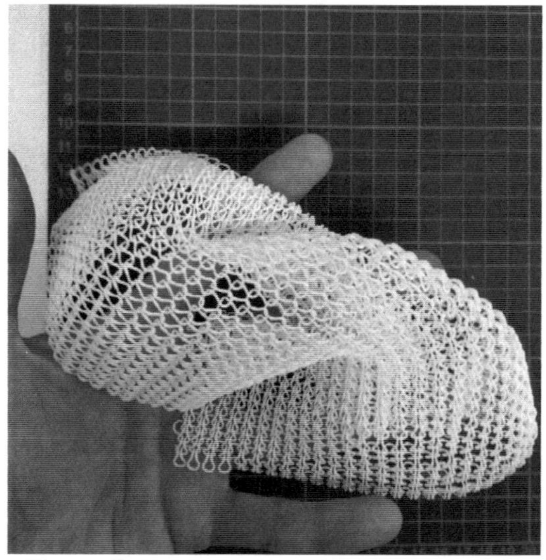

Figure 7. Bounce back of weft knitted piece.

3.2. 3D modelling interlock double-faced weft knit structure
A 3D CAD model was created in Rhinoceros to represent double-faced interlock weft knitted fabric structure. Fig. 8 shows the CAD line drawing (black and blue highlighting each of the interlocking layers). Due to the interlock structure a rib is created across the weft face. Fig. 9 shows the piped CAD model. The model was initially piped at 1.2mm thickness to test the structure and its flexibility, strength and stretch.

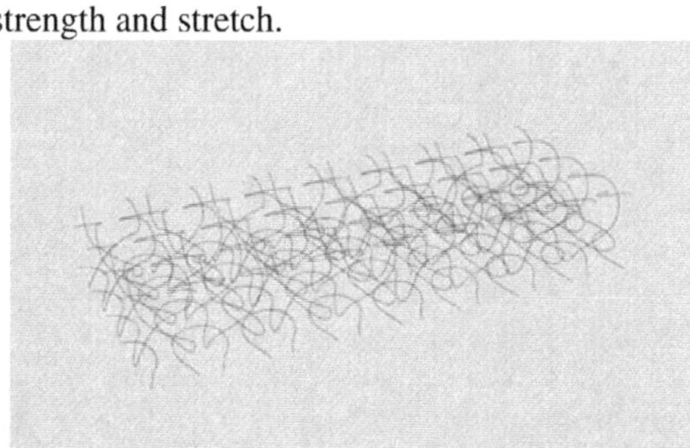

Figure 8. Rhino 3D line drawing of interlock structure.

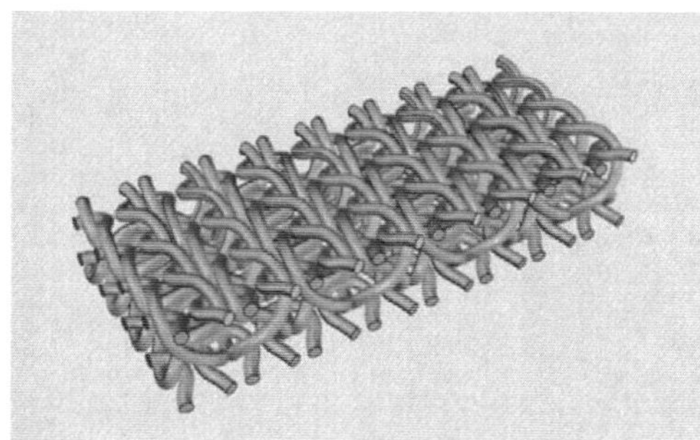

Figure 9. 3D piped model of interlock structure.

3.2.1. 3D printing interlock double-faced weft knitted structure 1.2mm. The model was initially printed at 1.2mm thickness to test the structure and its flexibility, strength and stretch (Fig. 10). Fig. 11 shows the flexibility of the piece. Both Fig. 12 and Fig. 13 show the interlock structure viewed from top to bottom. Like the single faced weft knit structure the interlock demonstrates great flexibility, bounces back to original form and stretches across the horizontal courses. The 1.2mm thickness is strong and resilient.

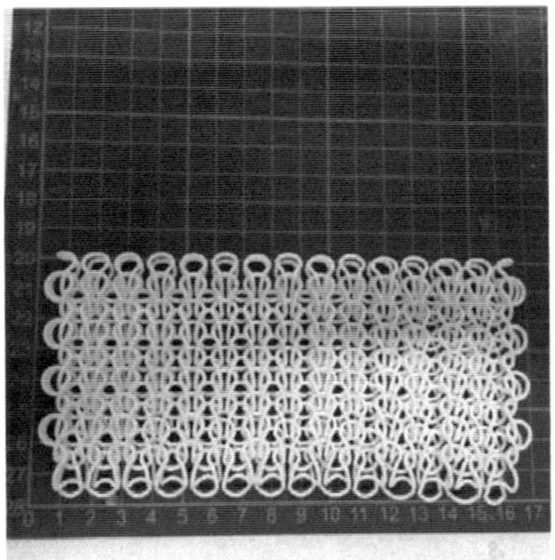

Figure 10. Contracted interlock structure.

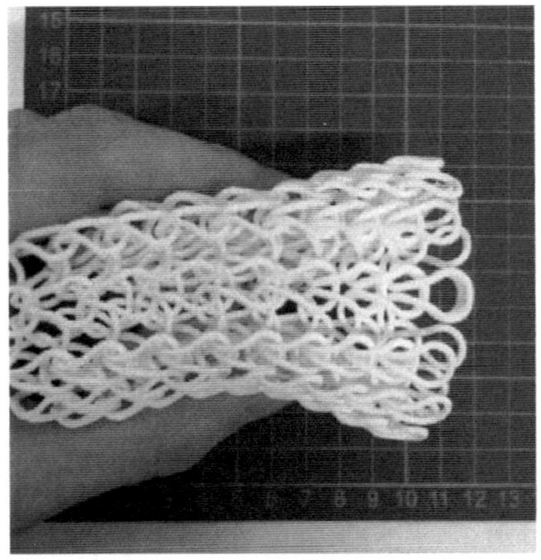

Figure 11. Flexibility of the interlock structure.

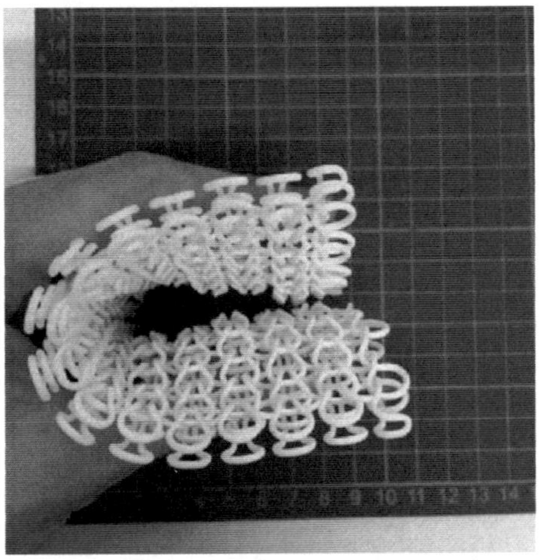

Figure 12. Interlock structure top view.

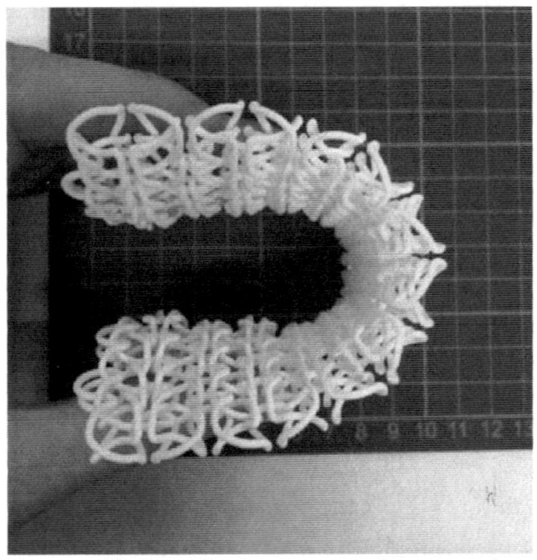

Figure 13. Interlock structure bottom view.

3.2.2. 3D printing interlock double-faced weft knitted structure 6mm. The model was subsequently printed at 0.6 mm thickness to test the structure and its flexibility, strength and stretch (Fig. 14). Again the printed sample exhibits flexibility (Fig. 15) and stretches across the horizontal courses with the structure returning back to its original form. The piece is strong but less so than the 1.2mm model.

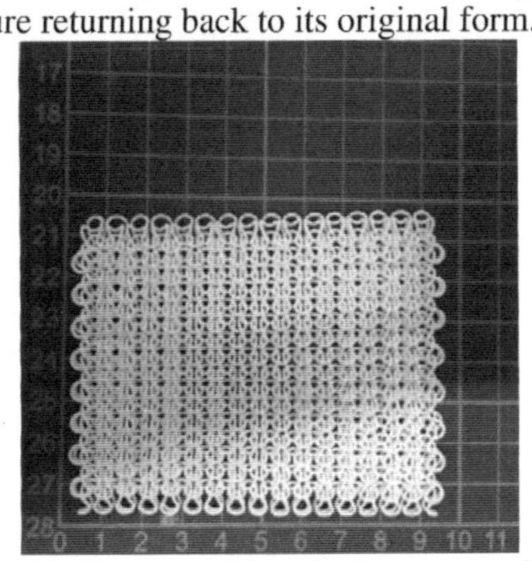

Figure 14. Contracted interlock structure 0.6mm.

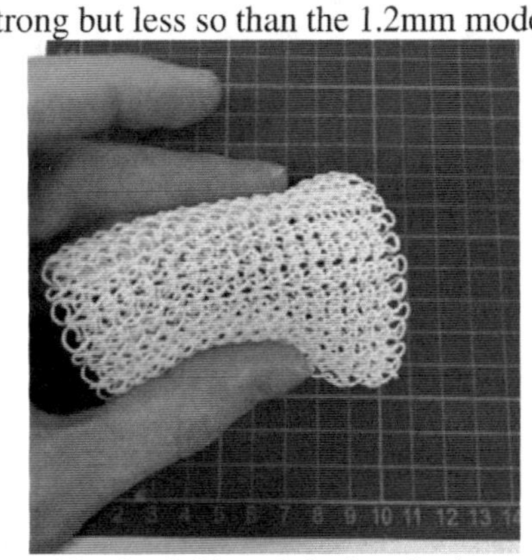

Figure 15. Flexibility of the interlock structure.

3.2.3. 3D printing interlock double-faced weft knitted structure 0.4mm. Finally, the model was piped at 0.4mm thickness to test the structure and its flexibility, strength and stretch (Fig. 16). Due to the finer resolution the rib structure across the weft face is more visible (Fig. 17 & Fig. 19).The printed sample exhibits flexibility and the piece bounces back to original form. The stretch across the horizontal courses is very limited, due to the 0.4mm thickness the loop structure is prone to break more easily whilst the close distance between the interlocking weft faces starts to fuse at intervals (Fig. 18& Fig. 19).

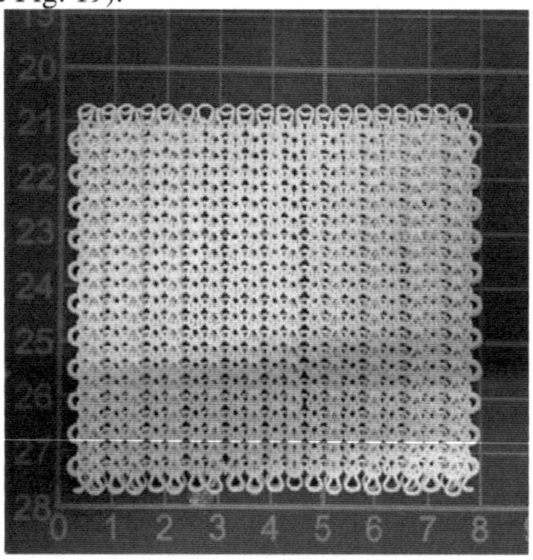

Figure 16. Interlock structure 0.4mm.

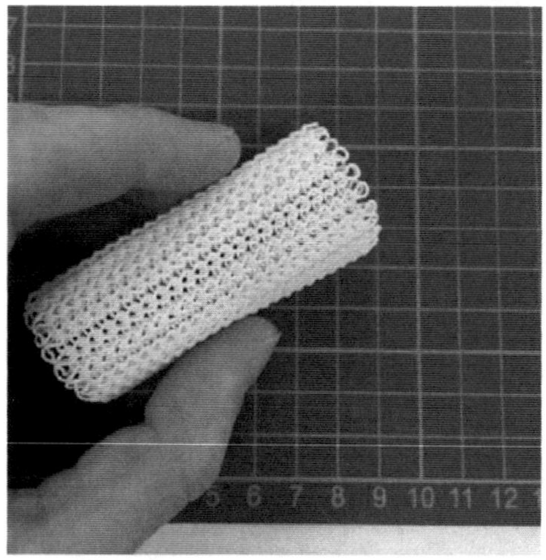

Figure 17. Flexibility of the interlock structure 0.4mm.

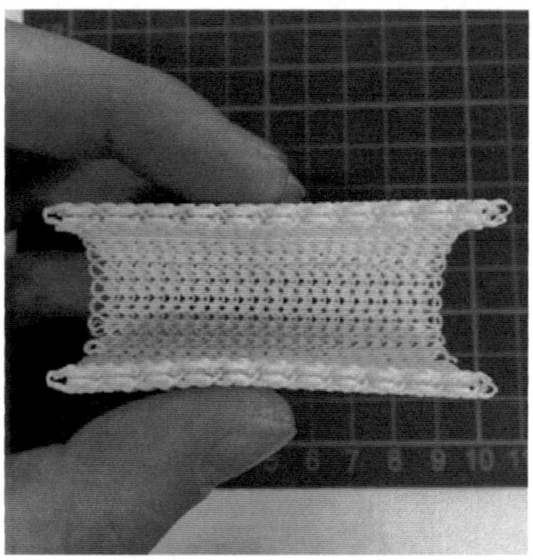

Figure 18. Interlock structure 0.4mm side view.

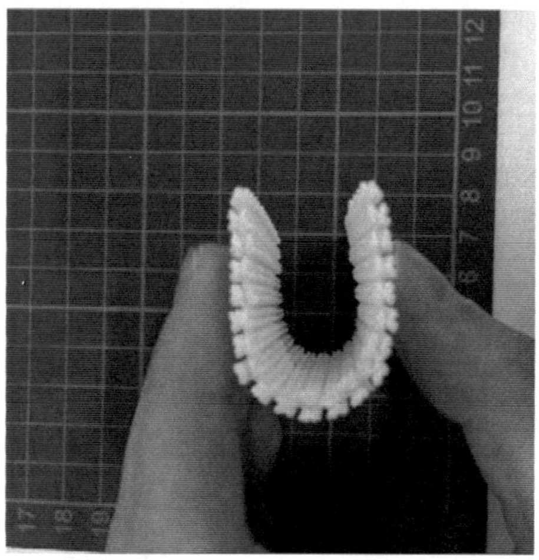

Figure 19. Interlock structure 0.4mm top view.

4. Conclusion/Further research

This paper has demonstrated the possibility of using Selective Laser Sintering of Nylon powder to print both single-faced and double-faced weft knitted structures. These structures when printed at various thicknesses demonstrate the mechanical properties of flexibility, strength and stretch which could make them viable solutions for use in the technical textile industry. Further research into different types of powdered materials i.e. Thermoplastic polyurethane (TPU) may produce softer fabrics, which may be more suitable for a fashion application.

Acknowledgments
This research has been funded by The Manchester Institute for Research and Innovation in Art and Design (MIRIAD). Product Designer Tom Mallinson of Digits2Widgets has assisted with the Rhino CAD work.

References
[1] Chalcraft, E 2013 *Voltage by Iris van Herpen with Neri Oxman and Julia Koerner* (http://www.dezeen.com/2013/01/22/voltage-3d-printed-clothes-by-iris-van-herpen-with-neri-oxman-and-julia-koerne/)
[2] PelegD 2015 (http://danitpeleg.com)
[3] Beckett R 2014 3D Printed Knitwear- *Pringle of Scotland AW 2014* (http://www.richard-beckett.com/pringle-3d-fabrics.html)
[4] Alec 2014 *English researchers use 3D printing to produce flexible and fine textile-like structures*(http://www.3ders.org/articles/20141109-english-researchers-use-3d-printing-to-produce-flexible-and-fine-textile-like-structures.html)
[5] Melnikova R, Ehrmann A and Finsterbusch K 2014 3D printing of textile-based structures by Fused Deposition Modelling (FDM) with different polymer materials *IOP Conf. Ser.: Mater. Sci. Eng.***62** 012018 IOP science
[6] Thompson R 2007 *Manufacturing Processes for Design Professionals* P.234. (Thames & Hudson)
[7] Millsaps B 2014 *3D Printing and Knitting Converge: Technical crafting brings new dimension to apparel*(http://3dprint.com/22901/knitting-3d-printing-textiles/)
[8] Digits2Widgets *Nylon SLS Design Guide*(http://www.digits2widgets.com/3d-printing-cad-services/nylon-sls_3d_printing/)
[9] http://www.performance-plastics.co.uk/product/classification/nylon-12-pa12/

Density functional theory studies on the nano-scaled composites consisted of graphene and acyl hydrazone molecules

J L Ren, L Zhou, Z C Lv, C H Ding, Y H Wu[1] and H C Bai

College of Chemistry and Chemical Engineering, Ningxia University, Yinchuan, Ningxia 750021, China

E-mail: yuhua236@foxmail.com

Abstract. Graphene, which is the first obtained single atomic layer 2D materials, has drawn a great of concern in nano biotechnology due to the unique property. On one hand, acyl hydrazone compounds belonging to the Schif bases have aroused considerable attention in medicine, pharmacy, and analytical reagent. However, few understanding about the interaction between graphene and acyl hydrazone molecules is now available. And such investigations are much crucial for the applications of these new nano-scaled composites. The current work revealed theoretical investigations on the nano-scaled composites built by acyl hydrazone molecules loaded on the surface of graphene. The relative energy, electronic property and the interaction between the counterparts of graphene/acyl hydrazone composites are investigated based on the density functional theory calculations. According to the obtained adsorption energy, the formation of the nano-scaled composite from the isolated graphene and acyl hydrazone molecule is exothermic, and thus it is energetically favorable to form these nano composites in viewpoint of total energy change. The frontier molecular orbital for the nano composite is mainly distributed at the graphene part, leading to that the energy levels of the frontier molecular orbital of the nano composites are very close to that of isolated graphene. Moreover, the counterpart interaction for the graphene/acyl hydrazone composites is also explored based on the discussions of orbital hybridization, charge redistribution and Van der Waals interaction.

1. Introduction

Acyl hydrazone compounds are organic matter that belongs to the Schif bases with great concern, which can be obtained by the condensation of hydrazide and ketone or aldehyde. Usually they exhibit strong coordination ability due to the various sites to coordinate. Also most Schiff bases possess good biological activities, such as anti-tumor and antibacterial [1]. With the aid of Schiff bases, the transamination can be achieved in biochemical reaction [2-4]. The C=N double bond in Schiff is found to be the effect group, and plays an important role in antibacterial activity. Therefore, Schiff bases currently have aroused considerable attention in medicine, pharmacy, and analytical reagent [1,5]. Compared with the ordinary Schiff bases, acyl hydrazone compounds are more stable, and exhibit even stronger biological activity. Besides, the oxygen and nitrogen atoms in acyl hydrazone molecules could be involved in the formation of hydrogen bonding networks of biochemical systems. In such a way, the acyl hydrazone compounds are good candidate in various physical chemical processes, such as resistance to high blood sugar, antitumor, antifungal and immune inhibitors [6-8].

On the other hand, graphene, which is the first obtained single atomic layer 2D materials, has drawn a great deal of attention due to the state-of-the-art potentials in physics, chemistry, bio-nanotechnology and pharmacy [9-14]. Currently graphene have been utilized in many practical fields. Shirzadmehr et. al developed graphene-based sensor composition. They demonstrate that the sensor is highly selective for heavy metal ions determination with very low limit, such as Hg^{2+} and Ag^{+} [11-12]. Along with a rise in the number of patents related to the graphene, its industrial applications have been dramatically increased, especially in pharmaceutical science and targeted drug delivery [15-19]. Because graphene possesses very large specific surface area which can be either the both edges and the top or bottom surfaces, therefore, its loading capacity is much higher than that of others. Dai et al. studied the polyethylene glycol loaded with nano-scaled graphene as anticancer drug carriers, and the drugs are interacted with the graphene by non-covalent bonding [17]. Zhang studied the functionalized graphene that can efficiently load two anticancer drugs, and has the specificity of breast cancer cells. The two anticancer drugs combined loading of the cancer cells have great toxicity, so as to more effectively to kill cancer cells [18]. Yang reported that oxide graphene can efficient load doxorubicin hydrochloride. And then the drug release can be controlled under different pH [19].

The current work presents a quantum chemical study on the nano-scaled composites made of acyl hydrazone molecules loaded on the surface of nanoscale graphene. The adsorption energy, electronic structure and the interaction nature between the counterparts of graphene/acyl hydrazone composites are investigated. And a solid understanding of the structure, energy, electronic property and interaction of the counterparts is much crucial for the construction new nano-scaled composite and also their applications. We expect that the current work here would be helpful to further studies of the acyl hydrazones drugs and graphene.

2. Models and Computational Details

As shown in Figure 1, several acyl hydrazone molecules (denoted as ***mol-1*** – ***mol-3***) are adopted as the model compounds in this work. These acyl hydrazone molecules can be obtained by the modification of pyridine and carboxylic acid. It is noticed that the three molecules studied here are isomers with different connecting sites for the pyridine rings. As for the nano-scaled graphene, a layer structure that contains 120 carbon atoms, is built with 8 carbon atoms and 15 carbon atoms along two axises of the surface. The surface area is 16.5 Å × 18.0 Å. The C–C bond length in the six-membered rings is 1.42 Å. The dangling bonds at the edges are terminated by hydrogen atoms in our model. It can be seen that the similar treatment has been used in previous theoretical studies [20-21].

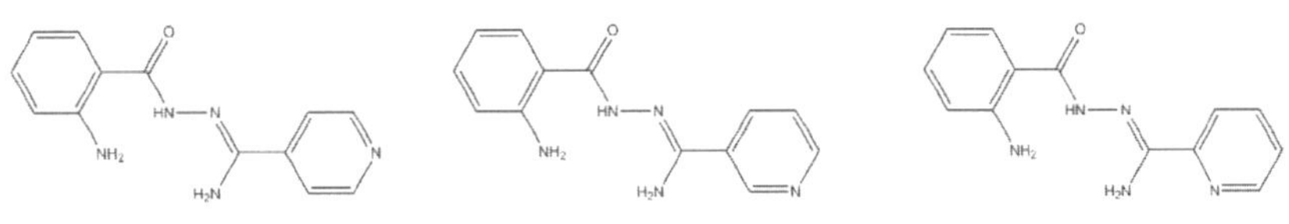

mol-1 ***mol-2*** ***mol-3***

Figure 1. Models of the acyl hydrazone molecules in this work.

We are aware that various possible configurations of acyl hydrazone molecules may exist when they are adsorbed on graphene surface. In order to get more reasonable configurations for the absorption models, molecular dynamics (MD) simulations are conducted with the aid of reactive force fields (ReaxFF) developed by Mattsson and co-authors [22]. We performed MD simulations at 300 K with each for 5 ps. The MD simulations are under the both constant temperature and volume condition, and the time step size is set as 1 fs. Different initial models are also considered to start the MD simulation. By using this strategy, the configurations are obtained as shown in Figure 2, and then computed by the quantum chemical methods.

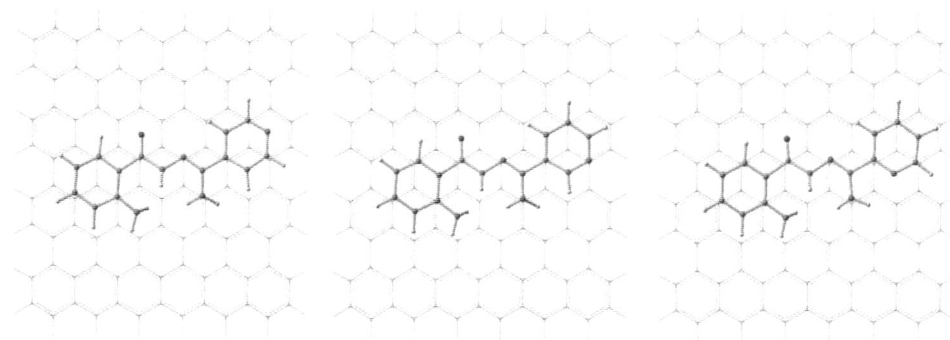

Figure 2. Structures of graphene/acyl hydrazone composites with fully geometrical optimization.

In this work the quantum chemical calculations are performed based on the Kohn-Sham density functional theory (DFT). M06-2X method [23] proposed by Truhlar and co-authors is selected as the exchange-correlation functional for all DFT computations. This selection is based on the fact that M06-2X method is more accurate to explore the system with non-covalent bonding, which may be very crucial for the nano-scaled composites in this work. All calculations are conducted by using the linear combination of atomic orbital (LCAO) technique with *Gaussioan09* code [24]. For all computations, self-consistent field molecular orbital (SCF-MO) method under the framework of DFT at the 6-31G (*d, p*) basis set theory level is adopted.

3. Results and Discussion

3.1. Relative Energy

In order to study the energetic effects in the formation of the graphene/acyl hydrazone composites, the adsorption energy (E_{ads}), which is the energy difference of the adsorption process, is calculated by:

$$E_{ads} = E\ (\text{composite}) - E\ (\text{graphene}) - E\ (\text{mol}) \quad (1)$$

where E (composite) represents the total energy of the composite, and E (graphene) and E (mol) are the energies of isolated graphene and acyl hydrazone molecule with their minimum energy structures respectively. A negative E_{ads} corresponds to a stable adsorption. The obtained results are listed in Table 1.

Table 1. Obtained adsorption energy (E_{ads}), the HOMO and the LUMO of the nano composites.

structures	E_{ads} (eV)	HOMO (eV)	LUMO (eV)
C1	-1.357	-4.175	-3.651
C2	-1.343	-4.175	-3.652
C3	-1.323	-4.144	-3.620
graphene	/	-4.121	-3.621
mol-1	/	-7.315	-0.510
mol-2	/	-7.220	-0.343
mol-3	/	-7.005	-0.382

It is observed that the values of E_{ads} are below zero for all three graphene/acyl hydrazone composites studied in the current work. Recall that the formation of the nanoscaled composite from the isolated graphene and acyl hydrazone molecule is exothermic when Eads<0. Thus it is energetically favourable to form these graphene/acyl hydrazone composites in viewpoint of total energy change. The calculated values of E_{ads} are -1.357, -1.342 and -1.323 eV for ***C1*** – ***C3***, respectively. Among them, the composite ***C1*** is the most exothermic to load the acyl hydrazone molecules on the surface of graphene due to the most negative Eads. Nevertheless, the differences between E_{ads} of the three composites are not evident (less than 0.1 eV). For a comparison, E_{ads} of the composites made of graphene and benzene or pyridine are also calculated with the same method. The obtained results for them are only -0.319 and -0.313 eV, respectively. These values are obviously smaller than those of ***C1*** – ***C3***. As the acyl hydrazone molecules studied in this work, they contain both benzene and pyridine structures. The summation of Eads for benzene and pyridine on graphene, however, is -0.632 eV, which is still only about 50% for those of ***C1*** – ***C3***. This fact implies that the interaction between graphene and other parts of acyl hydrazone (with the exclusion of benzene and pyridine rings) is also important to adsorption energy.

3.2. Electronic Property

It is well-known that the highest occupied molecular orbital (HOMO) and the lowest unoccupied molecular orbital (LUMO) are much crucial when the reactant is involved with chemical reactions [25], and thus the descriptions of the frontier molecular orbital (FMO) of the nano composites is necessary. The energy levels of the FMOs graphene/acyl hydrazone composites are listed in Table 1. It is quite clear here that HOMO and LUMO energy levels of ***C1*** – ***C3*** quite close to each other and vary very little (less than 0.03 eV). For a comparison, the isolated graphene flat and the acyl hydrazone molecules in this work are also calculated, and the obtained results are listed in Table 1. It is observed that both HOMO and LUMO energy levels of ***C1*** – ***C3*** are very close to that of isolated graphene, whereas far from those of the acyl hydrazone molecules. As a result, the FMO energy levels of composites change a little when the acyl hydrazone molecules are loaded on the surface of graphene. From this point, we may conclude that the FMO of the graphene/acyl hydrazone composites would be mainly derived from the graphene flat, other than from acyl hydrazone molecules.

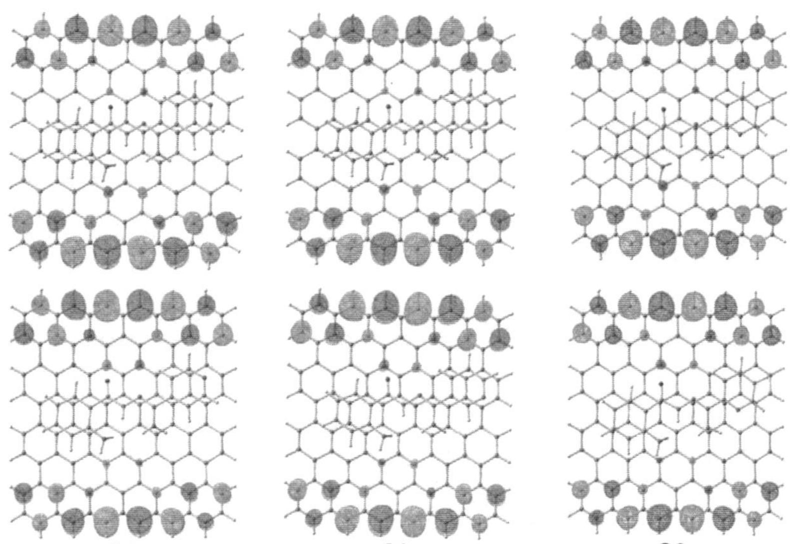

Figure 3. HOMO (top) and LUMO (bottom) of the composite ***C1*** – ***C3***.

To confirm this assumption, the distributions of HOMO and LUMO for composite ***C1*** – ***C3*** are sketched in Figure 3. It is obvious that FMOs of the pure graphene and composite ***C1*** – ***C3*** all exhibit much localization at the edge regions of graphene part. Nearly no distribution is observed at the acyl hydrazone molecule. This feather is nearly the same as that of the isolated graphene flat. According to the frontier molecular orbital theory, the graphene flat is thus the dominating part when the graphene/acyl hydrazone composites are involved in chemical reactions.

3.3. *Counterpart interaction*

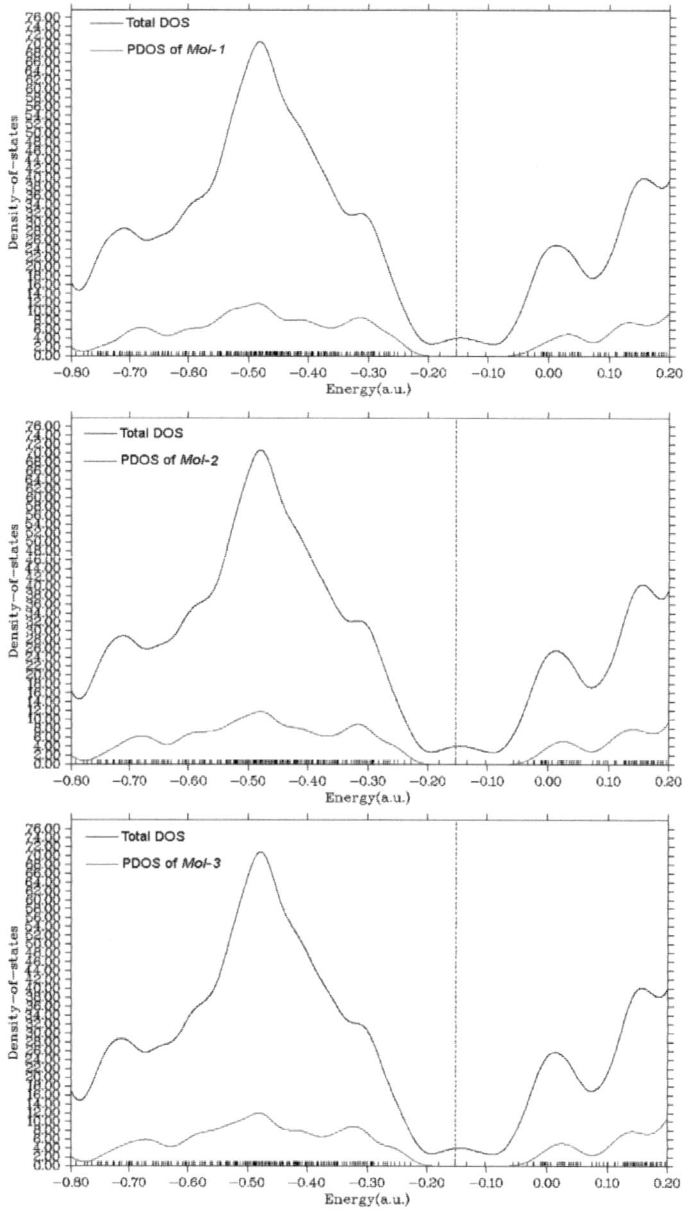

Figure 4. Total density of states and projected density of states of the composite ***C1*** – ***C3***.

In the current work, the graphene/acyl hydrazone composites can be considered as the acyl hydrazone molecules loaded on the graphene surface. And the understanding of the interaction of the counterparts is very crucial for the construction of the nano composites. Thus the investigation of the counterpart interaction for the graphene/acyl hydrazone composites is necessary. Firstly, we focus on the orbital hybridization for the counterpart. This can be analysed based on the projected density of states (PDOS) and total density of states (DOS), as shown in Figure 4. It is found that most parts of DOS are consisted with both graphene flat and acyl hydrazone molecules. This means that the orbital hybridization dose occur for the graphene/acyl hydrazone composites. Nevertheless, the DOS near the

FMOs are dominated by the graphene part, which agrees with the results of FMO analyses mentioned above.

Then we turn to the charge redistribution when the composites are formed by the nano graphene and acyl hydrazone molecules. It is observed that the electronic charge transfer from acyl hydrazone molecules to the graphene part can take place for these nano composites. The amounts of charger transfer are 0.027, 0.025 and 0.019 *e* respectively for ***C1*** – ***C3*** on the basis of the population based on Mulliken analysis. These values of charger transfer are much larger than 0.005 (or 0.007) *e* for that of the composites made of graphene and benzene (or pyridine). To give a videotext description for the charge redistribution, Figure 5 shows the electronic density difference of ***C1*** – ***C3*** relative to the isolated counterparts. It is observed that charge redistribution is mainly spreading at the interlayer space where oxygen and nitrogen atoms are located.

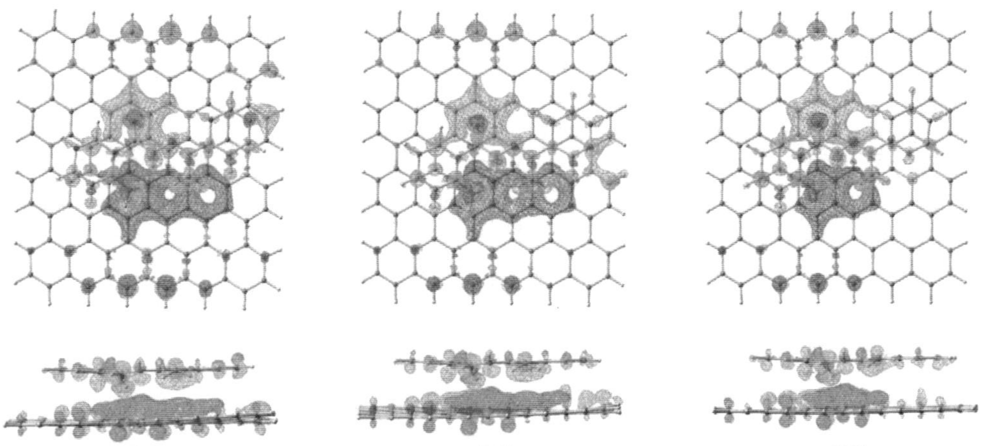

Figure 5. Top view (top) and side view (bottom) of the obtained electronic density difference of the composite ***C1*** – ***C3***.

We also noticed that the acyl hydrazone molecules are loaded on the graphene surface in the nano-scaled composites. Thus Van der Waals interaction may also be important for the counterpart interaction. The nearest distances between the counterparts are calculated to be 3.098, 3.095 and 3.081 Å, respectively for composite ***C1*** – ***C3***. These values are very close to the distances of Van der Waals interaction for carbon, nitrogen and oxygen atoms if their Van der Waals radii are considered to be 1.65, 1.55 and 1.60 Å, respectively. We are also aware that these distances are much longer than the covalent bond lengths between carbon, oxygen and nitrogen atoms. Thus these results indicate the Van der Waals interaction, not the covalent bonds, between the counterparts of the graphene/acyl hydrazone composite.

4. Conclusion

In summery, we have studied the nano-scaled composites composed of graphene and acyl hydrazone molecules based on DFT calculations. The absorption configurations were determined by means of the MD simulations. Then the relative energy, electronic property and the interaction between the counterparts of the nano-scaled composites are calculated and discussed. According to the obtained adsorption energy, the formations of the nano-scaled composites from the isolated graphene and acyl hydrazone molecules are always exothermic in view point of total energy change. This fact indicates that the formation of these nano composites is energetically favorable. Among them, composite ***C1*** is the most stable. The results of FMO analysis show that both HOMO and LUMO of these nano composites are mainly derived from the graphene flat, rather than from the acyl hydrazone molecule.

Thus graphene is the dominating part when the graphene/acyl hydrazone composites are involved in chemical reactions ~~pointing~~ in view point of FMO theory. The counterpart interaction for the graphene/acyl hydrazone composites is also investigated. It is observed that orbital hybridization, charge redistribution and Van der Waals force are important for the interaction between the counterparts of these nano-scaled composites.

Acknowledgements
The authors thank the financial support from Natural Science Foundation of Ningxia (No. NZ15015).

References

[1] Kumar S, Dhar D N, Saxena P N 2009 *J. Sci. Ind. Res.* 68 181.
[2] Casasnovas R *et al* 2013 *J. Phys. Chem. B* **117** 2339.
[3] Laila H *et al* 2013 *J. Mol. Struct.* **1040** 9.
[4] Martell A E, Taylor P 1984 *Inorg. Chem.* **23** 2734.
[5] Rosua T, *et al* 2012 *Polyhedron* **31** 352.
[6] Mintzer M A and Grinstaff M W 2011 *Chem. Soc. Rev.* **40** 173.
[7] Gromova A V *et al* 2012 *Chem. Commun.* **48** 2131.
[8] Müllera I A *et al* 2010 *Tetrahedron Lett.* **51** 4371.
[9] Novoselov K S *et al* 2004 *Science* **306** 666.
[10] Lee C, Wei X *et al* 2008 *Science* **321** 385.
[11] Afkhami A, Shirzadmehr A *et al* 2015 *Talanta* **131** 548.
[12] Shirzadmehr A, Afkhami A *et al* 2015 *J. Mol. Liq.* **204** 227.
[13] Bagheri H, Shirzadmehr A *et al* 2015 *J. Mol. Liq.* **212** 96.
[14] Shao Y, Wang J *et al* 2010 *Electroanal* **22** 1027.
[15] Qin W *et al* 2010 *Biomaterials* **31** 1007.
[16] Saikia N, Deka R C 2013 *J. Comput. Aided Mol. Des.* **27** 807.
[17] Sun X *et al* 2008 *Nnao Res.* **1** 203.
[18] Yang X *et al* 2011 *J. Mater. Chem.* **21** 3448.
[19] Zhang L *et al* 2010 *Small* **6** 537.
[20] Vovusha H *et al* 2013 *J. Phys. Chem. Lett.* **4** 3710.
[21] Saikia N and Deka R C 2014 *New J. Chem.* **38** 1116.
[22] Mattsson T R *et al* 2010 *Phys. Rev. B* **81** 054103.
[23] Zhao Y, Truhlar D G 2008 *Theor. Chem. Acc.* **120** 215.
[24] Frisch M J *et al* 2009 *Gaussian 09*, Gaussian, Inc., Wallingford CT.
[25] Fukui K 1982 *Science* **218** 747.

Tensile properties of bacterial cellulose nanofibers - polyester composites

H Abral[1] and M Mahardika
Mechanical Engineering, Andalas University, Padang, Indonesia

E-mail: abral@ft.unand.ac.id

Abstract. The paper shows tensile properties of bacterial cellulose (BC) nanofibers and polyester (PO) matrix composites. Tensile properties including tensile strength (TS), modulus elasticity (ME), and elongation (EL) were observed respectively. BC nanofibers exist in the form of a sheet that was then varied in matrix PO. The BC sheet was mounted by one, three, five and seven pieces respectively in the matrix PO. The tensile strength of the composites was conducted by using the tensile equipment. The results showed that the tensile strength of the composite with a single sheet of BC was lower than that of pure PO. The ST value achieved maximum level in the number of layers of BC three pieces, but then it decreased for the composites reinforced five and seven pieces of BC nanofiber, respectively. Scanning Electron Microscope (SEM) observation exhibits bad interface bonding between BC nanofibers and PO matrix.

1. Introduction

BC has been studied extensively for application a commercial product such as for the speaker diaphragm, diet food, high-grade paper [1, 2]. Other using that was very spectacular is that BC is applied in the health field for wound dressing materials [3]. Right now many researchers are interested in conducting research BC due to its unique properties. BC contains a pure cellulose, mostly crystallinity, and the nano-sized fibers [4].

Mechanical properties of BC nanofiber are very excellent. Due to the low density, the specific strength of BC is higher than that of steel. Therefore, it is very interesting to apply BC nanofibers to develop military products that require strong and lightweight. Such product is like body armour, helmet having high strength and toughness.

Polyester is one of kind of the polymer owning interesting properties, such as high tensile strength, rapid drying, low density. Because of the advantage properties, polyester is popularly used as a matrix in the composite. However, polyester is brittle due to thermosetting material and so easy to fracture into pieces due to the impact. Therefore, it is necessary to reduce the brittleness of polyester by using reinforcement. The aim of this study is to see the characteristics of BC nanofibers composite mixed with polyester. In this paper, results of tensile strength, modulus of elasticity, and strain, respectively are reported.

2. Methods

2.1. Preparation of BC nanofibers

The BC nanofiber used in this study came from local food product as a dessert. The product that consists of nanofiber was in the form of a sheet. Nanofiber sheet that contains a lot of water was compressed so that water out. Then the fiber is soaked with 5% NaOH solution over 120 minutes aiming to eliminate impurities to bind to the polyester better. Subsequently, the fiber is washed with water to remove NaOH solution. Then the fiber was dried by the sun, and lastly by using a hair dryer until constant weight is reached.

2.2. Manufacture of composites

The dried fibers are mixed with unsaturated polyester resin. The composite sample was varied in one, three, five and seven layers, respectively, of BC nanofibers as the matrix. Composites were made by hand lay-up method.Standard of the samples used for tensile testing was based on ASTM D638.

2.3. Tensile Testing

The tensile strength of composite sample were measured by using equipment Com-Ten testing machineof 95T Series with testing speed of 3 mm/minute at room temperature.

2.4. SEM observation

A kind of SEM equipment used for observation fracture surface of BC nanofibers PO matrix composite was Hitachi 3400 N series.

3. Results and Analysis

3.1. Tensile Strength

The tensile strength of the composites reinforced by different BC sheets in PO matrix is displayed in figure 1. TS value of pure PO is 38.3 MPa that is a little bit lower than other TS of PO from referees [5]. As shown in the figure 1 that strength of pure PO matrix is higher than that of all composite strength after being mixed by BC sheets of nanofibers. Decreases in TS may be due to defects in the composites appearing during the manufacturing the sample such as porosities from air trapped in PO matrix. Another defect in lowering of TS is a bad interface bonding between BC sheet and PO matrix.

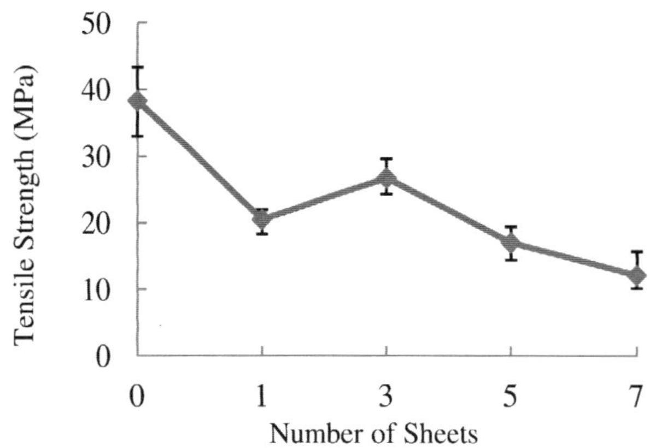

Figure 1. Tensile strength versus number of sheets.

Average TS value of the composite for one sheet BC was 20.5 MPa. Then, it increased for three sheets BC in the PO matrix composite. Average TS of three BC reinforced composites was 26.7MPa. The increment of TS values may be due to the fact that the BC sheet was effective to reinforce the applied load.

3.2. Modulus of Elasticity

Figure 2 shows modulus of elasticity not only for pure PO but also all composites. The ME value of pure PO was 576.7 MPa, and it was lower compared to last report [6]. As increasing the sheets in PO matrix, the ME of composite increased, especially for one and thee sheets. The ME of the composite for one and three BC sheets was 630.8, 669.4 MPa, respectively. However, it decreased as increases in the number of BC sheet in PO matrix. The ME for five and seven sheets of BC was 604.5, 482.6 MPa, respectively. The decreasing may be due to the fact that the sample was so fast fractured due to many defects like porosities and micro crack in PO matrix occurring in manufacturing composites. Also, a bad interface bonding between BC sheet and PO matrix provided lowering of ME [7].

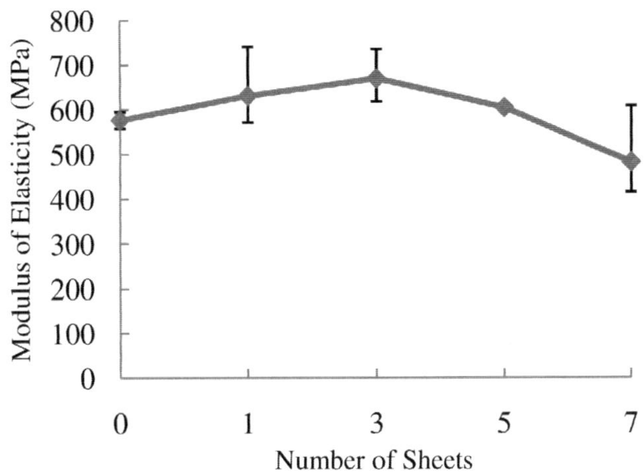

Figure 2. Tensile strength versus number of sheets.

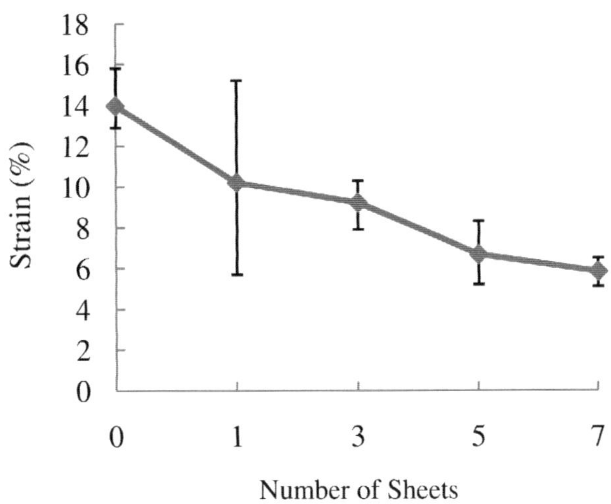

Figure 3. Tensile strength versus number of sheets.

3.3. Maximum Strain
Characterization of the composite strain of each different BC sheet was displayed in figure 3.The strain of pure polyester was 14% and the highest of whole strains of the composites. At the beginning of the composites for one sheet of BC, the elongation was still high. However, EL of the composites decreased continuously for further BC sheets in the PO matrix. This decreasing of the EL values may be due to increasing the number of defects in samples.

3.4. SEM photographs
Fracture surface of composites in SEM observation was displayed in figure 4. As shown in the figure 4 that micro crevice was observed on the fracture surface. Clearly, BC sheet was not bonded well with PO matrix resulting crevice between matrix and BC sheet. The difficulties of the BC bonding with PO matrix may be due to the fact that different characterisation of both materials. The behaviour of PO matrix is hydrophobic and the BC hydrophilic.

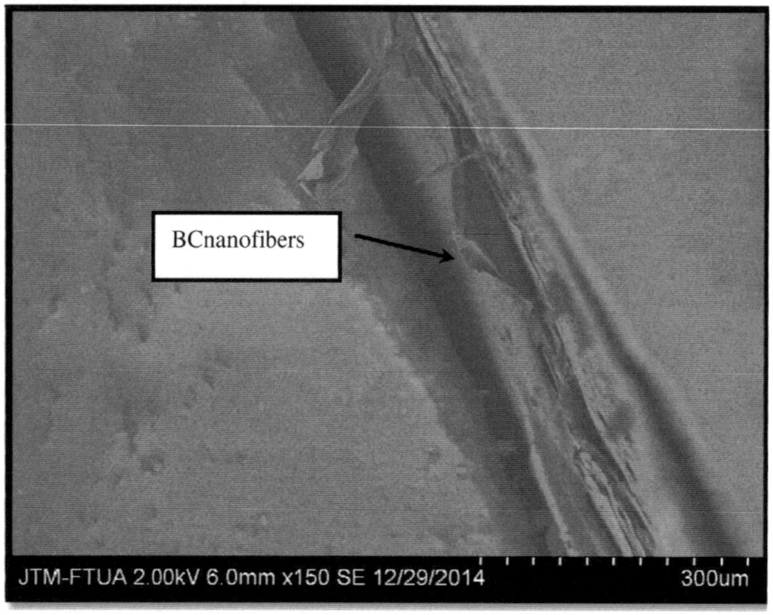

Figure 4. SEM fracture surface of tensile sample reinforced by one sheet of BC nanofibers.

4. Conclusion
The tensile properties of BC nanofibers polyester composite were presented in this paper. Some conclusions can be drawn as follows:

- TS value of pure PO matrix was higher than that of BC sheets reinforced composites. The maximal TS was achieved on 26.7 MPa by three layers of BC in the PO matrix.
- ME of the composites were increased as an increment of BC sheets in PO matrix. The maximum value of ME was 669.4 MPa obtained on number of 3 sheets.
- Fracture strain of the composites decreases continuously as number of sheets increases.
- SEM fracture surface revealed bad interfacial bonding between PO matrix and BC sheets due to different characterisation.

Acknowledgments

Acknowledgement is addressed from the authors to Directorate General of Higher Education, Ministry of National Education, Indonesia, with the project "International Research Collaboration and Publication, 2016", for supporting research finance.

References

[1] Cannon R E and Anderson S M 1991 *Microbiology* **17** 435-447.
[2] Evans B R, O'Neill H M, Malyvanh V P, Lee I and Woodward J 2003 *Biosensors and Bioelectronics* **18** 917-923.
[3] Czaja W, Krystynowicz A, Bielecki S and Brown R M 2006 *Biomaterials* **27** 145-151.
[4] Shah N, Ul-Islam M, KhattakW A and Park J K 2013 *Carbohydrate Polymers* **98** 1585-1598.
[5] Abral H, Gafar MF,Andriyanto H, Ilhamdi, Sapuan SM, Ishak MR and Evitayani 2012 *Polym-Plast Technol Eng* . **51** 12-18.
[6] Abral H, Andriyanto H, Samera R, Sapuan SM and Ishak MR 2012 *Polym-Plast Technol Eng.* **51** 500-506.
[7] Abral H, Putra H, Sapuan S M and Ishak M R 2013 *Polym-Plast Technol Eng.* **52** 446-451

The influence of g-C_3N_4 loading on the photocatalytic activity of $Bi_{12}O_{17}Br_2/Bi_2O_3$ composite in the phenol red degradation

X Lin[1,2], S S Jiang[2], Z Lin[2], M Wang[2], Y S Yan[1,3]

[1]School of Chemistry and Chemical Engineering, Jiangsu University, Zhenjiang Jiangsu 212013, P. R. China
[2]Key Laboratory of Preparation and Application Environmentally Friendly Materials of the Ministry of Education, Jilin Normal University, Siping, Jilin 136000, P. R. China

E-mail: jlsdlinxue@126.com

Abstract In this work, the influence of graphite-like C_3N_4 (g-C_3N_4) loading on the photocatalytic performance of $Bi_{12}O_{17}Br_2/Bi_2O_3$ composite was studied. The results indicated that g-C_3N_4-$Bi_{12}O_{17}Br_2/Bi_2O_3$ photocatalysts displayed much higher photocatalytic efficiency for the phenol red degradation than the bare Bi_2O_3 and $Bi_2O_3/Bi_{12}O_{17}Br_2$ composite under visible light illumination. The best photocatalytic performance of the composite sample with almost 100% phenol red degradation located at g-C_3N_4-$Bi_{12}O_{17}Br_2/Bi_2O_3$-50 under visible light illumination for 80 mins. This excellent photocatalytic performance was displayed according to the efficient separation and transportation of the photogenerated charges, which was resulted from the coupling of C_3N_4 and $Bi_{12}O_{17}Br_2/Bi_2O_3$.

1. Introduction

Photocatalysis has become an ideal method for producing clean sources of energy and environmental pollution treatment [1-3]. The crucial step for its application is to develop new, efficient, and clean catalysts [4-10]. Recent years, bismuth-based semiconductor photocatalysts such as Bi_2O_3 [11,12], $BiVO_4$ [13-15], Bi_2MoO_6 [16-18], BiOX (X = Cl、Br、I) [19], and Bi_2WO_6 [20] have attracted much attention because of their attractive photocatalytic activities under visible light irradiation. Bi_2O_3 has been broadly used in photocatalysis filed due to its perfect optical and photocatalitic properties [21-23]. Furthermore, Bi_2O_3 is also a material, which has a narrow band gap (2 to 3.9 eV) all of which make Bi_2O_3 to degrade dyes under UV and visible-light illumination [11,12]. Results come to that nanometer structures of Bi_2O_3 can improve the stability of metastable Bi_2O_3 at low temperature, which in return can also enhance the photocatalytic activities [24]. However the photodegradation activity isnot so active. To resolve these issus, great efforts have been made on heteronanostructures to enhance the separation of photogenerated charges, for further improving the photocatalytic performance [24-26].

Herein, the ternary heterostructured g-C_3N_4-$Bi_{12}O_{17}Br_2/Bi_2O_3$ composite photocatalysts were successfully synthesized for the first time. The phenol red was employed as a mode to study the photocatalytic performances of the ternary heterostructured composites. The results reveal that compared with pure Bi_2O_3 and $Bi_{12}O_{17}Br_2/Bi_2O_3$ composite, the ternary heterostructured g-C_3N_4-$Bi_{12}O_{17}Br_2/Bi_2O_3$ composite have a marked enhanced phenol red photodegradation efficiency under visible-light. The g-C_3N_4-$Bi_{12}O_{17}Br_2/Bi_2O_3$-50 catalyst performed the best in the degradation of phenol red. Moreover, the possible photocatalysis mechanism of the g-C_3N_4-$Bi_{12}O_{17}Br_2/Bi_2O_3$ nanoheterostructure was also studied.

2. Experimental

2.1. Preparation of samples

2.1.1 Preparation of g-C_3N_4. In a typical step, dicyandiamide (2.0 g) was heated to 550 °C (with a rate of 2.3 °C/min), after that, it was still heated at the tempered of 550 °C for 2 hs [26-28]. The product was collected for later use.

2.1.2 Preparation of g-C_3N_4-$Bi_2O_3/Bi_{12}O_{17}Br_2$ photocatalysts. Different mass ratios g-C_3N_4-$Bi_{12}O_{17}Br_2/Bi_2O_3$ composites were prepared via hydrothermal method. As 50% g-C_3N_4-$Bi_{12}O_{17}Br_2/Bi_2O_3$ for example, in a typical procedure, 100 mg g-C_3N_4, 100 mg $Bi(NO_3)_3{\cdot}5H_2O$, and 30 mg sodium dodecyl benzene sulfonate (SDBS) were mixed in 10 mL ethylene glycol. The mixture was vigorously stirred for 10 mins, and then 30 mL ethanol was added and vigorously stirred for around 10 mins, and heat treated at 160 °C for 12 hs in an oven. At last, the obtained 50wt% g-C_3N_4-$Bi_{12}O_{17}Br_2/Bi_2O_3$ powder was washed with distilled water \twice, and dried. Different concentrations of g-C_3N_4-$Bi_{12}O_{17}Br_2/Bi_2O_3$ composites at 10wt%, 30wt%, 70wt% and 90wt% were synthesized (denoted as g-C_3N_4-$Bi_{12}O_{17}Br_2/Bi_2O_3$-10, g-$C_3N_4$-$Bi_{12}O_{17}Br_2/Bi_2O_3$-30, g-$C_3N_4$-$Bi_{12}O_{17}Br_2/Bi_2O_3$-70, and g-$C_3N_4$-$Bi_{12}O_{17}Br_2/Bi_2O_3$-90, respectively).

2.2. Characterization of samples
The structures of the as-fabricated samples were tested by X-ray diffraction (XRD). A scanning electron microscopy (SEM, Japan JEOL, JSM-6510) was used to study the microstructure of the photocatalysts. The chemical compositions of the as-synthesized samples were tested through scanning electron microscope-X-ray energy dispersion spectra (SEM-EDX, JSM-6510). The diffuse reflectance spectra (DRS) were studied through a UV-vis spectrometer (UV-2550, Shimadzu).

2.3. Photocatalytic activities
The photocatalytic performances of the as-fabricated samples were studied by using phenol red photocatalytic degradation tests. The phenol red solution containing 50 mg of catalyst was mixed. A 300 W Xe lamp (λ > 420 nm) was used to offer visible light illumination. Before visible light illumination, the phenol red solution was vigorously stirred in the dark (60 mins). At 20 mins intervals, 4 mL of the phenol red solution was collected and measured.

3. Results and discussion
From Fig. 1a, it can be observed that the bare g-C_3N_4 has two peaks of around 13.0° and 27.3 ° respectively [27]. And the $Bi_{12}O_{17}Br_2/Bi_2O_3$ composite shows a coexistence of Bi_2O_3 phase (JCPDS No. 41-1449) and $Bi_{12}O_{17}Br_2$ phase (JCPDS No. 37-0701), revealing that the mixture of Bi_2O_3 and $Bi_{12}O_{17}Br_2$ are the main existing forms of the sample. For g-C_3N_4- $Bi_{12}O_{17}Br_2/Bi_2O_3$ composites, similar diffraction peaks to the $Bi_{12}O_{17}Br_2/Bi_2O_3$ are detected [28].

In the FT-IR spectrum (Fig. 1b) of bare g-C_3N_4, three absorption peaks in the region of 800–4000 cm^{-1} can be observed [27]. The broad absorption band at 3000– 3500 cm^{-1} is resulted from the stretching vibration of O–H and the stretching vibration of N–H [27]. The strong peaks of 1200-1700 cm^{-1} (centered at 1243, 1335, 1411, and 1649 cm^{-1}) are due to the vibration of CN heterocycle [27].

The peak of 802 cm^{-1} is ascribed to the mode of N≡H units [27]. All the peaks of g-C_3N_4 appear in the g-C_3N_4-$Bi_{12}O_{17}Br_2$/Bi_2O_3 samples, revealing the existence of g-C_3N_4.

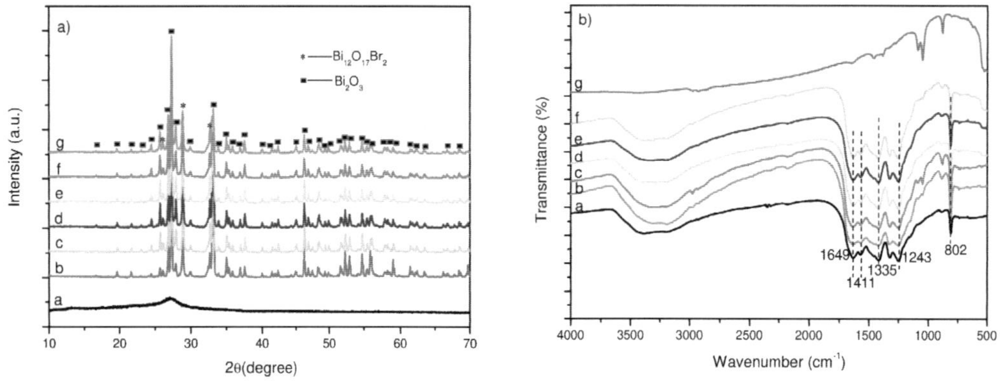

Figure. 1 (a) XRD patterns of the as-fabricated samples: (a) g-C_3N_4, (b) $Bi_{12}O_{17}Br_2$/Bi_2O_3, (c) g-C_3N_4-$Bi_{12}O_{17}Br_2$/Bi_2O_3-10, (d) g-C_3N_4-$Bi_{12}O_{17}Br_2$/Bi_2O_3-30, (e) g-C_3N_4-$Bi_{12}O_{17}Br_2$/Bi_2O_3-50, (f) g-C_3N_4-$Bi_{12}O_{17}Br_2$/Bi_2O_3-70, (g) g-C_3N_4-$Bi_{12}O_{17}Br_2$/Bi_2O_3-90. **(b)** IR spectra of the as-obtained photocatalysts: (a) g-C_3N_4, (b) 10% g-C_3N_4-$Bi_{12}O_{17}Br_2$/Bi_2O_3-10, (c) g-C_3N_4-$Bi_{12}O_{17}Br_2$/Bi_2O_3-30, (d) g-C_3N_4-$Bi_{12}O_{17}Br_2$/Bi_2O_3-50 (e) g-C_3N_4-$Bi_{12}O_{17}Br_2$/Bi_2O_3-70, (f) g-C_3N_4-$Bi_{12}O_{17}Br_2$/Bi_2O_3-90, (g) $Bi_{12}O_{17}Br_2$/Bi_2O_3.

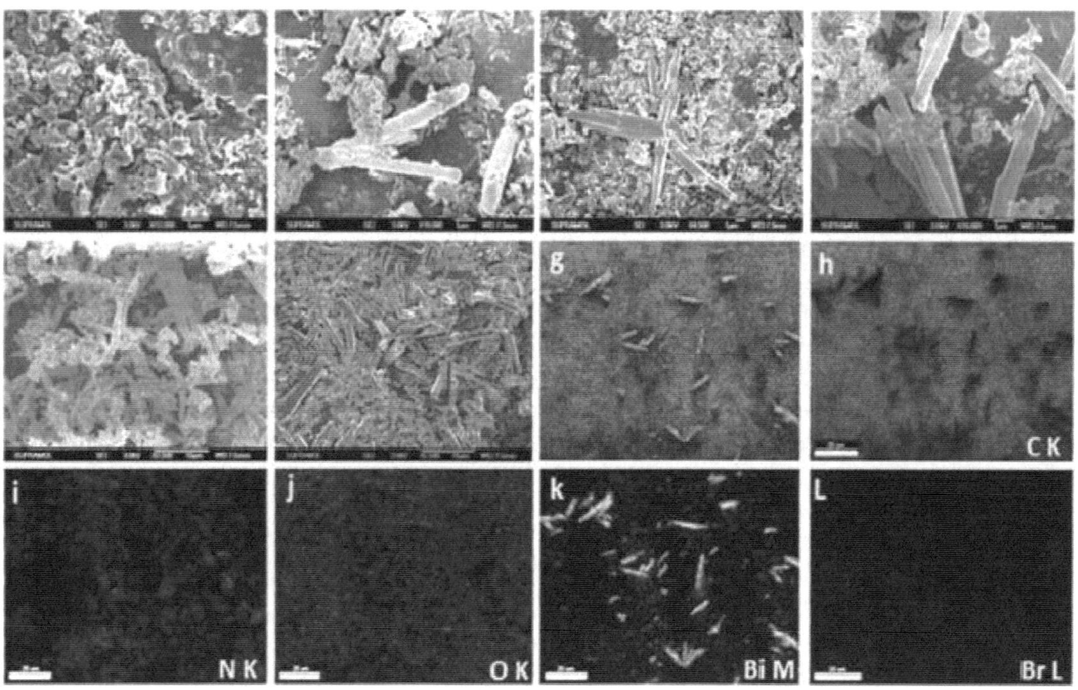

Figure. 2 SEM images of the as-fabricated photocatalysts: (a) g-C_3N_4, (b) g-C_3N_4-$Bi_{12}O_{17}Br_2$/Bi_2O_3-10, (c) g-C_3N_4-$Bi_{12}O_{17}Br_2$/Bi_2O_3-30, (d) g-C_3N_4-$Bi_{12}O_{17}Br_2$/Bi_2O_3-50, (e) g-C_3N_4-$Bi_{12}O_{17}Br_2$/Bi_2O_3-70, (f) g-C_3N_4-$Bi_{12}O_{17}Br_2$/Bi_2O_3-90, (g) $Bi_{12}O_{17}Br_2$/Bi_2O_3, and EDS elemental mapping maps of g-C_3N_4-$Bi_{12}O_{17}Br_2$/Bi_2O_3-50 sample (g-l).

The SEM images of bare g-C_3N_4 and g-C_3N_4-$Bi_{12}O_{17}Br_2$/Bi_2O_3 are revealed in Fig. 2. From Fig. 2(a), pure g-C_3N_4 particles are big irregular crystals (average diameter of 1–2 μm). After g-C_3N_4 loading, g-C_3N_4 particles were deposited on the $Bi_{12}O_{17}Br_2$/Bi_2O_3 samples, leading to the formation of g-C_3N_4-$Bi_{12}O_{17}Br_2$/Bi_2O_3 heterostructure (Fig. 2b-f). The formation of the g-C_3N_4-$Bi_{12}O_{17}Br_2$/Bi_2O_3 nanoheterostructure was illustrated through the elemental mapping of the as-fabricated g-C_3N_4-$Bi_{12}O_{17}Br_2$/Bi_2O_3-50 sample (Fig. 3g-l). Images of Bi–M, Br-L, O-K, N-K, and C-K show the Bi, O, N, C, and Br signals, respectively, further revealing the existence of g-C_3N_4, Bi_2O_3, and $Bi_{12}O_{17}Br_2$ in the g-C_3N_4-$Bi_{12}O_{17}Br_2$/Bi_2O_3-5 sample.

Fig. 3a shows the comparation of photocatalytic gradation rates of phenol red over different catalysts under visible light illumination. From the catalytic experiments, g-C_3N_4-$Bi_{12}O_{17}Br_2$/Bi_2O_3 samples were found to be more photoactive towards phenol red than bare g-C_3N_4, Bi_2O_3, and $Bi_{12}O_{17}Br_2$/Bi_2O_3 sample. As can be seen in Fig. 3b, visible light illumination of phenol red by the g-C_3N_4-$Bi_{12}O_{17}Br_2$/Bi_2O_3-50 sample resulted in the apparent decrease in absorption.

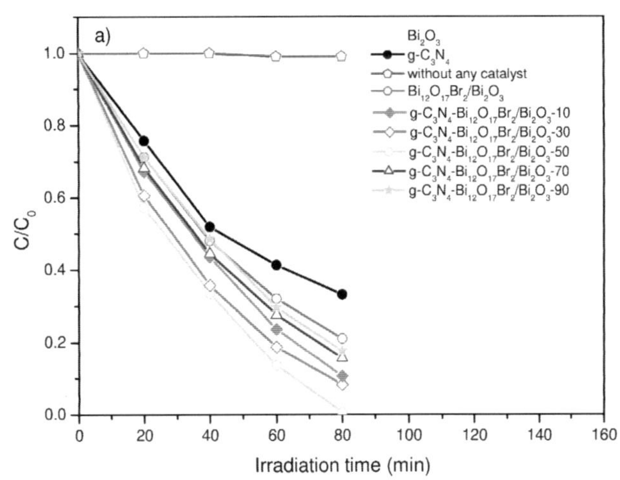

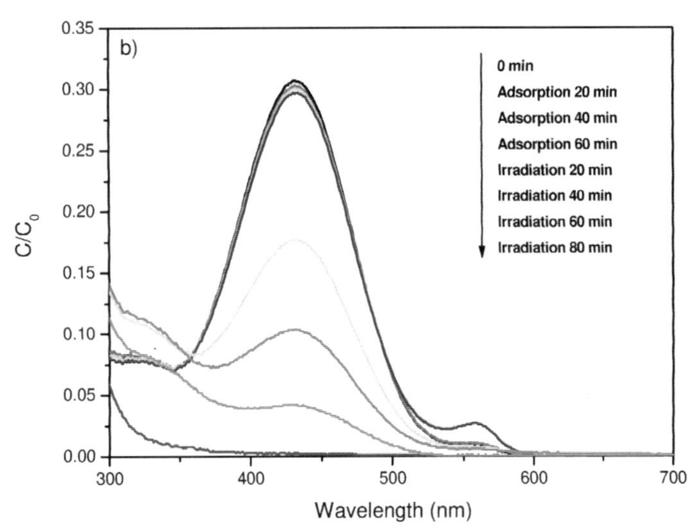

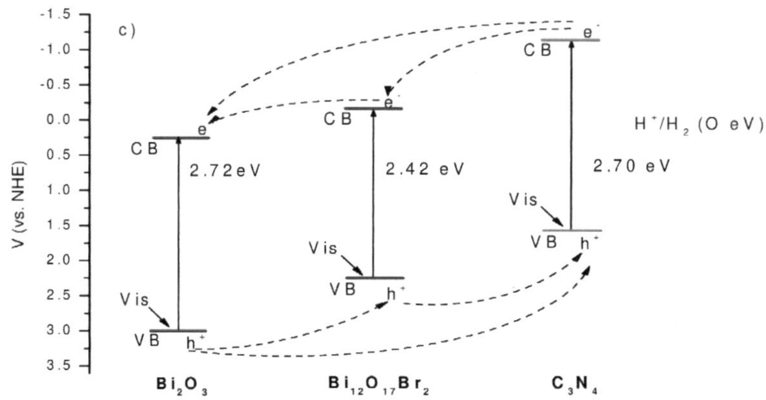

Figure. 3 (a) Photodegradation efficiencies of different samples. (b) UV-visible spectral changes of phenol red in an aqueous g-C_3N_4-$Bi_{12}O_{17}Br_2$/Bi_2O_3-50 dispersion. (c) Schematic diagram of the separation and transfer of photogenerated charges in the photocatalyst.

The photocatalysis mechanism of the g-C_3N_4-$Bi_{12}O_{17}Br_2$/Bi_2O_3 was studied, as revealed in Fig. 3c. The CB levels and VB levels of g-C_3N_4, $Bi_{12}O_{17}Br_2$, and Bi_2O_3 were obtained from the reported works [24,27]. The g-C_3N_4-$Bi_{12}O_{17}Br_2$/Bi_2O_3 hybrid catalysts demonstrate much higher photocatalytic performance than bare g-C_3N_4 and $Bi_{12}O_{17}Br_2$/Bi_2O_3, which shows that photogenerated charges separate more efficiently in the g-C_3N_4-$Bi_{12}O_{17}Br_2$/Bi_2O_3 composite. When the ternary heterostructured samples are irradiated, g-C_3N_4, $Bi_{12}O_{17}Br_2$, as well as Bi_2O_3 can be excited and produce photogenerated charges. Most of charges recombine easily, thus the bare sample has a respectively low photocatalytic effeciency [29-37]. For the ternary heterostructured g-C_3N_4-$Bi_{12}O_{17}Br_2$/Bi_2O_3 sample, at first, when the ternary heterostructured composite is excited by visible light illumination, the electrons in the CB of g-C_3N_4 can migrate to the CB of $Bi_{12}O_{17}Br_2$ because o the CB of g-C_3N_4 is more negative than that of $Bi_{12}O_{17}Br_2$. In addition, the holes on the $Bi_{12}O_{17}Br_2$ can move to g-C_3N_4. Second, the CB of $Bi_{12}O_{17}Br_2$ is more negative than that of Bi_2O_3 can account for the fact that the photogenerated electrons in the CB of $Bi_{12}O_{17}Br_2$ can migrate to the CB of Bi_2O_3 The holes on the Bi_2O_3 surface can transfer to $Bi_{12}O_{17}Br_2$. The electrons in the CB of g-C_3N_4 can directly migrate to the CB of $Bi_{12}O_{17}Br_2$. These charge transfer can improve the charges separation, thus enhancing the photocatalytic efficiency.

4. Conclusions

In short, ternary heterostructured g-C_3N_4-$Bi_{12}O_{17}Br_2$/Bi_2O_3 composite photocatalysts were successfully synthesized, and the composite samples showed excellent photocatalytic performance. And g-C_3N_4-$Bi_{12}O_{17}Br_2$/Bi_2O_3-50 composite performed best.

Acknowledgment

This work was supported by the China postdoctoral science foundation (2013M531286) and Science and Technology Research Project of Jilin Province Department of Education (2015220).

References

[1] Lu S Y, Wu D, Wang Q L, Yan J H, Buekens A G and Cen K F 2011 *Chemosphere* **82** 1215-1224

[2] Huang C X, Zhu K R, Qi M Y, Zhuang Y L and Cheng C 2012 *J. Phys. Chem. Solids* **73** 757-761

[3] Hasanpour A, Niyaifar M, Mohammadpour H and Amighian J 2012 *J. Phys. Chem. Solids* **73**

1066-1070
[4] Mao Y B and Wong S S 2006 *J. Am. Chem. Soc.* **128** 8217-8226
[5] Chen F, Zou W W, Qu W W and Zhang J L 2009 *Catal. Commun.* **10** 1510-1513
[6] Wang S, Yi L X, Halpert J E, Lai X Y, Liu Y Y, Cao H B, Yu R B, Wang D and Li Y L 2012 *Small* **8** 265-271
[7] Yang N L, Liu Y Y, Wen H, Tang Z Y, Zhao H J, Li Y L and Wang D 2013 *ACS Nano* **7** 1504-1512
[8] Lin Y, Geng Z G, Cai H B, Ma L, Chen J, Zeng J, Pan N and Wang X Q 2012 *Eur. J. Inorg. Chem.* **28** 4439-4444
[9] Wang L B, Wang Y, He C R, Zhuang A W, Wang X P, Zeng J and Hou J G 2013 *J. Am. Chem. Soc.* **135** 1272-1275
[10] Zeng J and Xia Y N 2012 *Nature Nanotech.***7** 415-416
[11] Zhou L, Wang W Z, Xu H L, Sun S M and Shang M 2009 *Chem. Eur. J.* **15** 1776-1782
[12] Brezesinski K, Ostermann R, Hartmann P, Perlich J and Brezesinski T 2010 *Chem. Mater.* **22** 3079-3085
[13] Lin X, Li H J, Yu L L, Zhao H, Yan Y S, Liu C B and Zhai H J 2013 *Mater. Res. Bull.* **48** 4424-4429
[14] Lin X, Yu L L, Yan L N, Li H J, Yan Y S, Liu C B and Zhai H J 2014 *Solid State Sci.* **32** 61-66
[15] Lin X, Wang Y S, Zheng J, Liu C, Yang Y and Che G B 2015 *Dalton Trans.* **44** 19185-19193
[16] Wang Q W, Guo X Y, Lin X, Liu C B and Zhai H J 2014 *Chinese J. Chem. Phys.* **27** 568-572
[17] Lin X, Guo X Y, Liu D, Wang Q W, Zhai H J and Chang L M 2015 *Mater. Res. Bull.* **63** 72-79
[18] Lin X, Liu D, Guo X Y, Sun N, Zhao S, Chang L M, Zhai H J and Wang Q W 2015 *J. Phys. Chem. Solids* **76** 170-177
[19] Yu C L, Cao F F, Shu Q, Bao Y L, Xie Z P, Yu J and Yang K 2012 *Acta Phys. -Chim. Sin.* **28** 647-653
[20] Lin X, Liu Z X, Guo X Y, Liu C B, Zhai H J, Wang Q W and Chang L M 2014 *Mater. Sci. Eng. B* **188** 35-42
[21] Drache M, Roussel P and Wignacourt J P 2007 *Chem. Rev.* **107** 80-96
[22] Cornei N, Tancret N, Abraham F and Mentre O 2006 *Inorg. Chem.* **45** 4886-4888
[23] Li W 2006 *Mater. Chem. Phys.* **99** 174-180
[24] Xu Y S, Zhang Z J and Zhang W D 2013 *Mater. Res. Bull.* **48** 1420-1427
[25] Xu J J, Chen M D and Fu D G 2011 *Trans. Nonferrous Met. Soc. China* **21** 340-345
[26] Guo F, Shi W L, Lin X and Che G B 2014 *J. Phys. Chem. Solids* **75** 1217-1222
[27] Shi W L, Guo F, Chen J B, Che G B and Lin X 2014 *J. Alloys Compd.* **612** 143-148
[28] Zhang S Q, Yang Y X, Guo Y N, Guo W, Wang M, Guo Y H and Huo M X 2013 *J. Hazard. Mater.* **261** 235-245
[29] Liu L, Qi Y H, Lu J R, Lin S L, An W J, Hu J S, Liang Y H and Cui W Q 2015 *RSC Adv.* **5** 99339-99346
[30] Tian Y L, Chang B B, Lu J L, Fu J, Xi F N and Dong X P 2013 *ACS Appl. Mater. Interfaces* **5** 7079-7085
[31] Lin X, Xu D, Lin Z, Jiang S S and Chang L M 2015 *RSC Adv.* **5** 84372-84380
[32] Guo F, Shi W L, Lin X, Yan X, Guo Y and Che G B 2015 *Sep. Purif. Techn.* **141** 246-255
[33] Jiang D L, Chen L L, Zhu J J, Chen M, Shi W D and Xie J M 2013 *Dalton T.* **42** 15726-15734
[34] Lin X, Guo X Y, Liu D, Wang Q W and Yan Y S 2015 *Nano* **10** (2) 1550020-1-550020-8
[35] Sun L M, Qi Y, Jia C J, Jin Z and Fan W L 2014 *Nanoscale* **6** 2649-2659
[36] Chen X, Zhou B H, Yang S L, Wu H S, Wu Y X, Wu L D, Pan J and Xiong X 2015 *RSC Adv.* **5** 68953-68963
[37] Wang W J, Cheng H F, Huang B B, Liu X L, Qin X Y, Zhang X Y and Dai Y 2015 *J Colloid. Interf. Sci.* **442** 97-102

IOP Conf. Series: Materials Science and Engineering **137** (2016) 012021 doi:10.1088/1757-899X/137/1/012021

First-principles study of ferromagnetism in Pd-doped and Pd-Cu-codoped BN

Q Wang[1,2,3], S Wang[1], J F Dai[1] and W X Li[1]

[1] School of Science, Lanzhou University of Technology, Lanzhou 730050, China
[2] State Key Laboratory of Advanced Processing and Recycling of Non-ferrous, Lanzhou 730050, China

E-mail: wangqing@lut.cn

Abstract. In this paper, we aimed at probing the ferromagnetism properties of Pd and Pd-Cu-codoped supercell BN based on the first-principles methods. The formation energy, lattice constants, energy band structures, spin density of state, energy difference between ferromagnetism (FM) and autiferromagnetism (AFM) orderings were calculated. Formation energy calculations showed that Pd atom tended to replace B atom in the supercell. Pd-doped BN exhibited a half-metallic ferromagnetic. And the ferromagnetism arised form the strong hybridization between the Pd4d and N2p state. Pd-Cu-codoped BN also displayed a half-metallic ferromagnetic. The incorporation of Pd and Pd-Cu induced some impurity energy differences between FM and AFM orderings. It also showed that FM state was the ground state, and room temperature ferromagnetism may be expected. These results pointed out the possibility of fabricating BN based on dilute magnetic semiconductors (DMS) by doping with Pd and Pd-Cu.

1. Introduction

Recently, people have expected to achieve signal processing and information storage capabilities in the same device. The dilute magnetic semiconductors (DMSs) were considered to be an ideal material for the manufactures of such semiconductors. At present, it has become a hot topic. Nanoscale materials exhibit a variety of unique properties that could be used in a wide of advanced technological applications. BN has a lot of similar characteristics with AlN which has many useful properties, such as physical and chemical properties, especially involving the electronic and magnetic properties. As we know, element doping is a promising approach to controllably modify the electronic structures and magnetic behaviors. Since the discovery of AlN, many theoretical and experimental studies about the doping of metal atoms have been conducted [1-3].

The BN is a wide band gap and large band energy of the semiconductor [4]. BN had four structures, which are named as cubic BN (c-BN), hexagonal BN (h-BN), rhombohedral BN (r-BN) and wurtzite BN (w-BN). The c-BN and h-BN are steady state structure, but r-BN and w-BN are not. Within reported cubic structure BN, similar to rhombus, the bonding manner of the cubic BN sheet is sp3 hybridization, and it has several properties including high melting point, high thermal conductivity, and high chemical hardness. Currently, the research was mainly focused on the mechanical properties and thermal properties of BN, while the researchers rarely focused on the optical and magnetic properties. Back in 1999, the experiments [5,6] showed that the deficient BN exhibited ferromagnetic at room temperature. Zhou and Liu [7,8] made some calculations for the ferromagnetism of the semi-metallic Cr-doped BN, and Pd doped BN had been reported [9] in the experiment, but the studies on DMS had not been reported. First-principles calculation method had been widely used in material simulation [10-12]. This paper carried out geometry optimization based on density functional theory plane wave pseudopotential (PWP) method for Pd-doped BN supercell system. And the results were analyzed to study electronic structure and optical properties of the Pd-doped BN. It explored the feasibility of obtaining Pd-doped BN based on the DMS.

2. Calculation model and methods

The space group of cubic BN semiconductor was F-43m. The experimental lattice constant was 0.3615nm [13]. $2 \times 2 \times 2$ of the supercell system containing 64 atoms was built. In all calculations, one or two Pd atoms replace the B atoms in the supercell system. Where x = 0.03125, 0.0625 represent two different doping concentrations ($Pd_xB_{1-x}N$). Pd atom, as a substitute, replaces bits in supercell high symmetry center point. Two Pd atoms, as an alternative, substitute bit option in the center of the high symmetry, and the other alternative place Pd at the four corners. For each configuration, it was optimized before calculating the average structure. Structural optimization was calculated by the CASTEP package [14]. The Perdew-Burke-Ernzerh (PBE) of the generalized gradient approximation (GGA) [15] was chosen to describe the effects of the electronic exchange-correlation interaction. In these calculations, the cutoff energy of the plane wave basis set was 380eV. The Monkhorst-Pack scheme K-point grid sampling was set as $6\times6\times1$ for the irreducible Brillouin zone. Structural optimization was continued until the energy converged to 1×10^{-5}eV/atom, the forces on the atoms had converged to less than 0.03eV/Å, the stress had converged to 0.05GPa and the atom displacement had converged to $1\text{x}10^{-3}$Å. Integral spin density and spin density absolute value integrator were analyzed. If the two numbers were very close and very small, the system was paramagnetic; provided that the two numbers were close and limited, the system was ferromagnetic; should the absolute value of the spin density integration was limited and the spin density integration itself was close to zero, the system was antiferromagnetic; when the two numbers were limited and the D-value was limited, the system was ferrimagnetic.

3. Results and discussion

3.1. Optimized Structures

To assess the accuracy of our computation method, we performed a series of calculations for BN supercell optimization. The equilibrium lattice constant after optimization is a=b=c=3.615, which is in good agreement with the experimental results (a=b=c=3.615) [16], hence it ensures the reliability of our method. We have calculated that the lattice constant of pure BN, the lattice constant is 0.3615 nm, and this value is exactly the same with the experiment. When using a Pd atom instead of one of the B atoms, the calculated lattice constants are 0.3615 nm, which indicates that Pd atoms do not affect the size of the lattice. However, the Pd-N bond increases compared with the B-N bond of the pure BN and the surrounding bond-length changes subsequently. This is because that the Pd atomic radius is larger than that of the B atoms leading to the redistribution of the electrons Under the situation where the direction of the electron spin was not set, we made analysis for the integral of the spin density and spin density absolute value integrator by calculation. We found that pure BN exhibits antiferromagnetic, and Pd doped BN system is calculated for the magnetic moment of ferromagnetic 2.62 μB. This shows that the Pd-doped BN can produce a magnetic spin polarization.

3.2. Electronic Structure

When the electronic calculations are done, the valence electron configuration of the main selection has B: $2s^22p^1$, and N: $2s^22p^3$. The pure BN was calculated, in order to compare with the case of doped BN. Fig. 1 is a band structure and a partial density of states (DOS) diagram of BN. Fig.1a shows that the calculated BN band gap is 4.492eV. This value is lower than the experimental values of 6.0eV. That is mainly attributed to the generalized gradient approximation (GGA) that cannot accurately calculate the band gap value [17]. But this does not affect the theoretical analysis of the electronic structure of BN. Fig.1b should be deviation, because the generalized gradient approximation (GGA) may calculate eigenstates in time, will produce a deviation. Fig. 1b shows that the valence bonds could be divided into the upper -22eV to -14eV and -11eV to 0eV band composition. It can be seen that valence band is mainly composed of B2s2p and N2p, the upper valence band is mainly composed of N2p, lower valence band is mainly composed of B2s2p and N2s which consist of a strong localized states at -15eV. The conduction band with a small amount of B2s and N2s is mainly composed of B2p and N2p.

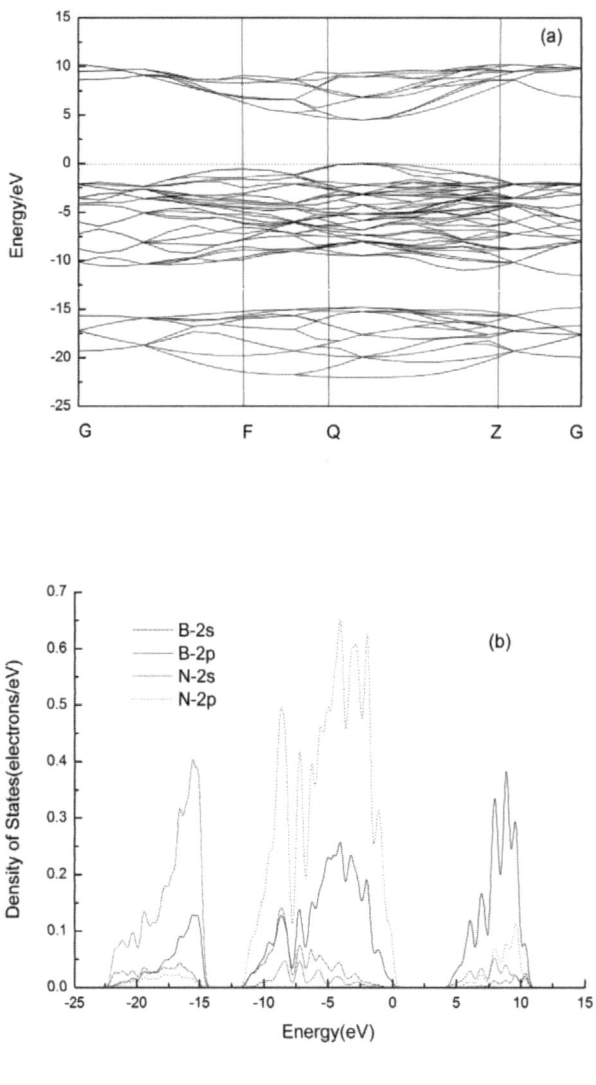

Figure 1. The band structure (a) and PDOS(b) of the pure BN

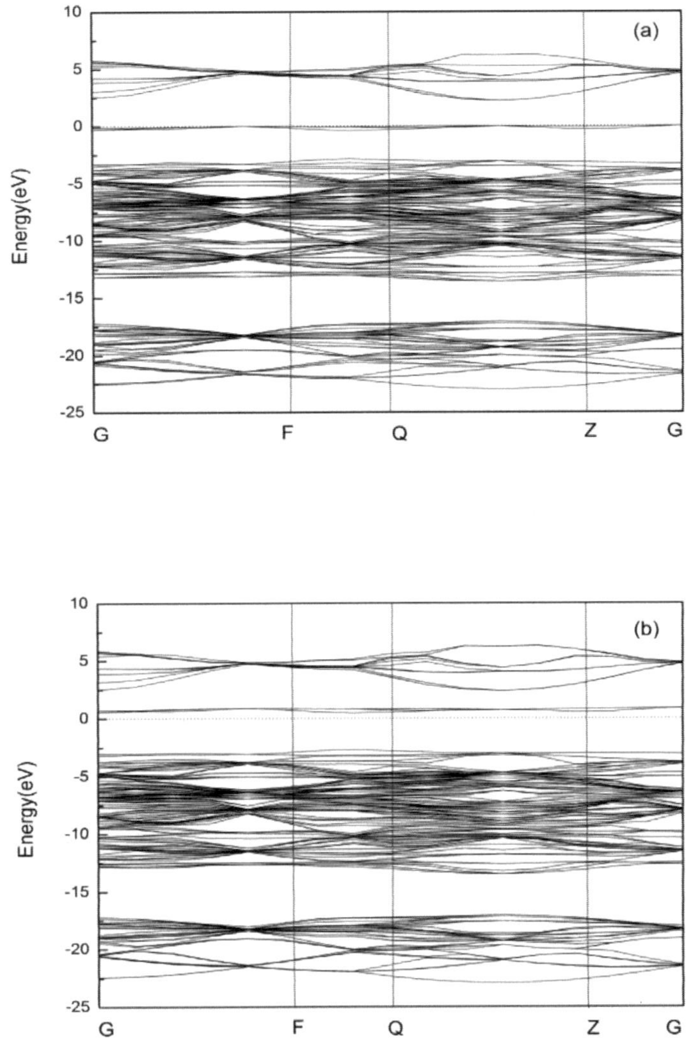

Figure 2. Spin-polarized down (a) and Spin-polarized up (b) band structures of BN doped with 3.125% of Pd-doped

Fig.2 shows a concentration of 3.125% Pd-doped BN spin polarization energy band diagram. It can be seen from the Fig. 2, whether the spin up or spin down of the energy band diagram, there is a band gap, which indicates that Pd doped BN does not destroy the overall properties of the semiconductor. In the energy band diagram of the spin, or the rendering semiconductor properties, the band gap reaches 2.436eV, it is smaller than the above calculated pure BN of 4.492eV. Under spin down, there are two full valence bands above the Fermi level, which exhibit a certain degree of metallic nature. Therefore, the whole (B, Pd) N exhibits semi-metallic nature. It is also consistent with the above calculated energy spin polarization conclusion and the experimental results. The role of these two discrete filled valence bands is like a free hole. Therefore, in the Pd-doped BN, we can theoretically achieve 100% spin polarization, which shows that (B, Pd) N possible spin injection is used. On the whole, the band gap is significantly reduced after the doping.

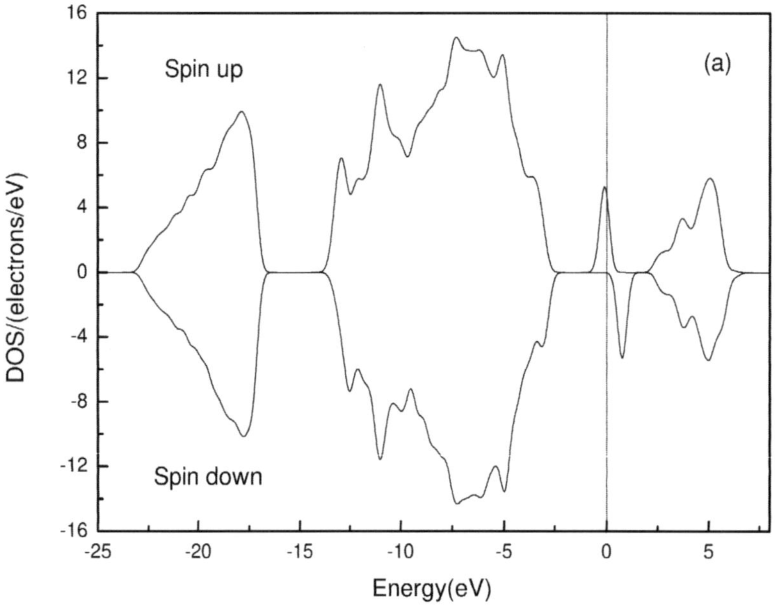

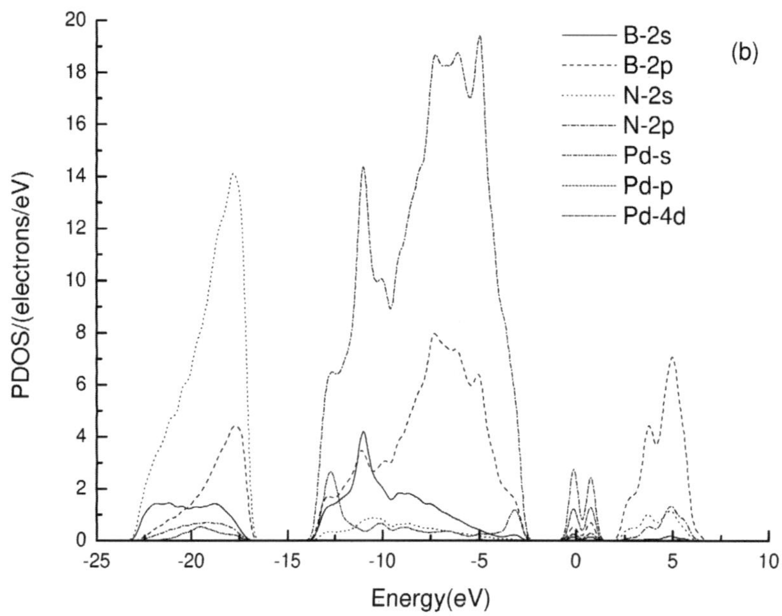

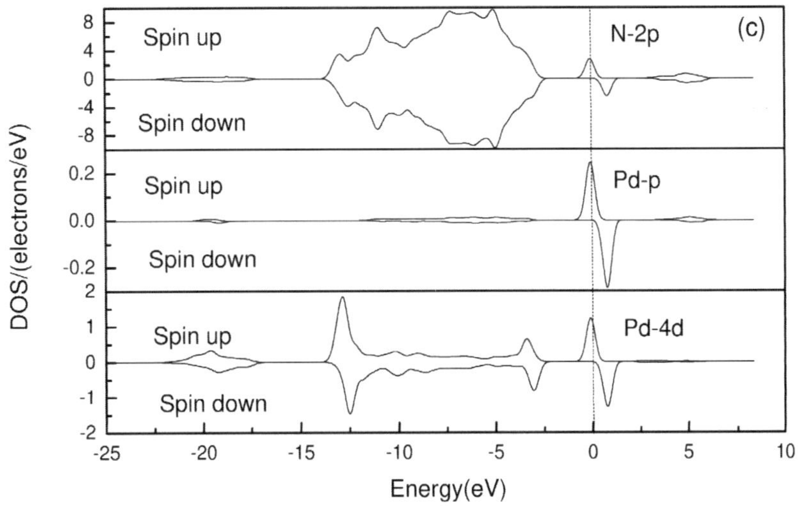

Figure 3. The total density of states (a), PDOS (b), DOS of N-2p , Pd-p and Pd-4d (c) of BN doped with 3.125% of Pd

Fig.3 shows the spin density of states of the total Pd-doped BN, partial wave density of states, and Pd, N spin density of states. Because the density of states near the Fermi level division determines the magnetic properties of matter, we focus on studying the density of states near the Fermi level. Fig.3(a) shows the splitting up of the electronic density of states of spin-up and spin-down state density distributions appearing obviously in the vicinity of the Fermi surface, which suggests that the system of electrons by spin exchange interactions have emerge arrange in order. Spin-up electrons across the Fermi surface show the metallic nature; and spin-down electrons in the Fermi surface open a gap show semiconducting, which is consistent with the above analysis of the energy band diagrams. Therefore, Pd-doped BN has a half-metallic nature. The total density of states distribution is different when the electron spin is up or down. It exhibits the asymmetry. In the occupied states under the Fermi level (vertical dashed line in Figure3a), we found that the number of the spin-up electrons is larger than that of the spin-down electrons after a calculation of the density of states by integral computation. The spin-up electrons exhibit a net magnetic moment in the outside, which shows ferromagnetism.

Fig.3(b) shows the density of states of B31PdN32, the peak of the lower valence band mainly coming from N2s electrons, the peak of the conduction band mainly from B2p electrons too, and the peak of the upper valence band mainly from B2p electrons, N2p electrons and Pd4d electrons. As can be seen from Fig. 3(c), Pd4d, B2p and N2p electrons below the Fermi level correspond to a few peaks, which indicates that Pd4d electronic and B2p, N2p have strong hybrid electronic effects. There are B31PdN32 B2p (a), N2p electron density of states (b) and Pd4d electron density of states (c) in the electronic density, as shown in Fig. 3. From the comparison, we can know that three peaks overlap each other in the plurality, strong hybrid peak can be found in Fig. 3a, the density of states near the Fermi surface is mainly composed of B2p, N2p and Pd3d electrons, all of which indicate the electronic states of Pd4d, B2p and N2p state electron mutual hybrid role. The calculations show that the number of occupy states below the Fermi level at the middle or on the electronic density of states Pd4d, B2p and N2p electronic integrating are much larger than that of spin-up electron spin-down electrons, suggesting that both Pd, B and N interaction are ferromagnetic state, nearest neighbor N atoms 2p electrons tend to be Pd4d electrons ferromagnetic order arrangement. These results indicate that Pd-doped BN ferromagnetic state can be obtained through the mechanism of the formation of hybrid p-d.

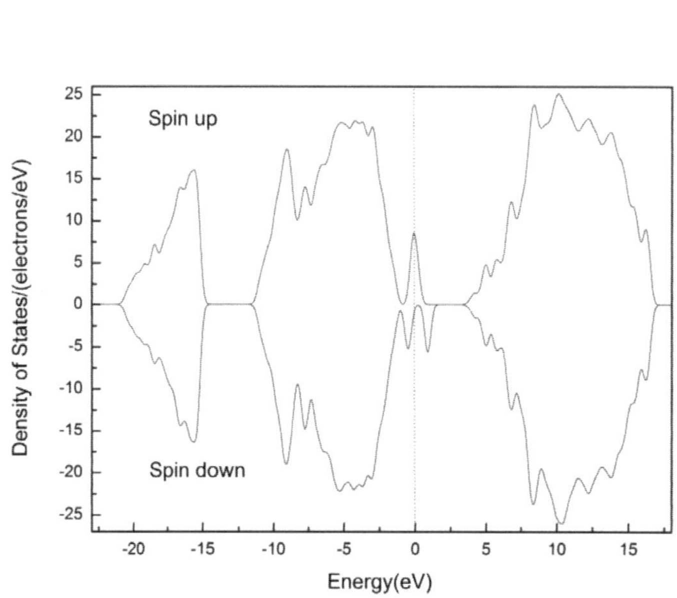

Figure 4. The total density of states of BN co-doped with Cu-Pd (Cu of 4.17% and Pd of 4.17%)

Many classes of computer simulation have two doped impurity element. In order to make the two elements doping together, we selected 3×2×2 supercell for co-doped Cu-Pd of BN. The density of states is shown in Fig. 4. Because of the density of states near the Fermi distribution determining the magnetic properties of matter, we focused on investigating the charges density of states near the Fermi distribution. Electronic density of states and splitting state density distribution of the spin-down can be seen from Fig.4,and in the vicinity of the Fermi surface, the spin-up is so obvious that it indicates the system in the electronic interaction taking place through the exchange of spin ordered. Spin-up electrons across the Fermi surface exhibit metallic nature; and spin-up electrons at the Fermi level lead to the opening of a gap, which exhibits semiconducting. That is consistent with the above analysis of the energy band diagram. Therefore, Cu -Pd co-doped BN has a half-metallic nature. There is an offset in spin-up and spin-down electron density of states. That also indicates the presence of magnetic ordering. There is a difference in the electronic density of states of spin-up and spin-down of the total distribution. It exhibits asymmetry. In the Fermi level (Figure 4) vertical dashed line occupying less state, we can know that spin-up number of points are larger than that of spin-down electrons after the electronic density of states, thus the net magnetic moment exhibits ferromagnetism.

4. Conclusion

In this paper, the first-principles study has been performed to evaluate the structural properties of the pure and Pd-doped BN, such as the band structure and the density of states. The calculations reveal that Pd-doped BN is a half-metallic ferromagnet with 100% carrier spin polarization, and the ferromagnetism arise from the strong hybridization between the Pd4d and N2p state. At BN doped with 3.125% Pd concentration, the different energy calculations show that the ground state is FM state. These results suggest that it is possible to fabricate BN based on DMS by doping with Pd. This DOS of BN compared with conventional dilute magnetic semiconductors, its doping concentration is smaller. And Pd doped BN is paramagnetic. Thus, Pd doped BN might be a kind of promising diluted magnetic semiconductors. Compared with conventional dilute magnetic semiconductors, doping concentration is much smaller. There is no doubt that you can find the ideal semiconductor. Thus, Pd doped BN might be a very promising diluted magnetic semiconductor. Pd-Cu codoped shows doped

BN group having two elements with BN group may be implemented, which is based on the latest proposed BN group.

References

[1] Zhang Z F, Zhou T G and Zhao H Y 2014 First-principles calcuations of 5d atoms doped hexagonal-AlN sheets: Geometry, magnetic property and the influence of symmetry and symmetry-breaking on the electronic structure *Chin.Phys.B* **23**(1) 2320

[2] Lei W W, Liu D and Zhu P W 2009 Ferromagnetic Sc-doped AlN sixfold-symmetrical hierarchical nanostructures *Appl Phys Lett* **95**(16) 162501

[3] Zhang Y, Liu W and Liang P 2008 Half-metallic ferromagnetism in Ca-doped AlN from first-principles study *Solid State Commun* **147**(7) 254

[4] R M Chrenko R M 1974 Diamond and c-BN *Solid State Commun* 511-515

[5] Wu Y F, Keller B P and Keller S 1996 Very high breakdown voltage and large transconductance realized on GaN heterojunction field effect transistors *Appl Phys Lett* **69** 1438

[6] Gorczyca I, Svane A and Christensen N E 1999 Theory of point defects in GaN, AlN and BN: Relaxation and pressure effects *Phys.Rev.B* **60**(15) 8147

[7] Zhou Q and Tian B 2009 Cr-doped BN First Principles Calculation of half-metallic ferromagnetism *J Sichuan Univ* **32** 226 (in Chinese)

[8] N Liu 2012 C-doped BN magnetic first-principles calculation *M.S. Thesis* (Univ Jilin) 2012. (in Chinese)

[9] Ji X H, Lau S P and Yu S F 2007 Ultraviolet photoluminescence from ferromagnetic Fe-doped AlN nanorods *Physics Letters* **90**(19) 193118

[10] Zhang J K, Deng S H and Jin H 2007 Optical properties of ZnO structure of first-principles study of rare earth doped *Acta. Phys.Sin* **56** 5371 (in Chinese)

[11] Meng F C, Ouyang B and Song J 2015 First-principles Study of Dislocation Slips in Impurity-doped raphene *J.Phys.Chem* **119**(6) 3418

[12] Si C, Liu Z and Duan W H 2013 First-principles Calulations on the Effect of Doping and Biaxial Tensile Starain on Electron-Phonon Coupling in Graphene *Phys.Rev.Lett* **111** 196802

[13] Jiang Q G,Ao Z M and Jiang Q2013 First principles study on the hydrophilic and conductive graphene doped with Al atoms *Phys Chem Chem Phys* **15** 10859

[14] Milman V, Winkler B and White J A 2007 Electronic structure, properties and phase stability of inorganic crystals: A pseudopotential plane-wave study *J Quant Chem* **77(5)** 895

[15] Perdew J P, Burke K and Emxerhof M1996 Generalized Gradient Approximation Made Simple *Phys Rev Lett* **77** 3865

[16] Mirkarimi P B, Mccarty K F and Medlin D L 1997 Medlin Review of advances in cubic boron nitride film synthesis *Mater. Sci. Eng* **R21** 47

[17] He K H, Zheng G and Kirtman B 2010 First principles study on the electronic structure and effect of vanadium doping of BN nanowrires *Solid State Commu* **150** 701

IOP Conf. Series: Materials Science and Engineering **137** (2016) 012022 doi:10.1088/1757-899X/137/1/012022

Research on the properties of dope-dyed bamboo pulp staple fibers

Y J Zhang[1]

Hebei University of Science and Technology, Shijiazhuang, Hebei

E-mail: zhangyongjiu1958@sohu.com

Abstract. In order to understand the properties of the dope-dyed bamboo pulp staple fibers, the moisture regains, tensile properties, friction properties and electrical conductivity of them and the white bamboo pulp fiber were tested, compared and analyzed. The results show that the moisture regains of the dope-dyed bamboo pulp fibers are smaller than the white bamboo pulp fiber's, whereas their friction coefficients, breaking strengths and elongations and mass ratio resistances are a bit larger.

1. Introduction

Bamboo fibers are classified into bamboo fiber directly from bamboo and bamboo pulp fiber according to their processing methods. The former is obtained to process bamboo by physical and mechanical methods. Its processing technology includes hydrolysis, steaming, boiling, softening, carding and so on [1-3]. The latter is made by chemical method and is a regenerated cellulose fiber with more than 93% of the regenerated cellulose which is made from bamboo as raw materials through special processes [4]. Currently, bamboo fiber used for textile processing generally refers to bamboo pulp one. Bamboo pulp fiber is not only purely spun, but also blended with other fibers such as cotton, hemp, silk, wool, viscose, Modal, Tencel and synthetic fiber. It can be produced by various styles of knitted fabrics, woven fabrics and so on. The fabrics are biodegradable, soft, cool and comfortable, and have many advantages like good drapability, high strength, good abrasion resistance and moisture permeability etc [5].

Dope dyeing is the green and environmental protection technology widely spread and applied in recent years and realizes the synchronous production of fiber and dyeing. The production of dope dyed fiber is to add an appropriate colorant into the dope during the polymerization or spinning of man made fiber process. After spinning, the colorant is uniformly dispersed in the fiber. Dope dyeing process compared with the conventional dyeing has a lot of advantages such as continuously spinning, more uniform color, better color fastness, higher dye uptake rate, shorter production cycle, lower cost, less pollution and so on[6-13]. Dope dyeing technology is widely used in man-made fibers and a lot of dope dyeing patents have been granted in the world [14-16]. China textile industry vigorously promotes dope dyeing technology and dope-dyed fiber accounts for 5%-10% of the total fiber during the twelfth five year period [17].

Because the dope-dyed bamboo pulp staple fibers are new and the study on their properties is now in its infancy, further tests need to be done. Since fiber properties can affect the textile processing and dyeing etc, the characteristics of the dope-dyed bamboo pulp staple fibers should be deeply studied.

2. Experimental materials and methods

2.1. Materials

Two dope-dyed bamboo pulp staple fibers and one white bamboo pulp staple fiber from Jilin Chemical Fiber Group CO., LTD. were prepared for testing whose specifications are shown in table 1

Table 1. Specifications of the fibers

Specimen No.	Fineness/dtex	Length/mm	color
1	2.11	38	White(no dope-dyed)
2	1.93	38	yellow
3	2.10	38	red

2.2. Test methods

2.2.1.Moisture absorption performance. Moisture absorption performance tests were performed on the apparatus—Model Y802A oven according to Chinese national standard "GB6503-2008 Testing method for moisture regain of man-made fibres". Before drying the samples, condition the fibers in the standard atmosphere of 20℃ and 60% RH (relative humidity) for a minimum of 24h, then weigh 10g fiber for each as W_1 (g) and put into the oven at 110℃ to expose. When the samples were exposed for 2h, weigh the specimens without removing them from the oven. After that, the samples were weighed once every 10 min until the constant weight is recorded as W_2 (g). Moisture regains (%) of the samples were calculated by the following equation:

$$\text{Moisture Regain } \% = \frac{W_1 - W_2}{W_2} \times 100 \tag{1}$$

2.2.2.Physical properties. Many forces such as friction, tensile and so forth can deform and do damage to fibers during textile processing. The mechanical properties of fibers are the abilities that the fibers bear all kinds of outer forces [18, 19]. Additionally, prior to testing, the samples were conditioned at the standard atmosphere (as mentioned above) for a minimum of 24h. The physical properties of the fibers were investigated as follows in this study:

Breaking Tenacity: Breaking tenacity was measured using a fiber strength tester YG004N according to Chinese national standard "GB/T14337-2008 Testing Method for Tensile Properties of Man-Made Staple Fibers".

Friction Performance: The friction coefficients of the fibers were investigated with the use of a friction tester Y151 by using the winch method [20]. The friction coefficients between one fiber and metal roller, one fiber and its own fiber roller, and one fiber and rubber roller were respectively measured at 30 rpm, 75 cm/min and 100 mg pre-tension. Every hanging single fiber was repeatedly measured for 2~3 times. 5 identical rollers were demanded to use and each kind of fiber was tested for 6 times using each roller, so 30 values were obtained. The friction coefficients were calculated based on the following equation:

$$\mu = \frac{\ln f_2 - \ln f_1}{\theta} \tag{2}$$

Where f_1 is the fiber tension of the loose end of the winch, f_2 is the fiber tension of the tight ends of the winch, μ is the angle between the fiber and the winch and θ is the friction coefficients between the fiber and the winch.

2.2.3. Electrical conductivity. Fiber mass ratio resistance was evaluated by using YG321 Fiber Mass Ratio Resistance Tester according to Chinese national standard "GB/T 14342-1993 Testing method for specific resistance of synthetic staple fibres". In addition, prior to testing, condition the samples at the standard atmosphere (as mentioned above) for a minimum of 24h. Fiber mass ratio resistance was evaluated by the following equation:

$$\rho m = R\frac{m}{l^2} \tag{3}$$

where R is the averaged resistance of the fiber (Ω), m is the mass (g) of the fiber and d is the density (g/cm^3) of the fiber.

3. Results and discussion

3.1. Evaluation of the moisture absorption of the fibers

Table 2 indicates that the moisture regains of 3 fibers are large, while the moisture regain of no dope-dyed (white) bamboo pulp fiber is larger than yellow fiber's and red fiber's. The reason the moisture regains of the dope-dyed bamboo pulp fibers are smaller is that some water-absorbent groups combining with the dyes, can reduce the number of fiber water-absorbent groups.

Table 2. Moisture regain of fibers

Specimen No	Moisture regain/%
1	13.91
2	13.43
3	11.06

3.2. Analysis of the physical properties of the fibers

The results of the breaking strength of the fibers are as in table 3.

Table 3. Tensile properties of the fibers.

Specimen No.	Breaking Tenacity/ $cN \cdot dtex^{-1}$		Breaking Elongation/ %		Dry breaking Tenacity/ CV
	Dry	Wet	Dry	Wet	
1	1.86	1.06	18.60	21.96	25.25
2	2.24	1.18	21.70	24.08	17.34
3	2.19	1.25	22.60	24.96	14.94

The dry and wet breaking strengths and elongations of the dope-dyed bamboo pulp fibers are larger than those of white bamboo pulp fiber as shown in table 3, whereas the coefficients of variation (CV) of the formers are smaller than the latter's. This shows that the strengths of the bamboo colored fibers are larger and more constant, which is helpful for the textile dyeing, finishing processing and fabric durability and quality.

The friction performance of fibers not only influences textile processing, but also affects the feeling and style of the products made from them. Additionally, friction can cause fibers to wear and deform, lead to mass transfer, and produce heat and static electricity [21, 22].

The friction coefficients of 3 fibers are shown in table 4. Generally speaking, the friction coefficients of two dope-dyed bamboo pulp staple fibers are a little larger than those of the white bamboo pulp staple fiber. The reason that the friction coefficients of two dope-dyed bamboo pulp fibers are larger is that their some groups combining with the dyes, can lead to their surfaces rough.

The larger fiber's dynamic friction coefficient, static friction coefficient and the differences between them are, the stiffer the fiber is, and vice versa. Therefore, two dope-dyed fibers are stiffer than the white bamboo pulp fiber.

Fiber's handle, spinning process and yarn qualities are all affected by its friction coefficients. The large friction coefficient among fibers, especially the static friction coefficient, can help the fiber aggregation in the spinning process, prevent fiber diffusion, and increase its yarn's evenness degree and strength [23]. The dynamic friction coefficients of two dope-dyed fibers are smaller than their corresponding static friction coefficients, which is helpful for their spinning process. The static and dynamic friction coefficients between each fiber and the rubber bar are the largest, and those between each fiber and its bar are the smallest. That the formers are large is helpful for the jaw to hold the card sliver during the drawing, but in the opening & carding, it is difficult to transfer fibers and it often happens to form the strikes. The difficult fiber transfer and strikes will do damage to fiber and will make yarns and machine parts abrase. The friction coefficient of two dope-dyeing fibers should be decreased with appropriate measures to ensure the spinning smoothly.

Table 4. Friction coefficients of three fibers.

Specimen No.	Static friction coefficient/ μ_s	Dynamic friction coefficient/ μ_d	$\mu_s-\mu_d$
1 to metal bar	0.2878	0.1904	0.0974
1 to its bar	0.2207	0.1846	0.0361
1 to rubber bar	0.4228	0.3578	0.0649
2 to metal bar	0.4052	0.2271	0.1781
2 to its bar	0.2578	0.2207	0.0371
2 to rubber bar	0.4894	0.3729	0.1165
3 to metal bar	0.3728	0.2337	0.1391
3 to its bar	0.2762	0.2578	0.0184
3 to rubber bar	0.4414	0.3628	0.0786

3.3. Electrical conductivity analysis

The mass ratio resistances of specimens 1-3 are 1.54×10^9, 3.83×10^9 and $6.36\times109\Omega\cdot g/cm^2$ respectively. The mass ratio resistances of the dope-dyed bamboo pulp fibers are obviously larger than the white fiber's. The moisture regains of 2 dope-dyed fibers are smaller than the white fiber's and their friction coefficients are larger with poor smooth, so they are easy to produce static electricity which affects production and processing. Even after adding some oil, provided that the temperature, humidity and spinning condition are not suitable, they can still produce static electricity easily which will lead to the fiber and the rubber roller or roller attracted to each other resulting in blocking the coiler, to make matter even worse, form hair on the surface of yarn, increase the flowers, even damage the rubber roller, influence carding machine output to a certain extent, and form yarn defections and so on. To make the dope-dyed fibers spin smoothly, anti-static measures should be taken to control their mass ratio resistance under $10^9\Omega\cdot g/cm^2$.

4. Conclusions

In this study, two dope-dyed bamboo pulp staple fibers and one white bamboo pulp staple fiber were tested. The moisture regains of the dope-dyed bamboo pulp staple fibers are smaller than the white bamboo pulp staple fiber's and their mass ratio resistances are larger, so it is necessary to keep relative humidity in the workshop to ensure spinning smoothly. Their breaking strengths and breaking elongations are higher, which make their durability better. Their friction coefficients are larger, so it is requisite to add some oil to decrease friction forces when they are spun and woven.

References

[1] Liu J F 2007 Green natural fiber—bamboo *Chin. Insp. Quara.* **8** 64

[2] Xue X C 2007 Properties and application of bamboo fiber *Chem. Fiber. Text. Tech.* **2** 30-33

[3] Le Y C and Wang G H 2004 Structure and properties of bamboo fiber and its product development *Sichuan Silk* **4** 10-12

[4] Song Y H and Zuo Z E 2009 Development of the bamboo fiber knitted loop transfer net fabric *Knit. Inds.* **7** 6-7

[5] Liang H X , Wang H Z and Zhang Y 2005 The exploitation and significance testing of the original bamboo fabric *Prog. Text. Sci. Tech.* **5** 39-40

[6] Xie S C, Qi Y H , Liu T, et al. 2013 Development of denim fabric with spun-dyed staple fibre yarn *Chin. Text. Leader* **11** 38-41

[7] Fu S H, Zhang K, Sun G H, et al. 2010 Development of dope dyeing technology for cellulose fibre fibre yarn *Chin. Text. Leader* **5** 73-75

[8] Chen L H, Kong F R, Xu R C 2014 Research on the tensile properties of the liquid coloring viscose fiber *J. Henan Inst. Eng.* **26(1)** 19-21

[9] Wu S Q 2015 Development and application of dope dyed polyester fiber used for automotive *Knit. Inds.* **11** 6-9

[10] Guo Q, LI J X and Ma Y J 2014 Study on the solvent carrier type PET before spinning dope colored fiber *PES Ind.* **27(3)** 27-29

[11] Gao Y , Liu J J , Du W H , et al. 2015 Production technology of dope-dyed PAN fiber *Synth. Fiber Chin.* **44(1)** 31-33

[12] Dobbelstein H 2004 Soiling behavior and processability of solution-dyed PA 6 BCF yarns Chem. *Fibers Int.*

[13] Franzen I and Masterbatches C. 2009 Solution dyed yarn vs. conventionally dyed yarn and fabric AFMA 1-38

[14] Mcintosh S A, H Y, Bailey B J et al. US Patent 6358458, 2002 Spinning and stability of solution-dyed nylon fibers

[15] Kobayashi S and Sugata Y US Patent Application 20020114917 Methods of coloring solution-dyed nylon

[16] Pacifici J A and Sims D G US Patent Application 20050144732 Process for providing dyed nylon fibers with resistance to staining and fading

[17] Zhu S 2013 The Countermeasures Probe into the Ecological Environment Rule of Dope-Dyed Polyester Fiber Chin. *Fiber Insp.* **5** 46-48

[18] Jiang Y X and Qin F. 2005 An Introduction to Textile (Beijing: China Textile & Apparel Press)

[19] Yu X F 2004 Experimental Technology of Textile Materials (Beijing: China Textile & Apparel Press)

[20] Zhao S J 2007 Textile Material Experiment Tutorial (Beijing: China Textile & Apparel Press)

[21] Bowden F P and Tabor D 1964 The Friction and Lubrication of Solids, Part II (London: Oxford University Press)

[22] Rabinowicz E Friction and Wear of Materials 1965 (New York: Wiley)

[23] Li R Z, Zheng Y S and Ao L M 2010 Research of silver plated fiber friction property and spinning oil *Cotton Text. Tech.* **38** 432-3

IOP Conf. Series: Materials Science and Engineering **137** (2016) 012023 doi:10.1088/1757-899X/137/1/012023

Properties of honeycomb polyester knitted fabrics

A F Feng

Hebei University of Science and Technology, Shijiazhuang, Hebei, China

E-mail: fengaf774@sohu.com

Abstract. The properties of honeycomb polyester weft-knitted fabrics were studied to understand their advantages. Seven honeycomb polyester weft-knitted fabrics and one common polyester weft-knitted fabric were selected for testing. Their bursting strengths, fuzzing and pilling, air permeability, abrasion resistance and moisture absorption and perspiration were studied. The results show that the honeycomb polyester weft–knitted fabrics have excellent moisture absorption and liberation. The smaller their thicknesses and area densities are, the better their moisture absorption and liberation will be. Their anti-fuzzing and anti-pilling is good, whereas their bursting strengths and abrasion resistance are poorer compared with common polyester fabric's. In order to improve the hygroscopic properties of the fabrics, the proportion of the honeycomb microporous structure modified polyester in the fabrics should not be less than 40%.

1. Introduction

Polyester fabrics have so many advantages like excellent elasticity, wrinkle resistance, durability etc, that they are widely used in the production of various types of clothing. However, in terms of moisture absorption, the clothing made from them is poor in comfort, and also easy to produce static electricity which brings about dust absorbing and clinging to body. Therefore, the main focus of this study on polyester fibers is to give hygroscopic polyester fiber with certain moisture absorption. It is well known that many polyester fibers with moisture wicking performance come into the world [1].

Honeycomb polyester fiber is developed through physical and chemical modification by "Shangyu Hongqiang color polyester Co. Ltd.". On its surface and in the interior, there are a lot of micropores which absorb and release moisture quickly and dissipate heat fast. What is more, cationic dyes can be used to dye the fiber at 100℃ and atmospheric pressure with a short time and a high dye-uptake, so that the energy can be saved and the production cost can be decreased. Because the honeycomb polyester fiber is softer and fluffier than common polyester fibers, its fabric has better drape and air permeability than common polyester fibers' fabric does [2-6]. It can be widely used in various kinds of clothing and accessories [7], and will also be used in other areas such as bedclothes, washcloth, military uniforms and so on. Since army training, exercises, and fights are usually performed in harsh natural conditions, especially in hot weather, one's mood and health will be affected by excessive sweating and overheating. Therefore, if he wears clothing that is cool and comfortable, the troubles would be solved easily and quickly [7-9].Water vapor mainly diffuses through the micropores in a fabric and capillary action, and it is absorbed by the fiber and evaporates in a lower vapor pressure inside of the fabric [10]. Domestic scholars have already extensively studied the moisture absorption and perspiration of fabrics [11-19]. Wicking behavior of fabrics was widely investigated [20-28].

The bursting strength, wear resistance, fuzzing and pilling, air permeability and moisture absorption and perspiration of honeycomb polyester knitted fabrics were investigated and compared with those of an ordinary polyester knitted fabric to provide theoretical basis for the product development, production and application of the honeycomb polyester.

2. Experimental

2.1. Materials

Seven weft-knitted fabrics with honeycomb polyester fiber and one common polyester knitted fabric were prepared and their parameters are shown in table 1.

Table 1. Parameters of weft-knitted fabrics

Sample No.	Fiber and yarn count	Structure	Thickness (mm)	Area density ($g \cdot m^{-2}$)	Density ($loops \cdot 25cm^{-2}$)
1	16.4 tex 80%H/20%B blended yarn	Jersey	0.804	Article I. 178.00	9 582.4
2	14.8 tex 80%H/20%B blended yarn， 2.2 tex lycra	Jersey	0.838	188.98	13 106.8
3	14.8 tex 65%H/35%B blended yarn	Jersey	0.609	130.66	6 663.0
4	9.8 tex 60%H/40%B blended yarn， 2.2 tex lycra	Jersey	1.045	185.30	14 481.0
5	11.8 tex 40%H/60%B blended yarn	Jersey	0.671	117.79	9 901.3
6	14.8 tex 100%H	Mesh	0.916	212.55	7 584.1
7	14.8 tex70%H/30%B blended yarn	Mesh	1.185	330.04	11 029.9
8	14.8 tex 100% common polyester	Interlock	0.801	269.89	10 732.8

*Note： where H= honeycomb polyester and B=bamboo pulp fiber.

2.2 Methods and instruments

2.2.1. Bursting strength. The bursting strengths of the fabrics were measured using YG065 Electronic Cylinder Bursting Strength Tester according to "GB/T 19976—2005 Textiles—Determination of bursting strength—Steel ball method".

2.2.2. Abrasion resistance. The damaged cycles of a fabric were measured using Y522 Fabric Abrader according to "ASTM D3884-2009 Standard Guide for Abrasion Resistance of Textile Fabrics (Rotary Platform, Double-Head Method)".

2.2.3. Fuzzing and pilling. The fuzzing and pilling grades of the fabrics were evaluated using YG502 Fuzzing and Pilling Tester according to "GB/T 4802.1—2008 Textiles—Determination of fabric propensity to surface fuzzing and to pilling—Part 1: Circular locus method".

2.2.4. Air permeability. The tests were carried out using an Digital Fabric Air Permeability Tester Y461E-II according to "GB/T 5453—1997 Textiles-Determination of the Permeability of Fabrics to Air".

2.2.5. Moisture absorption and perspiration.

(1) *Water absorption.* Water absorption was evaluated by the following equation according to "GB/T 21655.1—2008 Textiles—Evaluation of Absorption and Quick-Drying—Part 1: Method for Combination Tests":

$$A=(m-m_0)/m\times 100 \tag{1}$$

Where A is the water absorption rate of the sample (%), m is the original mass (g) of the sample and m_0 is the mass (g) of the sample after water soaked dropping.

(2) *Moisture permeability.* The moisture permeability was evaluated by the following equation according to "GB/T 12704—1991 Fabrics—Determination of water vapour transmission rate (WVT, g/(m^2·h)) —Dish method":

$$WVT = \frac{\Delta m}{S\cdot t} \tag{2}$$

where Δm is the difference (g) of the two masses of the same test assembly, S is the test area (m^2) of the sample and t is the test time (h).

(3) *Capillary effect.* The capillary effect of the sample was measured using YG (B) 871 Capillary Effect Tester according to "FZ/T 01071-2008 Textiles—Test method for capillary effect".

3. Results and discussion

The test results of the bursting strength, abrasion resistance, fuzzing and pilling resistance and air permeability of the samples are shown in table 2.

Table 2. Properties of the weft-knitted fabrics.

Sample No.	Bursting strength (cN)	Abrasion resistance (cycle)	Fuzzing and pilling (grade)	Air permeability rate (mm·s^{-1})
1	173.5	52	4	1114.36
2	160.7	126	3.5	580.32
3	165.9	54	4	2152.66
4	106.6	220	4	626.93
5	149.2	57	4.5	1997.88
6	238	127	4.5	989.13
7	352.9	241	4	490.55
8	612.1	788	4	575.71

3.1. Bursting strength and discussion

The bursting strengths of the knitted fabrics with the honeycomb microporous polyester are smaller than that of the pure polyester knitted fabric. This is because the strengths of the honeycomb microporous polyester and bamboo pulp fiber are both smaller than that of the ordinary polyester. What is more, the pure polyester fabric has interlock structure.

It has been found that there is a positive correlation between the bursting strengths and area densities of 7 fabrics with the honeycomb microporous polyester, while there are no correlation between the bursting strengths and the honeycomb polyester contents, or fabric thicknesses, or total densities after the influences of their honeycomb contents, thicknesses, area densities and total densities on the bursting strength were analyzed. The bursting strength of fabric 7 is the largest, the bursting strength of fabric 6(pure honeycomb polyester)is second larger, and others bursting strengths are smaller. Though both Fabrics 6 and 7 are mesh structures, the thickness, area density and total density of fabric 7 are larger than those of fabric 6. Fabrics 1 to 5 are all jersey and their area densities and thicknesses are basically smaller than those of fabrics 6 and 7, so their bursting strengths are also smaller. There is positive correlation between the bursting strength and yarn finenesses of fabrics 1 to 5 by regression analysis, as shown in figure 2. The finer the yarns are, the smaller their fabrics. Because the yarn of fabric 4 is the finest, its bursting strength is the smallest.

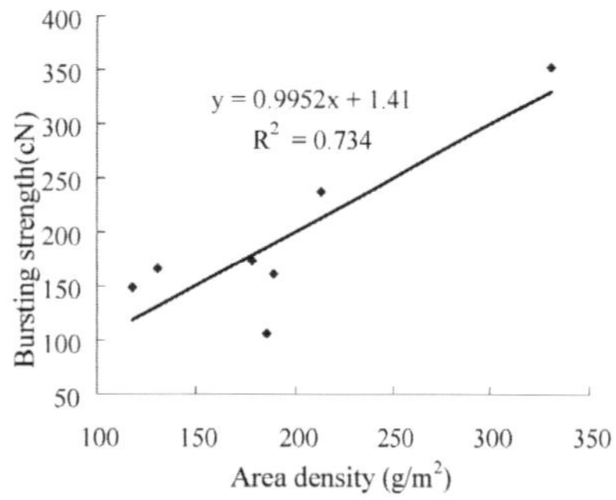

Figure 1. Relationship between bursting strengths and area densities of 7 honeycomb fabrics

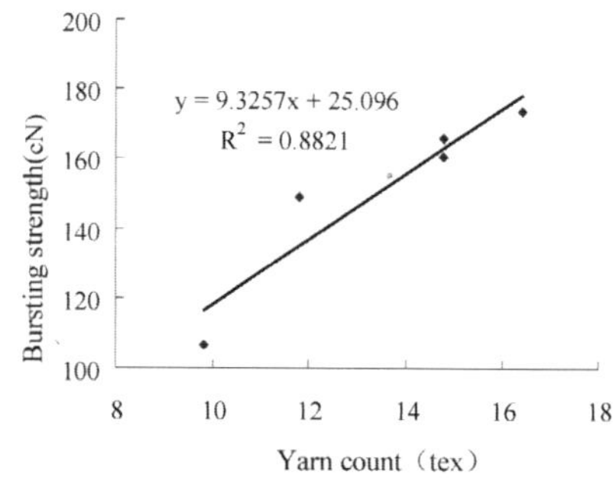

Figure 2. Relationship between bursting strengths and yarn counts of 5 honeycomb jersey fabrics

3.2. Abrasion resistance and discussion

The abrasion resistance of the pure common interlock polyester fabric is much better than that of the fabrics with honeycomb polyester, and the reasons are shown as in the fabric bursting strength. It has been found that there are some correlations between the abrasion cycles and the thicknesses and area densities of 7 fabrics with the honeycomb polyester, but there are no correlation between the abrasion cycles and the total densities, after the influences of their area densities, thicknesses and total densities on the abrasion resistance were analyzed. The regression equation between the abrasion cycles and area densities is $y=0.8911x-45.71$, $R^2=0.6143$ with certain relevance and the relationship between the abrasion cycles and the thicknesses is shown in figure 3. In general, by increasing the thicknesses of the fabrics, their abrasion resistance is also increased. The abrasion resistance of fabric 6 is the best whereas the abrasion resistance of fabric 1, 3 and 5 is poor. In addition, adding Lycra to the fabrics helps to improve the abrasion resistance. Fabrics 2 and 4 contain lira, so they have relatively good abrasion resistance, although the yarn of fabric 4 is the finest. This may be the fabrics with Lycra having good extensibility and elasticity to improve their abrasion resistance.

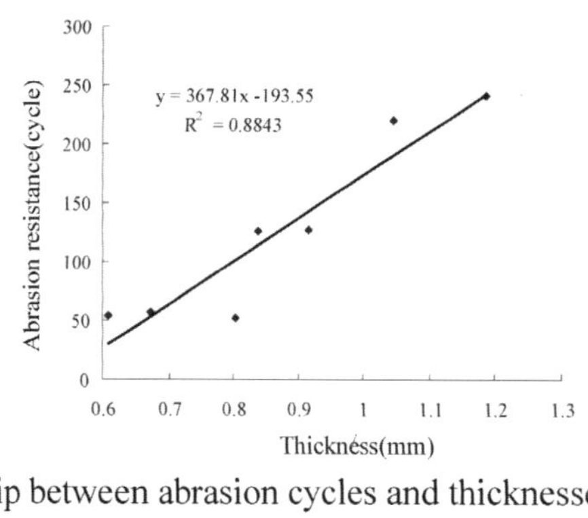

Figure 3. Relationship between abrasion cycles and thicknesses of 7 honeycomb fabrics

3.3. Fuzzing and pilling and discussion

During wearing and maintaining garments, they are often subjected to a variety of abrasions, which not only can make garments wear, but will make them fuzz and pill. The fabric fuzzing and pilling will not only affect the appearance surface gloss and wear resistance of clothing, but also influence its comfort. All fabrics in test have good fuzzing and pilling resistance (table 2). The fuzzing and pilling resistance of the fabrics with the exception of fabric 2 (grade 3.5) have reached 4 grade and above. The fuzzing and pilling of knitted fabrics is influenced by raw materials, and other factors like yarn count and twist, fabric density, area density, thickness, and spandex [29]. Theoretically, the larger the density of the weft-knitted jersey fabric is, the better its fuzzing and pilling resistance is. The density of fabric 4 is the largest, followed by fabric 2, however, because of them containing Lycra, the fuzzing and pilling of fabric 2 is the worst and fabric 4 is the second lowest.

3.4. Air permeability and discussion

The air permeability of the fabrics with the honeycomb polyester except-fabric 7 is better than the air permeability of the common polyester fabric. Among 7 fabrics with the honeycomb polyester, the air permeability of fabric 3 is the best while fabric 7 is the worst.

After the influences of the area densities, thicknesses and total densities of 7 honeycomb polyester fabrics on the abrasion resistance were analyzed, and it was found that there is negative correlation between the air permeability and the thicknesses, which is shown in figure 4, and that there are no relationships between the total densities and area densities. In addition, there is a negative correlation between the air permeability and the thickness of the fabrics. The thinner the fabric is and the smaller the resistance when the air passes through the fabric, the better its air permeability is. The air permeability for fabric 7 is the worst because its thickness and area density are the largest. The air permeability of fabric 6 with larger area density and thickness is good because of its mesh structure. Although the yarn of fabric 4 is the finest, Lycra in it leads to it having both larger density which makes gaps in the fabric smaller [12] and larger thickness which makes air pass through a longer distance, so its air permeability is poorer. Fabric 2 also contains spandex and its thickness is large, so its ventilation is poor.

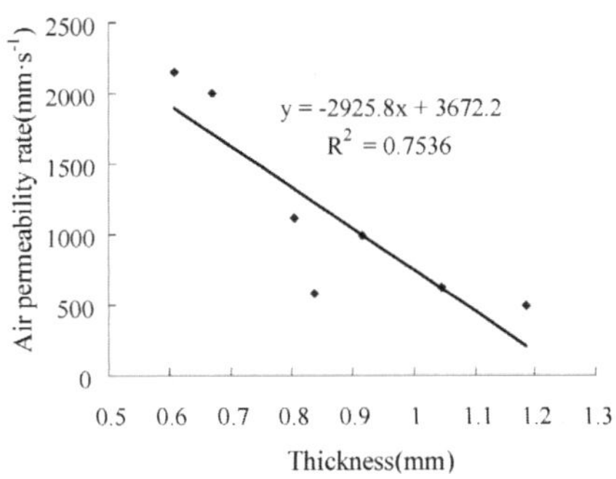

Figure 4. Relationship between air permeability rate and thicknesses of 7 honeycomb fabrics

Table 3. Results of moisture absorption and perspiration of the weft-knitted fabrics.

Sample No.	Water absorption rate (%)	Water vapour transmission rate($g \cdot m^{-2} \cdot d^{-1}$)	Capillary effect		
			Walewise wicking height(cm)	Coursewise Wicking height(cm)	Comprehensive value(cm^{-2})
1	185	279	11.8	12.4	146.32
2	197	223	12.5	13.5	168.75
3	252	290	13.2	11.0	145.20
4	257	275	13.2	12.8	168.96
5	258	309	13.6	13.8	187.68
6	190	306	12.2	10.6	129.32
7	209	289	17.1	16.6	283.86
8	134	211	4.9	6.4	31.36

3.5. Moisture absorption and perspiration and discussion

The results of the moisture absorption and perspiration of the fabrics are shown in table 3.

3.5.1. Water absorption and discussion. It can be seen from figure 3 that the water absorption rates (WAR) of fabrics 3, 4, 5 and 7 are more than 200% and meet the requirements of the moisture absorption rate for absorbent fabrics according to the standard "GB/T 21655.1-2008". The water absorption rate of the common polyester fabric is the worst. The other fabrics except for fabric 6 among 7 honeycomb polyester fabrics are honeycomb polyester and bamboo pulp fiber blended ones. Bamboo pulp fiber is regenerated cellulose fiber with excellent moisture absorption, and the moisture regain of the honeycomb polyester is higher than ordinary polyester's because of its microporous structure with large specific surface area. So the water absorption rates of 7 honeycomb polyester fabrics are higher.

The water absorption rate increases with the increase of honeycomb polyester fiber content. When the honeycomb polyester content reaches up to 40-65%, the water absorption rate is the maximum, and then it decreases with increasing honeycomb polyester content. The rate of the fabric is not only related to the water absorption rate of the fiber used, but also related to its structure, thickness, area density and so on.

When the fitting processes of the water absorption rate with the area density and thickness were done, it was found that fabric 4 had big deviation, therefore it was eliminated. The regression equations of the water absorption rates of 6 honeycomb polyester fabrics with their thicknesses and area densities are shown in figures 5 and 6.

The water absorption rates of the fabrics decreases with the increase of their area densities. That is why when the area densities of the fabrics are smaller, their structures are loose and their air rates are larger. Although the area density of fabric 8 is the largest, its structure is mesh and its thickness is the largest, which leads to looser and bigger air rate. Therefore fabric 8 has a bigger air rate.

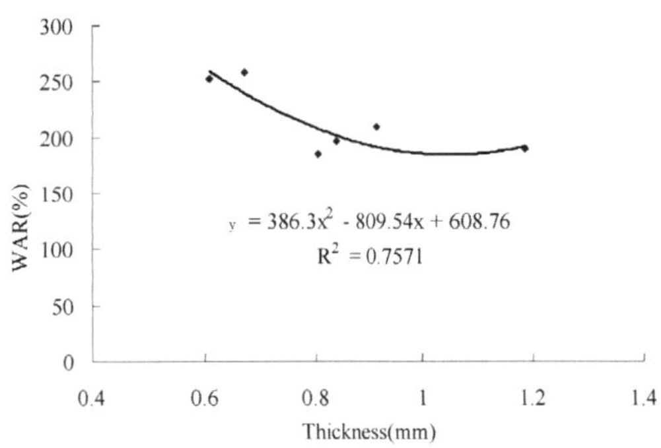

Figure 5. Relationship between water absorption rates and thicknesses of 6 honeycomb fabrics

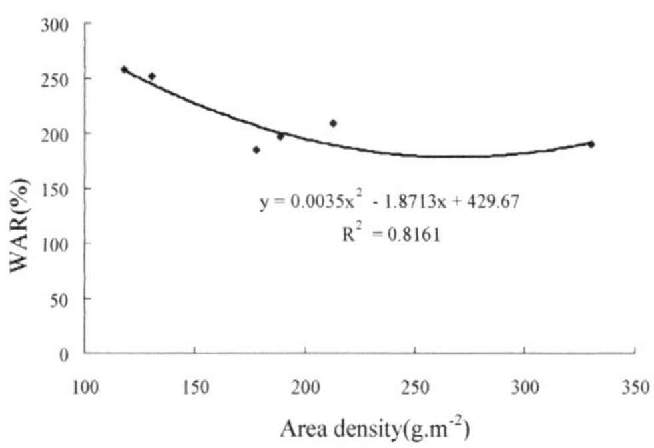

Figure 6. Relationship between water absorption rates and area densities of 6 honeycomb fabrics

3.5.2. Moisture permeability and discussion. It is known from figure 3 that the order of the water vapour transmission rates of the fabrics is 5 # >6 # >3 # >7 # >1 # >4 # >2 # >8 # . The common polyester fabric has the worst water vapour transmission rate. There are no significant correlations between the water vapour transmission rate and the honeycomb polyester content, density, area density and thickness. Therefore, the moisture permeability of the honeycomb polyester fabrics is affected by various factors.

Because fabric 5 with the finest yarn has the least density and area density, which makes it loose and big gaps, there is no wonder that its moisture permeability is the best. The moisture permeability of fabric 6 is very good because it is made from pure honeycomb polyester. Owing to its density being small, when the fiber absorbs moisture, its cross section does not expand, unlike other cellulosic fibers.

3.5.3. Capillary effect and discussion. On the basis of the standard "GB/T 21655.1-2008", when the wicking height is either 10cm or higher than 10cm, the fabric is hygroscopic. Table 3 shows that all honeycomb polyester fabrics have excellent capillary effects. Their walewise and coursewise wicking heights are larger than 10cm. The walewise and coursewise wicking heights of fabric 7 are the largest with 17.1cm and 16.6cm respectively. The fabric has excellent wicking property because of its mesh structure with good wicking effect and high proportion of the honeycomb polyester (70%). The capillary effect of fabric 5 among the jersey knitted fabrics is the best. The walewise and coursewise wicking heights of the common polyester fabric are the lowest.

Combined with the water absorption test results, bamboo fiber has good moisture absorption and water absorption, whereas its moisture permeability is worse than that of the honeycomb polyester fiber. Therefore, when two fibers are blended with their ratio about 50% to 50%, their jersey weft-knitted fabrics have good wicking effects.

4. Conclusions

The honeycomb polyester weft–knitted fabrics have excellent moisture absorption and liberation. The smaller their thicknesses and area densities are, the better their moisture absorption and liberation. Their anti-fuzzing and anti-pilling is as good as the common polyester fabric's, while their bursting strengths and abrasion resistance are poorer compared with than the common polyester fabric's. If the fabrics with good hygroscopic property are developed, the proportion of the honeycomb microporous structure modified polyester in the fabrics should not be less than 40%.

References

[1] Gu C Y 2005 Development and application of moisture absorption and liberation polyester fibers *Chin. Text. Leader* **7** 57-62
[2] Ma Y, Zhou Y K and Zhang H 2014 Union dyeing of polyester filaments /microporous polyester fibers with one-bath process *J. Text. Res.* **35(6)** 68-72
[3] Su D B, Li H X and Zhou H H 2014 Design and performance of honeycomb microporous PET/bamboo fiber blended flat knitting products *J Donghua Univ. (Nat. Sci.)* **40(4)** 451-455
[4] Yang S M and Dou H P 2014 Sizing practice of honeycomb porous structure photocatalyst fiber blended yarn *Co. Text. Tech.* **42(11)** 61-64
[5] Su D B and Li H X 2014 Evaluation on wearing performance of honeycomb PET blended knitted *W. Text. J.* **42(2)** 57-61
[6] Jiang J Q, Yang X Y and Zhang Y H 2014 Development of two kinds functional polyester and Model blended Yarn *Co. Text. Tech.***43(8)** 57-61
[7] Qi S M, Xu Y L and Ye Q L 2011 The effect of microstructure of honeycomb cationic modification polyester on the properties *J. Zhejiang Sci-Tech Uni.* **28** 44-9
[8] Huang Y L 2011 The performances of a new type of honeycomb micropore modified polyester fabric *Shanghai Text. Sci. & Tech.* **39** 52-3
[9] Liu Yulei and Meng Jiaguang 2009 Classification and application of moisture absorption and perspiration fabric *Progress in Textile Science & Technology* **5** 27-30
[10] Yu W D and Chu C Y 2009 *Textile Physcis* (Shanghai: Donghua University Press) 2nd Edi
[11] Zhao H Y, Wang Q and Yu J Y 2003 The study on wet permeability and vapor transmission of Coolst fiber and fabrics *Text. Sci. Res.* **3** 14-18
[12] Sun S P, An H and Lin S F 1993 The present situation and development trend of domestic and international comfortable knitted clothing *Knit. Inds.* **3** 30-36
[13] Zhang W Y 1987 Study on the mechanism of moisture permeability of fabrics *J. Chin. Text. Univ.* **13(4)** 75-79
[14] Wu J, Gao Z F and Xue F 2009 The comparison of moisture absorbing and quick drying textile

and its effect *Chin. Fiber Insp.* **9** 72-73

[15] Xue G Z 2002 Body's perspiration and the liquid wet-comfort of knitted apparel *Knit. Inds* **4** 122-124

[16] Guo C H and Li L 2005 Testing methods for moisture absorbing and quick drying textile *Insp. Quar. Sci.* **4** 15-17

[17] Li R Q and Wang Y 1991 Measuremnet of moisture absorption rate of textile materials J. *Text. Res.* **12** 35-7

[18] Li Y L and Lin S 1996 The moisture absorbing and quick drying properties of fabrics and their evaluation methods *J Beijing Inst. Clo. Tech.* **16(1)** 79-84

[19] Zhou Q 2006 Test methods and comprehensive evaluation of moisture comfort of fabrics *Knit. Inds* **4** 29-30

[20] Laughlin R and Davies J 1961 Some aspects of capillary absorption in fibrous textile wicking *Text. Res. J.* **31** 904-910

[22] Rita M and Randall J 1998 The interaction of water with fabric *Text. Res. J.* **68** 280-288

[23] Erik. K 1996 Wetting and wicking *Text. Res. J.* **66** 660-668

[24] Ramachandran T and Kesavaraja N 2004 A study on influencing factors for wetting and wicking behavior *IE (I) J., TX,* **84** 37-41

[25] Harzell M and Hsieh Y 1998 Enzymatic scouring to improve cotton fabric wettability *Text. Res. J.* **68**, 233–241

[26] You Lo H and Cram L 1998 Enzymatic hydrolysis to improve wetting and absorbency of polyester fabrics *Text. Res. J.* **68** 311-319

[27] Fangueiro R., Filgueiras A, Soutinho F, et al. 2010 Wicking behavior and drying capability of functional knitted fabrics *Text. Res. J.* **80(15)** 1522-1530

[28] Adler M M and Walsh W K 1984 Mechanisms of transient moisture transport between fabric *Text. Res. J.* **54**, 334-343

[29] Feng A F and Zhang Y J 2011Analysis of factors affecting fuzzing/pilling of bamboo fiber knitted fabrics *J Text. Res.* **32(9)** 53-58

Tearing analysis of a new airship envelope material under uniaxial tensile load

F X Wang, W Xu, Y L Chen and G Y Fu[1]

School of Naval Architecture, Ocean and Civil Engineering, Shanghai Jiao Tong Univ. Shanghai 200240, China
E-mail: gyfu@sjtu.edu.cn

Abstract. This paper experimentally investigated the tearing properties of a new kind of coated woven fabrics, GQ-6, made of ultra-high molecular weight polyethylene fiber. Such material can be used for the envelope materials of a stratospheric airship. First, the uniaxial tearing tests were carried out. Effects of the stretching rate, the initial crack length, and the initial crack orientation on the material's tearing tensile strength were investigated. Experimental results showed that the initial crack length and the initial crack orientation can be represented by the equivalent initial crack length while the stretching rate has a slight influence on tearing behavior of the uniaxial tensile specimens. Then analytical studies using three methods, i.e. Griffith energy theory, the stress intensity factor theory, and Thiele's empirical theory, among which, the stress intensity factor theory gives the best correlation with the test data. Finally, a 48mm threshold of the equivalent initial crack length was recommended to the envelope material in operation.

1. Introduction

There has been an increasing interest in stratospheric airships as an alternative to traditional air vehicles for aerial exploiting and environmental monitoring. The envelope of an airship is one of the most important structural components, which is usually made of flexible membranes perhaps with the following properties [1]: high strength, high tear resistance, light weight, low creep, low permeability, reliable joints and resistance to environmental degradation due to temperature, humidity, and ultraviolet light factors, etc. The envelope fabrics of airships are prone to be punctured by internal components or external sharp objects in manufacturing. There is no doubt that the presence of cracks can lead to either the reduction of mechanical fabric performance or catastrophic failure. Therefore,, investigations on the tearing behavior of fabrics can provide guidance for its design and industrial applications.

In recent decades, there have been some efforts made [2-5] to investigate the tearing behaviors of woven fabrics by carrying out experimental studies. As an airship envelope material, central crack tensile tearing test is the foremost method which is specified in the Airship Design Criteria FAA P-8110-2, [6]. In general, the uniaxial, bi-axial, multi-axial tensile tearing tests and pressurized cylinder tests are the most commonly applied approaches. For instance, Minami et al.[4] carried out uniaxial and bi-axial tearing tests on specimens containing a single crack, a bias crack or a circular defect. It was concluded that the tearing strength of uniaxial specimens with initial crack oriented in the yarn direction was the lowest. Rigaud et al. [7] investigated the effects of the load ratio and the crack orientation on the tearing strength of polyester fabrics under mono-axial and bi-axial tensile tests. They discovered that crack propagation in coated fabrics was triggered by specific mechanisms linked to the matrix softness. Meng et al. [8] studied the orientation influence on a certain stratospheric airship envelope subjected to a bi-axial load, then compared the experimental data with four empirical equations. They have concluded the various initial damages can be modeled as the effect of warp yarns. However, their conceptual design was lack of analyzing specimens with slit in the weft direction. Luo et al. [9] studied the mechanical behavior of a PVC coated warp fabric subjected to multi-axial load. It was found that the mechanical performance decreased with an increase of initial crack length for a given crack orientation, and the initial crack orientation perpendicular to a tensile direction had a maximal effect on the reduction of the mechanical performance. Topping et al. [10] performed an experimental investigation on crack growth in pressurized cylinders made of Dacron neoprene fabric. They employed the Griffith theory to predict the critical slit length, which had a good correlation with the experimental results. Maekawa et al. [11] carried out a bi-axial tensile test and a pressurized cylinder test to measure the tear propagation properties of the Zylon envelope material.

Additionally, a number of theoretical methods have been developed to analyze the mechanical properties of woven fabrics. Robert et al.[12] developed a probabilistic method to convert the distribution of applied failure load to the distribution of critical energy release rate. According to Minami's[4] test data, Timothy et al.[13] presented three fracture models to evaluate the failure load in coated fabrics containing cracks and the models coincided well with the experimental data. Evans et al. [14]improved the stress intensity factors which were determined by four key geometric configurations as well as uniaxial tearing tensile experimental data. Based on the uniaxial test data, Chen [15] derived a theoretical formula to evaluate the tearing strength of envelope materials with central cracks and the inclined orientation, etc.

Recently, the 46th Research Institute, 6th Academy of China Aerospace Science & Industry Corp. developed a new kind of woven fabrics, GQ-6, for stratospheric airship envelope. Such new textile is an ultra high molecular weight polyethylene fiber (UHMWPF). Its average thickness is approximately 264 μm and the specified areal density is 170 g/m^2. The ultimate tensile stresses on the warp and weft direction are 100.4kN/m and 80.44kN/m, respectively. And the elastic modulus in the warp and weft direction are 3625.55N/mm and 3652.61 N/mm, respectively. However, the tearing tensile mechanical properties of such ripped fabrics are yet to be explored and understood.

This paper aims to examine the tearing properties of GQ-6 through uniaxial tearing tensile tests. First, the effects of the stretching rate, crack orientation, and initial crack length on crack propagation were discussed thoroughly. Then, three analytical methods were utilized to predict the crack propagation. The fidelity of the analytical results were assessed by the test data. Finally, the allowable equivalent initial crack length of the airship envelope material, GQ-6, was recommended.

2. Experimental Program

2.1. Specimens and test set-up

The ripped specimen for uniaxial tests was prepared by introducing a center crack in accordance with the standard of American airship design FAA-8110-2[6]. Such strip specimen is 101.6mm wide and 152.4mm long. All specimens were stretched aligned to their long axis. For instance, the warp specimen is the warp yarns vertical to loading direction, and the intersection angle between the initial crack and the shorter edge is defined as the initial crack angle, see figure 1(a). The uniaxial tearing test was carried out by using a universal testing machine, MINEBEA-CHA-20KN (figure 1(b)), made in Shanghai Taiyang membrane structure co., LTD. Displacement-controlled testing was carried out on the uniaxial specimens.

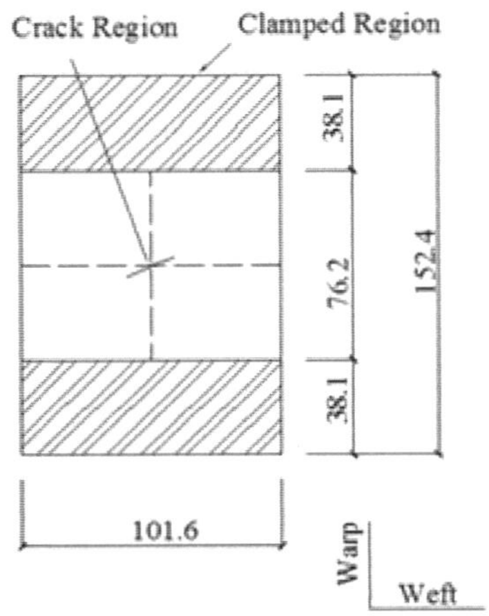

(a)

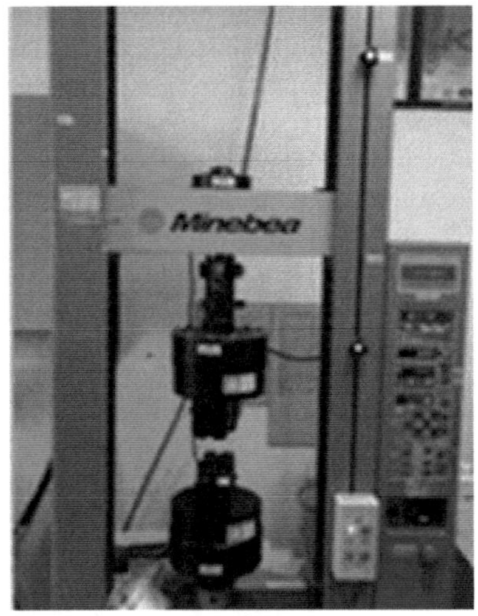

(b)

Figure 1. Test set-up. (a) schematic drawing of the warp specimen and (b) universal testing machine, MINEBEA-CHA-20KN.

2.2. Test procedure

Before the tearing specimens were installed on the testing machines, some preparations were made to facilitate the crack observation and the test data evaluation. All specimens were carefully made along the yarn directions. Orthogonal grid lines were so strictly drawn on the specimen surface that crack propagation can be identified easily. A specific crack was cut in a specimen according to the crack length and orientation determined by its test purpose.

In the uniaxial test, the variations of initial crack length, initial crack orientation and stretching rate were investigated. Hence three groups of specimens were used to reveal the effects of the three variables, respectively. For each group, three specimens were tested to examine their influence. During the test, a pre-tension of 5 N was first applied to the specimen. Then the load is increased at a specific loading rate as listed in table 1. The details of the crack length and crack orientations were summarized in table 1.

Table 1. Layout of the uniaxial experimental protocol.

Directions	Stretching rate (mm/min)	Initial crack length 2a (mm)	Crack orientation
Warp Weft	3,5,10	10,20,40,50	0°, 45°, 60°

3. Results and Discussions

3.1. Influence of Stretching rate

To aid the understanding of failure mode, a specified illustration on the crack initiation and propagation was offered first. Specifically, a tearing specimen was stretched slowly with the coating and the fabric extending together. Meanwhile, neither remarkable elongation nor crack expending occurred. However, the initial crack then translated from an approximate line to ellipse form with tiny gap emerging on the edge of the initial crack. The yarns behaved as a nonlinear material, and a plastic region on the crack adjacent area formed gradually. Such phenomenon indicated the second stage was coming. Then, as the elliptic crack increasing, one yarn on an arbitrary slit tip detached suddenly and adjacent yarns began to slip out alternately from the initial crack tips with the first detached yarn extracted. The above mentioned phenomena indicated the initial crack in the slit tip region was propagating. Finally, the extracted yarns appeared to be ruptured successively, and the specimen cannot sustain anymore load, then catastrophic failure came into being suddenly. The typical failure phenomena of specimens subjected to uniaxial tensile loading were shown in figure 2.

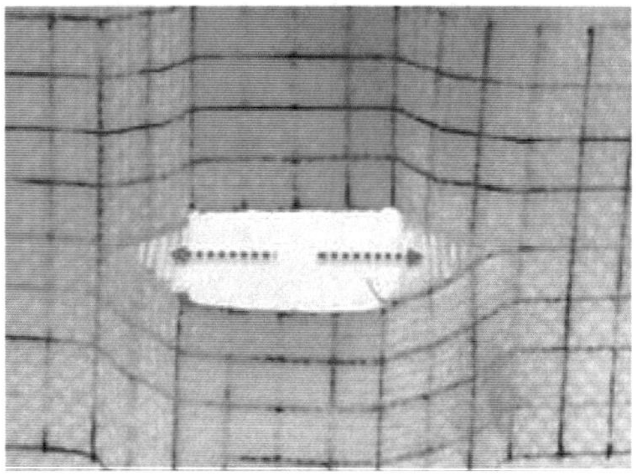

(a)

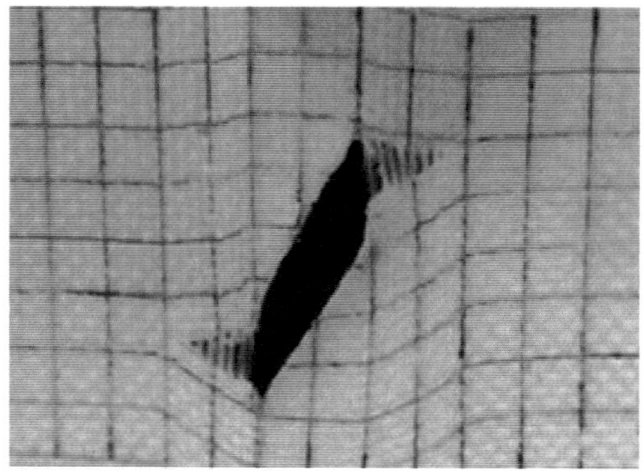

(b)

Figure 2. Typical failure modes for uniaxial tearing tensile specimens (20mm). (a) the transverse rupture and (b) the 'Z' shaped rupture.

From figure 2(a) and figure 2(b), it can be found all crack propagation of the uniaxial specimens were vertical to loading direction. Specifically, the specimens with in-axial initial cracks, which are the typical mode I crack, were ruptured without doubt by enlarging initial crack along its orientation. Such failure was defined as the transverse rupture. However, the initial cracks of the off-axial specimens, which were mixed between I and II crack modes, still propagated perpendicular to the loading direction. Such interesting phenomena were not similar to mental materials which could appear oblique crack or branching crack. That was because the application of the plain weave technology in the new textile. Consequently, the 'Z' shaped rupture was defined to the second failure mode.

In the uniaxial tearing tensile tests, three different stretching rates were carried out on the new textile to evaluate its sensitivity to such influence. First, prepare a set of specimens with 20mm initial crack length and oriented at 0° for an example. The corresponding force-elongation curves were revealed in figure 3. Then, to make a vivid illustration, the three maximum tearing tensile forces of specimens oriented at 45° were averaged and summarized in figure 4.

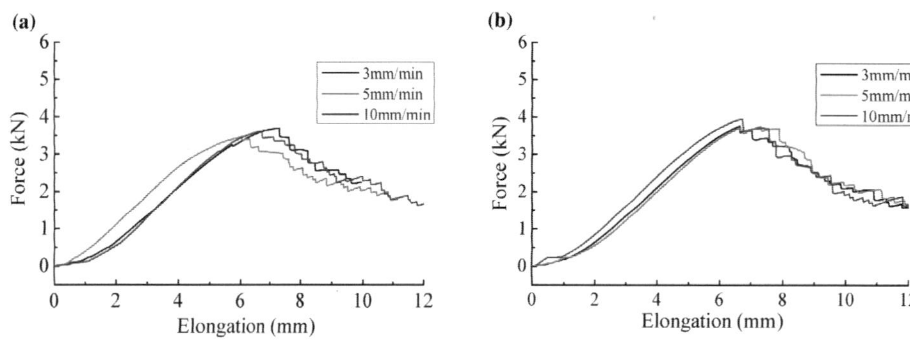

Figure 3. Stretching rate influence on uniaxial specimens. (20mm, 0^0) (a) the warp specimens and (b) the weft specimens.

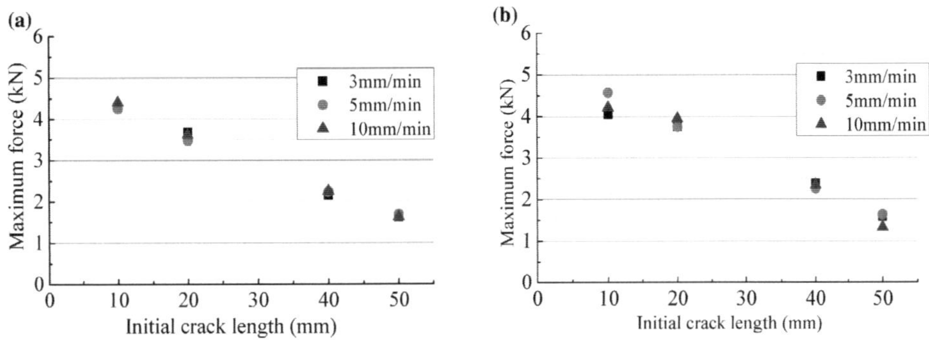

Figure 4. Stretching rate influence on maximum force of uniaxial specimens. (45°) (a) the warp specimens and (b) the weft specimens.

Figure 3 shows the rigid of each specimen was almost the same and increasing moderately. Meanwhile, few fluctuations were observed until the first yarn ruptured. Such phenomenon testifies that the inter-yarn friction of the new woven fabric could perform well on absorbing energy [18]. What is more, it is apparent that the stretching rate has no remarkable effect on the maximum tensile forces of specimens, which can be seen in figure 4. However, an interesting phenomenon is the maximum tensile forces of the warp and weft specimens are nearly identical for the same crack length with almost 20% discount on the weft direction for the intact specimens. That was due to the density of the warp and weft yarns almost the same for GQ-6. Therefore, the friction forces in the warp and weft tearing specimens are nearly identical. Meanwhile, the initial crack can break the weaving process exerting well on the woven fabric.

From the above analysis, consequently, the stretching rate has no remarkable effect on the new coated fabrics under uniaxial tension condition. Therefore, the stretching rate of 10mm/min was chosen arbitrarily at the following researches.

3.2. Influence of Initial crack length

To investigate the influence of the initial crack length on uniaxial specimens, the force-elongation curves corresponding to the warp, weft directions for different initial crack lengths oriented at 0°are shown in figure 5 as examples. All the maximum forces of specimens in warp and weft directions are concluded in figure 6. What should be noticed is the failure mode of uniaxial specimens with various initial crack lengths all ruptured in two modes. Generally, the in-axial specimens ruptured as the transverse mode, and the off-axial specimens ruptured as the "Z" shape mode.

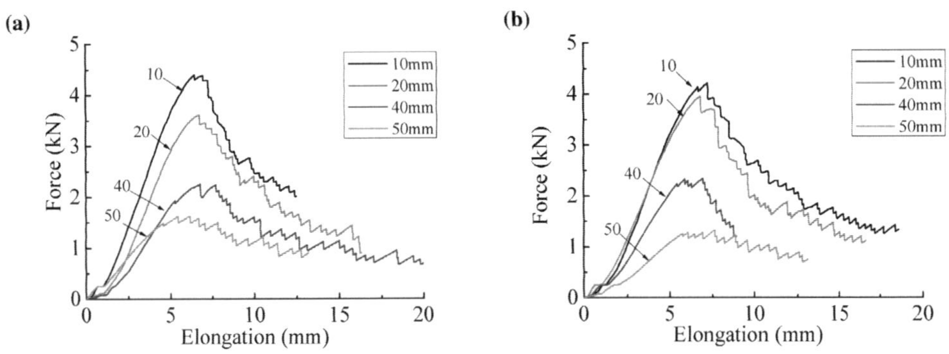

Figure 5. Initial crack length influence on uniaxial specimens (10mm/min, 0^0). (a) the warp specimens and (b) the weft specimens.

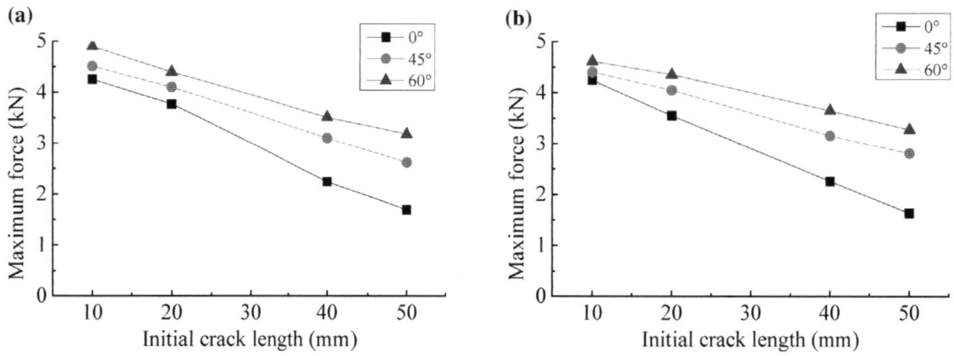

Figure 6. Maximum forces of uniaxial specimens (10mm/min). (a) Warp direction specimens and (b) Weft direction specimens.

From figure 5, it can be observed that the crack length has a significant influence on the new coated fabrics. In general, the maximum tensile force of the specimens decreased considerably with crack length increasing. A similar phenomenon has also been observed experimentally in Zylon airship envelope [11]. Slopes of the force-elongation curves were decreased while getting gentle with the initial crack length getting larger. That was due to fewer yarns left for a specimen with larger initial crack length. As for the maximum strain of such textile, around 4.2% to 5% was observed, which means tiny effect occurred on textile distortion for tearing specimens. From figure 6, the maximum tearing forces reduced rapidly with initial crack length increasing for specimens in various orientations, especially when the crack length was larger than 20mm. Therefore, we can conclude the larger initial crack can lead to less maximum tearing tensile force when subjected to uniaxial tension.

3.3. Influence of Crack orientation

In uniaxial tearing tensile tests, the specimens stretched at 10mm/min with 40mm initial crack length were taken as an example to show the crack orientation influence on mechanical properties of such textile. All failure modes of uniaxial specimens oriented at 0° were the transverse failure. However, the off-axial specimens, i.e., oriented at 45° or 60° were ruptured as "Z" shape failure mode. The force-elongation curves between the warp and the weft specimens are shown in figure 7. Meanwhile, all the maximum forces of specimens in warp and weft directions are plotted in figure 8, respectively.

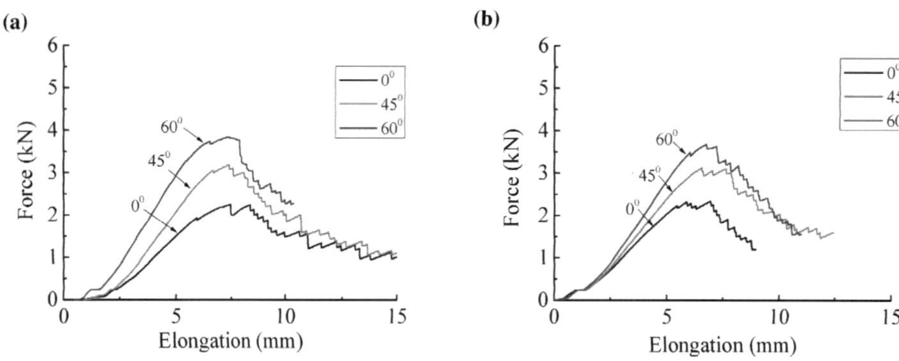

Figure 7. Crack orientation influence on uniaxial specimens(40mm, 10mm/min). (a) the warp specimens and (b) the weft specimens.

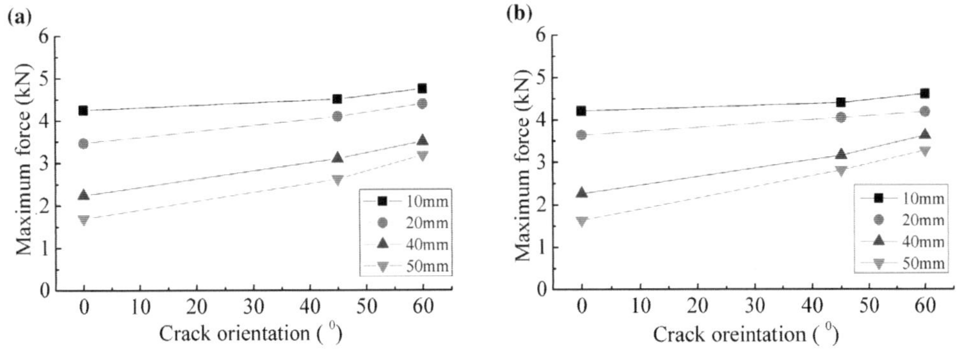

Figure 8. The maximum force of uniaxial specimens. (10mm/min) (a) the warp specimens and (b) the weft specimens.

In figure 7 and figure 8, it is obvious that the maximum tensile forces are increasing with the initial crack orientation getting larger. Theoretically, for a specific crack length, as the orientation getting larger, the number of cutting yarns in the projection region which were perpendicular to the loading direction got to be much smaller, i.e., there were more yarns sustaining stretching load, and the maximum force increased consequently. Meanwhile, it was generally known, for a specific airship, the crack may occur at a random orientation during working condition. Therefore, cracks inclined to the axis of the textile are less critical than those of equal length parallelled to a natural axis.

To cope with the aforementioned circumstances, Rigaud et al.[7] proposed a concept of equivalent initial crack length in their paper. Such length was obtained by projecting the initial crack length in an axis paralleled to the loading direction. For instance, a specimen oriented at 0° with 10mm initial crack length has an identical equivalent initial crack length to a specimen oriented at 60°with 20mm initial crack length .To obtain the relationship between maximum force and the equivalent initial crack length of such new coated fabric, table 2 is concluded.

Table 2. Maximum force of specimens with each equivalent initial crack length.

Specimens	Equivalent initial crack length (mm)	Warp specimens (kN)				Weft specimens (kN)			
		Maximum force			Average force	Maximum force			Average force
10-0°	10	4.29	4.26	4.40	4.32	4.06	4.57	4.21	4.28
20-60°	10	4.44	4.39	4.32	4.38	4.44	4.36	4.17	4.32
20-0°	20	3.67	3.77	3.62	3.69	3.66	3.55	3.45	3.55
40-60°	20	3.58	3.51	3.83	3.64	3.57	3.65	3.68	3.63

As can be seen from table 2, the maximum tensile stresses are approximate with each other for the specimens with identical equivalent initial crack length. For instance, the average maximum force of 20-60° specimen was 4.32kN, only 1.3% higher than that of 10-0° specimen. Accordingly, the crack orientation and the initial crack length can be combined as one factor -- equivalent initial crack length. To make a preparation for the next theoretical simulation, the critical stresses of all uniaxial specimens with effective critical length are listed in table 3 then.

Table 3. Critical stress of uniaxial tearing tensile specimens.

Specimens	Equivalent initial crack length (mm)	Warp (N/mm)				Weft (N/mm)			
		1	2	3	Average	1	2	3	Averag e
10-60°	5	58.80	63.36	65.76	62.64	58.03	61.53	61.60	60.39
10-45°	7.08	56.87	60.13	58.33	58.44	59.60	58.72	60.80	59.71
10-0°	10	57.23	56.73	58.69	57.55	54.07	56.55	56.16	55.59
20-60°	10	59.23	58.63	57.63	58.50	59.13	58.09	55.53	57.58
20-45°	14.16	56.60	54.69	55.80	55.70	53.88	54.01	54.58	54.16
20-0°	20	49.16	50.29	48.23	49.23	48.73	47.34	45.96	47.34
40-60°	20	47.70	46.80	51.07	48.52	47.56	48.67	48.42	48.22
50-60°	25	40.94	42.44	43.47	42.28	42.27	43.68	42.16	42.70
40-45°	28.32	43.03	41.38	42.43	42.28	42.54	42.08	41.67	42.10
50-45°	35.4	36.23	34.96	36.56	35.92	35.89	37.47	37.36	36.91
40-0°	40	28.73	29.89	30.07	29.56	31.73	30.13	31.23	31.03
50-0°	50	21.47	22.56	21.73	21.92	21.29	21.80	17.67	20.25

4. Theoretical Simulation

It is well known that an airship will undergo various pressures in different working conditions. The relationship between the threshold values of initial crack length and various critical stresses of envelope is very important. However, it is almost impossible to do experimental researches and unnecessary to study all the operating conditions. Therefore, precise descriptions on mechanical properties of membrane material by the numerical simulations are unquestionable and proper methods. Hence, three theoretical simulations, that is to say, the Griffith theory, the Stress intensity factor theory and the Thiele's empirical theory were made then.

4.1. Three theoretical simulations

The Griffith theory is based on the energy balance method according to Griffith's studies on the glass fracture. The fundamental of fracture mechanics is the definition of a function known as the stress intensity factor. Such function can describe the stress field concentration around the slit tip regarding the overall stress and crack length. Irwin [20] introduced the plastic work to modify the fracture mechanics theory. Further, he proposed a simplified model to analyze the stress redistribution in the plastic region. Isida et al. [21, 22]and Tada[23] studied the semi-empirical solutions according to a large amount of experimental data. Topping [10] researched the relationship between critical crack length and the stress with a pressurized cylinder. Thiele performed tests on two full-size airship envelopes to further Topping's investigation. He deduced an equation called Thiele's experimental formula to correlate cut slit tear stress with critical slit length.

The critical tearing tensile stress based on the Griffith theory is shown as follows:

$$\sigma_c = \left(\frac{G_{IC}E}{\pi a}\right)^{\frac{1}{2}} \quad (1)$$

Where E is the elastic modulus, and a is half of the center crack length, σ_c is tearing tensile stress, G_{IC} is the critical energy release rate.

The form of stress intensity factor is applied as:

$$K_I = F_T\sigma(\pi a)^{\frac{1}{2}} \quad (2)$$

Where σ is maximum tensile stress, a is half of the center crack length, and F_T stands for geometrical factor considering specimen dimension. The units of the above mentioned variables were N/mm, mm, mm, respectively. Based on the research output of Isida [22], when the ratio of length to width for specimen is 0.75, the fitting function of is written as

$$F_T = 2.08\left(\frac{a}{b}\right)^2 + 0.078\left(\frac{a}{b}\right) + 1 \quad (3)$$

Therefore, the stress intensity factor, K$_I$, is deduced to be $276.74N/mm\sqrt{mm}$ and $274.17N/mm\sqrt{mm}$ by warp and weft experimental data, respectively.

Thiele's experimental formula to correlate cut slit tear stress with critical slit length is written as follows:

$$\sigma_c = Pr = \frac{C_I C_s}{(2a)^n\left(1 + \frac{2a}{r}\right)} \quad (4)$$

Where σ_c is the tearing stress, P is the filling pressure, r is the diameter of cylinder, C_I is the tearing stress in accordance with the FAA-P-8110-2[6], and n is constant and can be deduced by the least square method.

4.2. Comparison of the stresses on the experimental data and theoretical simulations

The relationship between the critical stress and the equivalent initial crack length were simulated by the above mentioned theories. Then, evaluated by observing how well they correlate with the experimental results, we get the calculated results depicted in figure 9 for the uniaxial tests.

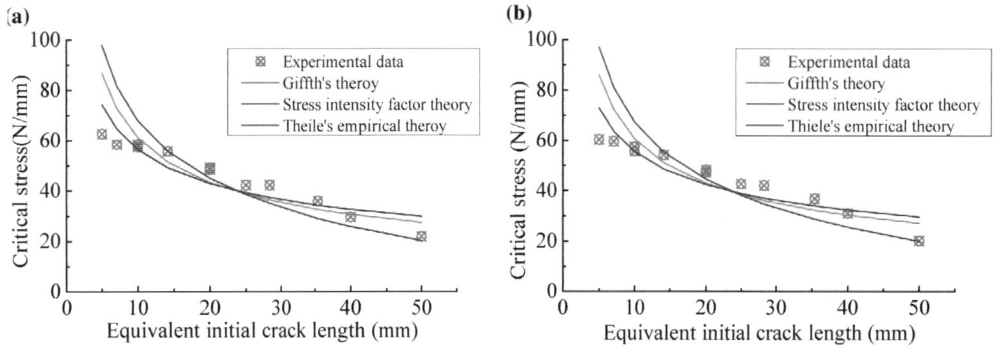

Figure 9. Comparison between experimental data and the theoretical values for uniaxial specimens (a) the warp specimens and (b) the weft specimens.

As for the uniaxial tearing tensile tests, theoretical data was higher than the experimental data when the equivalent initial crack length was less than 10mm. It should be noticed that theoretical simulations show unreasonable high values approaching infinity as equivalent initial crack length near zero. Such phenomenon has also been observed by other previous researchers. However, the theoretical resolutions were approximate to the experimental data during the equivalent initial crack length in the range of 10mm to 40mm. However, variation emerged when crack length was getting larger, as in figure 9. Among the three theory resolutions, the Stress intensity factor theory is the closest to the experimental data, whereas resolution of the Griffith theory and Thiele's empirical theory are higher than the experimental data. Theoretically, Griffith theory is based on the infinite plank and the Thiele's empirical theory based on the pneumatic cylinder tearing tensile tests in which irrespective of the dimension influence. Therefore, deviation comes to be higher during larger initial crack length. As for the stress intensity factor theory, the specimen's dimension is considered, hence such theory is the most suitable one for the uniaxial tearing tensile tests.

4.3. The allowable crack length

According to the standard of FAA-P-8110-2[6], there is no permission on the crack tear propagation under working condition for an airship envelope material. The limit stress of the envelope fabric is assumed to be one forth (4 is the safety factor specified in the FAA) of the ultimate tensile stress of the intact specimens. Consequently, the relevant minimum slit size can be calculated by the Stress intensity factor theory, as shown in figure 10. That is to say, the technology demonstrator can withstand up to around 48mm initial effective crack during the operation condition. Such dimension is higher than that of the Zylon envelope material Z2929T-AB whose technology demonstrator is 40mm slit[11].

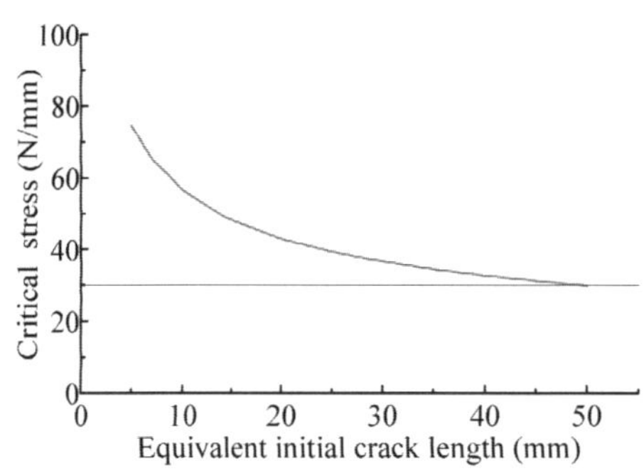

Figure 10. The allowable crack length of the technology demonstrator.

5. Conclusions

In the present work, the tear propagation characteristics of a new envelope fabric for airships, GQ-6, have been successfully identified by uniaxial tearing tensile tests. It is observed that the woven fabric has significant sensitivity to the crack orientation and initial crack length, whereas the stretching rate factor shows tiny sensitivity to specimens subjected to uniaxial tensile condition. After thoroughly researching the uniaxial tensile experimental data, we conclude that the initial crack length and the initial crack orientation can be combined with one factor of equivalent initial crack length. Then, by the experimental data, three theoretical simulations (Griffith energy theory, Stress intensity factor theory, Thiele's empirical theory) are deduced, and comparisons between the theoretical solutions and the experimental data are made. The stress intensity factor theory is selected to be the optimal one while analyzing the tearing mechanical properties of the GQ-6 textile. Finally, an equivalent initial crack length, i.e., 48mm, is testified to be safe during the operation condition.

The above conclusions can be directly applied in the preliminary design in practical engineering. Meanwhile, these work should be extended further to biaxial and multi-axial tearing tensile tests simulating stable crack growth by FEM analysis for arbitrary cracks.

Acknowledgments

The authors would like to acknowledge the 46th Research Institute, 6th Academy of China Aerospace Science & Industry Corp. for the GQ-6 materials support of the research work. Acknowledgements should also be extended tothe Shanghai Taiyang membrane structure co., LTD, where we carried out our uniaxial tearing tensile tests.

References

[1] Khoury G A and Gillett J D 1999 *Airship Technology* (Cambridge: Cambridge University Press)
[2] Chu CY and Chen F 1992 Tearing failure mechanism of woven fabrics and comparison between tear test methods *Text. Res. J.* **13** 196-200.
[3] Ueda Y 1983 Characteristics of brittle fracture under general combined modes including those under bi-axial tensile loads *Eng. Fract. Mech.* **6** 1131-58
[4] Minami H and Motobayashi S 1981 Experimental research on uniaxial and biaxial tensile strength of coated fabrics with variously shaped defects *J. Coated Fabrics* 24-40
[5] Williams J G and Ewing P D 1972 Fracture under complex stress - the angled crack problem *Int. J. Fract. Mech.* **8** 441-6
[6] FAA-P-8110-2-1995 Airship Design Criteria (USA Department of Transportation)
[7] Bigaud D, Szostkiewicz C and Hamelin P 2003 Tearing analysis for textile reinforced soft composites under mono-axial and bi-axial tensile stresses. *Compos. Struct.* **62** 129-37
[8] Meng J H, Cao S and Lv MY 2015 Analysis methods for tear properties of stratospheric airship envelope materials *J. Astronautics* **36** 230-5
[9] Luo Y and Hu H 2009 Mechanical properties of PVC coated bi-axial warp knitted fabric with and without initial cracks under multi-axial tensile loads *Compos. Struct.* **89** 536-42
[10] Topping A D 1973 The critical slit length of pressurized coated fabric cylinders *J. Coated Fabrics* 96-110
[11] Maekawa S and Yoshino T 2008 Tear propagation of a high-performance airship envelope material *J. Aircraft* **5** 1546-53
[12] Wetherhold R C and Park S 1986 Energy release rate calculations for the compact tension specimen using brittle orthotropic materials *Compos. Sci. Technol.* 219-30.
[13] Kenedy T C 1994 Application of composite fracture to coated fabrics *J. Coated Fabrics* 129-37
[14] Evans R, Clarke A, Gravina R, Heller M and Stewart R 2014 Improved stress intensity factors for selected configurations in cracked plates *Eng. Fract. Mech.* **127** 296-312

[15] Chen J W, Chen W J, Hou H Q and Hu Y 2015 Analysis on tearing strength of envelope materials under central crack tearing *Acata Materiae Compositae Sinica* **32** 1-9
[16] Wang Y, Chen X, Young R and Kinloch I 2016 Finite element analysis of effect of inter-yarn friction on ballistic impact response of woven fabrics *Compos. Struct.* **135** 8-16
[17] Irwin GR 1958 Fracture in encyclopedia of physics (New York: Springer-Verlag)
[18] Isida M 1970 Effect of width and length on stress intensity factors of internally cracked plates under various boundary conditions *Int. J. Fract. Mech.* **7** 301-16
[19] Isida M 1973 Analysis of stress intensity factors for the tension of a centrally cracked strip with stiffened edges *Eng. Fract. Mech.* **5** 647-65
[20] Hiroshi T 1971 A note on the finite width corrections to the stress intensity factor *Eng. Fract. Mech.* **3** 345-7

IOP Conf. Series: Materials Science and Engineering **137** (2016) 012025 doi:10.1088/1757-899X/137/1/012025

Tensile and fatigue behaviour of self-piercing rivets of CFRP to aluminium for automotive application

J Kang[1,3], H Rao[1], R Zhang[1], K Avery[2], X Su[2]

[1] CanmetMATERIALS, 183 Longwood Road South, Hamilton, L8P 0A5, Canada
[2] Ford Research and Innovation Center, Ford Motor Company, Dearborn, MI 48124, USA

E-mail: jidong.kang@canada.ca

Abstract. In this study, the tensile and fatigue behaviour of self-piercing rivets (SPRs) in carbon fibre reinforced plastic (CFRP) to aluminium 6111 T82 alloys were evaluated. An average maximum lap-shear tensile load capacity of 3858 N was achieved, which is comparable to metal-to-metal SPR lap-shear joints. The CFRP-Al SPRs failed in lap-shear tension due to pull-out of the rivet head from the CFRP upper sheet. The CFRP-Al SPR lap-shear specimens exhibited superior fatigue life compared to previously studied aluminium-to-aluminium SPR lap-shear joints. The SPR lap-shear joints under fatigue loads failed predominantly due to kinked crack growth along the width of the bottom aluminium sheet. The fatigue cracks initiated in the plastically deformed region of the aluminium sheet close to the rivet shank in the rivet-sheet interlock region. Scatter in fatigue life and failure modes was observed in SPR lap-shear specimens tested close to maximum tensile load.

1. Introduction

The mandatory fuel efficiency requirements established in the Corporate Average Fuel Economy (CAFE) standards compel every auto manufacturer in the US to build new cars and light trucks that meet the minimum fuel efficiency targets. One way to achieve improved fuel efficiency is vehicle light weighting, wherein aluminum and magnesium alloys, high strength steels, and polymer composites are used extensively in structural members. With the advancement in research, fiber reinforced polymer composite materials have shown a number of advantages over other lightweight alloys due to their high strength-to-weight and stiffness-to-weight ratios[1]. Studies have shown that the introduction of fiber reinforced plastics in automotive body structures could achieve weight savings up to 50-60% when compared to other lightweight metals, such as aluminum (which can only obtain weight savings up to 40-55% over the baseline steel structures) [2,3].

The use of lightweight fiber reinforced composites in automotive structural members yet faces a new challenge in terms of joining. Resistance spot welding (RSW), as a conventional joining method which has been widely used in automotive industry, is not feasible if one of the materials in use is a fiber-reinforced polymer composite. Instead, commonly-used joining process involving polymer and metallic materials include adhesive bonding and mechanical fastening [4]. Self-piercing riveting (SPR) is a high speed mechanical fastening technique where a tubular rivet is pierced through the top sheet and partially through the bottom overlapping sheet which are held against a die of asymmetrical cavity[5]. SPR is widely used by several automotive manufacturers as an economical and effective technique to join aluminum vehicle bodies [6–9]. SPR joints of aluminum to aluminum [6,10–14], magnesium to magnesium [15], and aluminum to steel [16–18] alloys have shown superior peel and fatigue strength compared to RSW joints. Unlike welding, SPR does not involve any heating or melting of material, nor does it require additional manufacturing processes, like pre-drilling holes, which could add additional manufacturing time and cost [5]. Therefore, it is feasible to use this technique when joining polymer-based fiber reinforced composite materials. Studies have shown that SPR technique can be effectively used in joining fiber reinforced composites to metals like steels [19] and aluminum alloys [1,20–24]. A majority of the studies on SPR in fiber-reinforced composites to metals focus on the influence of the process parameters on joint quality, including die geometry [25], distance between rivets [21], and oil pressure in the riveting system [1,26,27]. Very few studies have been conducted on the fatigue performance of SPR in composite to metal joints [1,21,23]. During the SPR process, the piercing of rivets through the composite panel causes localized damage by cutting the fibers; it also brings about deformation of the bottom aluminum sheet which results in a stress concentration and causes local changes in the material properties around the rivet shank. These changes in material properties can influence the mechanical property and the structural integrity of the SPR joint overall.

In the present study, the effort is focused on determining the mechanical properties, including the quasi-static and fatigue performance, of SPR joints in dissimilar CFRP-to-aluminum alloys for potential automotive applications.

2. Materials and experimental procedures

The CFRP to aluminium 6111 T82 lap-shear SPR specimens were fabricated on a Henrob servo-electronic riveting gun using steel rivets. Each sheet measured 2.5 mm in thickness and the rivets measured 7.8 mm in diameter. The CFRP panels used in this study were carbon fibre reinforced epoxy comprised of a three-layer braided fabric $(0^{o}/\pm 60^{o})_3$. The rivets were installed while the aluminium alloy was in the T4 condition. Then, after the joints were created, the entire specimen was subjected to a heat treatment similar to that experienced by structural members of the vehicle during the manufacturing process, resulting in a final heat treatment of T82. The final geometric configuration of the SPR joined test specimen is shown in Figure 1 and the cross-section of the SPR joint is shown in Figure 2. The macro image of cross-section (Figure 2(a)) shows the top CFRP sheet, the steel rivet, and the bottom aluminium sheet which is interlocked with the rivet. Both Figure 2(b) and 2(c) show the magnified regions closer to rivet shank indicating the presence of a thin sandwich layer of CFRP between the steel rivet surface and the aluminium sheet.

Quasi-static lap-shear testing of the specimens was performed on a MTS servo-hydraulic test frame in displacement control at a crosshead speed of 2 mm/min. The load-controlled fatigue test was performed on a MTS servo-hydraulic testing machine at a frequency of 20 Hz in a tension-tension mode at a load ratio of R=0.1. Shims of 2.5 mm thickness were used on both ends of the lap-shear specimens to ensure alignment during testing.

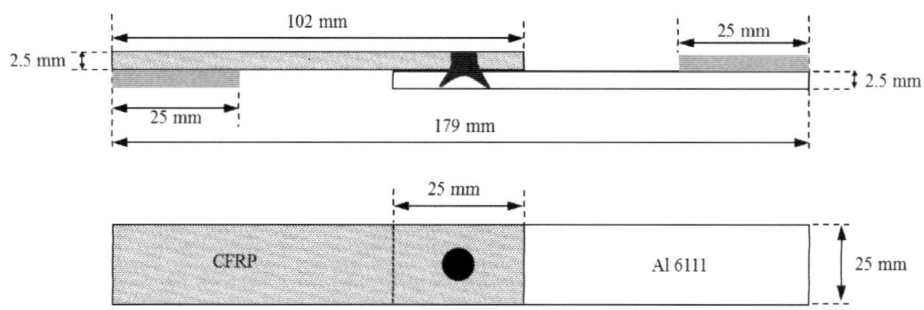

Figure 1. Geometrical dimensions of the representative SPR lap-shear CFRP-Al 6111 T82 test specimens.

After testing, fractography was performed on the SPR lap-shear specimens using an FEI Nova Nano 650 scanning electron microscope (SEM). Some of the tested and untested SPR lap-shear specimens were then cut parallel to the loading direction along the center of the specimen width. These were cold mounted, ground using standard metallurgical techniques, and finely polished using a 0.5 μm colloidal silica solution. The macrographs of the mounted samples were taken under a Zeiss digital optical microscope.

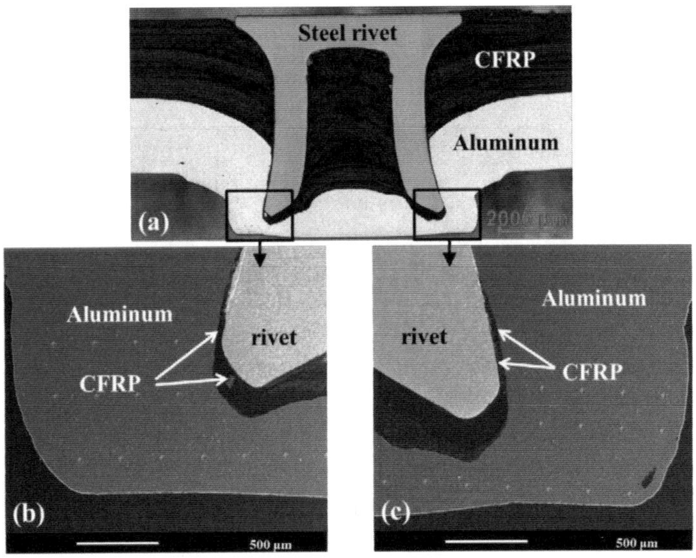

Figure 2. Representative cross-section of an untested SPR CFRP-Al test specimen; (a) macrograph showing the CFRP and aluminium sheet interlocked by a steel rivet; (b) and (c) magnified view of the rivet and aluminium surface interface under SEM showing the presence of a thin layer of CFRP.

3. Results and discussion

3.1. Tensile test

The lap-shear tensile test results of the SPR CFRP to aluminium 6111 T82 joints are shown in Figure 3. A maximum load of 3890 N was achieved with an average of 3858 N maximum load between the two specimens tested. Both of the tested lap-shear specimens failed due to pull-out of the rivet from the top CFRP facilitated by the damage to the CFRP around the rivet head. The inset in Figure 3 shows the resulting damage of the CFRP after the lap-shear tensile test. Note that the steel rivet is more rigid than the CFRP panel and thus a compressive force is applied at the interface of the rivet head and the CFRP panel. This results in penetration of the rivet head into the laminate forming a relief on the CFRP panel [1,20]. In this study, the rivet head entered the top CFRP laminate and damaged the top carbon fiber ply leading to the pull-out of the CFRP panel. This type of failure, where the top fiber reinforced composite panel fails due to the damage induced in the laminate by cutting of fibers, is generally referred to as the bearing failure [1,23].

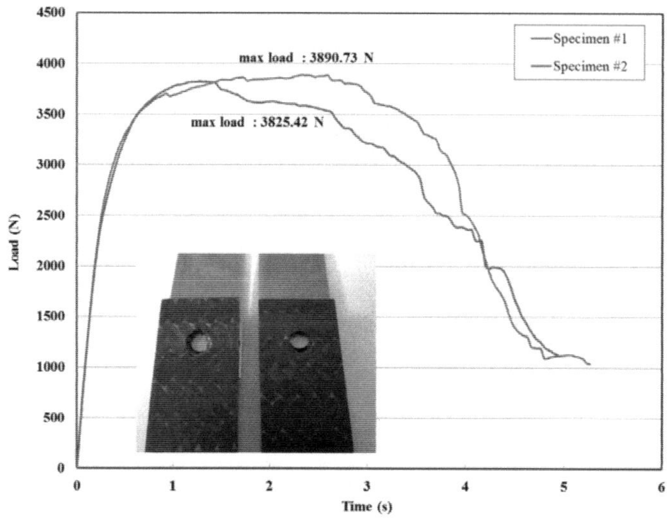

Figure 3. Lap-shear tensile test plot of the SPR CFRP to aluminium 6111 T82 joint. The inset picture shows the bearing failure of the CFRP in the two test specimens.

3.2. Fatigue test

The fatigue test results of the CFRP to aluminium 6111 T82 SPR lap-shear specimens are presented in Figure 4. The plot in Figure 4 also compares the fatigue life of AA6111-AA6111 (2 mm thickness) SPR lap-shear specimens [10]. From Figure 4, it is seen that the fatigue life of CFRP-Al SPR lap-shear specimens are higher than that of the Al-Al SPR lap-shear specimens of 2 mm thickness, despite the lower quasi-static load of the CFRP to aluminium 6111 T82 joints compared to AA6111-AA6111 SPR lap-shear specimens. The fatigue test results of the CFRP to aluminium 6111 T82 SPR lap-shear specimens also displayed a large scatter in fatigue life as the maximum fatigue load approached the maximum quasi-static load (3857 N). A similar scatter in fatigue life was also reported elsewhere in studies performed on SPR in composite to aluminum alloys [1,21,23]. This scatter in fatigue life close to maximum static load may be due to the scatter in the quasi-static behavior of the joints, which is suggested in Figure 3. This scatter in maximum load bearing capacity of the SPR joints may be due to the riveting process, the localized damage of the CFRP during the riveting process or other factors, which are yet to be fully understood. In general, the scatter in lap-shear tensile test is within the nominal range; however, the scatter in the fatigue data close to the maximum load level is significantly larger, as observed in this study.

The dominant mode of failure observed in this study was the fatigue cracks growing in a kinked manner along the width of the bottom aluminum sheet. The fracture surface of a representative tested aluminum coupon is shown in Figure 5 (loading direction is out of the plane of the page). Observation of the aluminum fracture surface in Figure 5(a) reveals the fatigue crack initiated at the periphery of the cold-worked region close to the aluminum-rivet interlock, as seen in Figure 5(b) and 5(c). The SEM micrographs indicate that the probable cause of crack initiation was the local stress concentration in the plastically deformed region. During the SPR process, the bottom aluminum sheet locally undergoes plastic deformation between the rivet shank and die, which results in cold working of the aluminum sheet [24]. Generally speaking, cold working not only increases the strength of the aluminum alloy but also introduces the high stress concentration in the region [1]. Outside the plastically deformed region, the material property of aluminum is that of the base metal. Hence, this plastically deformed region is small, and is localized to the area around the rivet and aluminum sheet interface. From the fatigue striations seen in Figures 5(d) and 5(e), it appears that the fatigue cracks grow steadily in the cold-worked plastically deformed region of the aluminum sheet. Once the fatigue crack reaches the base metal, catastrophic failure can occur in the bottom aluminum sheet.

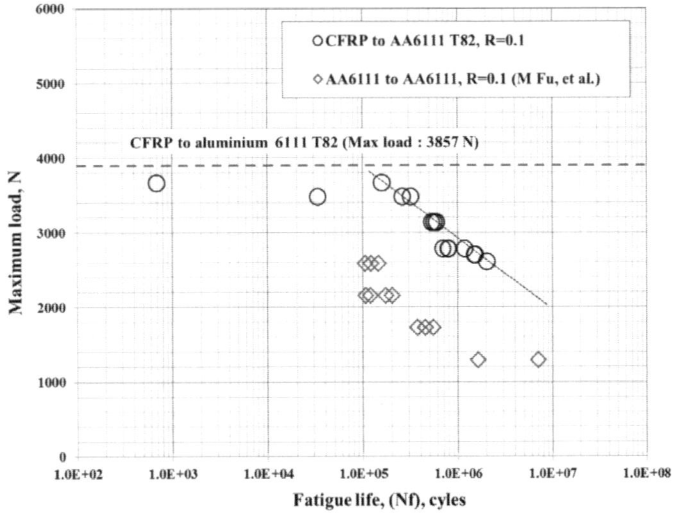

Figure 4. Fatigue test plot comparing the fatigue life of CFRP to aluminium 6111 T82, and AA6111 to AA611 [10] SPR lap-shear joint (R=0.1).

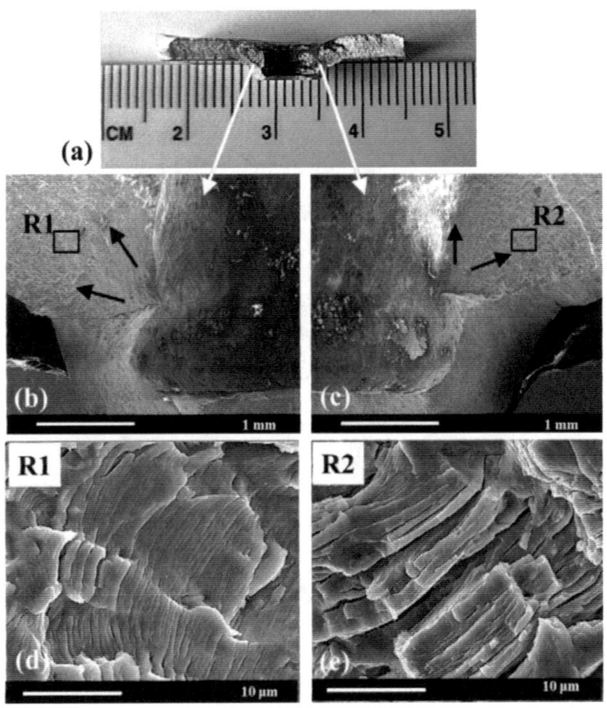

Figure 5. Fracture surface analysis of aluminum sheet that failed due to kinked cracks (a) macrograph of the fractured aluminum sheet (loading direction out of the page); (b) and (c) SEM images of the crack initiation region on the left and right side of the rivet respectively, the bold black arrows indicate the crack propagation direction; (d) and (e) magnified SEM images region R1 and R2 showing the fatigue striation in the crack propagation region.

The superior fatigue life performance of the CFRP to aluminium 6111 T82 SPR lap-shear joints compared to AA6111-to-AA6111 lap-shear joints may be attributed, in part, to the absence of fretting damage in the former. In several studies of metal-to-metal SPR lap-shear joints, fretting has been observed to be one of the dominant causes for crack initiation and significantly accelerated the crack growth in these SPR lap-shear joints [8,10,28–30]. Even though fretting was also observed in SPR lap-shear joints of aluminum to glass-fiber reinforced thermoplastic composite [23], no traces of fretting were observed in any of the SPR lap-shear test specimens tested in this study. Absence of fretting in this study may be attributed to the presence of a thin layer of CFRP sandwiched between the interfaces of the rivet and the aluminum sheet, as seen in Figure 2(b) and 2(c). The friction coefficient between the CFRP and the aluminum sheet is around 0.2, while the friction coefficient between two faying aluminum sheets is around 1.2 [31]. Hence the low friction coefficient between the CFRP and aluminum sheet produced no fretting which otherwise could reduce the fatigue life of the CFRP-Al SPR lap-shear specimens.

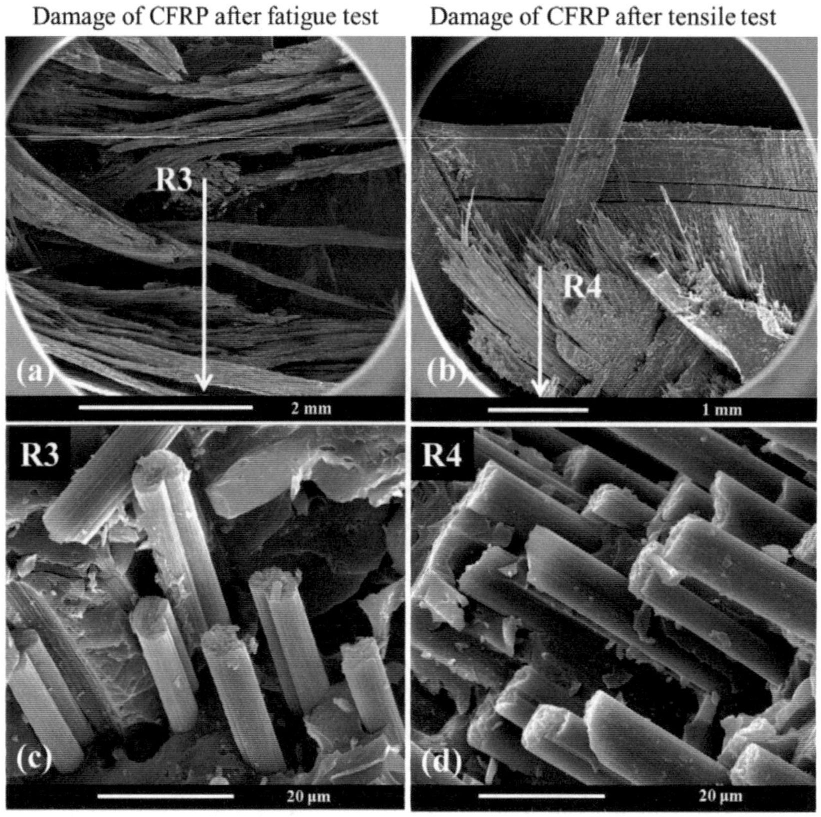

Figure 6. Top view of the damaged CFRP panel in (a) fatigue test; (b) tensile test; (c) and (d) are the magnified view of the region R3 and R4 showing the carbon fibre breakage and matrix damage in SPR lap-shear specimens tested under fatigue and quasi-static load respectively.

In quasi-static lap-shear tension, significant damage to the CFRP panel close to rivet head was observed which led to bearing failure. It is interesting to observe a similar failure mode also occurred in some SPR lap-shear specimens that were fatigue tested close to maximum static tensile load. In the two SPR lap-shear specimens that were fatigue tested at a maximum load of 3664 N (95% of the maximum tensile load), bearing failure was observed in one of the SPR lap-shear specimen that failed at 689 cycles, while the other suffered failure in the bottom aluminum sheet at 159,730 cycles. Figure

IOP Conf. Series: Materials Science and Engineering **137** (2016) 012025 doi:10.1088/1757-899X/137/1/012025

6 shows the SEM pictures comparing the bearing failure due to the damage in the CFRP panel in fatigue and tensile tested specimens. Comparison of Figure 6(a) and Figure 6(b) shows that the damage of the CFRP panel in an SPR lap-shear specimen tested close to the maximum quasi-static load is very similar to that observed in the quasi-static tensile specimens, respectively. Magnified views of the regions R3 and R4 show severe damage to polymer matrix and fiber breakage in both the fatigue and quasi-static tested SPR lap-shear specimens. In addition, once the top ply of carbon fibers and the polymer matrix are damaged, the compacting or locking mechanism between the rivet and CFRP panel can be released. The significant disparities in fatigue failure mode and fatigue life at this load level may be due to manufacturing variation of the joints. Further researches should be done to understand the impact of possible differences in the manufacturing process on the fatigue performance of the CFRP to aluminium 6111 T82 joints at both high and load low levels.

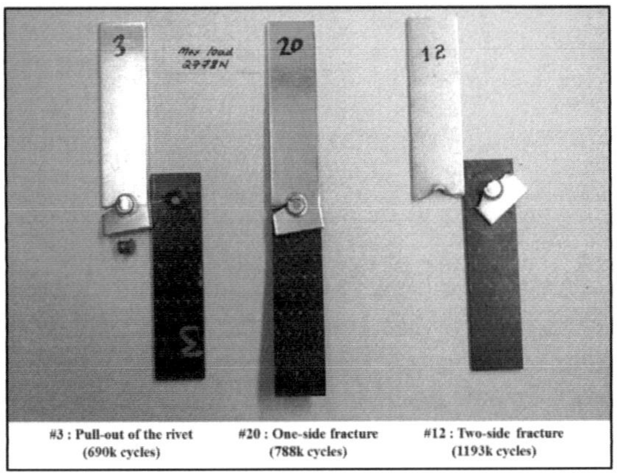

Figure 7. Failure modes observed in SPR joints tested at a load of 2778 N.

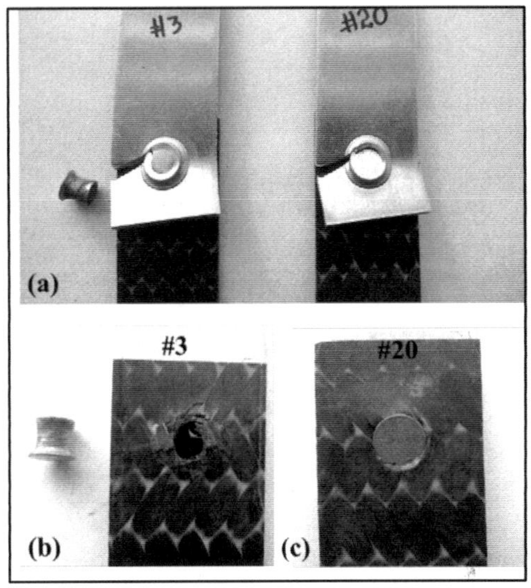

Figure 8. Comparison of the failure modes in SPR specimens #3 and #20; (a) one-sided crack on the aluminium sheet; (b) delamination of the CFRP in SPR specimen #3 and (c) rivet head and CFRP interface in SPR specimen #20.

In SPR lap-shear specimens that were fatigue tested at a maximum load of 2778 N (72% of the maximum tensile load), different failure modes were observed as well (Figure 7). Rivet pull-out resulted in the lowest fatigue life could count for the failure of the SPR lap-shear specimen, while the two sided fracture of the bottom aluminum sheet resulted in the highest fatigue life. Figure 8 shows the comparison of two SPR lap-shear specimens, i.e. #3 and #20 that failed due to rivet pull-out (#3) and one-sided fracture of the aluminum sheet (#20). Delamination of the top CFRP panel close to the rivet head interface was observed in SPR lap-shear specimen #3 (Figure 8(b)) and no delamination was observed in SPR lap-shear specimen #20 (Figure 8(c)). The probable reason for delamination and rivet pull-out failure still needs to find. It is expected that variations in the manufacturing process may lead to the observed scatter in fatigue life, as well as the failure mode.

4. Conclusions

Tensile and fatigue properties of dissimilar CFRP to aluminium 6111 T82 alloy SPR joints were evaluated in this study. A comprehensive failure analysis was performed to study the failure modes and factors influencing the mechanical properties. The following conclusions can be drawn from this study:

- The SPR technique can be used to effectively join dissimilar CFRP and AA6111 T82 alloy for automotive structural applications. The average maximum lap-shear tensile load capacity of the CFRP to aluminium 6111 T82 SPR lap-shear joints in this study is 3858 N, which is comparable to the metal-to-metal SPR lap-shear joints studied earlier.
- In quasi-static lap-shear tensile tests, damage to the top ply of the CFRP panel around the rivet head interface could lead to bearing failure. No further damage to CFRP panel was observed either in static or fatigue tests.
- In fatigue tests, the dominant mode of failure was crack growth along the width of the bottom aluminium sheet. Fatigue cracks initiated at the high stress concentration region in the plastically deformed area close to the rivet-aluminium interlock.
- The CFRP to aluminium 6111 T82 SPR lap-shear joints in this study exhibited superior fatigue life compared to AA6111 to AA6111 SPR lap-shear joints studied earlier. The absence of fretting may have contributed to the improved fatigue life.

References

[1] Di Franco G, Fratini L, Pasta A and Ruisi V F 2013 On the self-piercing riveting of aluminium blanks and carbon fibre composite panels *Int. J. Mater. Form.* **6** 137–44

[2] Eusebi E 1995 Composite intensive vehicles: past present and future *Struct. Mater. challenges next Gener. Veh. Washington, DC US Dep. Commer.*

[3] Gjostein N A 1995 Technology needs beyond PNGV *Basic Needs Veh. Futur. New Orleans, Louisiana (5 January 1995)*

[4] Kah P, Suoranta R, Martikainen J and Magnus C 2014 Techniques for joining dissimilar materials: Metals and polymers *Rev. Adv. Mater. Sci.* **36** 152–64

[5] Litherland H 1998 Joints in Aluminum: INALCO' 98 Proceedings *Joints in Aluminum: INALCO' 98 Proceedings* (Cambridge: Woodhead Publishing) pp 135–47

[6] Cai W, Wang P C and Yang W 2005 Assembly dimensional prediction for self-piercing riveted aluminum panels *Int. J. Mach. Tools Manuf.* **45** 695–704

[7] Mortimer J 2001 Jaguar uses X350 car to pioneer use of self-piercing rivets *Ind. Robot An Int. J.* **28** 192–8

[8] Huang L, Bonnen J, Lasecki J, Guo H and Su X 2016 Fatigue and fretting of mixed metal self-piercing riveted joint *Int. J. Fatigue* **83** 230–9

[9] Barnes T A and Pashby I R 2000 Joining techniques for aluminium spaceframes used in automobiles Part I - solid and liquid phase welding *J. Mater. Process. Technol.* **99** 62–71

[10] Fu M and Mallick P K 2003 Fatigue of self-piercing riveted joints in aluminum alloy 6111 *Int. J. Fatigue* **25** 183–9

[11] Lin P and Su Z 2013 Fatigue Behavior of Self-Piercing Rivets and Clinch Joints in Lap-Shear Specimens of Aluminum Sheets CLINCH JOINTS OF 6111-T4 *Int. J. Fatigue* **72** 293–8

[12] Zhao L, He X, Xing B, Lu Y, Gu F and Ball A 2015 Influence of sheet thickness on fatigue behavior and fretting of self-piercing riveted joints in aluminum alloy 5052 *Mater. Des.* **87** 1010–7

[13] Chen Y K, Han L, Chrysanthou a. and O'Sullivan J M 2003 Fretting wear in self-piercing riveted aluminium alloy sheet *Wear* **255** 1463–70

[14] Iyer K, Hu S J, Brittman F L, Wang P C, Hayden D B and Marin S P 2005 Fatigue of single-And double-rivet self-piercing riveted lap joints *Fatigue Fract. Eng. Mater. Struct.* **28** 997–1007

[15] Miyashita Y, Jack Teow Y C, Karasawa T, Aoyagi N, Otsuka Y and Mutoh Y 2011 Strength of adhesive aided SPR joint for AM50 magnesium alloy sheets *Procedia Eng.* **10** 2532–7

[16] Abe Y, Kato T and Mori K 2009 Self-piercing riveting of high tensile strength steel and aluminium alloy sheets using conventional rivet and die *J. Mater. Process. Technol.* **209** 3914–22

[17] Mori K, Kato T, Abe Y and Ravshanbek Y 2006 Plastic joining of ultra high strength steel and

aluminium alloy sheets by self piercing rivet *CIRP Ann. - Manuf. Technol.* **55** 283–6

[18] Han L and Chrysanthou a. 2008 Evaluation of quality and behaviour of self-piercing riveted aluminium to high strength low alloy sheets with different surface coatings *Mater. Des.* **29** 458–68

[19] Wagner J, Wilhelm M, Baier H, Füssel U and Richter T 2014 Experimental analysis of damage propagation in riveted CFRP-steel structures by thermal loads *Int. J. Adv. Manuf. Technol.* **75** 1103–13

[20] Di Franco G, Fratini L and Pasta a. 2013 Analysis of the mechanical performance of hybrid (SPR/bonded) single-lap joints between CFRP panels and aluminum blanks *Int. J. Adhes. Adhes.* **41** 24–32

[21] Di Franco G, Fratini L and Pasta A 2012 Influence of the distance between rivets in self-piercing riveting bonded joints made of carbon fiber panels and AA2024 blanks *Mater. Des.* **35** 342–9

[22] Zhang J and Yang S 2014 Self-piercing riveting of aluminum alloy and thermoplastic composites *J. Compos. Mater.*

[23] Gay A, Lefebvre F, Bergamo S, Valiorgue F, Chalandon P, Michel P and Bertrand P 2016 Fatigue performance of a self-piercing rivet joint between aluminum and glass fiber reinforced thermoplastic composite *Int. J. Fatigue* **83** 127–34

[24] Kroll L, Mueller S, Mauermann R, Gruetzner R, Technology F, Structures L and Gmbh E 2011 Strength of Self-Piercing Riveted Joints for Cfrp / Aluminium Sheets *18 Th Int. Conf. Compos. Mater.* 1–6

[25] Pickin C G, Young K and Tuersley I 2007 Joining of lightweight sandwich sheets to aluminium using self-pierce riveting *Mater. Des.* **28** 2361–5

[26] Fiore V, Bella G Di, Galtieri G, Alagna F, Borsellino C and Valenza A 2012 Mechanical Behaviour of Spr / Co-Cured Composite To Aluminium Joints *ECCM15-15th European Conference of Composite Materials* (Venice, Italy) pp 24–8

[27] Fratini L and Ruisi V F 2009 Self-piercing riveting for aluminium alloys-composites hybrid joints *Int. J. Adv. Manuf. Technol.* **43** 61–6

[28] Han L, Chrysanthou a. and O'Sullivan J M 2006 Fretting behaviour of self-piercing riveted aluminium alloy joints under different interfacial conditions *Mater. Des.* **27** 200–8

[29] Li D, Han L, Thornton M and Shergold M 2012 Influence of rivet to sheet edge distance on fatigue strength of self-piercing riveted aluminium joints *Mater. Sci. Eng. A* **558** 242–52

[30] Zhang Z, Yang X, Zhang J, Zhou G, Xu X and Zou B 2011 Effect of welding parameters on microstructure and mechanical properties of friction stir spot welded 5052 aluminum alloy *Mater. Des.* **32** 4461–70

[31] Schön J 2004 Coefficient of friction for aluminum in contact with a carbon fiber epoxy composite *Tribol. Int.* **37** 395–404

Microstructures of $A_4(BC)_4$ star copolymers and the influence of architecture

L Zhou[1], J Peng, M F Cheng and J H Fang

School of Science, Nantong University, Nantong 226019, China

E-mail: ZL7103@163.com

Abstract. The self-consistent field theory is used to investigate the microstructures of $A_4(BC)_4$ star copolymers. The phase diagram is mapped out to show different phase regions. Compared with $A(BC)_4$ star copolymers, several microstructures are lost, and more lamellar structures are found due to the stronger topological constraint of the junction. The results may be helpful to understand the influence of molecular architectures on microstructures and provide an effective way to design controllable microstructures.

1. Introduction

The self-assembly of block copolymers has attracted much attention in recent years for its promising applications in material science, nanotechnology and biology. For the simplest linear *AB* diblock copolymers, the fascinating periodic morphologies have been deeply understood both experimentally and theoretically [1-4]. As the distinct components increase, say *ABC* triblock copolymers, the variety and complexity of microstructures may increase dramatically [5, 6]. Especially, once the topological architecture changes from line shape to star, comb-like or other complicated shapes, more novel structures can be found [7-10]. The star architecture imposes a strong topological constraint on the polymer chain and may lead to new phenomena. In particular, the development of synthesis technology allows us to obtain more complex copolymers, which provides an opportunity to find new periodic microstructures due to the additional entropic effect. By far, systematic investigations on how the microstructures relate to the architectures are still desired. In this work, we use the self-consistent field theory (SCFT) to investigate the microstructures of $A_4(BC)_4$ star copolymers, where four *A* chains and four *BC* chains are attached to a junction point. Furthermore, we make a comparison between the microstructures of $A(BC)_4$ star copolymers, where one *A* chain and four *BC* chains are attached to a junction point. It is very helpful to explore the microstructure differences between $A_4(BC)_4$ and $A(BC)_4$ star copolymers to understand the influence of molecular architecture on microstructures.

2. Model and method

Figure 1 gives a schematic of $A_4(BC)_4$ and $A(BC)_4$ star copolymers. Green, red and blue curves represent *A*, *B* and *C* blocks, respectively. Four *A* chains and four *BC* copolymer chains are attached to the junction *O* by the *A* ends and *B* ends, respectively. Each chain *A* consists of N_A *A* segments, and each chain *BC* consists of N_B *B* segments and N_C *C* segments. All segments have the same volume ρ_0^{-1} and the statistical length *a*. The only difference between two kinds of copolymers is the connection mode of chain *A*. For $A(BC)_4$ copolymers, only one *A* chain is attached to the junction *O*. In both copolymers, the number of the segment component *A* is same. Therefore, the length of *A* chain in the

$A_4(BC)_4$ star copolymers is one fourth of that in the $A(BC)_4$ star copolymers, which means that two kinds of copolymers have same polymerization $N = 4N_A + 4N_B + 4N_C$ though their architectures are different.

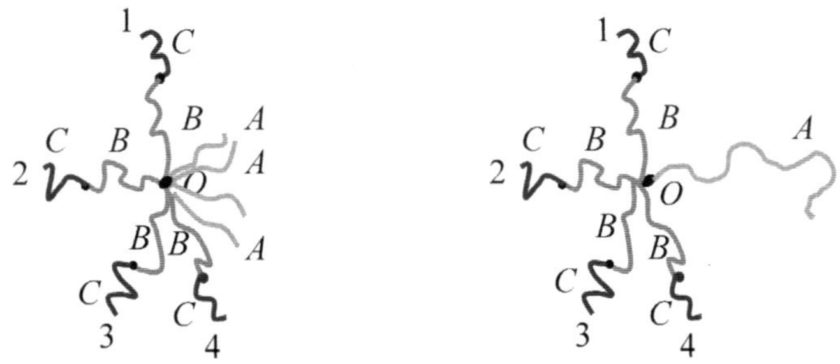

(a) $A_4(BC)_4$ (b) $A(BC)_4$

Figure 1. (Colour online) A schematic of (a) $A_4(BC)_4$ and (b) $A(BC)_4$ star copolymers. Green, red and blue curves represent A, B and C blocks, respectively.

As an important mean-field theory for polymers, the SCFT has been widely used in the prediction of phase behavior of block copolymers and other related systems [11, 12]. For the $A_4(BC)_4$ or $A(BC)_4$ copolymer melts in a volume V, the free energy F can be written as

$$\frac{NF}{\rho_0 k_B TV} = \frac{1}{V}\int d\vec{r}[\chi_{AB}N\varphi_A(\vec{r})\varphi_B(\vec{r}) + \chi_{BC}N\varphi_B(\vec{r})\varphi_C(\vec{r}) + \chi_{AC}N\varphi_A(\vec{r})\varphi_C(\vec{r}) - W_A(\vec{r})\varphi_A(\vec{r}) - W_B(\vec{r})\varphi_B(\vec{r})$$

$$-W_C(\vec{r})\varphi_C(\vec{r}) - \xi(\vec{r})(1-\varphi_A(\vec{r}) - \varphi_B(\vec{r}) - \varphi_C(\vec{r}))] - \ln(\frac{\Omega}{V}) \ . \tag{1}$$

In the equation (1), T is the temperature and k_B is the Boltzmann's constant. The variables φ_i, $W_i(i = A, B, C)$ and ζ are the functions of the position $\vec{r}$. φ_i is the local densities of different components. χ_{ij} $(= \chi_{ij}, i, j = A, B, C)$ is the Flory-Huggins interaction parameter between two different components. W_i is the effective chemical potential field acting on the different component. ζ is a Lagrange-multiplier potential field ensuring the system incompressibility. Ω is the partition function of a single star copolymer and is defined by

$$\Omega = \int d\vec{r} q_i(\vec{r}, s) q_i^+(\vec{r}, s) . \tag{2}$$

In the equation (2), q_i and $q_i^+(i = A$ or $BC)$ are the propagators and denote the probability that finds the segment s at the position $\vec{r}$. The propagators satisfy the modified diffusion equations:

$$\frac{\partial q_A(\vec{r},s)}{\partial s} = \frac{Na^2}{6}\nabla^2 q_A(\vec{r},s) - W_A(\vec{r})q_A(\vec{r},s) \quad 0 < s < N_A(\ A_4(BC)_4) \text{ or } 0 < s < 4N_A(A(BC)_4); \tag{3}$$

$$\frac{\partial q_A^+(\vec{r},s)}{\partial s} = -\frac{Na^2}{6}\nabla^2 q_A^+(\vec{r},s) + W_A(\vec{r})q_A^+(\vec{r},s) \quad 0 < s < N_A(A_4(BC)_4) \text{ or } 0 < s < 4N_A(A(BC)_4); \tag{4}$$

$$\frac{\partial q_{BC}(\vec{r},s)}{\partial s} = \frac{Na^2}{6}\nabla^2 q_{BC}(\vec{r},s) - W_i(\vec{r})q_{BC}(\vec{r},s) \quad if\ 0 < s < N_B, i = B; else\ \ if\ \ N_B < s < N_C, i = C; \tag{5}$$

$$\frac{\partial q_{BC}^+(\vec{r},s)}{\partial s} = -\frac{Na^2}{6}\nabla^2 q_{BC}^+(\vec{r},s) + W_i(\vec{r})q_{BC}^+(\vec{r},s) \quad if\ 0 < s < N_B, i = B; else\ \ if\ \ N_B < s < N_C, i = C. \tag{6}$$

Due to the different architectures of the $A_4(BC)_4$ and $A(BC)_4$ copolymers, the initial conditions of above equations are different.

For the $A_4(BC)_4$ copolymers, the initial conditions are given by

$$q_A(\vec{r},0) = 1,\ q_{BC}^+(\vec{r}, N_B + N_C) = 1,$$

$$q_A^+(\vec{r}, N_A) = [q_A(\vec{r}, N_A)]^3 * [q_{BC}^+(\vec{r},0)]^4,\ \ q_{BC}(\vec{r},0) = [q_A(\vec{r}, N_A)]^4 * [q_{BC}^+(\vec{r},0)]^3 . \tag{7}$$

For the $A(BC)_4$ copolymers, the initial conditions are given by

$$q_A(\bar{r},0)=1,\ q_{BC}^{+}(\bar{r},N_B+N_C)=1,\ q_A^{+}(\bar{r},N_A)=\ [q_{BC}^{+}(\bar{r},0)]^4,\ q_{BC}(\bar{r},0)=q_A(\bar{r},N_A)*[q_{BC}^{+}(\bar{r},0)]^3. \quad (8)$$

Minimizing the free energy F, we obtain a set of self-consistent equations. The field W_i and the local density φ_i at the equilibrium can be determined by using the combinatorial screening method [12] to solve the self-consistent equations with periodic boundary conditions.

3. Results and discussion

To compare the microstructures of two kinds of copolymers, we ensure all factors are same except for the architectures. The Flory-Huggins interaction parameters and the polymerization in both copolymers are set as $\chi_{AB}N = \chi_{BC}N = \chi_{AC}N = 140$ and $N = 600$. All the sizes are in units of a.

Table 1. (Colour online) Summary of observed microstructures of $A_4(BC)_4$ star copolymers.

Name (Abbreviation)	Graphic	Schematic representation
hexagonal microstructure (HEX)	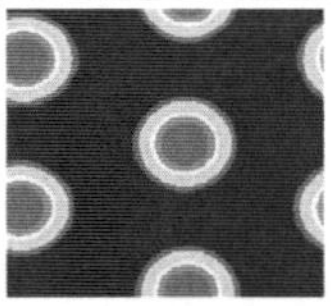	
core-shell hexagonal microstructure (CSH)		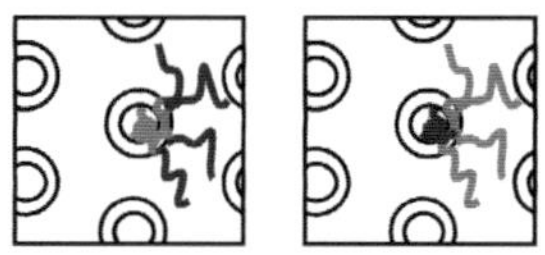
two-phase lamellar microstructure (L2)	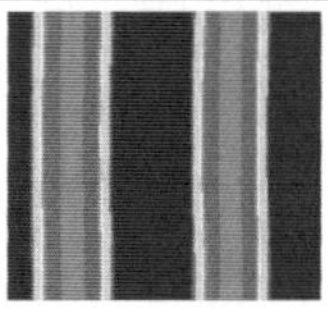	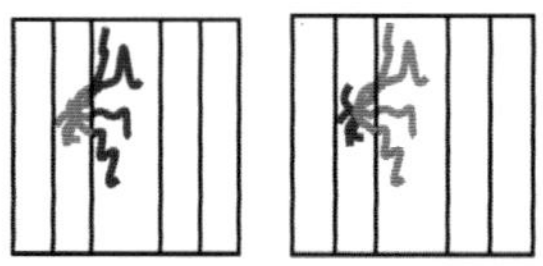
three-phase lamellar microstructure (L3)	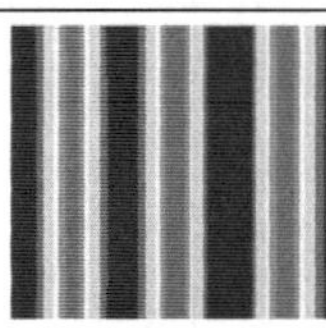	
three-phase lamellar microstructure with one mixed phase (L3M)	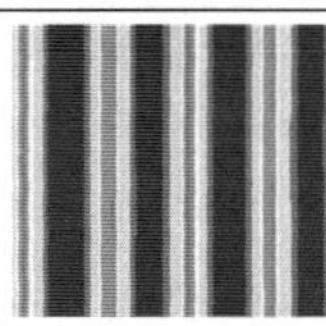	
interpenetrating tetragonal microstructure (TET)	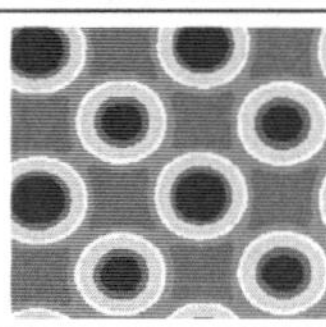	

We can introduce the average densities of different components: $D_A = 4N_A/N$, $D_B = 4N_B/N$, $D_C = 4N_C/N$. With the change of the average densities, the distinct block has different length. The corresponding copolymer may self-assemble into an ordered microstructure. All observed microstructures of the $A_4(BC)_4$ star copolymers are shown in table 1, including the hexagonal microstructure (HEX), the core-shell hexagonal microstructure (CSH), the two-phase lamellar microstructure (L2), the three-phase lamellar microstructure (L3), the three-phase lamellar microstructure with one mixed phase (L3M) and the interpenetrating tetragonal microstructure (TET). The phase diagram in figure 2(a) gives the microstructures varying with the average densities. Three sides of the triangle represent the average densities D_i ($i = A, B, C$). Each of them changes from 0 to 1. The increment is 0.1. Each grid point corresponds to a set of (D_A, D_B, D_C). For example, the grid point in the lower left of the phase diagram corresponds to (0.1, 0.1, 0.8). In this case, $N_A = 15$, $N_B = 60$ and $N_C = 480$, the length of chain A is very short whereas that of chain C is very long. The microstructure is obtained by using the method described in Section 2. The corresponding structural symbol is plotted on the grid point. In the three corners of the phase diagram, the systems can form the hexagonal microstructure. Three kinds of schematic representations of the composition are plotted in table 1. Taking the first schematic representation as an example, the cores consist of two kinds of short blocks A and B. The matrix consists of the long block C. The cores arrange hexagonally in the matrix. When (D_A, D_B, D_C) = (0.2, 0.1, 0.7) and (0.7, 0.1, 0.2), the core-shell hexagonal microstructures are observed. If (D_A, D_B, D_C) = (0.2, 0.1, 0.7), blocks A form the cores, the mixed blocks A and B form the shells, and C blocks form the matrix. Similarly, if (D_A, D_B, D_C) = (0.7, 0.1, 0.2), C blocks form the cores, the mixed blocks B and C form the shells, and A blocks form the matrix. When (D_A, D_B, D_C) = (0.1, 0.4, 0.5), (0.2, 0.2, 0.6), (0.6, 0.2, 0.2), (0.6, 0.3, 0.1) and (0.7, 0.2, 0.1), the two-phase lamellar microstructures are found. The schematic representations in table 1 show that the two short connected blocks form a mixed layer and the other long block forms another layer. With the separation of the mixed phase, L3M is observed, where most of the long blocks in the mixed layer is separated to form another individual layer, and the rest of the long blocks and the short blocks still form the mixed layer. The further separation can lead to the three-phase lamellar microstructure, where each different component forms an individual layer. When the density of B block is relativelyhigh, the system can form an interpenetrating tetragonal microstructure, where A and C blocks form tetragonal lattices respectively, and B blocks form the matrix. When (D_A, D_B, D_C) = (0.1, 0.8, 0.1), the system cannot be separated and keeps homogeneous (DIS).

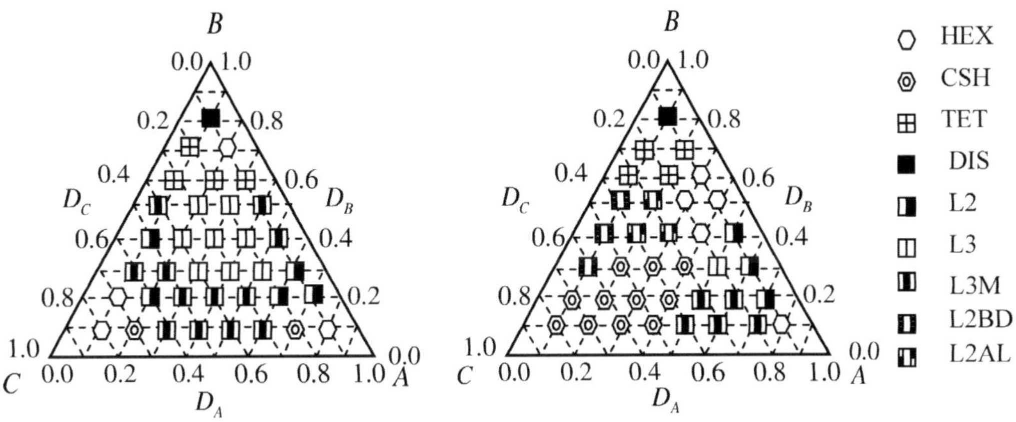

(a) $A_4(BC)_4$ (b) $A(BC)_4$

Figure 2. Phase diagrams of (a) $A_4(BC)_4$ and (b) $A(BC)_4$ star copolymers.

To further examine the influence of architecture, we compare the microstructures with those of the $A(BC)_4$ star copolymers. For clarity, the phase diagram of the $A(BC)_4$ star copolymers is shown in figure 2(b). Obviously, the microstructure number of the $A_4(BC)_4$ star copolymers is smaller than that of the $A(BC)_4$ star copolymers. Other structures such as the two-phase lamellar structure with alternating blocks inside a layer (L2AL), the two-phase lamellar structure with beads at the interface (L2BD) and the interpenetrating hexagonal microstructures (INHE) observed in the $A(BC)_4$ star copolymers(see Ref.[8]) cannot be found in the $A_4(BC)_4$ star copolymers. The first two lamellar microstructures show that the mixed blocks in the lamellar structures may further separate inside the mixed layer which can be attributed to the flexibility of one arm A. In addition, we find that most of the grid points in the phase diagram are occupied by the lamellar structures in the $A_4(BC)_4$ star copolymers. In comparison with the phase diagram of the $A(BC)_4$ star copolymers, most of the hexagonal microstructures and the core-shell hexagonal microstructures in the $A(BC)_4$ star copolymers are replaced by all kinds of the lamellar structures in the $A_4(BC)_4$ star copolymers. Due to the junction constraint of four chains A, the formation of the lamellar structures may help to lower the free energy by relieving the interfacial crowding. The symmetry of four blocks A and four blocks BC also favours the formation of symmetrical structures. Therefore more lamellar structures are observed in the $A_4(BC)_4$ star copolymers.

4. Conclusion

We studied the microstructures of the $A_4(BC)_4$ star copolymers by using self-consistent field theory. Six kinds of different microstructures are observed by varying the average densities of distinct components when $\chi_{AB}N = \chi_{BC}N = \chi_{AC}N = 140$ and $N = 600$. Compared with the microstructures of the $A(BC)_4$ star copolymers with the same parameters, the number of the microstructures decreases and more lamellar structures are found due to a stronger junction constraint of blocks A. It is helpful to understand the influence of architectures on microstructure and provide an effective way to design controllable microstructures with these results.

Acknowledgements

This work was supported by the National Natural Science Foundation of China (Nos. 61371057, 11204143), the Major National Scientific Instrument and Equipment Development Project of China (No.2011YQ03013403), Innovation Fund for the Combination of Production, Study and Research of Jiangsu Province-Prospective Project for Joint Research (No.BY2014081-06), and the Open Research Fund Program of Jiangsu Provincial Key Laboratory of Center for Photon Manufacturing Science and Technology (GZ201309).

References

[1] Bohbot-Raviv Y and Wang Z G 2000 *Phys. Rev. Lett.* **85** 3428-31
[2] Li S B, Chen P, Wang X H, Zhang L X and Liang H J 2009 *J. Chem. Phys.* **130** 014902
[3] Shin D O, Mun J H, Hwang G, Yoon J Moon, Kim J Y, Yun J M, Yang Y, Oh Y, Lee J Y, Shin J, Lee K J, Park S, Kim J U and Kim S O 2013 *ACSnano* **7** 8899-907
[4] Chen P, Liang H J, Xia R, Qian J S and Feng X S 2013 *Macromolecules* **46** 922-6
[5] Tang P, Qiu F, Zhang H D and Yang Y L 2004 *Phys. Rev. E* **69** 031803
[6] Zhu X M, Wang L Q and Ling J P 2011 *Macromolecules* **44** 8314-23
[7] Tang P, Qiu F, Zhang H D and Yang Y L 2004 *J. Phys. Chem. B* **108** 8434-8
[8] Zhou L, Liu Y, Peng J and Fang J H 2013 *J. Polym.Res.* **20** 102
[9] Zhou L, Zhang J G, Fang J H and Sun B H 2011 *J. Polym.Res.* **18** 1053-8
[10] Kirkensgaard J J K, Fragouli P, Hadjichristidis N and Mortensen K 2011 *Macromolecules* **44** 575-82
[11] Matsen M W 2002 *J. Phys.: Condens. Matter* **14** R21-47
[12] Drolet F and Fredrickson G H 2001 *Macromolecules* **34** 5317-24

IOP Conf. Series: Materials Science and Engineering **137** (2016) 012027 doi:10.1088/1757-899X/137/1/012027

Layered phases in the composites prepared using ZrC and LnB_6 (Ln = La, Ce, and Gd)

L Y Zhao[1] and K S Wang

Shanghai Academy of Spaceflight Technology, Shanghai, 201109, China

E-mail: zhaoliyou0617@163.com

Abstract. We found some layered phases in the composite prepared by reactive hot pressing at 1800°C for 30 mins, using ZrC and LnB_6 (Ln = La, Ce, and Gd) as raw materials. The layered phases are related to rare-earth metal borocarbides, but they are relatively and chemically stable in moisture environment. XRD patterns of these layered phases cannot be identified far away, indicating that the layered phases are possibly some new compounds.

1. Introduction

Ceramics with layered grains usually show many excellent properties. An important family of layered ceramics is $M_{N+1}AX_N$ (N = 1, 2 or 3). In the $M_{N+1}AX_N$, M is a transition metal, A is an A-group (mostly ⅢA and ⅣA) element, and X is either C or N. $M_{N+1}AX_N$shows low density, high elastic modulus, easy machinability, excellent thermal shock resistance, and damage tolerance [1-7]. This unusual combination of properties is traceable to their layered structure [8-11]. Another family of layered ceramics is $M_xB_yC_z$, where M is Al or rare-earth metal (RE). Metal borocarbides also display various attractive mechanical, physical, and chemical properties [12-17]. Ceramics with layered grains are emerging materials. Much is unknown about them by far, especially RE borocarbides. RE borocarbides are usually prepared by arc-melting technique [18-20]. The process is complex, but its yield is low. Because the products are usually powders and sensitive to moisture [21], most work focus on the calculation of crystal structure, therefore, reports on microstructure characterization are scarce and rare.

In our previous work [22-23], Lanthanum-containing layered phase was found in composites prepared by a simple reactive hot pressing using zirconium carbide and Lanthanum boride as raw materials. This finding is interesting, because there have been no similar reports before. In this research, another two lanthanide-containing layered phases were synthesized by the same method, which indicates we may obtain more layered ceramics by this method and the results can make a significant contribution to studying future layered ceramics.

2. Experimental

One starting powder is zirconium carbide (ZrC, D_{50} = 2.1μm, >98% purity, Changsha Wing High High-Tech New Materials Co., Ltd.,China). The other starting powder is lanthanide hexaboride (LnB_6, Ln = La, Ce, and Gd, -300 mesh, >99% purity, Northwest Institute for Non-ferrous Metal Research, China). The powder mixture of both ZrC and each $LnB_6$was first mixed in alcohol with SiC balls for 24 hours in a plastic bottle and then dried. The volume fraction of LnB_6 was 30 vol.% in this study. The mixed powders were sintered by reactive hot pressing in a graphite die with a BN coating at the

IOP Conf. Series: Materials Science and Engineering **137** (2016) 012027 doi:10.1088/1757-899X/137/1/012027

temperature of 1800°C under a pressure of 25 MPa for 30 mins in the vacuum. The heating rate was 10°C/min and the cooling was achieved by switching the power off. The resulting ceramics were cleaned, and the surface layers were ground off. Surfaces of the specimens were polished with SiC paper up to #1500 using water as polishing media. According to the adopted LnB_6, the specimens were referred as ZLa, ZCe, and ZGd, respectively. The phase composition of resulting ceramics was identified using X-ray diffractometer (XRD, Cu Kα radiation, D/max-2200VPC, Rigaku, Tokyo, Japan). The microstructure was characterized by scanning electron microscope (SEM, FEI Quanta 200F, Eindhoven, the Netherlands), which is equipped with an energy-dispersive spectroscopy (EDS, EDAX, USA) system. Before SEM examination, a thin layer of carbon was sputtered on the surfaces of samples.

3. Results and discussion

Figure 1 shows the XRD patterns of resulting ceramics. ZrB_2 is detected besides the major phase of ZrC. Some relatively weak peaks are found. Because LnB_6 diffraction peaks disappear, the weak peaks should be related to Ln-containing phases. These peaks cannot be identified based on the data in JCPDF database and literature. It is likely that we found some new phases.

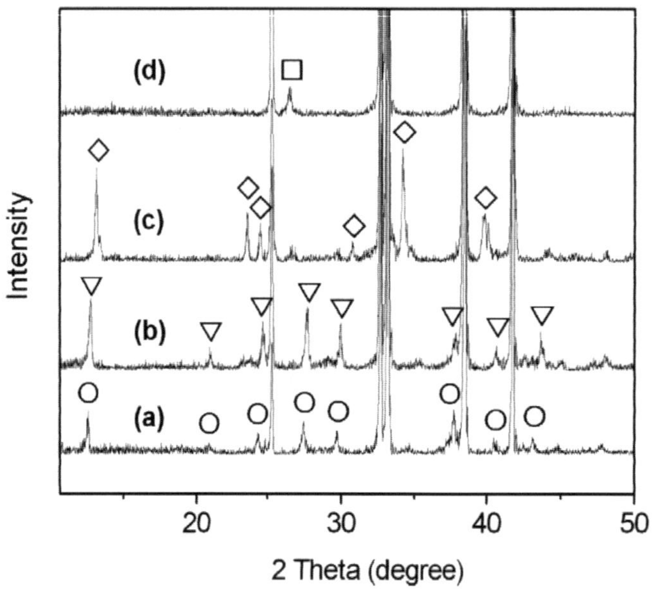

Figure 1. XRD patterns of (a) ZLa, (b) ZCe, (c) ZGd, and (d) the composite prepared using ZrC and amorphous B. Unidentifiable peaks are marked in (a), (b), and (c). "□" shows the graphite peak in (d).

Low-magnification surface SEM images (figure 2) show that lots of light phases distribute in the resulting ceramics. Surface and fracture surface SEM images exhibit a layered structure of the light phases. EDS analysis confirms the existence of Ln in the layered phases. Some oxygen is detected, which is also found in the raw LnB6 powders. We prepared a composite using mixed ZrC and amorphous B with the same molar ratio as that of ZrC to LnB6. Besides the peaks of ZrB2, graphite peak is found in the XRD pattern of this composite [figure 1(d)]. The graphite peak is stronger than that in the XRD pattern of the composite prepared using ZrC and LnB6. It can be inferred that the layered phases contain C element. Considering the terrible instability in the air and no reported layered structure [24-26], the formed layered phases should be not LnCx. In contrast, some work has shown that combination of boron and carbon elements with rare-earth metal can form ternary layered compounds [21,27,28]. Because boron-carbon substructures are abundant in variety, RExByCz has numerous different structural arrangements. Therefore, it is believed that the layered phases found in this study are related to boron-carbon-RE compounds. As mentioned above, most processing steps of samples are conducted in the rich water condition, and the samples are exposed to air for a long time before analysis. Almost no damage can be seen for the layered phases. This intactness of the layered phases is a novel phenomenon, because RE borocarbides are moisture sensitive and need no additional attention and effort, they would decompose to form metal hydroxides, and hydrocarbon and hydroboration species by themselves [18-21]. We have tried to characterize the crystal structure of the layered phases found in this study, but failed to do so because it is difficult to analyze the selected area electron diffraction.

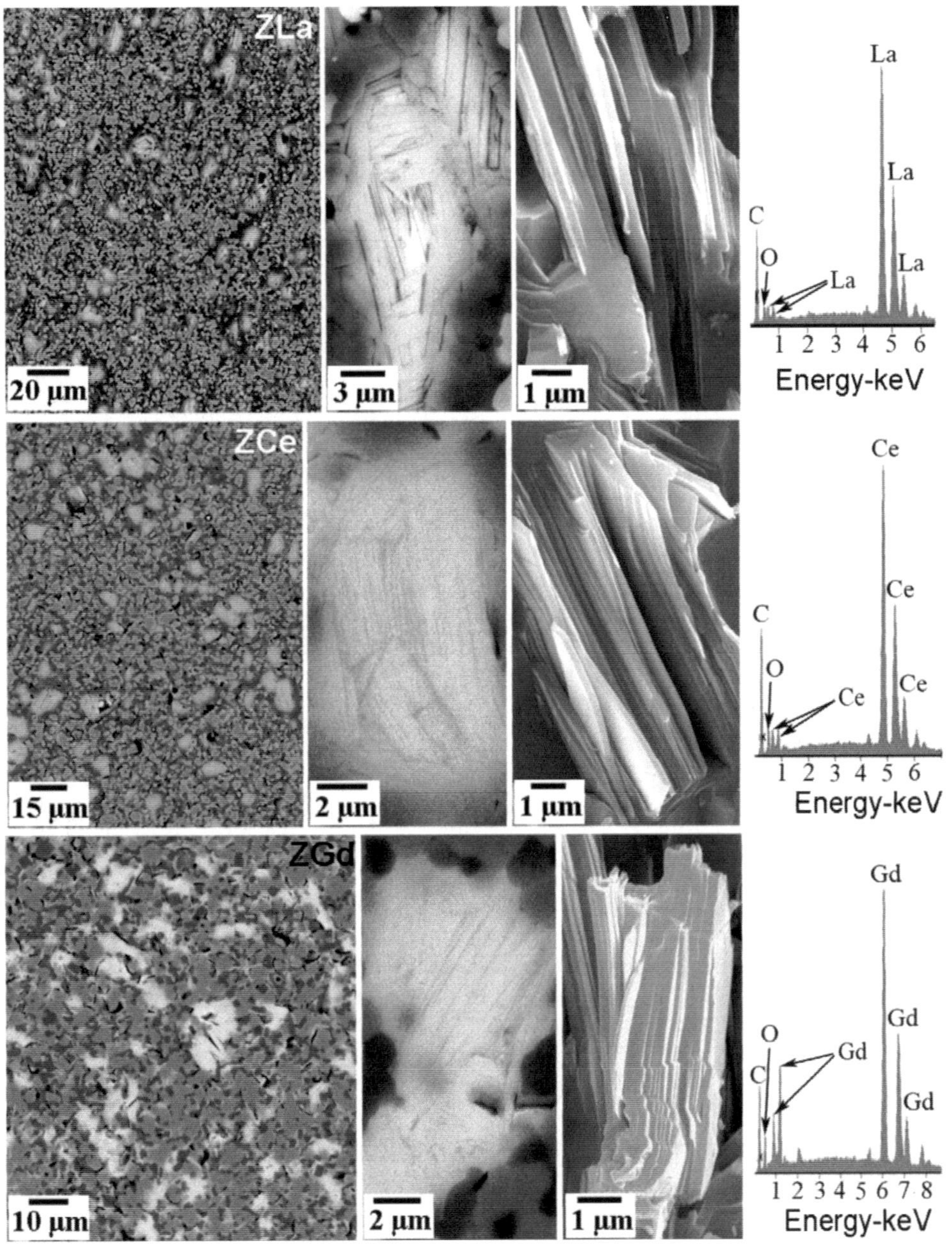

Figure 2. Low-magnification surface SEM images of the resulting ceramics, and the corresponding surface SEM images, fracture surface SEM images, and EDS analysis for the layered phase.

4. Conclusions
We found some layered phases in the composites prepared by reactive hot pressing using lnB_6 (Ln = La, Ce, Gd) and ZrC as raw materials. The layered phases are related to rare-earth metal borocarbides. Further work still needs to do. First is the crystal structure of these layered phases. Second is whether the rest of lanthanide hexaboride can react with ZrC to form layered phase. They are perhaps from a big family. Third is the reason why these layered phases are relatively and chemically stable in moisture environment.

References
[1] Barsoum M W, Yoo H I, Polushina I K, Rud V Y, Rud Y V and El-Raghy T 2000 *Phys. Rev. B* **62** 10194-8.
[2] El-Raghy T, Zavaliangos A, Barsoum M W and Kalidindi S R 1997 *J. Am. Ceram. Soc.* **80** 513-6.
[3] Barsoum M W, Farber L and El-Raghy T 1999 *Metall. Mater. Trans. A* **30** 1727-38.
[4] Grieseler R, Hähnlein B, Stubenrauch M, Kups T, Wilke M, Hopfeld M, Pezoldt J and Schaaf P, 2014 *Appl. Surf. Sci.* **292** 997-1001.
[5] Mauchamp V, Yu W, Gence L, Piraux L, Cabioc'h T, Gauthier V, Eklund P and Dubois S 2013 *Phys. Rev. B* **87** 235105.
[6] Ingason A S, Petruhins A, Dahlqvist M, Magnus F, Mockute A, Alling B, Hultman L, Abrikosov I A, Persson P O Å and Rosen J 2013 *Mater. Res. Lett.* **2** 89-93.
[7] Ingason A S, Mockute A, Dahlqvist M, Magnus F, Olafsson S, Arnalds U B, Alling B, Abrikosov I A, Jorvarsson B, Persson P O Å and Rosen J, 2013 **110** 195502.
[8] Barsoum M W. Prog 2000 *Solid State Chem.* vol **28** 201-81.
[9] Guitton A, Van P S, Tromas C, Joulain A and Van S 2014 *Appl. Phys. Lett.* **104** 241910.
[10] Wang C X, Yang T F, Xiao J R, Liu S S, Xue J M, Wang J Y, Huang Q and Wang Y G 2015 *Acta Mater.* **98** 197-205.
[11] Zhang Y H, Franke P, Seifert H J and Wang J Y 2015 *J. Am. Chem. Soc.* **98** 2570-2578.
[12] Bidaud E, Hiebl K, Bauer J, Rogl P 1998 *J. Alloys Compd.* **266** 1-6.
[13] Li F, Zhou Y, He L, Liu B, Wang J 2008 *J. Am. Ceram. Soc.* 91 2343-8.
[14] Wang J, Zhou Y, Liao T and Lin Z 2006 *Appl. Phys. Lett.* **89** 021917-1-3.
[15] Wiitkar F, Halet J F, Saillard J Y, Rogl P and Bauer J 1994 *Inorg. Chem.* **33** 1297-1305.
[16] Pyzik A J, Beaman D R 1995 *J. Am. Ceram. Soc.* **78** 305-12.
[17] Sea-Hoon L, Hidehiko T 2009 *J. Am. Ceram. Soc.* **92** 2172-4.
[18] Jardin C, Oeckler O, Mattausch H, Simon A, Halet J F, Saillard J Y and Bauer J 2000 *Inorg. Chem.* **39** 5895-900.
[19] Oeckler O, Jardin C, Mattausch H, Simon A, Halet J F, Saillard J Y and Bauer J Z 2001 *Anorg. Allg. Chem.* **627** 1389-94.
[20] Bidaud E, Hiebl K, Hoffmann R D and Pottgen R 2000 *J. Solid State Chem.* **154** 286-95.
[21] Yahia M B, Roger J, Rocquefelte X, Gautrer R, Bauer J, Guerin R, Saillard J Y and Halet J F 2006 *J. Solid State Chem.* **179** 2779-86.
[22] Zhao L, Jia D, Yang Z Duan X and Wang Y 2010 *Scripta Mater.* **63** 887-90.
[23] Zhao L, Jia D, Yang Z and Duan X 2011 *J. Am. Ceram. Soc.* **94** 3648-3650.
[24] Anderson J S, Clark N J and McColm I J 1968 J. *Inorg. Nucl. Chem.* **30** 105-11.
[25] Pelino M, Gingerich A K, Nappi B, Haque R 1992 *J. Phys. Chem.* **80** 839-46.
[26] Bowman A L, Krikorian N K, Arnold G P, Wallace T C and Nereson N G 1968 *Acta Crystallogr. Sect. B: Struct. Sci.* **24** 459-60.
[27] Shi Y, Leithe-Jasper A and Tanaka T 1999 *J. Solid State Chem.* **148** 250-9.
[28] Fang C M, Bauer J, Saillard J Y and Halet J F 2007 *J. Solid State Chem.* **180** 2465-70.

IOP Conf. Series: Materials Science and Engineering **137** (2016) 012028 doi:10.1088/1757-899X/137/1/012028

Synthesis and simultaneous self-assembly of novel antibacterial polyurethanes

J H Duan[1], F Yin[1], G C Jiang*[1, 2]

[1]Tianjin University of Science and Technology, Tianjin 300222, China
[2]Department of Chemistry, Tsinghua University, Beijing 100084, China

E-mail: wangziqing1978@163.com

Abstract. Novel physically crosslinked polyurethane (PUII) based on isophorone diisocyanates (IPDI) was prepared by a conventional two step method. The chemical structures of the PUII were characterized by fourier transform infrared (FTIR), proton nuclear magnetic resonance (1H-NMR), gel permeation chromatography (GPC) and scanning electron microscopy (SEM). The PUII hydrogels were subjected to solvent-induced self-assembly in THF + water to construct a variety of morphologies. The self-assembly morphology of the PUII was observed by scanning electron microscopy (SEM). Different amounts (0.2%, 0.4%, 0.6%, 0.8%, 1.0%) of 1,3,5-Tris(2-hydroxyethyl)hexahydro-1,3,5-triazine (TNO) was added as antibacterial agent to the polyurethane prepolymers. The inhibiting capacity of the antibacterial films to the Escherichia coli, Staphylococcus aureus, Bacillus subtilis and Gray mold has been studied. The inhibiting capacity of films for each strain effect became obvious with the increase of content of antibacterial agent and the sensitive degree to all kind of bacterial species was different.

1. Introduction

Polyurethanes (PUs) are one of the crucial classes of polymers with wide applications and properties, because their properties can easily be tailored by varying the components from which they are constructed [1]. Polyurethanes offer excellent versatility in terms of their mechanical properties and are widely applied in many fields. In general, the synthesis of polyurethanes is carried out in two stages. The first stage is the synthesis of a prepolymer containing free isocyanate groups and in the second stage there are reactions of chain extension or crosslinking of the polymer chains. PUs are composed of hard segments and soft segments. In the hard segment of polyurethane, ethyl urethane forms hydrogen bond, which is a kind of chemical bond. Hard segment structure determines the mechanical properties of polyurethane materials. However, the morphology and mechanical properties of polyurethane materials depend largely on the structure of the soft segment [2]. Self-assembled polyurethanes (PUs) are a special class of biodegradable and biocompatible materials which are currently used in tissue engineering and drug delivery applications [3, 4]. Because of the structural characteristics of polyurethane, it can be applied to many fields: for instance, coatings, adhesives, elastomers, sealants and plastics [5].

Traditionally, PUs are compound of diisocyanate monomers such as 4, 4-diphenylmethane diisocyanate (MDI) and polyatomic alcohol such as Polytetramethylene oxide (PTMG) [6-8]. The polyurethanes obtained with these diisocyanates are easily oxidized and this causes the color change of the products. Under ultraviolet light, a urethane bond is easily damaged and ethyleniminoquinone generates, so we always find that the color of PUs made by MDI turns yellow [9]. Compare with MDI, isophorone diisocyanates (IPDI) has excellent light stability and chemical resistance [10]. It is generally used in the manufacture of high-grade polyurethane resin.

Copolymer self-assembly behavior has been a hot research topic in recent years. By the structure or properties of different chain segments through chemical bond connected and block copolymers, microphase separation phenomenon will happen at certain conditions, and the formation of the various dimensions of nanometer or submicron structure will form. Self-assembly process and the formation of structure can be controlled by the chain segment of the chemical structure[11-13], chain length, the regulation of the connecting way [14,15], the category and function of the external force field[16,17]. In this study, a novel physically crosslinked polyurethane (PUII) by utilizing Isophorone diisocyanates was synthesized. In order to suppress interference of other substance, we synthesize PUII without dissolvent. The final products are characterized by FTIR, 1H NMR and SEM. Finally, the assembly behavior of PUII was investigated by SEM. The biocidal activity was examined by using the inhibition zone method.

2. Experimental

2.1. Materials

IPDI (Aladdin, Shanghai) was used without further purification. Polytetramethylene oxide (PTMG, Mw2000) (Aladdin, Shanghai) was dried at 90℃ in vacuum for 48 h. 1,4-Dihydroxybutane (BDO) (Chemicals, Tianjin, China) and tetrahydrofuran (THF) were used as received. 1,3,5-Tris(2-hydroxyethyl)hexahydro-1,3,5-triazine (TNO) was dewatered before use. All other reagents and solvents were used as received without further purification.

2.2. Preparation of PUII

PUII was prepared by a two-step addition polymerization. A 250 mL glass reactor equipped with a heating element, a mechanical stirrer, a charging and sampling port, and a nitrogen inlet and outlet was charged with polytetramethylene oxide 2000 (15 g, 0.0075 mol) and isophorone diisocyanates (4 g, 0.018 mol) . The reaction temperature was 60-70℃, and the reaction was continuous for 3 hours at this temperature. Then small amount of 1,4-Dihydroxybutane (BDO) as chain extender were added and allowed to react for 60 min. During the experiment, the characteristic peaks of the prepolymer were observed by using FTIR analyses until the peak (2265 cm^{-1}) disappeared. The PUII was injected with water methanol mixture precipitation (v/v 1:3) and dried in a vacuum oven at 30 ℃. Molecular weight: 7.9×104 g/mol (Mn). IR (figure 1): 3328 cm^{-1} (–NH), 2939 cm^{-1} (–CH), 1707 cm^{-1} (C=O), 1546 cm^{-1} (–CN), 1239 cm^{-1} (C–O–C, PUII), 1108 cm-1 (C–O–C, PTMG). 1H NMR (The spectrum was taken in $CDCl_3$, figure 2): δ=0.96-0.98 ppm (d, e, f , CH_3, IPDI)， 1.62-1.63 ppm (b, CH_2, PTMG), 3.06 ppm (g, CH2, IPDI), 3.40-3.41 ppm (c, CH_2O, PTMG), 4.06-4.07 ppm (a, OCH_2, PTMG), 7.29 ppm (h, NH, PUII).

2.3. Preparation of PUII with antibacterial agent

Different amounts(0.2%, 0.4%, 0.6%, 0.8%, 1.0%) of 1,3,5-Tris(2-hydroxyethyl)hexahydro-1,3,5-triazine (TNO) was added as antibacterial agent to the polyurethane prepolymers after BDO was superinduced and allowed to react for 60 min.

2.4. Characterization

IR spectra were recorded with a BRUKER-VECTOR22 FT-IR spectrophotometer using KBr disks. Nuclear magnetic resonance (1H NMR) spectra were recorded in $CDCl_3$ on a Bruker AVANCE III 400 MHz apparatus. By gel permeation chromatography (GPC) method for the determination of molecular weight and its distribution of polymer using water 515 pump and the 2410 type showed differential refractometer polystyrene gel column HT2, HT3, and 4 are connected in a serial way. THF was used as the eluent at a flow rate of 1.0 mL/min. Polystyrene standards with dispersity of 1.08-1.12

obtained from water were employed to calibrate the instrument. Scanning electron microscopy (SEM) measurements were done on a JEOL JSM-5600 LV SEM. The samples were fabricated by dropping one blob of the mixed solution of PUII, THF and water onto a glass patch, and then evaporated for 48 h under ambient temperature. Finally, the samples were coated with gold and observed by SEM.

3. Results and discussion

3.1. Synthesis of PUII

Novel physically crosslinked polyurethane (PUII) based on isophorone diisocyanates was prepared by a conventional two step method. The chemical structure of the final polymer (PUII) was characterized with FTIR in KBr matrices. The FT-IR spectra of PUII are shown in figure 1. As shown in figure 1, the peaks at 3328 cm^{-1} (–NH), 2939 cm^{-1} (–CH), 1707 cm-1 (C=O), 1546 cm^{-1} (–CN), 1239 cm^{-1} (C–O–C, PUII), 1108 cm^{-1} (C–O–C, PTMG) are observed, which demonstrate the synthesis is successful.

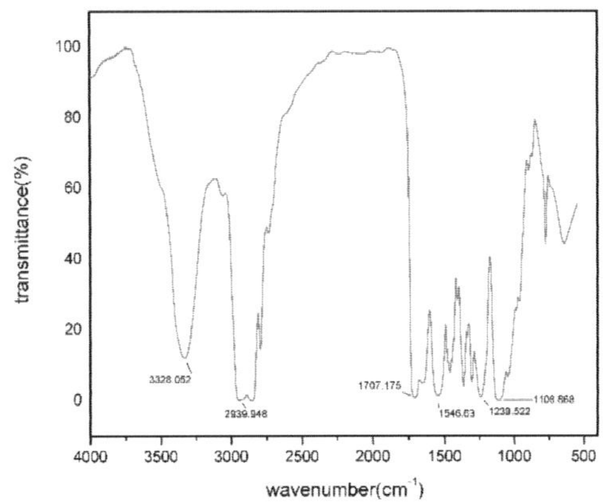

Figure 1. FTIR spectrum of the PUII.

1H NMR spectrum of the PUII was shown in figure 2. In the 1H NMR spectrum of PUII (figure 2), peak corresponding to the OCH2 (a) of the PTMG, was observed at 4.06-4.07 ppm. The peak related to the CH_2 (b) of PTMG appeared between 1.62 and 1.63 ppm. Other peaks were listed as follows: 3.40-3.41 ppm (c, CH_2O, PTMG), 0.96-0.98 ppm (d, e, f , CH_3, IPDI), 3.06 ppm (g, CH_2, IPDI), 7.29 ppm (h, NH, PUII). This confirms the successful incorporation of IPDI to the polyurethane backbone.

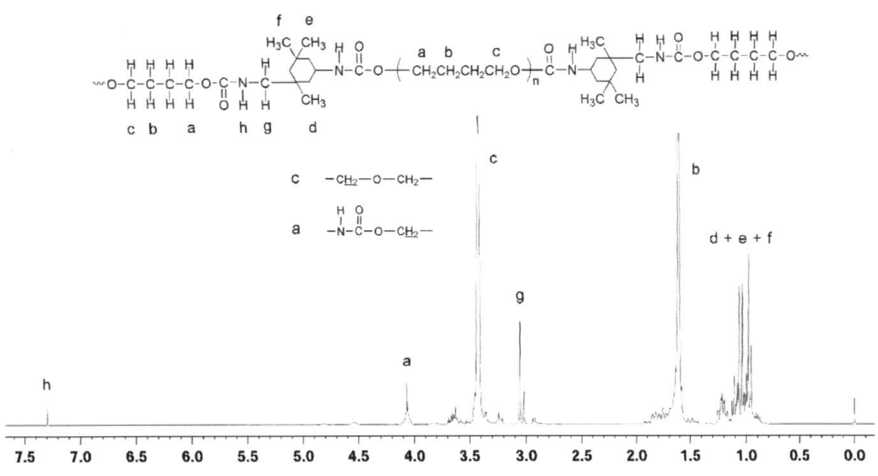

Figure 2. 1H NMR spectrum of the PUII.

GPC results indicate that the materials do not contain oligomers, and the PUII has molecular weight distributions (weight-average molecular weight of 72,000-79,000 and polydispersity of 1-2).

3.2. Self-assembly behavior of PUII

The self-assembly behavior of PUII was investigated by dissolving in the mixed solution of THF and water. Six samples (THF: water=90: 10/ 80: 20/ 70: 30/ 60: 40/ 50: 50/ 40: 60, v/v, corresponding serial number: 1-6) were prepared for SEM measurement. The samples were fabricated by dropping one blob of the mixed solution of PUII, THF and water onto a glass patch, and then evaporated for 48 h under ambient temperature. Finally, the samples were coated with gold and observed by SEM.

As shown in figure 3, (a)-(d), the SEM images of samples demonstrate that the polymer creates irregular globular morphologies. The average particle diameter was calculated by SEM images and the sizes range from 1μm to 7μm. The samples of number 6 and 1 (THF: water= 40: 60/ 90: 10) are correspondingly shown in figure 3(a) and figure 3(c). Figure 3(b) and figure 3(d) respectively reveal particulars of number 6 and 1. The space between irregular globular morphologies is gradually diminished when the quantity of water increases. So water quantity plays an important role in the self-assembly behavior. In this study, number 5 (THF: water= 50: 50) is critical value. When the ratio of THF/water is over this point, the aggregation becomes uncontrollable. By using scanning electron microscope techniques confirmed the good reproducibility of the microscopic material with solvent induced in the assembly process.

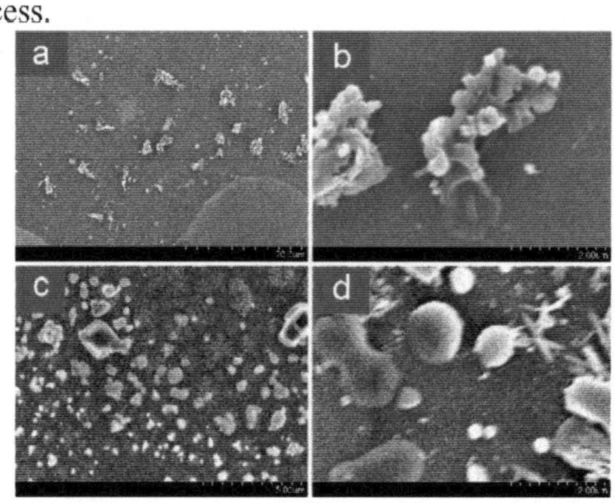

Figure 3. SEM images of PUII self-assembly behavior.

3.3. Antimicrobial properties

The inhibition zone method was employed. The PUII films with different amounts (0.2%, 0.4%, 0.6%, 0.8%, 1.0%) of 1,3,5-Tris (2-hydroxyethyl) hexahydro-1,3,5-triazine (TNO) were challenged with Escherichia coli, Staphylococcus aureus, Bacillus subtilis and Gray mold. As shown in Table 1. The results showed that when a small amount of antibacterial agent were added, the antibacterial effect of films on Gray mold was more obvious. When more antibacterial agent was added, the films showed better inhibitory effect on Staphylococcus aureus and Escherichia coli. In addition, the inhibitory effect on the Bacillus subtilis was not effective.

Table 1. The antibacterial circle diameter of different samples.

The content of antibacterial agent /%	The antibacterial circle diameter /mm			
	Escherichia coli	Staphylococcus aureus	Bacillus subtilis	Gray mold
0	0	0	0	0
0.2	14.1	10.2	10.3	17.8
0.4	18.1	11.7	11.9	18.5
0.6	19.3	18.3	13.8	19.1
0.8	20.9	21.9	15.5	19.9
1.0	23.2	22.1	17.1	22.2

4. Conclusions

In this work, a novel polyurethane (PUII) based on isophorone diisocyanates was synthesized with no dissolvant and characterized by FTIR, 1H NMR, GPC and SEM.

The self-assembly behavior of PUII also has been tested, and the results showed that the micron materials created irregular globular morphologies.

Antimicrobial property of the films contained antibacterial agent against Escherichia coli, Staphylococcus aureus, Bacillus subtilis and Gray mold was examined and showed different antibacterial properties.

Acknowledgements

The Project was supported by the Tribology Science Fund of State Key Laboratory of Tribology (SKLTKF14B04) and the China Postdoctoral Science Foundation (Nos. 20080430033 and 200902090).

References

[1] Xia Y, Zhang Z Y, Kessler M R, Stecher B, and Larock R C 2012 *J. Chem. Sus. Chem.* **5** 2221
[2] Li C Y, Chen J H, Chien P C, Chiu W Y, Chen R S and Don T M 2007 *J. Sci. Polym. Eng.* **47** 625
[3] Mishra A, Singh S K, Dash D, Aswal V K, Maiti B, Misra M and Maiti P 2014 *J. Acta. Biomaterialia* **10** 263
[4] Andreas L and Robert L 2002 *J. Sci.* **296** 1673
[5] Madbouly S A and Otaigbe J U 2009 *J. Prog. Polym. Sci.* **34** 1283
[6] Zhang Y, Wang L, Zhang Z, Zhang Y and Tuo X 2013 *J. Polym. Sci.* **51** 2161
[7] Rogulska M, Kultys A and Podkoscielny W 2007 *J. Eur. Polym.* **43** 1402
[8] Chang J G, Tuo X L and Wang D R 2012 *J. Mater. Sci: Mater. Med.* **23** 1877
[9] Singh R P, Tomer N S and Bhadraiah S V 2001 *Polym. Degrad. Stab.* **73** 443
[10] Thapliyal B P and Chandra R 1990 *Prog. Polym. Sci.* **15** 735
[11] Marsat J N, Heydenreich M, Kleinpeter E, Berlepsch H V, BooTtcher C and Laschewsky A 2011 *J. Macromol.* **44** 2092
[12] Jenekhe S A and Chen X L1999 *J. Sci.* **283** 372
[13] Li X M and He J D 2015 *J. Acta. Polym. Sinica* **1** 41
[14] Winey K I, Thomas E L and Fetters L J 1991 *J. Chem. Phys.* **95** 9367
[15] Li Z, Hillmyer M A and Lodge T P 2006 *J. Langmuir* **22** 9409
[16] Amundson K, Helfand E, Davis D D, Quan X, Patel S S and Smith S D 1991 *Macromol.* **24** 6546
[17] Xu T, Goldbach J T and Russell T P 2003 *Macromol.* **36** 7296

The effect of surface morphology on Model-I fracture toughness of carbon fiber reinforced titanium laminates

Z M Zheng, L Pan[1], L X Duan, Y Z Shen, Y B Hu, A Aamir, S Bhuwan and J Tao

College of Material Science and Technology, Nanjing University of Aeronautics and Astronautics, Nanjing 210016, China

E-mail: bettypan@nuaa.edu.cn

Abstract. The present study is focused on the relation between the microscopic sinusoidal surface morphology and model- I fracture toughness of carbon reinforced titanium laminates, based on cohesive elements. The interface toughness was computed as a function of geometric parameters of the interface texture. The results suggest that the toughness is increased when wavelength (λ) increase, which provides the need to design fracture/failure resistance materials by carefully selecting the suitable parameters of the interface texture.

1. Introduction

Fiber-metal laminates (FMLs) are commonly used in engineering field due to their interesting mechanical properties. However, for relatively weak interfaces, stiffness and strength of composite were significantly decreased after low velocity impacts. The damage may not be visible with optical observation, retaining detrimental consequences, unless proactive action is taken.

As a dissimilar material, the interfacial failure is an important issue in designing, manufacturing and applications of laminates. A common effective method is to modify the interface by sand-blasting, etching or laser ablation [1-4], which introduces a certain microstructure and increase fracture/failure resistance. However, the quantitative relation between dissimilar materials has not been reported since random textures were created by those methods. Additionally, the scientists, find that various creatures also make full use of the interface texture from their shell, bond to plant stem [5-10] in its body in order to natural environment. In fact, the texture, such as the sinusoidal found in pearls and shell, tend to decrease the interfacial energy, which proves an increase in fracture toughness [11-12]. Also, there are many methods, such as lithography, to control this micro-pattern on the metal-polymer surface. In addition, delamination based on numerical simulation developed for 30-years. The method named the cohesive zone model (CZM) is considered as the most effective one. A number of studies have been focused on simulating of initial crack and extension with cohesive interface elements. C. Balzani and W. Wagner [13], studying a computerized simulation, utilize a cohesive model included both exponential model and linear softening model and, the results suggest that the exponential model is especial fit for FMLs delamination numerical simulation due to the improved convergence behavior.

In the present study, the effect of geometrical parameters on a sinusoidal interface was investigated, between metal (Commercially Pure titanium, Ti) and carbon fiber reinforced PEEK composite (CF/PEEK), through Finite Element (FE) simulations.

2. Materials and methods

The grade TA2 at thickness of 2 mm was considered as mental-base and the CF/PEEK composite was used as the reinforcing material.

To study the pattern effect of interface toughness between metal and polymer, a kind of commercial photoresist (SUN-1150P from SUNTIFIC, CHINA) was covered on the polished Ti surfaces, and then exposed to Ultra Violet through a patterned mask. The part of photoresist film which was exposed to Ultra Violet was developed to have a number of line patterns, whose width include 100, 200 and 300 μm using different masks. After exposing, in order to make micro-patterns on the Ti surface (figure 1), the part of the un-covered Ti surfaces of the Ti, developing solution was chemically etched in a mixed solution nitric acid and hydrofluoric acid. After that, 10 piles of pregreg stacked on the Ti surface and put onto a hot press in a certain pressure and temperature together. The ASTM D5528-01 standard [14], which has been well established for double cantilever beam (DCB), gives the preparation and characterizations details of DCB test.

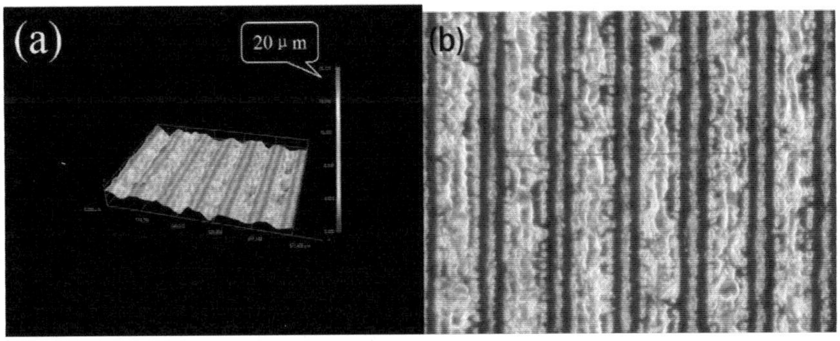

Figure 1. (a) 3D profile of the Ti interface after etching; (b) the surface coating with the photoresist.

3. Numerical modeling

A bi-linear cohesive zone law considered as the constitutive model of the cohesive elements describing the relationship between traction (σ) and displacement jump (δ) at the interface where crack develops, which has a fundamental assumption that the initiation and evolution of interfacial failure experiences an initially linearly elastic behavior is employed in the present work.

Two TA2 panels and a CF/PEEK are bonded a along sinusoidal interface. For the sinusoidal interface, variables, such as the material properties, the applied load and geometric parameters, have significant effect on the critical interface toughness (G_{IC}). However, for the TA2/CF/PEEK laminate, the geometrical parameters which describe the characteristic of the sinusoidal patterns on the interface of Ti determine G_{IC} as the others are constant. The sinusoidal interface is described by $y(x) = A\sin\left(\frac{2\pi x}{\lambda}\right)$. Here, the initial crack tip position defined by Cartesian coordinate system (x, y = 0). For $x < 0$, the interface suffers an artificial crack, while for $x \geq 0$, the interface is bonded and cohesive elements are inserted along sinusoidal interface while are used for simulating the interfacial delamination. In the case that the initial shape of the crack is defined by a straight line, having the initial artificial crack placed in the transition zone between a smooth and a sinusoidal part regarded this case is handled as the sinusoidal interface.

The specimens experience large deformation where geometrically and material non-linear finite elemental analysis should be considered. Otherwise, the materials experience both elastic deformation and plastic deformation. A von Misses isotropic plasticity model is used to control yielding.

Finite element software was employed for all simulations all simulation. The four-node is oparametric quadrilateral plane element, Plane182, was employed for solid materials. The relatively weak sinusoidal interface between Ti and CF/PEEK was represented by incorporating cohesive elements which are controlled by the bilinear softening model. These elements which carry the normal and shear separation force between two solids describe weak interface. The contact pair incorporated with cohesive elements was inserted along the sinusoidal.

4. Results and discussion

The way of crack development by gradual stable growth, showed in figure 2(a), has three stages generally. In the first stage, the crack maintains its original position while stress concentration occurs. In the second stage, the stable crack occurs along the sinusoidal interface. In the third stage, the crack propagates rapidly when the load increases. At the same time, a typical cure describing the relationship both loading and displacement which is from DCB test, showed in figure 2(b), suggests that the interface failure mechanism is a slow transition from linear-elastic to elastic-plastic behavior as the loading going on when the crack development from the stable stage to unstable stage.

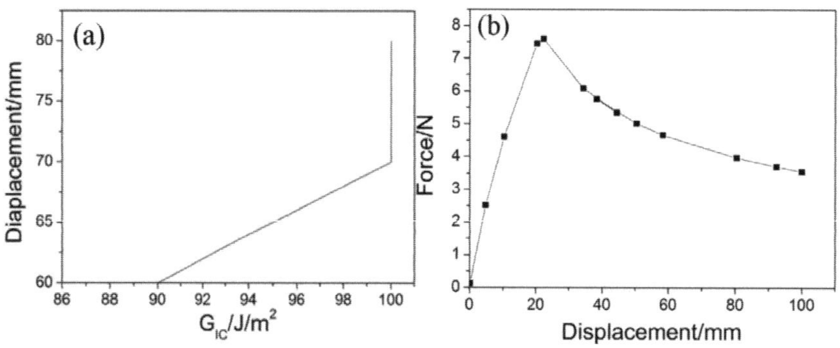

Figure 2. (a) Evolution of the crack propagation along the sinusoidal interface, showing increasing G_{IC}; (b) a typical cure describe the relationship both loading and displacement from DCB test.

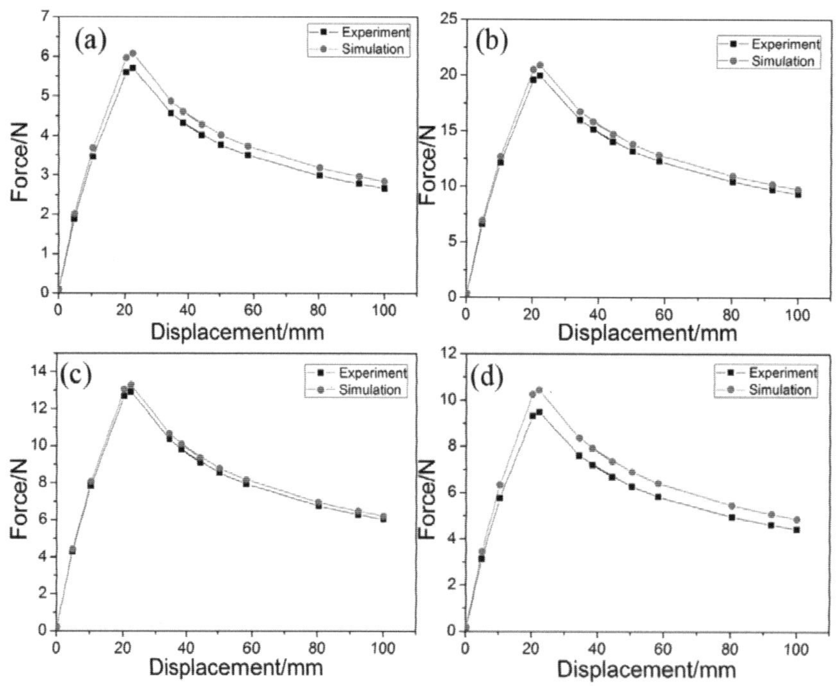

Figure 3. The results obtained from both experiment and numerical simulation of DCB of the specimens with sinusoidal interface: (a) λ =0 μm, (b) λ=100 μm, (c) λ =200 μm and (d) λ =300 μm.

The typical cure describing the relationship both loading and displacement, indicate that the critical load is increased when the λ is increased and are in a good agreement with the calculated curves, as presented in figure 3. Both results of the simulation and DCB test demonstrate that the sinusoidal interface plays an important role in the critical loading, which increases when the value of the λ is increased.

The highest toughness value(G) was obtained for a wavelength of 200μm, while the G values that were obtained for 100 μm or 300 μm, were above the one obtained for a smoother interface. In order to study the reasons of failure, the interfaces were inspected by optical microscope. As seen in figure 4(a), there is not residual adhesive, adjacent the excavated region. However, the preserved region on the surface of titanium has many residual adhesive, as showed in figure 4(b). For the wavelength of 200μm, there is not only residual adhesive but also the carbon fiber was failed, as presented in figure 4c. The experiments demonstrate that the interface toughness can be manipulated by sinusoidal interface, which change the mechanism of crack propagation from adhesive failure to cohesive failure.

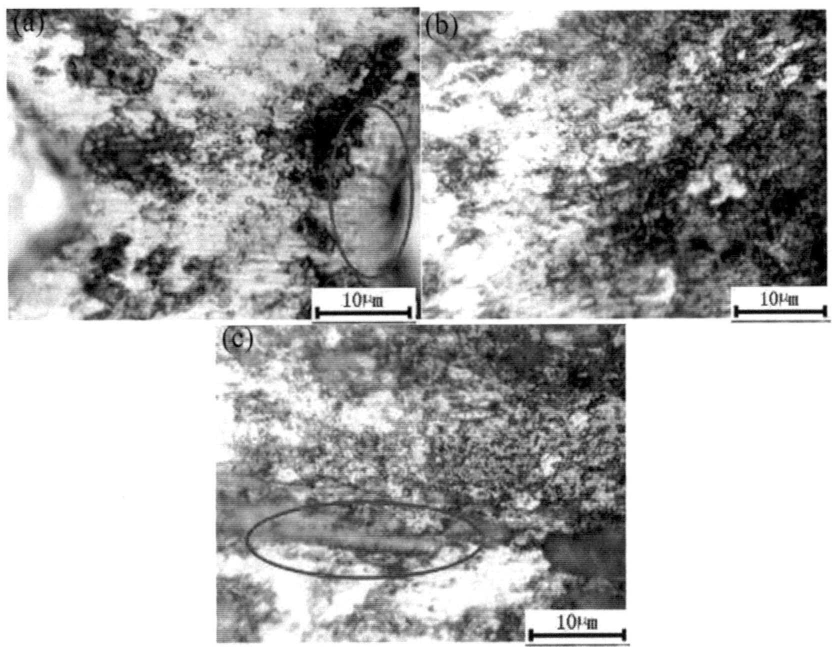

Figure 4. Surface of Ti at (a) wavelength with 300μm; (b) wavelength with 200μm.

Figure 5 shows the evolution of the stress distribution with the λ. For λ=100 μm, although many stress concentration fields were observed, the stress concentration field is relatively small, while the stress concentration field, adjacent to the peak of the sinusoidal interface, is relatively narrow, as can be observed in figure 5(b). The stress concentration field distribution presents a continuous morphology for λ = 0 μm and λ = 300 μm, as shown in figure 5(a) and 5(d), respectively. However, for a wavelength of 200 μm, the stress concentration field distribution demonstrated approximately seven discrete discontinuous peeks, while the total volume area is higher, compared to the area presented for a wavelength of 100 μm. Also, the stress distribution can affect polymer deformation near the interface, as showed in figure 4. The stress concentration field distributions indicate that the interface, in which a wavelength of 200 μm was employed, demonstrates a higher efficiency in energy dispersion when the crack propagation takes place along the sinusoidal interface.

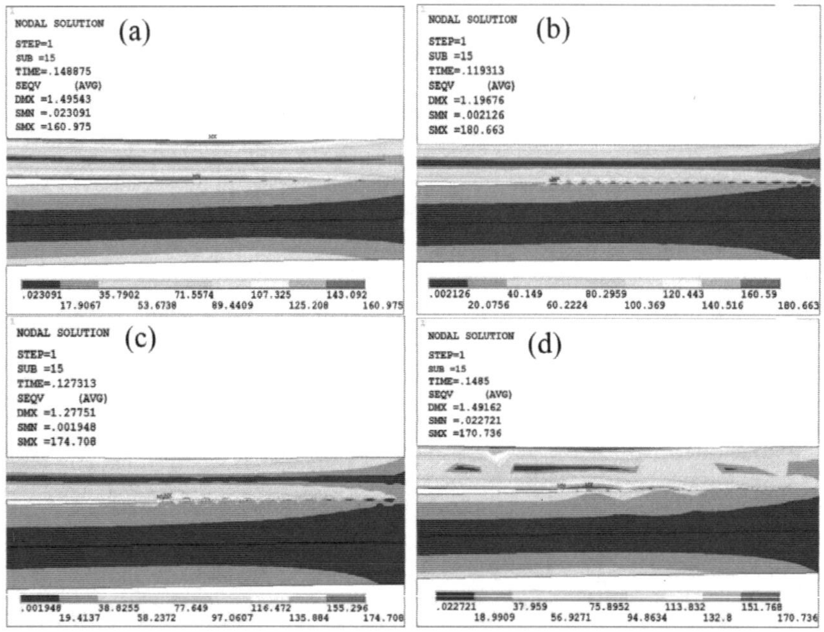

Figure 5. The stress distribution along with the interface for (a) λ=0μm, (b) λ=100 μm, (c) λ=200 μm and (d) λ=300μm.

5. Conclusions

The objective of the current work was to analyze the relationship between the interface toughness and the non-planar geometry characteristics. A sinusoidal interface was inserted between Ti and CF/PEEK. The major findings of the present study include that the interface toughness can be changed by selecting of the wavelength and amplitude, which supply suggestions for designing and manufacturing high interfacial fracture toughness FMLs.

Acknowledgments

This work was supported by the National Natural Science Foundation of China (Grant No. 51503099); the National Shipbuilding Project (New type of liquefied natural gas (LNG) ship cargo containment system research in advance) and a Project Funded by the Priority Academic Program Development of Jiangsu Higher Education Institutions.

References

[1] Jennings C W 1972 Surface roughness and bond strength of adhesives *J. Adhes.* **4** 25-38.
[2] Mulville D R and Vaishnav R N 1975 Interfacial crack propagation. *J. Adhes.* 7 15-233.
[3] Baburaj E G, Starikov D and Evans J 2007 Enhancement of adhesive joint strength by laser surface modification.2001 *Int. J. Adhes. Adhes.* **27** 268-276.
[4] Sancaktar E and Gomatam R 2001 A study on the effects of surface roughness on the strength of single lap joints *J. Adhes. Sci. Technol.* **15** 97-117.
[5] Barthelat F, Tang H, Zavattieri P D, Li C. M and Espinosa H D 2007 On the mechanics of mother-of-pearl: a key feature in the material hierarchical structure *J. Mech. Phys. Solids.* **55** 306-337.
[6] Wang R Z, Suo Z, Evans A G, Yao N and Aksay I A 2001 Deformation mechanisms in nacre. *J Mater Res.* **16** 2485-2493.
[7] Dunlop J W C, Weinkamer R and Fratzl P 2011 Artful interfaces within biological materials *Mater. Today* **14** 70-78.
[8] Espinosa H D, Juster A L, Latourte F J, Loh O Y, Gregoire D and Zavattieri P D 2011 Tablet-

IOP Conf. Series: Materials Science and Engineering **137** (2016) 012029 doi:10.1088/1757-899X/137/1/012029

level origin of toughening in abalone shells and translation to synthetic composite materials *Nature Commun.* **2** 509-527.
[9] Rafsanjani A, Derome D, Wittel F K and Carmeliet J 2012 Computational up-scaling of anisotropic swelling and mechanical behavior of hierarchical cellular material. *Compos. Sci. Technol.* 72 744-751.
[10] Gibson L J 2005 Biomechanics of cellular solids *J. Biomech.* **38** 377-99.
[11] Syn C J and Chen W W 2008 Surface morphology effects on high-rate fracture of an aluminum/epoxy interface *J. Compos. Mater.* **42** 1639-58.
[12] Kim W S, Yun I H, Lee J J and Jung H T 2010 Evaluation of mechanical interlock effect on adhesion strength of polymer–metal interfaces using micro-patterned surface topography. *Int. J. Adhes. Adhes.* **30** 408-417.
[13] Balzani C and Wagner W 2008 An interface element for the simulation of delamination in unidirectional fiber-reinforced composite laminates. *Eng. Fract. Mech.* **75** 2597-2615.
[14] D-A, A. 2001 Standard test method for mode i interlaminar fracture toughness of unidirectional fiber-reinforced polymer matrix composites *ASTM Standard #D5528-94a*.

IOP Conf. Series: Materials Science and Engineering **137** (2016) 012030 doi:10.1088/1757-899X/137/1/012030

The effects of thermally reversible agents on PVC stability properties

J Wang[1,4], J Yao[2], X H Xiong[1], C X Jia[1], R Ren[1], P Chen[1] and X M Liu[3]

[1] Liaoning Key Laboratory of Advanced Polymer Matrix Composites Manufacturing Technology, Faculty of Aerospace Engineering, Shenyang Aerospace University, Shenyang 110136, China

[2] Shenyang Rui Da Machinery Export and Import co., Ltd, Shenyang, 110000, China

[3] School of Textile & Material Engineering, Dalian Polytechnic University, Dalian 116034, China

E-mail: wally9932@126.com

Abstract. One kind of thermally reversible cross-linking agents for improving PVC thermally stability was synthesized. The chemical structure and thermally reversible characteristics of cross-linking agents were investigated by FTIR and DSC analysis, respectively. FTIR results confirmed that the cyclopentadienyl barium mercaptides ($(CPD\text{-}C_2H_4S)_2Ba$) were successfully synthesized. DSC results showed it has thermally reversible characteristics and the depolymerization temperature was between 170 ℃ and 205 ℃. The effects of cross-linking reaction time on gel content of Poly(vinyl chloride) compounds was evaluated. The gel content value arrived at 42% after being cross-linked for 25 min at 180 ℃. The static thermally stability measurement proved that the thermally stability of PVC compounds was improved.

1. Introduction

Polyvinyl chloride (PVC) is a kind of good comprehensive performance resin. The production is second only to polyethylene resin. It has been used in many fields with its excellent properties [1, 2]. However, since its poor heat resistance (softening point is 80℃), deformation resistance, abrasion resistance, electric breakdown resistance and poor mechanical strength is not enough higher shortcomings, limit its application.

Recently, many researchers have focused on PVC modification through physical (filling, blend) and chemical (copolymerization, grafting and cross-linking) modification methods [3-8]. It is well known that PVC thermal decomposition is from active chlorine structure began to emerge the HCl autocatalytic continuous degradation process, so try to remove or replace the active chlorine can effectively improve the strength of the PVC and heat resistance. The cross-linking of PVC is an effective method to enhance it [9-16]. Moreover, through the thermally reversible cross-linking method for modification of PVC, maybe improve the mechanical properties, environmental stress cracking resistance and reduce the permanent deformation products, moreover, the thermal stability performance of PVC may be enhanced through the introduction of sulfhydryl groups [17-19].

One kind of thermally reversible cross-linking agent for improving PVC thermal properties was synthesized in this work. Its chemical structure and thermally reversible characteristics were tested by FTIR and DSC, respectively. The influence of reaction time on gel content of flexible PVC and amount with cyclopentadienyl barium mercaptides ($(CPD-C_2H_4S)_2Ba$) on thermal stability of flexible PVC were evaluated.

2. Experimental

2.1. Materials
All the materials used in this work has been described by Wang and Fang[19].

2.2. Synthesis procedure

2.2.1. Synthesis of cyclopentadiene (CPD). Dicyclopentadiene (DCPD) (70 ml) was added in 100 ml three-mouth flask. The whole reaction system was filled with nitrogen gas for protection. Heated the reactants to boiling and then collected the boiling point at 41~42 ℃ fraction. 60 ml colorless transparent liquid, cyclopentadiene (CPD) was obtained for 2 h and kept in ice water bath during the splitting decomposition process. The anhydrous magnesium sulfate was added in the bottle.

2.2.2. Synthesis of cyclopentadienyl sodium (CPD-Na). NaOH (73 g, 1.8 mol) was separately added in two 250 ml suction flasks connected by Y type three. Then 135 ml THF was slowly added with magnetic stirring. The two suction flasks were placed in ice bath. The newly prepared 60 ml CPD was injected into the two suction flasks. Meanwhile, Y type three ways while the other end was connected to a magnetic rotor 500 ml suction flask, preparing for the next reaction. The purple red liquid, cyclopentadienyl sodium (CPD-Na) was obtained for 17 h. The whole reaction was under ice bath and with nitrogen protection conditions.

2.2.3. Synthesis of cyclopentadienyl ethyl bromide ($CPD-C_2H_4Br$). 1, 2- dibromoethane ($C_2H_4Br_2$) (63 ml) was added into a 500 ml suction flask. Then, the obtained CPD-Na was slowly poured into the suction bottle with magnetical stirring. After reacting at the room temperature for 0.5 h, the mixture was heated to 50 ℃ for 8 h. Then separated the product and washed oily liquid with deionized water for several times. After that added a small amount of ether oil, fractions were collected and then purified the oil liquid.

2.2.4. Synthesis of cyclopentadienyl ethyl mercaptan ($CPD-C_2H_4SH$). Thiourea (CH_4N_2S) (34 g, 0.45 mol) and Benzyltrim-ethyl-ammonium chloride ($C_{10}H_{16}N{\cdot}Cl$) (0.3 g, 0.0016 mol) were dissolved in 140 ml deionized water with heating and stirring. Cooled to 50~60 ℃ , slowly dripped cyclopentadienyl ethyl bromide ($CPD-C_2H_4Br$) with stirring. After reacting for 4 h, the compound was cooled and then added NaOH (30%) water solution. The product was washed with deionized water until neutral.

2.2.5. Synthesis of cyclopentadienyl barium mercaptides ($(CPD-C_2H_4S)_2Ba$). The cyclopentadienyl ethanethiol salt, cyclopentadienyl barium mercaptides ($(CPD-C_2H_4S)_2Ba$) was prepared as follows: a concentration of barium hydroxide aqueous solution was prepared in advance and then slowly dropped cyclopentadienyl ethyl mercaptan under stirring. Until the PH value was neutral, the final product was filtered off, and then cleaned using deionized water and dried under vacuum at 40 ℃ to a constant weight.

2.3. PVC compounds specimen preparation
The composition of flexible PVC compound was given in Wang and Fang [19]. It was pointed that there was 1 Phr organic tin in the compositions without thermally reversible cross-linking agents.

2.4. Chemical structure analysis by FTIR

The structure of cross-linking agent and derivatives were measured on a Perkin-Elmer FTIR (coating or KBr) at 4 cm^{-1} resolution and 32 times scanning.

2.5. Thermally reversible characteristics measured by DSC

DSC test was run on a TA 910 DSC from 23 °C to 300 °C at heating rate with 10 °C /min.

2.6. Gel content testing

The method has been described by Wang and Fang[19].

2.7. Static thermal stability

The Congo red method was used to test PVC thermal stability properties. The method was as follows, the cross-linked PVC composite specimen were cut into small pieces of 2 mm×2 mm at first, and then loaded into the test tube. When the glycerol was heated at 200 ℃, put the test tube in the oil bath. The time for the color of Congo red test paper changing to the equivalent of PH of 3 is recorded.

3. Results and discussion

3.1. FTIR

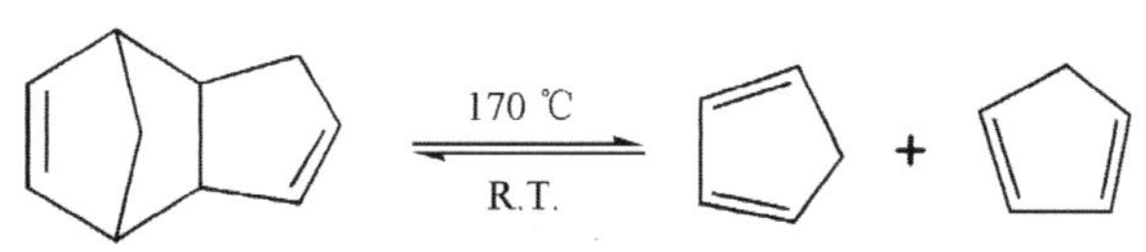

Figure 1. Diels-Alder reaction.

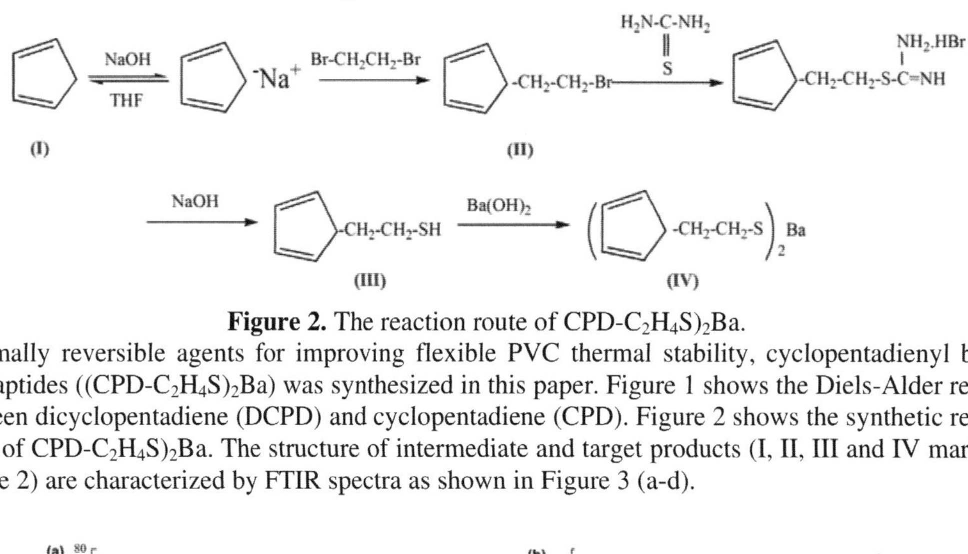

Figure 2. The reaction route of CPD-$C_2H_4S)_2Ba$.

Thermally reversible agents for improving flexible PVC thermal stability, cyclopentadienyl barium mercaptides ((CPD-$C_2H_4S)_2Ba$) was synthesized in this paper. Figure 1 shows the Diels-Alder reaction between dicyclopentadiene (DCPD) and cyclopentadiene (CPD). Figure 2 shows the synthetic reaction route of CPD-$C_2H_4S)_2Ba$. The structure of intermediate and target products (I, II, III and IV marked in Figure 2) are characterized by FTIR spectra as shown in Figure 3 (a-d).

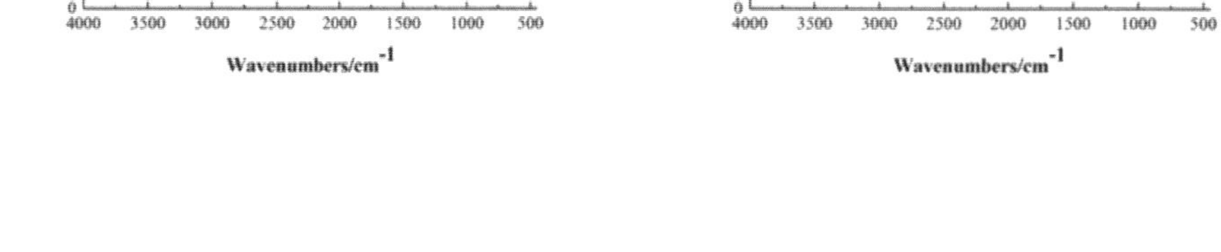

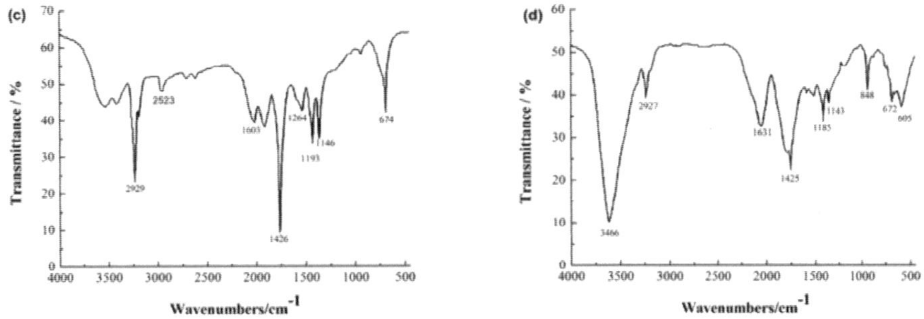

Figure 3. FTIR spectra of (a) CPD (b) CPD-C_2H_4Br (c) CPD-C_2H_4SH (d) (CPD-$C_2H_4S)_2Ba$.

From the FTIR spectra, it was found that the absorption peaks at about 3104 cm^{-1}, 2914 cm^{-1} and 1364 cm^{-1} in Figure 3(a) were attributed to -CH=CH- bond, -CH_2- bond and C-H bond of cyclopentadiene, respectively, which represent the major chemical structure of CPD. As can be seen from Figure 3(b), a new peak at 588 cm^{-1} which was associated with C-Br bond stretching vibration appeared. The other characteristic peaks of CPD still existed, which showed that CPD-C_2H_4Br was obtained. From Figure 3(c), it could be seen that a new peak at 674 cm^{-1} assigning to C-S bond and another new peak at 2523 cm^{-1} belonging to S-H bond appeared. It proved that CPD-C_2H_4SH was obtained. And these two peaks changed as seen from Figure 3(d), one peak at 2523 cm^{-1} disappeared and meantime the other peak at 674 cm^{-1} turned to be blue shift which proved S-Ba is forming. From the above FTIR analysis, it is confirmed that the target product was successfully synthesized.

3.2. DSC

Figure 4 illustrates the DSC curves of (CPD-$C_2H_4S)_2Ba$. The result shows that there exist three endothermic peaks in Figure 4. The peak at 90 ℃ belongs to the water in (CPD-$C_2H_4S)_2Ba$. The peak at 137 ℃ belongs to the melting points of (CPD-$C_2H_4S)_2Ba$. The endothermic peak between 170 ℃ and 205 ℃ is assigned to the depolymerization temperature of (CPD-$C_2H_4S)_2Ba$.

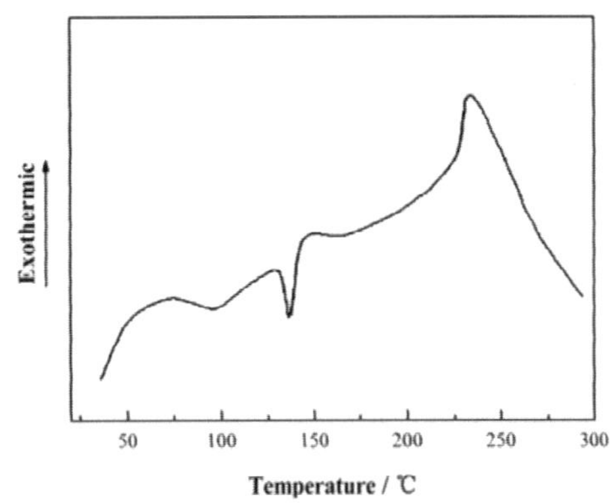

Figure 4. DSC curves of cyclopentadienyl barium mercaptides ((CPD-$C_2H_4S)_2Ba$).

3.3. Influence of reaction time on gel content for PVC

Figure 5 illustrates the results of gel content under different cross-linking time for flexible PVC with 3 Phr (CPD-$C_2H_4S)_2Ba$ at 180 ℃. It is found that the cross-linking time has obvious influence on gel content of PVC. The value is below 10% for the cross-linking time less than 15 min at this reaction temperature. The gel content value reaches at 15% for 20 min cross-linking reaction and it increases to 42% at maximum for 25 min. However, the gel content decreases for prolonging the reaction time for 35 min.

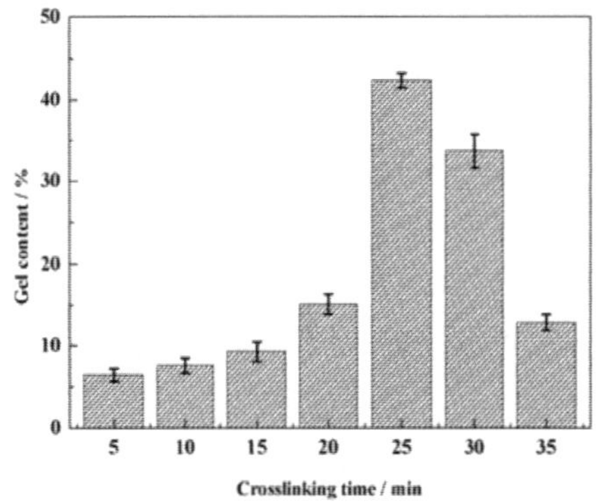

Figure 5. Gel content changes with different reaction time.

3.4. Influence of the amount of agents on thermal stability of flexible PVC

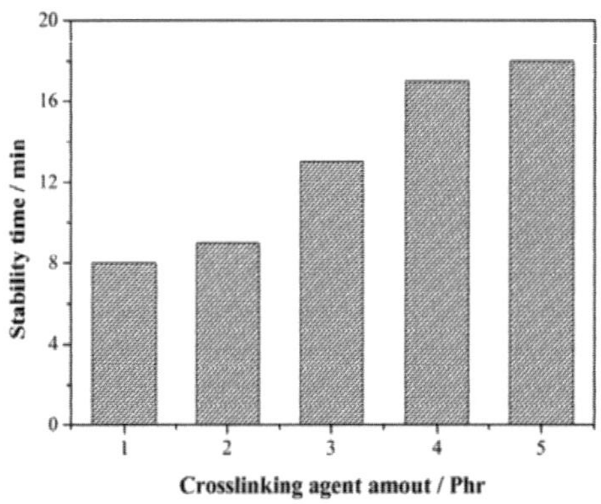

Figure 6. Static stability time with different agent amount for flexible PVC.

It is well known that the static thermal stability time is the reflection of PVC compounds thermal stability. Figure 6 shows the results of stable time of flexible PVC with different agent amount. The result shows that the stable time increases with adding agent amount from 1 Phr to 5 Phr. it is proved that the agent can react with the active chlorine in PVC structure and then enhance the thermal stability of PVC.

4. Conclusions

The cyclopentadiene derivatives, cyclopentadienyl barium mercaptides ($(CPD\text{-}C_2H_4S)_2Ba$) was successfully synthesized. The chemical structure of intermediate and target products were characterized by FTIR. It proved that the target products were obtained. From the DSC analysis, it is found that the cyclopentadiene derivatives have thermally reversible temperature. The gel content value arrived at 42% after being cross-linked for 25 min at 180 ℃. The cross-linking agent amount has strong effects on PVC compounds static stability.

References

[1] Parnell S and Min K 2005 *Polymer* **46** 3649
[2] Al-Ghamdi A A, Al-Salamy F, Al-Hartomy O A, Al-Ghamdi A A, Abdel Daiem A M and El-Tantawy F 2012 *J. Appl. Poly. Sci.* **124** 1144
[3] Kosuda T, Okada T, Nozaka S, Matsuzawa Y, Shimizu T, Hamanaka S and Mishima S 2012 *Polym. Degrad. Stabil.* **97** 584
[4] Xu J Z, Liu C H, Qu H Q, Ma H Y, Jiao Y H and Xie J X 2013 *Polym. Degrad. Stabil.* **98** 1506
[5] Pan M W, Xing S N, Yuan J F, Wu J W and Zhang L C 2010 *Polym. Eng. Sci.* **50** 1085
[6] Chantara T, Kamaruddin S and Abdullah Z 2011 *Polym. Plast. Technol. Eng.* **50** 833
[7] Wanderson C A, Honda R Y, Kayama M E, Schreiner W H Cruz N C and Rangel E C 2003 *Plasmas Polym.* **8** 1
[8] Samira T and Khalil A M 2012 *Polym. Adv. Technol.* **23** 1394
[9] Tsuchiya Y and Endo K 2011 *Polym. Degrad. Stabil.* **96** 1321
[10] Basfar A A 2002 *Polym. Degrad. Stabil.* **77** 221
[11] Chakrabarti R and Chakraborty D 2008 *J. Appl. Poly. Sci.* **110** 170
[12] Ambrogi V, Brostow W, Carfagna C, Pannico M and Persico P 2012 *Polym. Eng. Sci.* **52** 211
[13] Han Y, Su L, Mao L X, Zhang L Q and Yue D M 2014 *Polym. Plast. Technol. Eng.* **53** 306
[14] Esmizadeh E, Naderi G, Ghoreishy M H R and Bakhshandeh G R 2011 *J. Polym. Eng.* **31** 83
[15] Raj J M and Ranganathaiah C 2009 *Polym. Degrad. Stabil.* **94** 397
[16] García-Castanéda C, Benavides R, Martínez-Pardo M E, Uribe RM CarrascoÁbrego H and Martínez G R 2010 *Phys. Chem.* **79** 335
[17] Li Y B, Chen X N, Zhang Y and Jiao S K 2000 *Acta. Polym. Sin.* **6** 696
[18] Dong Y Q, Xia Y Z and Chen X N 2002 *Acta. Polym. Sin.* **3** 374
[19] Wang J Fang W Yuan Y and Liu X M 2014 *Advan. Mater. Res.* **900** 3

IOP Conf. Series: Materials Science and Engineering **137** (2016) 012031 doi:10.1088/1757-899X/137/1/012031

Preparation and electrochemical performance of manganese porphyrin /titanate intercalated nanocomposite

Y Sun[1], J P Deng[1,3], Z Y Tu[1] and J J Ma[2]

[1]Department of Materials Science and Engineering, Xi'an University of Science and Technology, Xi'an 710054, Shaanxi, China

[2]Department of Chemical Engineering, Huaihai Institute of Technology, Lianyungang 222005, Jiangsu, China

E-mail: jj_xiaoqian@sina.com

Abstract. A new nanocomposite of manganese porphyrin/titanate (MnPP-Ti_4O_9) was prepared successfully by delamination/reassembling (DR) method, with layered titanate as the host material, manganese porphyrin (MnPP) as the guest material. The microstucture of MnPP-Ti_4O_9 was characterized by XRD, UV-Vis, SEM and TEM. MnPP molecules were closely tilted against the host nanosheets at an inclined angle of 42.7°, by simple geometric calculation. The electrochemical performance of the nanocomposite was measured using cyclic voltammetry (CV). It was found that MnPP-Ti_4O_9 modified glassy carbon electrode (MTGCE) showed desired electrochemical performance and good catalytic activity for oxygen reduction.

1. Introduction

Inorganic layered compounds have a two-dimensional spatial structure, of which the distance between neighbor layers is at the molecular level. There is a strong covalent bond inside the layer, while the layers are stacked together by the weak force. If the layers are electrically neutral, such as graphite, the force is Van der Waals' force; if charged, it turns to electrostatic force [1]. According to the structural features of the layer, many nanocomposites can be fabricated by inserting different kinds of functional guest molecules into the layer space by appropriate methods [2-4]. Although some methods, such as direct-reaction method, ion-exchange method and direct-embedding method have been reported to accomplish successful intercalations, the uses of them are limited due to either the high charge density of the host layers or the size of guest molecules or the host layer spacing, which may lead to a long reaction period or failing intercalation [5-6]. Delamination/reassembling (DR) method, which is a relatively new method, has been proved feasible in fabricating intercalated nanocomposite recently [7]. This method can effectively overcome the steric hindrance of the layer spacing, thus the reaction time is remarkably reduced, compared with the previous methods which may have a reaction period of several months. What's more, intercalated materials prepared by this method seem to show a larger specific surface area than those prepared by other methods, which may lead to a better catalytic performance.

Intercalated nanocomposite not only exhibits some advantages of both layered host and functional guest molecules, but also show some differences in catalytic, adsorption, electrical and magnetic properties, compared with single component, which has potential applications in providing the new material system of building photoelectric devices, chemical and biological sensors, and in the field of molecular recognition and catalysis [8-9]. Titanate is a class of layered compounds, having regular layer structure and properties of layer-swellable and ion-exchangeable, which makes it an ideal material as good support for functional guest molecules [10]. Metalloporphyrins are a well-known kind of naturally-occurring compounds that represent many important enzymes. In this paper, the nanocomposite of MnPP-Ti_4O_9 is fabricated by DR method with layered potassium titanate ($K_2Ti_4O_9$) as the host body, MnPP as the guest, and the electrochemical performance of the nanocomposite is studied.

2. Experimental

$K_2Ti_4O_9$ was synthesized by solid-state reaction at 800℃ for 20h with K_2CO_3 and TiO_2 in the mole ratio of 1:1. The product was acidified 5 times with 3M HCl for 12h. The protonated product ($H_2Ti_4O_9$) was washed with distilled water to remove Cl^{-1} and dried in the vacuum. The resulting product was treated with tetrabutylammonium hydroxide (TBAOH) to exfoliate titanate. $H_2Ti_4O_9$ (0.2g) was firstly dispersed evenly into the distilled water, then a 10 wt% TBAOH aqueous solution was added slowly to the suspension until pH value reached 9.7. The exfoliated product was obtained after stirring for 2 weeks at room temperature. The colloidal suspension was centrifuged at 2000rpm for 30minutes to remove the particles that were not exfoliated, and the colloidal suspension containing titanate nanosheets was collected. The addition of MnPP aqueous solution to the colloidal suspension resulted in immediate aggregation of the nanosheets as precipitates. The final product MnPP-Ti_4O_9 was obtained after washed with distilled water 3 times and dried in the vacuum at 60℃ overnight.

Powder X-ray diffraction (XRD, XRD-6100), scanning electron microscope (SEM, JSM-5600), transmission electron microscope (TEM, JEM-2010) and ultraviolet-visible spectroscopy (UV-Vis, UV-2550) were employed to the characterization of the products. The electrochemical performance was performed on CHI 660C electrochemical workstation, with a saturated calomel electrode as the reference electrode, a platinum electrode as the auxiliary electrode, and a glassy carbon electrode (GCE) as the working electrode. The MTGCE was prepared by an absorption method, with 10 μL MnPP-Ti_4O_9 aqueous suspension coated on a GCE and dried for at least 24h. All electrochemical tests were performed at 25℃.

3. Results and Discussion

3.1. Microstructure characterization

Figure 1(a) shows the XRD patterns of $K_2Ti_4O_9$, protonated product ($H_2Ti_4O_9$), and intercalated product MnPP-Ti_4O_9, respectively. It can be found that the (200) peak of $K_2Ti_4O_9$ shifts slightly to the lower angle side after the acidification process, with 2θ angle changing from 10.14° to 10.07°, corresponding to the increase of layer spacing. This is due to that H^+ is inserted into the layers in hydrated state, of which the ion volume is bigger than K^+. The delamination occurs after the addition of the TBAOH molecules, driven by the force of neutralization reaction. The reassembling process is considerably fast, and the intercalated deposition can be observed immediately when adding metalloporphyrin aqueous solution to the nanosheets suspension. After the process, the (200) peak significantly shifts to lower angle after the reassembling process, which suggests MnPP is inserted successfully into the layers and the distance between neighbor layers is further expanded. The thickness of titanate, as described previously, is 0.5nm and the size of MnPP molecule is $1.80\times1.80\times0.75$ nm^3 (estimated by Chem 3D for an optimized structure) [10]. With these datum, it can be speculated that MnPP molecules are closely tilted against the host nanosheets at a specific inclined angle whose value is 42.7°, by simple geometric calculation.

Figure 1(b) shows the UV-Vis absorption spectrum of homogeneous MnPP and the nonocomposite of MnPP-Ti_4O_9. It can be found, both have two absorption bands in the wavelength range from 300 to 800 nm, where strong absorption bands correspond to the Soret absorption band and weak absorption bands are Q band absorption bands. The maximum absorption wavelength of homogeneous MnPP in Soret absorption band is at 462nm, while that of MnPP-Ti_4O_9 changes to 480nm. This phenomenon may be caused for that the terminal substituents of MnPP are distorted due to the restriction of the layer spacing in nanoscale. As a result, the coplanar effect between substituents and the porphyrin rings increases, leading to the formation of the big π Bond, thereby reducing the energy level difference [11]. This further indicates that the MnPP molecules are tilt-arranged between the titanate layers, because only in this way can MnPP molecules be closely packed.

TEM and SEM images of the intercalated nanocomposite are shown in figure 2(a) and figure 2(b), respectively. It is obvious titanate still keeps the layer structure and the layers are still orderly-arranged after acidizing, exfoliating and resembling processes. These suggest MnPP molecules can be successfully inserted into titanate layers, while the two-dimensional spatial structure of the host body is retained. The structural model of MnPP-Ti_4O_9 nanocomposite is shown in figure 3, according to the analyses above.

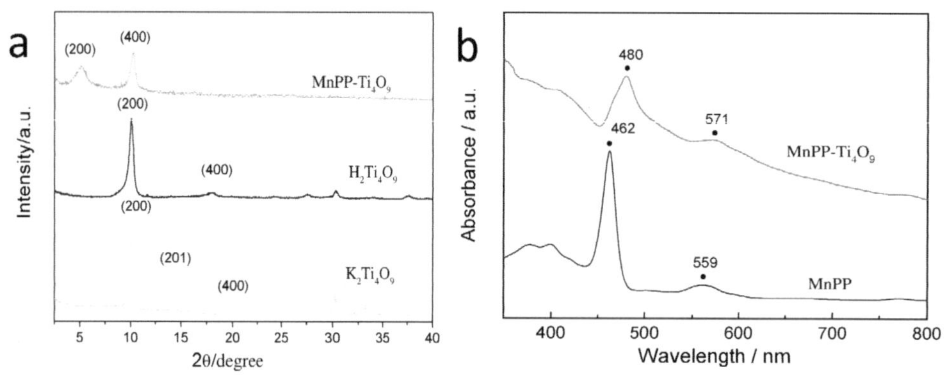

Figure 1. (a) XRD patterns of $K_2Ti_4O_9$, $H_2Ti_4O_9$ and MnPP-Ti_4O_9 and (b) UV-vis spectrum of pure MnPP and the nanocomposite of MnPP-Ti_4O_9.

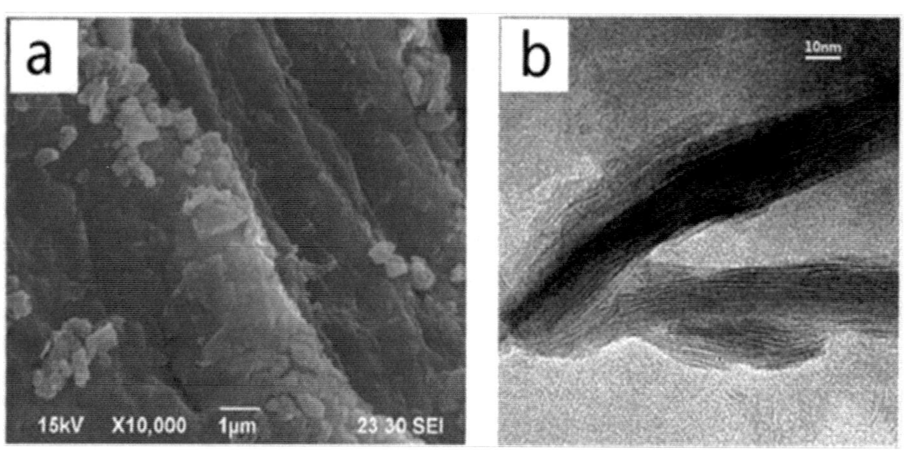

Figure 2. SEM (a), TEM(b) images of the nanocomposite of MnPP-Ti_4O_9.

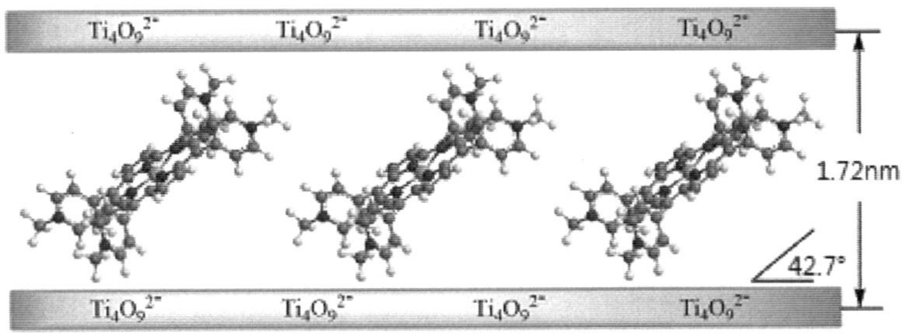

Figure 3. Model of the nanocomposite of MnPP-Ti_4O_9.

3.2. Electrochemical performance

The bare GCE is put into a 3×10^{-4} mol L^{-1} MnPP+0.1 mol L^{-1} KCl, N_2 saturated aqueous solution. CV tests are carried out in the voltage range of −0.5-0.5 V at the scanning rate of 50 mV s^{-1}. The results are shown in figure 4(a). The oxidation peak potential (E_{pa}) is located in −0.188V, while the reduction peak potential (E_{pc}) is in −0.258V, and the half maximum peak potential of E_m is −0.223V. The peak separation value of ΔE_p is 69 mV, which is greater than that of the single electron transfer theory value (59mV), suggesting the electrode reaction is quasi-reversible reaction, in accordance with the previously reported redox couple of $Mn^{III}PP^{5+}/Mn^{II}PP^{4+}$[12].

The MTGCE is fabricated by a absorption method. The CV curves of the MTGCE are shown in figure 4 (a). The electrode shows good redox activity, of which the waveform is similar to the bare GCE in homogeneous MnPP aqueous solution, suggesting it also has good redox properties. E_{pa} is located in −0.177V, while E_{pc} is in −0.256V which is almost identical with the bare GCE. The ratio of reduction peak current and oxidation peak current (I_{pc}/I_{pa}) is 1.9, showing a worse reversibility than homogeneous MnPP. This is due to the electron transfer process is somewhat hampered by host layers who possess semi-conductor properties. Both I_{pc} and I_{pa} of the MTGCE increase, which indicates that the concentration of MnPP between layer spacing is larger than that in the homogeneous MnPP aqueous solution, which results in a stronger electrochemical response current.

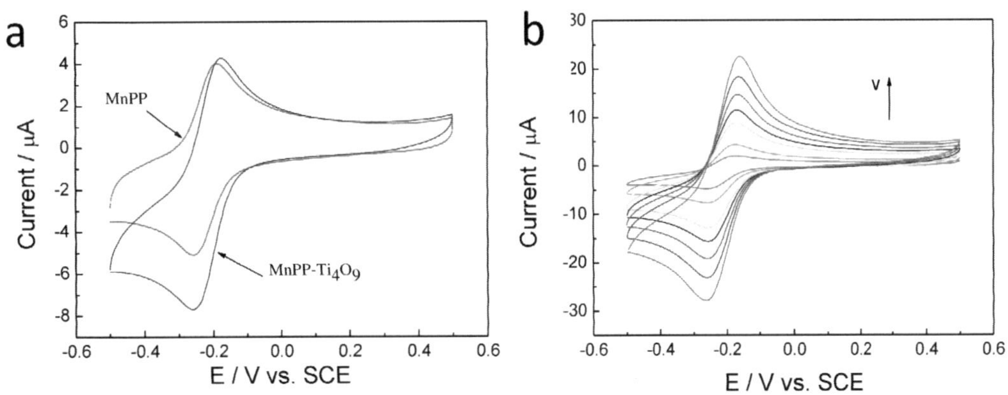

Figure 4. (a) CV grams of GCE in 3×10^{-4} mol L^{-1} MnPP + 0.1 mol L^{-1} KCl, N_2 saturated aqueous solution and MTGCE in 0.1 mol L^{-1} KCl, N_2 saturated aqueous solution, at a scanning rate of 50 mV s^{-1}, pH=7.0, T=25℃; (b) CV grams of MTGCE in 0.1 mol L^{-1} KCl, N_2 saturated aqueous solution, at the scanning rate: 50, 100, 150, 200, 250, 300, 350 and 400 mV s^{-1}, respectively, pH=7.0, T=25℃.

Further researches on the CV behaviors of the MTGCE at different scanning rates are shown in figure 4(b). With the increase of the scanning rate, the position of E_{pa} and E_{pc} gradually change to the positive and negative sides, respectively. Besides, both I_{pc} and I_{pa} are gradually increasing, and the value of ΔE_p has also shifted from 54 mV to 108 mV, which further indicates that the host layers with semi-conductor properties have a negative effect on the electron transfer process of the reaction, and the effect is extremely significant at a high scanning rate. Two straight lines can be obtained by drawing the relationship diagrams between the peak current（I_{pa} or I_{pc}）and the square root of the scanning rate. This means there is a linear relationship between peak current（I_{pa} or I_{pc}）and the square root of scanning rate, which can be described by two regression equations.

$$I_{pa}(\mu a) = -1.3941+0.05466v^{1/2}(mV\,s^{-1})^{1/2}) \quad (r=0.9832) \tag{1}$$

$$I_{pc}(\mu a) = -0.6327-0.05530v^{1/2}(mV\,s^{-1})^{1/2}) \quad (r=0.9921) \tag{2}$$

Where r is referred to correlation coefficient. This suggests that the electrochemical reaction of the active substance in the modified electrode is diffusion-controlled.

The stability of the MTGCE is studied on the same modified electrode for continuous 7 days, 40 times a day, using CV tests, in 0.1 mol L^{-1} KCl, N_2 saturated aqueous solution, at a scanning rate of 50 mV s^{-1}. Figure 5(a) is the relationship gram between peak potential and time while figure 5(b) is the relationship gram between peak current and time. It is obvious the peak potential almost has no change, while the peak current has only decreased by 8%, indicating that the nanocomposite can be well absorbed by the GCE and the nanocomposite has better stability than homogeneous metalloporphyrin which decompose easily in the oxidation process and has difficulty in recovery [13-15].

Based on the excellent electrochemical performance and stability of the MTGCE, the application of the MTGCE in the catalytic reduction of oxygen is researched using CV, in O_2 saturated phosphate buffer solution, the result is shown in figure 6. It is obvious that the reduction peak of the MTGCE has shifted positively by 260 mV, compared with the bare GCE in the same solution, what's more, it is observed that the peak has also sharpened and there is a upward trend in the reverse scanning curve. All of these suggest that the MTGCE has good catalytic activity for oxygen reduction. which results in a stronger electrochemical response current.

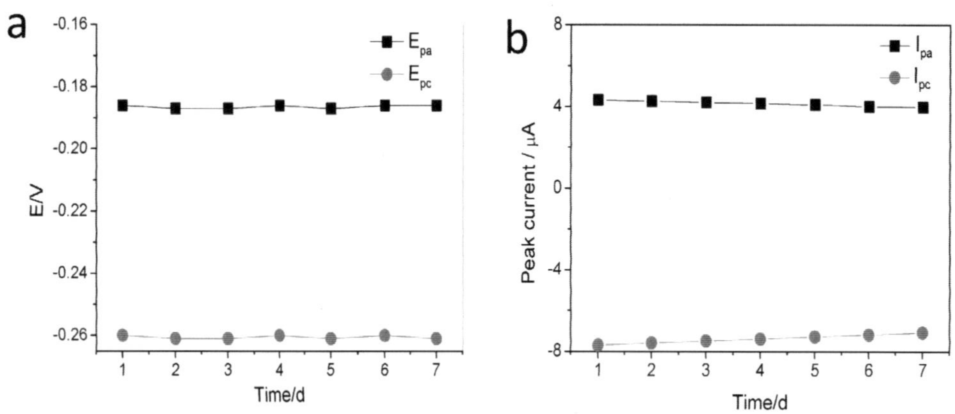

Figure 5 (a) The relationship gram between peak potential and time, and (b) the relationship gram between peak current and time

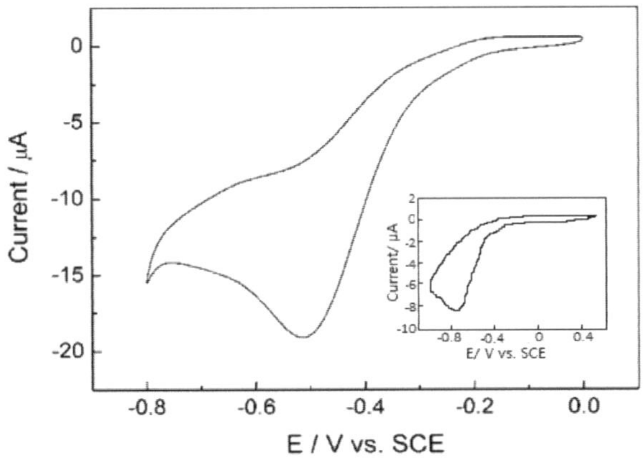

Figure 6. CV gram of MTGCE in O_2 saturated phosphate buffer solution, at a scanning rate of 50 mV s^{-1}, pH=7.0, T=25℃. Built-in gram is the CV test of bare GCE in O_2 saturated phosphate buffer solution, at a scanning rate of 50 mV s^{-1}, pH=7.0, T=25℃.

4. Conclusions

The nanocomposite of MnPP-Ti_4O_9 is fabricated by DR method. XRD, SEM and TEM images reveal that the insertion of MnPP into titanate layers is successful, and MnPP molecules show close tilt arrangement against the host nanosheets at an inclined angle of 42.7°, which is further supported by UV-Vis. CV tests show that titanate layer structure has litter impact on the electrochemical property of the nanocomposite and the electrochemical reaction is diffusion-controlled. What's more, the MTGCE has good catalytic activity for oxygen reduction and potential for applications in the field of electrochemical catalysis.

Acknowledgments

The research was financially supported by the Natural Science Foundation of Shanxi Province, China (2011JY007).

References

[1] Whittingham M S and Chianelli R R 1980 *J. Chem. Edu.* **57** 569.

[2] Takagi S, Eguchi M, Tryk D A and Inoue H 2006 *J. Photochem. Photobiol. C* 7 104.

[3] Demel J and Lang K 2012 *Eur. J. Inorg. Chem.* **2012** 5154.

[4] Takagi S, Shimada T, Ishida Y, Fujimura T Masui D and Tachibana H 2013 *Langmuir* **29** 2108.

[5] Chen Z J, Lin B Z and Chen Y L 2010 *J. Phys. Chem. Solids* **71** 841.

[6] Tong Z W, Takagi S and Shimada T 2006 *J. Am. Chem. Soc.* **128** 684.

[7] Ma J J, Yang M and Chen Y H 2015 *Mater. Lett.* **150** 122.

[8] Bizaia N, de Faria E H and Ricci G P 2009 *ACS Appl. Mater. Interfaces* **1** 2667.

[9] Soto-Oviedo M A, Araújo O A and Faez R 2006 *Synthetic Met.* **156** 1249.

[10] Ma J J, Wu J, Zheng J, Liu L, Zhang D E and Xu X Y 2012 *Mater. Lett.* **71** 4.

[11] Cady S S and Pinnavaia T 1978 *J. Inorg. Chem.* **17** 1501.

[12] Yin M Y, Yuan G L, Wu Y Q, Huang M Y and Jiang Y Y 1999 *J. J. Mol. Catal. A: Chem.***147** 93.

[13] Ferrand Y, Daviaud R, Le Maux P and Simonneaux G 2006 *Tetrahedron Asymmetry* **17** 952.

[14] Mirkhani V, Moghadam M, Tangestaninejad S, Mohammdpoor-Baltork I, Kargar H and Araghi M 2009 *Appl. Catal. A* **353** 61.

[15] Ghiaci M, Molaie F, Sedaghat M E and Dorostkar N 2010 *Catal. Commun.* **11** 694.

IOP Conf. Series: Materials Science and Engineering **137** (2016) 012032 doi:10.1088/1757-899X/137/1/012032

Computation of macro-fiber composite integrated thin-walled smart structures

S Q Zhang[1, 3], S Y Zhang[1], M Chen[2], J Bai[1] and J Li[1]

[1] School of Mechanical Engineering, Northwestern Polytechnical University, 710072 Xi'an, China
[2] Department of Industrial Design, Xi'an Jiaotong - Liverpool University, 215123 Suzhou, China

E-mail: sqzhang@nwpu.edu.cn

Abstract. Due to high flexibility, reliability, and strong actuation forces, piezo fiber based composite smart material, macro-fiber composite (MFC), is increasingly applied in various fields for vibration suppression, shape control, and health monitoring. The complexity arrangement of MFC materials makes them difficult in numerical simulations. This paper develops a linear electro-mechanically coupled finite element (FE) model for composite laminated thin-walled smart structures bonded with MFC patches considering arbitrary piezo fiber orientation. Two types of MFCs are considered, namely, MFC-d31 in which the d_{31} effect dominates the actuation forces, and MFC-d33 which mainly uses the d_{33} effect. The proposed FE model is validated by static analysis of an MFC bonded smart plate.

1. Introduction

Thin-walled metal structures integrated with smart materials, like piezoelectric, magnetostrictive, and shape memory alloys, called smart structures, have been attracted by many researchers and engineers in various fields for vibration suppression, acoustic control, and health monitoring. Due to high flexibility, reliability, and high actuation forces, macro-fiber composite (MFC) materials are more attractive in the applications of smart structures instead of piezoceramics and piezopolymers. MFC was first invented by NASA Langley Research Center [1] and now is produced by Smart Material Corporation [2]. The MFC material has very complex structural arrangement integrated with many different materials. MFC patches have five major layers with an active layer sandwiched between layers of adhesive, electrodes, and polyimide film. The active layer is manufactured by rectangular piezoceramic rods embedded in an epoxy matrix with interdigitated electrodes. Due to different polarization of piezo rods, there have two different types of MFCs, MFC-d31 and MFC-d33. For the patches of MFC-d31, the d_{31} effect dominates the actuation forces, while MFC-d33 patches mainly use the d_{33} effect.

The complexity of MFC materials increases the difficulty in modeling and computation of structures integrated with those materials. After the literature survey, it can be found that very few publications are available concerning the numerical analysis of MFC bonded smart structures. However, most of them were analyzed by commercial software, e.g. ANSYS [3, 4], ABAQUS [5, 6], with modified constitutive equations and using equivalent actuation forces. Beyond using the commercial software, Park and Kim [7], Bilgen et al. [8] developed analytical models respectively for MFC bonded active twist rotor blades and benders. Furthermore, based on the first-order shear deformation (FOSD) hypothesis, Azzouz and Hall [9] developed a von Kármán type nonlinear finite element model for frequency analysis of a rotating MFC actuator.

Based on the author's previous work [10], this paper presents a computational FE model using the FOSD hypothesis for MFC bonded structures with considering various piezo-fiber orientations. Using the proposed FE model, a thin plate integrated with either MFC-d31 or MFC-d33 patches is analyzed and compared.

2. Mathematical models

As mentioned before, there are two types of MFCs, in which MFC-d31 has the polarization in the thickness direction and MFC-d33 has the polarization parallel to the piezo fiber direction, as shown in Figure 1.

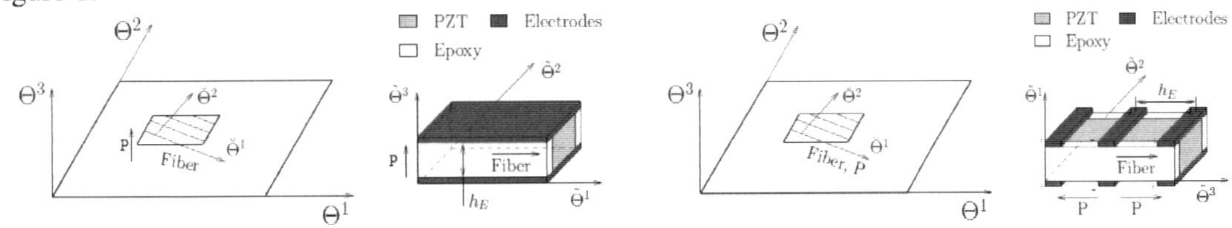

(a) MFC-d31 type (b) MFC-d33 type

Figure 1. Two types of MFC.

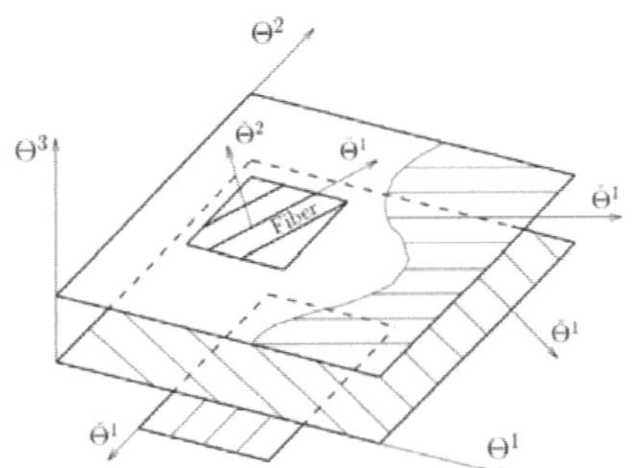

Figure 2. Multi-layer laminated composite structures with MFCs.

Using the homogenization method presented in [11, 12], MFC can be finalized as orthotropic materials including both elastic fiber and piezo fiber angles. Considering multi-layer laminated structures, as Figure 2 shown, the constitutive equations in the fiber coordinate system for each layer can be expressed as [10]

$$\breve{\boldsymbol{\sigma}} = \breve{c}\breve{\varepsilon} - \breve{e}^{\mathrm{T}}\breve{\boldsymbol{E}} \tag{1}$$

$$\breve{\boldsymbol{D}} = \breve{e}\breve{\varepsilon} + \breve{\chi}\breve{\boldsymbol{E}} \tag{2}$$

Here $\breve{\boldsymbol{\sigma}}$, $\breve{\varepsilon}$, $\breve{\boldsymbol{E}}$ and $\breve{\boldsymbol{D}}$ are respectively the stress, strain, electric field, and electric displacement vectors, the overhead □ represents the quantities in the fiber coordinate system. Furthermore, $\breve{c}$, $\breve{e}$ and $\breve{\chi}$ denote the elastic constant matrix, the piezoelectric constant matrix, and the dielectric constant matrix. The components in the piezoelectric constant matrix can be obtained by

$$\breve{e}_{p1} = \breve{d}_{p1}\frac{\breve{Y}_1}{1-\breve{v}_{12}\breve{v}_{21}} + \breve{d}_{p2}\frac{\breve{v}_{12}\breve{Y}_2}{1-\breve{v}_{12}\breve{v}_{21}} \tag{3}$$

$$\breve{e}_{p2} = \breve{d}_{p1}\frac{\breve{v}_{12}\breve{Y}_2}{1-\breve{v}_{12}\breve{v}_{21}} + \breve{d}_{p2}\frac{\breve{Y}_2}{1-\breve{v}_{12}\breve{v}_{21}} \tag{4}$$

where $\breve{Y}_1$ and $\breve{Y}_2$ are the Young's moduli, $\breve{v}_{12}$ and $\breve{v}_{21}$ are the Poisson's ratios. Furthermore, $\breve{d}_{p1}$ and $\breve{d}_{p2}$ are the piezoelectric coefficients, in which $p=1$ is for MFC-d33, $p=3$ is for MFC-d31.

In order to compute the elastic potential energy of each layer with various fiber orientations, the constitutive equations must be transformed to the structural coordinate system as [10]

$$\boldsymbol{\sigma} = \boldsymbol{c\varepsilon} - \boldsymbol{e}^{\mathrm{T}}\boldsymbol{E} \tag{5}$$

$$\boldsymbol{D} = \boldsymbol{e\varepsilon} + \boldsymbol{\chi E} \tag{6}$$

with

$$\boldsymbol{c} = \boldsymbol{T}^{\mathrm{T}}\breve{\boldsymbol{c}}\boldsymbol{T},\ \boldsymbol{e} = \breve{\boldsymbol{e}}\boldsymbol{T},\ \boldsymbol{\chi} = \breve{\boldsymbol{\chi}} \tag{7}$$

Due to the assumption of no thickness change after deformation, the stress, strain, electric field, and electric displacement vectors are obtained as

$$\boldsymbol{\sigma} = \begin{Bmatrix} \sigma_{11} \\ \sigma_{22} \\ \sigma_{12} \\ \sigma_{23} \\ \sigma_{13} \end{Bmatrix},\ \varepsilon = \begin{Bmatrix} \varepsilon_{11} \\ \varepsilon_{22} \\ 2\varepsilon_{12} \\ 2\varepsilon_{23} \\ 2\varepsilon_{13} \end{Bmatrix},\ \boldsymbol{E} = \begin{Bmatrix} E_p^{(1)} \\ E_p^{(2)} \\ \vdots \\ E_p^{(N)} \end{Bmatrix},\ \boldsymbol{D} = \begin{Bmatrix} D_p^{(1)} \\ D_p^{(2)} \\ \vdots \\ D_p^{(N)} \end{Bmatrix} \tag{8}$$

Assuming that an electric voltage is applied on two neighboring electrodes in the direction of polarization, the piezoelectric constant and the dielectric constant matrices are respectively expressed as [10]

$$\boldsymbol{e} = \begin{bmatrix} e_{p1}^{(1)} & e_{p2}^{(1)} & 0 & 0 & 0 \\ e_{p1}^{(2)} & e_{p2}^{(2)} & 0 & 0 & 0 \\ \vdots & \vdots & \vdots & \vdots & \vdots \\ e_{p1}^{(N)} & e_{p2}^{(N)} & 0 & 0 & 0 \end{bmatrix},\ \chi = \begin{bmatrix} \chi_{pp}^{(1)} & 0 & \cdots & 0 \\ & \chi_{pp}^{(2)} & \cdots & 0 \\ \vdots & \vdots & \ddots & \vdots \\ 0 & 0 & \cdots & \chi_{pp}^{(N)} \end{bmatrix} \tag{9}$$

Here superscript N stands for the number of layers.

Using the principle of virtual work one obtains the static equilibrium equations

$$\boldsymbol{K}_{\mathrm{uu}}\boldsymbol{q} + \boldsymbol{K}_{\mathrm{u}\phi}\boldsymbol{\phi}_{\mathrm{a}} = \boldsymbol{F}_{\mathrm{ue}} \tag{10}$$

in which $\boldsymbol{K}_{\mathrm{uu}}$ and $\boldsymbol{K}_{\mathrm{u}\phi}$ are the elastic stiffness and the piezoelectric coupled stiffness matrices, respectively, $\boldsymbol{q}$, $\boldsymbol{F}_{\mathrm{ue}}$, and $\boldsymbol{\phi}_{\mathrm{a}}$ represent the nodal displacement vector, the external force vector, and the actuation voltage vector. More details are referred to [10].

3. Numerical simulations

First, the present FE model is tested by a clamped plate bonded with MFC patches proposed by Bowen et al. [4]. The MFC plate is comprised of one aluminum layer in the middle as the host structure and two MFC-d33 (or MFC-d31) patches respectively bonded on the top and bottom surfaces acting as actuations, as shown in Figure 3. The dimensions of the aluminum plate are $300\times75\times2\ \mathrm{mm}^3$ and those of MFC patches (only active part) are $85\times57\times0.3\ \mathrm{mm}^3$. The material constants of the aluminum plate, MFC-d31, and MFC-d33 are shown in Table 1.

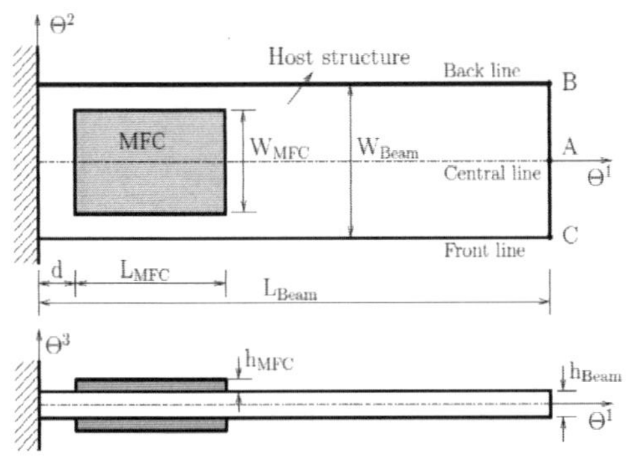

Figure 3. Clamped plate integrated with two MFC patches.

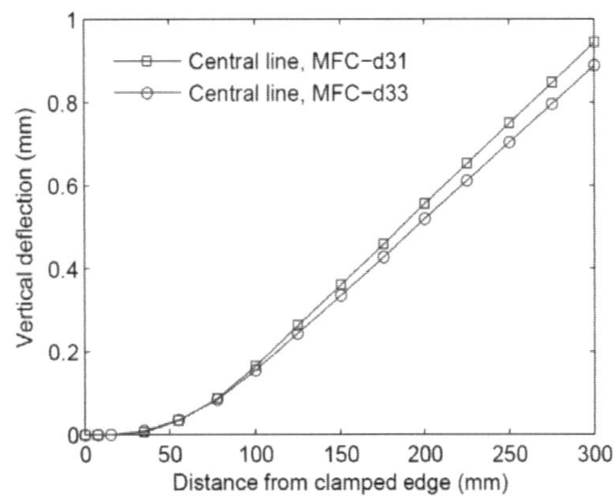

Figure 4. Central line displacements for MFC-d31 and MFCd33 bonded plates.

Table 1. Material Constants of aluminum and MFCs [2, 10].

Aluminum	MFC-d33	MFC-d31
$\breve{Y}$=70	$\breve{Y}_1 = 29.4$ GPa	$\breve{Y}_1 = 30.336$ GPa
$\breve{v} = 0.3$	$\breve{Y}_2 = \breve{Y}_3 = 15.2$ GPa	$\breve{Y}_2 = \breve{Y}_3 = 15.857$ GPa
	$\breve{G}_{12} = \breve{G}_{13} = 6.06$ GPa	$\breve{G}_{12} = \breve{G}_{13} = 5.515$ GPa
	$\breve{G}_{23} = 5.79$ GPa	$\breve{G}_{23} = 5.515$ GPa
	$\breve{v}_{12} = \breve{v}_{13}$=0.312	$\breve{v}_{12} = \breve{v}_{13}$=0.31
	$\breve{v}_{23}$=0.312	$\breve{v}_{23}$=0.438
	$\breve{d}_{11}$=467×10^{-12} m/V	$\breve{d}_{31}$=-170×10^{-12} m/V
	$\breve{d}_{12}$=-210×10^{-12} m/V	$\breve{d}_{32}$=-100×10^{-12} m/V
	$h_E = 0.5$ mm	$h_E = 0.18$ mm

An electric voltage of 400 V is applied on the top MFC patch for both cases of MFC-d31 and MFC-d33 bonded structures. The distances of two neighboring electrodes of MFC-d33 and MFC-d31 are respectively $h_E = 0.5$ mm and $h_E = 0.18$ mm, which leads to the electric fields along the polarization being equal to $E_1 = 400/0.5$ V/mm for MFC-d33 patches and $E_3 = 400/0.18$ V/mm for MFC-d31 patches. The vertical deflections at the central line of both cases are presented in Figure 4, where the piezo fiber angle is zero. From the figure, we can see that the central line deflection of the MFC-d31 bonded structure are quite close to those of the MFC-d33 bonded structure, even though the coefficients d_{11} and d_{31} are very different. This is because the elongation strains $\varepsilon_{11} = d_{11}E_1 = 3.736\times10^{-4}$ for MFC-33 and $\varepsilon_{11} = d_{31}E_3 = 3.778\times10^{-4}$ for MFC-31 are very close to each other, with the former one being slightly smaller than the latter one.

Under the same electric loading for the above two cases, considering MFC piezo fiber orientations of 0°, 30°, and 90°, results in different deformed shapes of the MFC plate, as can be seen in Figures 5-7. The results first show that the deformation of the MFC-d31 bonded plate does not change the sign, while the deformation of the MFC-d33 bonded plate changes the sign. This can be explained by the fact that the piezoelectric constants d_{31} and d_{32} of MFC-d31 have the same sign. Obviously, we can see that the twists of the MFC-d33 bonded plate are much larger than that of the MFCd31 bonded plate, due to the change of piezo fiber angles.

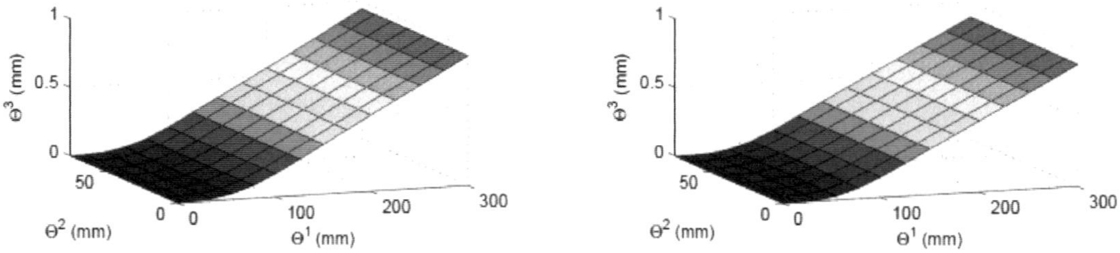

(a) MFC-d31 (b) MFC-d33

Figure 5. MFC piezo fiber orientation 0°.

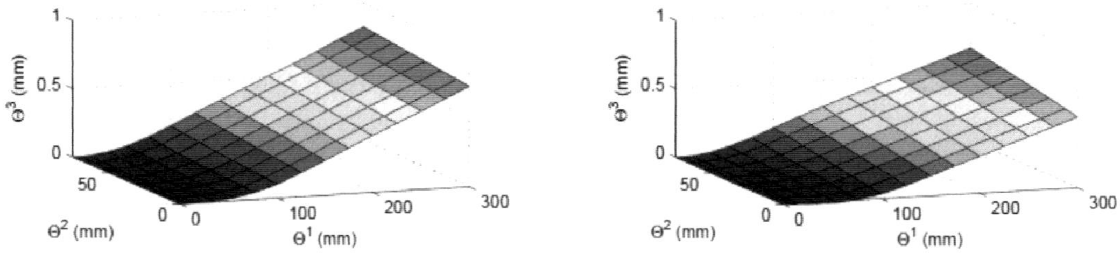

(a) MFC-d31 (b) MFC-d33

Figure 6. MFC piezo fiber orientation 30°.

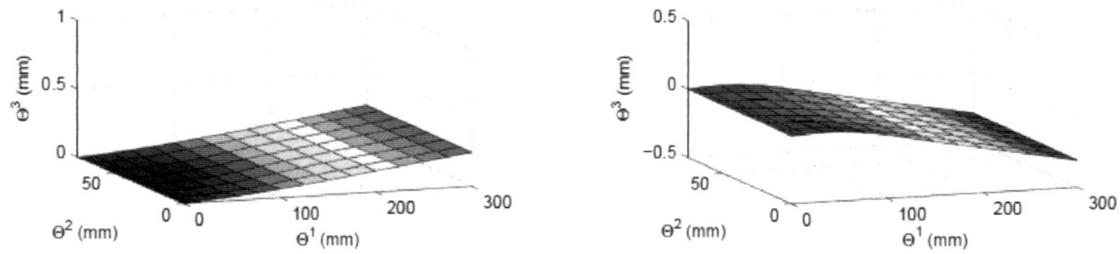

(a) MFC-d31 (b) MFC-d33

Figure 7. MFC piezo fiber orientation 90°.

4. Conclusions

This paper has developed a linear piezoelectric coupled finite element model for MFC smart structures using a 2-dimenssional FE method based on the FOSD hypothesis. An aluminum plate bonded with different MFCs is modeled and analyzed, namely MFC-d31 and MFC-d33. The present model has been applied to MFC-d31 or MFC-d33 patches bonded aluminum plates with different piezoelectric fiber orientations. The displacements and the deformation shapes have been simulated and compared with each other. The results showed that the MFC-d33 bonded plate has larger twist compared to the MFC-d31 plate. Furthermore, due to the different sign of $\breve{d}_{11}$ and $\breve{d}_{12}$ coefficients for MFC-d33, the aluminum plate bonded with MFC-d33 patches changes the sign of the displacements, while that with MFC-d31 does not.

Acknowledgements

This work was carried out under the financial support from the Research Foundation for Advanced Talents of Northwestern Polytechnical University - China, the Opening Fund of State Key Laboratory of Structural Analysis for Industrial Equipment, Dalian University of Technology - China (Grant No. GZ15212) and the "111 project" of China (Grant No. B13044).

References

[1] Wilkie W K, Bryant R G, High J W, Fox R L, Hellbaum R F, Jalink A, Little B D and Mirick P H 2000 *Proc. of SPIE's 7th Annual Int. Symp. on Smart Structures & Materials* **3991** 323-34 (Bellingham USA: SPIE).

[2] Smart Material Corporation: www.smart-material.com.

[3] Dano M L, Gakwaya M and Jullière B 2008 *J. Intel. Mat. Syst. Str.* **19** 225-33.

[4] Bowen C R, Giddings P F, Salo A I T and Kim H A 2011 *IEEE T. Ultrason. Ferr.* **58**(9) 1737-50.

[5] Ren L 2008 *Compos. Struct.* **83** 110-8.

[6] Binette P, Dano M L and Gendron G 2009 *Smart Mater. Struct.* **18** 025007.

[7] Park J S and Kim J H 2005 *Smart Mater. Struct.* **14** 745-53.

[8] Bilgen O, Erturk A and Inman D J 2010 *J. Vib. Acoust.* **132**(5) 051005.

[9] Azzouz M S and Hall C 2010 *Proc. of 51st AIAA/ASME/ASCE/AHS/ASC Structures, Structural Dynamics, and Materials Conf.* **AIAA 2010-2547**. (Reston USA: The American Institute of Aeronautics and Astronautics).

[10] Zhang S Q, Li Y X and Schmidt R 2015 *Compos. Struct.* **126** 89-100.

[11] Deraemaeker A, Nasser H, Benjeddou A and Preumont A 2009 *J. Intel. Mat. Syst. Str.* **20**(12) 1475-82.

[12] Li Y X, Zhang S Q, Schmidt R and Qin X S 2016 *J. Intel. Mat. Syst. Str.* (doi:10.1177/1045389X16633763).

TG-FTIR characterization of flame retardant polyurethane foams materials

W Liu[1], Y Tang, F Li, X G Ge and Z J Zhang

Sichuan Fire Research Institute of Ministry of Public Security, Chengdu 610036, Sichuan, China

E-mail: weiliu20141202@126.com

Abstract. Dimethyl methylphosphonate (DMMP) and trichloroethyl phosphtate (TCEP) have been used to enhance the flame retardancy of polyurethane foams materials (PUF). Flame retardancy and thermal degradation of PUF samples have been investigated by the LOI tests and thermal analysis. The results indicate that the excellent flame retardancy can be achieved due to the presence of the flame retardant system containing DMMP and TCEP. TG-FTIR reveals that the addition of DMMP/TCEP can not only improve the thermal stability of PUF samples but can also affect the gaseous phase at high temperature.

1. Introduction

Polyurethane foam (PUF) as an organic insulation material is of great interest to many consumers due to its excellent properties including good mechanical properties, low cost and effective thermal insulation [1-3]. However, like some other organic materials, the flammability of PUF limits its use in many areas. During the period of 2000-2010, many fire originated from PUF insulation materials has been reported in China. And the investigation from these accidents suggests that the hazard during combustion of the PUF can be caused by not only the high burning heat but also the release of the volatile products formed during its thermal degradation at high temperature, which could induce asphyxia, coma or death of people.

In our research, dimethyl methylphosphonate (DMMP) and trichloroethyl phosphtate (TCEP) have been used as the flame retardants containing phosphorus and halogen to improve the flame retardancy of the PUF samples. Many previous literatures have reported that DMMP and TCEP are the traditional flame retardants to enhance the thermal stability of materials at high temperature range. However, the influence of these flame retardants on the pyrolysis products in the gaseous phase during thermal degradation of PUF has been rarely mentioned. In this study, flame retardancy and thermal stability of PUF containing DMMP and TCEP have been investigated. TG-FTIR analysis has been used to characterize the change of the pyrolysis products in the gaseous phase during heating of the samples here. The aim of this work is to investigate the thermal degradation characteristics of the PUF samples and the influence of DMMP and TCEP on the gaseous phase at high temperature range.

2. Experimental

2.1. Materials

Polyether polyol (Model: 4110, 635) and polyaryl polymethylene isocyanate (PAPI) (Model: GR-4110G) were purchased from Shanghai Gaoqiao Petro Co, China. Cyclohexylamine used as the catalyst and dichlorofluoroethane (HCFC-141b) used as the foaming agent were obtained from Changzheng Chemical Reagent Company, China. Dimethyl methylphosphonate (DMMP) and trichloroethyl phosphtate (TCEP) were supplied from Jinan Enter Chemical Co., Ltd.

2.2 Preparation
Polyurethane foam samples were prepared by the cast molding method. Polyether polyol was mixed with cyclohexylamine, HCFC-141b, DMMP, TCEP, water together, and then stirred with an electric stirrer at the agitation speed of 1500 r/min until the solution was uniform. Then PAPI was added into the solution and stirred for 90s with the speed of 500 r/min. Thereafter, the solution was cast into a mold and then heated at 70℃for 48hr. Finally, polyurethane foam samples were taken out from the mold for using.

2.3 Measurements

2.3.1 LOI. The limiting oxygen index (LOI) tests were carried out on a JF-3 type instrument (Jiangning, China), according to GB/T 2406-1993.

2.3.2 TG-FTIR. Thermogravimetric analysis-infrared spectrometry (TG-FTIR) of the PUF samples was performed using DT-50 instrument that was interfaced to IRAffinity-1 FTIR spectrometer. About 5 mg of the PUF sample was put in an alumina crucible and heated from 25 to 800℃ in nitrogen atmosphere with the heating rate of 10℃/min.

3. Results and discussion

3.1 Flame retardancy
LOI values of polyurethane samples are presented in table 1. It indicates that LOI value of the PUF samples increase with flame retardant concentration, the % P value and the % Cl value. The higher the % P value and the % Cl value, the higher the LOI value. LOI values of all the flame retardant PUF samples are higher than that of the neat sample (23.0%). PUF/DMMP-TCEP (30%) has the highest LOI value of 30.6%, which is much higher than that of PUF sample without flame retardant containing phosphorus and chloride. It is suggested that the improved flame retardancy of PUF samples is attributed to the presence of DMMP and TCEP. And in the limiting oxygen index test, it can be observed that the intumescent residual char layer can be formed as a barrier between flame and the flame retardant PUF samples to inhibit the transmission of heat, energy and O_2 during combustion [4, 5]. The results reveal that the flame retardant system containing DMMP and TCEP could improve the flammability of polyurethane foam materials.

Table 1. LOI values of polyurethane samples.

Samples	DMMP Concentration (w/w %)	TCEP Concentration (w/w %)	P %	Cl %	LOI (%)
1	0	0	0	0	23.0
2	14	6	4.2%	2.2%	27.4
3	21	9	6.2%	3.3%	30.6

3.2 Thermal analysis (TGA)
Thermogravimetric (TG) and derivative thermogravimetric (DTG) were used to investigate the thermal stability of the PUF samples in detail. TG curves of the samples are shown in figure 1. It can be seen that starting decomposition temperature of the flame retardant samples are 223℃ (20% DMMP/TCEP) and 190℃ (30% DMMP/TCEP), respectively, which are lower than that of the PUF

IOP Conf. Series: Materials Science and Engineering **137** (2016) 012033 doi:10.1088/1757-899X/137/1/012033

sample. The higher % P value, the lower the starting decomposition temperature. This can be attributed to the earlier decomposition of P-O-C from DMMP and TCEP during the thermal degradation of the flame retardant PUF sample due to its lower bond energy, which could form some compounds containing phosphate to catalyze the formation of the residual layer at high temperature [6, 7]. Residue of the flame retardant PUF sample containing 20% DMMP/TCEP (Sample 2) at 800℃ is higher than that of the neat PUF sample. However, sample 3 containing 30% DMMP/TCEP has the lowest residual amount of 14.0% at 800℃ compared with the neat PUF sample (14.1%) and the PUF sample containing 20% flame retardants (16.7%). DTG curves from figure 2 show that the Sample 2 has three maximal weight loss peaks while the PUF sample has two maximal weight loss peaks. Compared the neat sample, the small new peak can be observed at the relative low temperature range for Sample 2, which could be explained by the thermal degradation of flame retardant containing phosphorus to catalyze the release of water and some volatile compounds during the decomposition of the sample. Meanwhile, Sample 3 containing 30% DMMP/TCEP has one more weight loss peak at the temperature range of 375-421℃ compared with Sample 2. This could be due to the gaseous flame retardant action of the higher content TCEP containing chloride at high temperature range. These results suggest that flame retardant system containing DMMP and TCEP could change the thermal degradation process of the PUF sample.

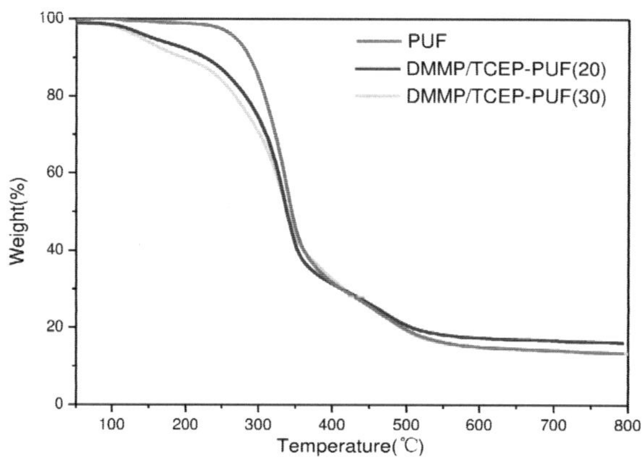

Figure 1. TG Curves of PUF samples

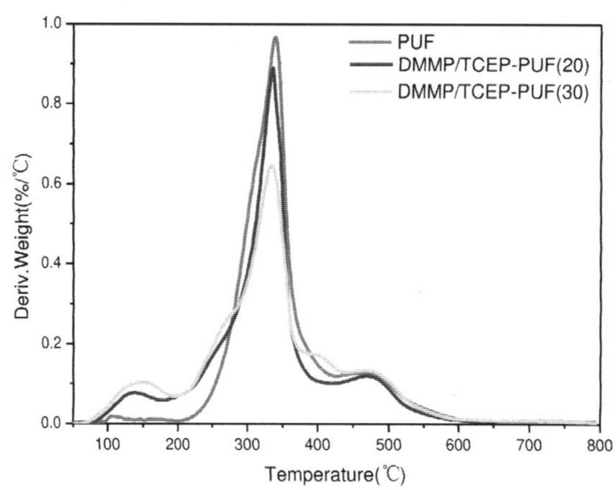

Figure 2. DTG curves of PUF samples

3.3 TG-FTIR

The gaseous products produced by pyrolysing various samples of the neat PUF and flame retardant PUF samples in nitrogen at different temperature (50-800℃) have been investigated by TG-FTIR. figure 3, 4 and 5 shows the 3D and 2D TG-FTIR spectra of neat PUF, PUF with 20% DMMP/TCEP and PUF with 30% DMMP/TCEP. As shown in figure 3, absorption peaks at 1114cm^{-1}, 1257cm^{-1}, 1220cm^{-1}-1320cm^{-1}, 3318cm^{-1}, 1622cm^{-1} and 2275cm^{-1} can be assigned to C-O-C, C-O from ester, C-N from amine group, N-H, -C=C and -NCO from neat PUF sample, respectively [8~10]. Sample 2 with 20% DMMP/TCEP has less and lower absorption peaks at high temperature range compared to neat PUF sample (figure 4), which indicates the lower evolution of the gaseous products during its thermal degradation process. This can be explained by the formation of the thick char layer covering the underlying PUF matrix due to the condensed flame retardant action of flame retardant containing phosphorus, which prevents the further thermal degradation of the sample in high temperature [11]. However, the more and higher absorption peak of Sample 3 with 30% DMMP/TCEP can be observed at high temperature range in figure 5, indicating the large release of the pyrolysis products to the gaseous phase. Combined with the high LOI value of Sample 3, this can be attributed to the gaseous flame retardant action of the flame retardant system containing DMMP and TCEP due to the higher content of chloride compared to Sample 2, which is in accordance with the results from thermal analysis of the PUF samples in Section 3.2.

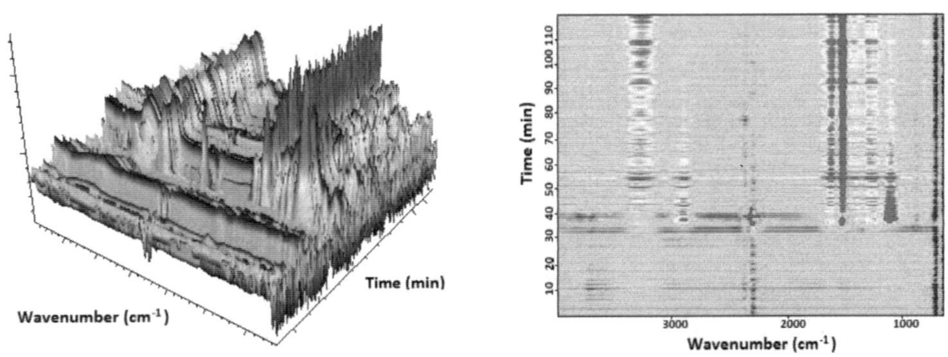

Figure 3. 3D and 2D TG-FTIR spectra of PUF without DMMP and TCEP

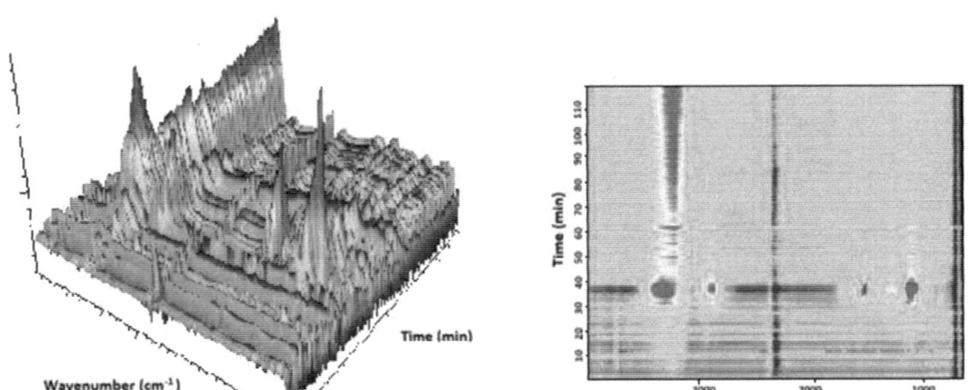

Figure 4. 3D and 2D TG-FTIR spectra of PUF/DMMP-TCEP (20)

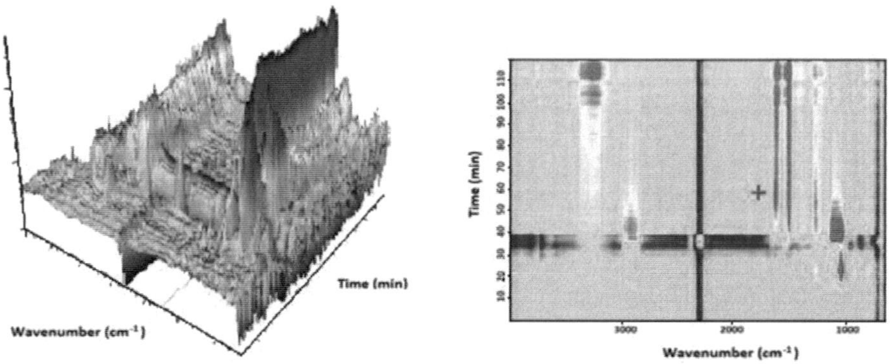

Figure 5. 3D and 2D TG-FTIR spectra of PUF/DMMP-TCEP (30)

FTIR spectra of neat PUF sample obtained at the maximum evolution rates (340 and 479℃) during heating are presented in figure 6, which can be used to get more information about the pyrolysis product in the gaseous phase. The characterized peaks of -NCO (2275cm^{-1}), CO_2 (2352cm^{-1}), H_2O (3579cm^{-1}) and -NH from -NHC=O (1530cm^{-1}) can be observed in spectrum (a) at 340℃, indicating that the depolymerisation of polyurethane and the break of its main chains [12]. In spectrum (b) at 479℃, the peaks at 1530cm^{-1} and 2275cm^{-1} have almost disappeared while some other peaks at 1513cm^{-1}, 2905cm^{-1}, 1465cm^{-1}, 1626cm^{-1} and 935cm^{-1}can be observed, which could be assigned to phenyl, $-CH_3$, $-CH_2-$ and C=C and $-CH_2-$ from alkene[13]. It suggests that thermal degradation of PUF materials can be described as two steps: the depolymerization of the main chains and the further pyrolysis of polyol and isocyanate [14].

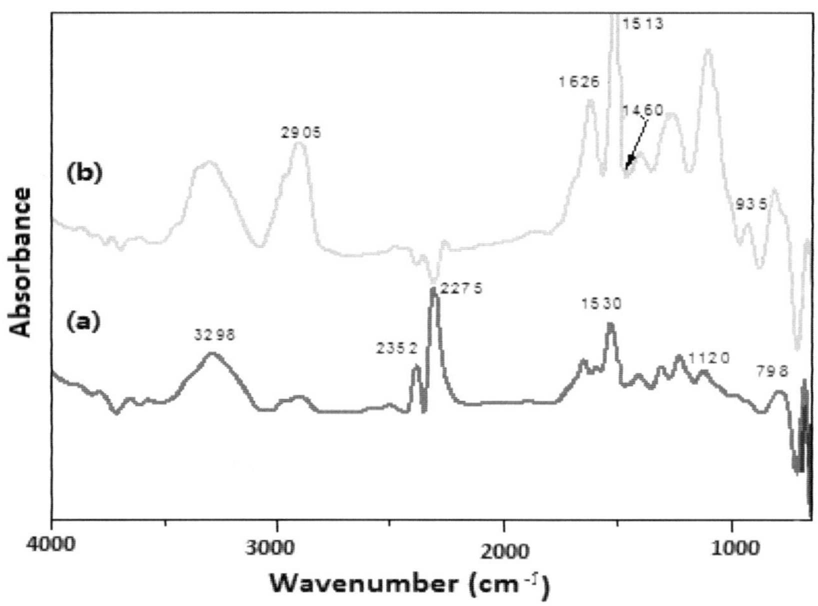

Figure 6. FTIR spectra of PUF at the maximal weight loss rates: (a) 340℃; (b) 579℃

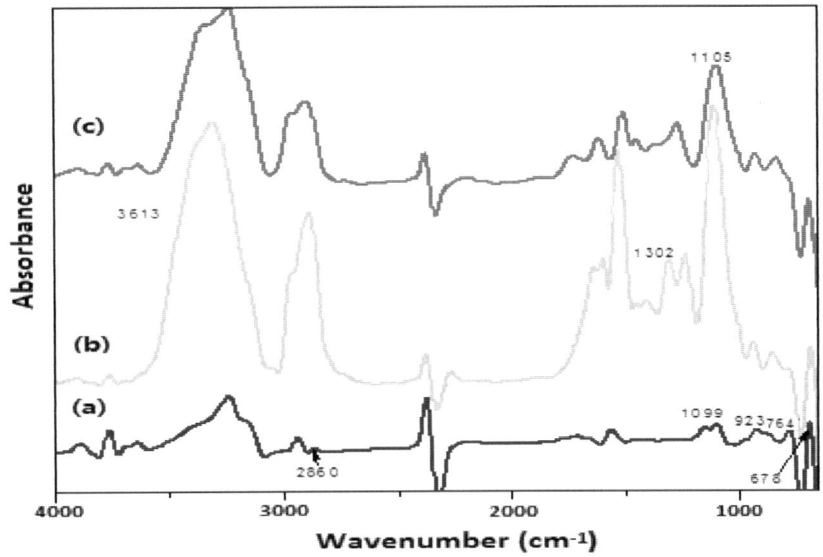

Figure 7. FTIR spectra of PUF/DMMP-TCEP (20) at the maximal weight loss rates: (a) 136℃; (b) 335℃ and (c) 474℃

FTIR spectra of Sample 2 at the maximum evolution rates (136, 335 and 474℃) during heating are presented in figure 7. In spectrum (a) at 136℃, the absorption peaks of 1099cm^{-1}, 923cm^{-1}, 764cm^{-1} and 678 cm^{-1} can be observed, which are assigned to C-O, P-C, P-O and C-Cl [15]. Absorption peak of HCl has appeared at 2860 cm^{-1}. These results confirm the decomposition of TCEP and DMMP at the initial decomposition stage, which is also supported by the result from TG and DTG presented in figure 2 and 3. Absorption peak of HCl at 2860 cm^{-1} has disappeared at 335℃. It may be due to the gaseous flame retardant action of TCEP containing chloride to capture the radical during combustion. Meanwhile, the peak of P=O at 1320 cm^{-1} can be observed in spectrum (b) [16]. The peak of P=O has disappeared at 474℃, indicating the flame retardant action of DMMP containing phosphorus in gaseous phase at high temperature range. However, the lower and less absorption peaks in spectrum (c) confirms that flame retardant activity of Sample 2 mainly takes effect in the condensed phase.

4. Conclusions

Dimethyl methylphosphonate (DMMP) and trichloroethyl phosphtate (TCEP) have been used as the flame retardants containing phosphorus and chloride to improve flame retardancy and thermal stability of the polyurethane foam material by the cast molding method. The LOI tests show that the LOI value of the PUF sample increases with the DMMP and TCEP concentration Thermal stability of the flame retardant sample can be enhanced at high temperature. Meanwhile, based on the information from TG-FTIR analysis, it can be concluded that although the flame retardant system containing DMMP and TCEP could improve the thermal stability of PUF samples by the formation of the intumescent residual char in the condensed phase, which covers the substrate to retard the thermal degradation of material, the release of the gaseous products can be also influenced due to the presence of DMMP and TCEP at high temperature.

Acknowledgement

The authors are grateful to the National Research Program of China (973 Program) (2012CB719701) and the Project-sponsored by Sichuan Fire Research Institute of Ministry of Public Security (20148806Z) for financial support.

References

[1] Grace T C, Salvador C N and Francisco V 2012 *Frontiers of Architectural Research* **1** 348
[2] Hu S J and Li Y B 2014 *Bioresour. Technol.* **161** 410
[3] Adel R K, Mustafa K, Kyle M, Mohini S and Oh S K 2015 *Ind. Crop. Prod.* **74** 1
[4] Wang D Y, Ge X G, Wang Y Z, Wang C, Qu M H and Zhou Q 2006 *Macromol. Mater. Eng.*, **291** 638
[5] Liu W, Chen D Q, Wang Y Z, Wang D Y and Qu M H 2007 *Polym. Degrad. Stab.* **92**1046
[6] Tian G X, Ji Q, Xu D M, Tan L W, Quan F Y and Xia Y Z 2013 *Fiber Polym.* **14** 767
[7] Chen D Q, Wang Y Z, Hu X P, Wang D Y, Qu M H and Yang B 2005 *Polym. Degrad. Stab.* **88** 349
[8] Liu W, Zhang S, Bourbigot S, Sun J, Yu L H, Feng Q L, Chen X S and Zhu XJ 2012 *Polym. Adv. Technol.* **23** 1550
[9] Hu J H and Zhen X F 2011 Bei Jing *Science Press* 286
[10] Wei L L, Wang D Y, Chen H B, Chen L, Wang X L and Wang Y Z 2011 *Polym. Degrad. Stab.* **96** 1557
[11] David W H, Gayani K and John M K 2005 *Polym. Degrad. Stab.* **87** 555
[12] Chen X L, Huo L L, Jiao C M and Li S X 2013 *J. Anal. Appl. Pyrol.* **100** 186
[13] Jiao L L, Xu G D, Wang, Q. S, Xu Q and Sun J H 2012 *Thermochim. Acta* **547**120
[14] Cao Z H, Zhang Y, Song P G, Cai Y Z, Guo Q, Fang Z P and Peng M 2011 *J. Anal. Appl. Pyrol.* **92** 339
[15] Liu W, Chen L and Wang Y Z 2012 *Polym. Degrad. Stab.* **97** 2487
[16] Gu J W, Zhang G C, Dong S L, Zhang Q Y and Kong J 2007 *Surf. Coat. Technol.* **201** 7835

IOP Conf. Series: Materials Science and Engineering **137** (2016) 012034 doi:10.1088/1757-899X/137/1/012034

Composites of Polyindole nanowires within Silicate and Aluminosilicate hosts with distinct conductive properties

J M Juárez, M B Gómez Costa, O A Anunziata[1]

Centre for Research in Nanoscience and Nanotechnology (NANOTEC), Córdoba Regional Faculty, National Technological University, Maestro López y Cruz Roja Argentina, 5016, Córdoba, Argentine.

[1]Corresponding author. E-mail: oanunziata@frc.utn.edu.ar

Abstract. Nanostructured silicate SBA-15 and aluminosilicate AlSBA-15 were synthesized in order to prepare polyindole composites. The Silica mesoporous materials were prepared by sol-gel method and alumination using post-synthesis technique and analysed by different methods (XRD, BET, TEM, and FTIR). Polyindole/host composites were prepared by *in situ* oxidative polymerization of pre-adsorbed indole, employing Cl_3Fe as oxidant. TG, FTIR, BET, XRD, SEM and TEM were used to characterize the resulting composites. These studies show that the porous structures of the materials are preserved after polymerization, and polyindole is found within the porous channels. The composites have an electrical conductivity range between values higher than those of the pure chemically synthesized polyindole, close to those of the pure electrochemically synthesized polymer and lower than those of the pure chemically synthesized polymer, in the order of 10^{-8} S/cm.

1. Introduction

Nanostructure materials through an unvarying nanopores array structure, with great precise surface area have been widely studied in the last decades because of its possible usage as catalytic materials, adsorbents of great carbon-based compounds or host as chemical supports. Among the nanoporous materials are found amorphous or polycrystalline solids and modified layered materials. A significant amount of work has been devoted to synthesizing, within the nanoporous range, materials characterized by a consistent and well-defined pore assembly [1, 2].

SBA-15 and Na-AlSBA-15 are crystalline mesoporous molecular frameworks with an ordered and precise mesoporous system. These materials possess regular networks and its pore diameter fluctuate thoroughly among 1.5 to 10 nm depending on the template, supplementary chemicals and different reaction environments. Because of their pore building and form, they have often been subject to study as a probe material as an absorbent for a variety of gases and vapours [3, 4].

In recent years they have aroused great scientific interest studies of electro activity of polypirrol (PPi) polyaniline (PANI), polycarbazol and its derivatives (conductive polymers with N_2 heterocycles), together with the great possibility of industrial applications [1, 3].

Polypyrrole displays good environmental stability and high electrical conductivity, and is good to prepare films of PPi by electrodeposition. In this way Poly (paraphenylene also shows suitable thermal stability. As Indole is a compound that contains aromatic linked with heterocycle no basic pyrrole ring, therefore, it is possible that PInd and the by-products have the characteristic that depict both polymers (poly(paraphenylene) and PPi). Nevertheless, from amongst the different characteristic of nanowire conducting polymers, research on PInd and its by-products has been sparse at best, in spite of close similarities observable with the other polymers [5].

The electroactive polymer polyindole (PInd) was synthesized electrochemically, or by oxidative process of indole (C_8NH_7).

Polyindole and derivatives are suitable for use in scientific and industrial areas including electrocatalysis, electronics and pharmacology. When doped, polyindole is green, its electrical conductivity lying between 5.10^{-4} and 8.10^{-2} S cm^{-1}, a variation range which depends slightly on the nature of the counterion [6]. Polyindole films have also been reported to show good thermal stability, great reduction/oxidation performance and slow degradation proportion compared to PPi and PANI [7].

The aim of this work is to synthesize PInd-SBA-15 and PInd-AlSBA-15 composites by means of in situ polymerization, to produce novel nanostructured composite showing both conductive behaviour as possible materials with catalytic properties for electrocatalytic developments. In this way, we expect to make an appreciated impact to the synthesis and applications of nanostructured composite materials.

2. Material and methods

2.1. Synthesis of Si-SBA-15 and Na-AlSBA-15

Well-organized nonporous SBA-15 silicates were produced employing P123, a triblock copolymer $EO_{20}PO_{70}EO_{20}$ as template and TEOS as silica basis. The procedure was the following: 20 g of template in solution on chloridric acid pH <1 (400 ml), stirring at 323 K. Later, TEOS (40 g) was incorporated under stirring at 323 K for a day. The product, white in appearance, was removed to a PP flask and subject to aging during 72 h and 373 K. The material was filtered, with Millipore water to achieve pH ~6.

The molecular mixture was Si: 0.018 $EO_{20}PO_{70}EO_{20}$: 2.08 HCl: 112 H_2O. The template was removed by first immersing the material under ethanol reflux for 6 h and dried at 363K [8], after being filtered and washed. To ensure the elimination of the structure-directing agent, the product was heated under N_2 flow at 573K, calcined at 823 K in air during 6 h. In order to prepare AlSBA-15, we used post-synthesis alumination with sodium-aluminate at room temperature; the procedure has already been described [9].

2.2. Indole adsorption and polymerization

The water contained in all the samples employed as hosts was eliminated at 673 K at 10^{-3}Torr during 1 h. Afterwards, indole was adsorbed into the host in gas phase for a day and at 333 K obtaining indole-SBA-15 following the technique described earlier [9]. In this way, the indole/host saturation ratio was attained.

The Polyindole were polymerizes whiting the host (PInd-AlSBA-15 and PInd-SBA-15) by *in situ* oxidation in a similar way reported by us [9], using $FeCl_3$ as oxidant. 30 ml of $FeCl_3$ 0.1M was added to the Indol/Hosts in a bottle during 24 h and static conditions. Then, the materials were extracted by filtration, cleaned using an ethanol /water solution. After this were dried at 333 K for 72 h.

3. Results and discussion

3.1. XRD, TEM and Structural studies of hosts and composites

Low-angle XRD analyses of samples employed as hosts was displays in Figure 1. The XRD analysis of SBA-15 and AlSBA-15 samples reveals a strong (1 0 0) reflection signal, as well as small signals, showing a high regularity at long range resultant from (1 1 0) and (2 0 0) diffraction planes, which are typical of SBA-15 material [8].

The aluminated sample suggests that the SBA-15 structure is preserved after Al has been incorporated by post-synthesis: the (1 0 0) reflection signals of Na-AlSBA-15 produced by post-synthesis broaden only slightly, and the (1 0 0) peak shifts marginally to higher 2θ values and lower d-spacing (Table 1). The uniformity of the pores (hexagonal arrangement) of these composites is indicated by the good resolution of individually XRD peaks.

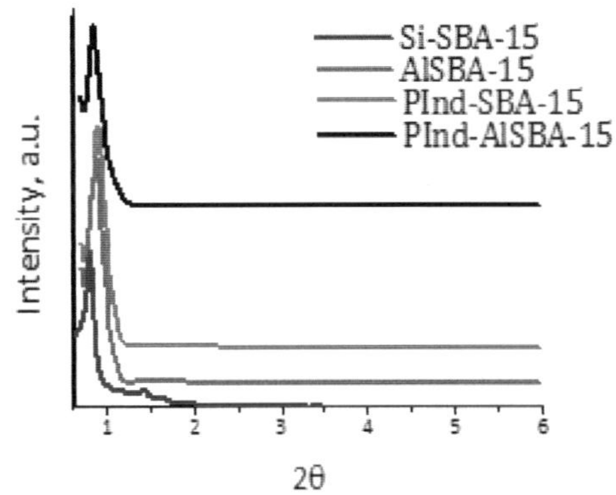

Figure 1. XRD of Si-SBA-15, AlSBA-15 and composite PInd-AlSBA-15 and PInd-SBA-15.

Moreover, the persistence of the hexagonal-ordered structure of the hosts after *in situ* polymerization is demonstrated for PInd-composites by X-ray results (Figure 1). A shift for different composites diffraction plane (1 0 0), distinctive of SBA-15, to higher 2θ angles can also be observed, which is an indicator of the successful integration of the polymer nanowire and the hosts, and has more orderly and uniform [10].

The composites surfaces are considerably minor in size than the hosts; nevertheless, its representative nanostructure is preserved with the PInd polymer within it, according to XRD analysis, (Table 1).

Table 1. Textural and structural properties.

Sample	**Area m^2/g**	V_P **(cm^3 g^{-1}) Total pore Volume**	**BJH Pore diameter (nm)**
SBA-15	1040	1.38	8.9
PInd-SBA15	550	1.14	5.1
AlSBA-15	960	1.05	8.6
PInd-AlSBA-15	736	0.89	5.0

TEM of composite images (Figures 2-3) show narrow pore size distribution indicating that the host structure is retained after the inclusion of the polymer. Images also show that no bulk polymer clusters are synthesized outside the channel hosts and we can observe the filling of pores with PInd.

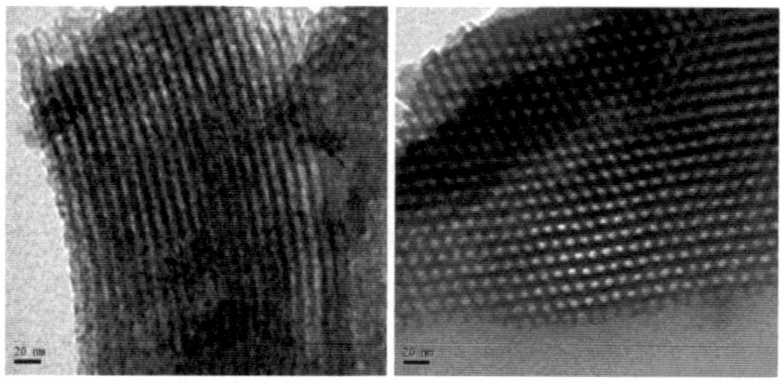

Figure 2. TEM images of PInd-Na-AlSBA-15 composite.

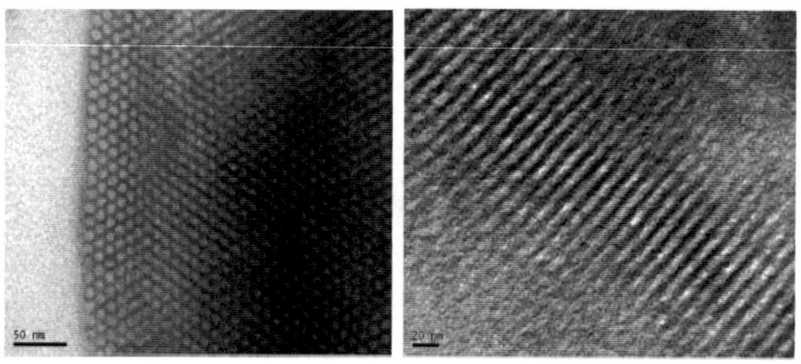

Figure 3. TEM images of PInd-SBA-15 composite.

3.2. FTIR characterization

FTIR spectra of polymer and composites can be seen in Figure 4. The FTIR absorption signals of pure PInd made it possible to detect detailed composite bands.

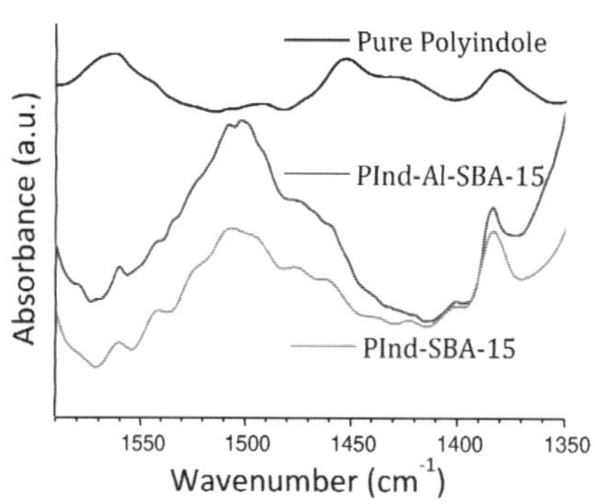

Figure 4. FTIR of pure polyindole and composites in the range of 1300-1700 cm^{-1}.

The bands observed at approximately 1574-1560, 1508, 1493, 1454-1460, 1383 cm^{-1} were assigned according to previous research [9].

The indole polymerization mechanism occurs by two main routes, 1,3 position or 2,3 position. The peak at 1564 cm^{-1} could be ascribed to N-H interactions in the pure PInd [11].The composites have a small band at about the same vibration of pure polymer (1560 cm^{-1}). These results suggest that the N-H bond is perturbed or diminishes in both composites. This indicates that polymerization occurs mainly in 1,3 position, marked by a small band of N-H bond vibrations. The polymerization process does not affect the benzene ring. In addition, new stretching signals due to the polymer chain appears at around 1460 and 1383 cm^{-1}. The bands observed at 1427 cm^{-1} in pure polymer are assumed to C-N stretching. The shift of this band in PInd-AlSBA-15 and PInd-SBA-15 (1420 cm^{-1}) shows that this bond is perturbed, confirming that polymerization in these composites occurs in 1,3 position. PInd - host interaction origins that PInd signals move to different absorption frequency than the pure PInd.

3.3. Thermogravimetric analysis

Figure 5 shows a TGA curve analysis for the samples. For pure PInd, the first mass elimination (300 - 373 K) was produced due to the water removal of polymer, which continue thermally constant nearby 430 K. Afterward, PInd begins to undergo a slow degradation until 700 K, when it commences to degrade quickly. MCM-41 and AlMCM-41 hosts, have a weight loss close 4% approximately 370 K and they become to stabilize; the complete mass loss being 7 % w/w at 770 K. SBA-15/AlSBA-15 hosts present a similar behavior to that of MCM-41, (total weight loss about 8 % w/w at 770 K).

The initial weight loss below 370 K of composites is likely to be caused by water loss and by adsorption of some gases. The composite is therefore fairly stable until around 770 K. Beyond this temperature, the weight loss rate of the composite materials (mainly polymer) increases.

Compared with the mass loss of PInd (16% w/w) and nanowire PInd whiting the composites about 770 K, the variance is extensive: 12% for Pind-AlMCM-41, 13% for Pind-MCM-41, 11% for Pind-SBA-15 and 9% for Pind-AlSBA-15 composites. The result is that PInd turn out to be stabilized when it is as nanowire within the hosts that make up the composites, preventing fast degradation of the polymer.

The mass loss ascribed to PInd in nanocomposites at 770 K is about 5% w/w for Pind-AlMCM-41, 6% for Pind-MCM-41, 3% for Pind-SBA-15 and 1% for Pind-AlSBA-15 composites. Note that the full mass loss for pure PInd was about 31%.

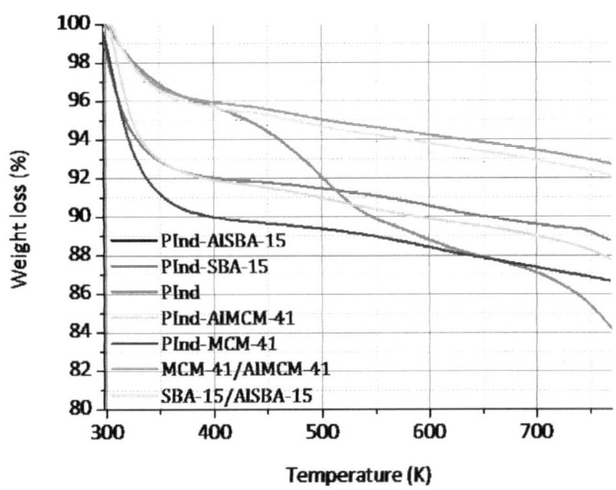

Figure 5. TGA analysis for: Pure polyindole, hosts and Composites.

3.4. Conductivity studies

Consistent with the results obtained by previous studies, chemically and electrochemically synthesized polyindole are characterized by an electrical conductivity of 1.10^{-4} and 5.10^{-2} S cm^{-1} respectively [12].

The electrical conductivity of PInd-nanocomposites at 298 K (RT conditions) lay between $8.7x10^{-4}$ (PInd-AlSBA-15) and $1.4x10^{-8}$ (Pind-SBA-15) (Figure 6). When the conducting polymers are placed within the channels of nanostructured Al-SBA or Al-MCM, the conductivity of the composite is able to alter from ionic conductors to semiconductors, because the changing influencing for conductive polymers exhibits conductivity more than 10 orders of level [13].

The electrical conductivity of conductive polymers are susceptible to be affected by a diversity of aspects: polaron, conjugation and complete chain length combined with the charge relocation to close molecules. Thus, if the polymers are inside the nanostructured channels, a variation of the conductivity can be observed.

The electrical conductivity of the PInd-AlMCM-41 composite is greater than pure polymer chemically synthesized, and approaches to the polymer produced by electrochemical methods [9].

The higher electrical conductivity of PInd-Al-hosts is probably because aluminosilicates, instead of silicates, impact the electrical conductivity. This performance can be ascribed because Al-hosts increase the transmission of the electrical charges from side to side of PInd chains in the nanopores. In this way, is possible that the host dopes polyindole.

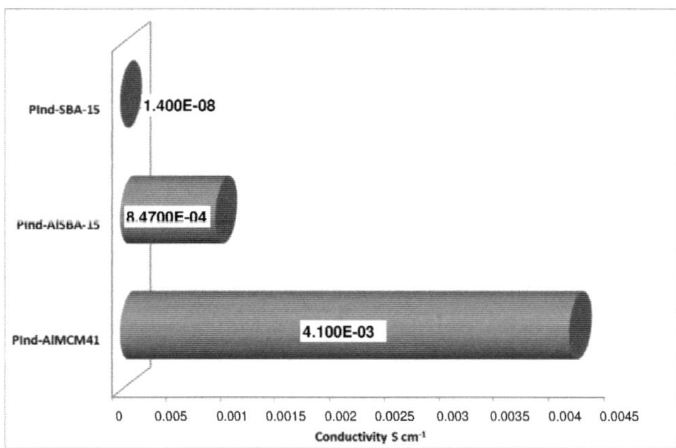

Figure 6. Conductive properties of Polymer/host composites (S/cm).

4. Conclusions

SBA-15 and sodium aluminosilicate as NaAlSBA-15 host materials with longitudinal channel arrangement were prepared and characterized. All the nanostructured materials employed as hosts were analysed using a variety of methods (XRD, BET, TEM, and FTIR).

FTIR analysis were achieved to analyse the interaction of indole adsorbed onto the mesoscopic materials. Infrared spectroscopy allowed us to corroborate that polymerization occurs mainly in 1,3 position indicated by a small band of N-H bond vibrations with the same electron-acceptor sites (Na^+ and silanol sites). Furthermore the spectrum indicated the PInd polymerization can be attained inside the hosts with hexagonal pore systems. XRD analysis of the resulting nanocomposites showed that the nanochannels structure of the hosts subsist after polymerization. Polyindole is originated inside the nanostructured channels. Moreover, according with our corroboration by BET and TEM analysis, PInd, is not on the external surface of the hosts.

These composites exhibit a wide range of desirable properties, with potential application in the field of electronics for developing e.g. electronic devices at a nanoscopic scale. The composites have higher electrical conductivity range than those PInd, obtainable by chemical synthesis, close to those of the pure electrochemically synthesized polymer, and lower than those of the pure chemically synthesized polymer.

Hybrid composites of organic molecular wires within inorganic reservoirs were successfully developed. The final conductivity of the composites can be modified according to the requirements of the final product by varying the hosts (which implies altering the anchor sites of the nanowires to the host), and nanowires with "tailor made" conductivity behavior could so be developed.

Acknowledgments

Juliana M. Juárez, Marcos Gómez Costa, Oscar A. Anunziata, NANOTEC, CONICET, National Technology University, Cordoba Faculty, Maestro Lopez y Cruz Roja Argentine. We recognize ICP-CSIC Madrid, Spain, by TEM analysis. We acknowledge the financial support of CONICET Argentina, PIP CONICET 11220120100218CO. 2014 -2016.

References

[1] Kresge C T, Leonowicz M E, Roth W, et al. 1992 Ordered mesoporous molecular sieves synthesized by a liquid crystal template mechanism *Nature* **359** 710–712

[2] Beck J S, Vartuli J C, Roth W J, et al. 1992 A new family of mesoporous molecular sieves prepared with liquid crystal templates *J. Am. Chem. Soc.* **114** 10834–10843

[3] Schmidt R, Stöcker M, Hansen E W, et al. 1995 MCM-41a model system for adsorption studies on mesoporous materials *Microporous Mater.* **3** 443-448

[4] Soler-Illia G, Sanchez C, Lebeau B, et al. 2002 Chemical Strategies To Design Textured Materials: from Microporous and Mesoporous Oxides to Nanonetworks and Hierarchical Structures *Chem. Rev.* **102** 4093–4138

[5] Cai Z and Yang G 2010 Synthesis of polyindole and its evaluation for Li-ion battery applications *Synth. Met.* **160** 1902-1905

[6] Billaud D, Maarouf E B and Hannecart E 1994 Electrochemical polymerization of indole *Polymer* **35** 2010-2011

[7] Abthagir P S and Saraswathi R 2004 Charge transport and thermal properties of polyindole, polycarbazole and their derivatives *Thermochim Acta* **424** 25–35

[8] Meynen V, Cool P and Vansant E F 2009 Verified syntheses of mesoporous materials *Microporous Mesoporous Mater.* **125** 170–223

[9] Gómez Costa M B, Juárez J M, Martínez M L et al. 2012 Synthesis and characterization of a novel composite: Polyindole included in nanostructured Al-MCM-41 material *Microporous Mesoporous Mater.* **153** 191–197

[10] Sun X, Ren J, Zhang L, et al 2010 Polyaniline nanocomposites via chemical oxidative polymerization in the presence of functional MCM-48 as in situ dopant *Synth. Met.* **160** 2244–2249

[11] An S, Abdiryim T, Ding Y, et al 2008 A comparative study of the microemulsion and interfacial polymerization for polyindole *Mater. Lett.* **62** 935–938

[12] Talbi H, Ghanbaja J and Billaud D 1997 Vibrational properties and structural studies of doped and dedoped polyindole by FTIR, Raman and EEL spectroscopies *Polymer* **38** 2099-2106

[13] Blinova N V, Stejskal J, Trchova M, et al 2007 Polyaniline and polypyrrole: A comparative study of the preparation *Eur. Polym. J.* **43** 2331–2341

IOP Conf. Series: Materials Science and Engineering **137** (2016) 012035 doi:10.1088/1757-899X/137/1/012035

Mechanical and thermal properties of promising polymer composites for food packaging applications

S F Abdellah Ali[1]

Institute Of Graduate Studies & Research, Alexandria University, Egypt & IRC in Polymer Science & School of Chemistry, University of Leeds, UK

E-mail: Salah15eg@yahoo.com

Abstract. Blending starches with biodegradable polycaprolactone (PCL) was used as a route to make processable thermoplastics. When developing biodegradable polymer composites it is important to use high concentrations of starch for legislative and cost reasons. The addition of starch has a significant effect on all physical properties including toughness, elongation at break and the rheological behaviour of the melt. To enhance the physical properties, we used cellulose acetate propionate (CAP) as a cellulose derivative with high amylase starch and PCL blends. It is suggested that the PCL/starch/CAP blends are partially miscible. It was found that the yield tensile strengths of most PCL/Starch/CAP blends were higher than that of pure PCL itself. There was a big difference between glass transition temperature values of PCL/Starch/CAP blends and the pure PCL glass transition temperature which indicates that no phase separation occurs. Addition of CAP to starch and PCL blends improved the mechanical and thermal properties even at high content of starch.

1. Introduction

Polycaprolactones (PCLs) are a particularly promising class of biodegradable polymers for blending with starches as they are readily available on an industrial scale. PCL is a biodegradable polymer with very high flexibility, and films of polycaprolactone have been reported to have an elongation and tensile strength at break more than 1000% and 32 MPa [1]. These values are very high compared with the elongation and tensile strength at break of a non biodegradable polymer such as low density polyethylene which has values in the range 500-725% and 9.7-17.2 MPa respectively [2]. PCL has melting point (~ 60 Celsius) which makes some processing routes difficult e.g. film production and the range of applications of PCL-based polymers and blends made using PCL is limited because of their low softening temperature [3]. Reducing the amount of expensive PCL employed through mixing with starch reduces its cost. Cellulose esters are thermoplastic materials used in films, moulded products, and fibres. In general cellulose esters are used in applications need good properties at temperatures above 60^0C e.g. modulus and tensile strength [4].To achieve good properties (modulus and tensile strength) in the temperature range 60 -120^0C requires the components of a blend to have appropriately high softening and melt temperatures.

This paper reports the results of the mechanical and thermal properties of a range of different compositions of polycaprolactone, starch and cellulose acetate propionate.

2. Experimental& characterization

2.1. *Materials&preparation*
Polycaprolactone (Mwt=80,000), (Aldrich), Hylon VII - a high amylose starch (National Starch) and cellulose acetate propionate (Acros Organics), were all used as supplied.PCL, CAP and starch were first mixed in the dry solid states in known proportions and the blends were prepared using Brabender mixer model PL2200 at 50 rpm for 15 minutes at 130 ^{0}C. Each component was compression moulded into 10cm x 10cm x 1mm sheets at a hydraulic pressure of 10 MPa for 10 minutes at 130 ^{0}C. All blend compositions were produced on a weight basis of the components. The different blends and their content of PCL, starch and CAP are shown as a percentage by dry weight in Table 1.

The percentage of PCL was 20, 40, 60 and 80% and the starch%being progressively replaced against CAP% until both have equal proportions. As the aim of this work was to study high-starch blends, compositions in which CAP had a higher percentage than starch were not investigated.

2.2. *Differential scanning calorimetery (DSC)*
Thermal analysis of the materials has been done using a Perkin Elmer DSC(Pyris1) equipped with a liquid nitrogen cooling accessory. The DSC was calibrated using indium. For each measurement, a sample weight of about 8 mg was placed in an aluminium pan and heated from 15 ^{0}C to 150 ^{0}C (10 ^{0}C/min) then quenched to 15 ^{0}C (20 ^{0}C/min) and heated again to 150 ^{0}C (10 ^{0}C/min). The melting points, the heat of fusion (the melting heat) and the degree of crystallinity were determined after two scan cycles by taking the average of the two values.

2.3. *Dynamic mechanical & thermal properties*
Dynamic Mechanical & Thermal Properties(DMTA) measurements were made using Rheometrics Solids Analyser, Model RSAII, in the rectangular torsional mode at 2^{0}C/min and 1H_Z (heating rate and frequency). The dimensions of specimens were 50.0 × 5.0 ×1.0 mm

Table 1. Blend Compositions (%) by weight

Blend codes	PCL	Starch	CAP	Blend codes	PCL	Starch	CAP
PCL	100	0	0	B-10	40	30	30
CAP	0	0	100	B-11	40	0	60
B-1	20	80	0	B-12	60	40	0
B-2	20	70	10	B-13	60	30	10
B-3	20	60	20	B-14	60	20	20
B-4	20	50	30	B-15	60	0	40
B-5	20	40	40	B-16	80	20	0
B-6	20	0	80	B-17	80	15	5
B-7	40	60	0	B-18	80	10	10
B-8	40	50	10	B-19	80	0	20
B-9	40	40	20				

2.4. Mechanical Properties

Tensile tests were made using an Instron, Model 5564 using dumb-bell shaped specimens in accordance to the recommended ASTM standard method (D638) [5]. The deformation rate employed was 10mm/min at 25^{0}C. Prior to the tensile tests, samples were adjusted at 50 ± 5% relative humidity for 72 h at room temperature in a closed chamber with hydrated calcium nitrate solution (in accord with American Society for Testing and Materials Standard E104).

3. Results &discussion

3.1. Thermal analysis properties

Due to the low melting point of PCL, It is difficult to process it by traditional techniques for thermoplastic polymers. For example, films produced using blowing technique aresticky during processing [6]. We can overcome this difficulty by blending it with starch and CAP. The melt temperature of starch is about 230 ^{o}C [6], so it helps to overcome PCL stickiness problems during extrusion when processed at high temperatures. For PCL, the recorded values of melting temperature (T_m) were 55.7 ^{o}C, 57.8 ^{o}C, 59.9 ^{o}C. These give mean equals to 57.8^{o}C and a standard deviation S = 1.69. To determine the confidence limits (error or uncertainty), we use the student's t-test parameters $t_{2,0.05,2}$with 95% confidence, such that T = mean ± $t_{2,0.05,2} \times S/(n)^{1/2}$ where n is the number of measurements (n=3 in this case and $t_{2,0.05,2}$ = 4.3).

Table 2. Thermal analysis data for the different blends

Code	T_m	ΔH_x(J/g)	X_c%	Code	T_m	ΔH_x(J/g)	X_c%
PCL	57.8±4	59.4	43.7	B-10	53.3±4	20.3	14.9
B-1	57.1±4	17.8	13.0	B-11	--------	--------	-------
B-2	55.9±4	17.6	12.9	B-12	57.7±4	37.3	27.4
B-3	54.3±4	16.7	12.3	B-13	56.9±4	36.8	27.1
B-4	-----	-------	------	B-14	56.5±4	35.5	26.1
B-5	--------	--------	------	B-15	55.3±4	33.6	24.7
B-6	--------	--------	------	B-16	57.8±4	47.5	34.9
CAP	--------	--------	------	B-17	57.7± 4	46.9	34.5
B-7	57.5±4	28.2	20.8	B-18	56.9±4	46.3	34.1
B-8	56.3±4	26.0	19.1	B-19	55.6±4	44.4	32.6
B-9	54.7±4	23.2	17.0				

Crystallinity percentage (X_c%) was calculated as a ratio of $\Delta H_x / \Delta H_m$ where ΔH_x is the fusion enthalpy of PCL in the blend and ΔHm for the hypothetical perfect crystal of PCL (ΔH_m = 136 J/g) [7].

The main thermal parameters, melting temperature (T_m), and percentage of crystallinity (X_c %) for the different samples tested are recorded in Table 2. The melt temperatures (shown in Table 2) of all the PCL/Starch blends were nearly similar to the melting temperature of pure PCL indicating a phase separation which normally found in non-miscible polymers. Similar results were also indicated by other researchers [8]. It was found that the melt temperature T_m decreased with increasing of CAP content. This is may be due to the amorphous nature of CAP which decrease the crystallinity content of PCL and lower the melting temperatures. Figures1 and 2 show the DSC thermograms for pure PCL and pure CAP and some blends after the second heating scan which shows typical outputs obtained with these mixtures. Figure 1a for pure PCL shows a single melting endotherm which indicates that the PCL sample is a semi-crystalline polymer. Figure 1b represents that CAP does not undergo a melting phase transition, and the thermogram is characteristic of an amorphous polymer with a high T_g (102.8°C). The peaks of some blends could not be determined due to the high content of amorphous CAP compared to PCL content.

The noticeable decrease in the melting point of all blends which contained CAP indicated that no phase separation occurred in those blends and CAP may act as a compatibilizer which lowers the melting temperatures, the melt viscosity and makes them easier to be processed. It will be useful in practice to study the compatibility between the components of these blends because it may control physical properties of such blends which affect their end use. Dynamic Mechanical Thermal Analysis (DMTA) was used to evaluate the compatibility between different components.

3.2. Dynamic Mechanical Thermal Properties

The glass transition temperatures for different composites with different CAP contents are shown in Table 3. There is a slight difference between $T_g$values of PCL/Starch blends and the PCL glass transition temperature indicating a phase separation found in all non-miscible polymers. Other researchers found same results [8]. On the other hand, there is a big difference between $T_g$values of PCL/Starch/CAP blends and the PCL glass transition temperature which indicates that no phase separation occurs in such blends. The presence of other components in the blend has a great effect on the primary nucleation of the crystallizing component [9]. The effect of the other polymers on the nucleation of the crystallizing component depends on the chemical propertiesof the polymer, its miscibility, its ability to crystallize and blend composition. When the polymers are miscible, the energy barrier to form a critical nucleus in the blend is greater than that of the homopolymer [10]. Therefore, the nucleation activity of heterogeneous nuclei in the blend decreases with increasing the concentration of the other components. The results of crystallinity percentage of PCL indicated that the crystallization of PCL-starch-CAP blends decreased by the addition of CAP. This is may be due to the amorphous nature of CAP as the CAP prohibited the movements of the polymer segments. The melting temperature and crystallinity percentage depressions are associated with miscibility [11]. These results suggested that the PCL/starch/CAP blends are partially miscible. Glass transition temperatures of all PCL/Starch blends values were nearly around PCL glass transition temperature which means that starch did not affect PCL glass transition temperature. On blending PCL/Starch with CAP, there was a change in the glass transition temperatures values which indicates that CAP acts as a compatibilizer between PCL and starch due to the existence of hydroxyl groups (in starch and CAP) and ester carbonyls (in PCL and CAP) [12].

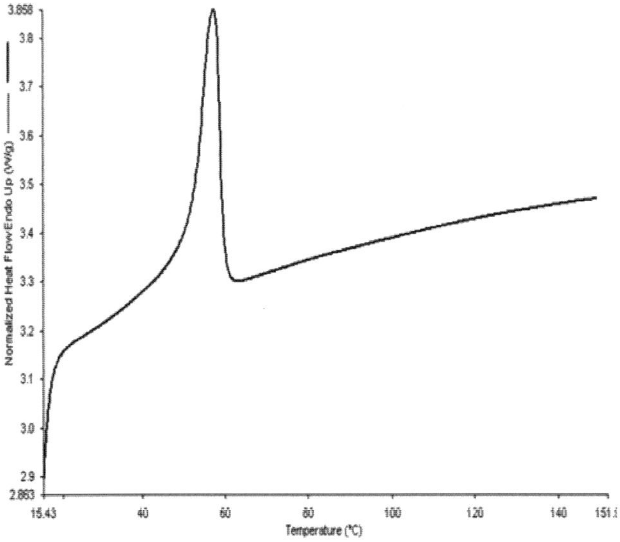

(a)

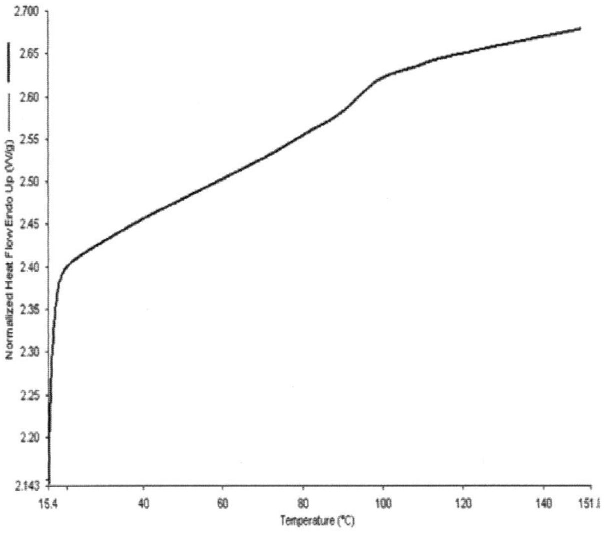

(b)

Figure 1.DSC curves for: (a) PCL (b) CAP

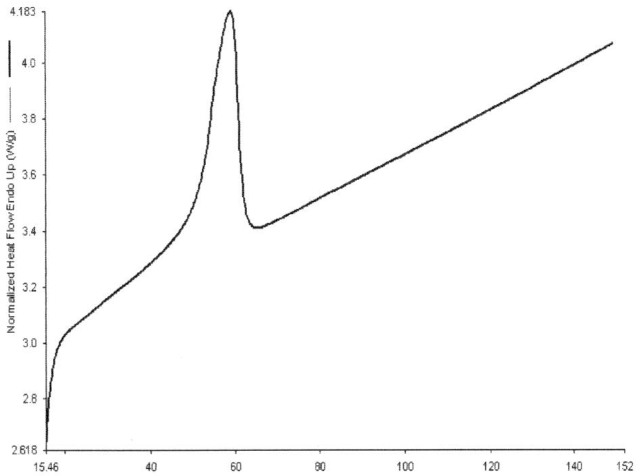

(a)

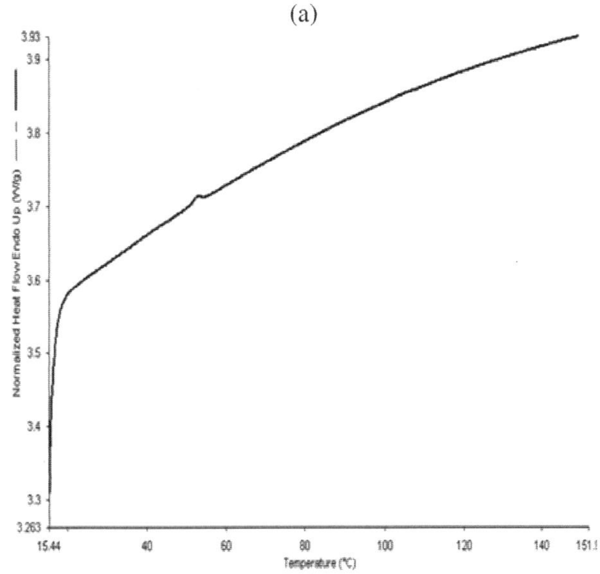

(b)

Figure 2. DSC curves for: (a) B13(b) B4

There is an interaction between PCL, CAP, and starch. This interaction binds them more strongly with each other than PCL/Starch blends components which give these blends better mechanical properties than PCL/Starch blends and can be used in some packaging and domestic applications such as plastic bags, trays, wash bottles, cutlery, flower pots and foodstorage.

Table 3. Glass-transitions temperatures

Blend	T_g	Blend	T_g
PCL	-53.9±4	B-10	-11.4±4
B-1	-53.5±4	B-11	-5.5±4
B-2	-22.3±4	B-12	-53.8±4
B-3	-12.3±4	B-13	-29±4
B-4	-9.4±4	B-14	-21.4±4
B-5	-7.9±4	B-15	-15.4±4
B-6	-1.4±4	B-16	-53.9±4
B-7	-53.7±4	B-17	-44.1±4
B-8	-25±4	B-18	-36.3±4
B-9	-15.9±4	B-19	-28.5±4

3.3. Mechanical Properties
Tensile strength at yield (σ_y), at break (σ_b) and elongation at break (ε_b) were calculated from stress-strain curves (Figure 3 as an example).By comparing the results obtained for binary blends of PCL/starch with the tensile properties data for PCL/starch/CAP blends, increases are noticed as well as significant increases in the elongation at break. In PCL/starch composites, the mechanical incompatibility of the two phases was great and increased with the starch content. As the starch content increased, the tensile strength and elongation at break of blends decreased. This decrease for the binary blends of polycaprolactone and starch could be due to the phase separation between the hydrophobic PCL and the hydrophilic starch. The higher the degree of phase separation, the lower the mechanical properties of the resulting blends [13-15]. It is suggested that CAP improvedthe compatibility and the interfacial adhesion between polycaprolactone and starch [16-18]. Mechanical properties of polymer blends depend on chain stiffness of the individual polymers and the intermolecular forces [19]. The mechanical properties of PCL/starch/CAP composites represented the polymer chains interactions during blending. Yield stress determines the maximum stress that could be borne without permanent deformation in the consideration of materials for packaging [19]. It was found that the tensile yield strengths of most PCL/Starch/CAP blends were higher than yield tensile strength of pure PCL itself. This is may be due to the existence of an interaction between these blend components. The values of tensile yield strength and tensile strength at break are quite high compared with non biodegradable material such as low-density polyethylene (LDPE) [2]. LDPE used in different applications such as trays and for general purpose containers such as those used in food storage. Many of these products require flexibility such as plastic bags, tubing and dispensing bottles. Most of these applications can be made by using these different blend compositions of PCL/Starch/CAP with consideration of their physical properties.

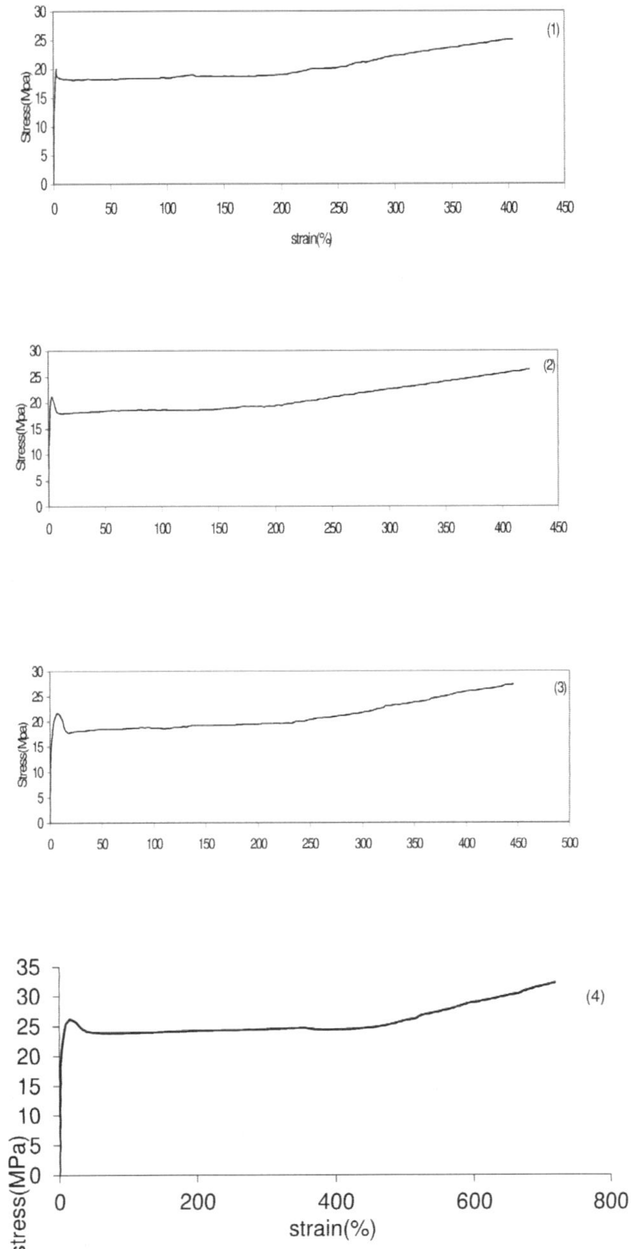

Figure 3. Stress-Strain curves of: (1) B 12 (2) B 13 (3) B 14 (4) B 15

4. Conclusion

A significant decrease in mechanical properties was noted at a starch high content with PCL; this is may be due to incompatibility between PCL and starch which has been reported by other workers. On blending CAP with starch and PCL, there was an improvement in most mechanical and thermal properties even at a high content of starch; this may be due to CAP acting as a compatibilizer between PCL and starch which prevents the phase separation between the components of the blends. The mechanical properties of the blends were enhanced by adding CAP, which could induce the best intermolecular adhesion between PCL and starch and govern the interaction between the components of such blends.

Acknowledgments

The author is grateful to all members of IRC in the Department of Physics for their help at all times and, in particular, since the instrumental facilities in IRC, Department of Chemistry, were destroyed by smoke emanating from a roof fire. Particular thanks go to the Egyptian Education and Culture Bureau for the financial support.

References

[1] Reddy M M, Vivekanandhan S, Misra M, Bhatia S K and Mohanty A K 2013 Biobased plastics and bionanocomposites: Current status and future opportunities *Progress in Polymer Science* **38**(10) 1653-1689

[2] Emblem A and Emblem H eds., 2012 Packaging technology: Fundamentals, materials and processes Elsevier: Amsterdam, the Netherlands

[3] Shimano M 2001 Biodegradation of plastics *Current Opinion in Biotechnology* **12** 242-247

[4] Jeon G W, An J E and Jeong Y G 2012 High-performance cellulose acetate propionate composites reinforced with exfoliated graphene *Composites Part B: Engineering* **43**(8) 3412-3418

[5] Gallagher F G, Shin H and Tietz R F inventors 1993 EI DuPont de Nemours, assignee; US **5**, 219, 646

[6] Bastioli C, Cerruti A, Guanella I, Romano G and Tosin M 1995 Physical state and biodegradation properties of starch – polycaprolactone systems *J. Environ Polym Degrad* **3**(2) 81-95

[7] Avella M, Errico M, Laurienzo P, Martuscelli E, Raimo M and Rimedio R 2000 Preparation and Characterization of compatibilised Polycaprolactone/Starch Composites *Polymer* **41** 3875-3881

[8] Kim C, Choi E and Park J 2000 Effect of PEG molecular weight on the tensile toughness of starch/PCL/PEG blends *Appl. Poly. Sci.* **77**(9) 2049-2056

[9] Wu T M and Chen E C 2006 Crystallization behavior of poly (ε-caprolactone) / multiwalled carbon nanotube composites *Journal of Polymer Science Part B: Polymer Physics* **44**(3) 598-606

[10] Averous L and Boquillon N 2004 Biocomposites based on plasticized starch: thermal and mechanical behaviours *Carbohydrate Polymers* **56**(2) 111-122

[11] COCK F, Cuadri A A and Partal P 2013 Thermal, rheological and microstructural characterisation of commercial biodegradable polyesters *Polymer Testing* **32**(4) 716-723

[12] Edgar J, Buchanan M, Debenham S, Rundquist A, Seiler D, Shelton C and Tindall D 2001 Advances in Cellulose Ester Performance and Application *Prog. Polym* **26**(9) 1605-1608

[13] Callister W and David G 2000 *Materials Science and Engineering: An Introduction*, Fifth Edition, John Wiley& sons, Inc.: Hoboken, NewJersey.

[14] Ishiaku U, Pang K, Lee W and Ishak Z 2002 Mechanical properties and enzymic degradation of thermoplastic and granular sago starch filled poly(ε-caprolactone) *European Polymer* **38**(2) 393-401

[15] Tsou C H, Lee H T, Tsai H A, Cheng H J and Suen M C 2013 Synthesis and properties of biodegradable polycaprolactone/polyurethanes by using 2, 6-pyridinedimethanol as a chain

extender *Polymer Degradation and Stability* **98**(2) 643-650
[16] Avella M, Errico M, Laurienzo P, Martuscelli E, Raimo M and Rimedio R 2000 Polymer **41** 3875
[17] Mohamed A, Sherald H G and Girma B 2007 Polycaprolactone/polystyrene bioblends characterized by thermogravimetry modulated differential scanning calorimetry and infrared photoacoustic spectroscopy *Polymer degradation and stability* **92**(7) 1177-1185
[18] Wu D, Zhang Y, Zhang M and Zhou W 2008 Phase behavior and its viscoelastic response of polylactide/poly (ε-caprolactone) blend *European Polymer Journal* **44**(7) 2171-2183
[19] Madi N K 2008 Thermal and mechanical properties of injection-molded recycled high-density polyethylene blends with virgin isotactic polypropylene *Materials & Design* **46** 435-441

IOP Conf. Series: Materials Science and Engineering **137** (2016) 012036 doi:10.1088/1757-899X/137/1/012036

Advances in the study of mechanical properties and constitutive law in the field of wood research

S Zhao[1], J X Zhao and G Z Han

Mechanical Engineering Department, Shijiazhuang mechanical Engineering College, Shi Jia-Zhuang 050003, Hebei, China

E-mail: jxzs163@163.com

Abstract. This paper presents an overview of mechanical properties and constitutive law for wood. Current research on the mechanical properties of wood have mostly focused on density, grain, moisture, and other natural factors. It has been established that high density, dense grain, and high moisture lead to higher strength. In most literature, wood has been regarded as an anisotropic material because of its fiber. A microscopic view is used in research of wood today, in this way, which has allowed for clear observation of anisotropy. In general, wood has higher strength under a dynamic load, and no densification. The constitutive model is the basis of numerical analysis. An anisotropic model of porous and composite materials has been used for wood, but results were poor, and new constitutions have been introduced. According to the literature, there is no single theory that is widely accepted for the dynamic load. Research has shown that grain and moisture are key factors in wood strength, but there has not been enough study on dynamic loads so far. Hill law has been the most common method of simulation. Models that consider high strain rate are attracting more and more attention.

1. Introduction

Wood is a natural polymer composite material that has been used throughout history in wide engineering applications. Many consider wood to be one of the oldest engineering materials [1]. Wood has many advantages, including its low weight, environmental friendliness, and renewability. Wood surrounds us in daily life—it is used in our chairs, cabinets, and other furniture. Recently, the mechanical properties of wood have been the subject of heightened interest. Knowledge about mechanical properties is the foundation of theoretical models and engineering analysis. Wood is a complex, porous, heterogeneous, and anisotropic material. The properties of wood are more uncertain than steel and other macromolecule materials. In wood, the density, grain, moisture, annual rings, and other natural factors are variable, yet, they must be accounted for. A reason for the growing interest in wood is the demand for theoretical models and engineering analysis. Engineering structures are becoming more and more complex, and it is harder to analyze complex structures than it was in the past, and precise results are required. Reliable theoretical models are needed in order to perform numerical analysis. Mechanical properties must be known in order to obtain proper models.

Another problem in studying wood is the constitutive model. Numerical simulation has become a new tool for engineering analysis, and can be used at times when simple experiments are not enough for an engineer. The constitutive model for wood has been studied for many years, and has advanced greatly; new models can consider temperature and strain rate, and are able to satisfy demands for wood structures most of the time. However, the complexities of wood mean that no single model can be used for all purposes, and many models must be used to solve problems.

In previous research into the properties of wood, some differences have been found between static and dynamic loads, such as destruction modes and strength change. Hill and other yield laws have not yet been able to simulate natural factors, but mental laws, such as J-C, are sometimes able to be used.

2. Mechanical property of wood

2.1. Static mechanical property

Forest products laboratory in the US has conducted a number of experiments regarding wood under compression loads [2], and in 1961, more detailed experiments were carried out that studied failures under compression load. It was found that compression failures are seen more frequently if the light, the piece, and the point of view are in a specific relationship to one other [3]; however, these studies did not consider the effects of direction and hardness. Vural and Ravichandran conducted an extensive study of axial responses and failures under quasi-static and dynamic axial compression [4-5]. Their study laid a foundation for future research. Recently, Da Silva observed the cellular microstructure shown in the micrograph of Balsa wood, and determined the causes of failure in the microstructure> Different failure patterns were found in different directions, and it was discovered that compared with hard wood, soft wood has more obvious softening and densification, shown in figure 1. Further, experiments about shear strength were conducted [6]. Researchers began to pay attention to the age of wood, and found that old wood has higher elasticity modulus in the radial and axial directions, and the shear modulus does not change in old wood [7].

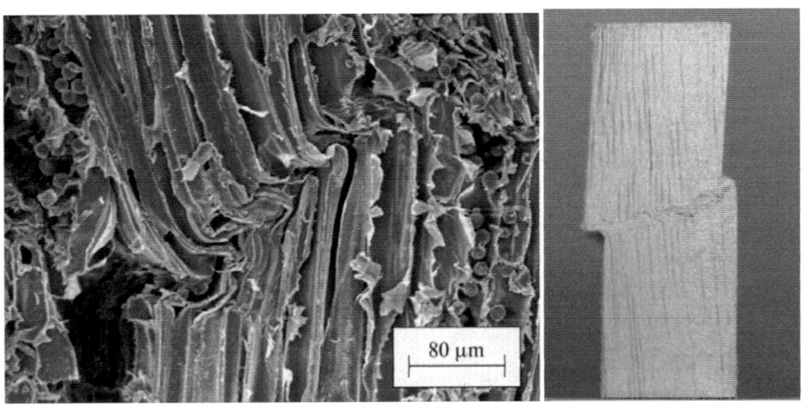

Figure 1. Photograph showing evidence of kink band broadening.

Sakai studied the effects of moisture content and incising on preservative penetration [8]. Manoochehr has investigated fracture energy and toughness of radial pine laminated veneer lumber (LVL), and detailed a method for making a sample. It was found that when the LVL fracture increases, the cumulative frequency increases. Simulation was conducted and the relationship between crack length and stress intensity was obtained [9]. Shear strength is another important parameter for wood. In many wood structures, shear load can lead to destruction. Shear strength is lower than compression for wood material. Wood's shear strength varies widely-- its orthotropic, shear plane, and density parameters are the most important factors. Shear strength is positively correlated to density and radial direction shear strength is higher than the remaining two directions. When axial strength is lowest, it softens under a shear load. The same is true for compression strength [10-12]. It appears that tensile strength is not as important as compression and shear strengths. In most circumstances, this parameter is ignored. The sample is difficult to make, and it is more complex to demand tensile strength [13]. Nonetheless, tensile strength is important for in depth study of complex structures. It had been found that tensile strength and density mass have a linear relationship [14]. Similarly, the wood's grain has a strong effect on strength. The failure mode is different though-- the crack always follows the grain and is not perpendicular to it [15-17].

Table 1. Tensile strength of pine at different grain angles and densities.

ρ (g/cm^3)	σ_0 (MPa)	ρ (g/cm^3)	σ_{30} (MPa)	σ_{45} (MPa)	σ_{90} (MPa)
0.434	64.3	0.451	10.9	7.0	3.7
0.442	57.6	0.471	8.7	4.4	0.9
0.511	66.5	0.482	15.9	10.1	2.1
0.527	91.4	0.483	13.2	3.0	1.2
0.534	89.4	0.500	12.8	8.5	3.3
0.537	115.2	0.510	11.6	6.4	1.5
0.591	82.4	0.577	11.8	6.0	0.8

Fabiane Salles proposed a new way to demand the elasticity modulus under a compression load. A sample was used as the standard, and other modulus were demanded though contrast to the standard. This method made the modulus easier to understand [18-19]. Moutou analyzed mixed-mode fractures of wood. The energy release rate in the wood material was evaluated by testing 2MCG specimens [20].

Grain angle, moisture, and density are the main factors in determining wood strength. The grain angle is especially important. Densification is obvious in the radial and tangential directions; however, the elasticity modulus in the axial direction is higher.

2.2. Dynamic property

Dynamic properties are more complex than static properties. Natural factors, rate effects, and load conditions should be taken into consideration when studying dynamic properties. Static properties were the subject of research in earlier literature, and the dynamic properties were considered as being static. The high strain rate response of end-grain balsa wood is very different from the quasi-static condition. There is about a 15% plateau in total, but the plateau is very few in the quasi-static state, which can be seen in figure 2 and 3 [21]. Tian Zhang studied viscoelastic properties of wood using nano-indentation experiments, and found that the storage modulus increases monotonically. When the damping decreases, the storage modulus and damping coefficient become dependent on the penetration depth [22]. It has been established that the age of wood affects the static mechanical properties and other physical properties of the wood. The dynamic mechanical properties also change with age. For older wood, a reduction of impact bending strength ocurrs [23-24]. The Hopkinson experiment device has played an important role in dynamic property testing. Nowadays, with the Split Hopkinson Pressure Bar (SHPB) device, researchers can test a material's strain rate range from 10^2s^{-1} to 10^4s^{-1}. At lower wood hardness levels, standard steel rods can't give a precise signal for the sample, so an aluminum stick should be used as an incident rod in the test. As differences between static mechanical properties and dynamic properties have become better known more attention has been given to dynamic mechanical properties. The main areas of interest are dynamic compression characteristics, as there is more change under that condition. Xu W found that birch exhibited high strain rate sensitivity, and radial-loading specimens are easier to separate than axial-loading specimens [25]. Zhong's test on spruce obtained similar results through using a theoretical model to analyze the mechanism of wood failure [26]. Many other researchers have conducted similar experiments. The SHPB device has advanced, which has allowed for testing of complex load conditions, such as twist and mixed-loading conditions [27-28].In their study, wood has higher yield strength under dynamic load, and would be higher with high strain rate, these works are very helpful to do simulate about wood.

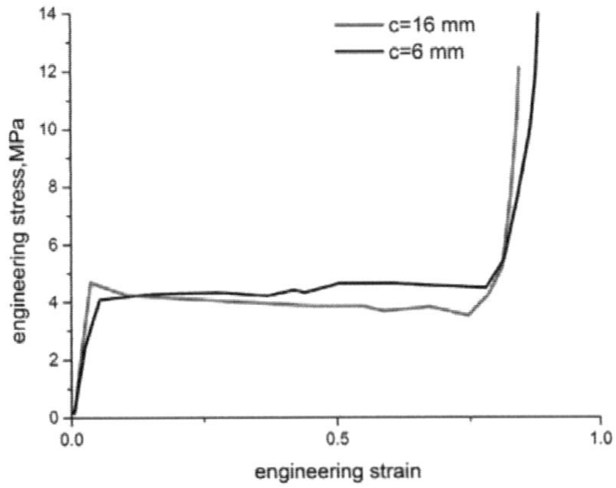

Figure 2. Quasi-static uniaxial compressive stress compared with strain responses in balsa wood.

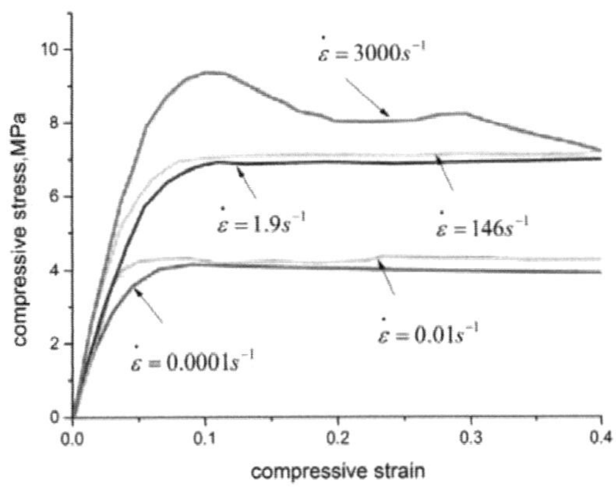

Figure 3. A comparison between the measured static, intermediate, and high strain rate compressive responses of balsa wood.

3. Constitutive model of wood

In general, wood is considered to be an elastoplastic orthotropic material; under load, it can be divided into three types: elasticity, plasticity, and densification. Sometimes the densification is not very important, but it can't be ignored in static mechanical analysis. Simple theories have been applied to solve early stage wood problems [29-31]. Recently, there have been great advances in wood modeling. Some of these models that have seen wide engineering use include maximum strain/stress criterion, Hill yield criterion, shear strength, and Tsai-Wu criterion [32-34]. Engineering demand drives the creation of advanced models.

Tom has applied a continuum failure criterion to wood, using modified Von Mises criterion, which can be seen in figure 4. The following formula can be used, where X is tensile strength and all constants of the general failure criterion can be determined from compression and tension [35-37].

$$\frac{\sigma^2_{x,or}}{XX'} + \frac{\sigma_{x,or}}{X} - \frac{\sigma_{x,or}}{X'} - 2F_{12}\sigma_{x,or}\sigma_{y,or} + \frac{\sigma^2_{y,or}}{YY'} + \frac{\sigma_{y,or}}{Y} - \frac{\sigma_{y,or}}{Y'} + \frac{\tau^2_{or}}{S^2} \tag{1}$$

Where S is the shear strength and

$$2F_{12} = 1/C'n_1 = 1/\sqrt{XX'YY'} \tag{2}$$

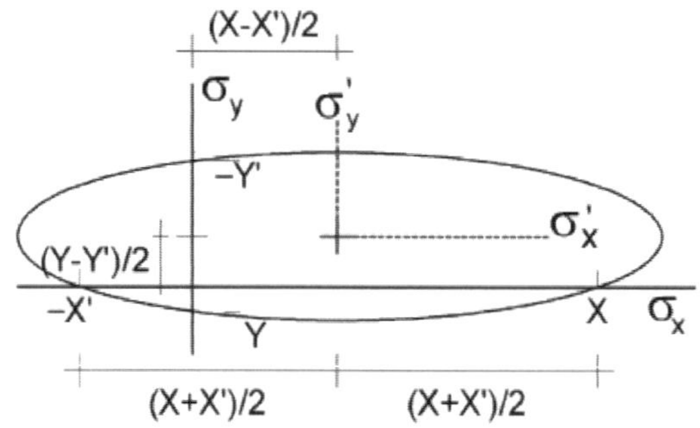

Figure 4. von Mises criterion for wood.

Adalian wrote a model based on hypoelastic formalism, which considered strain hardening. This was the first numerical model which used numerical modeling to design new containers [38]. Demand pushes theory forward. The strength of timber joints has attracted great attention [39]. Xu has developed the wood constitutive law for behavior of timber joints, Hoffman criterion was used in law, experiments and simulation got similar results [40-41]. A key factor in simulation is anisotropy evolution under finite strains. The strength of wood is strongly tied to its direction. In an attempt to obtain the damage of wood under a finite load, Houssem studied the plastic anisotropy problem, by examining a non-associative finite strain anisotropic elastoplastic model coupled with anisotropic ductile damage. It was shown that the model was suitable for large deformation kinematics [42-43]. Table has presented a three-dimensional (3D) constitutive equation for wood materials, which assumes that wood is an orthotropic material [44]. Recently, Oudjene and Khelifa have developed a coupled model for wood. Under this model the curve of the mechanical property of wood was described, shown in figure 5 [45-47].

Application research has also been done in other areas, including dowel-type fasteners, wood composites, crack, buckle, and impact analysis [48-52].

A simulation model for a static load has been widely studied, using a variety of laws, including the Hill and Hoffman yield, Tsai-wu, and van der Put laws. However, there have been few simulations of dynamic loads, and no model is widely accepted. There are two reasons for an upsurge in interest in dynamic models: the dynamic mechanical property is very different from the static model and many models are not suitable for the dynamic load; further, there are multiple factors affecting wood strength and it is difficult to consider all factors using one model.

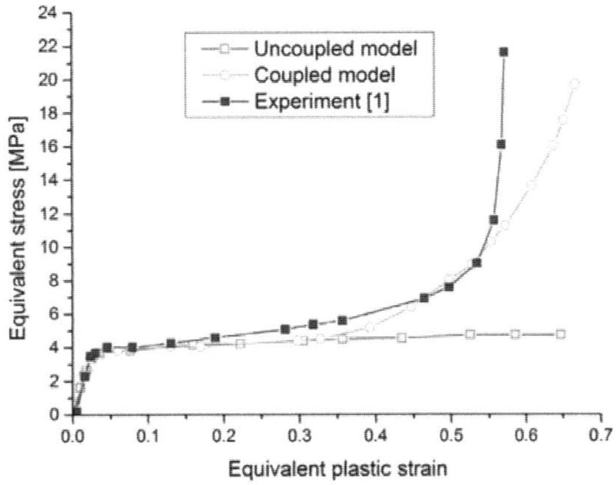

(a) radial direction (90^0)

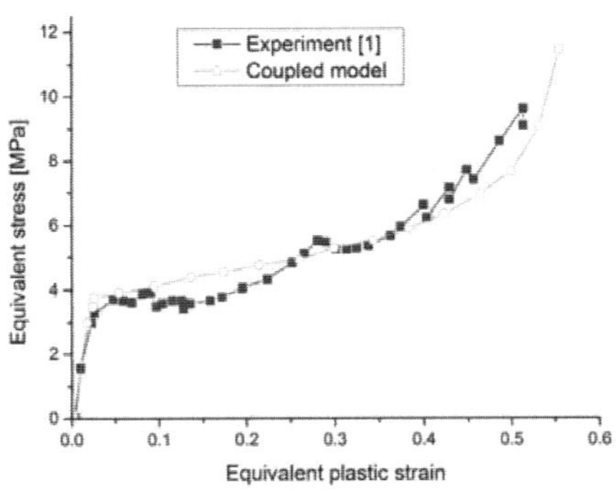

(b) longitudinal direction （70^0）

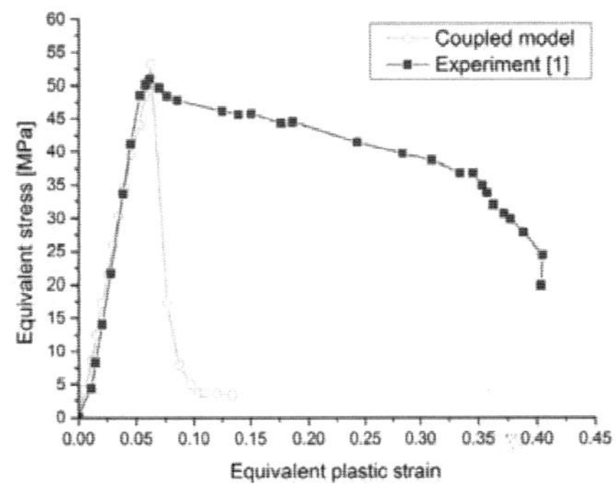

(c) parallel to the grain (0^0)

Figure 5. Stress – strain curves under loading in different directions: comparison between simulation and experiments.

4. Conclusion

This study reviewed advances in the study of the mechanical properties of wood. Compression, tension, twist, and mixed-loading conditions have been widely tested since their hardness, age, moisture, quality, grain angle, and moisture are the main factors for wood strength. Dynamic loads are more complex than the static condition. The strength of the dynamic condition is reliant on the strain rate. Typically, SHPB experiments are used to study high strain rate. While there have been many advances, more research is still needed on dynamic loads and higher rate experiments should be conducted. In order to satisfy developmental engineering demands, a suitable numerical model must be developed for complex structural analysis. Although many models have been proposed no single model can be adapted for all demands. Existing models are not able to effectively reflect the effects of natural factors like temperature, moisture, and grain. Because of the complexities of dynamic forces, and its differences from static forces, there have been very few simulations for dynamic loads and limited research on the properties of high strain rate. These are areas of study within the wood material field that are ripe for future research.

Reference

[1] Dinwoodie J 1975 Timber-a review of the structure-mechanical property relationship *J. Micro.* **104**(1) 3-32

[2] Zhong W Z, Song S C, Huang X C, et al 2011 Research on static and dynamic mechanical properties of spruce woody three loading directions *Chinese J. Theory Appl. Mech.* **43**(6) 1141-1150

[3] Kazuhiro O and Hiroshi Y 1997 Yield behavior of wood under compression-shear combined stress condition *Japan J. Soc. Mat. Sci* **46**(4) 385-389

[4] Vural M and Ravichandran G 2003a. Microstructural aspects and modeling of failure in naturally occurring porous composites *Mech. Mater.* **35** 523

[5] Vural M and Ravichandran G 2003b. Dynamic response and energy dissipation characteristics of balsa wood: experiment and analysis *Int. J. Solids Struct.* **40** 2147

[6] Andre D and Stelios K 2007 Compressive response and failure of balsa wood *Sol. Struc.* **44** 8685

[7] Kránitz K 2015 *Effect of natural aging on wood* (Dissertation: ETH Zurich) Zurich p145

[8] Haruko S, Yasuhiro T and Hajimu K 2011 Drying and impregnating for heavy timbers of Sugi heartwood (1)Effects of moisture content and incising on preservative penetration *Wood Preser.* **37**(1) 12-19

[9] Manoochehr A, Bruce D and Massimo F 2012 Experimental results of fracture energy and fracture toughness of Radiata Pine laminated veneer lumber (LVL) in mode I (opening) *Mater. Struct.* **45** 1189

[10] Michael O, Julia C and Anastasios P V 2013 Shear mechanical characterization of balsa wood as core material of composite sandwich panels *Constr. Build. Mater.* **41** 231

[11] ASTM Standard C 1994 Standard test method for shear properties of sandwich core materials *Am. Soc. Test Mater.* 273-94

[12] Soden P D and McLeish R D 1976 Variables affecting the strength of balsa wood *J. Strain Anal. Eng. Des.* 11(4) 225-34

[13] Sun H L and Lu K Y 2013 Development of curved surface profile modeling for specimen preparation of tensile strength perpendicular to wood grain *China Wood Industry* **27**(5) 49-51

[14] U.S Forest Service Research Paper 1967 Tension parallel-to-grain properties of southern pine dimension lumber

[15] Ashkenazi K 1966 Strength of Woods and Synthetics Materials (in Russian) (Moscow:*Lesnaja Promuslennosty*) pp 223

[16] Jan G and Michal C 2005 Tensile strength of softwood in LR orthotropy plane *Mech. Mater.* **37** 677

[17] Galicki J and Czech M 2001 Compression strength of pine wood in the main orthotropy plane *Int. J. Appl. Mech. Engrg.* **6** (4) 899-912

[18] Fabiane S F, Felipe H I 2013 Verification of test conditions to determine the compression modulus of elasticity of wood *Int. J. Agric. Forest.* **3**(2) 66-70

[19] Steven M and Anton D P 1987 Compact shear specimen for wood mode II fracture investigations *Int. J. Fract.* **35** 163

[20] Moutou P R and Chateauneuf A 2012 Statistical and reliability analysis for mixed-mode fracture tests applied to wood material *Wood Sci. Technol.* **46** 1099

[21] Tagarielli V L, Deshpande V S and Fleck N A 2008 The high strain rate response of PVC foams and end-grain balsa wood *Compos.: Part* B **39** 83

[22] Zhang T, Bai S L and Zhang Y F 2012 Viscoelastic properties of wood materials characterized by nanoindentation experiments *Wood Sci. Technol.* **46** 1003

[23] Walter S, Katalin K, Claus-Thomas B, et al 2015 Aging effects on physical and mechanical properties of spruce, fir and oak wood *J. Cultural Heritage* **16** 883

[24] Xu W, Hua J, Zhang S, et al 2015 Research on dynamic compression characteristics of birch

wood based on SHPB test *Scientia Silvae Scinicae* **51**(5) 95-101
[25] Erhardt D, Mecklenburg M F, Tumosa C S and Olstad T M 1996 New versus old wood: differences and similarities in physical, mechanical, and chemical proper-ties, in: J. Bridgeland (Ed.) *Int Council of Museums-Committeefor Conservation 11th Triennial Meeting* James&James, London pp 903-910
[26] Borst K, Jenkel C, Montero C et al 2013 Mechanical characterization of wood: An integrative approach ranging from nanoscale to structure *Comput. Struct.* **127** 53 - 67
[27] Wang Z, Gu L and Gao Z 2015 Experimental study on poisson's ratio of lumber by dynamic testing Scientia *Silvae Scinicae* **51**(5) 102-107
[28] Dou L, Wang X G and Liu Y C 2008 Dynamic mechanical behaviors of poplar wood *Explosion and Shock Waves* **28**(4) 368-3711
[29] van der Put TACM 1982 A general failure criterion for wood. *Proceedings of 15th Conseil Industrielle des Bois–International Union of Forestry Research Organizations Meeting* Boras Sweden
[30] Tsai S W and Wu E M 1971 A general theory of strength for anisotropic materials *J. Compos. Mater.* **5** 58-80
[31] Norris C B 1939 *The elastic theory of wood failure* Trans ASME **61** 259–261
[32] van der Put 2005 *The tensor polynomial failure criterion for wood* (Delft: Delft Wood Science Foundation)
[33] Hill R 1948 A theory of the yielding and plastic flow of anisotropic metals Proceedings of the Royal Society of London Series A *Math. Phys. Sci.* **193**(1033) 281-297
[34] Hoffman O 1967 The brittle strength of orthotropic materials *J. Compos. Mater.* **1**(2) 200-206
[35] van der Put TACM 2007 A new fracture mechanics theory of orthotropic materials like wood *Eng. Fract. Mech.* **74** 771
[36] Adalian C and Morlier P 2001 A model for the behavior of wood under dynamic multi-axial compression *Compos. Sci. Technol.* **61** 403-408
[37] Tom A C M and van der Put 2009 A continuum failure criterion applicable to wood *J. Wood Sci.* **55** 315
[38] Galicki J 2009 Application of the fields of objects formed on the basis of unit tensor objects to describe the tensile and compression failure stresses of pine wood in the principal orthotropy plane *Int. J. Solids Struct.* **46** 4298
[39] Johansen K W 1949 Theory of timber connections *Int. Assoc. Bridge Struc. Eng.* **9** 249
[40] Xu B H, Abdelhamid B 2014 Appropriate wood constitutive law for simulation of nonlinear behavior of timber joints *J. Mater. Civ. Eng.* 26
[41] Xu B H and Cai J 2015 State-of-the-art in strength criteria for wood *China Civil Eng. J.* **48**(1) 64-72
[42] Houssem B, Khemais S and Trong D N 2015 Damage anisotropy and its effect on the plastic anisotropy evolution under finite strains *Int. J. Solids Struct.* **63** 11-31
[43] Gao X, Zhang T, Hayden M and Roe C 2009 Effects of the stress state on plasticity and ductile failure of an aluminum 5083 alloy *Int. J. Plast.* **25** 2366
[44] Ala Tabiei and Jin Wu 2000 Three-dimensional nonlinear orthotropic finite element material model for wood *Compos. Struct.* **50** 143
[45] Blanco C, Cabrero J M and Martin-Meizoso A 2015 Design oriented failure model for wood accounting for different tensile and compressive behavior *Mech. Mater.* **83** 103-109
[46] Khelifa M, Oudjene M and Khennane A 2007 Fracture in sheet metal forming: effect of ductile damage evolution *Compos. Struct.* **85** 205
[47] Khelifa M and Oudjene M 2008 Numerical damage prediction in deep-drawing of sheet metals. *J. Mater. Process. Technol.* **200** 71
[48] Xu B H and Abdelhamid B 2013 Numerical simulation of embedding strength of glued laminated timber for dowel-type fasteners *J. Wood Sci.* **59** 17-23
[49] Moses D M and Prion L 2004 Stress and failure analysis of wood composites: a new model.

Composites Part B **35** 251

[50] Eckart R and Michael K 2010 Three-dimensional numerical analyses of load-bearing behavior and failure of multiple double-shear dowel-type connections in timber engineering *Comput. Struct.* **88** 165

[51] Li S, Ma W L and Zeng J F 2013 Study on Simulation Experiment of Wood Ability of Resisting High Speed Impact Based on Abaqus *Modern Scientific Instruments* **1** 105-107

[52] Sun Y L, Zhang Q and Zhao D 2010 Stress and strain field at wood crack tip *Journal of Beijing Forestry University* **32**(1) 103-107

IOP Conf. Series: Materials Science and Engineering **137** (2016) 012037 doi:10.1088/1757-899X/137/1/012037

Effect of DMMP on the pyrolysis products of polyurethane foam materials in the gaseous phase

W Liu[1], F Li, X G Ge, Z J Zhang, J He and N Gao
Sichuan Fire Research Institute of Ministry of Public Security, Chengdu, 610036, China

E-mail: weiliu20141202@126.com

Abstract. Dimethyl methylphosphonate (DMMP) has been used as a flame retardant containing phosphorus to decrease the flammability of the polyurethane foam material (PUF). Flame retardancy and thermal degradation of PUF samples have been investigated by the LOI tests and thermal analysis. The results show that LOI values of all PUF/DMMP samples are higher than that of the neat PUF sample and the LOI value of the samples increases with both DMMP concentration and the %P value. Thermal analysis indicates that flame retardant PUF shows a dominant condensed flame retardant activity during combustion. Thermogravimetric analysis-infrared spectrometry (TG-FTIR) has been used to study the influence of DMMP on the pyrolysis products in the gaseous phase during the thermal degradation of the PUF sample. Fourier transform infrared spectrometry (FTIR) spectra of the PUF sample at the maximum evolution rates and the generated trends of water and the products containing –NCO have been examined to obtain more information about the pyrolysis product evolutions of the samples at high temperature. These results reveal that although DMMP could improve the thermal stability of PUF samples through the formation of the residual char layer between fire and the decomposed materials, the influence of DMMP on the gaseous phase can be also observed during the thermal degradation process of materials.

1. Introduction

Recently, insulation materials have been widely used in construction to meet the energy-saving and emission-reduction requirements from the national building code in many countries. Polyurethane foam as an organic insulation material is of great interest to many consumers due to its excellent properties including good mechanical properties, low cost and effective thermal insulation [1-3]. However, like some other organic materials, the flammability of PUF limits its use in many areas. During the period of 2000-2010, many fire originated from PUF insulation materials has been reported in China. In November 2010, the big fire occurred at residential high-rise in Shanghai Jingan district was caused by the ignition of PUF used in the insulation external wall, which made 58 people killed and more than 70 people injured. Another fire in Jilin leaded to the direct economic loss of 5 million and the burned area of 7900 m^2, which resulted from the combustion of the PUF insulation material used in the roof. Meanwhile, investigation from these accidents suggests that the hazard during combustion of the PUF can be caused by not only the high burning heat but also the release of the volatile products formed during its thermal degradation at high temperature, which could induce asphyxia, coma or death of people.

According to the previous literature, DMMP as the flame retardant containing phosphorus could exhibit a good condensed phase flame retardant during combustion. Many researches have revealed that the flame retardant containing phosphorus could promote the carbonization reaction of the materials and increase its char yield at high temperature, indicating that the condensed phase activity is dominant during burning [4, 5]. However, the influence of DMMP on the gaseous phase during thermal degradation of PUF has been examined in this study. Flame retardancy and thermal stability of PUF containing DMMP have been investigated here. TG-FTIR analysis has been used to characterize the change of the pyrolysis products in the gaseous phase during heating of the samples. The aim of this work is to investigate the thermal degradation characteristics and the pyrolysis product evolutions of the PUF samples at high temperature range.

2. Experimental

2.1. Materials

Polyether polyol (Model: 4110, 635) and polyaryl polymethylene isocyanate (PAPI) (Model: GR-4110G) were purchased from Shanghai Gaoqiao Petro Co, China. Cyclohexylamine used as the catalyst and dichlorofluoroethane (HCFC-141b) used as the foaming agent were obtained from Changzheng Chemical Reagent Company, China. Dimethyl methylphosphonate (DMMP) was supplied from Jinan Enter Chemical Co., Ltd.

2.2 Preparation of samples

Polyurethane foam samples were prepared by the cast molding method. Polyether polyol was mixed with cyclohexylamine, HCFC-14b, DMMP, water together, and stirred with an electric stirrer at the agitation speed of 1500 r/min until a uniform mixture was formed. Then PAPI was added into the mixture and stirred for 90s with the speed of 500r/min. Thereafter, the mixture was cast into a mold and then heated at 70℃for 48hr. Finally, polyurethane foam samples were taken out from the mold for using..

2.3 Measurements

The limiting oxygen index (LOI) tests were carried out on a JF-3 type instrument (Jiangning, China), according to GB/T 2406-1993. Thermogravimetric analysis-fourier transform infrared spectrometry (TG-FTIR) of the neat PUF and the flame retardant PUF was performed using DT-50 (Setaram, France) instrument that was interfaced to IRAffinity-1 FTIR spectrometer. About 5 mg of the sample was put in an alumina crucible and heated from 25 to 800℃ in nitrogen atmosphere with the heating rate of 10℃/min.

3. Results and discussion

3.1 Flame retardancy

LOI test has been used to evaluate the flame retardancy of the PUF samples. The LOI value is the minimum concentration of oxygen required to maintain combustion of the material, and is the simplest method to measure flammability of materials [6]. LOI values of the PUF samples are presented in table 1. It indicates that LOI values of all PUF/DMMP samples are higher than that of the neat PUF sample. And LOI value of the PUF samples increases with both DMMP concentration and the %P value. The higher the %P value, the higher the LOI value. PUF/DMMP (30%) has the highest LOI value of 31.0%, which is 8.0% higher than that of the neat sample. It is suggested that the improved flame retardancy of the PUF samples can be attributed to the presence of DMMP. And in the LOI test, the residual char layer (figure 1) can be formed as a barrier between fire and the samples to inhibit the transmission of heat, energy and O2 during combustion of the flame retardant PUF. This phenomenon reveals that DMMP as the flame retardant containing phosphorus could enhance the flame retardancy of the PUF samples based on the condensed flame retardant mechanism.

Table 1. LOI values of PUF samples.

Samples	DMMP concentration (w/w %)	P%	LOI (%)
1	0	0	23.0
2	20	5.0	27.7
3	30	7.5	31.0

Figure 1. Photograph of sample 3 after the LOI tests.

3.2 Thermal analysis

Thermogravimetric (TG) and derivative thermogravimetric (DTG) were used to study the thermal stability of the PUF samples. TG curves of the samples are shown in figure 2. It can be seen that starting decomposition temperature of the PUF/DMMP samples are 252℃ (20% DMMP) and 216℃ (30% DMMP), respectively, which are lower than that of the neat PUF sample. The higher %P value, the lower the starting decomposition temperature. This may be attributed to the earlier decomposition of DMMP at the initial stage of the whole thermal degradation [7]. And the residual char of the flame retardant samples at 800℃ are higher than that of the neat sample. Compared to the other samples, Sample 3 with the DMMP content of 30% has the highest residual char of 16.8%. It indicates the thermal degradation of the flame retardant sample could be inhibited due to the incorporation of DMMP. DTG curves in figure 3 show that Sample 2 and 3 have three maximal weight loss peaks while the neat PUF sample has two maximal weight loss peaks. It is suggested that the presence of DMMP has changed the thermal degradation process of the PUF samples. Meanwhile, it also shows the maximal weight loss rate temperature of flame retardant PUF is lower than that of the neat sample. Among all the samples, sample 3 with the highest %P value of 7.5% has the smallest maximal weight loss rate. These results indicate that the thermal stability of PUF samples can be enhanced with the content of DMMP [8].

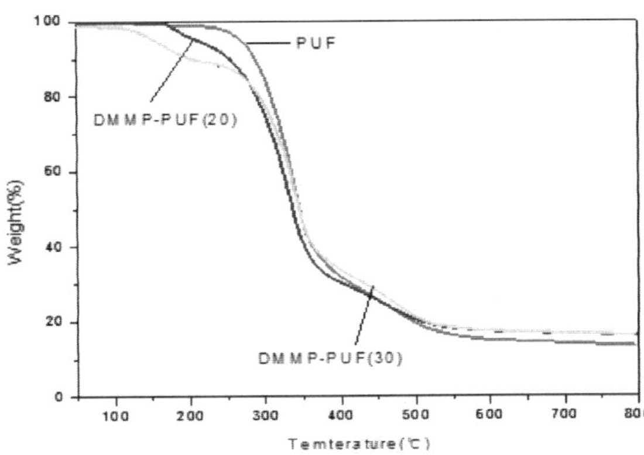

Figure 2. TG Curves of PU samples.

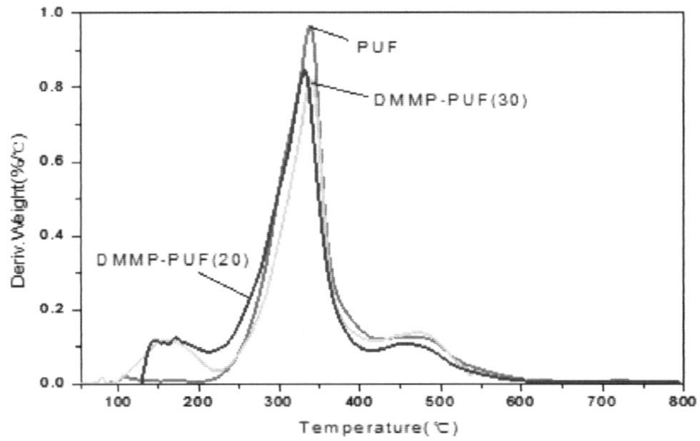

Figure 3. DTG curves of PU samples.

3.3 TG-FTIR

TG-FTIR analysis has been used to investigate the influence of DMMP on the pyrolysis products in the gaseous phase during the thermal degradation of the PUF sample. figure 4 and 5 show the 3D and 2D TG-FTIR spectra of the neat PUF sample and the flame retardant PUF sample with the 30% DMMP concentration. As shown in figure 4, absorption peaks at 1114cm^{-1}, 1257cm^{-1}, 1220cm^{-1}-1320cm^{-1}, 3318cm^{-1}, 1622cm^{-1} and 2275cm^{-1} can be observed for the neat PUF sample, which could be assigned to C-O-C, C-O from ester, C-N from amine group, N-H, -C=C and -NCO, respectively [9]. However, it can be seen in figure 5 the flame retardant PUF sample containing DMMP has less and lower absorption peaks at high temperature range compared with the neat sample. This can be explained by the formation of the thick char layer covering the decomposed PUF substrate after the incorporation of the flame retardant containing phosphorus, which prevents the further thermal degradation of the sample in the high temperature and inhibits the release of some low molecular decomposed compounds into the gaseous phase during thermal degradation process of the flame retardant sample [10].

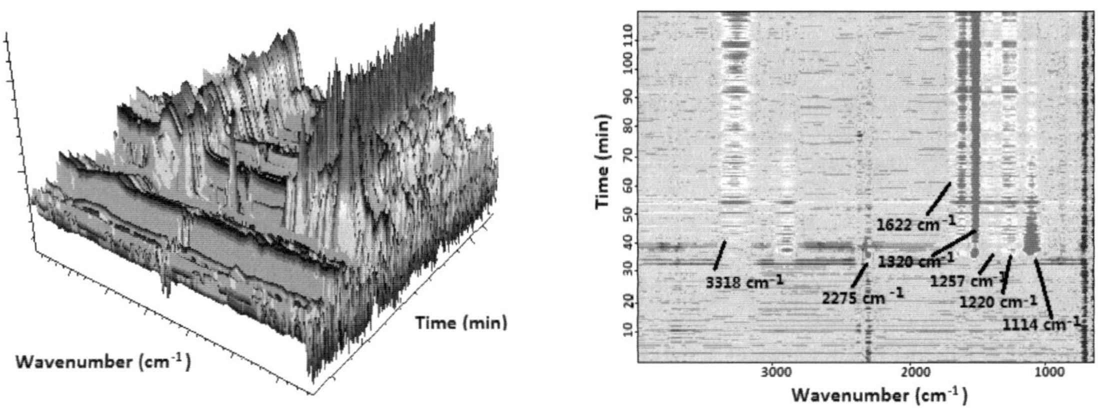

Figure 4. 3D and 2D TG-FTIR spectra of PU without DMMP.

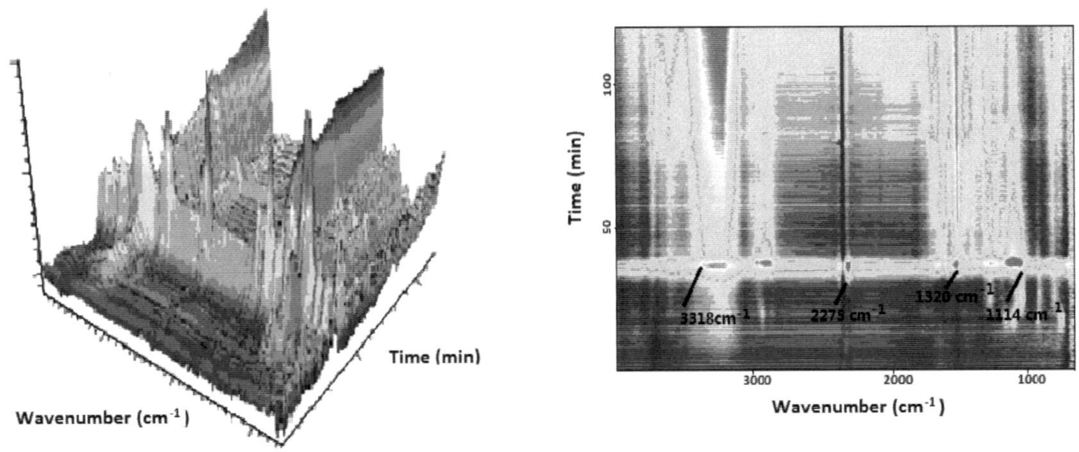

Figure 5. 3D and 2D TG-FTIR spectra of the flame retardant PUF sample with the DMMP concentration of 30%.

In order to obtain more information about the pyrolysis product evolutions of the samples during heating, FTIR spectra of the neat PUF sample at the maximum evolution rates has been examined. FTIR spectra of the neat sample at 340 and 479℃ are presented in figure 6. The characterized peaks at 2275cm^{-1}, 2352cm^{-1}, 3579cm^{-1} and 1530cm^{-1} can be observed in spectrum (a) [11], indicating the formation of some compounds containing -NCO, CO_2, H_2O and some products containing -NH from -NHC=O. It suggests the process of depolymerization at this temperature range, which could be caused by the break of the main chains in the PUF. In spectrum (b), the peaks at 1530cm^{-1} and 2275cm-1 have almost disappeared while some other peaks at 1513cm^{-1}, 2905cm^{-1}, 1465cm^{-1}, 1626cm-1 and 935cm-1can be observed, which could be assigned to phenyl, $-CH_3$, $-CH_2-$, C=C and $-CH_2-$ from alkene [12]. This indicates the thermal degradation of polyol and isocyanate, which are formed during the depolymerization of the PUF main chains. These results reveal that thermal degradation of the neat PUF can be described as two steps: the depolymerization of the main chains and the further pyrolysis of polyol and isocyanate. And thermal degradation of the neat PUF is presented in figure 7.

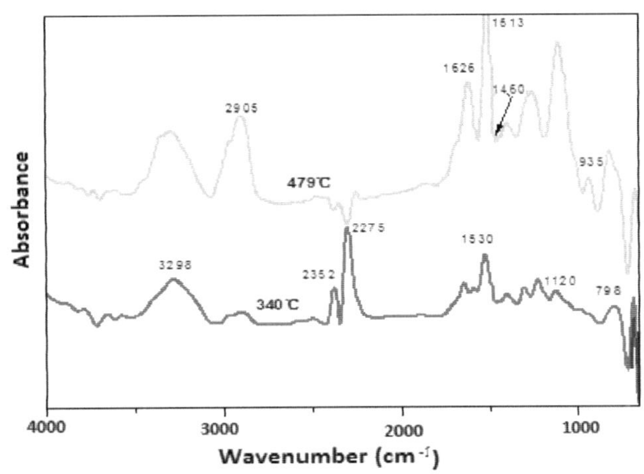

Figure 6. FTIR spectra of the neat PUF sample obtained at the maximal weight loss rates.

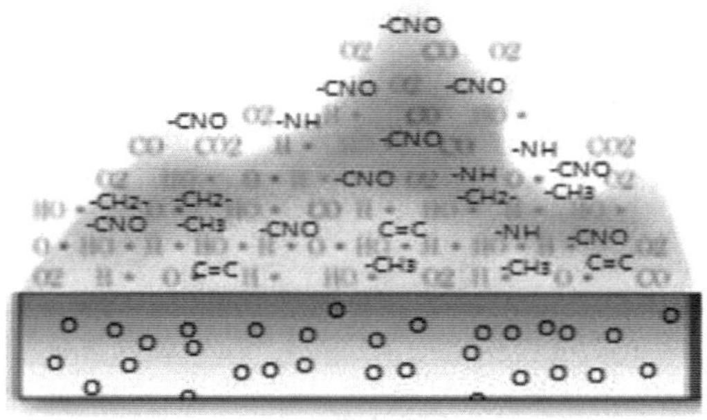

Figure 7. Thermal degradation of the neat PUF.

FTIR spectra of the flame retardant PUF sample at the maximum evolution rates of 163, 331 and 467℃ are shown in figure 8. The absorption peaks of 2314cm^{-1} and 1321cm^{-1} can be observed for the flame retardant PUF in spectrum (a), which can be assigned to P-H and P=O [13]. It suggests that the compounds containing phosphorous groups could be released into the gaseous phase during the decomposition of flame retardant PUF. And the derivatives of these compounds could terminate the free radicals existing in the gaseous phase during combustion and reduce the flammability of materials. It also indicates that the flame retardant action of DMMP can be also detected in the gaseous phase at relative high temperature range [14]. In addition, as shown in figure 8, spectrum (b) and (c) are almost similar with spectrum (a). This phenomenon proves the inhibited thermal degradation of the underlying matrix, which may be due to the covering mechanism of the intumescent residual layer in high temperature.

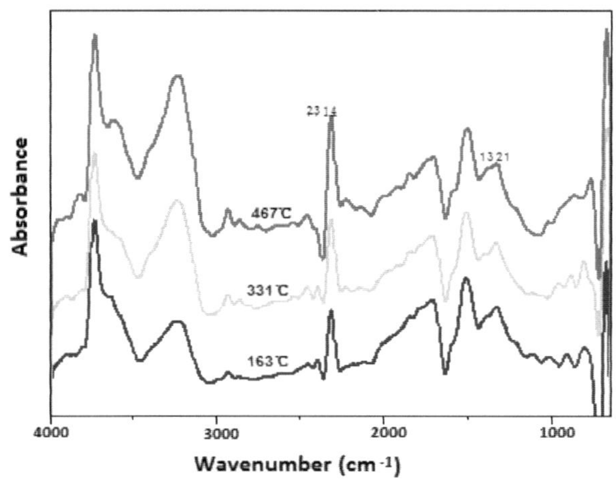

Figure 8. FTIR spectra of PUF/DMMP (30) obtained at the maximal weight loss rates.

The generated trends of water are shown in figure 9. It can be seen that the temperature of the maximum intensity peak is 364℃ for the flame retardant PUF, which is 48℃ lower than that for the neat PUF. This may be due to the decomposition of DMMP at the initial stage of combustion to form some phosphorus oxide and polyphosphorus acid, which could catalyze the dehydration reaction among polymer chains at the relative low temperature [15, 16]. The release of water could dilute the flammable gas in the gaseous phase and absorb the burning heat when subjected to the flame, which retards the thermal degradation of PUF and enhances its flame retardancy. However, the release of water for flame retardant PUF is inhibited obviously above 480℃ while that for the near PUF remains until 800℃. It is possibly because the intumescent residual char layer covering the decomposed flame retardant PUF could inhibit the release of water and prevent the thermal degradation of the material further at higher temperature range, which can be also confirmed by the constant residual amount at the temperature range of 500-800 ℃ in TG curve of the sample (30% DMMP concentration) (Figure 2).

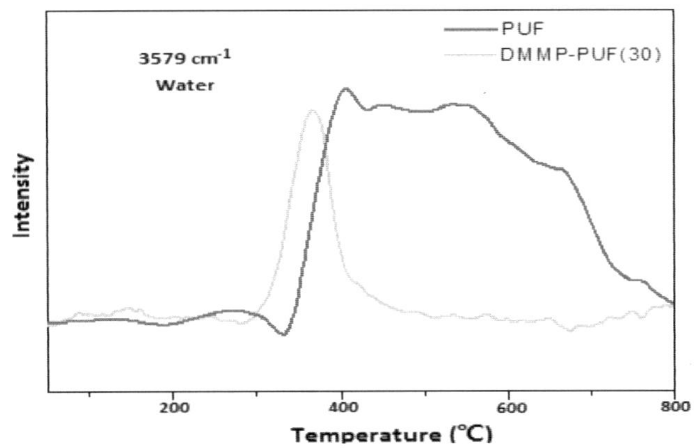

Figure 9. Relationship between intensity and temperature for water.

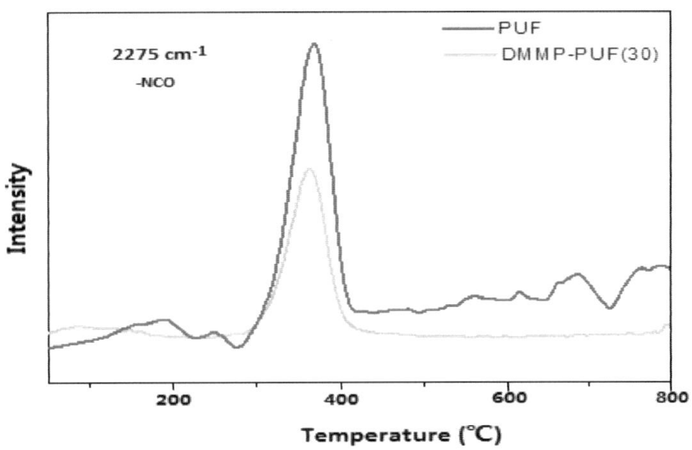

Figure 10. Relationship between intensity and temperature for -NCO containing products.

The generated trends of -NCO are shown in figure 10. The maximum intensity peak of -NCO for the flame retardant PUF sample can be observed in temperature range of 300-405℃. And the highest maximum peak of -NCO for the neat PUF is at the same temperature range. However, it shows that the intensity of the maximum peak for the neat PUF is relatively higher than that for the flame retardant PUF sample. And the -NCO release from the neat sample could be detected until 800℃ compared to that from the flame retardant PUF. This implies that the thermal stability of the flame retardant PUF could be improved due to the covering mechanism of the intumescent residual layer on the substrate at higher temperature range. And the result reveals that the incorporation of DMMP could inhibit the release of some decomposed compounds containing -NCO in high temperature range, which could finally reduce the toxic hazard on people during the combustion of the PUF material.

4. Conclusions

Dimethyl methylphosphonate (DMMP) has been used as a flame retardant containing phosphorus to improve flame retardancy and thermal stability of the polyurethane foam material by the cast molding method. The LOI tests show that the LOI value of the PUF sample increases with the DMMP concentration and the P%. Thermal stability of the flame retardant sample can be enhanced at high temperature by forming the higher residual char covered on the decomposed PUF substrate at high temperature. Results from the LOI test and thermal analysis confirm the condensed flame retardant activity of DMMP for the flame retardant PUF sample. However, based on the information from TG-FTIR analysis, it can be concluded that although DMMP could improve the thermal stability of PUF samples by the formation of the residual char in the condensed phase, which covers the substrate to retard the thermal degradation of material, the release of the gaseous products can be also inhibited due to the presence of DMMP at high temperature.

Acknowledgement

The authors are grateful to the National Research Program of China (973 Program) (2012CB719701) and the Project-sponsored by Sichuan Fire Research Institute of Ministry of Public Security (20148806Z) for financial support.

References

[1] Grace T C, Salvador C N and Francisco V 2012 *Frontiers of Architectural Research* **1** 348
[2] Hu S J and Li Y B 2014 *Bioresource Technol.* **161** 410
[3] Adel R K, Mustafa K, Kyle M, Mohini S and Oh S K 2015 *Ind. Crop. Prod.* **74** 1
[4] Wang D Y, Ge X G, Wang Y Z, Wang C, Qu M H and Zhou Q 2006 *Macromol. Mater. Eng.*, **291** 638
[5] Liu W, Chen D Q, Wang Y Z, Wang D Y and Qu M H 2007 *Polym. Degrad. Stab.* **92**1046
[6] Tian G X, Ji Q, Xu D M, Tan L W, Quan F Y and Xia Y Z 2013 *Fiber Polym.* **14** 767
[7] Chen D Q, Wang Y Z, Hu X P, Wang D Y, Qu M H and Yang B 2005 *Polym. Degrad. Stab.* **88** 349
[8] Liu W, Zhang S, Bourbigot S, Sun J, Yu L H, Feng Q L, Chen X S and Zhu XJ 2012 *Polym. Adv. Technol.* **23** 1550
[9] Hu J H and Zhen X F 2011 Bei Jing *Science Press* 286
[10] Wei L L, Wang D Y, Chen H B, Chen L, Wang X L and Wang Y Z 2011 *Polym. Degrad. Stab.* **96** 1557
[11] David W H, Gayani K and John M K 2005 *Polym. Degrad. Stab.* **87** 555
[12] Chen X L, Huo L L, Jiao C M and Li S X 2013 *J Anal. Appl. Pyrol.* **100** 186
[13] Jiao L L, Xu G D, Wang, Q S, Xu Q and Sun J H 2012 *Thermochim. Acta* **547**120
[14] Cao Z H, Zhang Y, Song P G, Cai Y Z, Guo Q, Fang Z P and Peng M 2011 *J. Anal. Appl. Pyrol.* **92** 339
[15] Liu W, Chen L and Wang Y Z 2012 *Polym. Degrad. Stab.* **97** 2487
[16] Gu J W, Zhang G C, Dong S L, Zhang Q Y and Kong J 2007 *Surf. Coat. Tech.* **201** 7835

IOP Conf. Series: Materials Science and Engineering **137** (2016) 012038 doi:10.1088/1757-899X/137/1/012038

Stabilizing effects of estertins mercaptide (methyl acrylate) for PVC degradation

S H Zhang[1], T M Liu, J L Li, C R Wang, C Li and Z Q Wang

College of Chemistry and Chemical Engineering, Shanghai University of Engineering Science, Shanghai 201620, China

E-mail: zhangsh@sues.edu.cn

Abstract. The thermal and UV light (ultraviolet light) stability of PVC films with estertins mercaptide (methyl acrylate), methyltins mercaptide and the compound consisted of estertins mercaptide (methyl acrylate) and hydrotalcite (2:2.5) were investigated by ageing in a circulation oven at 190℃ and irradiating with 72W UV light for 96h, respectively, and then the yellowness and transmission rate were tested by Color Quest XE. Hydrotalcite was proved to have good synergies with estertins mercaptide (methyl acrylate) on improving the thermal stability and UV light stability. The retarding effects of the heat stabilizers to PVC degradation were tested by TGA from 50℃ to 600℃. The results show that temperature of HCl evolution from PVC film was improved obviously by compounding with estertins mercaptide(methyl acrylate) and hydrotalcite and estertins mercaptide(methyl acrylate) was found to have a better long term stability. Sn^{4+} consistence of water and seawater in which films before and after UV light irradiation were soaked for 60 days was analyzed by ICP; the results indicate that the Sn^{4+} consistence from the films with estertins mercaptide(methyl acrylate) as thermal stabilizer was lower than that from the film with methyltins mercaptide. The crosslink moderately by UV irradiation for PVC films can hold back the dissolution of organotin heat stabilizers from PVC products into water and seawater.

1. Introduction

Organotin has been used for pesticides, wood preservatives and heat stabilizers for polyvinyl chloride (PVC) and antifoulant for several decades [1]. They are mainly classified into two categories, estertins and alkyltins. Dimethyltins bis-(isooctyl thioglycollate)(methyltins mercaptide) are widely used as thermal stabilizer in the production of sheets, profiles, general purpose and potable water pipes, sidings, films, foils, bottles and articles meant for paper packaging [2,3]. However, the byproducts with which they react such as labile chlorine atoms or HCl exhibit greater toxicity [4-8], and cause a serious pollution to the water when they are used as the stabilizer in PVC water pipe [9-11]. According to the literature [12], estertins, such as di-β-carboalkoxyethyltins bis (isooctyl thioglycollate) [estertins mercaptide (methyl acrylate)] are lower toxic, cheaper and safer than alkyltins, and the byproducts of the reaction between estertins mercaptide (methyl acrylate) and the polymeric chains, mainly accumulated are much lower toxic than those of methyltins mercaptide[12], but preliminary researches showed that the retarding effects to PVC degradation at early stage were not so effective as methyltins mercaptide.

Hydrotalcite [$Mg_6Al_2(OH)_{16}CO_3 \cdot 4H_2O$] which has been known for an extended period of time has good synergies with many kinds of stabilizers such as Ca/Zn compounds and lead salts [13]. The objective of the present study is to investigate the retarding effects of estertins mercaptide (methyl acrylate) mercaptide (methyl acrylate) and hydrotalcite(2:2.5) on degradation of PVC.

2. Experimental

2.1 Materials, equipments and samples preparation

PVC paste resin (P440) from Shanghai Chlor-Alkali Chemical Co., Ltd. , dioctyl phthalate (DOP) and stearic acid from Shanghai Sinopharm Chemical Reagent Co., Ltd., hydrotalcite from Xiamen Xindakang Inorganic Materials Co., Ltd., methyltins mercaptide from Shanghai Zhiqiang Plastic Auxiliary Co., Ltd., and estertins mercaptide (methyl acrylate) made according to the literature [14], seawater was from Yellow Sea located in Fengxian District of Shanghai, circulating water was from laboratory. Films(about 30mm diameter×1.5mm thickness) were prepared from PVC (100phr), DOP (60phr), stearic acid(1phr) and heat stabilizer(2phr) by mixing at 25°C and gelating at 170°C for 15 min. The thermal stability, the UV light stability and the stability in water and seawater of three film samples were tested. The films $1^{\#}$, $2^{\#}$ and $3^{\#}$ are compounded with different thermal stabilizers, namely estertins mercaptide (methyl acrylate), methyltins mercaptide and estertins mercaptide (methyl acrylate)/ hydrotalcite (2:2.5), respectively.

2.2 Test methods

Each film sample was cut into nine circular specimens (30mm diameter×1.5mm thickness) for testing thermal stability for 45 minutes in a forced air convection oven, and took out one piece every 5 minutes. The thermal stability of film samples was investigated both by comparing the color changes of every specimen aged for 5 min, 10 min, 15min, …, 45 minutes directly and by measuring yellowness and transmission rate of the specimens by a Color Quest XE from American Hunterlab Company. Mass loss during heating of the film samples was measured by thermogravimetry using a Germany Linseis Thermogravimetric Analyzer (TGA) with heating at a rate of 10℃ min^{-1} from 50℃ to 600℃ under nitrogen at 40 cm^3 min^{-1}.

The film samples were irradiated during 96 h at room temperature, using 72W, T8 G13, 365nm, UV lamps and the effect was measured by yellowness before and after UV aging.

About 0.5g films $1^{\#}$-$3^{\#}$ before and after UV aging were cut into 2mm ×2mm size particles and immersed into 150mL water or seawater for 60 days at room temperature. Sn^{4+} consistence of the filtered water and seawater was analyzed by means of VARIAN inductively coupled plasma atomic emission spectrometer (ICP OMS) to determine the stability of three kinds of stabilizers both in water and seawater.

3. Results and discussions

3.1 Thermal stability and transparence of the film samples

The air oven ageing results of the samples were shown in figure 1.

When PVC is under heating, HCl which is evolved as a result of PVC degradation induces the formation of conjugated double bonds. The stability of the samples can be characterized by yellowness because the colors of PVC films turn from yellow to black as conjugated structures increase.

The experimental results from Fig.1 showed that significant difference wasn't observed of all the film samples within 25 minutes at 190℃, the colors of film $1^{\#}$ and $2^{\#}$ became brown in 30 minutes and 40 minutes, respectively, and turned to dark brown in 45minutes. However, film $3^{\#}$ sample was still light brown till 45minutes. The results above indicated that the thermal stability of methyltins mercaptide was better than that of estertins mercaptide (methyl acrylate), and that is the compound of estertin mercaptide (methyl acrylate) and hydrotalcite that shows the best stability under thermal aging.

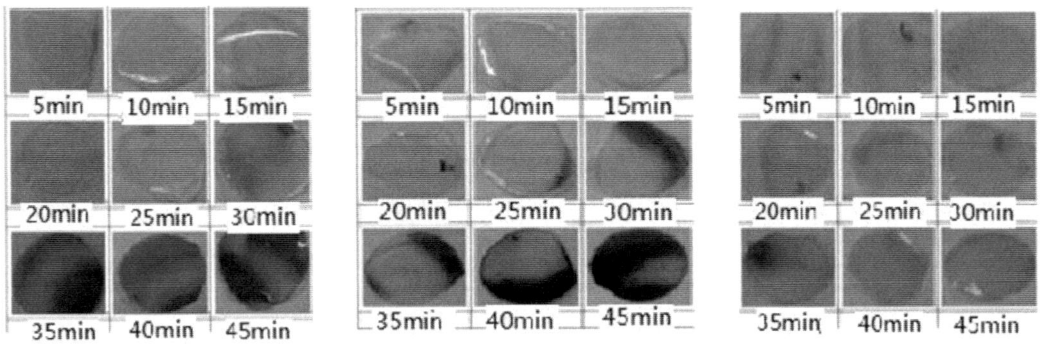

(a) film 1# (b) film 2# (c) film 3#

Figure 1. Pictures of three films during 45 minutes thermal ageing.

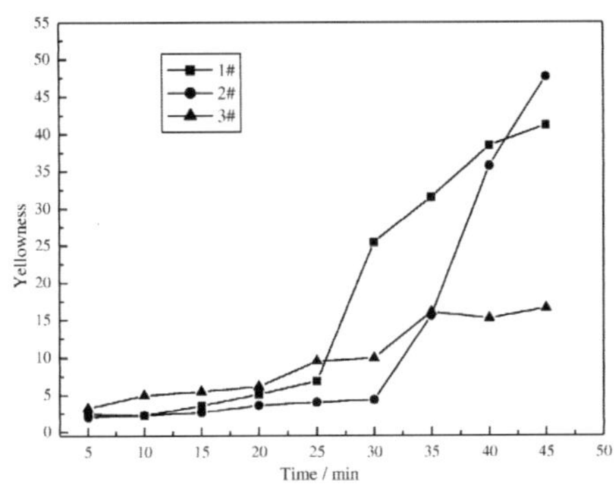

Figure 2. Curves of yellowness and time

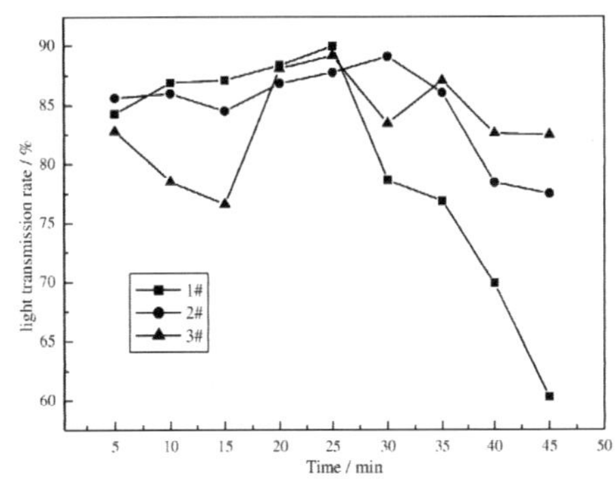

Figure 3. Curves of transmission rate and time

The yellowness from films 1# -3# given in figure 2 increased from 2.22, 1.99 and 3.24 to 6.85, 4.02 and 9.52 within 25 minutes, respectively. The films 1# and 2# were over 40 after 45 minutes, while film 3# attainonly 16.57 after the same time of aging. The results told that hydrotalcite shows synergistic effects with estertins mercaptide (methyl acrylate) .

Figure 3 shows the light transmission rate difference of film 1#-3# The films 1# and 2# increased from 84.26% and 85.60 to 89.97% and 87.74%, respectively within 25minutes, and film 1# had excellent transparence, probably because the ester group in estertins mercaptide (methyl acrylate) had good plasticizing action, but the transparence decreased quickly with the increase of conjugated structures after 25 minutes. The results that the light transmission rate of film 3# increased from 82.77% to 89.16% in the first 25 minutes and stayed at 82.45% till 45 minutes indicating the hydrotalcite could be well dispersed in PVC under the action of the plasticizer.

3.2 Analysis to UV light irradiation

The yellowness and the light transmission rates of films which were irradiated by 72 KWh UV light for 96h were reported in table 1 giving the quantitative information about the resistance to UV light aging of the samples.

Table1. Test results of UV light irradiation.

Characteristic parameters	Experimental samples		
	film 1#	film 2#	film 3#
Before aging:			
Yellowness	4.52	5.02	3.19
Transmission rate (%)	89.97	87.74	89.16
After aging:			
Yellowness	7.45	6.78	3.65
Transmission rate (%)	88.29	87.22	88.24

It was observed from table 1, before UV light irradiation, the transparence and yellowness of film 2# is not so good as film 1#, however, the transparence of the film 3# with 2.5phr hydrotalcite is 89.16, which proves hydrotalcites can enhance the thermal stability to PVC. Comparing the results of before and after UV irradiation, the transparence of the film 1# decreases for 1.87%, film 2# for 0.59% and film 3# for 1.03%, which indicates methyltins mercaptide can keep the film better transparence after UV irradiation. The yellowness of film 1# increases for 65%, film 2# for 35% and film 3# for 14%. The yellowness of film 3# changes less, which proves that the compound of estertins mercaptide (methyl acrylate) and hydrotalcite can increase the UV irradiation stability of PVC film.

3.3 PVC degradation analysis

When PVC was heated, DOP evaporated from the surface and HCI evolved from thermal degradation released from film as gaseous products [5]. The mass loss caused by DOP evaporation was neglected because its content in the three film samples were the same. TGA curves presented in figure 4 and the respective results presented in table 2 indicate that when the remaining mass rate was 99%, the temperatures were 138.1℃, 163.2℃, and 156.4℃ from films 1#, 2# and 3#, respectively. At the remaining mass rate was 95%, the temperatures were 198.1℃, 204.2℃, and 200.5℃, respectively. This result proves film 2# is more stable than the other two films at the initial stage of PVC degradation, which means methyltins mercaptide gives the better stability to PVC. However, with the increase of temperature, films 2# and 3# are more stable indicates estertins mercaptide (methyl acrylate) and the compound stabilizer have the better stability for PVC at higher temperature.

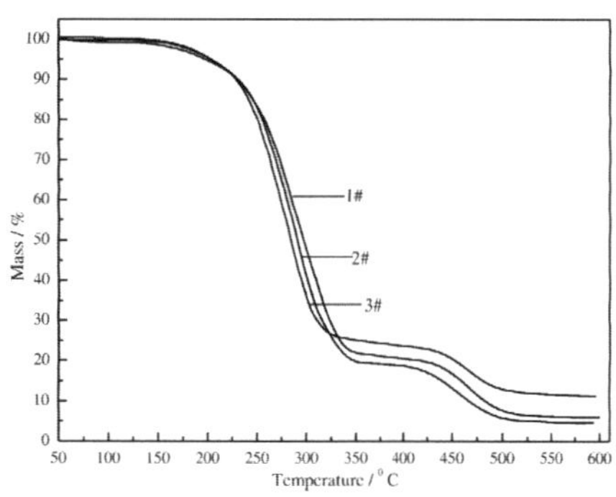

Figure 4. TG curves

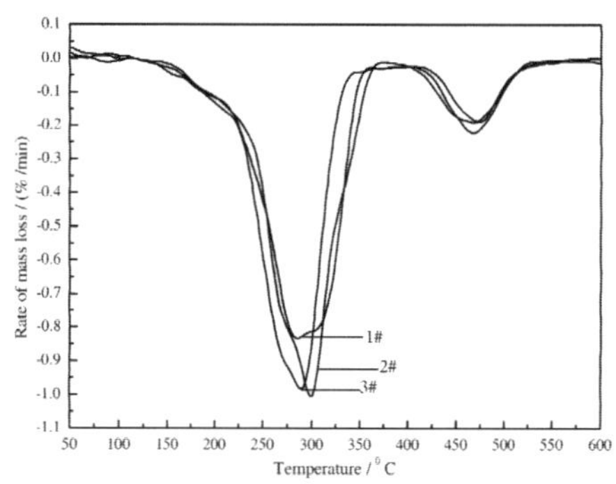

Figure 5. DTG curves

DTG curves in figure 5 showed that the degradation speed and the mass loss rate from film 2# was the biggest confirming that the results in table 2 furthermore. The degradation speed of film 3# as shown in figure 5 was faster than that of film 1# because the crystalwater, OH^- and CO_3^{2-} also evaporated from the layers of hydrotalcite accompanied with the increasing of temperature. But OH^- and CO_3^{2-} can react with HCl, so it has the best thermal stability even if the degradation speed is bigger when the temperature is over 200℃ [15].

Table 2. TGA analysis of thermal degradation.

Characteristic parameters	Temperature at different stage (℃)		
	film1#	film2#	film3#
$T_{1\%}$	138.1	163.2	156.4
$T_{5\%}$	198.1	204.2	200.5
$T_{10\%}$	230.1	231.1	227.2
$T_{16\%}$	248.6	248.4	242.7
$T_{20\%}$	257.3	255.3	249.4
Δm_1/ %	79.47	81.20	76.80

note: Tm% is temperature at m% remaining mass rate;
Δm_1 is the rate of mass loss from 50 to 400℃.

3.4 Analysis of Sn^{4+} consistence in water

Table 3 gave the Sn^{4+} consistence of the water in which films 1#-3# before and after UV irradiation had been soaked for 60 days was analyzed by means of ICP. At first, The standard liquid is prepared with 20mg/L Sn^{4+} consistence. The Sn^{4+} consistence of fresh water and seawater is analyzed by ICP, in fresh water no Sn^{4+} trace is found and Sn^{4+} consistence of seawater is 0.0722mg/L.

Table 3. The analysis results of ICP test.

Samples	Sn^{4+} consistence of Water (mg/L)	Sn^{4+} consistence of Seawater (mg/L)
film 1#	0.823	0.954
1#UV aging	0.383	0.416
film 2#	1.066	1.930
2# UV aging	0.482	0.779
film 3#	2.568	4.029
3# UV aging	1.348	1.547

Organotins and organotins chlorides produced by organotin reacting with HCl or liable chloride atoms are digested during the analysis by ICP and exist in the form of Sn^{4+} in freshwater, so organotin's consistence dissolved into the water from PVC film samples can be characterized by means of detecting Sn^{4+} consistence of the freshwater. The results from table 3 indicated that Sn^{4+} consistence from film 1# was lower than those from film 2# both in water and seawater, it meant estertins mercaptide (methyl acrylate) were more stable. It is interesting to find that Sn^{4+} consistence from film 3# was far higher than those from film 1# and 2# assigned to organotin hydroxides and organotin carbonate compounds formed by OH^- and CO_3^{2-} evaporated from the interlayers of hydrotalcite reacted with organotin chlorides dissolved in water, but the effects of these byproducts to the freshwater need to be investigated furthermore [16]. We also find that Sn^{4+} consistence of every film after UV irradiation was higher than that of before UV irradiation, proves UV irradiation accelerates the solution of organotin from PVC film into water

The information that Sn^{4+} consistence in seawater was more than those in water showed salt compounds and microorganisms in seawater can accelerate the solution of organotin from PVC films. It's interesting to find Sn^{4+} consistence of the water and seawater which films after UV irradiation soaked were lower than that without being irradiated. This result showed that the crosslink moderately by UV irradiation for PVC films can hold back the dissolution of organotin heat stabilizers from PVC products into water and seawater.

4. Conclusions

The thermal stability of PVC films with estertins mercaptide (methyl acrylate), methyltins mercaptide and estertins mercaptide (methyl acrylate)/ hydrotalcite (2:2.5) compound was tested by air oven aging and TGA, respectively. The results showed that the property of inhibiting to PVC degradation under heating and UV light irridiation of estertins mercaptide (methyl acrylate) which didn't retard HC1evolution as well as methyltins mercaptide is improved by compounding with hydrotalcite $[Mg_6Al_2(OH)_{16}CO_3 \cdot 4H_2O]$. Compared to methyltins mercaptide, estertins mercaptide (methyl acrylate) kept a better stability both in water and seawater. The crosslink moderately by UV irradiation for PVC films can hold back the dissolution of organotin heat stabilizers from PVC products into water and seawater.

Acknowledgments
The work financially supported by science and technology development project of Yanbian state (2015-GX11) and university students' research and innovation project of Shanghai (Nos. cs1504007).

References

[1] Jindra P and Kim F D 2005 *Marcromolecules* **38(15)** 6352-6366
[2] Hakkarainen M 2008 *Berlin: VerFag Berlin Heidelberg* 11-50
[3] Duan R Z, Jiang P P and Zhang X Y 2007 *Fine and Speciality Chemicals* **15(6)** 1-5
[4] Cai J I, Wang M W and Li B W 2009 *Chemistry Research Toxicol* **22(9)** 1582-1587
[5] Fang Z Y, Wu B and Yong X J 2008 *Zhejiang Preventive Medical Association* **6(2)** 36-37
[6] Arkis E and Balkose D 2005 *Polym. Degrad. Stab.* **88(1)** 46-51
[7] Gao E J, Tang W and Meng J X 2011 *Plastic Additives* **2** 46-48
[8] Zhang X Q, Zhou J Y and Zhou E G 2006 *Industrial Health and Occupational Diseases* **32(5)** 304-306
[9] Madoka O, Takaomi A and Sayaka M 2007 *Water Air Soil Poll.* **178(6)** 255-265
[10] Xie W, Wang X H and Zhang J S 2010 *Environmental Science* **31(4)** 1072-1078
[11] Wu W Q, Zhou B and Yao Z W 2009 *Journal of Dalian Maritime University* **35(3)** 115-118
[12] Ronald E H and Oakes V 1976 *England: Akzo Chemie, Liverpool*: 122-133
[13] Zhang S H 2010 *Harbin: Harbin University of Science and Technology* 75-82
[14] Zhang S H, Liu B, Li H S and Hui Zhu 2011 *Journal of Chemical Engineering of Chinese Universities*, **25(4)** 662-669
[15] Ahmet N A, Birgül Z K and Abidin T 2011 *Appl. Clay Sci.* **51(1)** 308-316.

IOP Conf. Series: Materials Science and Engineering **137** (2016) 012039 doi:10.1088/1757-899X/137/1/012039

Polymeric efficiency in remove impurities during cottonseed biodiesel production

H L Lin[1], Y H Liang[2], J Yan[1], H D Lin[3] and A R Espinosa[4,5]

[1]Chemistry & Chemical Engineering Department, Zhongkai University of Agriculture and Technology, Guangzhou 510225, China.
[2]Mechanical Engineering Department, University of Wisconsin-Madison, WI 53706, USA.
[3]Guangdong Institute for Veterinary Drug and Feedstuffs Control, Guangzhou 510230, China.
[4]Soil Net LLC,WI 53508, USA

E-mail: aicardoroa@soilnetllc.com

Abstract: This paper describes a new process for developing biodiesel by polymer from crude cottonseed oil. The study was conducted to examine the effectiveness of the alkali transesterification-flocculation-sedimentation process on fast glycerol and other impurities in the separation from biodiesel by using quaternary polyamine-based cationic polymers SL2700 and polyacylamide cationic polymer SAL1100. The settling velocity of glycerol and other impurities in biodiesel was investigated through settling test experiments; the quality of the biodiesel was investigated by evaluating the viscosity and density. The results revealed that SL2700, SAL1100 and their combination dramatically improved the settling velocity of glycerol and other impurities materials than traditional method. SL 2700 with molecular weight of 0.2 million Da and charge density of 50% then plus SAL1100 with molecular weight of 11 million Da and charge density of 10% induced observable particle aggregation with the best settling performance.

1. Introduction

The most common method of biodiesel production is by transesterification of vegetable oils or animal fats with methanol in the presence of an alkaline catalyst [1-3]. The primary concern for biodiesel producers is the steady availability of low-cost oil feedstock, which yields high quality biodiesel product that is competitive in price with diesel fuels. Unfortunately, the efforts expended to obtain and maintain high purity refined oil make the resulting biodiesel less economically viable alternative to diesel under the prevailing market conditions [4]. Therefore, there is considerable interest and economic benefit in using unrefined (crude) oil directly for biodiesel production. However, compared to the refined vegetable oils, their crude oil counterparts contain higher free fatty acids (FFA) and form higher amounts of soap during transesterification, which inhibits the separation of the glycerol from methyl esters using the traditional method [5]. During biodiesel production, the cottonseed oil is de-gammed following complicated steps of adding water and acid first to reduce the phosphorous content to a very low level and rewashing the oil with water. The glycerol produced by transesterification is separated and removed by gravity or centrifugation because glycerol is insoluble in the biodiesel. The gravity method takes several hours, and the centrifugation process is very expensive [6]. Therefore, the developing alternative methods to efficiently remove glycerol, gossypol and other impurities from methyl esters are essential when crude cottonseed oil is used directly for biodiesel production. If this can be performed rapidly and inexpensively, the economic competitiveness of the biodiesel will improve significantly leading to widespread use of biodiesel in place of diesel. In addition, the phosphorous content of crude cottonseed oil adversely affects the biodiesel production process [7-9].

Herein we describe a polymeric flocculation method to improve the biodiesel production process by directly transesterifying crude cottonseed oil. The polymeric flocculants destabilize the particles through charge neutralization, electrical double layers and subsequent formation of particle-polymer-particle bridges, which generate a small volume of sludge output [10-13]. Previous studies have focused on thickening operations in mineral processing, with flocculant [14-18]. However, flocculation technology practice in making cottonseed biodiesel has not performed and little work has been reported. If phosphorus and other impurities can be coagulate and flocculate by polymers then sink from the biodiesel, it will save energy and time of the making process. Our objectives are: 1) investigate the effectiveness of using polymeric flocculants as a sedimentation aid during biodiesel production from crude cottonseed oil ,2) characterize the biodiesel properties and 3) molecular weight, charge density of polymer are analyzed to gain a better understanding of the flocculation mechanism. Cottonseed oil was chosen because transesterification of cottonseed oil produces methyl ester whose properties are comparable to those of the conventional diesel fuel [19-20].

2. Materials and methods

2.1. Materials

Crude cottonseed oil was obtained from Wisconsin, Madison. Methanol of 99.9% purity was purchased from Fisher Scientific (Fair Lawn, NJ). Sodium hydroxide of 98.2% purity was purchased from Wausau Chemical, China. Potassium hydroxide of 99.0% purity was purchased from Fisher Scientific (Fair Lawn, NJ). Two commercial grade cationic polymers of different molecular weight (Mw) and charge density (CD) were obtained (Soil Net LLC, WI). The polymers were designated as SL2700 (Mw=0.5×10^5 Da; CD=50 %) and SAL1100 (Mw=11×10^6; CD=10%). SL2700 is a quaternary polyamines-based polymer and SAL1100 is a polyacrylamide.

2.2. Transesterification of crude cottonseed oil

Free fatty acid (FFA) content of the biodiesel was determined by titration [21] to be 2.51 mg KOH/g; phosphorus content of crude cottonseed oil was between 200-330 mg/L [22]. Crude cottonseed oil was transesterified by methanol (molar ratio 1:6) with a basic catalyst (0.8 wt-% of NaOH to oil) in beaker. The reaction was carried out for 90 min at 60 °C while mixing thoroughly with a magnetic stirrer (~1000 rpm).

2.3. Flocculation and Sedimentation

The polymers SL2700, SAL1100, and their 1:1 (w/w) mixture (SL2700 added first, then SAL1100 the second by step) were added following transesterification as above. The following four polymer concentrations were used: 1.0, 2.0, 3.0, and 4.0 mg/L of oil. The polymers were mixed in thoroughly using magnetic stirrer (~300 rpm) under 60 ºC for 10 min.

The samples were held at 30ºC undisturbed for settling of the flocs of gums, gossypol and other impurities formed. The settling rates of the flocs were determined by measuring the mud-line heights by taking the samples in 100 mL graduated cylinders. The cylinders were stoppered and carefully inverted

IOP Conf. Series: Materials Science and Engineering **137** (2016) 012039 doi:10.1088/1757-899X/137/1/012039

back-and-forth 8-10 times for thorough mixing and then held upright. Non-polymer amended samples were concurrently analyzed for transesterification using the same techniques. Results were then read, calculated, recorded and reported in units of cm / min. The measurements were conducted three times for each sample and the results were averaged. Unreacted methanol and water were removed in oven at 100 °C until the constant weight was observed.

2.4. Effect of polymers on sample characterization
Free fatty acid (FFA) content of the biodiesel was determined by titration; Gossypol content of biodiesel was tested by UV–Vis absorption spectra in 1 cm cells using a UV- visible spectrophotometer (UV-1601PC, Japan); Trace elements were measured by wavelength dispersive X-ray fluorescence (WDXRF) spectrometry (S4 Pionees, Bruker AXS, Inc., Madison, WI) all the samples were analyzed applying a measurement time of 1200 s; specific gravity of the biodiesel was precisely measured to investigate the glycerol settling in the biodiesel with polymer types and dosage, the measurements were done from 20 °C to 100 °C by using specific gravity meter (Precision hydrometer, Chase, USA, 233743), the measurements were conducted three times for each sample and the results were averaged; dynamic viscosity of the biodiesel samples were measured at 15-100°C over a shear rate range of 1 to $100s^{-1}$ using rheometer (Anton Paar Instruments Inc., Physical MCR301) equipped with cup-and-bob geometry, three replicates were produced for each type of samples and the results were averaged.

3. Results and discussion

3.1. Flocculant dose and settling behavior
Figure 1 (a)-(d) shows the biodiesel/impurities settling velocity with different dosages of SL2700, SAL1100, and combination of SL2700+SAL1100 (1/1, w/w), respectively. From the settling graphs, it could be pointed out that impurities settling could be finished in around 90, 75, 50, and 50 min at 30 °C when polymer dosage was 1.0, 2.0, 3.0, and 4.0 mg/L, respectively. At polymer dosage of 1.0-3.0 mg/L, the higher velocity of solid settling is occurred by higher polymer dosages, but the settling velocity has no significant difference when different kinds of these polymers and dosages are used in this range. It can be seen in figure 3 that the addition amount of polymers (4.0 mg/L) does not help in enhancing the settling velocity. It is also observed that impurities settling could be faster by using polymers comparing no polymer is used at the same settling condition. When polymer is not added, the time to reach the stable compact heights of impurities is near 5 h at 30 °C. In addition, the sequence of compact efficiency of impurities in canola blend by polymers is SL2700 + SAL1100 > SAL 1100 > SL2700 when same dosage polymers are used. The stable compact height of impurities in cylinder is 22.1, 23.6 and 24.3 cm after 90 min when SL2700+SAL1100, SAL1100, and SL2700 are used respectively.

The clarification process of crude cottonseed oil using polyamines-based flocculant polymers has been discussed by Aicardo et al. [23]. In the transesterification process of cottonseed crude oil, glycerin can form a uniform phase at any mixing ratio. Therefore, the glycerin layer from the biodiesel system has a higher density than the biodiesel resulting in a separation between the biodiesel layer and glycerin layer due to the difference in the density of both layers. At the same time, the use of the polymers causes the agglomeration soap formed in the process of neutralization or bridging. Soap can help to combine glycerin and gossypol, which greatly enhances the separation of the glycerin fraction from the biodiesel. A suitable amount of soap also reduces the interfacial tension between the biodiesel and the impurities aqueous phase (phosphorus and salts). This improves the reaction between polymer and the impurities.

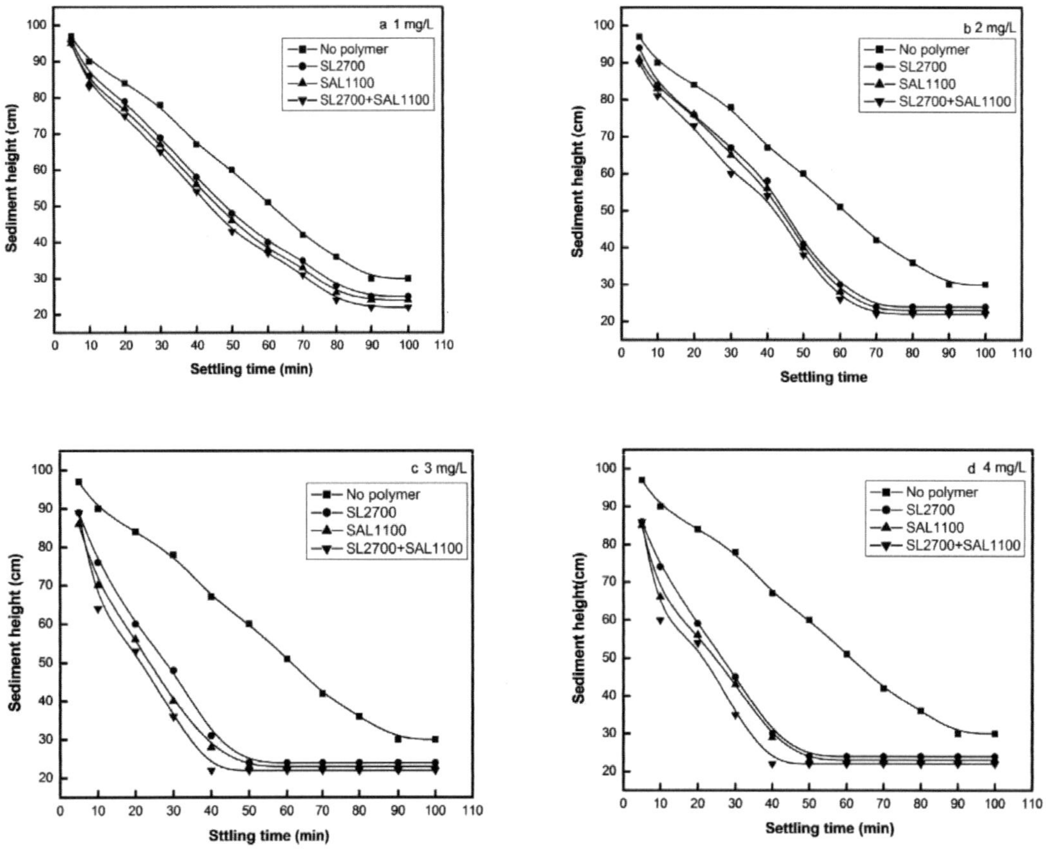

Figure 1. Settling behavior of cottonseed biodiesel suspension with polymers dosages of a,1.0; b,2.0; c,3.0; and d,4.0 mg/L at 30 °C.

3.2. *Effect of polymer molecular weight and charge density on the flocculation process*

Figure 1 also shows that the combination of polymers SL2700 and SAL1100 induce observable particle aggregation with the best compact efficiency of impurities in canola biodiesel blend; SAL1100 induces particle aggregation with the second compact efficiency of impurities; and SL2700 induces particle aggregation with the third compact efficiency of impurities. SAL1100 is polyacylamide cationic polymer with lower charges and higher molecular weight, which can be ascribed to the strong bonds of polymer chains due to bridging. Some of the segments of SAL1100 chain can adsorb onto the surface of particles through electrostatic interactions, some of the reminder of the polymer extending into solution phase. As other particles approach the SAL1100 chain, they interact with SAL1100 chain, thus a bridge is formed between particles then flocculate. SL2700 is quaternary polyamine-based cationic polymer with higher charges and lower molecular weight, which can adsorb onto the surface of the anionic particles in cottonseed biodiesel through electrostatic interactions, most of the charged sites on the polymer attach to the particle surface with oppositely charges then flocculate. The combination of SL2700 and SAL1100 function by charge neutralization and bridging, this adds the function of flocculation of particles then sedimentation.

3.3. Effect of polymers on gossypol, FFA and trace elements

It is established that the maximum absorption wavelength of gossypol on a spectrophotometer is 375 nm. Figure 2 showed that the clarified cottonseed biodiesel exhibited an absorption peak at 375 nm on a UV-vis spectrophotometer when no polymer was used. The cottonseed biodiesel clarified with various polymers didn't exhibit any absorption peak at this wavelength. The results indicate that gossypol contained in the biodiesel with polymer was sufficiently removed.

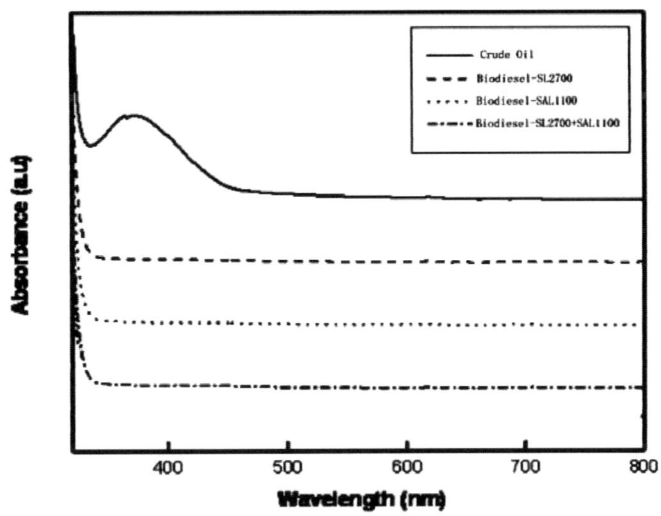

Figure 2. Changes in light absorbance of the crude cottonseed oil and cottonseed biodiesels at different wavelength.

Table 1. FA content in crude cottonseed oil and biodiesel (treated with polymer and no polymer). Values marked with the same superscript letters for the same test parameter indicate no significant differences at α=0.05.

	FFA (%)					
	1.0 mg/L	2.0 mg/L	3.0 mg/L	4.0 mg/L	0.0 mg/L	Crude oil
AL2700	0.33 ± 0.01^{a}	0.34 ± 0.01^{a}	0.33 ± 0.01^{a}	0.32 ± 0.01^{a}		
SAL1100	0.35 ± 0.01^{a}	0.34 ± 0.02^{a}	0.36 ± 0.04^{a}	0.33 ± 0.00^{a}	0.35 ± 0.04^{b}	2.51 ± 0.03^{c}
1:1 mixture	0.30 ± 0.02^{a}	0.35 ± 0.03^{a}	0.34 ± 0.02^{a}	0.32 ± 0.03^{a}		

FFA must be reduced to very low levels to improve the appearance and the oxidative stability of the vegetable biodiesel. According to EN14213 standard, the standard level of FFA in biodiesel is less than 0.5%. The effect of polymers types and polymer dosage on cottonseed biodiesel was reported in table 1.

There were no significance changes ($p > 0.05$) when different types and dosages (0.0-4.0 mg/L) polymers were used, FFA content in all cottonseed biodiesel was between in 0.30-0.36%, which met the biodiesel standard of FFA. The result also means that FFA content in biodiesel from crude cottonseed oil is not influenced by polymers treatment.

Trace concentration of elements can influence emission and flow properties of biodiesel. Table 2 lists the mineral contents of cottonseed oil and biodiesel samples. P, S, C and Na + K were the most abundant minerals found in crude cottonseed oil. The around level of P (310 mg/L), S (100 mg/L), C (1.0 mg/L), and Na+K (119 mg/L) in crude cottonseed oil, which was reduced to 0.0, 0.0, 0.0 and <3.0 mg/L respectively after transesterification-flocculation-sedimentation, and reduced to 3.00, 0.0, 0.3 and 2.0 mg/L respectively after the process of transesterification-sedimentation (no polymer treatment). It is evident that elemental concentration existing in cottonseed biodiesel treated by polymers and no polymers meet the EN14213 standard, but the the transesterification-flocculation-sedimentation combined process were more effective in reducing the level of P, C, and Na+K than the process of transesterification-sedimentation. Both of these methods could reduce S to 0.0 mg/L. The other concentrations of major element like sulfur, carbon, sodium and potassium existing in cottonseed biodiesel was higher than concentration of samples treated by polymer.

Table 2. DXRF measurement results (3.0 mg/L polymer was used). Values marked with the same superscript letters for the same elemental test parameter indicate no significant differences at α=0.05.

concentration (mg/L)	SL2700	SAL1100	SL2700+SAL1100	No polymer	Crude oil	EN14213 standard
P	0.0 ± 0.00^{a}	0.0 ± 0.00^{a}	0.0 ± 0.00^{a}	3.0 ± 0.00^{b}	310.0 ± 0.94^{c}	<10
S	0.0 ± 0.00^{a}	0.0 ± 0.00^{a}	0.0 ± 0.00^{a}	0.0 ± 0.00^{a}	100.0 ± 0.97^{b}	<10
C	0.0 ± 0.00^{a}	0.0 ± 0.00^{a}	0.0 ± 0.00^{a}	0.3 ± 0.00^{a}	1.0 ± 0.12^{b}	<0.3(%)
Na+K	2.0 ± 0.00^{a}	3.0 ± 0.00^{a}	1.0 ± 0.00^{a}	2.0 ± 0.00^{a}	119.0 ± 0.23^{b}	<5

3.4. Effect of polymers on specific gravity

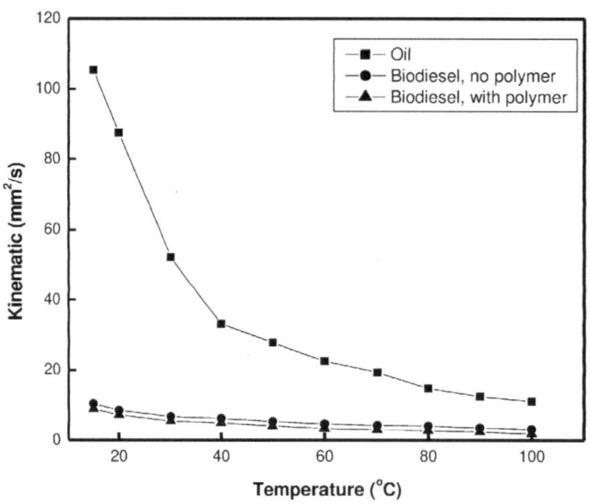

Figure 3. Specific gravity of the cottonseed biodiesel with different temperature.

Specific gravity plays an important role of diesel combustion and exhaust emissions.The specific gravity of biodiesel depends on the fatty acid composition. The specific gravity of biodiesel usually varies between 0.86 and 0.90 [24]. In this study, the specific gravity of biodiesel obtained by three polymers and no polymer are measured from 20-100 °C (figure 3). The testing result is approximately between 0.86-0.89. Regression analysis shows that the temperature depends of specific gravity to be linear over the measured temperature range. The regression line is in the form of

$$\rho = \mathrm{m}T + \mathrm{b}$$

where r is Specific gravity, T is temperature (°C), and m and b are given in Table 3 for all biodiesels. The average m value is -3.0×10^{-4} g/ml T. The straight line correlation fits the data very well, and R^2 is 0.99. Biodiesel has a density of 0.86-0.89. The specific gravity of biodiesel with no polymer and polymer treatment has no significant differences.

Table 3. Specific gravity correlation constants for cottonseed biodiesel.

Biodiesel type	m (g/ml °C)	b (g/ml)	R^2
Biodiesel-no polymer	-3E-04	0.892	0.99
Biodiesel-SL2700	-3E-04	0.890	0.99
Biodiesel-SAL1100	-3E-04	0.888	0.99
Biodiesel-SL2700 + SAL1100	-3E-04	0.887	0.99

3.5. Effect of polymers on viscosity and flow behavior

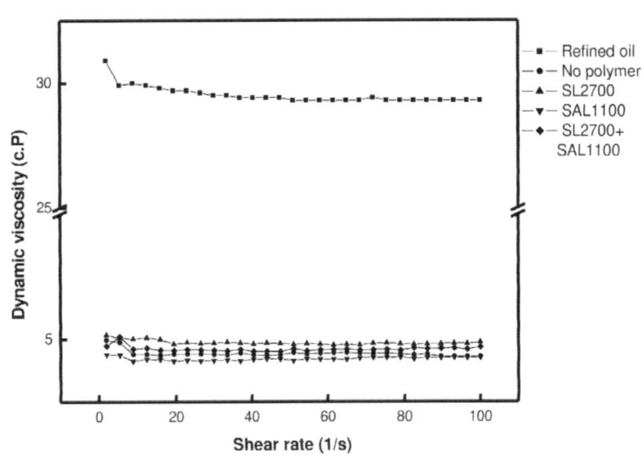

Figure 4. Viscosity curves of canola oil and biodiesel with different polymers treatment (settling 90 min) and no polymer treatment (settling 5 h)

Viscosity also plays an important role of diesel combustion and exhaust emissions. Accordingly, vegetable oil is converted to biodiesel primarily to reduce fuel viscosity. Figure 4 is the dynamic viscosity of refined cottonseed oil and biodiesel treated with SL2700, SAL1100 and SL2700+SAL1100 settling after 90 min and no polymers treatment after settling 5 h (methanol was removed also) at 40 ºC. These rheograms show approximately Newtonian behavior for cottonseed oil and biodiesel at 40 ºC because of the evidenced of the linear curves. They show Newtonian flow behaviors approximately because of double bonds [25]. Cottonseed oil has a relatively high feedstock dynamic viscosity (around 30 mPa s). It was observed that the viscosity of biodiesel obtained from no polymer is around 8.8 mPa s after settling 90 min and 5.0 mPa s after settling 5 h. Biodiesel obtained from the three polymers has much lower and similar dynamic viscosity after settling 90 min (around 4.8 mPa s). There was no significant difference for the performance of three polymers to viscosity. There was no significant difference for the performance of three polymers and no polymer treatment to viscosity after the systems are stable.

4. Conclusions

The use of cationic polymers as flocculation aid can hasten the cottonseed biodiesel to separate from glycerin and other impure materials after transesterification process. At this moderate polymer dosage of 1.0-3.0 mg/L, the efficiency of solid settling occurred by higher polymer dosage is higher. The impurities settling could be finished in around 50 min at 30 ºC when 3 mg/L of SL2700, SAL1100 and SL2700+SAL1100 (1:1, w/w) is used respectively. When polymer is not added, it needs at least 5 h to separate the impure materials from biodiesel. The use of cationic polymers could quicken the process of making cottonseed biodiesel.

Acknowledgments

This work is supported by the project of Guangdong Science and Technology (20140503, 2015B020215012) and the project of Guangzhou Science and Technology (201508030039)

Reference

[1] Issariyakul T, Kulkarni M, Meher L and Dalai A K 2008 Biodiesel production from mixtures of canola oil and used cooking oil *Chem. Eng. J.* **140** 77-85
[2] Atabani A E, Silitonga A S and Badrudin I A 2012 A comprehensive review on biodiesel as an alternative energy resource and its characteristics *Renew. Sust. Energ. Rev.* **16** 2000-2093
[3] Dmytryshyn S N, Dalai A K, Chaudhari S T and Misha H K 2004 Synthesis and characterization of vegetable oil drived ester: evaluation for their diesel additive properties *Bioresour. Technol.* **92** 55-64
[4] Leung D Y C, Wu X and Leung M K H 2010 A review on biodiesel production using catalyzed transesterification *Appl. Energ.* **87** 1083-1095
[5] Ojolo S J, Adelaja A O, and Sobamowo 2012G M Production of bio-diesel from palm kernel oil and groundnut oil *Adv. Mater. Res.* **36** 501-506
[6] Knothe G, Gerpen J V and Lrahl J 2004 *The biodiesel handbook* Acos Press, Illinois p70-87
[7] Smith E G, Janzen H H and Newlands N K 2007 Energy balances of biodiesel production from soybean and canola in Canada *Canadian J. of Plant Sci.* **87** 793-801
[8] Boey P L ,Maniam G P and Hamid S A 2009 Utilization of waste crab shell (Scylla serrata) as a catalyst in palm olein transesterification *J. Oleo Sci.* **58** 499-502
[9] Chu F F, Chu P N and Shen X F 2014 Effect of phosphorus on biodiesel production from scenedesmus obliquus under nitrogen-deficiency stress *Bioresour. Technol.* **152** 241-246
[10] Yeap K Y, Addai-Mensah J and Beattie D A 2010 The pivotal role of polymer adsorption and

flocculation conditions on dewaterability of talcaceous dispersions *Chem. Eng. J.* **162** 457-465
[11] Avci D, Mol N and Dagasan L 2002 Newcationic polyelectrolytes for flocculation process of bakers, yeast waste water *Polym. Bul.l* **48** 353-359
[12] Li H H, O'Shea J P and Franks V 2009 Effect of molecular weight of poly(N-isopropyl acrylamide) temperature-sensitive flocculants on dewatering *AIChE J.* **55** 2070-2080
[13] Yoon S Y and Deng Y 2004 Flocculation and reflocculation of clay suspension by different polymer systems under turbulent conditions *J. Colloid Interface Sci.* **278** 139-145
[14] Wong S S, Teng T T and Ahmad A L 2006 Treatment of pulp and paper mill wastewater by polyacrylamide (PAM) in polymer induced flocculation *J. Hazard. Mater.* **135** 378-388
[15] Gregory J and Barany S 2011 Adsorption and flocculation by polymers and polymer mixtures *Adv. Colloid Interface Sci.* **169** 1-12
[16] Yan Y D, Glover S M and Jameson G J 2004 The flocculation efficiency of polydisperse polymerflocculants *Int. J. Miner. Process.* **73** 161-175
[17] Zhou Y and Franks G V 2006 Flocculation mechanism induced by cationic polymers investigated by light scattering *Langmuir* **22** 6775-6786
[18] Yue Q Y, Gao B Y and Wang Y 2008 H. Zhang, Synthesis of polyamine flocculants and their potential use in treating dye wastewater *J. Hazard. Mater.* **152** 221–227
[19] Espinosa A R and Lin H L 2014 Enhanced biodiesel process. U.S. Patent No: 8907113
[20] Warabi Y, Kusdiana D and Saka S 2004 Biodiesel fuel from vegetable oil by various supercritical alcohol *Appl. Biochem. Biotechnol.* **115** 793-801
[21] Mahesar S A, Sherazi S.T.H and Khaskheli A R 2014 Analytical approaches for the assessment of free fatty acids in oils and fats *Anal. Methods* **6** 4956-4963
[22] Gupta M K 2007 *Pratical Guide to Vegetable Oil Processing* AOCS Press, Illinois, P 6-12
[23] Lin H L, Wedegaertner T C, and Espinosa A R 2015 A method to refine crude cottonseed oil using non-toxic polyamine-based cationic polymers, *Chinese Journal of Chemical Engineering* 23 379-383
[24] Stanciu I 2012 A new viscosity-temperature relationship for vegetable oil *J. Pet. Technol. Altern. Fuels* **3** 19-22
[25] Maskan M 2003 Change in color and rheological behavior of sunflower seed oil during frying and after adsorbent treatment of used oil *Eur. Food Res. Technol.* **218** 20-25

IOP Conf. Series: Materials Science and Engineering **137** (2016) 012040 doi:10.1088/1757-899X/137/1/012040

Internal stress determination in a polymer composite by Coda wave interferometry

Q Zhu[1], C Binetruy, C Burtin

GeM (Institut de recherche en génie civil et mécanique), CNRS UMR 6183, Ecole Centrale de Nantes, 44321, Nantes, France

E-mail: qzhu.nantes@gmail.com

Abstract. Coda wave interferometry (CWI) is largely employed in geotechnical applications to monitor changes due to cracks in materials but it is still not used for composite materials. In this paper, the technique is proposed to study internal stresses in a composite laminate $[0°/90°]_{63}$ and was compared with the traditional acoustic technique. It is shown that the Coda wave interferometry has better precision and sensibility than the method based on the first arriving time of flight (TOF) measurement, especially when the fiber orientation is normal to the wave propagation. This method is found to be promising for residual stress evaluation in composite materials.

1. Introduction

Polymer based composites are thought to have a great advantage for industry application because of their high specific weight, stiffness and strength. Internal stresses, especially residual stresses generated during the composite fabrication remains a problem and can cause premature failure of parts and/or warpage[1]. Due to their high sensitivity, FBG (Fiber Bragg Grating) is now often integrated in mold for deformation prediction [2, 3]. It is embedded in the host material and considerable calibration is needed for this 1D information [4]. Ultrasonic method is thought to be a powerful tool for the non-destructive detection and has been widely used for chemical reaction inspection [5] and crack detection in composites [6]. Referring to acoustoelastic property [7], the wave velocity is sensible to stresses imposed to materials. This is found for all types of materials although based on different physical reasons with a general form under small strains [8,15]:

$$V_{ij} = V_{ij}^{0}(1+\beta_{ij}\bar{\sigma}) \tag{1}$$

Where V_{ij} is the ultrasonic wave velocity that depends on averaged stress $\bar{\sigma}$ where i is the wave propagation direction and j is the wave polarization direction. The superscript '0' indicates for the velocity without stress. β_{ij} is the acoustoelastic constants. Such relation gives an access to the stresses generated during processing under averaged stress and isotropic material assumption.

Few works are devoted to polymer composites related to acoustoelasticity. Mase et al. [9] measured the acoustoelastic coefficients in Kevlar 49/SP 328 composites with different ply sequences by longitudinal and shear waves. Such coefficients vary according to different laminated structures. Santos et al. [10] used critical refractive longitudinal waves (L_{cr}) to address the stress influence in carbon/epoxy composite. This technique is characterized by the wave propagation parallel to and right below the surface. It can inspect the body from the surface because of the difficulty of access to the thickness of the composite. The time of flight (TOF) [8] is inversely proportional to the stress applied with 0° fiber direction sample. In 45° and 90° fiber direction samples, this proportionality is 3-4 magnitude order lower. However, when the composite becomes thicker, the scattering phenomenon comes to be strong and makes the TOF measurement more challenging.

Coda Wave Interferometry (CWI) [11, 12] or seismic doublet method [13] is a technique for monitoring changes in media using acoustic waves. It is widely used in geophysics and civil engineering [14, 15], but not yet for composite materials. In the latter case, the waves will be multiple scattered by heterogeneities (fibers) and Coda waves will be generated. This technique can be promising for composite manufacturing inspection in solving problems as in quality control with different structures.

The changes in the waveforms can be quantified by computing the time shifted cross-correlation parameter $R(t_s)$ over a time window with center time t and width 2T [12]:

$$R(t_s) = \frac{\int_{t-T}^{t+T} u(t')u(t'+t_s)dt'}{\sqrt{\int_{t-T}^{t+T} u^2(t')dt' \int_{t-T}^{t+T} u^2(t')dt'}} \tag{2}$$

where $u(t')$ is the unperturbed wave and $u(t')$ is the perturbed wave, t_s is the time shift of the perturbed waveform relative to the unperturbed waveform. If the waves are not perturbed, then:

$$u(t') = u(t') \text{ and } R(t_s=0)=1. \tag{3}$$

When the perturbed wave is a time shifted version of the original wave, then:

$u(t') = u(t'-\delta t)$ and $R(t_s)$ attains its maximum Rmax for $\delta t = t_s$.

The travel time perturbation can be expressed as

$$\delta t = -\frac{\delta V}{V} t \tag{4}$$

where the relative velocity perturbation $\frac{\delta V}{V}$ can be obtained after the determination of δt, from which the influence of the stress could be known. Recently Larose et al. [16] and Niederleithinger et al. [17] have used Coda wave interferometry to study the weak stress effect in heterogeneous medium, as for example testing of concrete samples. They have found that it was possible to link the weak stress change to the velocity variation when analyzing the Coda waves [16], according to:

$$V - V^0 = \frac{\partial V}{\partial \sigma} \Delta\sigma + o(\sigma) \tag{5}$$

where V^0 is the velocity at stress-free state, $\Delta\sigma$ is the uniaxial stress and

$$\beta = \frac{1}{V^0} \frac{\partial V}{\partial \sigma} \tag{6}$$

is defined as an effective acoustoelastic coefficient.

Here the Coda wave interferometry is proposed to study internal stresses in another type of heterogeneous medium like polymer composites, based on the fiber displacement.

2. Experiments

A $[0°/90°]_{63}$ glass/ epoxy composite sample of 34.62×35.42×87.47mm3 in dimension, as shown in figure 1 fabricated from our laboratory is used for acoustic tests. It is a thick composite made of crossed unidirectional plies. It is scanned with tomography (Micro XCT400, Xradia) as shown in figure 2. The average thickness of each ply is 273.5μm. The Young's modulus in each direction is evaluated first by compression test with an INSTRON machine: in-plane moduli E1=E3=27.9GPa, out-of-plane modulus E2=3.4GPa.

The composite specimen was subjected to uniaxial loading during the ultrasonic measurement. The applied uniaxial load was increased from 0kN to 100kN in five steps of 20kN with ultrasonic acquisition at each force increment. The loading speed is 0.002mm/s. The sample is always under elastic state with strain less than 1%. Kaiser effect should be considered in composites in case of defects or damage existence, which is often included during CWI measurement in concrete. Longitudinal sensors (MSW-QC 2.25MHz, General Electric) were used to generate Coda in transmission mode, fixed by a laboratory made system. A universal couplant ZG-F (General Electric) is used here under ambient temperature. Waveforms were registered by a commercial ultrasonic recorder (Krautkramer USN60, General Electric). Two compression schemes (a) and (b) shown in figure 3 were tested and analyzed.

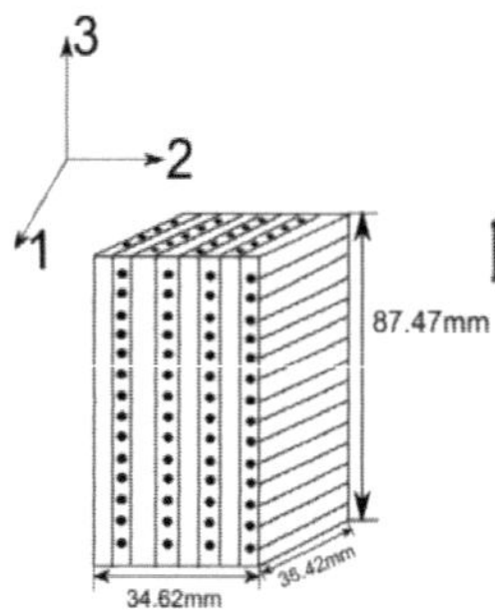

Figure 1. Composite sample

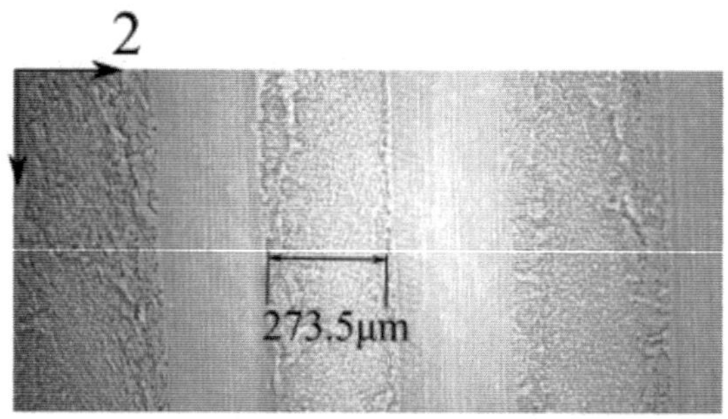

Figure 2. Tomography image in plane12

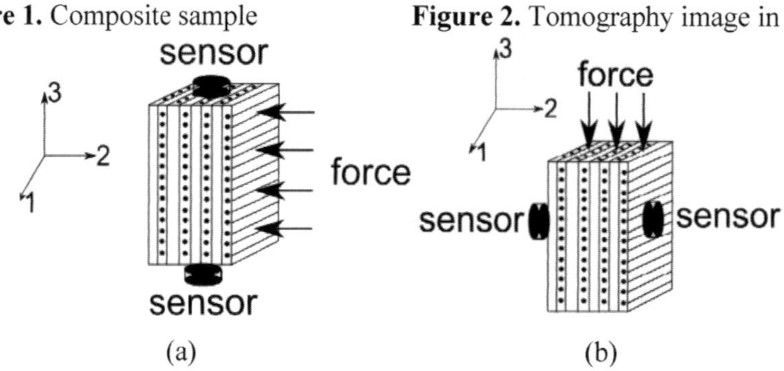

Figure 3. Two compression schemes (a) Measurement in direction 3, compression in direction 2 (b) Measurement in direction 2, compression in direction 3

3. Results and discussion

3.1. Force applying in a perpendicular direction to the composite ply

Force was applied along direction 2 and sensors were positioned perpendicularly direction 3 as shown in compression scheme (a) in figure 3. All the raw signals (time interval Δt=0.076μs, total length T_t=22μs) are stored and treated with matlab. Interpolation was done on the signals with 'cubic' type by ×20 zoom to improve the signal resolution. A correlation window length equal to $T=T_t/10$ is chosen. The center time t is shifted each Δt. The Coda waves at 0kN and at 100kN in figure 4(a) are compared at first. For the interval early in time, the waveforms are very similar. But for the later interval (Coda), the waveforms are distinctly different, where one of the waveforms is a time shifted version of the other. The same analysis can be done for the other loads.

According to equations (2)-(4), the travel time perturbation δt in figure 4(b) and cross-correlation parameter R in figure 4(c) can be calculated when shifting the center time in the interval [21μs, 30μs].

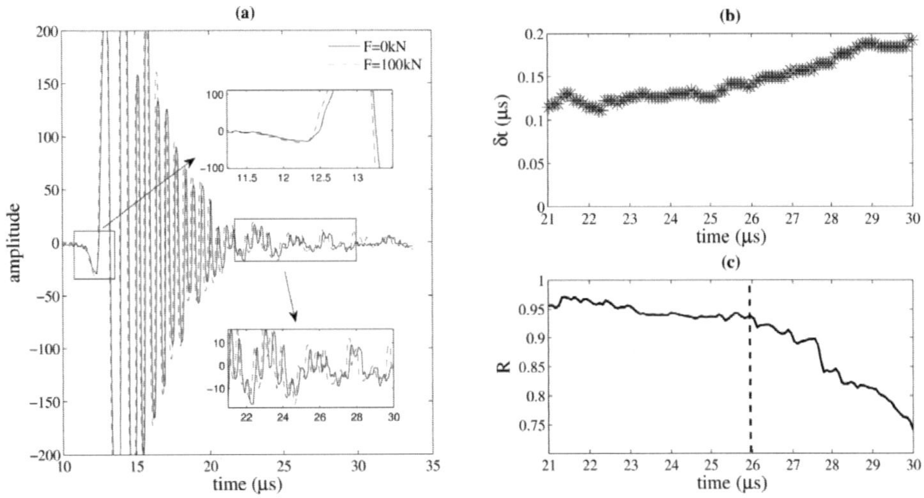

Figure 4. (a) The variation of the amplitude versus the time at 0kN and 100kN (b) Travel time perturbation versus the time (c) Cross-correlation parameter value versus the time (Equation (2))

The curve in figure 4(c) can be decomposed into two parts: the first part where the Coda can be used and the second part where R decreases sharply indicating a failure correlation. Only the first part in the range [21μs, 26μs] is used, and the mean of R_{max} and δt are computed in this domain. The whole procedure is repeated for every applied load, and the results are listed in table 1.

Table 1. The average values of R_{max} and δt at each applied load and stress

F(kN)	σ(MPa)	R_{max}	$-\delta t/t(\times 10^{-3})$
20	6.5	0.93	1.0
40	12.9	0.94	2.2
60	19.4	0.93	3.2
80	25.8	0.93	4.1
100	32.3	0.93	5.7

From that, the average velocity variation was computed and varied linearly with the applied stress as shown in figure 5. A linear regression of the data yields β=1.75×$10^{-2}$$MPa^{-1}$according to equation (6). A further comparison between CWI (Equation (4)) and TOF (Equation (1)) presented in figure 5 demonstrates that CWI gives a better linear behavior with a higher precision. It has a strong ability for the minor strain detection. TOF method becomes reliable only after 40kN with a similar slope of that of CWI.

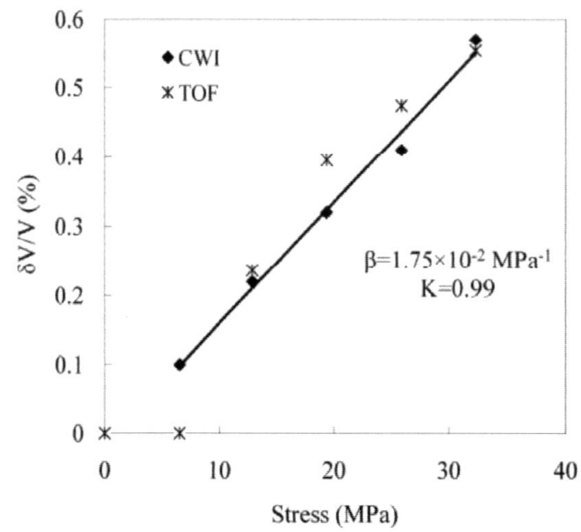

Figure 5. Velocity variation with external force applied (K is the coefficient of determination)

3.2. Force applying parallel to the direction of the ply

A much more severe attenuation is encountered in measurement along direction 2 when the wave travels through the thickness of the plies as shown in compression scheme (b) in figure 3. The amplitude decreases quickly and sharply after the maximum peak which makes the use of Coda wave more challenging. A similar procedure is used as done in section 3.1. The results for 0kN and 100kN are presented in figure 6.

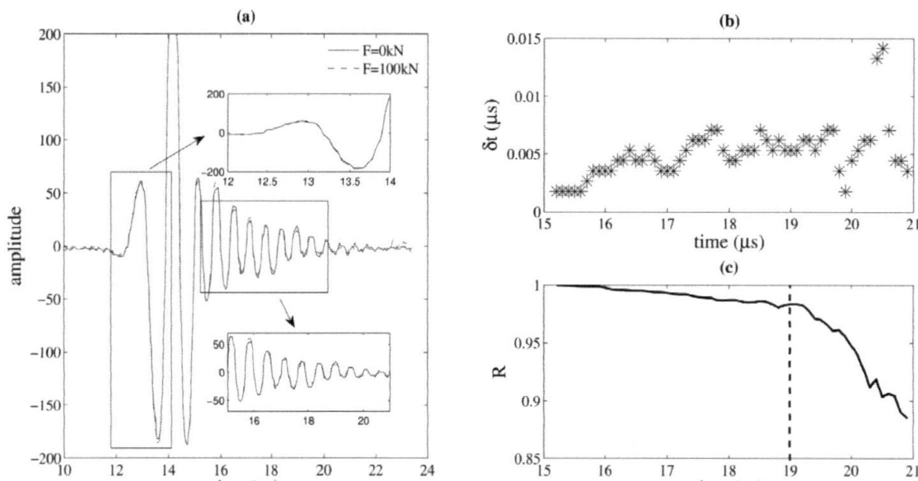

Figure 6. (a)The variation of the amplitude versus the time at 0kN and 100kN (b) Travel time perturbation versus the time (c) Cross-correlation parameter value versus the time (Equation (2))

The cross-correlation parameter R drops quickly after 19μs and the time domain between [15μs 19μs] is set for the R_{max} computation. The value of δt/t from TOF is found much larger than that from CWI, which is shown in table 2.

Table 2. The average values of R_{max} and δt at each applied load and stress

F(kN)	σ(MPa)	R_{max}	$-\delta t/t(\times 10^{-4})$
20	16.3	0.99	1.1
40	32.6	0.99	1.7
60	48.9	0.98	1.9
80	65.2	0.98	2.0
100	81.6	0.99	2.3

Such large discrepancy is due to the too small velocity variation in this situation and a high precision is required for velocity calculation when the TOF technique is used. Based on a statistical time shift, CWI can prevent such problem, and a linear relationship is again found for this method with $\beta=2\times10^{-4}MPa^{-1}$.

The fiber orientation shows a strong effect on the detection of acoustic signals depended on the fiber displacement (deformation) and signal attenuation. In figure 7(a), CWI gives a good precision even at low stress level. In figure 7(b), the small deformation from the fiber dominated direction and strong attenuation makes the TOF measurement difficult. In this situation, TOF becomes not reliable, but CWI still works.

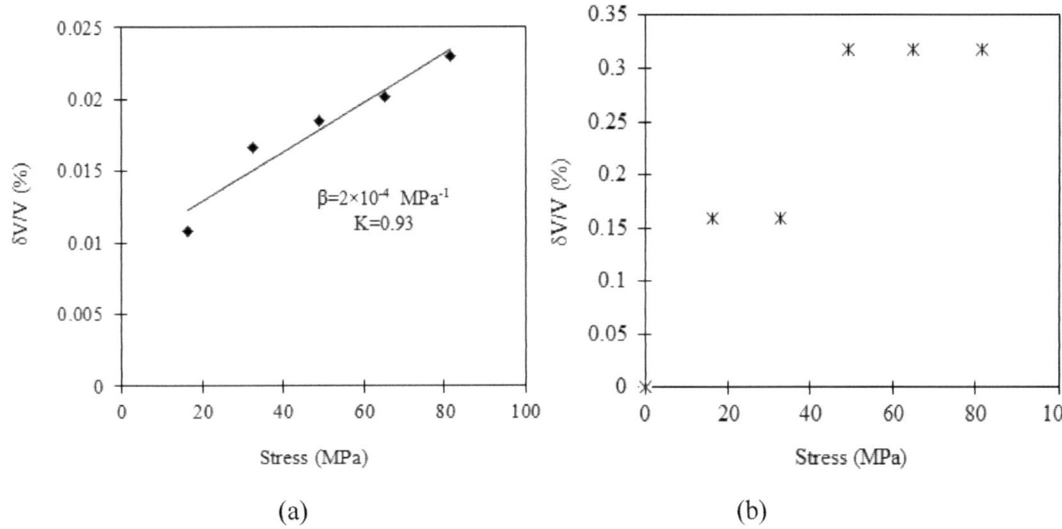

Figure 7. Velocity variation with external force applied (a) CWI (K is the coefficient of determination) (b) TOF

4. Conclusion

This study reports the first set of results from CWI technique for internal stress determination in a polymer composite subjected to uniaxial compression. The velocity variation can be obtained according to the travel time perturbation calculated from this technique. And the internal stresses can be estimated with a linear approximation to the velocity variation.

When the wave propagates parallel to direction of the ply, strong scattering waveforms are observed. Waveform shifts are found during loading due to the fiber displacement. The CWI result is comparable to that from the TOF technique but shows a better precision and sensitivity. A strong attenuation is noticed when the wave propagation direction are normal to the ply, although there is still distinguishable Coda wave shift for the stress determination. TOF measurement under this situation needs a higher precision measurement for both strain and travelling time. The fiber displacement can be expected not to be evident under this situation because the loading direction is along the high stiffness fibers. Factors as temperature [13], humidity and matrix type need to be further investigated which will influence CWI measurements.

This method is applicable to composites with complex fibrous geometry such as fabric. It seems also promising for in-line inspection in composite processing: such as RTM (Resin Transfer Molding), autoclave, etc... The early part of ultrasonic wave (TOF measurement) could predict the major modulus variation during the polymerization of the resin and the later part (CWI) can give an indication when internal stress establishes and other factors take place.

References

[1] Parlevliet P, Bersee H and Beukers a 2007 Residual stresses in thermoplastic composites – a study of the literature. Part III: Effects of thermal residual stresses *Compos. Part A Appl. Sci. Manuf.* **38** 1581-96

[2] Okabe Y, Yashiro S, Tsuji R, Mizutani T and Takeda N 2002 Effect of thermal residual stress on the reflection spectrum from fiber Bragg grating sensors embedded in CFRP laminates *Compos. Part A Appl. Sci. Manuf.* **33** 991-9

[3] Khoun L, Oliveira R De, Michaud V and Hubert P 2010 Investigation of process-induced strains development by fibre Bragg grating sensors in resin transfer moulded composites *Compos. Part A Appl. Sci. Manuf.* **42** 274-82

[4] Luyckx G, Voet E, Lammens N and Degrieck J 2011 Strain measurements of composite laminates with embedded fibre bragg gratings: criticism and opportunities for research. *Sensors (Basel).* **11** 384-408

[5] Schmachtenberg E, Heide J S zur and Töpker J 2005 Application of ultrasonics for the process control of Resin Transfer Moulding (RTM) *Polym. Test.* **24** 330-8

[6] Kinra V K, Ganpatye A S and Maslov K 2006 Ultrasonic Ply-by-Ply Detection of Matrix Cracks in Laminated Composites *J. Nondestruct. Eval.* **25** 37-49

[7] Pao Y and Sachse W 1984 Acoustoelasticity and ultrasonic measurement of residual stress *Physical acoustics* ed W P Mason and R N Thurston (Orlando: Academic Press) pp. 62-140

[8] Zhu Q, Burtin C and Binetruy C 2014 Acoustoelastic effect in polyamide 6: Linear and nonlinear behaviour *Polym. Test.* **40** 178-86

[9] Mase G T, Wong T E and Johnson G C 1990 Acoustoelasticity in Layered Composites *J. Eng. Mater. Technol.* **112** 266

[10] Santos A A D, Ambiel L B, Garcia R H and Rodovalho T G 2013 Stress analysis in carbon/epoxy composites using Lcr waves *J. Compos. Mater.* **48** 3425-34

[11] AKI K and CHOUET B 1975 Origin of Coda Waves : Source , Attenuation , and Scattering *Geophys. Res.* **80** 3322-42

[12] Snieder R 2006 The Theory of Coda Wave Interferometry *Pure Appl. Geophys.* **163** 455–73

[13] Grêt A, Snieder R and Özbay U 2006 Monitoring in situ stress changes in a mining environment with coda wave interferometry *Geophys. J. Int.* **167** 504-8

[14] Stähler S C, Sens-Schönfelder C and Niederleithinger E 2011 Monitoring stress changes in a concrete bridge with coda wave interferometry. *J. Acoust. Soc. Am.* **129** 1945-52

[15] Larose E and Hall S 2009 Monitoring stress related velocity variation in concrete with a 2× 10–5 relative resolution using diffuse ultrasound *J. Acoust. Soc. Am.* **125** 1853-7

[16] Niederleithinger E 2014 Coda wave interferometry used to localize compressional load effects in a concrete specimen *EWSHM-7th European workshop on structural health monitoring* (Nantes: Inria) pp. 1427-33

Optimization of the contents of hollow glass microsphere and sodium hexametaphosphate for glass fiber vacuum insulation panel

C D Li[1, 2, 3], Z F Chen[1, 2, 3] and J M Zhou[3]

[1]International Laboratory for Insulation and Energy Efficiency Materials, College of Materials Science and Technology, Nanjing University of Aeronautics and Astronautics, Nanjing 210016, P.R. China
[2]Jiangsu Collaborative Innovation Center for Advanced Inorganic Function Composites, Nanjing University of Aeronautics and Astronautics, Nanjing 210016, P.R. China
[3]Suzhou V.I.P. New Material Co., Ltd, Hong Da Fang Yuan Group, Suzhou, 215400, P.R. China

E-mail: li690612564@126.com

Abstract. In this paper, various additive amounts of hollow glass microspheres (HGMs) and sodium hexametaphosphate (SHMP) powders were blended with flame attenuated glass wool (FAGW) to form hybrid core materials (HCMs) through the wet method. Among them, the SHMP was dissolved in the glass fiber suspension and coated on the surface of glass fibers while the HGMs were insoluble in the glass fiber suspension and filled in the fiber-fiber pores. The average pore diameter of the FAGW/HGM HCMs was 8-11 μm which was near the same as that of flame attenuated glass fiber mats (FAGMs, i.e., 10.5 μm). The tensile strength of the SHMP coated FAGMs was enhanced from 160 N/m to 370 N/m when SHMP content increased from 0 wt.% to 0.2 wt.%. By contrast, the tensile strength of the FAGW/HGM HCMs decreased from 160 N/m to 40 N/m when HGM content increased from 0 wt.% to 50 wt.%. Both the FAGW/HGM HCMs and SHMP coated FAGMs were vacuumed completely to form vacuum insulation panels (VIPs). The results showed that both the addition of SHMP and HGM led a slight increase in the thermal conductivity of the corresponding VIPs. To obtain a high-quality VIP, the optimal SHMP content and HGM content in glass fiber suspension was 0.12-0.2 wt.% and 0 wt.%.

1. Introduction

Thermal insulation is an indispensable part of current strategies to lower the total energy consumption worldwide. Conventional insulation materials, like expanded polystyrene, glass wool, and polyurethane, feature high thermal conductivities between 20 to 40 mW/(m·K) and poor thermal insulation performance [1]. Thick layers (tens of cm) of these conventional insulation materials are necessary to achieve a low overall thermal conductivity for their applications in structures and buildings [2].

Vacuum insulation panels (VIPs) are super insulation materials with a low thermal conductivity of < 8 mW/(m·K). In recent years, VIPs have been widely applied in buildings sectors, refrigerators and aeronautics and astronautics [3-5]. Core material is the skeleton of a VIP and greatly influences the thermal insulating performance of the VIP. Core materials with a low thermal conductivity, low outgassing rate, and high mechanical properties are preferred in the high-quality VIPs. In Asia countries, glass fibers are commonly used as core materials due to their excellent thermal insulation properties, low density and superior chemical and thermal stability (> 1000 °C). The resulting VIPs are known as glass fiber VIPs, i.e., GF-VIPs [3]. Compared with fumed silica VIPs and foam based VIPs, GF-VIPs possess a lower initial thermal conductivity; and thus, they are more suitable for the applications in the household appliances like refrigerators, freezers, and vending machines.

In general, polymer binders are not added in the glassfiber core materials (GF-CMs) due to their high outgassing rate. But, the mechanical properties of GF-CMs that do not contain binders were very low because of the weak interlocking force among glass fibers. Thus, in order to get a better GF-CM with enhanced overall properties, extensive works have been carried out. Li et al. [6] proposed a novel core material structure of for VIPs, i.e. fiber pore packing with different-size powders structure, and stated that it is the development direction of future VIP core materials. Di et al. [7] used the glass fiber chopped strand and flame attenuated glass wool (FAGW) as the core materials for VIPs and found that the service life of GF-VIPs with getters was above 15 years. Kim et al. [8] studied the relation between the density of glass wool and external pressing load and found that GF-VIPs were superior when the pressing load was less than 25 kPa. Regarding enhancing the tensile strength of GF-CMs, Li et al. [9, 10] optimized the number of beating revolutions for centrifuged glass wool mat (CGM), flame attenuated glass wool mat (FAGM), and hybrid glass wool mat (HGM).

In current technology, there are two different fabricating techniques for GF-CMs: (1) wet method (like the papermaking technique); and (2) dry method (like the hot pressing technique and centrifugal-spinneret-blow technique). The essential of wet method is to form wet glass fiber mats through the removal of the solvent in glass fiber suspension by the vacuum filtration. In the current wet method, the solvent for glass fiber suspension is recycled white water (tape water with some small fibers). The addition of chemical additives in the white water is beneficial to the bonding strength among the GF and fiber dispersion in the GF suspension. However, there are few kinds of literature on this aspect. The knowledge thereof is of vital importance to obtain a high-strength GF-CM and low-thermal-conductivity GF-VIPs.

In this paper, various amounts of sodium hexametaphosphate (SHMP) and hollow glass microspheres (HGMs) was added in the white water to form wet-laid mats, i.e. the SHMP coated FAGMs and FAGW/HGM HCMs. A total of 20 pieces of the SHMP coated FAGMs and FAGW/HGM HCMs were horizontally put together to get the GF-CMs. Microstructural morphologies, pore diameter distribution and tensile strength of the GF-CMs were investigated. The aim of this paper was to investigate the SHMP and HGM addition on the tensile strength of the GF-CMs and thermal conductivity of the associated GF-VIPs.

2. Experimental

All the raw materials, including FAGW, HGMs, and SHMP, were provided by Suzhou V.I.P. New Material Co., Ltd. (Taicang, P.R. China). The SHMP content in the glass fiber (FAGW) suspension was 0 wt.%, 0.04 wt.%, 0.12 wt.%, and 0.2 wt.%, respectively. A total of 0 g, 2.02 g, and 4.71 g of HGMs, 4.71 g of FAGW, and 1500 ml of the above white water were stirred in a cylindrical tank for 30 min and then dispersed in the GBJ-A fiber dissociation device, respectively. The rotation frequency and the beating revolution of impeller were $48.3\pm1.65\ s^{-1}$ and 5000 r, respectively. Each associated FAGW suspension was dispersed by mechanical crushing and abrasion on fibers through contacting with edge and faces of rapidly moving metal blades in white water. Each dispersed FAGW suspension was then transported to a papermaking machine (ZCX-200A) and stirred by a blender for 5 min. After blending, white water was drained out from the suspensions by the vacuum pump through an 80 mesh copper net while FAGW and HGMs were aligned at the surface of the copper net and then formed a wet mat at the surface of the copper net. Finally, the wet mat was dried at 150 °C for 60min.

Surface morphologies of the as-prepared mats were observed by the field emission scanning electron microscopy (FE-SEM, Hitachi S-4800). The pore diameter distribution of the mats was determined by the mercury intrusion porosimeter (MIP, AutoPore IV 9510). Tensile strength tests of the rectangular SHMP coated FAGMs and FAGW/HGM HCMs with a length of 100 mm and width of 15 mm were carried out by the electronic diaphragm tension machine (XH-018A), according to the JB/T 7630.1-2008 standard where the clamping space and velocity of the grips were 50 mm and 100 mm/min, respectively. Thermal conductivities of the as-prepared VIPs were evaluated by the heat flow meter (Netzsch HFM 436).

3. Results and discussion

3.1. Microstructure

Figure 1 shows the SEM micrographs of the FAGM, SHMP coated FAGM and FAGW/HGM HCM. As shown in Figure 1(a) and (b), the glass fibers were slightly curled, randomly oriented and possessed a great aspect ratio. The mean diameter of glass fibers was 3.5 μm. The glass fibers built the pores with various diameters ranging from 1 μm to 100 μm. As for the SHMP coated FAGM, there was a thin SHMP coating on the surface of glass fibers. But the SHMP coating was not continuous; SHMP coating was attached at only a part of glass fibers. As shown in Figure 1(c), the FAGW/HGM HCMs were densely packed with HGMs and glass fibers. The HGMs were filled in the fiber-fiber pores, making the pores more compact.

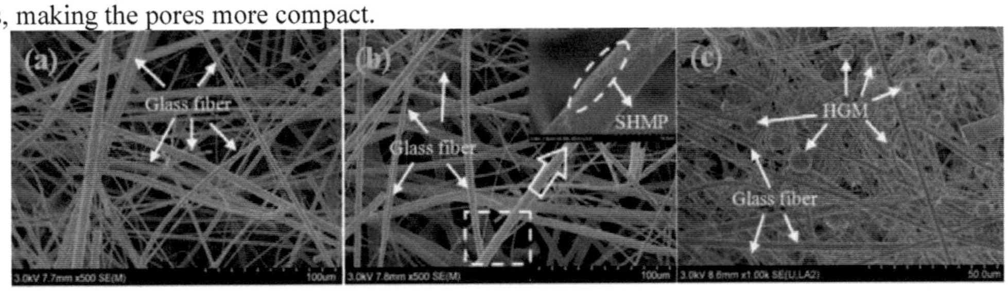

Figure 1. SEM micrographs: (a) FAGM; (b) SHMP coated FAGM; and (c) FAGW/HGM HCM.

3.2. Pore structure of core material

Table 1 shows the pore parameters of various core materials. Each core material possessed a high porosity of 84-91%. Also, the average pore diameter of Sample 1-3 was 8-11 μm while that of Sample 4 was only 0.3 μm.

Table 1. Pore parameters of various core materials

	Composition, %			Total intrusion volume V, mL/g	Total pore area A, m^2/g	Average pore diameter (4V/A), μm	Porosity, %
	FAGW	HGM	SHMP				
Sample 1	100	0	0	6.9	2.6	10.5	86.7
Sample 2	70	30	0	4.8	2.3	8.3	85.8
Sample 3	50	50	0	3.9	1.5	10.6	84.5
Sample 4	0	100	0	8.2	116.2	0.3	90.8

Figure 2 shows the distribution of cumulative intrusion and differential intrusion in the core material calculated by MIP technique. As shown in Figure 2(a), around 41% (Sample 1), 70% (Sample 2), 57% (Sample 3), and 41% (Sample 4) of pore volume was occupied by pores with diameter ≥ 10 μm while 54% (Sample 1), 30% (Sample 2), 43% (Sample 3), 10% (Sample 4) of pore volume was taken over by pores with diameters of 1-10 μm. As shown in Figure 2(b), the pore diameter distribution of Sample 1-3 was similar with the highest peak of ≈ 9 μm and two short peaks at different pore diameters. By contrast, the highest peak of Sample 4 was around at 0.07 μm and the remaining two peaks were at the range of 0.01-0.05 μm. Thus, the presence of HGMs in the FAGW made the pore distribution in the core material moving into a multi-peak pattern.

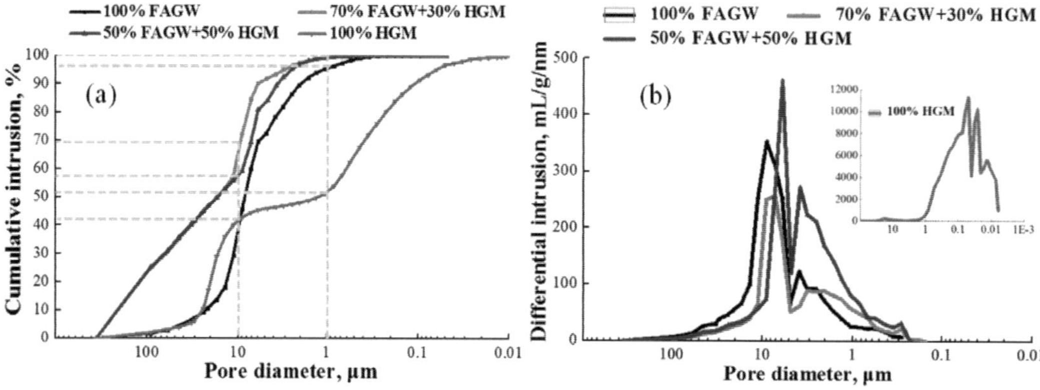

Figure 2. Distribution of cumulative intrusion and differential intrusion in the core material calculated by MIP technique.

3.3. Tensile strength of core material

SHMP is a mixture of polymeric metaphosphates and a hexamer of composition $(NaPO_3)_6$. SHMP is usually used as the deflocculant in wet method for nonwoven mats. Also, it can be a dispersing agent that helps to break down the fiber aggregations.

Figure 3 shows the effect of SHMP addition on tensile strength and density of core material. The tensile strength significantly enhanced from 160 N/m to 370 N/m while the density slightly decreased from 115 kg/m^3 to 99 kg/m^3 when the SHMP addition increased from 0 wt.% to 0.2 wt.%. It was noted that the tensile strength at SHMP content of 0.12 wt.% and 0.2 wt.% was near the same, one was 360 N/m while the other was 370 N/m. Hence, the addition of SHMP not only had a great effect on enhancing the bonding between glass fibers but also made the fiber mat fluffier.

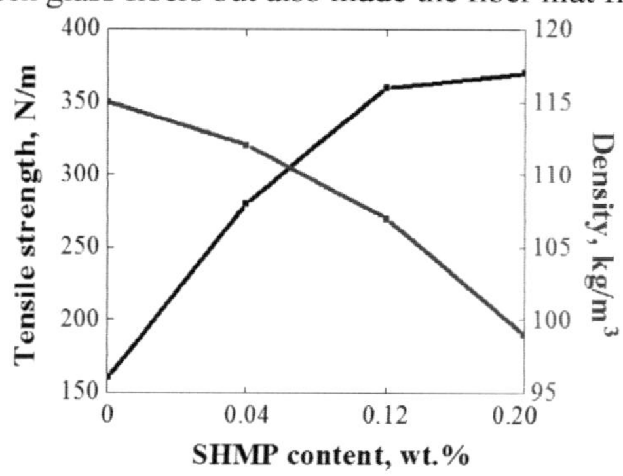

Figure 3. Effect of SHMP addition on tensile strength and density of core material.

HGMs were microscopic glass spheres and acted as lightweight fillers in the composite core materials. In this paper, the diameter of HGMs was 15-135 μm. Figure 4 shows the effect of HGM content on tensile strength and density of core material. The tensile strength of the core material significantly decreased with the increase of HGMs content while the density increased rapidly with HGMs content. The maximum density arrived at 229 kg/m^3 when the HGMs content was 50 wt.%.

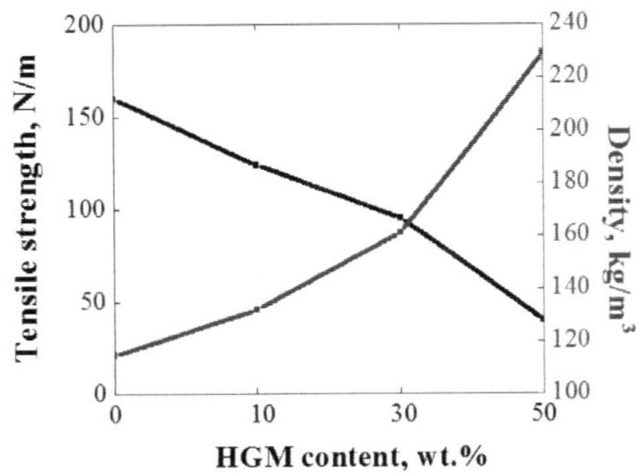

Figure 4. Effect of HGM content on tensile strength and density of core material.

3.4. Thermal conductivity of VIP

Figure 5 shows the thermal conductivity of VIPs. Compared with VIPs with pure FAGMs, the VIPs possessed a higher thermal conductivity. Both the addition of SHMP and HGM increased the thermal conductivity of the VIPs. As shown in Figure 5 (a), the thermal conductivity of VIPs roughly linearly increased from 2.039 mW/(m·K) to 2.432 mW/(m·K) when SHMP content increased from 0 wt.% to 0.2 wt.%. As shown in Figure 5 (b), the thermal conductivity of VIPs roughly monotonously increased from 2.039 mW/(m·K) to 2.356 mW/(m·K) when HGM content increased from 0 wt.% to 50 wt.%.

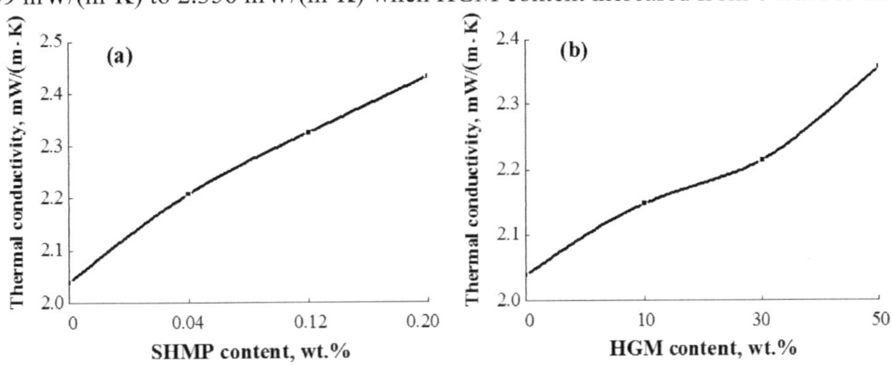

Figure 5. Thermal conductivity of VIPs: (a) core material: SHMP coated FAGMs; (b) core material: FAGW/HGM HCMs.

4. Conclusion

Adding the SHMP in the glass fiber suspension was very beneficial to the enhancement of tensile strength of the FAGMs while adding the HGMs in the glass fiber suspension increased the density and decreased the tensile strength of the GF-CMs and increased the thermal conductivity of the corresponding GF-VIPs. The optimal SHMP content and HGM content in the glass fiber suspension were 0.12-0.2 wt.% and 0 wt.%.

Acknowledgments

This work was supported by Funding for Outstanding Doctoral Dissertation in NUAA (BCXJ13-10), Funding of Jiangsu Innovation Program for Graduate Education (the Fundamental Research Funds for the Central Universities, CXLX13_149), Major Achievements Transfer Projects of Jiangsu province (BA2013097), and a Project Funded by the Priority Academic Program Development of Jiangsu Higher Education Institutions.

References

[1] Bouquerel M, Duforestel T, Baillis D and Rusaouen G 2012 *Energ Buildings* **54** 320-336

[2] Ruckdeschel P, Kemnitzer T W, Nutz F A, J. Senker and Retsch M 2015 *Nanoscale* **22** 10059-10070
[3] Li C D, Duan Z C, Chen Q, Chen Z F, Boafo F E, Wu W P and Zhou J M 2013 *Mater. Design* **50** 1030-1037
[4] Li C D, Chen Z F, Boafo F E, Chen Q, Zhang J, Zhou J M, Ye X L and Li C Y 2013 *Dry Technol.* **31** 1084-1090
[5] Li C D, Chen Z F, Boafo F E, Xu T Z and Wang L 2014 *J Mater Process Tech.* **214** 539-543
[6] Li C D, Chen Z F, Wu W P, Chen Z, Zhou J M, Cheng X Y and Su D 2012 *Applied Mechanics and Materials* **174-177** 1437-1440
[7] Di X B, Gao Y, Bao C, Hu Y and Xie Z 2013 *Vacuum* **97** 55-59
[8] Kim J and Song T 2013 *Int J Heat Mass Tran* **64** 783–791
[9] Li C D, Chen Z F and Saeed M 2016 *Mater Struct.* **49** 1861-1869
[10] Li C D and Chen Z F 2015 *Powder Technol.* **279** 221-227

IOP Conf. Series: Materials Science and Engineering **137** (2016) 012042 doi:10.1088/1757-899X/137/1/012042

Microwave extinction characteristics of nanoparticle aggregates

Y P Wu[1], J X Cheng[1,2,4], X X Liu[3], H X Wang[3], F T Zhao[1] and W W Wen[1]

[1]Kadinuo Science and Technology (Beijing) Co., Ltd., Beijing, 100013, China
[2]Institute of Nuclear and New Energy Technology, Tsinghua University, Beijing, 100084, China
[3]Xi'an Research Institute of High Technology, Xi'an, 710025, China

E-mail: chengjx@tsinghua.edu.cn

Abstract. Structure of nanoparticle aggregates plays an important role in microwave extinction capacity. The diffusion-limited aggregation model (DLA) for fractal growth is utilized to explore the possible structures of nanoparticle aggregates by computer simulation. Based on the discrete dipole approximation (DDA) method, the microwave extinction performance by different nano-carborundum aggregates is numerically analyzed. The effects of the particle quantity, original diameter, fractal structure, as well as orientation on microwave extinction are investigated, and also the extinction characteristics of aggregates are compared with the spherical nanoparticle in the same volume. Numerical results give out that proper aggregation of nanoparticle is beneficial to microwave extinction capacity, and the microwave extinction cross section by aggregated granules is better than that of the spherical solid one in the same volume.

1. Introduction

Absorbing composite coating, prepared by mixing the nano-absorbent and matrix harmoniously, is one of the effective ways to reduce the radar cross section (RCS) of object. While fabricating composite coating, filler granules inevitably collide with each other, and then congregate to aggregates with fractal structure [1, 2]. The scattering and absorption characteristics of nanoparticle aggregates are of great significance for researches on microwave absorption mechanism. Because of nanoparticles aggregating randomly, the actual fractal structure is complicated and changeable, which has a significant impact on the extinction performance of aggregates. It is obvious that the performance cannot be investigated according to the monomer theory [3]. There are lots of researches on clusters of given fractal structure [4-6]. Nevertheless, Studies on extinction characteristics associated with the actual complex structures have been carried on infrequently [7, 8].

The fractal theory is a promising approach for study on extinction performance of nanoparticle clusters. The diffusion-limited aggregation (DLA) is easy and feasible to simulate the self-similar fractal growth by the way of solving simple kinematics and kinetic equations. The fractal structures simulated by this model are in good agreement with the experimental results [9, 10]. According to the rule of three dimensional DLA, under the conditions of nearest neighbour and subordinate neighbour, the fractal aggregates are all satisfied with the scale invariance and self similarity, and have the same fractal dimension (about 2.48) despite their growth in different nearby conditions and different appearance. It is shown that the fractal dimension of cluster is not associated with the structure of lattice in the small range of particle number. Thus DLA is a kind of potent method for exploring the structure and growth process of clusters. However the interaction between electromagnetic wave and particle aggregates is very complex, the Maxwell Equations often possessing analytic solution can only be solved numerically. The discrete dipole approximation (DDA) is a general method to calculate electromagnetic scattering and absorption of particles of arbitrary geometry and composition. As a good method for numerical solution, the DDA possesses several features, such as simple procedure, high efficient iterative algorithm, fast computing speed, and so on. Therefore, it has been widely utilized to investigate scattering and absorption performance of particles [11-14].

In this paper, the fractal aggregation model DLA is utilized to simulate the possible structures of spherical nanoparticle clusters. Then the impacts of the aggregation states on microwave extinction performance are numerically analyzed based on the DDA. The results would provide profitable theoretical references for the application of nanoparticles in microwave absorbing composites.

2. Theory of discrete dipole approximation

The way of calculating extinction performance of nanoparticle with arbitrary shape by DDA method is as present. The DDA replaces the solid particle by a set of *N* point dipoles, with the spacing between the dipoles far small than the wavelength. Thus the research on the actual particle is displaced to that on these small dipole groups. It is supposed that the relative permittivity of the solid particle is $\varepsilon_r(\bar{r})$. The position vector of the *j* electric dipole is $\bar{r}_j$, and the polarization α_j. The electric dipole moment can be expressed as follow.

$$\bar{P}_j = \alpha_j \bar{E}_j(\bar{r}) \tag{1}$$

Where $\bar{E}_j(\bar{r})$ is the total domain electric field on the dipole at $\bar{r}_j$, including the electric field of the incident wave and the dipole electric field formed by other dipoles on the location.

$$\bar{E}_j = \bar{E}_0 \exp(i\bar{k}\cdot\bar{r}_j - i\omega t) - \sum_{i\neq j} \bar{A}_{ji}\bar{P}_i \tag{2}$$

Here $\bar{E}_0$ denotes the field of the incident wave, and $k=\omega/c$ the incident wave vector. The interaction matrix is $\bar{A}_{ji}$ [11].

$$\bar{A}_{ji}\bar{P}_i = \frac{\exp(ikr_{ji})}{r_{ji}^3}\left\{k^2\bar{r}_{ji}\times(\bar{r}_{ji}\times\bar{P}_i) + \frac{(1-jkr_{ji})}{r_{ji}^2}\left[r_{ji}^2\bar{P}_i - 3r_{ji}(\bar{r}_{ji}\cdot\bar{P}_i)\right]\right\} \quad (j\neq i) \tag{3}$$

Where $\bar{r}_{ji} = \bar{r}_j - \bar{r}_i$. For a system containing *N* point dipoles, $\bar{A}_{ji}$ is a $3N\times3N$ symmetric complex matrix. While $j=i$ and $\bar{A}_{jj} = \alpha_j^{-1}$, Equation of electric dipole moment $\bar{P}_j$ can be gained according to above formulas.

$$\sum_{i=1}^{N} \bar{A}_{ji}\bar{P}_i = \bar{E}_{inc,j} \quad (j = 1, 2, \cdots, N) \tag{4}$$

This is a $3N$-dimension linear complex vector equation, which can be solved by the fast Fourier transform and the complex conjugate gradient technique. For the sake of calculating $\bar{P}_j$ accurately, the appropriate dipolar polarizability α_j should be determined according to the dielectric parameters of the solid particle. In the case of $|m|kd < 1$, the calculating results obtained from the DDA are good agreement with that by other theory based on lattice dispersion relation (LDR) [15]. Thereby the LDR expression is selected as the type of polarization.

$$\alpha_{LDR} = \frac{\alpha_{CMR}}{1 + \frac{\alpha_{CMR}}{d^3}\left[(b_1 + m^2 b_2 + m^2 b_3 \sum_{j=1}^{3} (\hat{a}_j \hat{e}_j)^2)(kd)^2 - \frac{2}{3} i (kd)^3 \right]} \tag{5}$$

Where b_1=-1.8915316, b_2=0.1648469, b_3=-1.7700004 and $\alpha_{CMR} = 3d^3(\varepsilon_r - 1)/[4\pi(\varepsilon_r + 2)]$. $\hat{a}$ and $\hat{e}$ presents the unit vector of the incident direction and the direction of the electric field vector, respectively. Here m denotes the complex refractive index of the particle, and the square of m is equal to ε_r .

Once Eq. (4) has been solved for the unknown polarizations $\vec{P}_j$, the extinction, absorption and scattering cross sections C_{ext}, C_{abs} and C_{sca} can be evaluated as follows [11].

$$C_{ext} = \frac{4\pi k}{|E_0|^2} \sum_{j=1}^{N} \mathrm{Im}\left[\vec{E}^*_{inc,j} \cdot \vec{P}_j \right] \tag{6}$$

$$C_{abs} = \frac{4\pi k}{|E_0|^2} \sum_{j=1}^{N} \left\{ \mathrm{Im}\left[\vec{P}_j \cdot (\alpha_j^{-1})^* \cdot \vec{P}_j^* \right] - \frac{2}{3} k^3 |\vec{P}_j|^2 \right\} \tag{7}$$

$$C_{sca} = \frac{k^4}{|E_{inc}|^2} \int d\Omega \left| \sum_{j=1}^{N} \left[\vec{P}_j - \hat{n}(\hat{n} \cdot \vec{P}_j) \right] \exp(-ik\hat{n} \cdot \vec{r}_j) \right| \tag{8}$$

Where $\hat{n}$ denotes the unit vector of the scattering direction, and $d\Omega$ the solid angle.

Based on aggregation structure generated by the DLA, the extinction capacity of radar wave by nanoparticle aggregates could be investigated by the available computer code DDSCAT [16]. Here the source program has been re-compiled and added the parts of results processing and output.

3. Results and discussion

Taking the nano-carborundum for an example, the influences of the particle number, original diameter, fractal structure and orientation on microwave extinction characteristics are investigated. Here the relative permittivity of SiC is cited from reference [17], namely $\varepsilon_r = \varepsilon_\infty + (\varepsilon_0 - \varepsilon_\infty) \cdot \left[1 + (j2\pi f \tau)^a \right]^{-1}$. Where f presents electromagnetic frequency, ε_0=82.4, ε_∞=4.1, τ=5.9×10^{-11} and a=0.52. Because the way of attenuating electromagnetic wave is due to the joint action by plenty of granules in composite, the calculated extinction parameters are replaced by the corresponding ones per unit volume, which makes the calculated results more reasonable.

3.1. Different extinction performance between aggregates and solid particle with the equal volume

Extinction performance of two different nano-carborundum aggregates has been compared with that of spherical solid particle in the equal volume. Wherein the number of granules is N=100 with particle diameter a=80nm in one aggregates, and N=50 with a=80nm in another ones. Fractal structure of aggregates is generated by the DLA. Relationship between extinction performance of the three scatters and the incident wave frequency has been investigated by the DDA. The numerical results are shown in figure 1. As can be seen from the graphs, extinction performance of nanoparticle aggregates is much better than that of the solid one in the same volume.

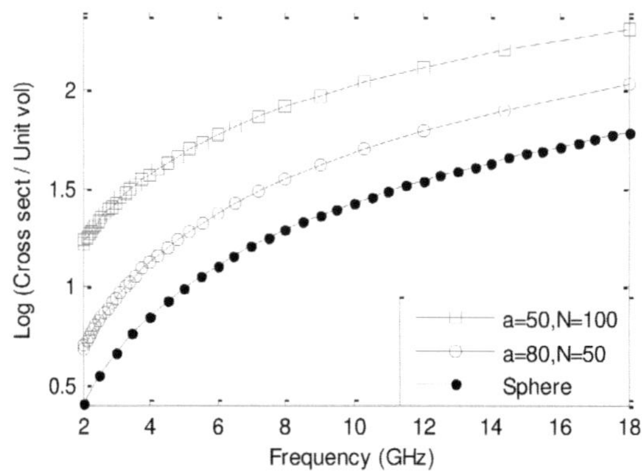

Figure 1. Extinction performance of aggregates and solid particles with the equal volume

3.2. Impact of original particle size on extinction performance

No matter what method is utilized to fabricate nanometer granules, the particle size of original granules is hardly single, but a certain distribution. So it is necessary to investigate the impact of original diameter on microwave extinction. Here the quantity of nano-carborundum in clusters is supposed as N=100. Figure 2 presents the relationship between extinction performance and original particle size in aggregates at different frequency. It is obvious that extinction performance of aggregates has nothing to do with the original particle size in the range of radar wave band. This conclusion is similar to that of single nanoparticle. It is worth to note that the result is drawn from the case of unit volume of particle.

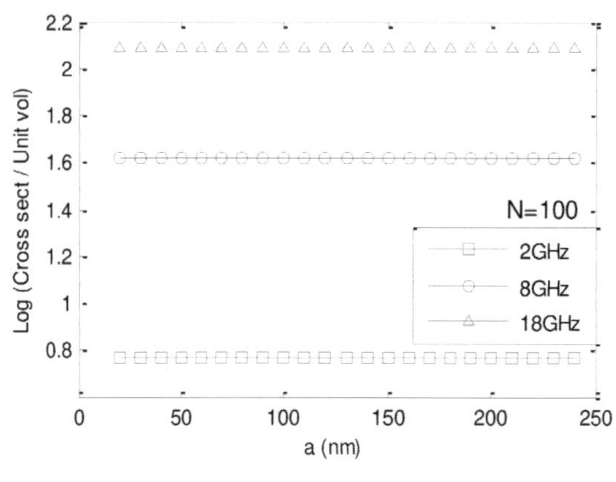

Figure 2. Relationship between extinction performance and original particle size in aggregates

3.3. Impact of particle quantity in aggregates on extinction performance

In general, particle quantity in aggregates is random, ranging from a few to more than thousands. Here the particle diameter is supposed as a=50nm. Figure 3 displays the relationship between extinction performance and particle quantity in aggregates. When the number of aggregated granules is less than 60, extinction cross section increases along with enhancing the particle quantity. When the number of aggregated granules is more than 70, extinction cross section has little attenuation while increasing the particle quantity. Too much number of aggregated particles would cause that aggregates develop to excessive large volume and the absorbent particles distribute unsymmetrical in composite coating, which may induce high conductivity and reflectivity locally. Therefore, proper aggregation of nanometer granules is beneficial to microwave extinction performance.

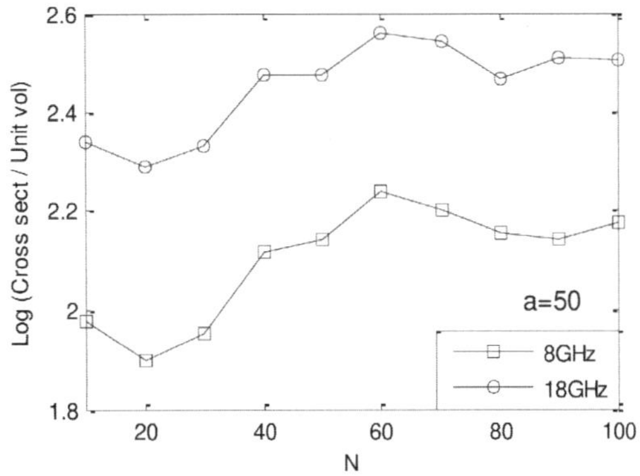

Figure 3. Relationship between extinction performance and particle quantity in aggregates

3.4. Impact of aggregate fractal structure and orientation on extinction performance

Figure 4 presents three different fractal structures with 100 aggregated nano-carborundum particles. Here it is supposed that Θ denotes the angle between the wave vector and the unit vector along with the principal axis of inertia. Figure 5 displays the relationship between extinction cross section and the angle at 12GHz. Numerical results indicate that diverse fractal structures lead to different extinction capacity even though these aggregates contains equal number of particles. With the increase of the angle between the incident wave and the principal axis of inertia of aggregates, extinction performance of each aggregates exhibits similar trend. When the direction of the principal axis of inertia of aggregates approaches parallels to that of incident wave, extinction capacity falls into decline. And the vertical relationship between the two directions would lead to the best extinction performance.

(a) 100-1 (b) 100-2 (c) 100-3

Figure 4. Fractal structures of nanoparticle aggregates at N=100

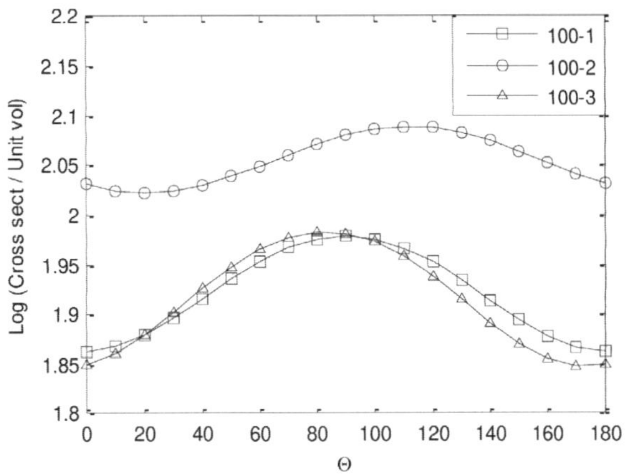

Figure 5. Relationship between extinction performance and the angle of incident wave

4. Conclusions

The DLA model for fractal growth is utilized to investigate nanoparticle aggregates. Possible structures of nanoparticle clusters are generated by computer simulation. Based on the DDA method, microwave extinction characteristics of different nano-carborundum aggregates are numerically analyzed. The analysis of calculated data shows that microwave extinction capacity of nanoparticle aggregates is better than that of spherical solid one in the same volume. Extinction performance of aggregates is associated with the quantity of granules but independent on the original particle size in it. Proper aggregation of nanoparticle is beneficial to microwave extinction capacity. Given identical original particle size and granule quantity, different fractal structure and orientation relative to the incident wave would result in different extinction performance. When the direction of the principal axis of inertia of aggregates approaches vertical to that of incident wave, extinction capacity is the best. And the parallel relationship between the two directions will lead to the worst extinction performance.

References

[1] Du T and Ilegbusi O J 2004 *J. Mater. Sci.* **39** 6105
[2] Nie P 2009 *Journal of Beijing University of Aeronautics and Astronautics* **35** 852
[3] Galyamov B S, Lushnikov A A and Maximenko V V 1996 *J. Aerosol. Sci.* **27** 551
[4] Mackowski D W 1994 *J. Opt. Soc. Am. A* **11** 2851
[5] Mackowski D W and Mishchenko M I 1996 *J. Opt. Soc. Am. A* **13** 266
[6] Stognienko R, Henning T and Ossenkopf V 1995 *Astron. Strophys.* **296** 797
[7] Huang C J, Liu Y F and Wu Z S 2007 *Acta Physica Sinaca* **56** 4068
[8] Lei C X and Li F L 2006 *Ludong University Journal (Natural Science Edition)* **22** 303
[9] Witten T A and Sander L M 1983 *Phys. Rev. B* **27** 5686
[10] Pang S Q, Chen L and Chen J 2007 *Journal of Guangxi Academy of Science* **23** 73
[11] Draine B T 1988 *Astrophys. J.* **333** 848
[12] Collinge M J and Draine B T 2004 *J. Opt. Soc. Am. A* **21** 2023
[13] Wriedt T, Hellmers J and Eremina E 2006 *J. Quant. Spectrosc. Ra.* **100** 444
[14] Muinonen K and Zubko E 2006 *J. Quant. Spectrosc. Ra.* **100** 288
[15] Draine B T and Goodman J J 1993 *Astrophys. J.* **405** 685
[16] Draine B T and Flatau P J http://arxiv.org/abs/astro-ph/0409262
[17] Sun G L 2004 *Doctoral Dissertation of Tsinghua University*

IOP Conf. Series: Materials Science and Engineering **137** (2016) 012043 doi:10.1088/1757-899X/137/1/012043

Electrochemical properties and microstructure of Al/Pb-Sn anodes during copper electrowinning

Y C Zhang[1] and S J Qin

Department of Mechanical Engineering, Baoji University of Arts and Science, Shaanxi Baoji 721013, China

E-mail: 1040130198@qq.com

Abstract. Al/Pb-Sn composite anodes used in copper electrowinning are prepared by electro-deposition onto the aluminum Substrates in lead methanesulfonate electrolyte. Using potentiodynamic investigations, the information about corrosion resistance and oxygen reaction (OER) potential of these anodes in the lead dioxide region have been measured. The results show that with an increase of Sn content the OER potential and corrosion rate decrease. The SEM observations show that with an increase of Sn content the lead grain size increases. The X-ray diffraction (XRD) analysis show that with an increase of Sn content the phase of β-PbO_2 with lower corrosion rate and OER potential increases.

1. Introduction

To search for a suitable anode alternative to conventional Pb-Ca-Sn anodes which operate at relatively high OER over-potential, resulting in significant energy consumption and corrosion during copper electrowinning, Pb-Sb and Pb-Co alloy anodes and Pb/Pb-Sb, Pb/Pb-Co and Pb/Pb-Co_3O_4 composite anodes have been prepared.

The literature [1-2] pointed that Sb can decrease OER overpotential of anodes for copper electrowinning. As the Sb content is sufficiently low, there is no interconnecting network of Sb and corrosion is reduced. But higher Sb content reduces corrosion resistance of Pb-Sb anodes used in copper electrowinning [3]. Co into Pb (in a concentration of 0.02–0.1%) has long been known to reduce notably the OER over-potential and the corrosion rate of Pb. But the components of Pb and Co cannot mix in liquid-casting condition even in a large temperature range [4]. Co^{2+} into electrolyte increases the electrocatalytic activity and corrosion resistance of anodes but decreases the current efficiency in zinc electrowinning for the difficult formation of lead persulphate [5-6]. For the reason, experiments have been performed to include metallic Co in a anode. Lead-cobalt anodes are obtained by electrodeposition from a lead sulphamate electrolyte onto Pb-Ca [7] or Pb-Ca-Sn [8] plates. A comprehensive study on the behavior of Pb-Co_3O_4 [9-10] composite anodes showed that compared to commonly use metallurgical Pb-5.85Sb and Pb-0.08%Ca-0.74%Sn anodes the Pb-Co_3O_4 composite anode shows a considerable depolarizing effect on the OER. The corrosion rate of Pb-Co_3O_4 composite anode during prolonged polarization 96h under galvanostatic conditions is lower than that of Pb-Sb and Pb-Ca-Sn anode anodes. But Pb-Co_3O_4 composite anodes are still in the laboratory. Literature [11] points Al/Pb-PANI-WC anode has lower OER potential and corrosion rate, but it only give the electrochemical properties of the fruit surface which is not the best method for characterizing the properties of anodes during metal electrowinning.

Aluminum has a lower cost, density, and electrical resistivity than lead which is probably reduce the bath voltage and weight used as the substrate of Al/Pb-Sn anodes.

In this paper, Al/Pb-Sn composite anodes are obtained by electrodeposition onto aluminum plates in lead methanesulfonate electrolyte. Electrochemical properties and microstructure of Al/Pb-Sn anodes have been researching by potentiodynamic investigations, SEM observations and XRD analysis.

2. Experimental details

2.1. Process method

The Al/Pb-Sn composite anodes materials were obtained by electrodeposition onto 40mm×50mm×3mm aluminum plates in the lead methanesulfonate electrolyte consisted of $HCH_3SO_3$80-120mldm^{-3}, and Pb^{2+}150-200gdm^{-3} from $Pb(CH_3SO_3)_2$, Sn^{2+} 2-50gdm^{-3}. The electrodeposition was carried out at a pH=1-3 and a cathodic current density of Jc=3Adm^{-2}, and with mechanical bubbling for 5h at 50℃.

The aluminum substrate were pretreated by buffing(mechanical buffing with fibre grinding wheel), degreasing (soak in fatty alcohol-polyoxyethylene ether degreaser at 40℃ for 4 to 12h), caustic wash (soak in 12gdm^{-3} NaOH solution at 20-40℃ for 120-180 seconds), acid pickling(38gdm^{-3} H_2SO_4 at 20-40℃ for 10-30 seconds), zinc immersion (soak in 12gdm^{-3} NaOH, 15gdm^{-3} ZnO, 10gdm^{-3} $KNaC_4H_4O_6 \cdot 4H_2O$, 2gdm^{-3} $FeCl_3 \cdot 6H_2O$ solution at 20-40℃ for 60 seconds), acid pickling(3-10s), zinc immersion (at 20-40℃ for 50 seconds) and nickel electroplating for 1 minute.

2.2. Measurement and analysis

The electrochemical workstation (CHI760C) with three electrodes were used for measuring CV for Al/Pb-Sn composite anodes in the copper electrowinning electrolyte consisted of Cu^{2+}(50gdm^{-3}) and H_2SO_4 (200gdm^{-3}) at 40℃. The counter electrode was the platinum plate, and the reference electrode was MSE. The Al/Pb-Sn composite anodes were used as the working electrode. Also, SEM and XRD were used for determining the surface microstructures and anodic layer phase for Al/Pb-Sn composite anodes.

3. Results and discussions

Figure 1 shows the cyclic voltammograms of Al/Pb-0.32%Sn, Al/Pb-0.41%Sn, Al/Pb-0.49%Sn, Al/Pb-0.62%Sn composite anodes after 24h of anodic polarization. It's obvious from Figure 1 that with an increase of Sn content in Pb-Sn composite the branch is shifted further negatively showing lower OER potential, and the cathode peak d shifted slightly positively, leading to a bigger depolarization region which means that with an increase of Sn content in Pb-Sn composite the corrosion rate of Al/Pb-Sn anodes decrease.

Figure 2 shows the SEM of Al/Pb-0.32%Sn(a), Al/Pb-0.41%Sn(b), Al/Pb-0.49%Sn(c) and Al/Pb-0.62%Sn(d) anodes before anodic polarization. As shown in Figure 2, with an increase of Sn content the Pb grain size increases, showing a higher corrosion rate [10]. The result is contradiction with the corrosion information by CV analysis which is probably due to the content β-PbO_2 phase on anodic layer increase after 24h of anodic polarization (as shown in Figure 3). SEM observations on the fruit surface of Al/Pb-0.32%Sn shows that the Al/Pb-0.32%Sn anode has dense and cubic crystals of lead, possessing a chaotic orientation, but changing to a well-organized orientation gradually when the Sn content increase.

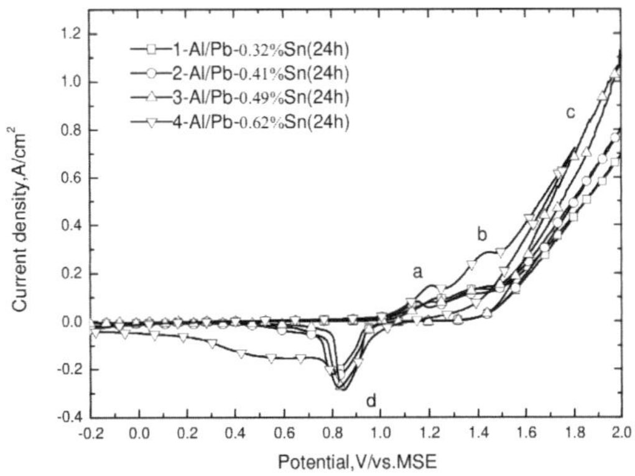

Figure 1. Cyclic voltammograms of Al/Pb-0.32%Sn, Al/Pb-0.41%Sn, Al/Pb-0.49%Sn, Al/Pb-0.62%Sn composite anodes after 24h of anodic polarization.

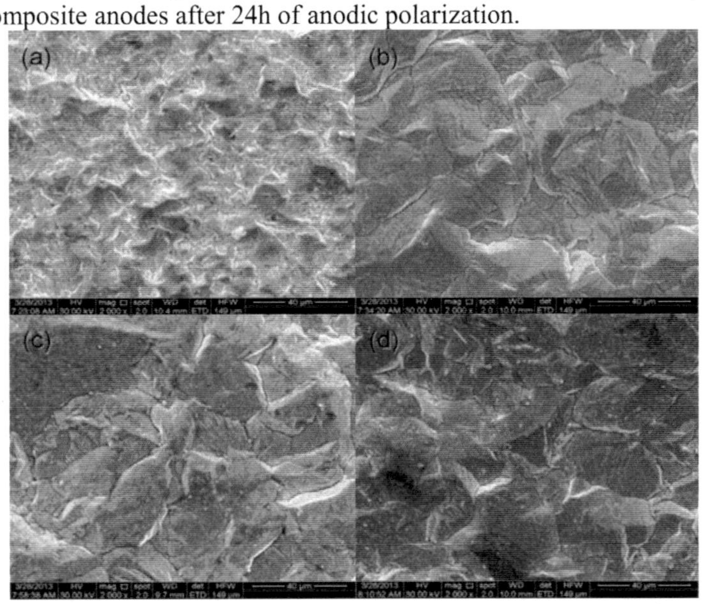

Figure 2. SEM of the Al/Pb-0.32%Sn(a), Al/Pb-0.41%Sn(b), Al/Pb-0.49%Sn(c) and Al/Pb-0.62%Sn(d) anodes before anodic polarization

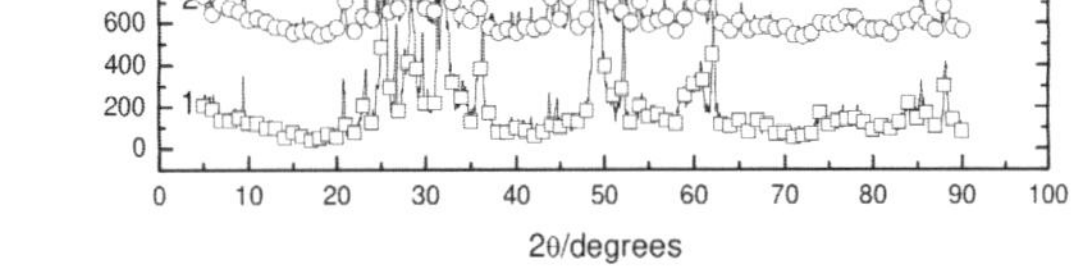

Figure 3. XRD of the Al/Pb-0.32%Sn, Al/Pb-0.41%Sn, Al/Pb-0.49%Sn and Al/Pb-0.62%Sn anodes after 24h of anodic polarization.

Figure 3 shows the XRD of Al/Pb-0.32%Sn(1), Al/Pb-0.41%Sn(2), Al/Pb-0.49%Sn(3) and Al/Pb-0.62%Sn(4) anodes after 24h of anodic polarization. The XRD analysis indicates that these anodic layers of Al/Pb-Sn composite anodes are composed of α-PbO_2, β-PbO_2, and Pb phase. The anodic layer of Pb anode presents a high content of α-PbO_2 and lower of $PbSO_4$. The existence of Pb phase reveals that the layers are relatively thin. The major phases of Al/Pb-Sn anodes are α-PbO_2, $PbSO_4$, β-PbO_2, and Pb. With an increase of Sn content, $PbSO_4$ first increases and then deceases while β-PbO_2 increases at the range of 2θ from 43 to 50.

4. Conclusions

(1) The result of potentiodynamic investigations deduced that with an increase of Sn content, the corrosion rate and OER potential of Al/Pb-Sn anodes decrease during copper electrowinning.

(2) With an increase of Sn content of Pb-Sn composite the grain size of Pb increases, showing a higher corrosion rate which is contradiction with the corrosion information by CV analysis which is probably due to the content β-PbO_2 phase on anodic layer increase after 24h of anodic polarization.

Acknowledgments

Financial supports for this work from the Baoji science and technology project (14GYGG-2-4).

References

[1] Laihonen S, Laitinen T, Sundholm G and Yli-Pentti A 1989 The Anodic behavior of Sb and Pb-Sb eutectic in sulphuric acid solution *Electrochim. Acta* **35**(1) 229-238

[2] Metikos-pekovic M, Babic R and Brignac S 1997 Influence of antimony on the properties of the anodic oxide layer formed on Pb-Sb alloys *Journal of Power Source* **64** 13-19

[3] Hrussanova A, Markova L and Dobrev TS 2002 Electrochemical properties of Pb-Sb, Pb-Ca-Sn and Pb-Co_3O_4 anodes in copper electrowinning *J. Appl. Electrochem.* **32**(5) 505-512

[4] Koch D F A 1959 The effect of cobalt on a lead anode in sulphuric acid *Electrochim Acta* **1**(1) 32-38

[5] Ivanov I, Stefanov Y, Noncheva Z, Petrova M, Dobrev Ts, Mirkova L, Vermeersch R and J P Demaerel 2000 Insoluble anodes used in hydrometallurgy Part II Anodic behaviour of lead and lead-alloy anodes *Hydrometallurgy* **57**(2) 125-139

[6] Zhang Y, Chen B, Yang H, Huang H and Guo Z 2014 Anodic behavior and microstructure of Al/Pb-Ag-Co anode during zinc electrowinning *J. Cent. Sou. Univ.* **21**(1) 83-88

[7] Dobrev T, Valchanova I, Stefanov Y and Magaeva S 2009 Investigations of new anodic materials for zinc electrowinning *Trans. Inst. Met. Finish.* **87** 136-140

[8] Maryam J B and Aleksandar N N 2012 Electrodeposition of lead-cobalt composite coatings electrocatalytic for oxygen evolution and the properties of composite coated anodes for copper electrowinning *Hydrometallurgy* **129** 59-66

[9] Hrussanova A, Markova L, Dobrev T and Vasilev S 2004 Influence of temperature and current density on oxygen overpotential and corrosion rate of Pb-Co_3O_4, Pb-Ca-Sn, and Pb-Sb anodes for copper electrowinning: Part II *Hydrometallurgy* **72** 215-224

[10] Xu R, Huang L, Zhou J, Zhan P, Guan Y and Kong Y 2012 Effects of tungsten carbide on electrochemical properties and microstructural feature of Al/Pb-PANI-WC composite inert anodes used in zinc electrowinning *Hydrometallurgy* **125** 8-15

[11] Han Q, Liu K, Chen J and Wei X 2003 A study on the electrodeposited Ni-S alloys as hydrogen evolution reaction cathodes *Int. J. Hydrogen Energy* **28** 1207-1212

IOP Conf. Series: Materials Science and Engineering **137** (2016) 012044 doi:10.1088/1757-899X/137/1/012044

The effect of magnetic field intensity and treatment time on graphene / epoxy composites' fracture toughness

Z Q Tian[1,3], L Zhang[1], S Fu[1], R H Yuan[1], Z W Dong[1] and X M Ren[2]

[1]Wuhan mechanical college, Wuhan, Hubei, 430075, China
[2]Naval Medicine Research Institute, Shanghai, 200433, China

E-mail: tianzq2009@hotmail.com

Abstract. The effect of the intensity of the magnetic field and the treatment time on the fracture toughness of graphene/epoxy composites is researched. Also, the mechanism of the effect of the magnetic field on the fracture toughness of graphene/epoxy composites and a method to improve the impact resistance is explored. Then, three-point bending tests are employed to characterize the fracture toughness of graphene/epoxy composite. The results show that the intervention of magnetic field could induce GNS to generated orientation arrangement, improving the fracture toughness of the graphene/epoxy composite. When the intensity of the magnetic field was increased, the growth rate of the fracture toughness slowed. However, when 2T magnetic was used to synthetically process the material, and when the processing time was less than 50 min, the fracture toughness of the composite material increased significantly.

1. Introduction

Epoxy resin is a high strength polymer material that is widely used in aerospace, machinery manufacturing, chemical building materials, shipping and other industries. Although epoxy resin has a high strength, it also has a lower impact resistance than other high-strength materials. In the epoxy resin toughening process, the preparation of composite materials, especially nano-composites, improve its impact resistance; this has become a hot topic for research. GNS sheets have thickness of a nanometer order. Also, the high modulus and high strength features of GNS make it an ideal component for the formation of nanocomposites [1]. Research related to GNS and epoxy composites has shown that that, in the case of direct recombination, the strength of the composite material is not increased significantly with an increase in the content of GNS [2,3]. In order to improve the strength of the composite material, the treatment of GNS must be modified. The intervention of a magnetic field can effectively change the orientation arrangement of GNS [4-15]. Based on the compound of the GNS and the epoxy resin, the toughening effect on the epoxy resin is studied. Toward this end, a constant magnetic field was used to intervene in the GNS and epoxy resin composite process. The orientation and arrangement of GNS in the epoxy matrix was changed by the magnetic field. This research can be used to deduce the effect of a magnetic field on composites.

2. Preparation and characterization method of the test material

2.1. Test material preparation

The GNS is dispersed in 100ml of acetone, at a frequency of 40KHz, for two hours in the power of 185W ultrasonic. After the GNS is uniformly dispersed, 6.6g epoxy is added, and the ultrasound is continued for two hours. The dispersed solution is stirred in a 75℃ environment by the use of a magnetic stirrer, then the solution is removed in acetone. While being stirred, the mixture is subjected to a degassing treatment for 30 min in order to remove any air bubbles. Finally, the degassed epoxy resin is poured into a homemade completed mold. After curing at 100℃ for twelve hours, the composite material is completely cured.

GNS / epoxy composites synthesis in a magnetic field is completed by using a constant magnetic field generated by an electromagnetic coil. Adjusting the magnetic field intensity to 0.5T, 1T, 1.5T, and 2T, the processing time is set at 10min, 30min, 50min, and 70min. After the magnetic treatment is completed, the mold is removed and placed in a curing oven, where it is cured for 12h at 100℃.

2.2. Test material characterization methods

The measure of fracture toughness is conducted with the ASTM D5045 for reference. The three-point bending method [16] is used. In order to reduce error, five splines from each sample are tested. Each sample is cut to length (S): Width (w): Thickness (B)=8:2:1. The prefabricated depth of sample surface is the 0.5W natural crack. Japan Shimadzu company's AG-10TA electronic universal testing machine is adopted. The indenter radius is 2mm, the bearing fillet radius is 2mm, and the loading speed is 1mm/min. The Inspect. F scanning electron microscope of Dutch FEI company is adopted to observe the structure of the sample section.

3. Results and analysis of test

3.1. The effect of magnetic field intensity

As the composite materials were synthesized, magnetic fields of different intensities (0.5T, 1T, 1.5T and 2T) were exposed to the sample for 30 min. The composite sample which included GNS content of 1wt.% was acquired. The samples were then tested by three-point bending. The results of KIC and GIC are shown in figure 1.

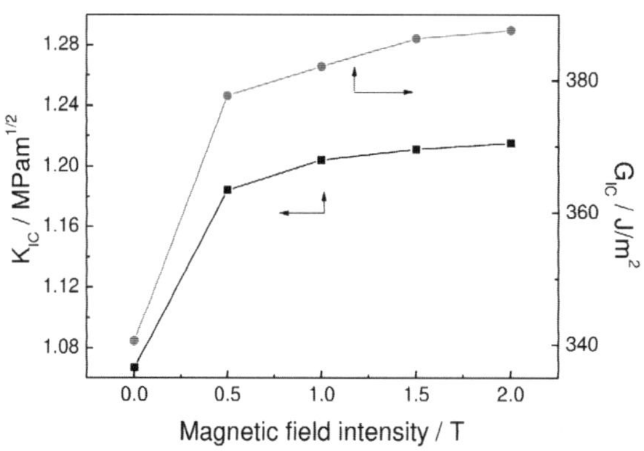

Figure 1. Relationship between K_{IC}, G_{IC} and magnetic field intensity

As can be seen from figure 1, as the strength of the magnetic field increased, KIC and GIC presented basically the same trend. When the magnetic field was 0.5T, KIC was 1.184MPam0.5. This is an 11.0% increase over the KIC without the presence of a magnetic field, 1.067MPam0.5. At the same time, the GIC value increased from 340.7J/m2 to 377.7J/m2, which is a rate of increase of about 10.9%. When these sets of data are contrasted, it can be clearly seen that the addition of a magnetic field significantly improved the fracture toughness of the composite. However, as the strength of the magnetic field continued to increase, the growth rate of KIC and GIC gradually slowed. When the magnetic field was enhanced 1T, 1.5T, or 2T, the growth of KIC and GIC was respectively 12.8%, 13.5%, 13.9% and12.3%, 13.5%, 13.8%. This implies that a lower magnetic field strength enables the GNS to achieve a better toughening effect. In contrast, the high intensity magnetic field only made a small increase of effect. The reason is that GNS has a good response for lower intensity magnetic fields, so that the GNS will have a certain orientation arrangement. The strength of the magnetic field can be enhanced to improve the degree of orientation of GNS; however, due to the restriction of the epoxy molecular chain, then its degree of orientation cannot raise too much, so the growth rate of the fracture toughness of the composite material gradually slows down.

3.2. The effect of treatment on time

In the process of composite material synthesis, the composite material was disposed of by using a 2T magnetic field for 10min, 30min, 50min, and 70min. The samples were tested to obtain the results shown in figure 2.

As can be seen from the figure, the fracture toughness values of KIC and GIC increased for the sample that was processed for 10min. Precisely, the increase was from 1.067MPam0 and 340.7J/m2 to 1.154MPam0.5 and 368.7J/m2. The growth rates are 7.5% and 8.2%. The sample that was processed for 30min also showed a similar growth trend. Although the slope efficiency of the curve indicates a slowing of the growth rate, the KIC and GIC values still appeared to increase significantly. However, when the time was extended to 50min and 70min, both values remained basically unchanged and the slope efficiency of the curve was approximately zero. This result could have been caused by two factors. First, the introduction of GNS could have increased the complexity and difficulty of the orientation. In a free state, due to the steric hindrance of epoxy molecular chains around GNS, the rotation of GNS becomes more difficult. As the degree of orientation with GNS improves, the epoxy molecular chain is closed to its entanglement, and GNS motion becomes more difficult. Therefore, even if the processing of the magnetic field is extended, GNS orientation movement does not proceed further. The second is to consider the epoxy curing time. The epoxy system solidified for 28min at 80℃ to fully achieve gelling. At this time, the viscosity of the system is very large, and the degree of crosslinking is already high. The system presents the basic curing characteristics. GNS is therefore fixed to the epoxy matrix, which can no longer produce orientation-induced motion.

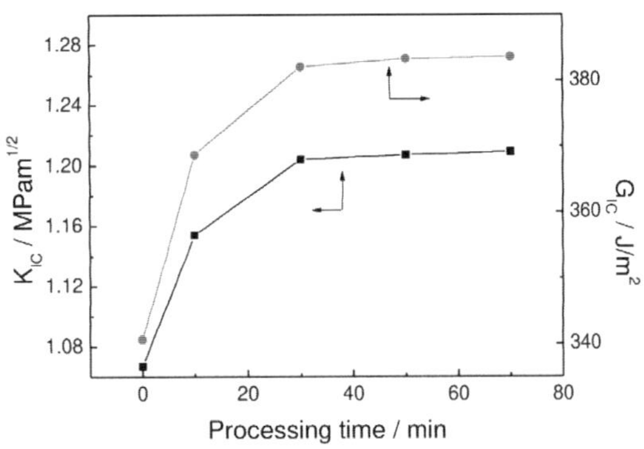

Figure 2. Relationship between K_{IC}, G_{IC} and magnetic field processing time

3.3. Mechanism analysis of the impact of a magnetic field on the fracture toughness of graphene/epoxy composite material

The above test results indicate that treatment with a magnetic field can further improve the toughness of GNS/epoxy composites. This effect is due to GNS being formed into a magnetic field-induced orientation. The schematic principle can be described by figure 3.

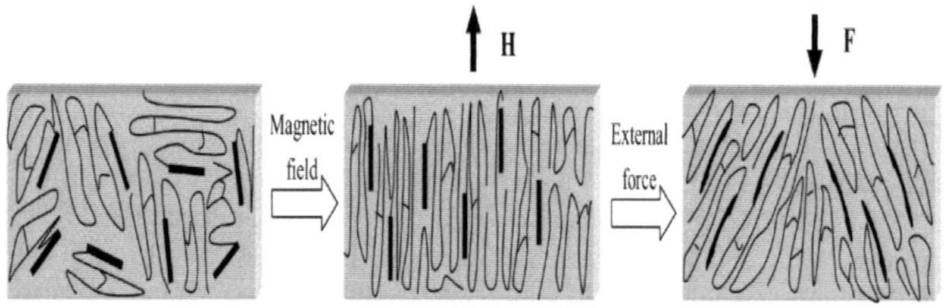

Figure 3. Schematic diagram of toughening mechanism in magnetic field

When composites were processed in the presence of a magnetic field, because GNS itself is strongly magnetic, its orientation can be manipulated by a magnetic field. With the GNS arranged in the same direction, the nearby molecular chain of epoxy will create an orderly arrangement. A study of thermotropic liquid crystal toughened epoxy composites found that the orientation of liquid crystal molecules can be aligned in the direction of the external force when it is subjected to external force because of the mutual traction LCD and epoxy matrix, and the epoxy matrix is also present in the orientation of a certain trend. The regular arrangement of the epoxy molecules results in a greater amount of deformation in a parallel direction. When the composite material is subjected to external force, the epoxy molecular chain segments can deform to absorb the energy of the outside force, which reduces molecular chain fracture. At the same time, the high flexibility of GNS can also produce bending deformation and absorb some of the energy. This creates a synergistic effect of GNS and the epoxy molecular chain, so that material damage caused by an external force decreases; this achieves the goal of improving toughness. According to the theory of polymer mechanics, the increase of fracture toughness must be accompanied by changes of breakage elongation. From the tensile property tests for the above two samples, the changes of breakage elongation are shown in figure 4. As can be seen from the figure, the breakage elongation with the trend of the magnetic field strength and magnetic treatment time was consistent with the KIC and GIC. This confirms that the orientation of GNS in composites by use of a magnetic field produces a toughening effect.

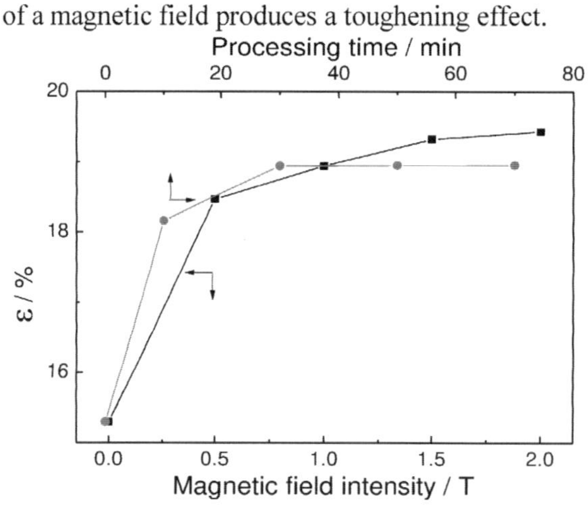

Figure 4. Relationship between magnetic field and elongation

3.4. Electron microscopy analysis of graphene/epoxy composites with magnetic field

The most direct and effective method of characterizing samples is to use an electronic microscope to directly observe them. Figure 5 shows images of two samples, one treated with a magnetic field and one not.

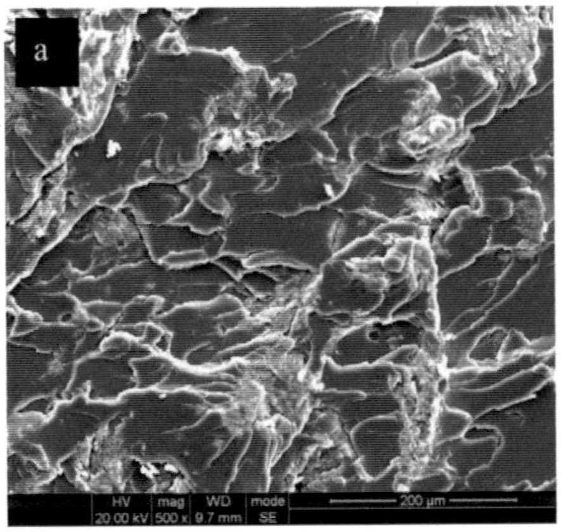

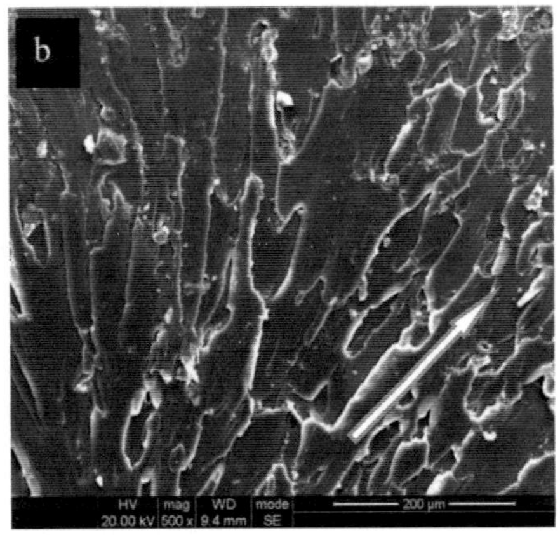

Figure 5. SEM images of different samples (a)untreated (b)1T

The SEM image shows that the sheet of sample (a) without the treatment of magnetic field is arranged in a disorderly fashion, and the layered stacked state appears irregular. However, the treatment sample of 1T magnetic field induced a clear orientation structure, and the sheet was orderly arranged in the same directions. This indicates that the orientation of GNS was induced by the magnetic field. By observation of the fracture surface, it becomes clear that the orientation of GNS caused the ordered ethylene molecule segments. In contrast with figure 4, stress concentration points in figure 5(a) become obvious from observing both 5(a) and 5(b), which may be caused by maldistribution of GNS in the matrix. However, in figure 5(b) stress concentration points are rare. This indicates that the magnetic field promotes proper distribution of GNS in the matrix. It was also shown that the magnetic field is the reason behind the improvement of characteristics of the composites.

4. Conclusion

In the synthesis process of the GSN/epoxy composite, the magnetic field was shown to induce directional arrangement of GSN. Within a certain range, the degree of orientation will increase as the strength and length of the magnetic field also increase.

Based on the above testing results, the magnetic field is an effective method to improve the performance of GSN/epoxy composites.

References

[1] Mohammad A R, Javad R, Iti S, et al. 2010 Fracture and fatigue in graphene nanocomposites. *Small* **6**(2) 179-183.

[2] Mohammad A R, Javad R and Zhou W, et al. 2009 Enhanced mechanical properties of nanicomposites at low graphene content. *Nano* **12**(3) 3884-3890.

[3] Bryan W D and Khalid L 2007 Use of exfoliated graphite filler to enhance polymer physical properties *Carbon* **5**(9) 1727-1734.

[4] Afzana A and Alan H W 1991 Orientation kinetics of thermotropic main-chain liquid-crystalline polymers in a magnetic field *Polymer* **32**(1) 103-108.

[5] Witold B, Elana A F and Mihail G K, et al. 1999 Orientation of a longitudinal polymer liquid crystal in a constant magnetic field *Polymer* **40**(6) 1441-1449.

[6] Altoiz B A, Naroditskaya T V and Popovskii Y M 2003 Influence of magnetic field on the molecular orientation in epitropic mesophase of nitrobenzene *Adv. Colloid Interfac.* **104**(1-3) 239-243.

[7] Sata H, Kimura T and Ogawa S, et al. 1996 Magnetic orientation of poly(ethylene-2,6-naphthalate) *Polymer* **37**(10):1879-1882.

[8] Janssen J, Lobert M and Muller-Warmuth W, et al. 1993 Magnetic field-induced orientation of crytallites in powdered samples of layered compounds: explanation and application *Solid State Nucl. Mag.* **2**(4) 201-204.

[9] Takahiko K, Ryota I and Yuzo Y, et al. Crystal orientation of β-phase isotactic polypropylene induced by magnetic orientation of N,N'-dicyclohexyl-2,6-naphthalenedicarboxamide 2002 *Polymer* **43**(26) 7301-7306.

[10] Kitamura N, Fukumi K and Takahashi K, et al. 2011 Orientation of Carbon Nano-fiber in Carbon/Silica Composite Prepared Under high Magnetic Field *Mat. Sci. Eng. A* **18**(5) 4-8.

[11] Nakano Y and Matsuo M. 2008 Orientation behavior of carbon fiber axes in polymer solutions under magnetic field estimated in terms of orientation distribution function *J. Phys. Chem. C* **112**(40) 15611-15622.

[12] Hiroaki Y, Koichi Y and Yuuichi Y, et al. 2009 Orientation of nanowires consisting of poly(3-hexylthiophene) using strong magnetic field *Synthetic Met.* **159**(9-10) 955-960.

[13] Hiroaki Y,Yuuichi Y and Sunao Y, et al. 2008 Magnetic orientation of single-walled carbon nanotubes or their composites using polymer wrapping *Sci. Technol. Adv. Mat.* **9**(2) 1-6.

[14] Chilerio A, Musso S, Sangermano M, et al. 2008 Preparation of polymer-based composite with magnetic anisotropy by oriented carbon nanotube dispersion *Diam. Relat. Mater.* **17**(7-10) 1590-1595.

[15] Erin C, Richard V and Marwan A H, et al. 2007 Properties of carbon nanotube-polymer composites aligned in a magnetic field *Carbon* **45**(10) 2037-2046.

[16] Sun C F, Xue Y P, Hu P 2005 Composite foam sandwich structure and mechanical properties of test methods *Fiber Reinforced Plastics/composite* 2: 3-6.

The effect of acrylic latex-based polymer on cow blood adhesive resins for wood composites

J Yan[1], H L Lin[1], G Z Feng[1] and S Gunasekaran[2, 3]

[1]Chemistry & Chemical Engineering Department, Zhongkai University of Agriculture and Technology, Guangzhou 510225, P.R. China
[2]Biological Systems Engineering Department, University of Wisconsin-Madison, WI, 53706, U.S.A

E-mail: guna@wisc.edu

Abstract. In this paper, alkali-modified cow blood adhesive (BA) and blood adhesive/acrylic latex-based adhesive (BA/ALB) were prepared. The physicochemical and adhesion properties of cow blood adhesive such as UV- visible spectra, particle size, viscosity were evaluated; share strength, water resistance were tested. UV- visible spectra indicates that the strong bonding strength of BA/ALB appeared after incorporating; the particle size of adhesive decreased with the increase of ALB concentration, by mixing ALB and BA, hydrophilic polymer tends locate or extand the protein chains and provide stability of the particles; viscosity decreased as shear rate increased in concordance with a pseudoplastic behavior; both at dry and soak conditions, BA and ALB/BA show significant difference changes when mass fraction of ALB in blend adhesive was over 30% ($p < 0.05$). ALB/ BA (ALB30%) is not significant different than that of phenol formaldehyde which was used as control. A combination of cow blood and acrylic latex-based adhesive significantly increased the strength and water resistance of the resulting wood.

1. Introduction

Wood adhesives in use today are normally synthetic polymers from petroleum chemicals.Toxic emissions from wood adhesives are of major concern [1-3]. One of the goals of the wood-products industry is to use natural materials as adhesives and reduce dependency on petrochemical [4-9]. Proteins from both plant and animal sources have been used to prepare adhesives with varying degrees of success. Weakley and Mehltretter [10] developed moisture-resistant plywood adhesives by cross-linking casein with dialdehyde. Ash and Lumbuth [11] prepared plywood glues containing high blood solids blended with phenol-formaldehyde (PF) resins that required a special mixing procedure to avoid high viscosity. Golick and Dike [12] formulated exterior phenolic plywood glues containing up to 70% dried blood. Thames, Cook and Mendon [13] formulated an adhesive by mixing of soy protein isolate; a polyolplasticizer and a vegetable oil derivative was used to impart water resistance. Trocino [14] prepared adhesives by copolymerizing hydrolyzed vegetable protein that has been functionalized with methylol groups, the adhesives were able to maintain long-term strength and water resistance. Wang, Wang and Sun [15] studied the effects of hexadecyltrimethyl ammonium bromide, ethylhexadecyldimethyl ammonium bromide and benzydimethylhexadecyl ammonium chloride, results indicate that electronic and hydrophobic interactions between soy protein isolate and detergent help improve adhesive performance. Jiang, Qin, Hse et al. [16] replaced phenol in the synthesis of phenol-formaldehyde resin with feather protein, performed as well as the PF resin. Qi, Li and Wang [17] found $NaHSO_3$-modified soy protein adhesives had many advantages over traditional soy protein isolate adhesive. Li, Qi, Sun et al. [18] proved that canola protein could be used as bio-degradable wood adhesives. $NaHSO_3$ could cleave disulfide bonds in the protein, induce extra charges ($RS\text{-}SO_3^-$) on the protein surface, reduce adhesives' apparent viscosity and improve the flow ability properties of canola protein adhesives. It had insignificant effects on the adhesion performance of the wood adhesives. Although different methods had been used to reduce viscosity, the viscosity of these protein-based glues were still very high.

Blood adhesives suffer from relatively lower adhesive strength and higher viscosity than chemical synthetic adhesives. The high viscosity of the modified blood protein adhesives is a result of increased intermolecular interactions of the unfolded protein molecules. Adhesive with high viscosity has poor flow ability and is not easy to spread in application. Possible approaches to solving the high viscosity problem are to minimize intermolecular interactions and enhance physicochemical factors that weaken or disrupt these forces without adversely affecting adhesive strength. Improved adhesive strength and water resistance have been observed for adhesives prepared from alkali-modified blood protein in our previous study [19]. Ionic environments have been known to weaken the electrostatic interactions between protein molecules. Thus the viscosity of protein can be varied by treating protein with salts or by using reducing agents without affecting the adhesive strength or water resistance. However, relatively little is known about acrylic latex-based adhesive blending with blood adhesive.In most latex applications, acrylic acid is often used for coating and adhesive formulations [20-21]. The carboxylic acid groups of acrylic provide colloidal stability through electrosteric stabilization [22]. It is an adhesive with superior bonding and water resistance properties, it is non-flammable, non-toxic and low viscosity. Cow-blood has higher protein ratio compared to other types of animal blood that could reach better adhesive properties through modifing. The objective of this study was to investigate the adhesive and water resistance properties and viscosity of blending acrylic latex-based adhesive and cow-blood adhesive.

2. Materials and methods

2.1. Materials

Fresh cow blood was purchased from Muscle Science Department, UW-Madison. Sodium hydroxide ($NaOH$), sodium silicate (Na_2SiO_3), calcium oxide (CaO) and ammonia ($NH_3.H_2O$) (density =0.9 g/cc) were purchased from Fisher Scientific (Fair Lawn, NJ). Sodium azide (NaN_3), EDTA disodium salt ($C_{10}H_{14}N_2Na_2O_8$. $2H_2O$), were purchased from Sigma Chemical Company (St, Louis, MO). Acrylic latex-based synthetic adhesive (ALB) was obtained from Soil Net Inc., (Madison, WI, USA).

2.2. Preparation of blood adhesive

Adhesive of alkaline modified cow blood adhesive (BA) was made as follows [19]: 300 g of cow blood was mixed with 0.5 mL sodium azide (1.0%, w/w)and 5.5 mL EDTA (10%, w/w) by stirring. Calcium hydroxide (1: 3.5 w/w) about 20 g, caustic soda (30% w/w) 10 to10.5 g was added in stages under stirring at room temperature (25 to 28 °C). Then about 3 g ammonia solution (10%) and 12 g sodium silicate were added in stages while stirring. The resulting product is adhesive and the pHs of the adhesives are around 10.0.

2.3. Preparation of blend adhesive

The BA adhesive was blended with ALB adhesive. The mass fraction of ALB in the blend adhesives was 0.00, 0.21, 0.33, 0.52, 0.75, 0.85 and 1.00, respectively. The mixture was stirred for about 20 minutes at room temperature under 300 rpm.

2.4. UV testing
The UV-Vis spectra of the samples were recorded for the wavelength range of 200 to 800 nm at room temperature using a UV-Vis spectrophotometer (UV-1601PC, Japan).

2.5. Rheological analysis
Rheological analysis was performed at 25℃ within 12 h after preparation. A standard rotational rheometer equipped with plate-plate geometry (CP 4/40) / (Bohlin CVOR, Malvern Instruments Inc., Southborough, MA) was used in steady state by varying the shear rate from 10 to 250 s^{-1}. The distance between plate and plate was set to 150 μm for all measurements. Three replicate measurements were performed for each type of adhesive.

2.6. Particle size testing
Surface-area-average diameter of the adhesives was determined by particle size analyzer (90Plus, Brookhaven Instruments Corporation, New York, USA). In all experiments, adhesives were diluted with distilled water prior to particle size measurements, to keep the concentration typically between 10^{-5} and 10^{-2} volume fraction to prevent multiple scattering effects. The particle size was calculated by the autocorrelation function of the ALV Sizer software. Each measurement was done in triplicate.

2.7. Dry and wet bonding shear strength
Bonding shear strength was measured using an Instron testing systems (Model 1000, Canton, MA). Two hard maple wood strips (10.0 × 1.2 × 0.08 cm) were bonded over a bonding area of 2.0× 1.0 cm by applying the adhesive (Figure 1). The wood strips were heated to 120℃ and held compressed at 220 N for two minutes. Bonding shear strength of dry specimens (moisture content 6 to 10%) was performed at a crosshead speed of 10 mm/min. Five replications were performed for each measurement. The bonding shear strength was also measured after soaking the specimens in water for 24 h at room temperature and then drying at room temperature in a fume hood for 24 h. Both in dry and soak tests, phenol formaldehyde (PF) was used as control.

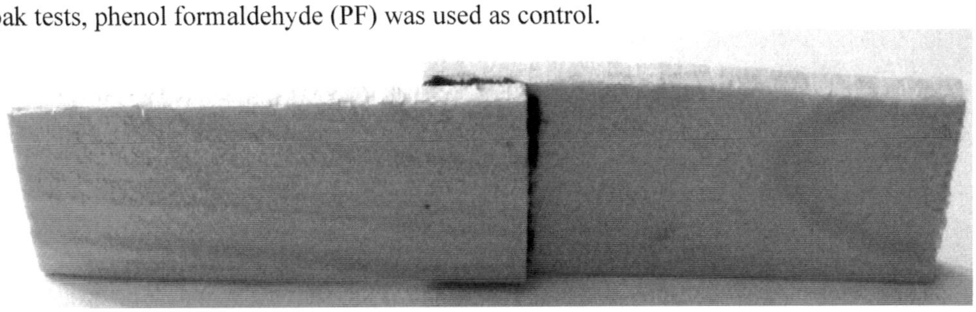

Figure 1. Maple wood strips bonded by adhesive.

3. Results

3.1. UV-Vis spectrophotometer
UV-Vis spectroscopy can provide information on possible denaturation of blood protein, particularly that of conformational change. UV-Vis spectra of BA, ALB and ALB/BA (w/w) are shown in Figure 2. The UV-Vis spectrum of BA exhibits a strong Soret absorption at 421 nm.

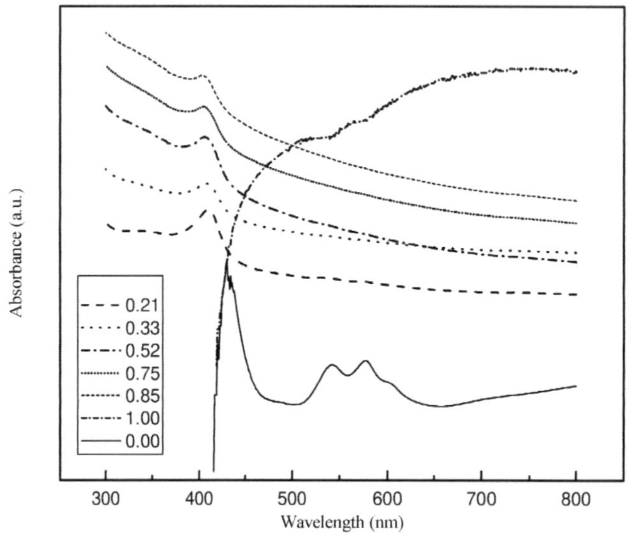

Figure 2. UV spectra of ALB in BA solution with different concentration.

ALB doesn't exhibit absorption peak. When ALB is combined with BA, Soret band absorption at 421 nm exhibits a shifts in the wavenumber. These results indicate that interactions between BA and ALB occurred. On the other hand, the shift phenomenon is more significant with higher ALB content. This indicates that stronger bonding strength of ALB/BA appeared after incorporating more amount of ALB because of more lateral hydrogen bonds formed.

3.2. *Effects of ALB amount on viscosity*

Adhesive of moderate viscosity is needed for wood products applications. To evaluate the behavior of the binary formulations of adhesive, rheological flow properties of the binary systems were evaluated. The flow curve of viscosity vs. shear rate of the adhesive system was reported in Figure 3. It shows all the BA has higher viscosities at lower shear rates, but has lower viscosity at higher shear rates, exhibiting a shear thinning behavior. At low shear rates, high molecular weight BA chains are subjected to Brownian randomization. The molecular chains become more ordered along the flow field hence exhibit lower viscosity as shear rate sufficiently increases to overcome the Brownian motion, The high viscosity at low shear rate gives BA good suspension property. It is also evident in Figure 3 that viscosity trended downward with an increase in ALB content in the adhesive. The situation can be attributed to osmotic pressure change in the system due to the change in charge density and structure from BA and ALB. The viscosity curves were fit to power-law model (equation 1).

$$\eta_\alpha = K(\gamma)^{n-1} \tag{1}$$

Where, η_α is apparent viscosity (Pa.s), γ is shear rate (s^{-1}), K is consistency coefficient (Pa s^n), and n is flow behavior index. K and n values are listed in Table 1. Based on these, the blood adhesives examined are classified as pseudoplastic fluids ($K > 0$, $0 < n < 1$). At the same time, we can see from Table 1 that K appeared to decrease with increasing mass fraction of ALB. K showed significant decline when mass fraction of ALB increased from 0.00 to 0.21, it means that viscosity of blood adhesive decreased apparently when ALB was added, more ALB was added, lower viscosity was obtained. Shear thinning character decreased gradually as judged from the increase in n value from 0.70 to 0.94 when mass fraction of ALB varied from 0.00 to 0.85.

Table 1. Power-law parameter K and n of BA with addition of ALB (25℃)

Mass fraction of ALB	K(Pa s^n)	n	R^2
0.00	9.81	0.70	0.97
0.21	4.38	0.82	0.96
0.33	3.30	0.83	0.96
0.52	1.88	0.92	0.95
0.75	1.49	0.93	0.94
0.85	0.84	0.94	0.81
1.0	0.82	0.95	0.80

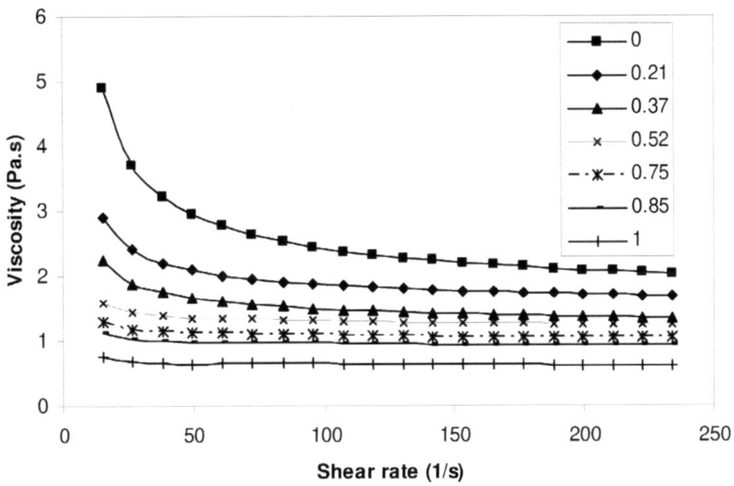

Figure 3. Viscosity curves of adhesives (pH = 10.2) at different mass fraction of ALB.

3.3. Effects of ALB amount on particle size changing and stability

Figure 4 shows the mean particle size decreased with the increasing of ALB content. This phenomenon is probably due to the decreased of viscosity of the internal phase, which increases the net shear stress and weakens the formation of large droplets. In addition, the decreasing viscosity could promote rapid dispersion of ALB into the blood adhesive aqueous phase, resulting in hindering aggregation of blood adhesive and decreasing smaller droplets. The carboxylate groups at basic condition resulted in dipoles and metal ions can joint in these dipolar interactions, which lead to a type of cross-linking [22]. It means that ALB was sufficient to cover the protein emulsion droplets which avoid the coalescence of protein droplets and then cause the formation of adhesive with smaller size. More ALB can be incorporated onto the protein surface, a large number of carboxylate groups extending into the continuous phase then could be cross-linked, forming a layer at the surface to avoid aggregation of protein molecules.

At the same time, more ALB can be oriented to reduce efficiently the interfacial tension which result in significant increase in the net shear stress at a constant energy density [23, 24] during emulsification and promoted the formation of smaller emulsion droplets. The mean particle size of adhesive decreased with the increase of ALB concentration, by mixing ALB and BA, hydrophilic polymer tends locate or extend the protein chains and therefore, provide stability of the particles.

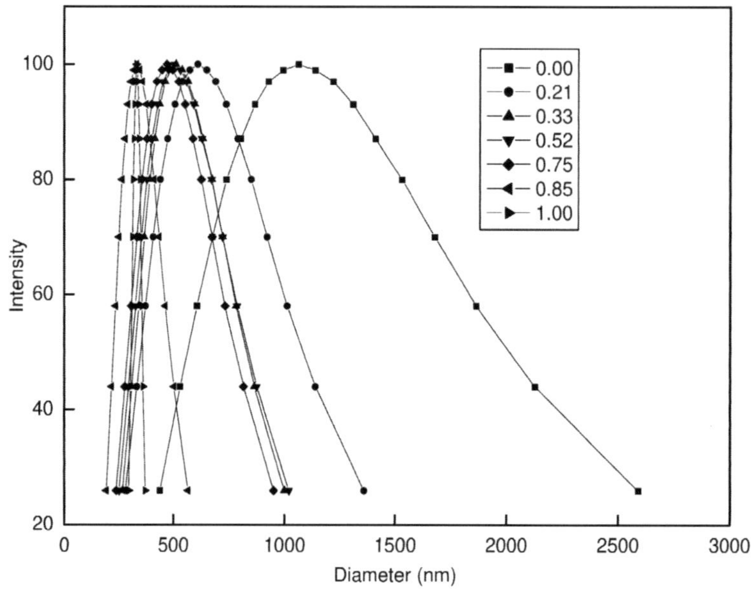

Figure 4. Particle size recorded at various mass fraction of ALB for blood protein-based adhesive.

3.4. Dry and soak bonding shear strength

Bonding shear strengths of blend adhesive at different mass fraction of ALB for both dry and soak tests are shown in Figure 5. It is clear that content of ALB greatly increased the bonding shear strength and water-resistances of plywood samples. In dry condition, the average shear strengths were 1110, 1241, 1368, 1598, 1601 and 1621 N when mass fraction of ALB were 0.00, 0.10, 0.20, 0.30, 0.40 and 0.50, respectively. BA adhesive and ALB/BA adhesive do not show significant difference changes ($p > 0.05$) when mass fraction of ALB was less than 20% at dry condition, but BA adhesive and ALB/BA show significant difference changes when mass fraction of ALB is over 30% ($p < 0.05$). ALB/BA (30%) is not significantly different than that of control phenol formaldehyde (PF, 1611 N).

Bonding shear strength of the samples soaked in water shows the trends like in dry condition. The average shear strength of blood adhesive was 229, 398, 643, 695, 703 and 721 N at soak condition when mass fraction of ALB are 0.00, 0.10, 0.20, 0.30, 0.40 and 0.50, respectively. BA and ALB/BA adhesive show significant difference changes when mass fraction of ALB are over 30% ($p < 0.05$). ALB/BA (30%) is not significantly different from that of control phenol formaldehyde (PF, 760 N). This suggests that the mixed adhesive are competitive with phenol formaldehyde for bonding shear strength in dry and soak conditions.

The bonding between blood protein molecules links mainly by hydrogen bonds, all kinds of salt-type keys and disulfide bonds are chemical cross-linking part. However, this is weak. The formation of protein-polymer (BA/ALB) complexes can be employed to improve functional properties of adhesive. Because ALB is easy to form hydrogen bonds, it also has high chemical activity and easy through grafting or cross - alliance to get a variety of branched-chain or mesh modifier system which has water resistance capability. ALB with certain chain length can form limited tangles points with BA, because there are many amino acids polar groups in BA. When BA adhesive blends with ALB, they will form a number of physical-chemical cross-linking points which guarantees reinforcement, bond strength and water resistance.

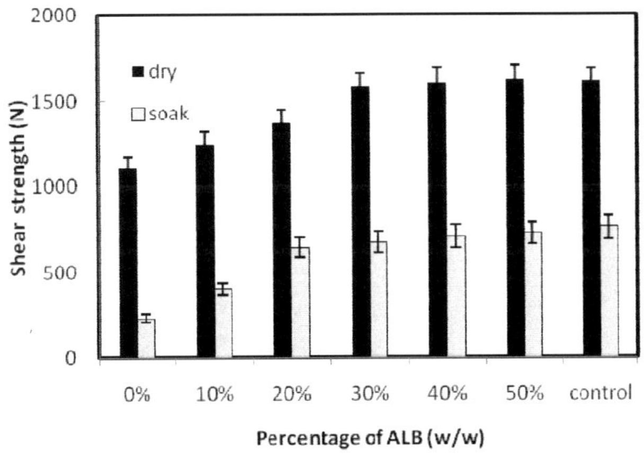

Figure 5. Dry and soak bonding shear strength at various mass fraction of ALB for blood protein-based adhesive and that of control phenol formaldehyde (PF).

4. Conclusions

ALB adhesive was successfully combined with blood adhesive. ALB adhesive increased the net shear stress and decreased viscosity of internal phase of blood adhesive, at the same time, provide stability of the blood protein particles. When BA adhesive blend with ALB adhesive, they will form a number of physical-chemical cross-linking points which guarantees the reinforcement, bond strength and water resistance.

Acknowledgments

Supported by the project of Guangdong Science and Technology (20140503, 2015B020215012), Chinese National Natural Science Foundation (31371880).

References

[1] Henderson J T 1979 *Tappi J* **62** 9396
[2] Heck H D Casanova M and Starr T B 1990 *Crit. Rev. Toxicol* **20** 397-426
[3] Edoga M O and Kovo A S 2006 *J Pract Technol* **8** 41-48
[4] Liu Z B Zhang Y L and Wang X M 2015 *Mater. Sci. App.* **6** 567-575
[5] Bekhta P Ortynska G and Sedliacik J 2014 *Drvna Industrija* **65** 293-301
[6] Hemmilä V Trischler J and Sandberg D 2013 *Pro Ligno* **9** 118-125
[7] Vazquez G Santos J Freire M S Antorrena G and Álvarez J G 2012 *J Therm Anal Calorim* **108** 605-611
[8] Vazquez G Pizzi A Freire M S Santos J and Antorrena G 2013 *Wood Sci Technol* **47** 523-535
[9] Garcia-Becerra F Y Acosta E J and Allen D G 2012 *J. Am. Oil Chem. Soc.* **89** 1315-1323
[10] Weakley F R and Mehltretter C I 1964 U.S. Patent 3153597
[11] Ash J R and Lumbuth A L 1954 U. S. Patent 2817639
[12] Golick A J and Dike T W 1941 U.S. Patent 2368466
[13] Thames S F Cook R C and Mendon S K 2004 U.S. Patent 6790271
[14] Trocino F S 2008 U.S. Patent 20080255333.
[15] Wang Y Wang D and Sun X S 2005 *J. Am. Oil Chem. Soc.* **82** 357-363
[16] Jiang Z H Qin D C Hse C L Lao Z H Wang G and Yu Y 2008 *J. Wood Chem. Technol.* **28** 240-246
[17] Qi G Y Li N B and Wang D H 2013 *J. Am. Oil Chem. Soc.* **90** 1917-1926
[18] Li N B Qi G Y Sun X S and Wang D H 2012 *J Polym. Environ.* **20** 905-915
[19] Lin H L and Gunasekaran S 2010 *Int. J. Adhes. Adhes.* **30** 139-144

[20] Ghim D and Kim J H 2016 *Korean J. Chem. Eng.* **33** 707-710
[21] Xiang A S Yan W L and Koel B E 2013 *J Nanopart Res* **15** 1705-1714
[22] Wang T Canetta E and Weerakkody T G 2009 *ACS Appl. Mat. Interfaces* **1** 631-639
[23] Mainardes R M and Evangelista R C 2005 *Int. J. Phytorem* **290** 137-144
[24] Nandi A Khakhar D V and Mehra A 2001 *Langmuir* **17** 2647-2655

IOP Conf. Series: Materials Science and Engineering **137** (2016) 012046 doi:10.1088/1757-899X/137/1/012046

Prediction of thermal strains in fibre reinforced plastic matrix by discretisation of the temperature exposure history

E K Ngoy[1]

Department of Chemical Engineering, Vaal University of Technology, South Africa

E-mail: etiennen@vut.ac.za; ek_yosh@yahoo.fr

Abstract. Prediction of environmental effects on fibre reinforced plastics habitually is made difficult due to the complex variability of the natural service environment. This paper suggests a method to predict thermal strain distribution over the material lifetime by discretisation of the exposure history. Laboratory results show a high correlation between predicted and experimentally measured strain distribution

1. Introduction

Due to the need for energy efficient materials, there has been a considerable increase of interest in light and green materials such as fibre reinforced plastics (FRP) in the aerospace industry. However, environmental effects have shown to be a hindrance to extensive use of FRP. A reliable method to quantify and predict environmental effects on FRP is required for safe and efficient use of these materials.

Stresses resulting from environmental factors such as cycling temperature and humidity have been recognised as one of the most deleterious effect on FRP materials [1]. Hygrothermal stresses are known to be the cause for crack initiation and cause the material to become brittle or to deform [2, 3]. They can approach or exceed the design load and acceptable material limits [4].

A number of works on the modelling of environmental stress effect on FRP material have been reported [5-12]. These reports deal mostly with the characterisation of the effect of temperature and humidity on the stress state of composite laminates. In this regard, models and computation algorithms of various performance levels have been made available. However, most of the time complication results from the material heterogeneity and the complex variability of the environment. In addition, only the effect of the physical degradation is taken into account explicitly, while the material undergoes a chemical degradation also.

This analysis focuses on the determination of environmental stresses without involving the resulting modification on the stress state of the material. Attention is on the prediction of a long-term effect, comprehensively taking into account the physical and the chemical degradation, as well as the viscoelastic, time-dependent effect. In the present paper, the thermal strain history is predicted based on the knowledge of temperature exposure. In a subsequent paper, strains will be used to compute stresses taking into account the time dependent relaxation and environmental degradation.

For the purpose of this paper, only a one-dimensional problem in plate laminate is considered. This is because it is demonstrated experimentally that the degradation of the mechanical strength of materials used in aviation varies in a layered way after exposure to a natural environment [13]. The degradation process can be considered as a progressing damaged front that moves from the outer surface through the laminate [14]. A gradient is generated through the laminate thickness and because of such a gradient effects on mechanical strength depend strongly on the laminate thickness [13, 15].

The suggested method is based on the environmental resistance of the matrix. This is because in practical applications, the environmental resistance of a laminate is determined by that of its barrier coat, made of a resin rich layer. It is also observed, in general, that the degradation mechanism of FRP initiates in the matrix and primarily affects the matrix dominated properties [14, 16-17]. Therefore, the investigation focuses on the matrix as the latter appears to be the most sensitive part of the composite laminate structure.

In order to resolve its complexity, the natural environment is discretised into a succession of constant environmental states over time. The time intervals of a constant environment can be minimised sufficiently in such a way as to obtain a discrete pattern as illustrated in Figure 1.

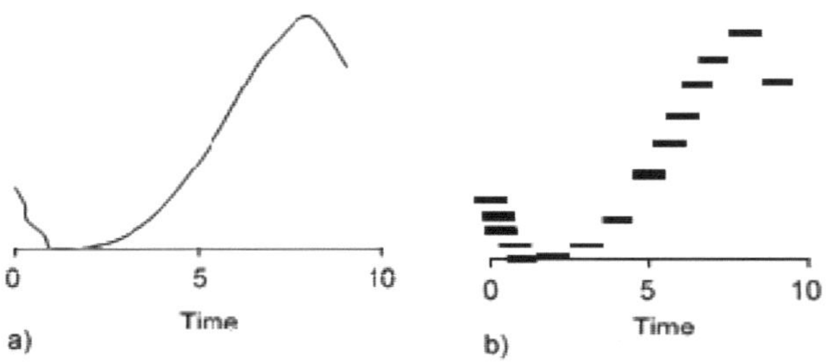

Figure 1. Schematic representation of a transformation from a continuous variable environment (a) to a succession of constant environments (b).

2. Mathematical Formulation

2.1. *Assumptions*

From all the above the following assumptions are introduced:

- The material is homogeneous and isotropic. Thermal strains result only from through the thickness gradient of the temperature.
- The environment to be considered is constant in function of time.
- The material structure is a plate laminate.

2.2. *Thermal strains*

Figure 2 represents a plate laminate of thickness l exposed to environmental temperature $f(x,t)$, which is a function of time and the material thickness.

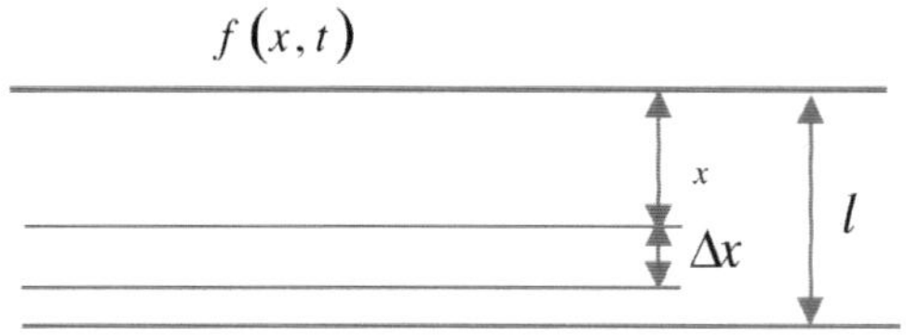

Figure 2. Schematic representation of a plate laminate.

At a given distance x from the material surface and fixed time t_i, the gradient of environmental temperature can be expressed as

$$gradf\left(x,t_i\right)=\frac{\partial f\left(x,t_i\right)}{\partial x}dx \tag{1}$$

Then, based on thermal expansion laws, the strain between two layers respectively situated at distances x and $x + \Delta x$ in the material can be evaluated as

$$\Delta\varepsilon(x,t_i) = \lambda \int_{x}^{x+\Delta x} \frac{\partial f(x,t_i)}{\partial x} dx \qquad (2)$$

where, ε and λ represent the strain and the material coefficient of thermal expansion respectively.

In equation (2), the determination of strain is straightforward if the temperature field inside the material is known. However, Fourier's law does not always allow for the handling of the complex variability of the external environment, frequently resulting in the solution for complex variation not being available. The method using constant environment, hereby, is introduced to determine thermal strain distribution over time.

2.3. Resultant strain at time t

Considering the succession of constant environmental states as in Figure 1b, the strain history in the laminate is given by a succession of strains such as the one represented in equation (2). Then, based on the Boltzmann superposition principle [18], the resultant strain over a period of time t, comprising a finite number of constant environmental states, can be estimated as:

$$\varepsilon(x,t) = \sum_{i=0}^{n} J(t-t_i)\Delta\sigma(x,t_i) \qquad (3a)$$

or

$$\varepsilon(x,t) = \sum_{i=0}^{n} \Delta\varepsilon(x,t_i) \qquad \forall t_i \prec t \qquad (3b)$$

In equation (3a), σ and $J(t)$ respectively represent the stress and the material compliance.

3. Experimental Method

Strain gauges installed inside the material at specific positions measure strains that develop in the material. The strain gauges used are made of constantan grid with one direction configuration, embedded in a phenolic resin. An epoxy adhesive is used to stick the strain gauge to the testing material. The outside temperature is recorded and used to compute and predict internal strains. Then the computed strains are compared to the measured strains.

3.1. Experimental material and test specimens

A 100% epoxy resin laminate is used. The formulation is shown in Table 1.

Table 1. Experimental material.

Epoxy resin: Ampreg 21	
Resin phr	100
Hardener	33
Curing cycle	48h/ 18°C

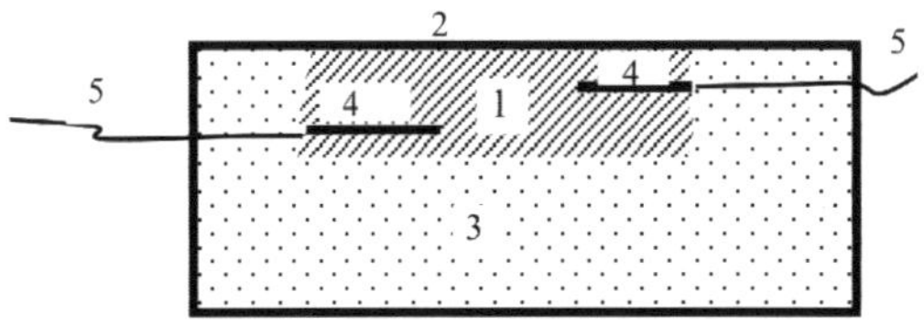

Figure 3. Test specimen.

The test specimen is constructed in such a way as to obtain exposure on one face only, and strain gauges are enclosed at specific distances (1 mm and 2 mm) of the external surface (Figure 3).

3.2. Experimental setup
The experimental set up is shown in Figure 4 here below.

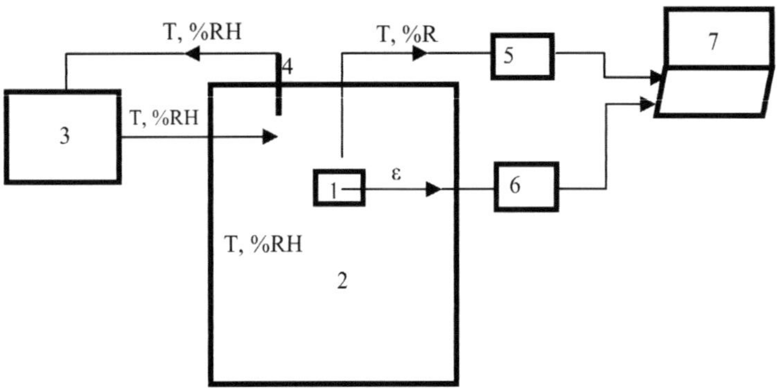

Figure 4. Experimental set up.

3.3. Experimental procedure and results

3.3.1. A set of three material samples was exposed in the weather simulator following a predetermined temperature and humidity pattern. The chamber temperature was recorded with a specialised data logger and the exposure duration was set at 3800 seconds. Figure 5 shows the exposure pattern. The temperature distribution is than discretised in a 10-second constant temperature interval, as shown in Figure 6.

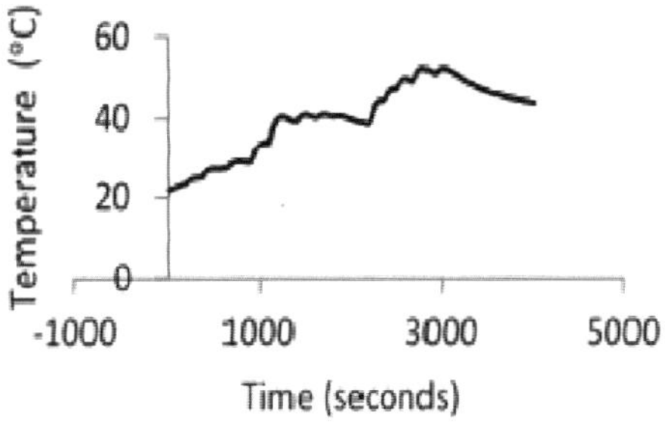

Figure 5. Temperature pattern.

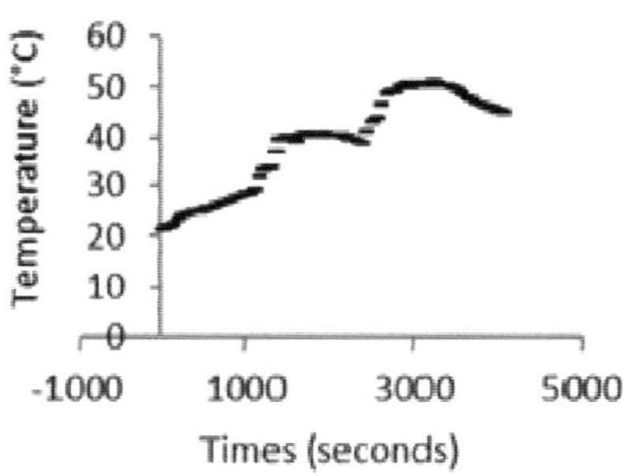

Figure 6. Temperature discretization.

3.3.2. Internal free strains at 1 mm and 2 mm distances from the external surface of the laminate were measured using strain gauges and recorded with the strain data logger. Figure 7 presents the results.

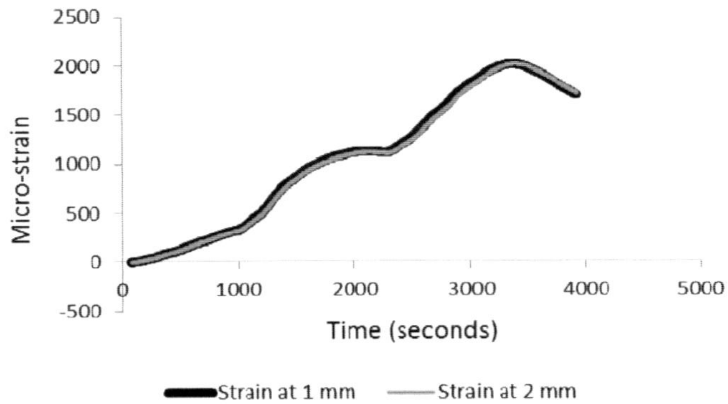

Figure 7. Measured free strain inside the laminate.

3.3.3. The temperature field at 1 mm and 2 mm distances were computed for the successive constant environmental temperature using Fourier's law. The specimens were considered to have semi-infinite, solid structure, for which an analytical solution is provided; see equation (4) where T_s and T_i respectively represent the constant temperature at the exposed surface and the initial temperature inside the material [19]. Symbols ρ , C, and k respectively, stand for the material density, specific heat capacity, and coefficient of thermal conductivity. The computed temperature profile is shown in Figure 8.

$$T(x,t)=(T_i-T_s)erf\left(\frac{x}{2}\sqrt{\frac{\rho c}{kt}}\right)+T_s \tag{4}$$

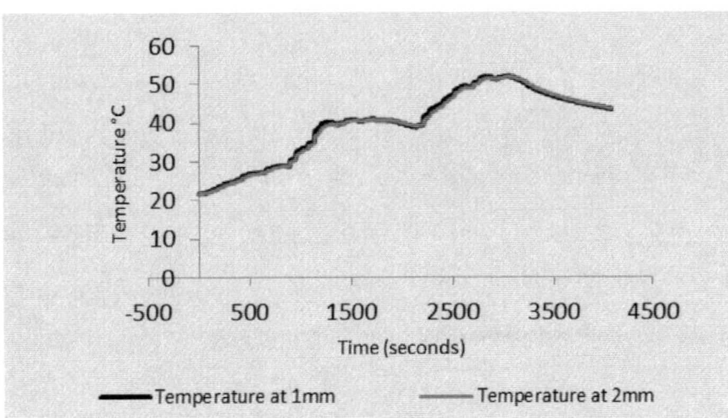

Figure 8. Computed internal temperature.

3.3.4. The temperature gradient between 1 mm and 2 mm distances from the outer surface was determined as per equation (1) and strains were computed as per equation (2). To this end, the material characteristics that were used are shown in Table 2. The exact value of the coefficient of thermal expansion for the material used in this experiment could not be found. However, a couple of values for the coefficient of thermal expansion of some epoxy resins were collected from various sources and averaged to provide the value displayed in Table 2. The use of that value is discussed in Section 4.

Table 2. Material thermal properties.

Epoxy resin Ampreg 21	
Density [Kg/m^3]	1100
Specific heat capacity [J/Kg. °K]	1670
Coefficient of thermal conductivity [W/m.°K]	0.17
Coefficient of thermal expansion [10^{-6}m/m]	30

Computation results are shown in Figure 9. Two different time intervals where used for the discretisation of the exposure history. At first, a 60-second time interval was used, and then the time interval was reduced to 10 seconds in order to improve the precision.

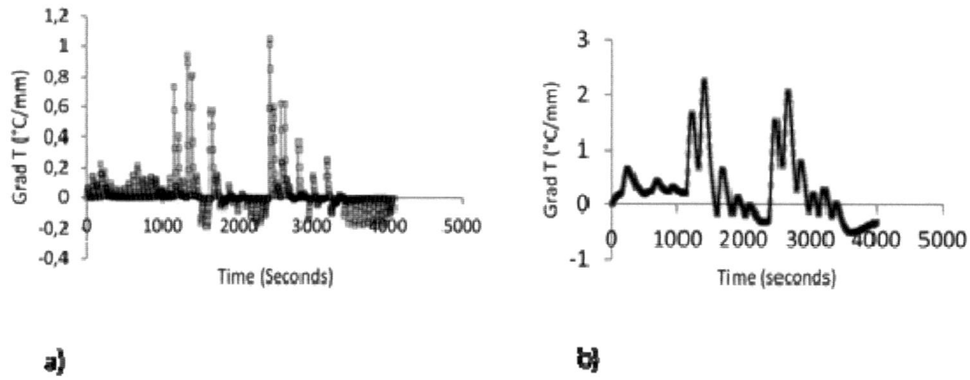

Figure 9. Distribution of temperature gradients computed a) per 60 seconds time interval, and b) per 10 seconds time interval.

3.3.5. Computed strains were then compared graphically to experimentally measured strains. The latter were obtained at each instant by differences in free strain of the material layers at 1 mm and 2 mm from the exposed surface; a comparison is shown in Figure 10.

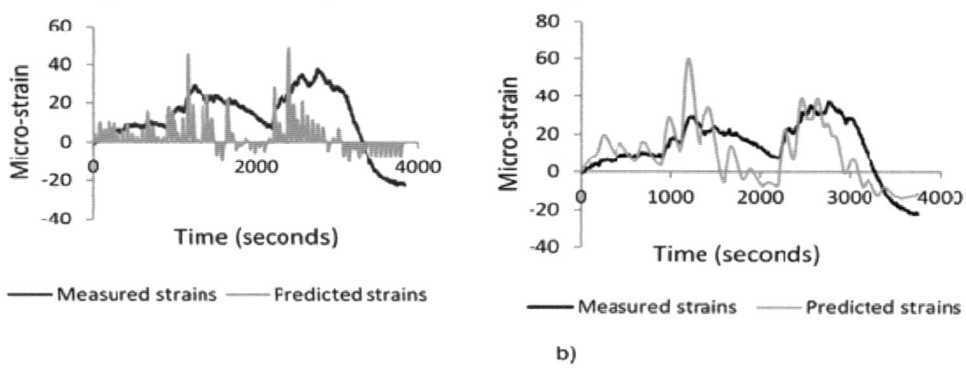

Figure 10. Measured strains compared to strains computed a) per 60 seconds time interval, and b) per 10 seconds time interval.

4. Discussion

The predicted strain history shows two major positive peaks, the first between 1000 and 1500 seconds and the second between 2000 and 2800 seconds. A negative peak is shown between 3100 and 3500 seconds. The positive peaks correspond to tension strain and the negative peak corresponds to a compression of the material. The pattern of predicted strain history consistently reflects what was experimentally recorded (Figure 10). However, as previously mentioned, the exact value of the coefficient of thermal expansion was not found. There is uncertainty on the exact values of these peaks. Nevertheless, it can be seen from equation (2) that the computed strains should be linearly correlated to the real strains and the correlation curve should pass through the axe origin. Hence, resultant strains in function of time were computed as per the Boltzmann superposition principle for each loading period (tension and compression) and the linear correlation was assessed between the predicted strains and the measured strains. The results of the linear correlation assessment are presented in Table 3.

It is shown that for the 60-second time interval, the correlation is quite poor, while the reduction of the time interval to 10 seconds leads to a considerable improvement of the prediction. At the 10-second time interval, the correlation is excellent and the correlation equation shows that obtained values of peaks need to be corrected by a multiplicative factor of 1.4. Therefore, it can be concluded that the predicted strain history excellently reflects the real strain history. However, considerable differences can be observed right after tension periods. These differences are recorded more or less from 1500 to 2200 seconds and between 2800 to 3300 seconds. These periods correspond to tension release and it is shown that the recovery of measured strains lags behind, while the recovery for predicted strains is relatively fast. In reality, the computed strain history is reflective of the thermal loading only while the measured strain includes the material response. The observed differences show the material response to the release of the thermal load. The observed delayed response is typical of viscoelastic material.

Table 3. Correlation between measured strains and predicted strains.

Interval	60 sec time interval		10 sec time interval	
	R^2	Correlation equation	R^2	Correlation equation
0 to 1500 sec	0.920	0.3λ	0.981	1.5λ
2200 to 2800 sec	0.561	0.3λ	0.975	0.9λ
3300 3500 sec	0.120	0.2λ	0.991	1.7λ
Average		0.3λ		1.4λ

5. Conclusion

The analysis has demonstrated a possible method to predict the thermal strain history in a laminate matrix. It indicated that a variable temperature environment could be discretised in a succession of fixed-time intervals of constant temperature. Each constant temperature environment is treated individually as an incremental thermal load, to calculate the temperature field inside the laminate, based on Fourier's law for heat transmission. This leads to the determination of the internal strain distribution, over time, by using the law of thermal expansion. The analysis has shown sufficiently reducing the time interval of constant temperature leads to an acceptable prediction that excellently reflects the real thermal strain history.

Acknowledgements

The author wishes to acknowledge the valuable support received from the National Aerospace Centre of Excellence toward this research. The support received from the Faculty of Engineering and the Built Environment of the University of Johannesburg hereby is acknowledged.

References

[1] Sachin Kishrsagar, Roberto A, Lopez-Anido and Gupta R K 2000 ACI Mater. J., title 97 M80, p 703
[2] Mills N J 1993 *Environmental Effects*. In: 2nd ed. Plastics, Microstructure and Engineering Applications. (Edward Arnold, UK) p 228
[3] Adam D S, Bowles D E and Herakovich C T 1988 *Thermally Induced Cracking in Graphite-Epoxy Cross-ply Laminate*. In: GS Springer ed. Environmental Effect on Composite Materials, vol.3. (Technomic, Westport CT)
[4] Stone M A, Schwartz I F and Chandler H D 1997 Compos. Sci. Technol. 57, p 47

[5] Benkheda A., Tounsi A., and Adda bedia E.A. 2008 Compos. Struct. 82, p 629
[6] Gigliotti M, Jacquemin F, Moliard J, and Vautrin A, 2007 Macromol. Symp. 247, p 199
[7] Pavankiran Vaddadi, Toshio Nakamura, and Raman P Singh, 2003 Comp. Part A 34, p 719
[8] Upadhyay P C, Wayne W S, Prucz J, and Lyons D W 2009 J. Reinf. Plast. Compos.
[9] Springer G S 1988 *Environmental Effects*. In: G S Springer ed. Environmental Effect on Composite Materials, vol. 3. (Technomic, Westport CT)
[10] Guglieno Carra and Valter Carvelli 2014 Compos. Struct. 108, p 1019
[11] Guzman E, Cugnoni J, and Gmur T 2014 Compos. Struct. 111, p 179
[12] Silva M A G, Sena Da Fonseca B, Hugo Biscaia 2014 Compos. Struct. 116, p 377
[13] Startsev O V, Krotov A S, and Startseva L T 1999 Polym. Degrad. Stab. 63, p 183
[14] Sevostianov I, Sookay N K, Von Klemperer C J, and Verijenko V E 2003 Compos. Struct. 62, p 417
[15] Larson F 1988 *The Effect of Ultraviolet Light on Mechanical Properties of Kevlar 49 Composites*. In: GS Springer ed. Environmental Effect on Composite Materials, vol. 3. (Technomic, Westport CT) p 132
[16] Kumar B G, Singh R P, and Nakamura T 2002 J. Compos. Mater. 36(24), p 2713
[17] Nakamura T, Singh R P, Vaddadi P 2006 Exp. Mech. 36, p 257
[18] McCrum N G, Buckley C P, and Bucknal C B 1997 *Principles of Polymer Engineering* (Oxford University Press, Oxford)
[19] Kreith F, Bohn M S, 2001 *Principles of heat transfer, 6th ed.*, (Thomson Learning, Australia)

Preparation of thin film nanofibrous composite NF membrane based on EDC/NHS modified PAN-AA nanofibrous substrate

Y Yang[1], X Wang[2,4] and B S Hsiao[3,4]

[1]Doctoral Student, State Key Laboratory for Modification of Chemical Fibers and Polymer Materials, Donghua University, No. 2999 Renmin Road (N), Songjiang District, Shanghai, 201620, P. R. China
[2]Professor, State Key Laboratory for Modification of Chemical Fibers and Polymer Materials, Donghua University, No. 2999 Renmin Road (N), Songjiang District, Shanghai, 201620, P. R. China
[3]Professor, Department of Chemistry, Stony Brook University, Stony Brook, NY 11794-3400, USA

E-mail: wangxf@dhu.edu.cn; bhsiao@notes.cc.sunysb.edu

Abstract. A novel kind of thin-film nanofibrous composite (TFNC) nanofiltration (NF) membranes consisting of a polyamide (PA) barrier layer were successfully fabricated by interfacial polymerization (IFP) based on electrospun double-layer nanofibrous substrates, which have an ultrathin poly (acrylonitrile-co-acrylic acid) (PAN-AA) nanofibrous layer as top layer and a thicker polyacrylonitrile (PAN) nanofiber layer as bottom porous support layer. Immersing PAN/PAN-AA nanofibrous substrates into 1-ethyl-(3-3-dimethylaminopropyl) carbodiimide hydrochloride/N-hydroxysuccinimide (EDC/NHS) aqueous solution and piperazine (PIP) aqueous solution (0.20 wt%) sequentially for a period of time, the carboxyl groups on PAN-AA nanofibers were activated by carbodiimide and then reacted with the amide groups. The as prepared composite membrane has an integrated structure with high rejection rate (98.0%); high permeate flux (40.4 L/m^2h) for $MgSO_4$ aqueous solution (2 g/L).

1. Introduction

TFNC membranes have been proven to be an effective media for nanofiltration (NF) and reverse osmosis (RO), which structure with a thin and dense functional barrier layer supported on a reinforced porous substrate [1-4]. The ultra-thin dense barrier layer is usually obtained through interfacial polymerization (IFP), by which the filtration performance of TFNC membranes is mainly determined. The conventional reinforced porous matrix is a less porous dense asymmetric membrane made by phase inversion methods. In recent decades, electrospun nanofibrous membranes have been used as supporting materials in composite membranes due to its high porosity (up to over 80%), interconnectivity, micro scale interstitial space, and a large surface-to-volume ratio [5-7]. But it is difficult to fabricate a structure-integrated nanofibrous composite NF membrane because the pore size of electrospun nanofibrous support is relatively large (submicron level) to support strugglingly the ultrathin barrier layer (~ 100 nm thickness) prepared IFP method [8].

In this study, a new kind of TFNC membrane was fabricated by introducing a PAN-AA intermediate layer between PA barrier layer and PAN nanofibrous substrate. Carboxyl groups on PAN-AA nanofibers were activated by EDC/NHS and reacted with secondary amine group on PIP, which could form covalent linkage with PA barrier layer. The nanofiltration performance of the resultant TFNC membrane was evaluated by the rejection of $MgSO_4$ from salt water using a cross-flow filtration device.

2. Experimental

2.1. Materials
PAN (Mw = 1.36×10^5) 2-Morpholinoethanesulfonic acid (MES), 1-Ethyl-3-(3-dimethylaminopropyl) carbodiimide hydrochloride (EDC), and N-hydroxysuccinimide (NHS) were purchased from J & K Scientific Ltd. N,N'-dimethyl formamide (DMF), n-hexane, piperazine (PIP), trimesoyl chloride (TMC) and magnesium sulfate ($MgSO_4$) were purchased from China State Medicinal Group Chemical Reagent Co., Ltd. PAN-AA with a viscosity-averaged molecular weight ($M\eta$) of 2.4×10^5 was synthesized by radical polymerization [9].

2.2. Preparation of PAN-AA/PAN nanofibrous substrate
Dried PAN and PAN-AA powder were dissolved in DMF respectively with gentle stirring at 50 °C in oil bath for 12 h to get homogeneous solutions respectively. PAN solution (10 wt%) was electrospun on a grounded rotating metal drum until the thickness of the PAN nanofibrous substrate reached to 40 μm (electric voltage: 20 kV, solution feed rate: 16.6 μL/min, spinneret diameter: 0.7 mm, distance between the spinneret and the grounded drum: 15-20 cm, environment temperature: 40 °C, humidity ~40%, diameter of drum: 10 cm, length of drum: 30 cm and rotating speed: 300rpm) [6]. Then an ultrathin layer of PAN-AA nanofibers was electrospun above the PAN nanofibrous substrate for 20 min (Electric voltage: 24kV, solution feed rate: 8.0 μL/min). The distance between the spinneret and the grounded drum was 15 cm. The as-prepared nanofibrous membranes were dried at 50 °C in vacuum for 24 h to remove the remained solvent.

2.3. Preparation of TFNC membrane
TFNC membrane was prepared through IFP technique. To begin with, the PAN/PAN-AA nanofibrous substrate was clamped with a Teflon holder, and activated in an MES buffer solution (50 mmol, pH 6.0) containing EDC and NHS (concentration of EDC range from 50-200 mmol, concentration ratio fixed at 2:1, w/w) for 4 h at room temperature [10]. Then the activated membranes were washed and immersed in PIP aqueous solution (0.2 wt%) for 1 h to impregnate the nanofibrous substrates. The surface was rolled with a glass rod to eliminate any little bubbles formed in the process of soaking. Excess solution was drained off the surface by air knife (EXAIR, USA). The n-hexane solution of TMC (0.2 wt%) was poured into the holder. After a certain period of reacting time (60 s), the excess organic solution was removed off the surface and the membrane were held in oven for 1.0 h at 60 °C, so that a dense PA barrier layer was formed on the support membrane. Finally, the resultant membranes were successively washed and stored in deionized water for testing [11].

2.4. Characterization
The morphology of TFNC membrane was investigated by scanning electron microscopy (SEM) system (Phenom G2 pro, FEI, USA). The nanofiltration (NF) performance of the resulting membranes was measured by using a cross-flow membrane filtration system with 5 parallel filtration cells (effective area: 3.8 cm^2). Membranes were pressured at 0.2 MPa for 30 min with distilled water before the NF tests to obtain stable permeate flux. After that, the permeate flux and salt rejection of TFNC membranes were tested with $MgSO_4$ aqueous solution (2 g/L) under certain operating pressure at 100 psi; The permeate flux was calculated using the equation: J (L/m^{2}h) = $V/(A\cdot\Delta t)$, where V is the volume of permeate flow during the test time, A is the effective membrane area, Δt is the collecting time of permeate flow. The $MgSO_4$ rejection rate was calculated using the following equation: R (%) = $(Cf\text{-}Cp)/Cf\times100\%$, where Cf and Cp were conductivity of feed and permeate

solution, respectively, which were measured by using a conductivity detector (FE30, Mettler Toledo) [9].

3. Results and discussion

3.1. Membrane morphology characterization

The prepared PAN/PAN-AA double-layer nanofibrous substrates were fabricated via electrospinning technique. The typical SEM image of PAN nanofibers (average diameter ~ 342 nm) and PAN-AA nanofibers (average diameter ~ 173 nm) was shown in figure 1(A) and 1(B) respectively. The combined PAN/PAN-AA double-layer nanofibrous substrates were fabricated by depositing an ultrathin PAN-AA nanofibrous layer in 20min (thickness < 0.2 μm) on the surface of PAN nanofibrous substrate (thickness ~ 40μm) as shown in figure 1C.

Figure 1. SEM images of electrospun nanofibers: A) PAN, B) PAN-AA and C) PAN/PAN-AA double-layer nanofibrous substrate

In figure 2(A) and 2(B), the surface and cross section morphologies of PAN-AA/EDC100/PATFNC membranes show that a PA barrier layer (thickness ~150 nm) was coated on PAN/PAN-AA nanofibrous substrate modified with EDC/NHS solution (100/50 mmol).

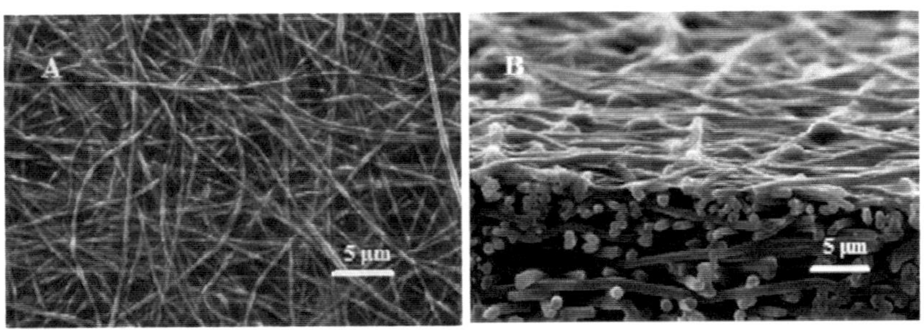

Figure 2. SEM images of surface (A) and cross section (B) of PAN-AA/EDC100/PA TFNC membrane

3.2. Filtration performance of PAN-AA/PA membrane

A series of TFNC membranes were prepared by interfacial polymerization (IFP) based on PAN-AA nanofibrous substrate modified with different concentration of EDC/NHS solutions (50-200 mol), while other IFP conditions is fixed (TMC 0.2 wt%, reacting time was kept at 60s, heat-treated in oven at 60℃ for 1h) as shown in figure 3. As the concentration of EDC/NHS solutions increased, the permeate flux of the PAN-AA/EDC/PA TFNC membrane decreased from 43.5 L/m^{2}h to17.3 L/m^{2}h, and the rejection of TFNC membrane increased from 96.3 % to 98.5 %/. When the concentration of EDC was 100 mmol, the rejection of TFNC membrane was above 98.3% and the permeate flux could obtain 40.4 L/m^{2}h. Compared with the TFNC membrane prepared by same IFP conditions (PIP 0.2

wt%, TMC 0.2 wt%, 60s reacting time, 60 °C for 1h) based on single-layer PAN nanofibrous substrate, the rejection of PAN/PA TFNC membrane is just 97.8 % and the permeate flux of that is 41.5 L/m^{2}h.

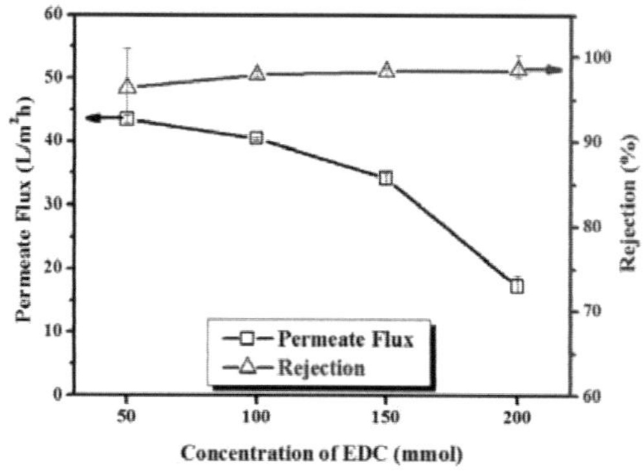

Figure 3. Filtration performance of PAN-AA/EDC/PA TFNC membrane effect with concentration of EDC

4. Conclusions

In this study, a novel kind of TFNC membranes was prepared by interfacial polymerization on electrospun double-layer PAN-AA/PAN nanofibrous substrates, which were activated with EDC/NHS to react with aqueous monomer PIP. The EDC/NHS modified PAN-AA med-layer generated an interconnection linkage between PA barrier layer and PAN nanofibrous substrate. The resulting TFNC membranes were evaluated for nanofiltration using $MgSO_4$ aqueous solution at 2 g/L. The PAN-AA nanofibrous layer modified with different concentration of EDC/NHS solution (50-200 mmol) played a critical role in interfacial polymerization, which changed the permeate flux decreasing from 43.5 L/m^{2}h to 17.3 L/m^{2}h. When the EDC concentration was 100 mmol, the filtration properties of TFNC membrane based on EDC/NHS modified double-layer PAN/PAN-AA nanofibrous substrate were better than that of TFNC membrane based on PAN nanofibrous substrate.

Acknowledgments

This work was supported by National Science Foundation of China (51273042, 21174028), Program for New Century Excellent Talents in University (NCET-13-0725).

References

[1] Kaur S, Sundarrajan S, Rana D, et al. 2014 Review: The characterization of electrospun nanofibrous liquid filtration membranes *J. Mater. Sci.* **49**(18) 6143-6159.

[2] Mokhena T C, Jacobs V and Luyt A S 2015 A review on electrospun bio-based polymers for water treatment *Express Polym. Lett.* **9**(10) 839-880.

[3] Subramanian S and Seeram R 2013 New directions in nanofiltration applications - Are nanofibers the right materials as membranes in desalination *Desalination* **308** 198-208.

[4] Sundarrajan S, Balamurugan R, Kaur S, et al. 2013 Potential of Engineered Electrospun Nanofiber Membranes for Nanofiltration Applications *Dry Technol.* **31**(2) 163-169.

[5] Li X, Wang L, Yang Y, et al. 2013 Thin film nanofibrous composite ultrafiltration membrane for oil removal from oilfield wastewater *Advanced Materials Research* **669**(669) 99-102.

[6] Shen L, Yang Y, Zhao J, et al. 2016 High-performance nanofiltration membrane prepared by dopamine-assisted interfacial polymerization on PES nanofibrous scaffolds *Desalin. Water Treat.* **57**(21) 9549-9557.

[7] Wang X, Yeh T-M, Wang Z, et al. 2014 Nanofiltration membranes prepared by interfacial polymerization on thin-film nanofibrous composite scaffold *Polymer* **55**(6) 1358-1366.

[8] Kaur S, Sundarrajan S, Rana D, et al. 2012 Influence of electrospun fiber size on the separation efficiency of thin film nanofiltration composite membrane *J. Membrane Sci.* **392-393** 101-111.
[9] Takaomi K, Hong Y W and Fujii N. 1995 Molecular Imprinting of Theophylline in Acrylonitrile-acrylic Acid Copolymer Membrane *Chem. Lett.* **10** 927-928.
[10] Peng J, Su Y, Chen W, et al. 2013 Polyamide nanofiltration membrane with high separation performance prepared by EDC/NHS mediated interfacial polymerization *J. Membrane Sci.* **427** 92-100.
[11] Zhao K 2014 Fabrication of a poly(piperazine-amide)/polysulfone hollow fiber membrane for effective removal of heavy metals in industrial wastewater *7th International Conference on BioMedical Engineering and Informatics* (Institute of Electrical and Electronics Engineers Inc., New York) p. 61-65

Reinforcement of SBR/waste rubber powder vulcanizate with in situ generated zinc dimethacrylate

X P Wang[1], B K Cheng, X Zhang and D M Jia

School of Materials Science and Engineering, South China University of Technology, Guangzhou, Guangdong, 510641, China

E-mail: wangxp@scut.edu.cn

Abstract. Methyl acrylic acid/zinc oxide (MAA/ZnO) was introduced to modify styrene-butadiene rubber/waste rubber powder (SBR/WRP) composites by blending. The enhanced mechanical properties and processing ability were presumably originated from improved compatibility and interfacial interaction between WRP and the SBR matrix by the in situ polymerization of zinc dimethacrylate (ZDMA). A refined interface of the modified SBR/WRP composite was observed by scanning electron microscopy. The formation of ZDMA significantly increased the ionic bond content in the vulcanizate, resulting in exceptional mechanical performance. The comprehensive mechanical properties including tensile strength, tear strength and dynamic heat-building performance reached optimum values with 16 phr MAA.

1. Introduction

Waste rubber products are the second largest waste polymer material, following waste plastics. Waste rubber products primarily consist of destroyed tires, rubber shoes and belts, sheets of sealing rubber, etc., which exhibit a solid network structure that makes them undissolved and decomposed before melt. The disposal of waste rubber has become a major environmental issue worldwide. Waste rubber powder (WRP) prepared by grinding could be served as a filler or substitute for rubber makes recycling processes for the material of great importance [1-5]. The development of an environmentally friendly means of waste rubber disposal will help prevent environmental pollution.

However, the incompatibility between WRP and polymer matrices caused by its dense crosslinked network structure and the inert surface of WRP makes the adding of WRP in a raw rubber often disrupt the properties of the mixture.

Unsaturated carboxylic acid salts such as methyl methacrylate acrylic acid zinc, magnesium, etc. produced by the reaction between unsaturated carboxylic acid and metal oxide, hydroxide or carbonate reaction proceeds, can be made by "in situ" preparation in a polymer matrix [6]. Since zinc dimethacrylate (ZDMA) reinforced by hydrogenated nitrile rubber (HNBR) which demonstrates enhanced tensile strength and excellent abrasion resistance was developed by Zeon Chemicals to produce golf balls, unsaturated carboxylic acid salts have been widely investigated as a means of rubber reinforcement by applying unsaturated carboxylic acid and metal oxides directly to the rubber matrix. The in situ generation of unsaturated carboxylic acid salt can be obtained during the mixing and vulcanization process [7-8]. Unsaturated carboxylic acid salt contains a large number of positive and negative ions, which gather together to form an ionic cluster as a result of electrostatic attraction, resulting in the reinforcement of the material. Moreover, the unsaturated groups in unsaturated carboxylic acid salt can copolymerize with one another as well as undergo a grafting reaction with the polymer matrix to enhance the rubber material [9-10]. The micron-scale dispersion of residual ZDMA particles and the nanoscale dispersion of poly-ZDMA aggregates was observed in the vulcanizates [11].

In this paper, methyl acrylic acid and zinc oxide (MAA/ZnO) were incorporated into styrene butadiene rubber/waste rubber powder (SBR/WRP) composites to increase the compatibleness of the rubber matrix and WRP as well as to subsequently reinforce the composites. The mechanical properties such as wear resistance and fatigue properties were investigated to verify the in situ reinforcement of ZDMA in the rubber matrix. The crosslinking density, processability of the vulcanizate, mechanical and dynamic mechanical properties, dynamic compression fatigue of the sample were measured and analyzed. Scanning electron microscopy (SEM) was used to observe the microstructure.

2. Experimental

2.1. Materials

Methyl acrylic acid（MAA）and zinc oxide (ZnO) were obtained from Tianjin Kemiou Chemical Reagent Co., Ltd. The WRP used in this research was offered by Guangzhou Aiqi technical Company. The mean particle diameter was 1000 mesh. SBR 1502 was purchased from China Sonipec Cro. The utilized ingredients of the composite including accelerator 2,2-dibenzothiazoledisulfide (DM), accelerator N-cyclohexyl-2-benzothiazolesulfenamide (CZ), and, activator stearic acid (St) were bought commercially. Carbon black N234 was provided by Shanghai Cabot Chemical Co., Ltd. The fumed silica was supplied by Guangzhou Hongchuang Trading Co., Ltd. Pure dicumyl peroxide (DCP), was purchased from Aladdin Reagent Co., Ltd.

2.2. Sample preparation

The rubber compounds was mixing by a two-roll mill, and the recipe was listed in table 1. Raw SBR was masticated with WRP to obtain a masterbatch, into which various contents of MAA and ZnO were mixed. The proportion of MAA was 0, 4, 8, 12, 16, and 20 phr, respectively. The ratio of MAA to ZnO was fixed in 2.1:1 (wt%:wt%) and other additives were loaded subsequently. The time and roller temperature for mixing was 10 minutes and 50℃ respectively. The compound rubber was placed one night before vulcanization. Thenceforth, the samples were vulcanized at 150℃ under a pressure of 10MPa for top optimum cure time T_{90}, which was measured by the UCAN UR-2030 vulcameter (Taipei, Taiwan). Kinds of vulcanized specimens were prepared for tensile angle tear and compress tests, respectively.

Table 1. SBR/WRP/filler composite contents.

Ingredient	SBR	RP	MAA	ZnO	St	N234	DCP
phr	100	10	0~20	0~9.5	2	34	2

2.3. Measurements

Tensile properties and tear strength were investigated according to ASTM D412 and ASTM D3185, respectively, on an Electron Omnipotence Experiment Machine U-CAN UT-2060 (Taipei, Taiwan) instrument with a strain rate of 500 mm/min according to standard descriptions. Dynamic mechanical analysis (DMA) spectras of modified and unmodified SBR/WRP composites were acquired by using a DMA243 dynamic mechanical spectrometer (Nestal, Germany) , on the tensile mode with 0.5% strain and 10 Hz frequency. The surface morphologies of the metal-sprayed samples were observed by SEM, Quanta 200 (FEI, Netherlands).

A crosslinking density test was conducted by swelling experiments. Weighted samples were immersed in toluene to achieve balance. The crosslinking density (n) was calculated from volume fraction of the rubber (V_r) in a swollen sample. The values of V_r and n were attainted from the equations provided by Valentín and Nie [12-13], as follows:

$$V_r = \frac{\dfrac{w_2 - f_{ins} w_0}{\rho_2}}{\dfrac{w_2 - f_{ins} w_0}{\rho_2} + \dfrac{w_1 - w_2}{\rho_1}} \tag{1}$$

$$-[\ln(1-V_r) + V_r + \chi V_r^2] = V_0 n (V_r^{1/2} - \frac{V_r}{2}) \tag{2}$$

where w_0 is the weight of dry sample originally; w_1 and w_2 represent the weights of the swollen samples and the unswollen samples, respectively; f_{ins} is the weight fraction of the components which was undissolved in toluene solvent; ρ_1 and ρ_2 are the densities of the solvent and the rubber, respectively; the Flory-Huggins polymer-solvent interaction term is represented as χ, which is assumed to be $0.29+0.90V_r$ for SBR and the toluene pair at 25℃;the toluene molar volume V_0=106.2 ml/mol. In our experiment, samples were immersed in toluene solvent for seven days at 25℃. Then, w_1 was determined from the surface-cleaned swollen rubber, and w_2 was determined as the constant weight of the swollen sample after being dried for five hours at 80℃. Accordingly, V_r and n can be obtained through equation (1) and (2). For measuring the ionic and covalent crosslink density separately,the samples were previously swelled in a mixed solvent of acetone/hydrochloric acid for three days at 60℃ to destroy the ionic network, and then immersing in toluene for seven days. The total crosslink density is a combination of the ionic crosslink density and the covalent crosslink density.

3. Results and discussion

3.1. Microstructure of the SBR/WRP blends

The microstructure of SBR/WRP and MAA/ZnO modified SBR/WRP were investigated by SEM at a magnification of 500X; results are depicted in Figure 1. The SEM micrographs show that the brittle fracture surface of the modified SBR/WRP is different from that of SBR/WRP. Hole formation as a result of the detachment of WRP from the SBR matrix can be seen in the micrographs, in which it can be seen that the MAA/ZnO modified SBR/WRP in figure 1(b) demonstrates more irregular fine shapes and a rough surface as compared to the SBR/WRP sample depicted in figure 1(a). This pattern indicates the well distribution of WRP in the rubber matrix and strong interfacial actions between the WRP and rubber matrix after modification. The improved bonding at the interface of the composite materials and the rubber powder dispersed in the matrix can be ascribed to the in situ generation of ZDMA by the reaction of ZnO and MAA, and by the grafting of MAA to the rubber matrix during the mixing and curing process. This is demonstrated by the increased hardness and maximum torque of the sample.

The effective wetting of WRP by the matrix and strong interfacial interactions are required on order to obtain an SBR/WRP compound with satisfactory mechanical properties and well distribution of WRP in the rubber matrix.

3.2. Vulcanization characteristics, mechanical properties and crosslink density of SBR/WRP compounds

As shown in table 2, increasing the rubber powder loading in SBR compounds reduces the optimum cure time T_{90} in the presence of unreacted accelerators in the rubber powder, as reported by Mathew

[14]. The elevated maximum torque M_H and the reduced minimum torque M_L were observed as the amount of MAA and ZnO increased. At the beginning of the curing process, the liquid MAA mixed in the gum can temporarily act as a plasticizer in order to reduce the minimum torque M_L.

(a) (b)

Figure 1. SEM photomicrograph of SBR/WRP blends, (a) unmodified SBR/WRP and (b) modified SBR/WRP.

Subsequently, the in-situ generation of ZDMA and the grafting of MAA to the rubber matrix triggered by dicumyl peroxide at the curing temperature lead to an increase in the maximum torque M_H.

The tensile strength of a 10 phr loading of unmodified WRP in an SBR matrix was reduced by approximately 15% as a result of the weak interaction between WRP and the rubber matrix. This resulted in the concentration of stress on the interface between the WRP and the rubber, causing failure of the material.

The mechanical properties gradually increased to a maximum at 16 phr of MAA, after which the further increase of MAA resulted in decreasing mechanical properties. As compared to the unmodified compound, the tensile strength and impact strength of the modified compound increased by 30% and 40%, respectively, while the modulus increased by 300% with an optimum loading of 16 phr MAA. This can be accounted for by the uniform dispersion of MAA and ZnO in rubber compounds below 16 phr of MAA loading, which contributes to the homogeneous dispensability of ZDMA and the grafting degree of MAA to the rubber matrix as reinforcement.

The total crosslinking density, ionic bond crosslinking density and covalent bond crosslinking density of the vulcanized rubber were all improved after modification. This is because MAA and ZnO reacted to generate an unsaturated carboxylic acid salt containing ionic bonds, and the composite network between WRP and the rubber matrix was created by the homopolymerisation of liquid MAA in swelled rubber powder and the rubber matrix with ZnO [15]. Additionally, the increased covalent bond crosslinking density was related to the double bond in the unsaturated MAA, which partially initiated the grafting on the rubber chain.

However, when more than 16 phr MAA was applied, the agglomeration of ZDMA became primarily responsible for the weaker interaction and bonding between the rubber substrate and the WRP, resulting in the deterioration of strength.

3.3. Dynamic compression fatigue of SBR/WRP composites

During the dynamic compression process, the generation of high dynamic compression was caused by the weak interface between the rubber powder and the rubber substrate. The addition of MAA/ZnO enhanced the interface between the rubber powder and the rubber matrix by the in situ reaction, thus reducing the temperature increase induced by dynamic compression fatigue of the SBR/WRP composites. As shown in table 2, the increasing fatigue temperature of SBR with 10 phr unmodified WRP was 40.80 ℃, and was significantly decreased after modification by the addition of the MAA/ZnO system. When the amount of MAA reached 16 phr, the fatigue temperature increased reached a minimum value of 30.25℃, with a decreasing rate of 25%. The demonstrated variation tendency of the dynamic compression fatigue properties was consistent with that of the mechanical properties.

Table 2. Vulcanization characteristics, mechanical properties, crosslink density and compression fatigue properties of SBR/WRP/MAA compounds.

Sample	SBR	SBR/WRP 10					
		MAA 0	MAA 4	MAA 8	MAA 12	MAA 16	MAA 20
Cure Characteristic							
T_{S1}/min	0.85	0.93	0.65	0.53	0.48	0.50	0.48
T_{90}/min	22.02	20.58	19.50	20.20	18.58	17.83	16.37
M_L/dN·m	5.4	4.04	2.44	2.03	1.99	1.87	1.66
M_H/dN·m	20.12	19.57	23.62	33.19	38.27	44.98	40.36
Mechanical Properties							
Hardness (shoreA)/ °	65	64	67	74	76	78	78
300% Modulus/MPa	8.21	7.98	11.26	17.76	19.48	23.70	21.42
Tensile Strength/MPa	24.75	20.88	23.37	24.66	26.06	27.42	25.02
Elongation at Break /%	591	535	532	410	403	406	391
Permanent Deformation /%	10	8	10	12	16	16	16
Tear Strength/KN·m^{-1}	37.66	37.96	50.29	51.98	52.31	53.94	51.30
Crosslink Density/10^{-5}mol·ml^{-1}							
Covalent Crosslink Density	8.92	7.67	5.74	7.36	7.76	8.33	7.27
Ionic Crosslink Density	1.13	1.29	4.41	3.90	3.65	3.09	4.10
Total Crosslink Density	10.05	8.96	10.15	11.26	11.41	11.42	11.37
Fatigue heat buildup/℃	37.05	40.80	38.40	35.40	33.27	30.25	40.50

3.4. Dynamic mechanical property of SBR/WRP composites

The dynamic mechanical properties of SBR/WRP composites with and without MAA/ZnO modification were investigated. The tan δ versus temperature curves as shown in figure 2 depict the presence of sole tan δ matching with the glass transition temperature of the modified and unmodified SBR/WRP composites. For MAA/ZnO modified SBR/WRP, the storage modulus greatly increased, which can be attributed to the improved crosslinking density and corresponds to the higher hardness. The glass transition temperature (Tg) of SBR/WRP at -35.58℃ was near that of the modified SBR/WRP composite, which occurred at -37.40℃. However, due to the restriction of the rubber chain movement induced by the improved crosslink density and the increased interaction between the rubber molecular chains and WRP particles, the value of the damping peak decreased in the MAA/ZnO modified SBR/WRP composite.

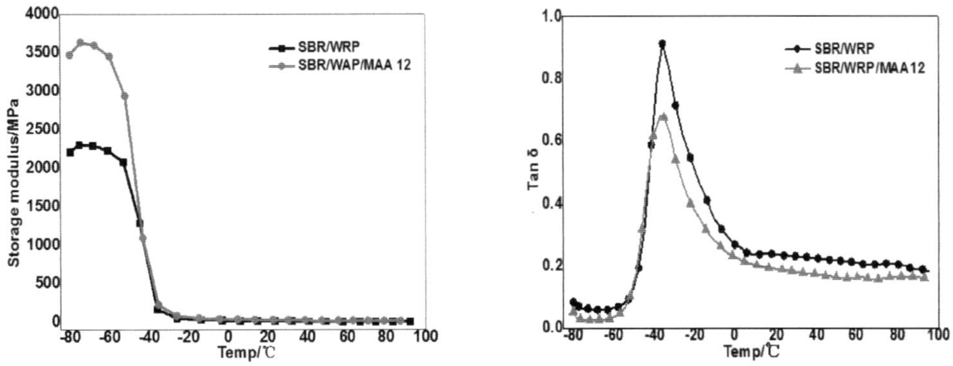

(a) (b)

Figure 2. Dynamic mechanical property of SBR/WRP composites, (a) storage modulus and (b) loss tangent with temperature.

4. Conclusion

MAA/ZnO was employed to effectively modify SBR/WRP composites. The in situ generation of zinc dimethacrylate (ZDMA) and the grafting of MAA to the rubber chains improved the interfacial bonding of WRP and the SBR matrix. The total crosslinking density, ionic bond crosslinking density and covalent cross-linking density of vulcanizate were all increased, resulting in enhanced static mechanical performance and in the reduction of heat generated by dynamic compression. The glass transition temperature remained relatively unaltered, while the value of the damping peak decreased in MAA/ZnO modified SBR/WRP composites.

References

[1] Formela K and Haponiuk J T 2014 Curing characteristics, mechanical properties and morphology of butyl rubber filled with ground tire rubber (GTR) *Iran. Polym. J.* **23** 185-194

[2] Yu R, Gong Z, Guo W, Zhang H, Liu C 2016 A novel grafting - modified waste rubber powder as filler in natural rubber vulcanizates *J. Appl. Polym. Sci.* **133**

[3] Zhu J, Zhang X, Liang M and Lu C 2011 Enhancement of processability and foamability of ground tire rubber powder and LDPE blends through solid state shear milling *J. Polym. Res.* **18** 533-539

[4] Marković G, Veljković O, Marinović-Cincović M, Jovanović V, Samaržija-Jovanović S and Budinski-Simendić J 2013 Composites based on waste rubber powder and rubber blends: BR/CSM *Composites Part B: Engineering* **45** 178-184

[5] Liu J, Liu P, Zhang X, Lu P, Zhang X and Zhang M 2016 Fabrication of magnetic rubber composites by recycling waste rubber powders via a microwave-assisted in situ surface modification and semi-devulcanization process *Chemical Engineering Journal* **295** 73-79

[6] Lin Y, Chen Y, Zhang Y, Jia D, Luo Y and Liu L 2015 The use of zinc dimethacrylate functionalized graphene as a reinforcement in rubber composites *Polym. Adv. Technol.* **26** 423-431

[7] Nie Y, Huang G, Qu L, Zhang P, Weng G and Wu J 2010 Cure kinetics and morphology of natural rubber reinforced by the in situ polymerization of zinc dimethacrylate *J. Appl. Polym. Sci.* **115** 99-106

[8] Guo B C, Chen F, Chen W W, Lei Y D and Jia D M 2010 Reinforcement of nitrile rubber by in situ formed zinc disorbate *Express. Polym. Lett.* **4** 529-538

[9] Yuan X, Peng Z and Zhang Y 2000 In situ preparation of zinc salts of unsaturated carboxylic acids to reinforce NB *J. Appl. Polym. Sci.* **77** 2740-2748

[10] Wen S, Zhou Y, Yao L, Zhang L, Chan T W, Liang Y and Liu L 2013 In situ self-

polymerization of unsaturated metal methacrylate and its dispersion mechanism in rubber-based composites *Thermochimica Acta* **571** 15-20

[11] Huang A, Wang X and Jia D 2012 Preparation of HNBR/PMgMA/OMT nanocomposites and their aging resistance in different media *Acta Polymerica Sinica* **12** 660-665

[12] Valentín J L, Carretero-González J, Mora-Barrantes I, Chassé W and Saalwächter K 2008 Uncertainties in the determination of cross-link density by equilibrium swelling experiments in natural rubber *Macromolecules* **41** 4717-4729

[13] Nie Y, Huang G, Liu Z, Qu L, Zhang P, Weng G and Wu J 2010 Improved mechanical properties and special reinforcement mechanism of natural rubber reinforced by in situ polymerization of zinc dimethacrylate *J. Appl. Polym. Sci.* **116** 920-928

[14] Mathew G, Singh R P, Lakshminarayanan R and Thomas S 1996 Use of natural rubber prophylactics waste as a potential filler in styrene-butadiene rubber compounds *J. Appl. Polym. Sci.* **61** 2035-2050

[15] Xu C, Chen Y, Cao L, Wang Y and Zeng X 2013 Study of the Crosslinking Evolution of Styrene-Butadiene Rubber/Zinc Dimethacrylate Based on Dissolution/Swelling Experiments *Journal of Macromolecular Science, Part B* **52** 319-333

Research of glass fibre used in the electromagnetic wave shielding and absorption composite material

M Xu[1], F Jia, H Q Bao, K Cui and F Zhang

China Building Material Academy, Beijing, 100024, China

E-mail: xumingcbma@163.com

Abstract. Electromagnetic shielding and absorption composite material plays an important role in the defence and economic field. Comparing with other filler, Glass fibre and its processed product—metal-coated glass fibre can greatly reduce the material's weight and costs, while it still remains the high strength and the electromagnetic shielding effectiveness. In this paper, the electromagnetic absorption mechanism and the reflection mechanism have been investigated as a whole, and the shielding effectiveness of the double-layer glass fibre composite material is mainly focused. The relationship between the shielding effectiveness and the filled glass fibre as well as its metal-coated product's parameters has also been studied. From the subsequent coaxial flange and anechoic chamber analysis, it can be confirmed that the peak electromagnetic shielding effectiveness of this double-layer material can reach -78dB while the bandwidth is from 2GHz to 18GHz.

1. Introduction

With the development of modern science and technology, electric equipment is widely used and the electromagnetic environment becomes much more complex. So it is pressing to reduce the electromagnetic interference between equipments and to make them work properly. Moreover, the widely used electric equipment has been presenting a great challenge to national economy and defense. The electromagnetic shielding (EMS) technology has been used to prevent important information from leaking out, and research of the basic material used for EMS becomes much more important [1]. At present, electromagnetic shielding composite materials are always filled with metal fibers and conductive powders, which make them heavy and highly conductive. Unlike this kind of EMS composite materials, glass fiber and its metal-coated product composed with resin, would have higher shielding effectiveness and strength with less mass density, which has been regarded as the potential replacement of them [2, 3]. In this work, the relationship between fiber parameters and shielding effectiveness of material were discussed; the relationship between absorbing properties and fiber filling volume were analyzed; and the declination between metrical result and calculated result were explained. In the end, it can be pointed out that glass fiber and its processed product could play an important role in material's electromagnetic shielding effect, and it also had significant meaning in achieving the integration of function and structure of material.

2. Experiment

In the experiment, phenolic resin and metal-coated glass fiber were used as raw materials, and the composite material was prepared by a hot-pressing method. The process mainly contains three parts as follows.

2.1. Raw Materials

Metal-coated glass fiber: Aluminum was coated on the E-glass fiber. The thickness of the Aluminized layer: 2.5 μ m; the diameter of the fiber: 20 μ m. Common E-glass fiber: the diameter was 17 μ m, insulated. Electromagnetic absorber: W-type ferrite and alloy powder, the particle size were 30 μ m and 8 μ m. Resin: Magnesium-Phenolic resin, trademark 2122.

2.2. Preparation of the Material

The material was made up of two different kinds of layers: the electromagnetic shielding layer and the microwave absorbing layer. All the raw materials should be mixed in the solution. Experimental process was: Firstly, cut Aluminized glass fiber into different kinds of length: 1.5mm, 10mm and 16mm, and the E-glass fiber were cut into 4mm. Then the Aluminized fiber and the resin were mixed together using acetone as thinner—which was the shielding layer; in the same way, E-glass fiber, absorber and resin were mixed together—which was the microwave absorbing layer. Both layers should be preheated in a mold at 110° C in 120 seconds, to form a plate shape. Secondly, previous two kinds of plates were pressed with a mold at 180° C, and the final sample was obtained with a thickness of 3 mm (electromagnetic shielding layer was 2mm and absorbing layer was 1mm). In this experiment, the viscosity of the system can be adjusted.

2.3. Shielding Effectiveness test of Material

Electromagnetic absorbing effectiveness of the material were obtained from methods for measuring the reflectivity of radar absorbing materials with anechoic chamber, while vertical polarization and shielding effectiveness were obtained from the coaxial guide method, which is a near field experiment, with the effectiveness represented by dB. Conductivity was analyzed by means of a 4 probe resistivity measuring instrument. The inner structure of the material was observed by a scanning electron microscope.

3. Result and Discussion

In the Aluminized glass fiber layer, fiber's filling volume and its size is crucial to the conductivity and shielding effectiveness of material. In addition to this, the filling volume of E-glass fiber in the absorbing layer has something to do with the microwave absorbing effectiveness. Through the research, we find that glass fiber and its processed product play an important role in the control of electromagnetic parameters.

3.1. The Relationship between Conductivity, Electromagnetic Reflection Properties and Aluminized Glass Fiber of Material

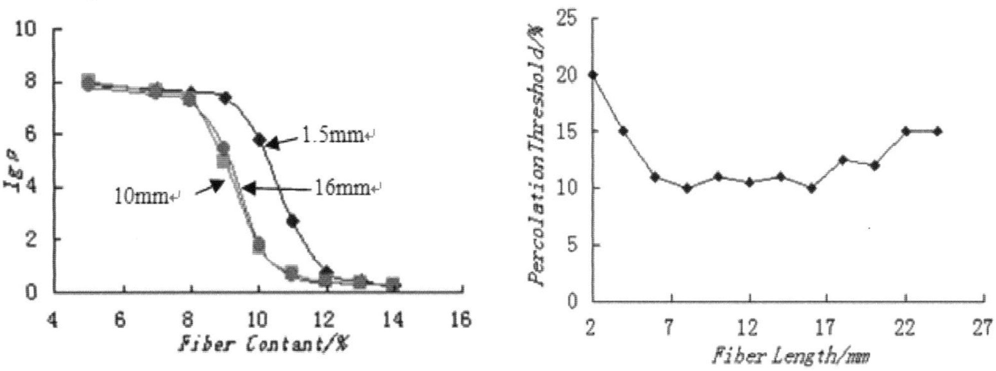

(a) (b)

Figure 1. Relationship between Percolation Threshold and Fibre Content.

An important parameter of the composite material is the percolation threshold [4], which will occur when the conductivity(lgρ) of the material changes suddenly with the increase of a fiber content. However, this is relative to aluminized fiber's length and diameter. In this experiment, the fiber's diameter was constant, while the relationship between the aluminized fiber's length and its electrical as well as electromagnetic properties were mainly investigated. The 1.5mm, 8mm and 16mm length aluminized fibers were used as the research object, as shown in figure 1(a).

When the fiber length is 1.5mm, its percolation threshold is about 12%(wt), so if the fiber content is higher than this demarcation point, the material will suddenly become a conductor. However, this demarcation point is changeable depending on the fiber length. If a longer fiber length was used instead, for example: 10mm or 16mm, the percolation threshold will decrease to 11%(wt) correspondingly. So elongating the fiber length can possibly decrease the fiber content which is necessarily needed. However, when fibers of longer length were used as the research object, it has been found that the above conclusion doesn't work, shown in figure 1(b). From figure 1, it can be seen that, if the fiber length was continually elongated, the percolation threshold will increase instead. It may be caused by technological factors in mixing and dispersing process. While the fiber length continually become longer, the fiber is hard to be dispersed. Some fiber will cluster as one, and make large insulating area, which was confirmed from the SEM photograph, as shown in figure 2.

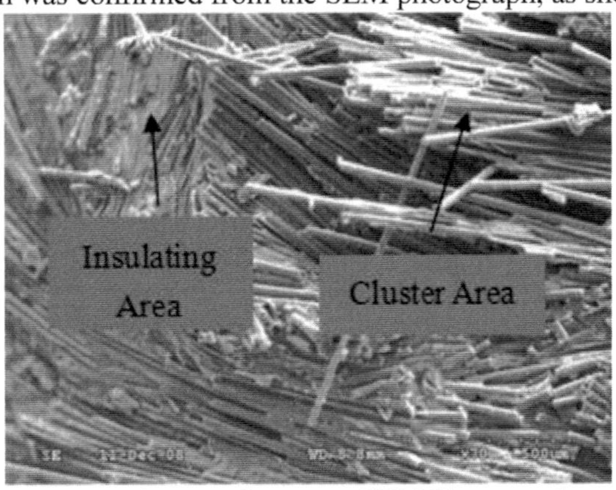

Figure 2. Inner Structure SEM Photograph.

Material may acquire its highest shielding effectiveness when its fiber length is in the appropriate range [5]. In this experiment, the best fiber length range for the percolation threshold is from 7mm to 16mm. Besides, the shielding effectiveness of material (fiber content 60wt %) at above fiber length range when the frequency is 2GHz to 18GHz, was also tested, as shown in figure 3.

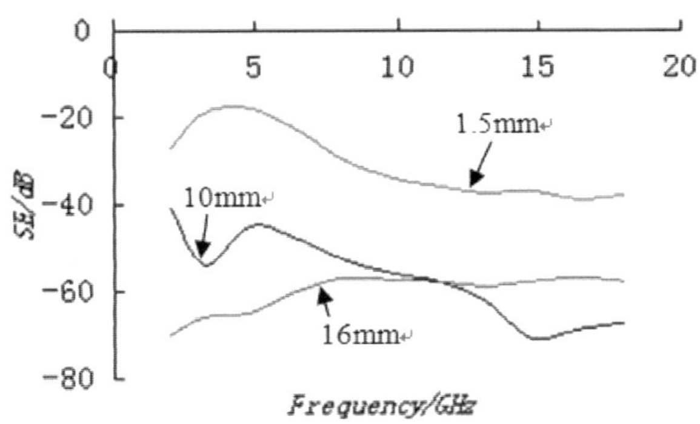

Figure 3. Shielding Effectiveness with the corresponding Fibre Length.

From figure 3, it can be concluded that the material with a longer length can possibly gain a higher shielding effectiveness comparing to that with shorter fiber length. It can also be observed that in higher frequency areas, 10mm length sample is even better than the 16mm sample. The half-wave resonance may be the reason for this, for 10mm fiber length just meeting the requirement of 2cm electromagnetic wave's resonance—the frequency of 15GHz, which conforms to the tendency in figure 3.

3.2. The Relationship between Electromagnetic Absorbing Properties and E-glass Fibre of Material

Single electromagnetic shielding layer of material could cause some new problems such as second clutter pollution. In order to solve this problem effectively, an electromagnetic absorbing layer which composed of E-glass fiber and absorbent was attached on the surface of reflection layer. Here, the usage of conventional reinforcing E-glass fiber in the electromagnetic absorbing material as well as the correlationship between E-glass fiber content and material's absorbing and shielding effectiveness, were mainly studied, with the result shown in figure 4.

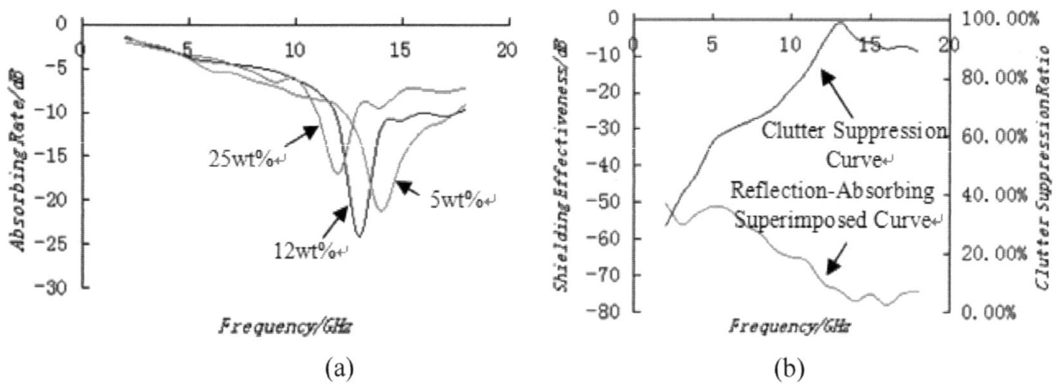

Figure 4. Absorbing Rate and Clutter Suppression Ratio.

From figure 4, it can be concluded that: conventional E-glass fibre can change electromagnetic parameters of material, thereby causing the different absorbing rate, like in figure 4(a). The relationship between fibre content and absorbing rate were studied, and it can be found that when the fibre content is 5wt%, the absorbing rate of material is close to the intrinsic absorption rate of 21.3dB. When the fibre content was continually increased to 12wt%, the maximum absorbing rate would rise to 24.2dB, and the peak of curve would move to the left. However, too much glass fibre will result in the reducing of absorbing rate. When fibre content was increased to 25wt%, there was an evident decrease of the peak with only 17dB. It may be caused by the E-glass fibre's multiple interface structure and the changing of electromagnetic parameters because appropriate fibre's multiple interfaces would cause electromagnetic wave's strongly scattering and attenuation, while at the same time which would also change the dielectric factors [6] and cause the peak to move to the left. However, too much fibre will lead to both decrease of permittivity and permeability and increase of the match thickness, so fibre content is important to material's absorbing ability. In this experiment, it has been found that the fibre content of 10wt% to 12wt% is the right amount.

The total shielding effectiveness when reflection layer and absorbing layer being superimposed was investigated in figure 4(b). It can be found that not only the total SE is higher than single layer, but also the clutter is perfectly suppressed. When the frequency is in the range of 11GHz to 18GHz, the clutter suppression ratio can reach up to about 90%, and the total SE can still maintain -70dB.

4. Conclusions

Conventional glass fibre and its processed product were used in the electromagnetic shielding material. Compared with other electromagnetic shielding materials, this material has a lower density and a higher shielding effectiveness; more important, it could perfectly suppress the clutter and maintain FRP's mechanical strength.

- The percolation threshold of material would decline with increase of the aluminized glass fibre length; besides, it would also be controlled by the technology factors, and the percolation threshold would increase when fibre length is longer than 16mm;
- Longer fibre length could possibly gain a higher shielding effectiveness, however, resonance should also be taken into account, especially in the high frequency area;
- Conventional glass fibre could efficiently adjust the absorbing curve's intensity and peak position, but excessive glass fibre would result in the decline of absorbing, and the content of 10wt% to 12wt% is found to be appropriate;

Reflection-absorbing superimposed material could efficiently shield electromagnetic wave, and it would be helpful to the electromagnetic compatibility of the composite material. Compared with conventional shielding material, the prepared material could finally attenuate 90% microwave's energy when the frequency range is from 11GHz to 18GHz.

References

[1] Yang S Y *Electromagnetic Shielding Theory and Practice* National Defense Industry Press, Beijing, China, 2006 pp. 96.

[2] Tan S T and Zhang M Q 1999 Conductivity and electromagnetic shielding properties of metal fiber filled polymer composite *Mater. Eng.* **12** 3.

[3] Xia Y, 2002 Research of carbon black filled polyethylene composites *Plastic* **31** 48.

[4] He D, Ekere N N 2004 Effect of particle size ratio on the conducting percolation threshold of granular conductive-insulating composites *J. Phys. D: Appl. Phys.* **37** 1848-152.

[5] Hochberg A and Versieck J 2001 Shielding for EMI and antistatic plastic resins with stainless steel fibres *Plastics Additives* **3** 24

[6] Sun J, Jiang T 2005 Research of nano-ferromagnetic thin films plating on glass fiber technology. *Journal of Yunnan University* **S1** 38

IOP Conf. Series: Materials Science and Engineering **137** (2016) 012050 doi:10.1088/1757-899X/137/1/012050

A novel preparation of milk protein/polyethylene terephthalate fabric

J F Zhou[1,4], D D Zheng[1,4], L Zhong[2], F X Zhang[3] and G X Zhang[1,4,5]

[1]College of Textile & Garments, Southwest University, Chongqing, China
[2]Chongqing Municipality Fibre Inspection Bureau, Chongqing, China
[3]School of Chemistry and Chemical Engineering, Southwest University, Chongqing, China
[4]Chongqing Engineering Research Center of Biomaterial Fiber and Modern Textile, Chongqing, China

E-mail: zgx656472@sina.com

Abstract. In this work, -NH_2 groups were introduced to polyethylene terephthalate (PET) fibers by nitration and reduction method, and then milk protein was grafted on the nitrated and reduced PET (NR PET) fibers by sucrose glycidyl ether crosslinking agent. FTIR suggested the milk protein was successfully grafted on PET fiber surface. SEM images showed a layer of substance covered on the PET fiber surface. DSC demonstrated an excellent thermal stability of milk protein/PET fiber. The moisture regain was improved by milk protein/PET fiber. Moreover, the crease recovery angle and stiffness were retained by the milk protein/PET fabric.

1. Introduction

PET fabrics are widely used in the textile industry due to its excellent mechanical property, dimensional stability and wash-and-wear property [1]. However, the poor skin-friendly property of PET fabric limits its application in other fields, such as underwear clothes, bedding and sports wearing. The milk protein is skin-friendly due to its wonderful biocompatibility [2]. Therefore, grafting milk protein on the PET fiber surface can improve the wearing comfort of PET fabric and retain the excellent mechanical properties.

Then, the first step is to endow PET fibers reactive species. Various modification technologies were developed, including chemical hydrolysis [3, 4], UV irradiation [5], plasma treatment [6, 7], surface oxidation [8] and aminolysis methods [9, 10]. Nevertheless, these modification methods would result in a decrease of mechanical property or produce waste in the effluent.

In this study, reactive –NH_2 groups were introduced to the PET fiber by nitration with gradually concentrating method and reduction process. And then the milk protein could be grafted on the NR PET fibers surface at the presence of sucrose glycidyl ether to produce a milk protein/PET fabric with high skin-friendly and excellent mechanical property.

2. Experimental

2.1. Materials
PET fabrics (27.0 tex × 22.2 tex/480 × 320,210 g/m^2), sucrose and milk protein were purchased from Chaotianmen market of Chongqing, China. Nitric acid, sodium sulfide, epoxychloropropane, sodiumhydroxide and tetrabutylammonium bromide were purchased from Chengdu Kelong, China.

2.2. The nitration and reduction of PET fibers
The PET fabric (20 cm × 20 cm) was immersed in 20 g/L HNO_3 solution for 10 min at room temperature with a bath ratio of 1:30. The PET fabric was then cured at 70 °C for 20 min in a drying oven. After that, the nitrated PET fabric was washed with distilled water and reduced in a 25 g/L Na_2S solution at 98 °C for 30 min. Finally, the nitrated and reduced PET fabric with reactive $-NH_2$ groups was washed with distilled water and dried at 110 °C (NR PET).

2.3. Synthesis of sucrose glycidyl ether crosslinking agent
Sucrose (20 g) was firstly dissolved in 200 ml distilled water, and then tetrabutylammonium bromide (0.1 g) and epoxychloropropane (27 mL) were added to the solution. After 30 min reaction at room temperature, NaOH (11 g) was gradually added into the solution, the temperature was kept at 40 °C for 4h with a mild agitation.

2.4. Grafting milk protein onto the NR PET fibers
NR PET fabrics were immersed into a mixed solution of sucrose glycidyl ether and milk protein (sucrose glycidyl ether/milk protein: 1/5) ranging from 10 g/L to 50 g/L for 10 min at ambient temperature with a bath ratio of 1:30. Then the immersed PET fabrics were padded to reach a wet pickup of 100% followed by parching at 60 °C for 30 min and curing at 110 °C for 30 min. Finally, the milk protein/PET fabric was produced. The weight gain rate (WGR) of milk protein/PET fabric was determined by weighing.

2.5. Test methods

2.5.1 Characterization. FTIR measurement of PET fabric samples was carried out by a Spectrum GX spectrometer (Perkin-Elmer Co., USA).SEM images of PET fabric samples were obtained using an S-4800 cold field-emission scanning electron microscope (Hitachi Limited, Japan).DSC of PET fabric samples was measured using a PerkinElmer DSC 200 F3 Maia instrument (Waltham, MA, USA).

2.5.2. Moisture regain. The moisture regains of PET fabric samples were evaluated according to the GB/T 6503-2008 standard test method.

2.5.3. Mechanical properties. The crease recovery angles of PET fabric samples were measured by a YG(B)541D-II automatic digital fabric crease elasticity measuring instrument (Wenzhou DaRong-Textile Instruments Co., Ltd.) following the AATCC method 66-2008 method.AnLLY-01 computer-controlled stiffness measurement apparatus(Universal Textile Instruments Co., Ltd., Changzhou, China)was used to determine the stiffness of the PET samples following the ASTM D 1388-96(2002) standard test method.

3. Results and discussion

3.1. The FTIR of PET fabric samples
Figure 1 shows the FTIR spectra of PET fibers. After the grafting of milk protein, the spectrum of milk protein/PET fiber has the same main peaks as that of original PET fibers. However, the spectrum of milk protein/PET fiber exhibits characteristic absorptions of amide group at 1558 cm^{-1} (amide II)

and 1650 cm^{-1} (amide I). Moreover, the absorption peak at 3540 cm^{-1} was assigned to the –NH stretching vibration. These indicated that the successful grafting of milk protein onto the PET fiber.

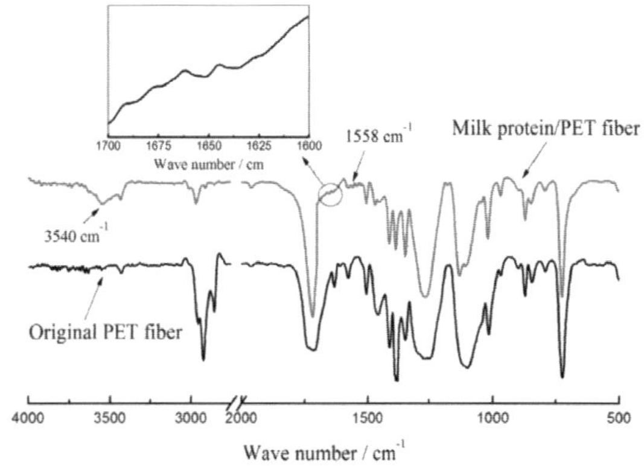

Figure 1. FTIR of PET fabric samples.

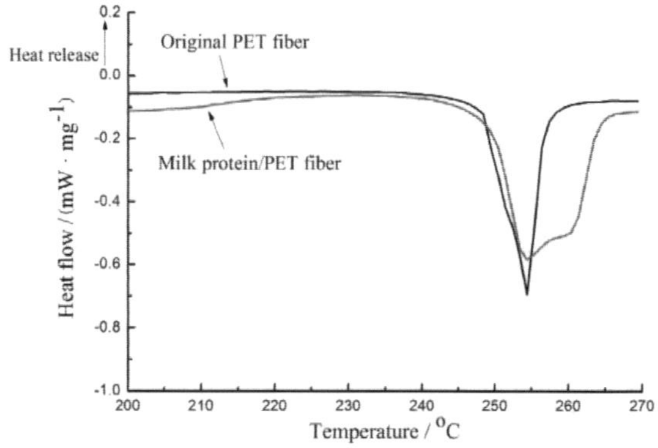

Figure 2. The DSC of PET fabric samples.

3.2. The DSC of PET fabric samples
The DSC thermograms of original PET fibers and milk protein/PET fibers are exhibited in figure 2. The endothermic peak of the original PET fibers is sharp while that of the milk protein/PET fibers is blunt, indicating that the melt of original PET fibers is faster than that of the milk protein/PET fibers. This was because the strong interactions between the $-NH_3^+$, $-COO^-$, and –OH groups in the milk protein and the crosslinking actions prevented the melt of PET fibers.

3.3. The SEM of PET fabric samples
The surface morphology of original PET fibers and milk protein/PET fibers are showed in figure 3. From figure 3, it can be seen that the initial PET fibers are smooth while the milk protein/PET fibers are covered with a layer of substance. These SEM images manifest the milk protein has been successfully grafted on the PET fibers surface.

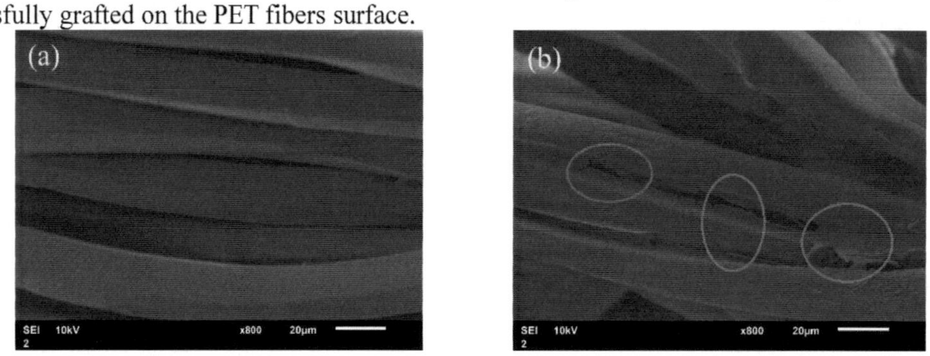

Figure 3. The SEM images of (a) original PET fibers and (b) milk protein/PET fibers.

3.4. The moisture regains of PET fabric samples
Figure 4 shows the relationship between the moisture regain and the weight gain rate. From figure 4, the moisture regain increases with the growth of WGR. The moisture regain of the original PET fabric is 0.45%, whereas that of the milk protein/PET fabric with 3.34% WGR is 1.14%, 153% higher. This was because the –OH, $-NH_2$, –COOH and –CONH– groups of milk protein on the surface of milk protein/PET fabric have high ability to absorb water vapor.

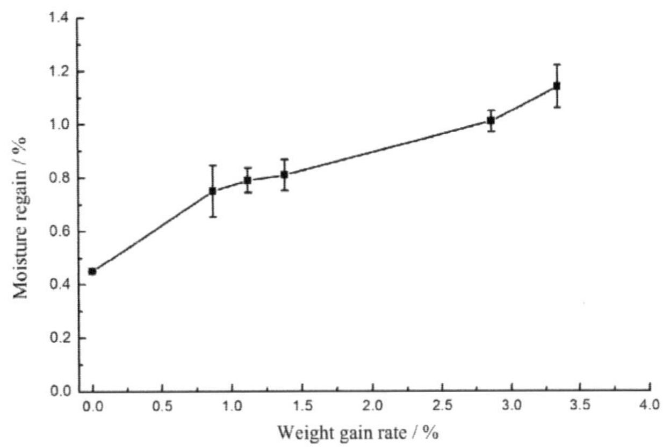

Figure 4. The relationship between the moisture regain and WGR.

3.5. The crease elasticity of PET fabric samples

The relationship between the crease recovery angle and weight gain rate is shown in figure 5. The crease immediate and delay-recovery angles all decreased slightly with the increase of WGR. This is because the grafting of milk protein has introduced some polar groups (–OH, $-NH_2$, –COOH) on the PET fabric surface. The polar groups would interact with each other when folding the milk protein/PET fabric, and the interactions would not disappear thoroughly after the vanished of external force. The crease recovery angle only decreased a little and retained well by the milk protein/PET fabric.

3.6. The stiffness of PET fabric samples

Figure 6 exhibits an increment of the stiffness after grafting milk protein on the PET fabric. The more milk protein was grafted, the higher the flexural rigidity of PET fabric would be. The flexural rigidity of original PET fabric is 0.013 cN·cm, whereas the flexural rigidity of milk protein/PET fabric with 2.34% WGR is 0.048 cN·cm. In general, the stiffness of fabric grafted macromolecule will increase. This is because many polar groups of milk protein/PET fabric interact with each other, resulting an increase of stiffness. However, the increment is not so much, and the hand feeling of milk protein/PET is still very excellent.

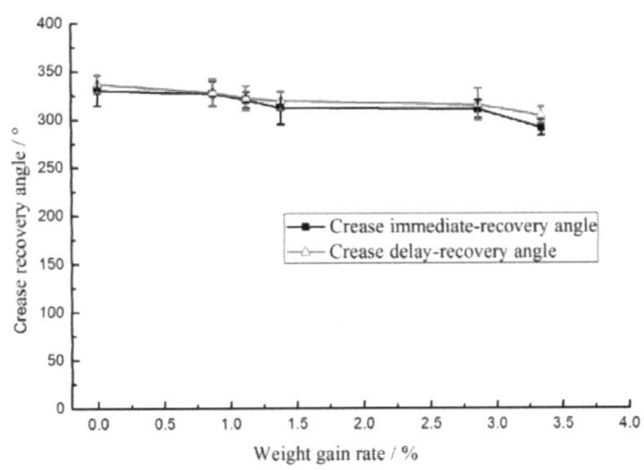

Figure 5. The creasability of PET fabric samples.

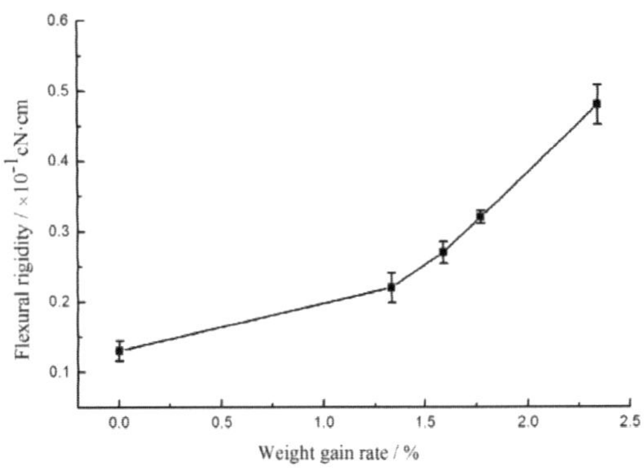

Figure 6. The stiffness of PET fabric samples.

4. Conclusions

Nitration and reduction method endowed the PET fibers reactive $-NH_2$ groups and then the milk protein could be grafted on the PET fiber surface with sucrose glycidyl ether to produce a high functional milk protein/PET fabric. The moisture regain of milk protein/PET fabric increased significantly. The crease elasticity and hand feeling were retained well by the milk protein/PET fabrics.

Acknowledgment

This study was supported by the Key Application and Development Project of the Chongqing Science & Technology Commission (Grant number CSTC2014yykfB50002) and the Fundamental Research Funds for the Central Universities (Project numbers XDJK2013A021 and XDJK2013D002).

References

[1] Zhang F X, Liang H and Zhang G X 2015 Colorant-free coloration and superhydrophilic modification of poly(ethylene terephthalate) fabric surface by H_2O_2 and nano-TiO_2 ultraviolet photocatalysis *TextRes J* DOI: 10.1177/0040517515603800.

[2] Yang L F 2007 Estination of its biological function of milk protein fiber and its product development Shanghai *Text. Sci. Technol.* **35** 46-48.

[3] Liu Y X, He T and Gao C Y 2005 Surface modification of poly(ethylene terephthalate) viahydrolysis and layer-by-layer assembly of chitosan and chondroitin sulfate to construct cytocompatible layer for human endothelial cells *Colloid Surface B* **46** 117-126.

[4] Eberl A, Heumann S and Brückner T, et al. 2009 Enzymatic surface hydrolysis of poly(ethylene terephthalate)and bis(benzoyloxyethyl) terephthalate by lipase and cutinase in thepresence of surface active molecules *J. Biotechnol.* **143** 207-212.

[5] Zhu Z M and Kelley M J 2004 Poly (ethylene terephthalate) surface modification by deep UV (172 nm) irradiation *Appl. Surf. Sci.* **236** 46-425.

[6] Gupta B, Hilborn J G and Bisson I et al. 2001 Plasma-induced graft polymerization of acrylic acid onto poly(ethylene terephthalate) films *J. Appl. Polym. Sci.* **81** 2993-3001.

[7] Tompkins B D, Dennison J M and Fisher E R 2013 H_2O plasma modification of track-etched polymer membranes for increased wettability and improve performance *J. Membrane. Sci.* **428** 576-588.

[8] Fávaro S L, Rubira A F and Muniz E C et al. 2007 Surface modification of HDPE, PP, and PET films with KmnO4/ HCl solutions *Polym. Degrad. Stabil.* **92** 1219-1226.

[9] Croll T I, O'Connor A J and Stevens G W, et al. 2004 Controllable surface modification of poly(lactic-co-glycolic acid) (PLGA) by hydrolysis or aminolysis I: physical, chemical, and theoretical aspects *Biomacromolecules* **5** 463-473.

[10] Bech L, Meylheuc T, Lepoittevin B and Roger P, et al. 2007 Chemical surface modification of poly(ethylene terephthalate) fibers by aminolysis and grafting of carbohydrates *J. Polym. Sci. Pol. Chem.* **45** 2138-72.

IOP Conf. Series: Materials Science and Engineering **137** (2016) 012051 doi:10.1088/1757-899X/137/1/012051

Preparation and dyeing of super hydrophilic polyethylene terephthalate fabric

D D Zheng[1,2], J F Zhou[1,2], F Xu[1,2], F X Zhang[3] and G X Zhang[1,2,4]

[1]College of Textile & Garments, Southwest University, Chongqing, China
[2]Chongqing Engineering Research Center of Biomaterial Fiber and Modern Textile, Chongqing, China
[3]School of Chemistry and Chemical Engineering, Southwest University, Chongqing, China

E-mail: zgx656472@sina.com

Abstract: In this study, the dyeing properties of PET fabrics modified with sulfuric acid was investigated using disperse red E-4B and disperse blue 2BLNG-L at high temperature and high pressure. The results revealed that the sulfuric acid modification improved the K/S value of dyeing PET fabrics, and the modified PET fabric could be dyed uniformly. The a, b, C, L and H of modified PET fabric were almost the same as that of original PET fabric. The water contact angles were still 0° after 10s, indicating that the hydrophilic property of modified PET fabrics still kept excellent. The wash fastness of dyed PET fabrics after modification was generally good.

1. Introduction

Polyethylene terephthalate (PET) fabric was widely used in apparel, home furnishings and various industrial products due to its excellent properties such as high strength, wear-resistant, washable [1,2]. However, PET fabric has poor hydrophilicity and hygroscopicity, which restricts its application [3]. Thus it is very important to improve the hydrophilicity and comfort of PET fabric, and it has become a hotspot in research over the decades.

To date, numerous hydrophilic modification methods for PET fabrics have been proposed, For example, the surface oxidation [4], surface grafting method [5, 6], coating. Our group has prepared super hydrophilic PET fabric by dilute sulfuric acid with gradually concentrating method under microwave irradiation [8].

PET fabric is generally dyed under the severe condition of high temperature and high pressure, where the disperse dyes penetrated into internal space of PET fibers. Introduction of sulfonic acid groups in PET macromolecules may affect the dyeing properties of modified PET fabric. So this paper study the dyeing properties of PET fabrics modified with sulfuric acid. In this paper, the K/S value, levelling property ΔE, wash fastness, lightness(L), chroma(C), hue(H) of the modified PET fabric were analyzed and compared with that of the original PET fabric.

2. Experimental

2.1. Material

PET fabric (27.0tex×22.2tex/500×360) was bought from Chaotianmen market, Chongqing China. Sulfuric acid was bought from the ChengDu Kelong Chemical Reagent Company, Sichuan China. Disperse red E-4B and disperse blue 2BLNG-L were purchased from Dyeing Plants Wenling, Zhejiang China.

2.2. Sulfuric acid modification and dyeing of PET fabrics

PET fabric was soaked in a certain concentration of sulfuric acid solution, padded through the nip to reach wet pickup of about 100%, then irradiated for 6 minutes with microwave at 600 W. Finally after washing and drying, super hydrophilic PET fabric was obtained.

The modified PET fabric was putting into the disperse dye solution with mass concentration of 1.0% o.w.f at a ratio of 1:30 (w/w), heated up to 120 C at a rate of 2 C/min and kept for 50 min. Subsequently, the dyed PET fabric was washed and dried at room temperature.

2.3. Characterization techniques

The K/S values represent the dyeing depth of the fabric under the maximum absorption wavelength. The test using D65 light source with perspective angle of 10 . Five different points on the fabric samples were tested, and the average values of which were calculated. The larger the K/S value is, the deeper the dyeing is. The calculation formula is as follows:

$$K/S = \frac{(1 - R)^2}{2R} \tag{1}$$

Where K is the light absorption coefficient of the fabric; S is the light scattering coefficient of the fabric; and R is the light reflectance of the fabric at the maximum absorption wavelength.

Levelling property of the fabric was measured with Data color 650 instrument. One position on the dyed PET fabric was selected as comparison. Then ΔE value of PET fabric at different position was measured.

The lightness(L), chroma(C), hue(H) of PET fabric were measured by Data color 650 instrument. These data could show the color of fabric based on Munsell color system.

The water contact angle(WCA) of the dyeing PET fabrics was measured with horizontal single-shot method using JGW-360 contact determination instrument. The WCA of the samples was calculated as following formula.

$$\theta = \arctan\frac{4dh}{d^2 - 4h^2} \tag{2}$$

where h is the height from top of the water droplet to the fabric surface; d is the contact diameter of water droplet on fabric.

The wash fastness of the dyeing PET fabrics was evaluated by the AATCC 61-2006 standard method using a soaping fastness tester (Roaches Co., England). The color change was rated according to the appropriate ISO grey scale.

3. Results and discussion

3.1. The K/S values of the samples

The super hydrophilic PET fabrics were dyed with disperse blue 2BLNG-L and disperse red E-4B. Figure 1 shows the relationship between the dilute sulfuric acid concentration and the K/S values of the dyed fabrics modified with gradually concentrating dilute sulfuric acid method. The results showed that the K/S value increased obviously with the increase of the concentration of sulfuric acid. When using disperse blue 2BLNG-L to dye fabrics, the K/S value of the original PET fabric was 6.68, and increased to 8.62 after modified with 8.0 g/L sulfuric acid. When dyed with disperse red E-4B, the K/S value of the unmodified PET fabric was 6.25, after modified with 8.0 g/L sulfuric acid, the K/S value of the modified PET fabric increased to 8.37. The higher the K/S value was, the higher the degree of exhaustion was, and the better the dyeing performance of PET fabric was. These show that the dyeing property of sulfuric acid modified PET fabrics is better than that of the unmodified PET fabrics. This should be because PET fabric modified with sulfuric acid has sulfonic acid groups on the PET macromolecule of the surface layer. The sulfonic acid groups are negative charges, which mutually exclusive on the PET fiber surface, maybe resulting in the interspaces among the PET macromolecule chains increasing greatly. Then the dyestuff molecules can entry the PET fibers inner space more easily. Thus the dyeing properties of the sulfuric acid modified PET fabric are better than that of unmodified PET fabric.

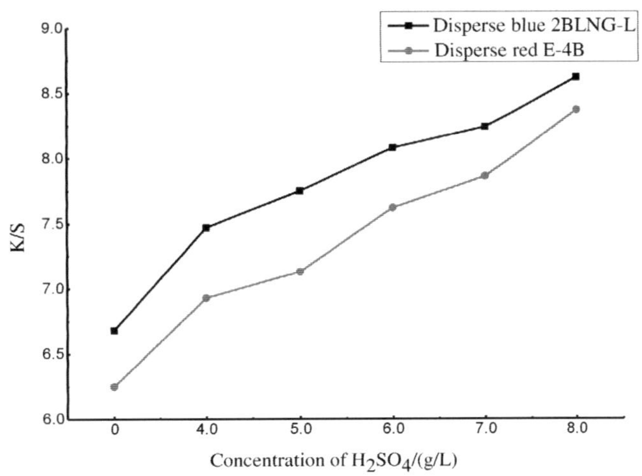

Figure 1. The K/S values of original PET fabric and PET fabrics modified by H_2SO_4.

3.2. The levelling property of sulfuric acid modified PET fabric

Table 1 shows the color value ΔE of PET fabrics before and after modified by 8.0g/L H_2SO_4. From table 1 it can be seen that the color values ΔE of different position on the dyeing modified PET fabrics dyed with disperse red E-4B and disperse blue 2BLNG-L only changed a little, and the average values were less than 0.6. When using disperse red E-4B and disperse blue 2BLNG-L to dye, the average ΔE values of modified PET fabric of 0.362, 0.304, respectively, are slightly higher than that of original PET fabric. Thus the levelling property of the sulfuric acid modified PET fabric is extremely excellent. The modification of PET fabric with sulfuric acid as modifying agent does not affect the levelling property of PET fabric.

Table 1. Color value Δ E of PET fabrics before and after modified by H_2SO_4.

		Color value ΔE of different point on PET fabrics										Average
		1	2	3	4	5	6	7	8	9	10	ΔE
Disperse red E-4B	Original PET	0.12	0.43	0.27	0.33	0.16	0.23	0.61	0.37	0.11	0.22	0.285
	Modified PET	0.46	0.19	0.32	0.53	0.59	0.66	0.32	0.30	0.11	0.14	0.362
Disperse blue 2BLNG-L	Original PET	0.42	0.15	0.43	0.26	0.22	0.23	0.17	0.31	0.20	0.35	0.274
	Modified PET	0.19	0.41	0.34	0.25	0.26	0.13	0.56	0.34	0.12	0.44	0.304

3.3. Relative dyeing values of sulfuric acid modified PET fabrics

Table 2 lists the relative dyeing values of PET fabrics before and after modified by 8.0 g/L H_2SO_4. PET fabrics were dyed with disperse blue 2BLNG-L and disperse red E-4B respectively. The color of the modified PET fabric was almost the same as that of the original dyeing PET fabric. The L and C value of modified dyeing PET fabric both decreased slightly, which corresponded to the increasing of K/S value of modified PET fabric. The Hunter a value of modified dyeing PET slightly decreased, the color was a little greener than that of original PET fabric; the Hunter b value of modified dyeing PET increased slightly, the color tended to yellow. The change of a, b value resulted that the H value of modified PET fabric dyed with disperse red E-4B changed from 1.20° to 1.66°, the H value of modified PET fabric dyed with disperse blue 2BLNG-L changed from 248.86° to 246.14°, the color of modified PET fabrics tended to be yellow-green. But the degree was very small, which could not be seen by naked eyes.

Table 2. Dyeing properties of PET fabrics before and after modified by H_2SO_4.

		a	b	C	L	H/(°)	%R corresponding wavelength（nm）
Disperse blue 2BLNG-L	Original PET	-14.09	-36.44	39.07	49.54	248.86	640
	Modified PET	-15.26	-34.51	37.73	47.28	246.14	640
Disperse red E-4B	Original PET	59.36	1.23	59.37	52.13	1.20	520
	Modified PET	58.87	1.73	58.90	49.37	1.66	520

3.4. The hydrophilic property of modified PET fabrics after dyeing

Figure 2 shows the hydrophilicity of modified and dyed PET fabrics. From figure 2, it can be seen that the water droplet could disappeared thoroughly. This directly proved that the dyed PET fabric modified by sulfuric acid still has super hydrophilicity.

Table 3 shows the hydrophilicity of the PET fabric modified by H_2SO_4 after dyeing. The hydrophilic property of modified PET fabrics after dyeing was characterized by the water contact angles when water droplet was on the PET fabric surface at 10s. It is shown that with the WCA of 0° at 10s, the PET fabrics maintain excellent hydrophilicity after dyed with disperse dyestuff.. It can also be seen that the hydrophilic property of the modified PET fabric dyed with disperse dyeing was not affected. The dyeing performance of the fabric was improved greatly, while the hydrophilic property kept well.

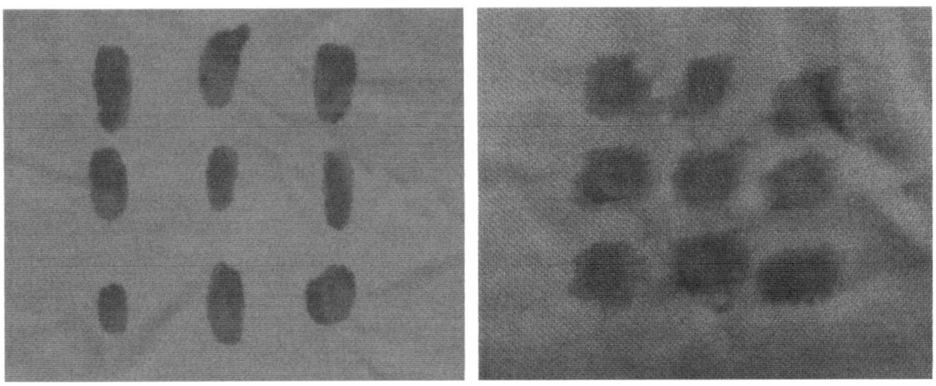

a b

Figure 2. The hydrophilicity of modified and dyed PET fabrics a. dyed by disperse red E-4B; b. dyed by disperse blue 2BLNG-L.

Table 3. Hydrophilicity of the PET fabric modified by H_2SO_4 after dyeing.

WCA at 10s /($^{\circ}$)	Concentration of H_2SO_4/(g/L)					
	0	4.0	5.0	6.0	7.0	8.0
Modified PET fabric	157.36	0	0	0	0	0
Modified PET fabric dyed by disperse blue 2BLNG-L	157.31	0	0	0	0	0
Modified PET fabric dyed by disperse red E-4B	157.43	0	0	0	0	0

3.5. Wash fastness of fabrics

Table 4 lists the wash fastness of original and modified PET fabrics after dyeing. In wash fastness of red and blue dyes, staining of original PET fabrics was very good (grade 5). Compared to the original PET fabrics, the Wash fastness of modified PET fabrics and dyed by disperse blue 2BLNG-l only decreased a little to grade 4, and the grade of wash fastness for modified PET fabrics dyed by disperse red E-4B was 4-5. During the washing process, dye molecules would migrate from interior to surface of the fibers.

Table 4. Color fastness to washing of original and modified PET fabrics after dyeing.

Samples	Disperse blue 2BLNG-l		Disperse red E-4B	
	Original PET	Modified PET	Original PET	Modified PET
Wash fastness	5	4	5	4-5

4. Conclusion

In this work, the effect of sulfuric acid modified on the dyeing properties of PET fabrics was discussed. The dyeing property of PET fabric modified by sulfuric acid is better than that of original PET fabric, and the value of K/S increased with the increase of the modifying concentration of sulfuric acid. The levelling property and wash fastness of PET fabrics maintains well after the modification, without significant change in the color of the fabric after dyeing. And the hydrophilic property of modified PET fabrics keeps excellent after dyeing.

Acknowledgments

This work was supported by the Key Application and Development Project of the Chongqing Science & Technology Commission (Grant No. CSTC2014yykfB50002) and the Chongqing City Graduate Research and Innovation Project (Grant No. CYS14059).

References

[1] Chuang M J and Chu A K 2011 Formation of a high hydrophilic/hydrophobic contrast surface on PET substrates by ECR generated sulfur hexafluoride plasma *Appl. Surf. Sci.* **257** 3943-47.

[2] Zemlji L F, Tkav T, Vesel A and Sauperl O 2013 Chitosan coatings onto polyethylene terephthalate for the development of potential active packaging material *Appl. Surf. Sci.* **265** 697-703.

[3] Mckeen LW 2010 *Fatigue and Tribological Properties of Plastics and Elastomers (Second Edition)* (Amsterdam, Holland: Elsevier) p 99-147.

[4] Liu X D, Sheng D K, Gao X M, et al. 2013 UV-assisted surface modification of PET fiber for adhesion improvement *Appl. Surf. Sci.* **264** 61-69.

[5] Liang M N, Yao J R, Chen X, Huang L and Shao Z Z 2013. Silk fibroin immobilization on poly(ethylene terephthalate) films: Comparison of two surface modification methods and their effect on mesenchymal stem cells culture *Mat. Sci. Eng. C* **33** 1409-1416.

[6] Zaman M, Liu H B and Xiao H N, et al. 2013 Hydrophilic modification of polyester fabric by applying nanocrystalline cellulose containing surface finish *Carbohydr. Polym.* **91** 560-567.

[7] Xu F, Zhang G X, Zhang F X, Zhang Y S 2015 Facile preparation of super-hydrophilic poly(ethylene terephthalate) fabric using dilute sulfuric acid under microwave irradiation *Appl. Surf. Sci.* **349** 437-444.

Fabrication of sisal fibers/epoxy composites with liquid crystals polymer grafted on sisal fibers

Q Y Luo, S R Lu[1], L F Song and Y Q Li

Key Laboratory of New Processing Technology for Nonferrous Metals and Materials, Ministry of Education, School of Material Science and Engineering, Guilin University of Technology, Guilin 541004, P.R. China

E-mail: lushaor@163.com

Abstract. In this word, microcrystalline cellulose fibers (MCFs), extracted from sisal fibers, were treated with function end-group hyperbranched liquid crystals (HLP). This work brought some insights into the successful surface modification in epoxy composite with HLP. The HLP-MCFs/epoxy composites are studied systematically. The HLP - MCFs/epoxy composites were studied by Fourier transform infrared spectroscopy (FT-IR), polarizing microscope (POM), X-ray photoelectron spectroscopy (XPS) and mechanical properties analysis. The results reveal that the reinforcement of EP composites was carried out by adding HLP-MCFs. In particular, with 1.0 wt% filler loading, the flexural strength, tensile strength, impact strength and flexural modulus of the HLP-MCFs/EP composites were increased by 60 %, 69%, 130 %, and 192 %, respectively. It anticipates that our current work exploits more efficient methods to overcome the few nature fiber/polymer (NPC) adhesion in the interface region and provides implications for the engineering applications of the development of NPC.

1. Introduction

The tendency of natural fibers as a sustainable materials have attracted many researchers because dwindled fossil resources.[1] It is particularly desirable that composites modified by natural fibers for the development of environmentally sustainable materials, because natural fibers offer more advantages than synthetic fibers such as its low cost, biodegradability, and renewability.[2] However, the main drawback are the little compatibility and poor interfacial adhesion between natural fibers with the polymer matrix. Accordingly, the affection of composites with natural fibers is unsatisfactory, and its modifications by specific treatments are certainly necessary. To date, several method of modified the surface of natural fibers are proposed, such as alkalization, acetylation, and cyanoethylation. Additionally, hyperbranched polymer is another efficient way to improve the interfacial bonding with matrix and graphene.[3,4] The end-groups of hyperbranched polymer with three-dimensional space structures interact with functional groups of epoxy resins(EP), which can effectively improve the compatibility between fillers and EP, and enhance the toughness of the EP.[5]

In the present work, the MCFs, extracted from sisal fibers, were modified by HLP to improve the compatibility and interfacial adhesion with epoxy matrix. Because HLP has dendritic architecture and abundant functional groups which can react to epoxy matrix via covalently incorporation, furthermore, it contain liquid crystal characters. The introducing of HLP is hoped to improve MCFs/EP composites with excellent mechanical properties. Furthermore, the MCFs derived from the eco-friendly and renewable of sisal fibers, would replace synthetic fibers to apply in polymer composite reinforcement and endow the NPC with some unexpected performances.

2. Materials

MCFs, as a reinforcement, were extracted from Sisal fibers. Yueyang Chemical Plant, China afford epoxy resin (DGEBA, E-44, epoxy value ¼ 0.44) to complete the research. triphenylphosphine (TPP) use as catalyst in this study. g-Aminopropyl-triethoxysilane (g-APTEOS) was coupling agent in this study. We also purchased drugs from Aladdin Chemistry Co., Ltd, such as The hardener (4,4-Diaminodiphenylsulfone), 3,5-diaminobenzoic acid (DABA). 4,4-Biphenol was obtained from Tokyo Chemical Industry Co., Ltd. N,N-dimethylacetamide (DMAc), acetone, Piperidine, N,N-dimethylformamide (DMF), N-methyl-2-pyrrolidone (NMP), used as a solvent, were obtained from Sinopharm Chemical Reagent Co., Ltd., China. furthermore, Sinopharm Chemical Reagent Co., Ltd., China also furnished necessary laboratory reagents to complete the study. such as, methanol, Lithium chloride (LiCl). The rest of the sample were provided by Guang Zhou Jin hua du Chemical Reagent Co., Ltd. in China. Such as xylene (Xyle), NaOH, 36%HCl, disodium tetraborate decahydrate, 30%H_2O_2, 65–68% HNO_3, 99.5% CH3COOH.

3. Experimental details

As shown in scheme1, the HLP-MCFs were synthesized as follows. Firstly, the MCFs were modified by KH550. Then the hyperbranched aromatic polyamides modified MCFs (BHAP-MCFs) were obtained. The last, the liquid crystal BP_2 were grafted on the surface of the BHAP-MCFs, so the hyperbranched liquid crystals modified MCFs (HLP-MCFs) were obtained. Afterward, HLP-MCFs were used to modified epoxy resin.

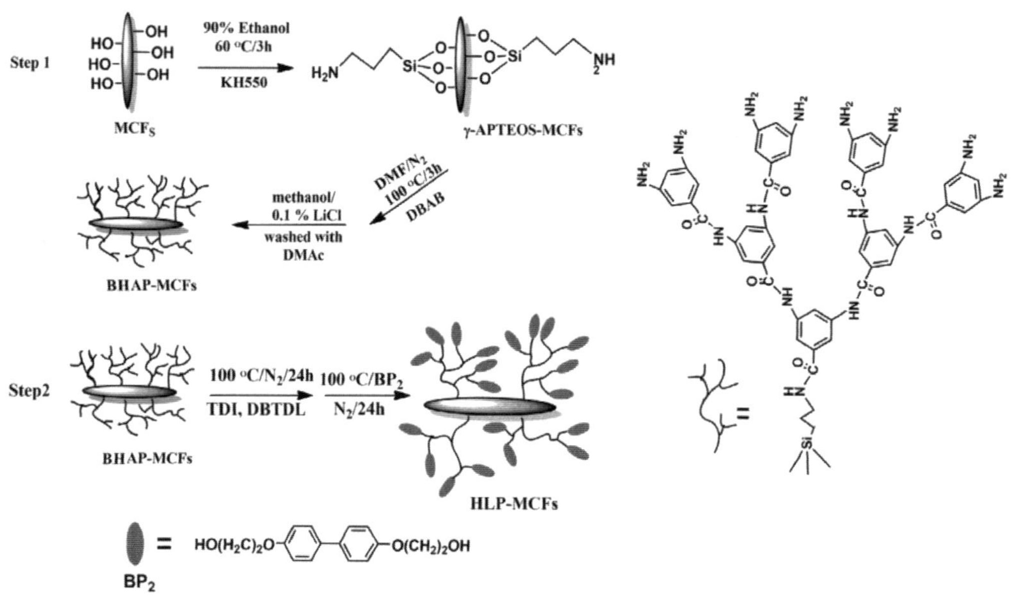

Scheme 1. Preparation process of HLP-MCFs

4. Results and discussion

As shown in Fig.1, FTIR of MCFs shows the peaks at 3343 cm^{-1}(due to -OH stretching) and 2920 cm^{-1}(arised from alkane C-H stretching) which show the composition of MCFs. The bands present at 1731 cm^{-1} and 1638 cm^{-1} pointed out that the cellulose is hygroscopic and forms strong hydrogen bonds. FTIR of BHAP-MCFs exhibits bands at 1548 cm^{-1}, 1344 cm^{-1} owing to -NHCO- and amine stretching, respectively. 1604 cm^{-1} and 1434 cm^{-1} caused by benzene ring, which indicates the reaction between BHAP and MCFs. Furthermore, in the spectrum of HLP-MCFs, the peak at 1605 cm^{-1}, 1498 cm^{-1}, 1249 cm^{-1} is the characteristic of the amine N-H deformation, aromatic hydrocarbon C=C stretching and the -NHCOO-, respectively. The absorptions at 1093 cm^{-1}, 1054 cm^{-1} are arising from the amine C-N stretching. These changes in the characteristic peaks prove that BP_2 have successfully grafted on the surface of BHAP-MCFs.

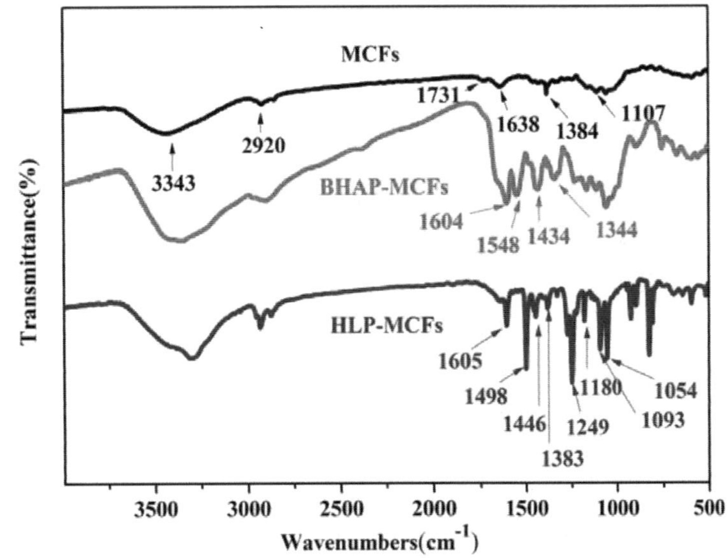

Figure 1. FTIR spectrum of the untreated and treated MCFs

The surface compositional analysis of HLP-MCFs have been performed by the X-rays photoelectron spectroscopy (XPS) in Figure 2. The detected elements were carbon (C), oxygen (O), nitrogen (N), and silicon (Si) which also indicates the HLP grafted on the surface of MCFs successfully.

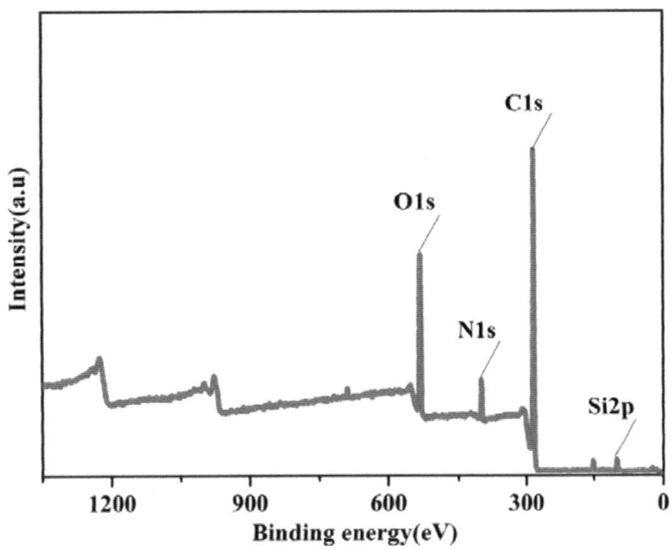

Figure 2. XPS analysis of HLP-MCFs

As shown in Fig. 3, the stress-strain curves of the pure epoxy and HLP-MCFs/epoxy composites were carried out. Compared with the pure epoxy, it was found that the HLP-MCFs/epoxy composites exhibit a higher elongation at break and higher stress, because of the higher cross-linking densities of the HLP-MCFs/epoxy composites and the more efficient interfacial interactions between the

HLP-MCFs and epoxy.[6] With the increment of HLP-MCFs, the stress of these composites was gradually improved. When HLP-MCFs content is 1.0 wt%, reached its highest value. Additionally, there is a maximum elongation at break for HLP-MCFs/EP composites with 0.5 wt% HLP-MCFs.

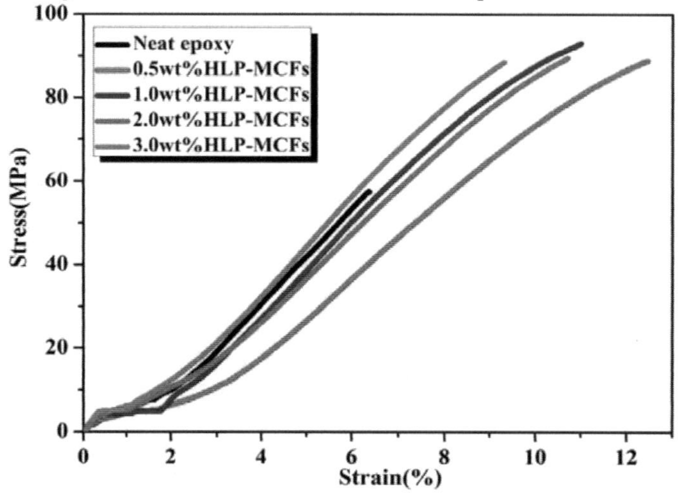

Figure 3. Tensile test of pure epoxy and HLP-MCFs/epoxy composites

As shown in Fig.4 and Fig.5, the mechanical property of the HLP-MCFs/epoxy composites such as flexural, impact, tensile strength and flexural modulus have increased significantly, and then decrease slightly with the increment of HLP-MCFs contents, because of the cross-linking densities of HLP-MCFs/EP composites. The higher cross-linking densities of the HLP-MCFs/EP composites lead to the more efficient interfacial interactions between the HLP-MCFs and EP matrix and stronger network structure of HLP-MCFs/EP composites, which can dramatically improve the mechanical properties of HLP-MCFs/EP composites.[7] Also, the improvement of various strength are attributed to the rigid benzene ring structure of HLP-MCFs[8] and the orientation of the mesogenic units, which hindered the development of the silver pattern. Therefore, a small amount of HLP-MCFs (1.0 wt%), as a filler, can dramatically affect the mechanical properties of HLP-MCFs/EP composites.[7] In short, a small amount of HLP-MCFs can significantly improve the strength and toughness of the EP.

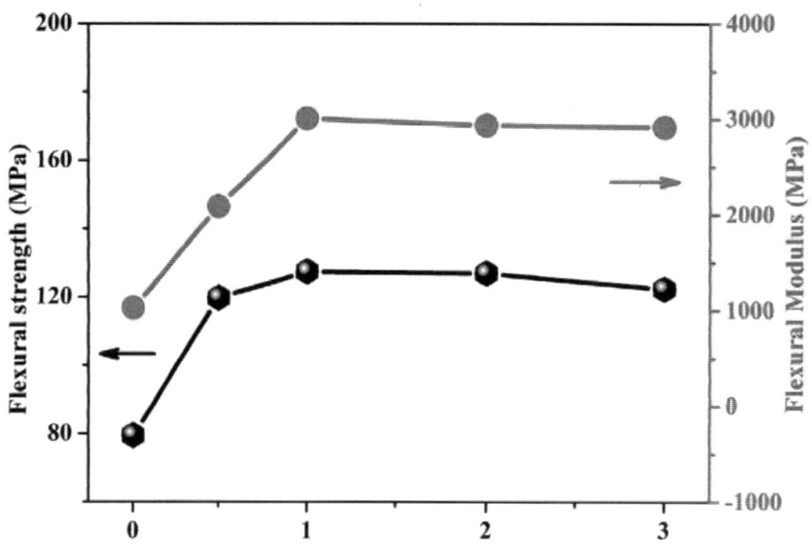

Figure 4. Flexural strength and modulus of pure epoxy and HLP-MCFs/epoxy composites

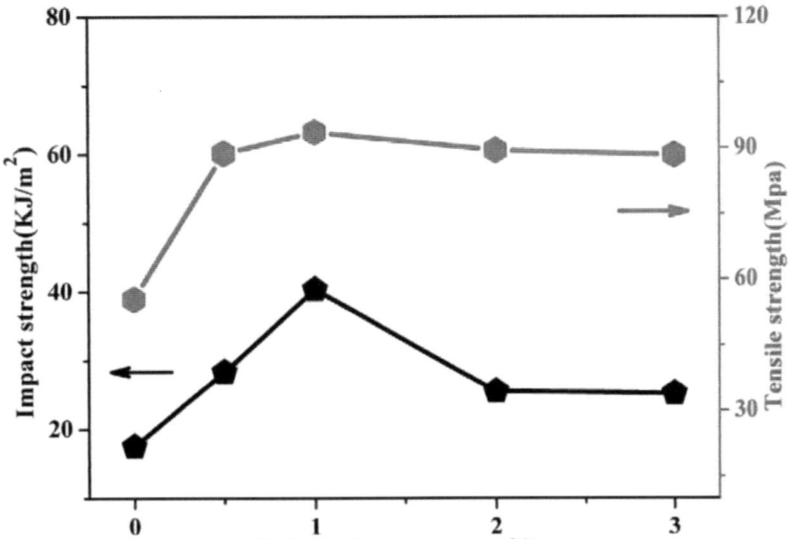

Figure 5. Impact, tensile strength of pure epoxy and HLP-MCFs/epoxy composites

5. Conclusion

This work brought some insights into the successful surface modification in epoxy composite with HLP. The HLP-MCFs/epoxy composites are studied systematically. The results reveal that the reinforcement of EP composites was carried out by adding HLP-MCFs. For instance, with 1.0 wt% filler loading, the flexural strength, tensile strength, impact strength and flexural modulus of the HLP-MCFs/EP composites were increased by 60 %, 69%, 130 %, and 192 %, respectively.

Acknowledgment

The authors are thankful to the National Natural Science Foundation of China (51463007 and 51163004), the Natural Science Foundation of Guangxi Province, China (2014GXNSFDA118006) which supply sufficient financial resource to complete the research.

References

[1] Sreenivasan V S, Ravindran D, Manikandan V and Narayanasamy R 2012 *Mater. Des.* **37** 111
[2] Saheb D N and Jog J P 1999 *Adv. Polym. Technol.* **18** 351
[3] Xiao X E, Lu S R, Qi B, Zeng C, Yuan Z K and Yu J H 2014 *RSC Adv.* **4** 14928
[4] Zeng C, Lu S R, Xiao X E, Gao J, Pan L L, He Z H and Yu J H 2014 *Polymer Bulletin.* **72** 453
[5] Lu S R, Ling R H, Luo C X, Li S R and Huang B 2013 *Plast. Rubber Compos.* **42** 361
[6] Drechsler P and Federle W 2006 *J. Comp. Physiol. A -Neuroethol. Sens. Neural Behav. Physiol.* **192** 1213
[7] Maeda K, Bismarck A and Briscoe B J 2005 *Wear.* **259** 651
[8] Coleman J N, Khan U and Gun'ko Y K 2006 *Adv. Mater.* **18** 689

Research on the melt impregnation of continuous carbon fiber reinforced nylon 66 composites

M Y Jia[1], C X Li, P Xue, K Chen and T H Chen

Institute of Plastics Machinery and Engineering, Beijing University of Chemical Technology, Beijing 100029, China

E-mail: jiamy@mail.buct.edu.cn

Abstract. Impregnation mold of continuous carbon fiber reinforced thermoplastic composites was designed and built in the article. Based on the theory of fluid mechanics and Darcy's law, a model of the melt impregnation was also established. The influences of fiber bundle width and impregnation pins' diameter on the impregnation degree were studied by numerical simulation. Continuous carbon fiber reinforced nylon 66 composites were prepared. The effects of coated angle and impregnation mold temperature on the mechanical properties of the composites were also described.The agreement between the experimental data and prediction by the model was found to be satisfactory.

1. Introduction

Demanding for advanced materials in aerospace and automobile industry applications has led to the emerge of composites. Composites were a kind of heterogeneous material, which consisted of two or more substances of different physical or chemical properties. In the 1960s, the short fiber reinforced thermoplastic composites had realized industrial production. However, the composites can not be applied in the force bearing component and structural use. Therefore, continuous fiber reinforced thermoplastic composites (CFRP) were developed, which was regarded as a novel material consisted of carbon or glass fibers and polymer matrix. It was proved that thermoplastic composites were superior to thermosetting counterpart for the superior mechanical properties in practical application, such as short cycle-time, high strength, high elastic modulus, light weight and so on. However, the thermoplastic composites were too difficult to be impregnated because of the high viscosity. The traditional impregnation processes of manufacturing continuous fiber reinforced thermoplastic composites were powder impregnation, solvent impregnation as well as melt impregnation [1]. Among these methods, the process which had got some attention was melt impregnation process. During the impregnation process, the fiber bundles were pulled through the guide pins, which were under the molten thermoplastic. Due to the high viscosity of the molten thermoplastic, it was difficult for the molten polymer completely penetrate into the fiber bundle. To solve the problem, many studies had been done to research how to reduce the viscosity and realize the better impregnation. Gaymans [2] studied that the polypropylene impregnated the glass fiber roving with pins and the results showed that the diameter of the pin, tension and melt viscosity were the key factors during the impregnation. Peltonen [3] studied the effects of the temperature, time, pressure, the length of the contact surface, tension force and MFR-value of PP on the impregnation. Nygard [4] pointed out that using the radial slit die, which included different vibration techniques and a silt die, was better than the cross-head impregnation mold. Meanwhile, they found that with the use of the radial slit die, the high volume fraction was obtained and the fiber-matrix content was controlled. Kumar [5] prepared the PP/LGF reinforced particles by melt impregnation device designed by themselves. Asai [6] used the screw to take the molten resin into the mold, then take the use of impregnation unit by specific design to achieve the full impregnation of fiber. Hatano [7] manufactured continuous fiber reinforced thermoplastic composites with the use of a combination of pull extrusion molding and weaving techniques. Cogswell [8] studied the impregnation technology of thermoplastic composites and the impregnation effect was best when the viscosity of melt was less than $30Pa{\cdot}s$. Törmälä [3] discussed the effect of temperature on the impregnation degree by the experiment of glass fiber reinforced polypropylene, the results showed that the impregnation degree of the polymer was improved by 20% when the temperature of the polymer increased from 180℃ to 230℃. Wevers [2] studied the effect of traction on impregnation effect by the experiment of continuous glass fiber reinforced polypropylene. It was shown that the impregnation degree increased with the increase of traction force. Kim [9] came to the conclusion from the experiment of carbon fiber reinforced polyetheretherketone that the impregnation effect improved with the increase of pressure.

In this paper, a new impregnation model was developed not only discussed the variety between the carbon fiber roving and the pins but also analyzed the influence on the mechanical properties of composites, process parameters and structure parameters on the impregnation efficiency.

2. Materials and equipment

Carbon fiber (T700SC) with Number of Filament being 12000 was supplied by Toray Industries, Inc. Tokyo, Japan. Nylon 66 (A-1) was supplied by Rhodia polyamide (Shanghai) Co., Ltd. Twin-screw extruder (KS20) with diameter of single screw being 20mm was supplied by Kunshan Kesun Plastic & Rubber Machinery Co., Ltd. Injection molding machine (PT-130) was supplied by L.K. Group, Hong Kong, China.

3. Establishment and numerical simulation of impregnation model

3.1. Establishment of impregnation model

The impregnation process of carbon fiber bundle was displayed in Figure 1. There were a series of parallel pins in the impregnation device. The carbon fiber bundle can be fully impregnated by the molten polymer due to the tension action when the carbon fiber bundle pulled out of the impregnation device. After fully impregnated, the carbon fiber bundle pulled through an exit mold of a diameter of 3mm. The fiber passed through a series of parallel pins, the impregnation would occur on the contact surface, as shown in Figure 2. The exit mold could control the fiber content. Impregnation was accomplished by the polymeric matrix penetrate into surrounding fiber bundles.

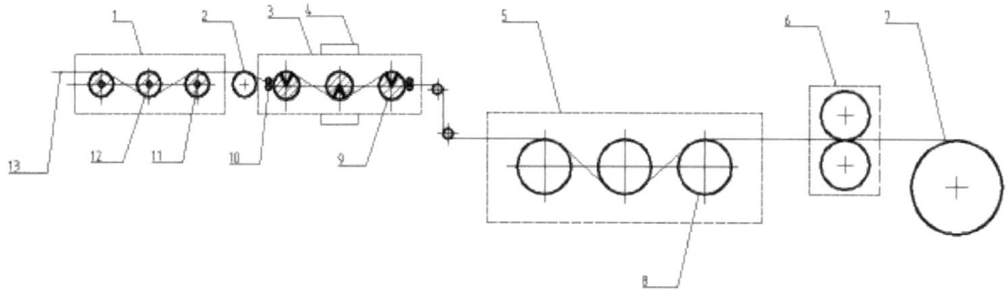

Figure 1. Melt impregnation process.

(1) dispersion device; (2) eliminate electrostatic device; (3) the radial impregnating mold; (4) twin-screw extruder; (5) secondary impregnating device; (6) double roller traction; (7) winding device; (8) the second tension roller; (9) the first tension roller; (10) floating roller; (11) heating rod; (12) dispersion rollers; (13) continuous fiber.

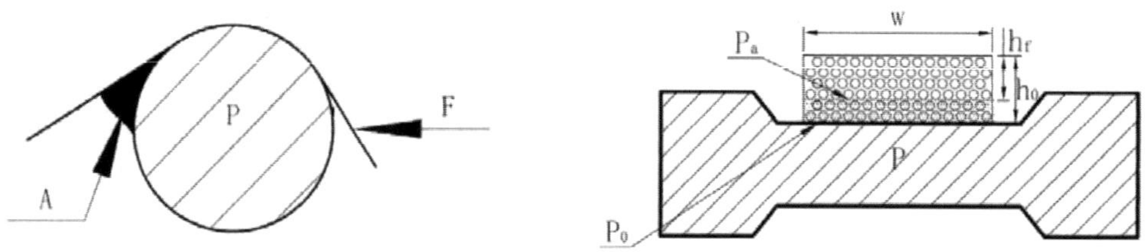

(a) The radial cutaway view of pin (b) The axial cutaway view of pin

Figure 2. The geometric model of melt impregnation.

According to the previous research [10, 11], the melting rate of polymer that penetrated into the fiber bundle was considered as the most important factors in the process of melt impregnation. The rate that the molten polymer penetrated into the fibers had been analyzed by Darcy's law [12], as shown in the equation (1), which was an empirical relation from experiments on water flow through sand beds. We assumed that Darcy's law could be used in the vertical flow of the thermoplastic composites,

$$V_h = -\frac{K}{\mu}\frac{dP}{dh} \tag{1}$$

Where V_h is the vertical velocity of the thermoplastic composites, P is the applied pressure, h is the designates the vertical position. The mass flow rate through the vertical position h is,

$$m = \rho\left(-V_h\right)W\varepsilon \tag{2}$$

And K is the permeability coefficient of the fiber structure, which can be calculated by the Kozeny-Carman equation [13],

$$K = \frac{D_f^2\varphi^3}{16k\left(1-\varphi\right)^2} \tag{3}$$

Where D_f is the fiber diameter, φ is the porosity of the carbon fiber bundle and k is a constant related to polymer viscosity, fiber structure and fiber type, which can be calculated from the following equation [3],

$$k=\frac{2\varphi^3}{(1-\varphi)\left\{\ln\left(\frac{1}{1-\varphi}\right)\left[\frac{1-(1-\varphi)^2}{1-(1-\varphi)^2}\right]\right\}} \tag{4}$$

Equations (1) and (2) are combined to get the following equation,

$$\frac{m}{\rho W\varepsilon}=\frac{K}{\mu}\frac{dP}{dh} \tag{5}$$

The pressure at the $h=h_0$ is the applied pressure, P_0. The following equation is obtained by the integration of equation (5) from h_0 to h_f,

$$\frac{m}{\rho W\varepsilon}\left(h_f-h_0\right)=\frac{K}{\mu}\left(P_a-P_0\right)\frac{dh_f}{dt} \tag{6}$$

Where hf is the location of the thermoplastic composites where the pressure is the atmospheric pressure P_a. The mass flow rate can be expressed in term of dh_f/dt as,

$$m=\rho W\varepsilon\left(-\frac{dh_f}{dt}\right) \tag{7}$$

Equations (6) and (7) are combined to get the following equation,

$$-\left(h_f-h_0\right)dh_f=\frac{K}{\mu}\left(P_a-P_0\right)dt \tag{8}$$

The following equation is obtained by the integration of equation (8) along with the initial condition ($h=h_0$ at $t=0$),

$$t=\frac{\mu h_0^2}{2K\left(P_0-P_a\right)}\left[1+\left(\frac{h_f}{h_0}\right)^2-2\frac{h_f}{h_0}\right] \tag{9}$$

The impregnation degree D_{imp} can be defined as the ratio between the number of impregnated fibers and the total number of fibers. If we assume that the fibers are uniformly distributed, the impregnation degree can be expressed as,

$$D_{imp}=\frac{area\ impregnated}{total\ area}=\frac{Wh_0-Wh_f}{Wh_0}=1-\frac{h_f}{h_0} \tag{10}$$

Substituting equation (10) into equation (9). The impregnation degree is obtained as a function of time, applied pressure, viscosity, initial bundle vertical and permeability,

$$t=\frac{\mu h_0^2}{2K\left(P_0-P_a\right)}\left[1+\left(1-D_{imp}\right)^2-2\left(1-D_{imp}\right)\right] \tag{11}$$

We assumed that the cross-sectional is the rectangular shape, the total cross-sectional area equal to the cross-sectional area of the mono-filaments and the cross-sectional area impregnated by thermoplastics composites. The relationship between the impregnation thickness (h_0) and the total number of carbon fibers (n_f) and the diameter of the carbon fiber (D_f) and the fiber bundle width (W) and the porosity (φ) can be calculated from geometric considerations. This is shown schematically in Figure 2 and expressed mathematically in equation (12),

$$h_0=\frac{\pi}{4}\frac{n_f D_f^2}{W\left(1-\varphi\right)} \tag{12}$$

The total coated time (t) has relations with the coated angle (θ), the number of the pin (N), the pin diameter (D), the traction speed (U). This is expressed mathematically in equation (13),

$$t=\frac{\theta D}{2U} \tag{13}$$

Bates [10] had studied the melt pressure while a glass fiber pulled over a pin in polyamide 66 or polypropylene polymers melt pool. The results showed that the melt pressure is,

$$P_0=\frac{2T}{WD} \tag{14}$$

Equations (11) to (14) are combined to get the following equation,

$$t=\frac{\theta D}{U}=\frac{\mu}{K\left(\frac{2T}{WD}-P_a\right)}\left[\frac{\pi}{4}\frac{n_f D_f^2}{W\left(1-\varphi\right)}\right]^2\left[1+\left(1-D_{imp}\right)^2-2\left(1-D_{imp}\right)\right] \tag{15}$$

The configuration of pins is shown in Figure 3.

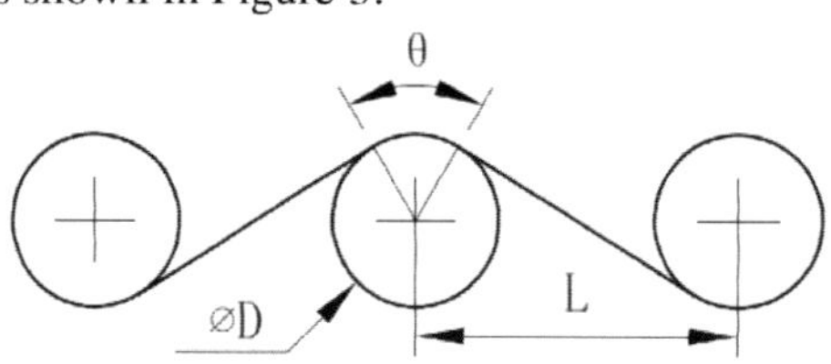

Figure 3. The configuration of pins

According to geometric knowledge, we get the relationship among the coated angle (θ), the pin diameter (D) and the center distance between the two pins (L), written as,

$$\theta = 2\arcsin\left(\frac{D}{L}\right) \tag{16}$$

Substituting equation (16) into equation (15), the following equation can get,

$$D_{imp} = \frac{4W(1-\varphi)}{\pi n_f D_f^2}\sqrt{\frac{2KD}{\mu U}\left(\frac{2T}{WD} - P_a\right)\arcsin\left(\frac{D}{L}\right)} \tag{17}$$

Equation (17) provides a relationship between the impregnation degree and the process variables in the impregnation process, namely, physical parameters, process parameters and structure parameters.

Equations (1) and (17) showed that the impregnation degree is determined by numbers of factors, including the permeability coefficient of fiber structure, the fiber bundle width, the pin diameter, the distance between two pins, the line speed of the fiber, the fiber pretension and melt viscosity. Among these factors, the permeability coefficient and Kozeny-Carman constant can be calculate based on the porosity and the melt viscosity, which regarded as constants for a Newton fluid. Therefore, in the following section, the influence of other impregnation parameters would be mainly discussed.

To quantitatively analyzed the influence of the line speed of fiber, a T700 carbon fiber through a high-viscosity melt was impregnated. The melt viscosity was assumed to be 50*Pa·s*, the fiber bundle width was 6*mm*, the pretension was 20*N*, the pin diameter was 8*mm*. These parameters were used for the calculation below unless otherwise specified.

3.2. Numerical simulation analysis

3.2.1. Effect of fiber bundle width on impregnation degree

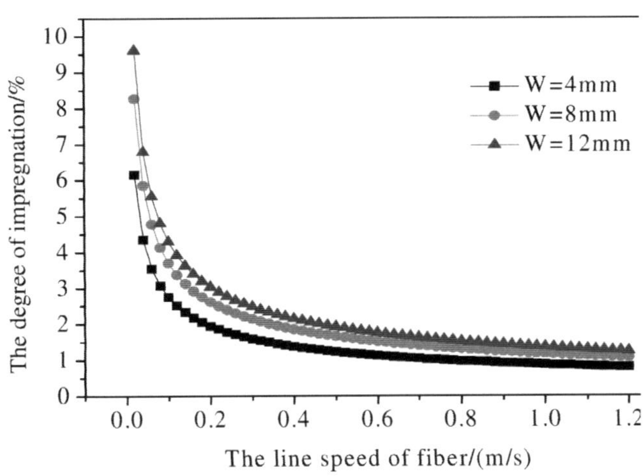

Figure 4. Degree of impregnation as a function of the fiber line speed and the fiber bundle width.

Figure 4 was the impregnation degree calculated as a function of the line speed of the fiber at different fiber bundle width of 4*mm*, 8*mm* and 12*mm*. For a single curve, such as the 4*mm* one, the impregnation degree decreased sharply at a low speed, reached the steady state at about 0.1*m/s*. The figure showed when the line speed of the fiber was greater than 0.1*m/s*, the degree of impregnation had a little changes. This can be explained as the melt flow. There are two factors affected the impregnation process, the amount of the melt and the impregnation duration. With the increase of the line speed of the fiber, the amount of the melt flowing into the fiber bundle decreased, while the duration shortening of impregnation process. For a low line speed of the fiber, more and more polymer penetrated into the fiber bundle, resulting in a high degree of impregnation due to the sufficient time. On the contrary, the duration of impregnation decreased sharply resulting in a relatively low degree of impregnation, but in the actual production, the relationship between the performance of the product, the production efficiency and the line speed of the fiber should be considered. As a result, in the premise of ensuring the performance of the product and the production efficiency, the maximum speed could be chosen.

With the increase of the fiber width, the distance between the mono-filament fibers became wider, resulting in an enhancement of the impregnation process. The degree of the impregnation increased, while the steady state was almost unchanged.

3.2.2. Effect of tension on impregnation degree

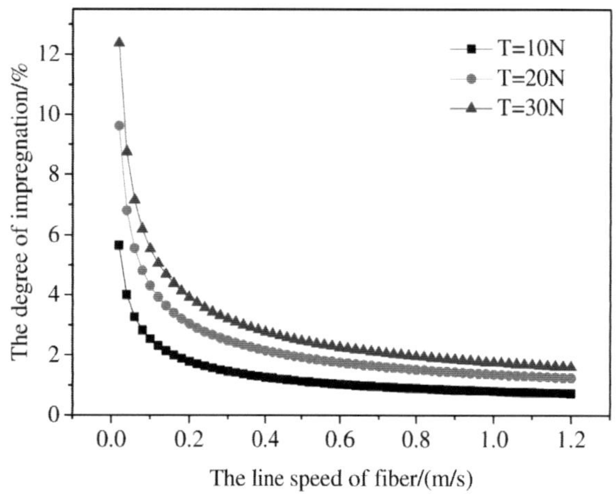

Figure 5. Degree of impregnation as a function of the fiber line speed and the fiber tension.

Figure 5 showed the influence of different tensions on impregnation degree. The calculation was done by the different tensions of 10*N*, 20*N* and 30*N*, while fixing other parameters mentioned above. Increasing the value of tension, degree of impregnation was improved, which could be proved in the experimental result reported by Peltonen [3] and Gaymans [2]. The starting value of the steady state was unchanged. It was indicated that the value of the stable state did not depend on the fiber bundle and the tension. Also, with the increase of fiber tension, the pressure between the fiber bundle and the pins increased according to equations (1) and (2), which could also enhance the degree of impregnation according to the experimental results reported by Kim [9].

3.2.3. *Effect of pin diameter on impregnation degree*

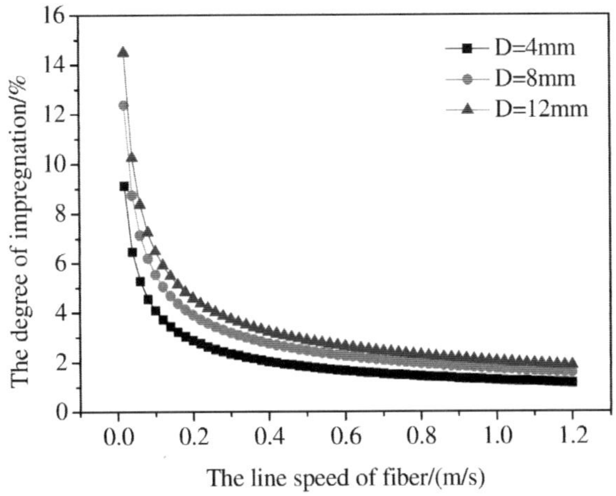

Figure 6. Degree of impregnation as a function of the fiber line speed and the fiber diameter.

The influence of different pin diameter on the impregnation degree was shown in Figure 6. The calculation was done by the different pin diameters of 4mm, 8mm and 12mm. By increasing the value of the pin diameter, the contact length of the fiber bundle and the pins increased resulting in enhancing the degree of the impregnation, which could be proved in the experimental results reported by Peltonen [14]. The amount of polymer impregnated with one pin was limited as we have seen above. Several pins were often used to strengthen the impregnation. According to the model in this paper, the impregnation degree of each pin and the total impregnation degree of multiple pins could be calculated.

The analyses discussed above was only based on the ideal boundary conditions. For a real case, there existed a minimum pressure between the fiber bundle and the pins. When the pressure was lower than the atmospheric pressure (P_0), the penetration could not be realized and the equation (15) would have little practical effect, so the impregnation mold should meet the following condition,

$$\frac{2T}{WD} - P_a > 0 \tag{18}$$

Consequently, it is necessary to optimize the fiber bundle width in practical application.

4. Results and discussion

4.1. *Effect of coated angle on the mechanical properties of composites*

By the melt impregnation mold and impregnation mechanism, continuous carbon fiber reinforced nylon 66 prepreg was prepared. The effect of configuration of impregnation pins on coated angle was studied by theoretical analysis. The effect of coated angle on the mechanical properties of fiber reinforced thermoplastic composites was also researched experimentally and compared with the theory of melt impregnation.

4.1.1. Effect of configuration of impregnation pins on coated angle. According to the different configurations of impregnation pins, the total coated angle and average coated angle could be obtained. The analysis results were shown in Table 1 with different the configuration of impregnation pins in Figure 7.

Table 1. Effect of configuration of impregnation pins on coated angle

Configuration of impregnation pins	a	b	c	d	e	f
Total coated angle/°	0	43	58	92	276	416
Average coated angle/°	0	11	19	31	40	60

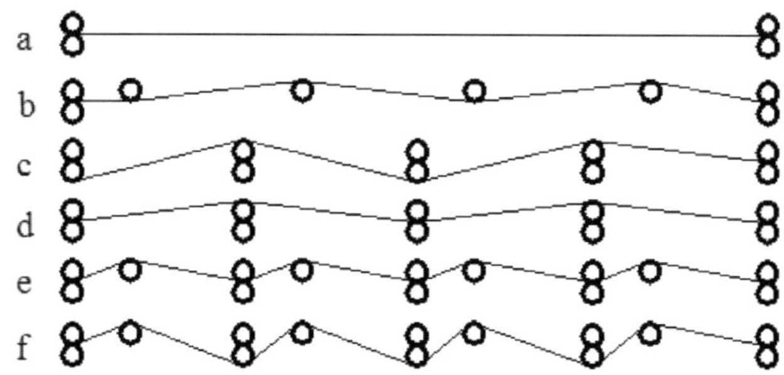

Figure 7. The configuration of impregnation pins

Under the action of a certain tension, the fiber bundles would repeatedly collapse when it passed through the impregnation pins so that the fiber bundles could be dispersed on the surface of impregnation pins. The larger the coated angle between fiber bundles and impregnation pins, the greater the dispersion effect. When the coated angle was zero, there would be a fiber accumulation phenomenon, which resulted in a poor impregnation effect.

4.1.2. Effect of coating angle on mechanical properties of composites

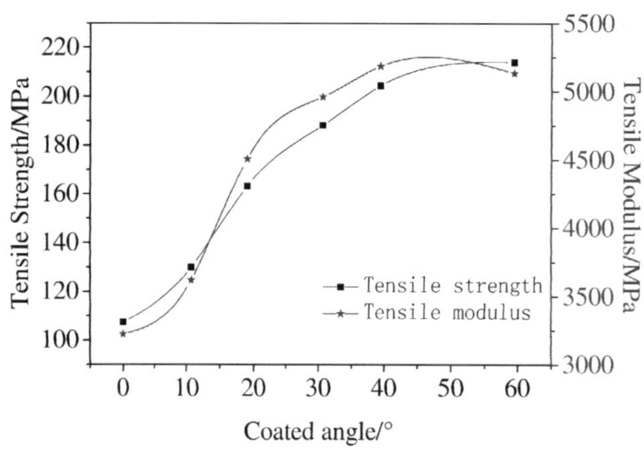

Figure 8. The influence of coating angle on tensile properties

Figure 8 was the influence of coated angle on tensile properties of composites. It can be seen from the figure that the tensile properties of carbon fiber reinforced nylon 66 composites improved with the increase of the coated angle. When the coated angle raised from 0°to 11°, the tensile strength increased from 107*MPa* to 130*MPa*, which improved by 21.5%. When the coated angle was increased to 60°, the corresponding tensile strength was 214*MPa*, which improved by 100%.

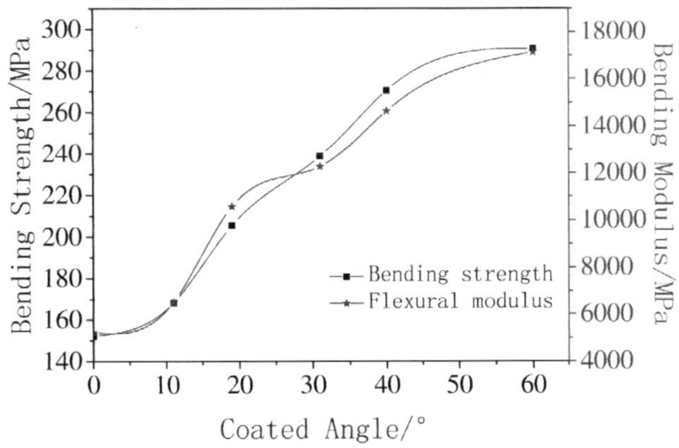

Figure 9. The influence of coated angle on flexural properties

According to the Figure 9, it can be found that the flexural properties of carbon fiber reinforced nylon 66 composites improved with the increase of the coated angle. When the coated angle raised from 0° to 11°, the bending strength increased from 107*MPa* to 130*MPa*, which improved by 10.5%. When the coated angle was increased to 60°, the bending strength was 290*MPa*, which improved by 91%.

With the increase of the value of coated angle and the configuration changes of impregnation pins, the fiber bundles across the surface of the impregnation pins were dispersed, which improved the mechanical properties of composites.

4.1.3. Theoretical model validation of coated angle. The line speed of the fiber was 0.09m/s and the pin diameter was 8mm in the experiment.

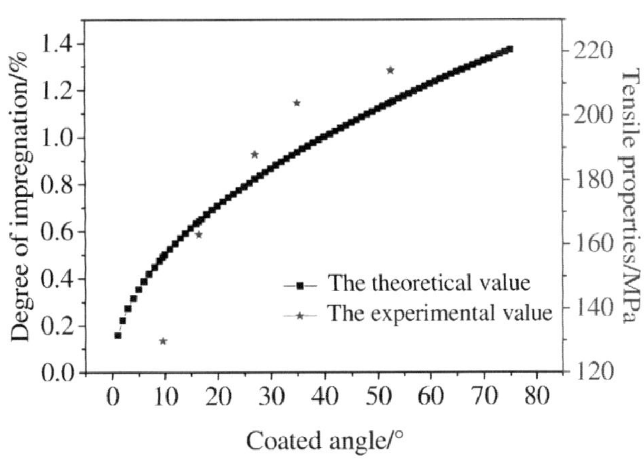

Figure 10. The influence of the coated angle for the degree of impregnation and tensile properties.

As shown in Figure 10, the influence of the coated angle on the degree of impregnation and tensile properties was described. It can be seen that the degree of impregnation increased with high coated angle. Corresponding, there was also a continual increase in tensile properties as the degree of impregnation. Through the comparison between theoretical analysis and contrast test, it was proved that the theoretical analysis and theoretical model had rationality.

4.2. Effect of impregnation mold temperature on mechanical properties of composites

To study the influence of the temperature of impregnation mold on the mechanical properties of carbon fiber reinforced nylon 66 composites (CF/PA66), screw speed with 75*r/min*, the feed rate with 12*r/min*, traction speed with 5.4*r/min* and pressure maintaining time with 30*s* were set in the experiment. Five groups of different temperature of twin-screw (area 1 to area 4) and impregnation mold (area 5 to area 7) were selected, as shown in Table 2:

Table 2. The temperature of twin screw and impregnation mold/℃.

	Area 1	Area 2	Area 3	Area 4	Area 5	Area 6	Area 7
1	250	255	260	265	270	270	270
2	250	260	270	275	280	280	280
3	260	270	275	285	290	290	290
4	260	280	290	295	300	300	300
5	260	280	290	300	310	310	310

Figure 11 illustrated the influence of impregnation mold temperature on the tensile properties of composites. The figure revealed that the tensile properties of the carbon fiber reinforced nylon 66 increased with the enhance of the temperature of the impregnation mold firstly and then tend to be stable. When the mold temperature increased from 270℃ to 290℃, the tensile strength increased from 136*MPa* to 214*MPa*, which improved by 57%. When the impregnation mold temperature was 300℃ and 310℃, its corresponding tensile strength was 213*MPa* and 215*MPa* respectively, tensile strength almost no changes. The tensile strength of the composites reached the maximum and it was in a stable state when the temperature of the impregnation mold was 290℃, just as shown in Figure 11.

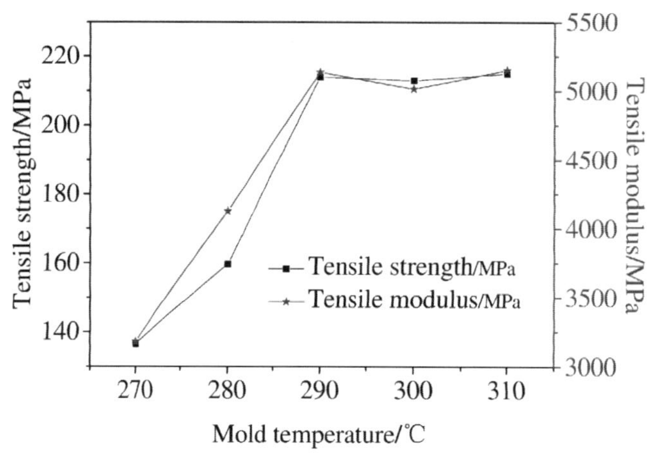

Figure 11. The influence of the impregnation mold temperature on tensile properties.

The influence of impregnation mold temperature on the flexural properties of composites was shown in Figure 12. The figure showed that the bending strength of the carbon fiber reinforced nylon 66 increased with the rose of the temperature of the impregnation mold and then fluctuation within a certain range. The bending strength increased from 241*MPa* to 317*MPa*, which improved by 48% when the impregnation mold temperature enhanced from 270℃ to 300℃. When the impregnation mold temperature was 310℃, the corresponding bending strength was 310*MPa*, bending strength almost no changes. It can be known from Figure 8 that the bending strength of the composites reached the maximum when the impregnation mold temperature was 300℃.

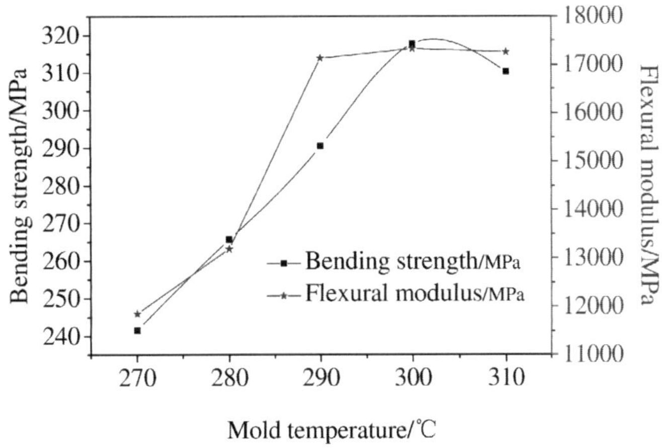

Figure 12. The influence of the impregnation mold temperature on flexural properties.

Figure 13 revealed the influence of impregnation mold temperature on the impact properties of the composites. As shown in the Figure 13, the impact strength of carbon fiber reinforced nylon 66 increased firstly and then decreased with the rose of the temperature of impregnation mold and finally tended to be gentle. When the mold temperature increased from 270℃ to 290℃, the impact strength increased from 35.6KJ/m^2 to 46.51KJ/m^2, which improved by 30.6%. Continue to improve the impregnation mold temperature, the impact strength increased on the contrary. It can be concluded from Figure 13 that the impact strength of the composites reached the maximum when the impregnation mold temperature was 290℃.

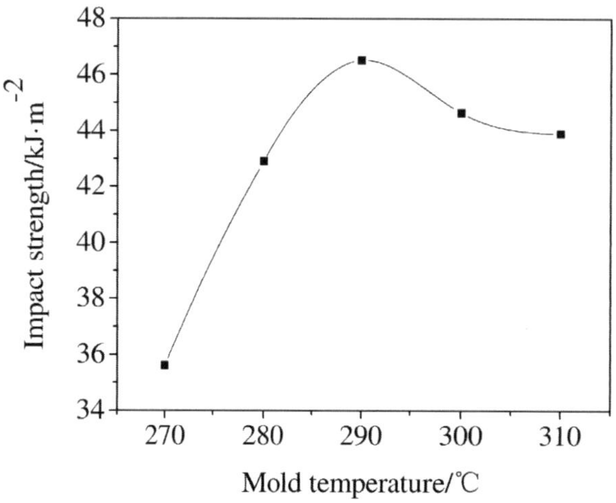

Figure 13. The influence of the impregnation mold temperature on impact properties.

Using differential scanning calorimetry measured nylon 66, the melting point was 262.576℃, so the melt impregnation mold temperature should be higher than the melting point of nylon 66. The comprehensive compared the effects of the impregnation mold temperature on tensile properties, flexural properties and impact properties of composites, the optimum temperature of melt impregnation mold should be between 290℃ and 300℃.

5. Conclusions

This paper studied the process of multi-rollers impregnation of fiber reinforced composites and a model was developed to estimate the degree of the impregnation when the fiber symmetrically pulled over a stable pin. Many factors were took into consideration, including the pin diameter, the fiber line speed, the fiber bundle width, fiber pretension, polymer viscosity and the permeability coefficient of fiber structure. The results showed that the maximum impregnation degree was not related to the fiber pretension and the pin diameter but highly depended on the fiber bundle width. The optimum mold temperature was in the range of 290℃~300℃, while continuous carbon fiber reinforced nylon 66composites obtained by melt impregnation method had better mechanical properties. The results had satisfactorily corroborated by experiment.

Acknowledgements

The authors would like to acknowledge the support of the Special Fund of the Central University Research of China (XK1518).

References

[1] Chang I Y and Lees J K 1988 Recent Development in Thermoplastic Composites: A Review of Matrix Systems and Processing Methods *J Thermoplast Compos* **1**(3) 277-296
[2] Gaymans R J and Wevers E 1998 Impregnation of a glass fiber roving with a polypropylene melt in a pin assisted process *Compos Part A-appl S* **29**(5) 663-670
[3] Peltonen P and Törmälä P 1994 Melt impregnation parameters *Compos Struct* **27**(1) 149-155
[4] Nygard P and Gustafson C G 2004 Continuous Glass Fiber-Polypropylene Composites Made by Melt Impregnation: Influence of Processing Method *J Thermoplast Compos* **17**(2) 167-184

[5] Kumar K S Bhatnagar N and Ghosh A K 2007 Development of Long Glass Fiber Reinforced Polypropylene Composites: Mechanical and Morphological Characteristics *J Reinf Plast Comp* **26**(3) 239-249

[6] Asai T Ohara T and Tanaka T 1996 Method of manufacturing continuous fiber-reinforced thermoplastic prepregs *US Patent* 5529652

[7] Hatano D Tanaka Y and Takagi M 2011 *26th Annu Tech Conf AM Soc Compos* vol 2 (Montreal QC Canada) p 1331

[8] Cogswell F N Hezzell D J and Williams P J 1992 Fiber-reinforced compositions and methods for producing such compositions *US Patent* 5213889

[9] Kim T W Jun E J and Um M K 1989 Effect of pressure on the impregnation of thermoplastic resin into a unidirectional fiber bundle *Adv Polym Tech* **9**(4) 275-279

[10] Bates P J and Charrier J M 2000 Pulling tension monitoring during the melt impregnation of glass roving *Polym Composite* **21**(1) 104-113

[11] Bijsterbosch H and Gaymans R J Impregnation of glass rovings with a polyamide melt. Part 1: impregnation bath *Compos Manuf* **4**(2) 85-92

[12] Happel J 1959 Viscous flow relative to arrays of cylinders *AM Inst Chem Eng* **5**(2) 174-177

[13] Ali R Iannace S and Nicolais L 2003 Effects of processing conditions on the impregnation of glass fiber mat in extrusion/calendering and film stacking operations *Compos SCI Technol* **63**(15) 2217–2222

[14] Peltonen P Lahteenkorva K and Paakkonen E J 1992 The influence of melt impregnation parameters on the degree of impregnation of a polypropylene/glass fiber prepreg *J Thermoplast Compos* **5**(4) 318-343

IOP Conf. Series: Materials Science and Engineering **137** (2016) 012054 doi:10.1088/1757-899X/137/1/012054

The gravity effect on spin coating glucose thin film

P Han and H Wang[1]

Department of Physics, Capital Normal University, Beijing Key Laboratory of Metamaterials and Devices, Key Laboratory of Terahertz Optoelectronics, Ministry of Education, and Beijing Advanced Innovation Center for Imaging Technology, Beijing, 100048, P.R. China

E-mail: wanghai@cnu.edu.cn

Abstract. We demonstrate the gravity effect on spin-coated glucose film by comparing properties of films fabricated by two different methods: erected spin coating (ESC) and inverted spin coating (ISC). A "Bi-Layer" model, *i.e.* substrate effect layer and free layer, for spin coating is proposed to analyze the gravity effect on spin coating process which offers a more accurate prediction on film thickness. The thickness and deposition pattern are different in samples fabricated by ESC and ISC methods in titrating-repeated spin coating experiment. Finally, in a glucose coating/ferromagnetic composite structure, an enhanced Kerr rotation with the enhancement factor of 2.11 is achieved by ISC method.

1. Introduction

Spin coating is an important method for polymer and organic thin films fabrication [1, 2]. In the spin coating process, the liquid is first dropped on the substrate, and then the substrate is accelerated to the designed spinning speed. The solution flows radially outward owing to the action of centrifugal force. However, a small amount is adsorbed on the substrate, and forms film [3, 4]. The first mathematic analysis proposed by Emslie, Bonner and Peck (EBP) that established a model describing the flow characteristics of Newtonian liquid on a rotating substrate [5]. Furthermore, Meyerhofer developed their model by taking the evaporation into account [6]. On the basis of these pioneer works, Bornside *et al.* modified the related parameters on evaporation rate in detail [7, 8]. Subsequent works focused on the exploration of the formation mechanism and properties of films combining liquid flow and evaporation [3, 9, 10].

In all these works, gravity effect on spin coating is neglected traditionally, for the substrate adsorption dominates the formation of thin film under spinning. In the case of erected spin coating (ESC), gravity force and the substrate adsorption are in the same direction in ESC, as shown in Figure 1a. Gravity exerts pressure on liquids and enhances the substrate effect. For the inverted spin coating (ISC), gravity force and the substrate adsorption are in opposite direction, as shown in Figure 1b. In this case, gravity stretches liquids and weakens the substrate adsorption. Therefore, films fabricated by these two methods are different. However, as far as is known, few people uses ISC method, and the differences between ESC and ISC methods have not been investigated in detail [11, 12].

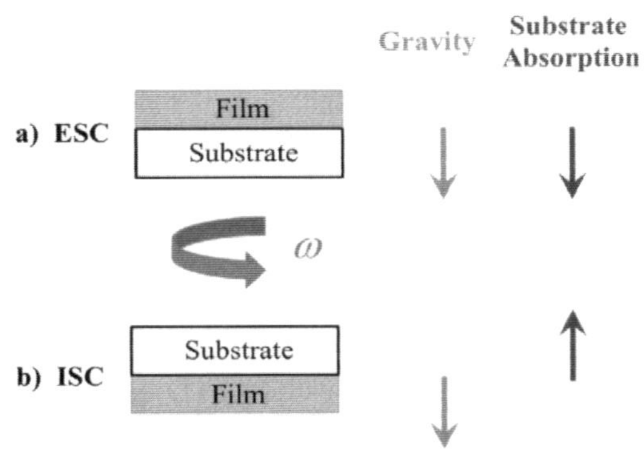

Figure 1. Schematic plot for the experiment, a).for ESC process, the gravity and the substrate adsorption are in the same direction; b). for ISC process, they are in the opposite direction. ω is the spin speed.

Here, we demonstrate that ESC and ISC share same fundamental physics approximately by the similar senses. But the gravity may exert its influences on the film formation resulting in the modifications of microstructures. A titrating-repeated spin coating experiment is performed to demonstrate the differences in the thickness variation with the increase of coating times. To understand the competition between gravity force and substrate adsorption, a "Bi-Layer" model, *i.e.* the substrate effect layer and the free layer, is proposed and applied for our data analysis. Also, we fabricate glucose films on the top of $Ni_{80}Fe_{20}$ substrates which can be used to enhance Kerr rotation of the $Ni_{80}Fe_{20}$ films.

2. Materials and methods

2.1. Experimental procedure

The experiments were performed using a spin coater (Model: WS-650-23NPP) with the maximum spin speed 12000Rpm. Glass substrates of size $25 \times 25\text{mm}^2$ were used for all experiments. The substrates were cleaned by electronic cleaning agent in an ultrasonic bath for ten minutes followed by de-ionized water for five minutes. Dispensed solution (0.5mL) onto the center of the substrate. For ISC, inverted the spin coater immediately after the substrate was fixed on the vacuum chuck. For both methods, the spinning time was sixty seconds. We examined the thicknesses of the full-dried glucose films employing a surface profile tester (Alpha-Step IQ). $Ni_{80}Fe_{20}$ and Fe thin films were deposited using ultra high vacuum magnetron sputtering system (Model: JGP600) with a base pressure 6.3×10^{-5}Pa. The thicknesses were fixed at 100nm ($Ni_{80}Fe_{20}$) and 30nm (Fe) respectively. The deposition rate was 3.18 Å/s ($Ni_{80}Fe_{20}$) and 2.64Å/s (Fe) in experiments. We measured the Kerr rotation θ_K employing the NanoMOKE-III, the longitude MOKE configuration was used. The wavelength of the incident linear polarized light was 660nm, and the incident angle was 45°.

3. "Bi-Layer" model

By the continuity equation of Newtonian liquid, EBP regarded the relationship of film-thinning-rate as a function of processing parameters under the assumption that the centrifugal force equals to viscous force [5]. Furthermore, taking the solvent evaporation during spin coating process into consideration, Meyerhofer revised the relationship between the final film thickness and spin speed assuming that the evaporation rate equals to the viscous flow rate [6]. The relationship between the film thickness h and spin speed ω is $h \propto \omega^{-1/2}$ [3, 4, 6, 7]. Using a mass transfer coefficient to describe vapor-liquid equilibrium process, Bornside *et al.* expressed the evaporation rate with specific parameters [7, 8]. However, some recent works revealed that a bottom-anchored layer formed on the substrate exhibits non-Newtonian behavior due to the substrate effect, which caused a deviation of the index for the spin speed [4]. Here we simplify the laminar flow model for liquid dynamics as "Bi-layer", *i.e.* the substrate effect layer and the free layer. The substrate effect layer represents the thin liquid layer attached to the substrate that exhibits non-Newtonian nature due to the substrate absorption. The free layer is the liquid layer on the top of the substrate layer. The interaction between the free layer and its "substrate" is the internal interaction in liquid, and therefore the free layer can be considered as the ideal Newtonian liquid. In our model, the film thicknesses of substrate effect layer and free layer are h_1 and h_2 respectively. The total film thickness h_f can be described by:

$$h_f = h_1 + h_2 = m_1\omega^{-\beta_1} + m_2\omega^{-\beta_2} \tag{1}$$

$$m_1 = \left(1 - x_A^0\right)\left[\frac{3\eta_0}{2\rho} \cdot k_1 \cdot \left(x_A^0 - x_{A\infty}\right)\right]^{1/3} \tag{2}$$

$$m_2 = \left(1 - x_A^0\right)\left[\frac{3\eta_0}{2\rho} \cdot k_2 \cdot \left(x_A^0 - x_{A\infty}\right)\right]^{1/3} \tag{3}$$

Here η_0 is the initial viscosity of the coating liquid; ρ is the density of the pure solvent, β_1, and β_2 are the indexes for the spin speed. The evaporation rate for the substrate effect layer and free layer are $k_1 \cdot \left(x_A^0 - x_{A\infty}\right)$ and $k_2 \cdot \left(x_A^0 - x_{A\infty}\right)$ respectively, where k_1 and k_2 are the mass transfer coefficients for the substrate effect layer and the free layer. x_A^0 is the initial concentration of solvent in the coating liquid, $x_{A\infty}$, is the mass fraction of solvent in the overlying gas [7]. The mass transfer coefficients for substrate effect layer and free layer are given by the following expressions:

$$k_1 = \gamma_1 \cdot \frac{1}{\rho} \cdot \frac{P_A M_A}{RT} \tag{4}$$

$$k_2 = \gamma_2 \cdot \frac{1}{\rho} \cdot \frac{P_A M_A}{RT} \tag{5}$$

P_A is the vapor pressure of pure solvent A at temperature T, R is the ideal gas constant, and M_A is the molecular weight of solvent A. γ_1, and γ_2 are parameters for the overlying gas for substrate effect layer and free layer respectively. According to Bornside's work, γ is defined as:

$$\gamma = \frac{cD_g}{v_g^{1/2}} \tag{6}$$

The constant c depends on the Schmidt number of the overlying gas. D_g is the binary diffusivity of the solvent in the overlying gas. v_g is the kinematic viscosity of the overlying gas. All of the parameters and expressions are well defined in the literature.

4. Result and discussion

4.1. Simulation results by "Bi-Layer model" and discussion

Figure 2a demonstrates the film thicknesses h_f as a function of spin speed ω. The solid black line is the fitting results by "Bi-layer" model. As shown in Figure 2a, the film thicknesses of ESC (circle points) and ISC (blue star symbol) series obey the same relationship versus spin speed. We list the fitting parameters in Table 1. The present work estimates the initial viscosity of aqueous glucose solution employing the literature [13].

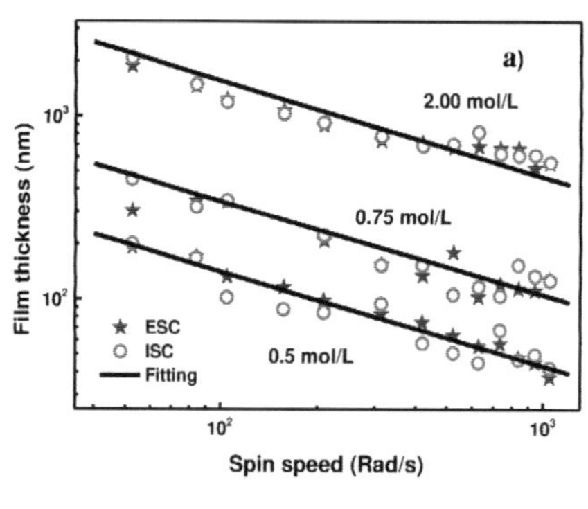

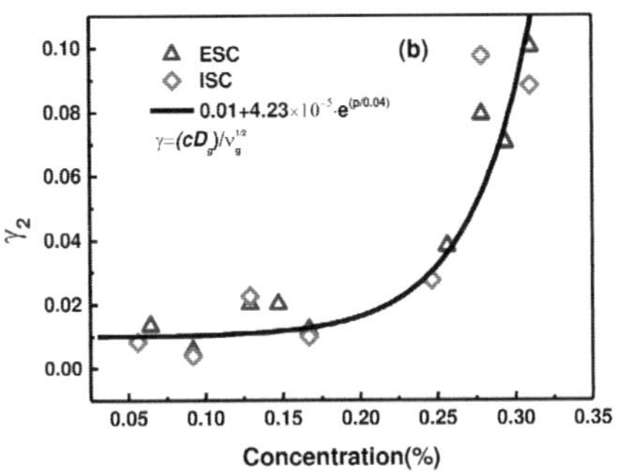

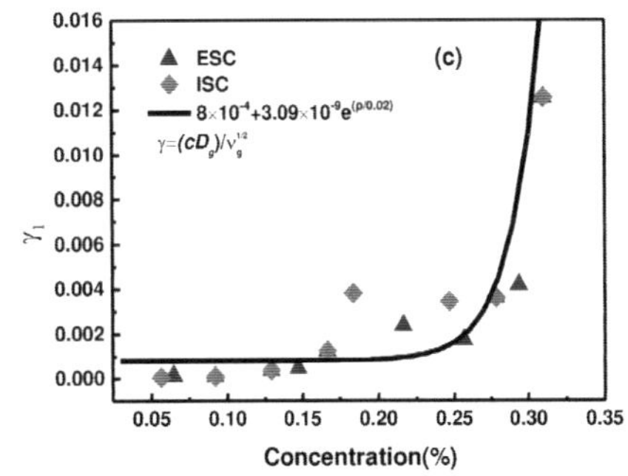

Figure 2. a) The film thickness as a function of spin speed ω, the concentration used is 0.5mol/L, 0.75mol/L, and 2.00mol/L. The blue and red points are the experimental data employing ESC and ISC methods respectively. The solid black lines represent the simulation results by "Bi-Layer model"; (b-c) Relevant parameters γ_2 and γ_1 for overlying gas as a function of the initial concentration of glucose aqueous solution. Erected spin coating and inverted spin coating methods are represented by the blue triangle and the red rhombus respectively. The solid black lines represent the fitting lines, the formula is given in the figure, p is the mass fraction of solute.

β_1 and β_2 are 0.64 and 0.5 respectively, the index for substrate effect layer β_1, deviates from $h_f \propto \omega^{-1/2}$ law due to the non-Newtonian fluid rheology, and the free layer suppresses the evaporation rate. This result is quite similar to the previous works on strong substrate effect, and $\beta_1 = 0.64$ is reasonable in experiments [4]. The latter one, β_2, for free layer term, is consistent with the one in theories for Newtonian liquid, owing to its liquid-to-liquid interface with substrate layer.

Table 1. Constant parameters used for predicting the glucose film thicknesses in Figure 2a.

Parameters for predicted film thickness in Fig.2a, Coating glucose aqueous solution
0.50mol/L-x_A^0-0.908; η_0 =0.0137 Poise;
0.75mol/L-x_A^0-0.871; η_0 =0.0151 Poise;
2.00mol/L-x_A^0-0.690; η_0 =0.0243 Poise;
$x_{A\infty}$=0;
$\rho = 1g/cm^3$;
P_A=0.03128atm;
$M_A = 18g/mol$
$R = 82.06atm \cdot cm^3/mol \cdot K$
$T = 298.15K$

Although substrate effect layer has less contribution on film thickness, its contribution cannot be neglected. Otherwise, the predicted film thickness is smaller than the experimental data as reported in the previous research. And the deviation becomes significant with the increase of concentration [7]. Figure 2b and 2c demonstrate when the mass fraction of solute is less than 20%, γ_2, and γ_1 values are almost constant which consists of previous works [7]. As the mass fraction of solvent is very high, the content of solute can not completely affect the diffusion, the dynamic viscosity of the overlying gas tends to be steady, further lead to the evaporation rate is almost invariant. However, as the mass fraction of solute is greater than 20%, the solute gradually suppress the solvent diffusion with the increase of the mass fraction. The dynamic viscosity of the overlying gas of free layer decreases which causes rapid growth of γ_2 as shown in Figure 2b. With the increase of concentration, γ_1 varies accordingly as shown in Figure 2c. In the case of glass substrate, the adsorption is significant and the gravity effect on substrate adsorption is ineffective for both ESC and ISC series. Hence, the differences between ESC and ISC is not obvious in our experiments.

Taking viscosity of 1mol/L glucose solution (1.586cP) for the free layer and varying the equivalent viscosity of substrate layer from 0.01 to 100 times than that of the free layer, we can analyze the substrate layer effect on the evolution of film thickness versus spin speed as shown in Figure 3. Compared with the initial mass loss due to centrifugal force at low spin speed the film thickness variation at higher spin speed reflects the liquid spreading process on the substrate as discussed by Meyerhofer. For a comparison, all data are normalized to the thickness obtained at 200rad/s. As shown in Figure 3, with the decrease of equivalent viscosity of the substrate effect layers, the curve gradually tends to flatten up, and the film thinning rate tends to be smaller.

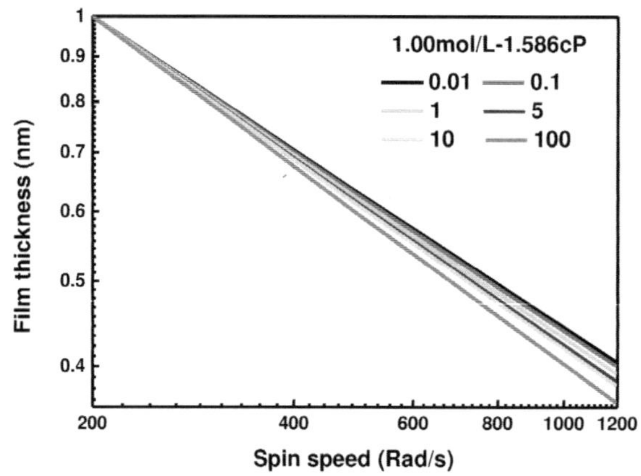

Figure 3. Calculated normalized film thickness versus spin speed using "Bi-Layer" model. Varying the equivalent viscosity ratio of substrate effect layer and free layer ranges from 0.01 to 100: 0.01(black), 0.1(red), 1(green), 5(blue), 10(cyan); 100(magenta). Taking viscosity of 1mol/L glucose solution (1.586cP) for the free layer.

4.2. Titrating-repeated spin coating experiment

We further demonstrate the differences between ESC and ISC through titrating-repeated spin coating method. The initial concentration of aqueous glucose solution is 1.00mol/L; ω is 1046rad/s for each experiment, and the substrates we used was glass. The sample preparation is similar to the above experimental procedure of this work. After drying, we measured the film thicknesses for the first time. Then, another 0.5mL solution was titrated on the surface of dried films for the second time. And the same process was repeated five times for both ESC and ISC methods. Figure 4 shows the relationship between film thickness variation and the number of the experiment.

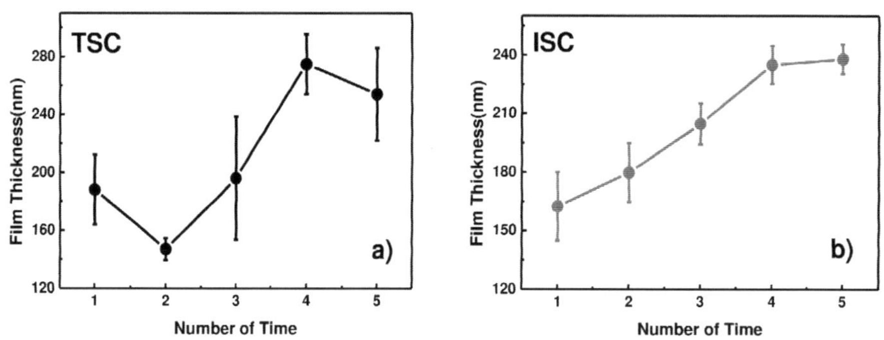

Figure 4. Film thickness versus experimental number: a) ESC method and b) ISC method.

For ESC, as shown in Figure 4a, the film thickness is 188nm after the first spin coating, and then decreases down to 146nm after the second coating and finally rises to 274nm after the fourth spin coating. As shown in previous works, spin-coated thin film is porous after being dried [14, 15]. Therefore, as we titrated liquid on the dried film surface, the liquid fills in the pores under gravity effece. Thus, the thickness of film reduces. Once the internal pores are all filled up, the film thickness rises. However, for ISC, gravity pulls the liquid out from the pores. Hence, the film becomes thicker, as shown in Figure 4b, in which film thickness increases from 163nm to 234nm after the fourth spin coating.

4.3. The gravity effect on deposition patterns
Not only the film thickness but also the deposition patterns may also affected by gravity due to the different kinetic process [16, 17]. We measured the wetting angles of films fabricated by ESC and ISC methods. The initial concentration of aqueous glucose solution is 1.00mol/L, and the spin speed ranged from 209 to 1046rad/s. After drying, another 0.5mL solution was titrated on the surface of dried films and the same drying process was repeated. Finally, a 30nm metal film was deposited on the film surface. Then, we titrated $3\mu L$ of ethanol on the sample surface and measured the wetting angles. Figure 5a and 5b illustrate the results for ESC and ISC respectively. In Figure 5c, the wetting angles are plotted as a function of ω. In this plot, for ESC, the wetting angle reduces from 16.2° to 12.1° with the increase of ω. However, for ISC, with the increase of ω, the wetting angle first decreases from 16.0° to 11.7° and then rises to 16.9°. A minimal value appears when $\omega = 628\text{rad/s}$. In the case of ESC, since gravity and substrate adsorption is in the same direction. The wetting angle reduces monotonically with the increase of ω due to the centrifugal force and substrate adsorption in common sense. However, for ISC, gravity and substrate adsorption are in the opposite direction, the centrifugal force breaks the continuous film at the high spinning rate resulting in a rougher surface morphology under weak substrate adsorption.

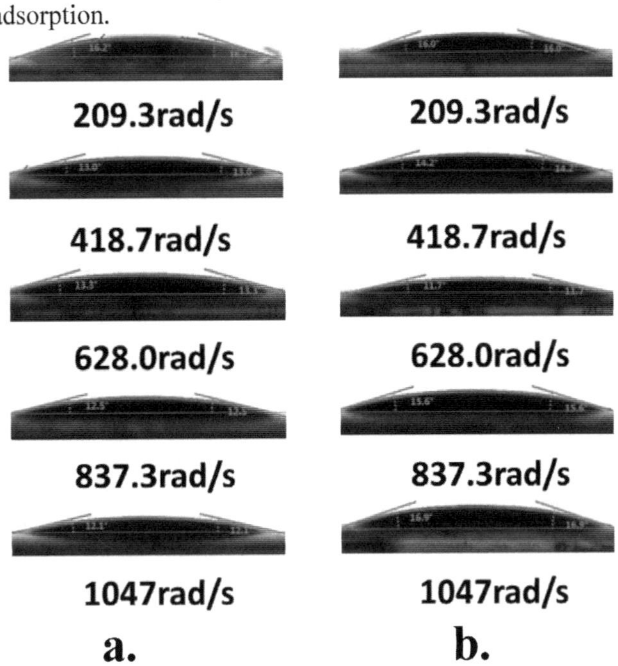

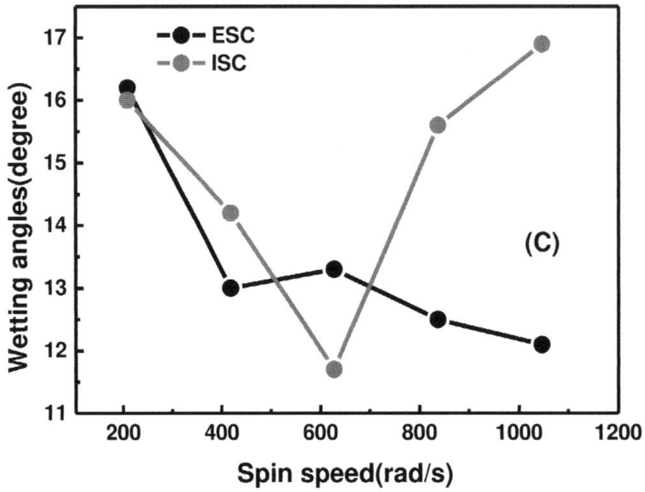

Figure 5. a) The wetting angles. a): ESC method and b): ISC method. c): Wetting angles as a function of spin speed, the black data points correspond to ESC, and the red data points correspond to ISC.

4.4. Application

As an application, we fabricate the glucose film on the top of the $Ni_{80}Fe_{20}$ film. The structure of the glucose film coated on both glass and $Ni_{80}Fe_{20}$ substrates is amorphous as shown in Figure 6a and 6b. The (111) and (200) peaks in Figure 6b are the diffraction peaks of $Ni_{80}Fe_{20}$. As shown in Figure 6c, with the increase of ω, the Kerr rotation rises from 65.75mdeg to 74.78mdeg and then reduces to 34.92mdeg for ESC, the enhancement factor is 1.95. While for ISC, the maximum θ_K is 80.29mdeg, and the enhancement factor is 2.11. The solid line (black) in Figure 6c represents the average Kerr rotation of the bare $Ni_{80}Fe_{20}$ films. The use of dielectric coatings for enhancing the Kerr rotation is a commonly used method [18, 19]. Also, the glucose film fabricated on the top of the $Ni_{80}Fe_{20}$ film equivalent to an organic Fabry-Perot cavity. Linear polarized light propagates within the cavity, which gives rise to multiple reflections off the magnetic layer, the sum of phase and polarization rotation are increased [18, 19].

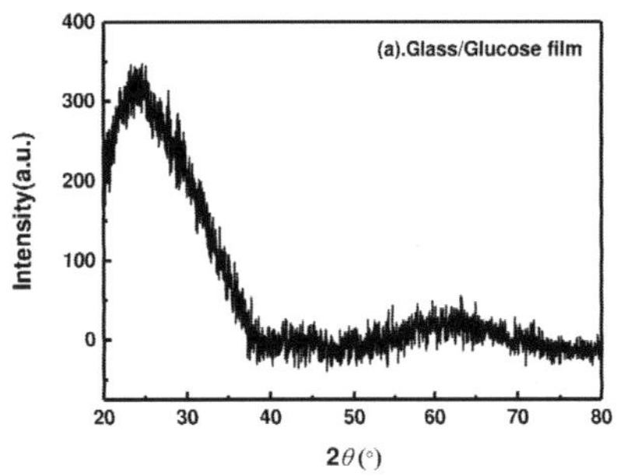

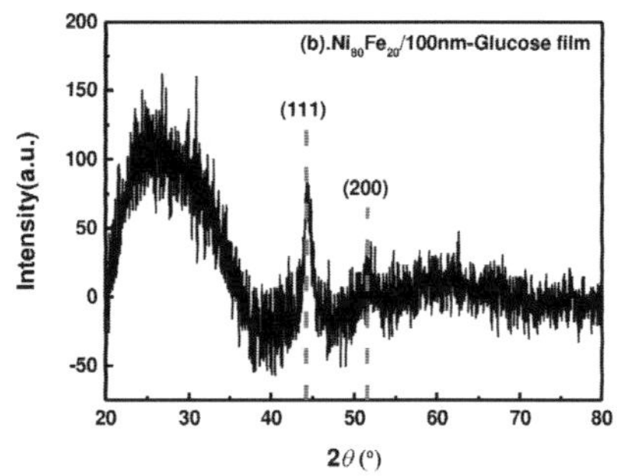

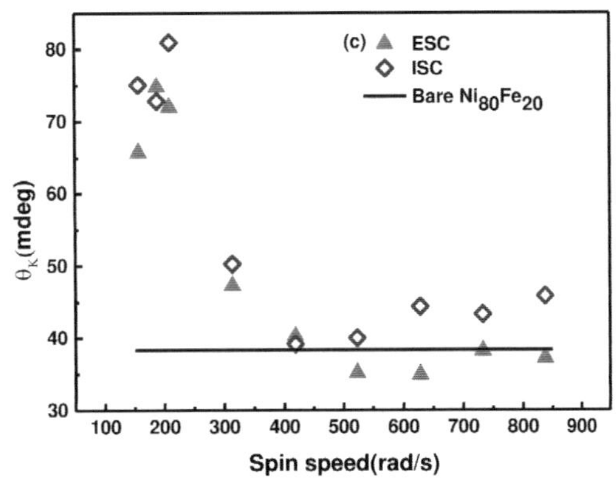

Figure 6. XRD high angle scan in the range $2\theta = 20^{\circ}\sim80^{\circ}$ for glucose film/glass (a) and glucose film/$Ni_{80}Fe_{20}$ film (b). (c). Kerr rotation as a function of spin speed, the glucose film was fabricated employing ESC and ISC method. The concentration of the aqueous glucose solution is 1.00mol/L and the ω ranges from 209.3 to 837.3rad/s. We measure the longitude Kerr rotation of the composite structure including organic films and ferromagnetic materials. The applied magnetic field is 80Oe. The red triangle and blue rhombus symbols represent the Kerr rotation. The solid black line represents the average Kerr rotation of $Ni_{80}Fe_{20}$ films.

5. Summary

In conclusion, we demonstrate the gravity effect on spin coating process through compared the properties of films fabricated employing ESC and ISC methods in detail. ISC shows a unique microcosmic evolution and could be useful for fabrication of films of organic and polymer films. Also, a new "Bi-Layer" model is proposed to understand the spin coating process and simulates the experiment data accurately. Although aqueous glucose solution is selected on purpose in the present work since it is cheap and environmental-friendly, the conclusion has universality.

Acknowledgments
This work was financially supported by the National Natural Science Foundation of China (NSFC) with grant No.11274233 and Beijing education committee under grant KM201610028004.

References

[1] Schmidt R H, Mosbach K and Haupt K 2004 *Adv. Mater.* **8** 719-722
[2] Ikawa M, Yamada T, Matsui H, Minemawari H, Tsutsumi J, Horii Y, Chikamatsu M, Azumi R, Kumai R and Hasegawa T 2012 *Nat. Commun.* **3** 1176
[3] Hall D B, Underhill P and Torkelson J M 1998 *Polym. Eng. Sci.* **38** 2039-2045
[4] Cheung K P, Grover R, Wang Y, Gurkovich C, Wang G and Scheinbeim J 2005 *Appl. Phys. Lett.* **87** 214103
[5] Emslie A G, Bonner F T and Peck L G 1958 *J. Appl. Phys.* **29** 858-862
[6] Meyerhofer D 1978 *J. Appl. Phys.* **49** 3993-3997
[7] Bornside D E, Macosko C W and Scriven L E 1991 *J. Electrochem. Soc.* **38** 317-320
[8] Bornside D E, Brown R A, Ackmann P W, Frank J R, Tryba A A and Geyling F T 1993 *J. Appl. Phys.* **73** 585-600
[9] Birnie D P and Manley M 1996 *Phys. Fluids* **9** 870-875
[10] Mouhamad Y, Mokarian-Tabari P, Clarke N, Jones R A L and Geoghegan M 2014 *J. Appl. Phys.* **116** 123513
[11] Mairaj A K, Curry R J and Hewak D W 2004 *Electron. Lett.* **40** 421
[12] Mairaj A K, Curry R J and Hewak D W 2005 *Appl. Phys. Lett.* **86** 094102
[13] Bui A V and Nguyen M H 2004 *J. Food Eng.* **62** 345-349

[14] Lai J H 1979 *Polym. Eng. Sci.* **19** 1117-1121
[15] Park M S and Kim J K 2004 *Langmuir* **20** 5347-5352
[16] Man X K and Doi M 2016 *Phys. Rev. Lett.* **116** 066101
[17] Bonn D, Eggers J, Indekeu J, Meunier J and Rolley E 2009 *Rev. Mod. Phys.* **81** 739-805
[18] Cantwell P R, Gibson U J, Allwood D A and Macleod H A M 2006 *J. Appl. Phys.* **100** 093901
[19] Qureshi N, Schmidt H and Hawkins A R. 2004 *Appl. Phys. Lett.* **85** 431-433

Influences of different parameters on the microstructure of magnetic-field-induced self-assembled film

X Dan[1] and J J Yang

Nanjing Xiaozhuang University, School of Environmental Science, Nanjing, P.R. China

E-mail: xiaozhuangxxc2010@126.com

Abstract. Self-assembled films with needle-like microarrays were fabricated using a mixture of cobalt and fluorocarbon resin under a magnetic field. The various influences of magnetic powder content, viscosity and size distribution on the structure of the self-assembled films were investigated. The self-assembled film morphologies were characterized by stereomicroscope and scanning electron microscopy. Experimental results indicate that an increase in magnetic powder content results in greater unit height and diameter, and that a reduction in viscosity results in increasing array density and decreasing unit width. Additionally, particles with narrow size distribution were able to attain more regular microarray structures. The structural alterations were closely related to numerous effects such as van der Waals forces, dipole-dipole interactions, and external-dipole interactions. The self-assembled film demonstrated magnetic anisotropy, as identified by vibrating sample magnetometry (VSM).

1. Introduction

As a special physical field, the magnetic field demonstrates marked effects on the movement of magnetic particles. In general, magnetic particles as induced by a magnetic field can self-assemble, resulting in a 1D, 2D or 3D structure and morphology [1-4]. It has been reported that self-assembled materials with 3D structures demonstrate more remarkable application prospects in magnetic recording devices, optoelectronic devices, biological sensors, etc. [5-7]. Many researchers have successfully assembled arrays with 3D structures by inducing magnetic fields. However, these structures are typically produced in the magnetic liquid, and are very difficult to maintain once the magnetic field is removed [8-10]. In order to obtain 3D arrays which can be maintained without a magnetic field, needle-like microarrays were fabricated on film surfaces by combining cobalt particles with fluorocarbon (FC) resin [11].

Previous studies have reported that the alteration of parameters such as the strength of magnetic interaction, field strength, or magnetic powder content in a magnetic fluid will result in direct variation of the structural arrangements under a magnetic field [12-15]. Additionally, the present experiment demonstrates that the regulation of parameters may obtain various self-assembled structures. The influence of magnetic field strength on the resulting structure has been previously investigated [11]. In the present research, magnetic powder content, viscosity and particle size distribution were altered in order to investigate the influence of the van der Waals forces, dipole-dipole interactions, and external-dipole interactions on a mesoscopic scale on the microstructure of self-assembled films. The present investigation provided both fundamental understanding and necessary contributions to potential practical applications. The magnetic anisotropy behavior of films with different applied directions of magnetic field were also studied.

2. Experiment

2.1. Materials

Cobalt (Co) powder was synthesized by chemical vapor deposition. Fluorocarbon (FC) resin with 50% solid content was prepared by the copolymerization of fluorinated monomers. The aliphatic hexamethylene-isocyanate allophanamide (HDI) and cyclohexanone (CYC) were selected for the curing agent and solvent of FC resin, respectively.

2.2. Preparation of self-assembled films

The detailed preprocess was explained in previous literature [11]. The strength of the magnetic field throughout the entire process was controlled at 0.38 T. In the experiment, the viscosities of the assembled solutions were adjusted to 7 mPa·s, 14 mPa·s, 24 mPa·s, 38 mPa·s, 63 mPa·s and 139 mPa·s, respectively, as controlled by the amount of cyclohexanone with a powder content of 0.0072 g/mL. The powder contents were 0.0024 g/mL, 0.0048 g/mL, 0.0072 g/mL and 0.0096 g/mL, respectively, and viscosity was maintained at 38 mPa·s. The powder content was equal to the amount of powder divided by the amount of resin. A narrower particle size distribution was obtained by centrifuging the mixture of Co powders, FC resin and cyclohexanone at a rate of 300 r/min.

2.3. Characterization of magnetic powders and self-assembled films

Scanning electron microscopy (SEM, JSM-5900, JEOL) was employed to observe the size and morphology of the Co powder and self-assembled film structures. A stereoscopic microscope (Olympus, BH-2) was also used to observe the self-assembled film structures. The magnetization curve of the magnetic powders and the magnetic anisotropy of the self-assembled films were characterized at room temperature by vibrating sample magnetometry (VSM, ADE-EV7, ADE).

3. Results and discussion

3.1. Influence of magnetic powder content on the structure of self-assembled films

Figure 1 depicts stereoscopic microscopic photographs of the self-assembled films with different Co powder contents, demonstrating the formation of numerous black, needle-like microarrays on the surface of the films, formed by the agglomeration of Co particles. In the present experiment, magnetic particles vibrate, fluctuate, attract one another and form small magnetic clusters under the influence of van der Waals forces during ultrasonic treatment and solvent evaporation. Each magnetic cluster can be regarded as a magnetic dipole, which exhibits larger dipole moments than any single particle. When an external magnetic field is applied, the magnetic clusters in the mixtures tend to arrange in the direction of the magnetic field and aggregate in a chain-like formation, thus forming the microarray structure.

As the content of Co powder increases, the height of the array increases, as well as the mean diameter of single needle structures. When the content reaches 0.0096 g/mL, the microarray disappears and a wall-like aggregation forms. In the system composed of magnetic particles, resin and solvent, the dominant magnetic dipole forces consist of van der Waals attractions, dipole-dipole interactions and external-dipole interactions [16]. The dipole-dipole interaction energy between two dipoles is given by the following:

$$U = r^{-3}[\boldsymbol{\mu_1} \cdot \boldsymbol{\mu_2} - 3(\boldsymbol{\mu_1} \cdot r)(\cdot \boldsymbol{\mu_2} \cdot r)/\mathrm{r}^2] \tag{1}$$

where $\boldsymbol{\mu_1}$ and $\boldsymbol{\mu_2}$ represent the dipole vectors of dipoles 1 and 2. The dipole moment $\boldsymbol{\mu}= M_s V$, M_s, which represents the saturation magnetization of magnetic particles, and where V is the volume of a single particle, and r is the distance between dipole centers [13].

Alteration of the magnetic powder content results in the variation of r. As the magnetic powder content increases, the distance between particles decreases and the dipole-dipole interactions become stronger. The enhanced resultant force creates longer chains of magnetic particles, which result in thicker, taller needles. Array height increases at a decreasing rate to a certain point, at which the number of chains begins to increase. The chains become so dense that aggregation occurs, as a result of the intensive force between chains.

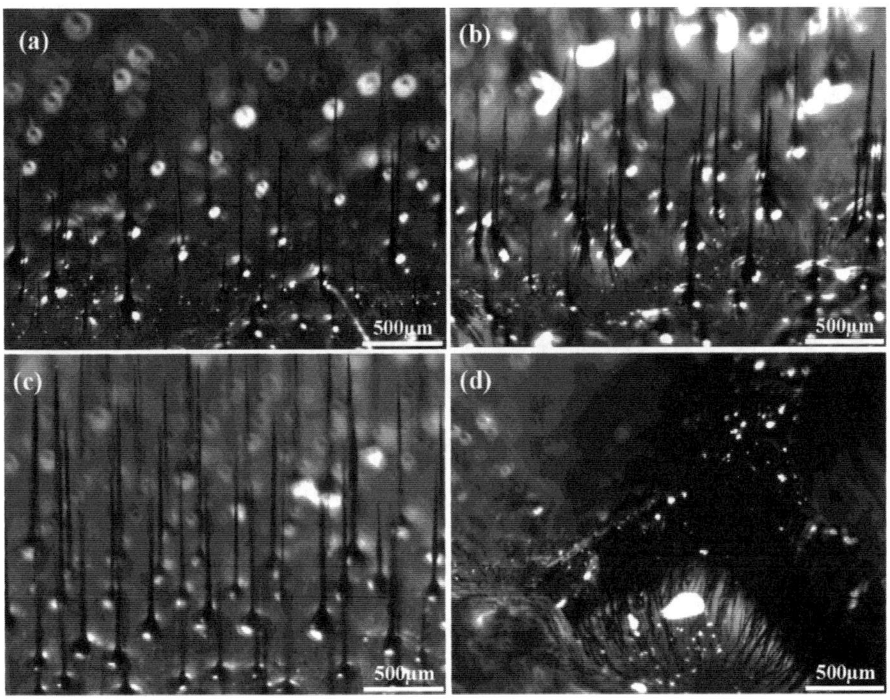

Figure 1. Stereoscopic microscope photographs of Co/FC magnetic-field-induced self-assembled films with Co powder content of (a) 0.0024g/mL (b) 0.0048g/mL (c) 0.0072g/mL (b) 0.0096g/mL.

3.2. Influence of viscosity on the structure of self-assembled films

Figure 2 depicts stereoscopic microscope images of Co/FC self-assembled films. Results indicate that lower viscosity results in denser arrays and smaller needle diameter. As the viscosity increases to 63 mPa·s, regular arrays no longer form, resulting in mounded structures with few needles. Further increasing the viscosity results in the elimination of all needles and the appearance of random structural bumps.

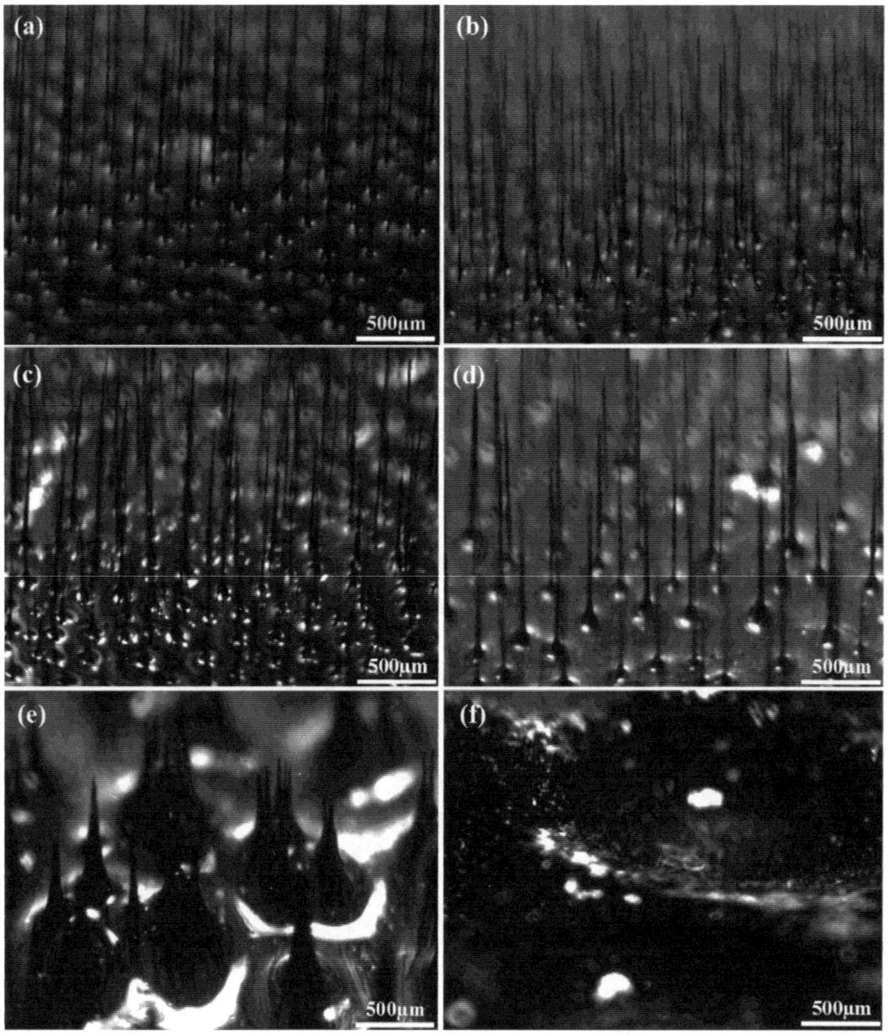

Figure 2. Stereoscopic microscope photographs of Co/FC magnetic-field-induced self-assembled films with viscosity of (a) 7 mPa·s (b) 14 mPa·s (c) 24 mPa·s (d) 38 mPa·s (e) 63 mPa·s (f) 139 mPa·s.

It is generally known that as a result of high molecular weight and viscosity, the presence of polymers will hinder the movement of magnetic particles, clusters and chains. However, the Co particles maintain chain-like arrangements at high viscosities (strong dipole-dipole interactions) under the magnetic field [17]. At higher viscosities, the chains move with greater difficultly and become unevenly distributed, which may result in further aggregation, therefore resulting in larger units. During the self-assembly process, greater repulsion between units results in greater distances between units [18]. The final microarray structures take shape as the result of the balance between the interactions and repulsions between chains. Therefore, viscosity reduction effectively achieves denser arrays and thinner units.

3.3. Influence of particle size distribution on the structure of self-assembled films

The influences of particles size distribution on the microstructural arrays were also studied. The centrifugation of a mixture of Co powder, resin and solvent resulted in narrow particle size distributions. As demonstrated by a comparison of Figure 3(a) and Figure 3(b), the particles demonstrate more uniform size after centrifugation, with identical viscosities and powder contents. The array structures with a narrower size distribution are more regular and the needles are thinner and demonstrate a broader size distribution, as demonstrated by the stereoscopic microscopy and SEM photographs.

Uniform particle volume can result in a uniform dipole-dipole force, resulting in regular structures. Some larger parties exist within the system with broader size distribution, and produce stronger dipole-dipole forces and chain-chain attractions, resulting in thicker arrays.

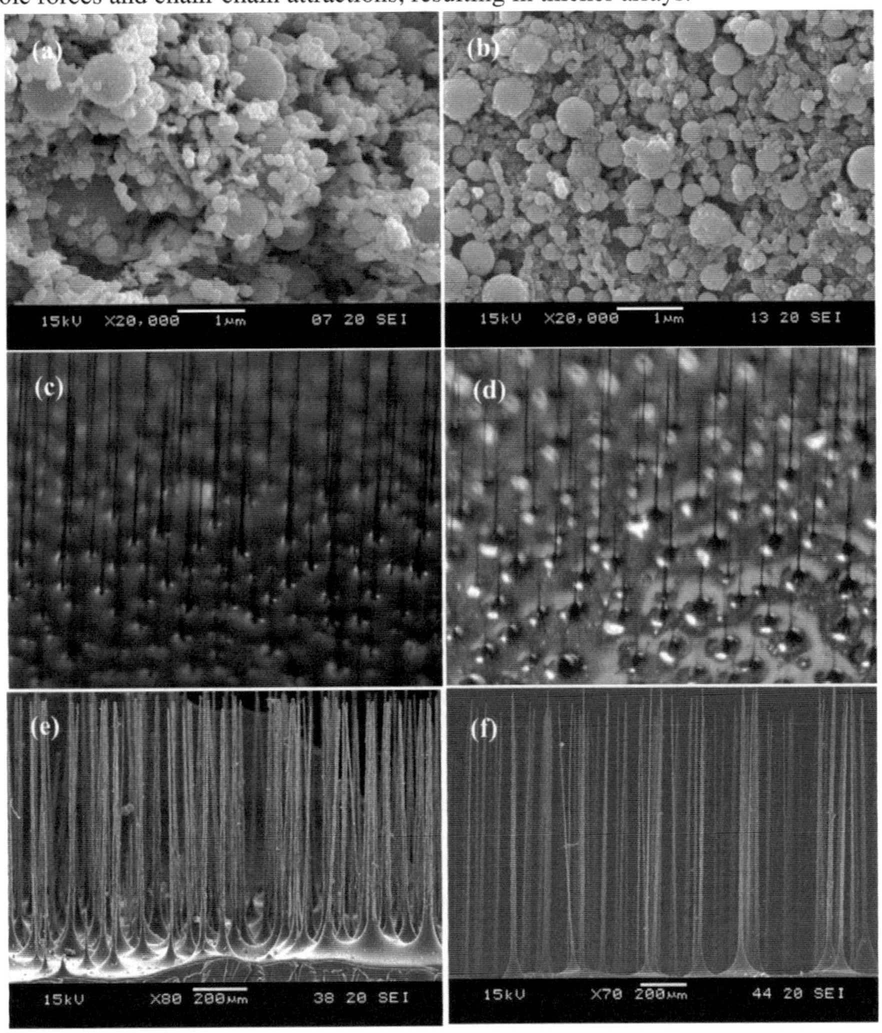

Figure 3. SEM photographs of Co powder (a) without centrifugation; and (b) with centrifugation. Stereoscopic microscopy photographs of magnetic-field-induced self-assembled films formed by Co powder (c) without centrifugation; and (d) with centrifugation. SEM photographs of magnetic-field-induced self-assembled films formed by Co powder (e) without centrifugation; and (f) with centrifugation.

3.4. Magnetic properties

The hysteresis loops of the self-assembled films are depicted in Figure 4. The shapes of the two hysteresis loops presented in the figure differ from one another when the magnetic field is applied perpendicular or parallel to the plane of the film. The film with a perpendicular applied field demonstrates greater saturation magnetization than that under the influence of a parallel applied filed. The reduced remanence (M_r/M_s) of Co is 0.074 when the applied field is perpendicular to the film plane, whereas the reduced remanence is 0.033 when the applied field is parallel to the film plane. Based on the results described above, the arrays growing in the direction perpendicular to the film plane can be easily magnetized. During the self-assembly progress, partial magnetic particles rotate to arrange with the field applied along the easy axis, resulting in the magnetic anisotropy of the self-assembled films [10].

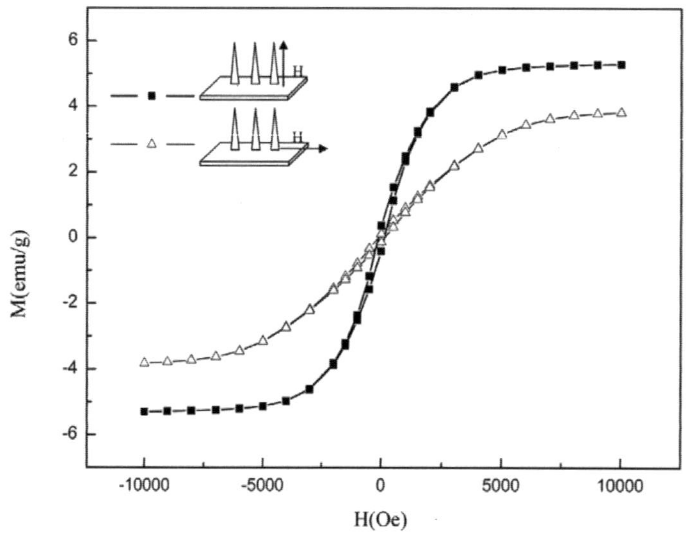

Figure 4. Magnetic hysteresis loop of self-assembled films under magnetic fields applied in different directions.

4. Conclusion

Cobalt was used to assemble microarrays under magnetic fields. Experimental results indicate that the particle volume and saturation magnetization of the magnetic particles had great influence on the self-assembled film structures. Increasing magnetic powder contents resulted in taller microarrays comprised of larger unites, while reduction in viscosity resulted in denser arrays comprised of thinner units. The narrower size distribution of particles was easily to get more regular microarray structures. The different shapes of magnetic hysteresis loops and the alteration between reduced remanences (M_r/M_s) measured under different magnetic field directions demonstrated the magnetic anisotropy of the self-assembled films.

Acknowledgments

The authors appreciate the support from the National Natural Science Foundation of China (21506102), Natural Science Foundation of the Higher Education Institutions of Jiangsu Province,

China (14KJD150007).

References

[1] Pileni M P 2001 Magnetic fluids: fabrication, magnetic properties, and organization of nanocrystals *Adv. Funct. Mater.* **11** 323

[2] Wang H, Chen Q W, Sun X, Qi H P, Yang X, Zhou S and Xiong J 2009 Magnetic-field-induced formation of one-dimensional magnetite nanochains *Langmuir* **25** 7135

[3] Wang M, He L, Xu, W, Wang X and Yin Y 2015 Magnetic assembly and field-tuning of ellipsoidal-nanoparticle-based colloidal photonic crystals *Angew. Chem. Int. Edit.* **54** 7077

[4] Yu X, Feng X, Hu J, Zhang Z L and Pang D W 2011 Controlling the magnetic field distribution on the micrometer scale and generation of magnetic bead patterns for microfluidic applications *Langmuir* **27** 5147

[5] Martın J I, Nogues J, Liu K, Vicent J L and Schuller I K 2003 Ordered magnetic nanostructures: fabrication and properties *J. Magn. Magn. Mater.* **256** 449

[6] Lin H, Xiu F, Fang M, Yip S, Cheung H Y, Wang F and Ho J C 2014 Rational design of inverted nanopencil arrays for cost-effective, broadband, and omnidirectional light harvesting *ACS nano* **8** 3752

[7] Sciacca B, van de Groep J, Polman A and Garnett E C 2016 Nanowires: Solution-grown silver nanowire ordered arrays as transparent electrodes *Adv. Mater.* **28** 976

[8] Skjeltorp A T 1985 Ordering phenomena of particles dispersed in magnetic fluids *J. Appl. Phys.* **57** 3285

[9] Černák J 1994 Aggregation of needle-like macro-clusters in thin layers of magnetic fluid *J. Magn. Magn. Mater.* **132** 258

[10] Ngo A T and Pileni M P 2003 Assemblies of cigar-shaped ferrite nanocrystals: orientation of the easy magnetization axes *Colloid Surf. A-Physicochem. Eng. Asp.* **228** 107

[11] Xu D, Lu C H, Zhang D P, Song J B, Ni Y R and Xu Z Z 2012 Fabrication and optical properties of self-assembled films with micro needlelike arrays under magnetic field *Mater. Lett.* **71** 94

[12] Holm C and Weis J J 2005 The structure of ferrofluids: A status report *Curr. Opin. Colloid Interface Sci* **10** 133.

[13] Yusuf N A 1989 Field and concentration dependence of chain formation in magnetic fluids *J. Phys. D: Appl. Phys.* **22** 1916

[14] Butter K, Bomans P H, Frederik P M, Vroege G J and Philipse A P 2003 Direct observation of dipolar chains in iron ferrofluids by cryogenic electron microscopy *Nat. Mater.* **2** 88

[15] Tang Y, Chen Q and Chen R 2015 Magnetic field induced controllable self-assembly of maghemite nanocrystals: From 3D arrays to 1D nanochains *Appl. Surf. Sci.* **347** 202

[16] Lalatonne Y, Richardi J and Pileni M P 2004 Van der Waals versus dipolar forces controlling mesoscopic organizations of magnetic nanocrystals *Nat. Mater.* **3** 121

[17] Jordan P C 1979 Field dependent chain formation by ferromagnetic colloids *Molecular Physics* **38** 769

[18] Ivey M, Liu J, Zhu Y and Cutillas S 2000 Magnetic-field-induced structural transitions in a ferrofluid emulsion *Phys. Rev. E.* **63** 011403

Preparation and radar absorptive properties of $BaFe_{12}O_{19}$ -coated glass fiber

F Jia[1], M Xu, H Q Bao, K Cui, F Zhang

China Building Material Academy, Beijing 100024, China

E-mail: jiafei870316@163.com

Abstract: Traditional passive jamming materials such as chaff and foil showed some limitations in use because they can only reflect the electromagnetic wave. Therefore, to develop a kind of absorptive passive jamming material to make up for deficiencies of traditional passive jamming materials and improve the jamming efficiency is of great significance. In this paper, the $BaFe_{12}O_{19}$-coated glass fiber, used as a kind of radar absorptive chaff, was prepared by sol-gel dip-coating method. The effects of heat treatment temperature, heat treatment time and coating times on film quality, tensile strength and attenuation efficiency of the samples were discussed. The study shows that an increase of the heat treatment temperature and an extension of the heat treatment time is conducive to the growth of barium ferrite grain, while they would introduce the loss of chaff strength at the same time. In addition, multi-coating process can improve the film quality and attenuation efficiency of the sample. Data show that the 10 times coated samples have a best reflectivity of (15GHz, -6.65dB) and the bandwidth of reflectivity lower than -5dB is11.8 GHz. According to the test results, the prepared material has certain attenuation efficiency in the range of 2GHz-18GHz, having a high practical value.

1. Introduction

Since World War II, the passive radar jamming technology has been rapidly developed. A series of passive jamming devices, mainly including chaff, infrared bombs, and smoke bombs, have been developed and equipped to the sea, land, air and space combatant platforms[1]. Meanwhile with the development of radar anti-jamming technology, especially the anti-chaff technology, traditional passive jamming materials are easily identified by radar and their operational efficiency has been reduced, which would limit their use. Therefore, to develop an efficient, practical, inexpensive absorptive passive jamming material, make up of the chaff, the smoke, and other conventional passive jamming materials to counter radar detector and multi-mode precision guided weapons, has an important practical significance. William[2] formed a millimeter wave-screening cloud comprised with an aerosol of fine fibers of a carbon composition, in which the particles are of micron diameter and millimeter length. Xu Yun[3] modified the surface properties of carbon fiber by nitric acid oxidation and in-situ PANI coating methods. And the polyaniline-coated carbon fiber has certain attenuation efficiency in the 3mm and 8mm wave. Wang Bei[4] prepared the carbon fiber/retinyl Schiff base salts composite materials which have an absorption and interference effect for 8mm wave. Chen Xin[5] prepared the nickel coated bamboo fibers as the millimeter wave smoke interfering materials. Han Chao[6] covered the nano wave absorption material on aluminum glass fiber and obtained the millimeter wave absorption chaff. The jamming materials prepared by the above research have a certain jamming effect on the millimeter wave, but the effect is not obvious in the range of 2GHz-18GHz. In this paper, a layer of barium ferrite film was coated on the surface of glass fiber by sol-gel[7-9] method. And the effects of heat treatment temperature, heat treatment time and coating times on film quality, tensile strength and attenuation efficiency of the $BaFe_{12}O_{19}$-coated glass fiber were discussed. According to the test results, the absorptive chaff with certain attenuation efficiency in the range of 2GHz-18GHz has been obtained, and from this work a kind of feasible method to prepare the absorptive passive jamming material has been provided.

2. Experimental procedure

2.1. Synthesis of Sol

First, appropriate amount of citric acid was dissolved in the minimum amount of deionized water. Next, $Fe(NO_3)_3 \cdot 9H_2O$ and $Ba(NO_3)_2$ were added into the prepared aqueous solution, in a Fe/Ba molar ratio of 11.5, as shown in Table 1. It should be mentioned that the molar ratio of $C_6H_8O_7/(Fe^{3+} + Ba^{2+})$ should be 1.5. Under the condition of rapid mixing, the mixed solution was neutralized to pH 7 by adding aqueous ammonia. Then, appropriate amount of polyethylene glycol was added, stirred evenly and kept there for 1h. After that, the neutralized solution was evaporated by heating at 90℃ for 3h in water bath with stirring. As the water evaporated, the solution became viscous and finally formed a transparent sol with a certain fluidity and viscosity.

Table1. Preparation parameters of sol

Fe^{3+}/Ba^{2+}	$C_6H_8O_7/M^{n+}$	polyethylene glycol(g/L)	PH	T(℃)	T(h)
11.5	1.5	6.6	7	90	3

2.2. Preparation of $BaFe_{12}O_{19}$-coated glass fiber

The concentrated sulfuric acid and hydrofluoric acid were mixed in a volume ratio of 1:3 as a coursing solution. The glass fibers (diameter of 20μm) were immersed into the coursing solution for 3 minutes at 30℃~40℃ under ultrasonic environment first, next washed with deionized water until the pH value of the washing water reached 7, and then washed with absolute ethyl alcohol 3 or 4 times or so before drying. After that, the cleaned glass fibers were immersed into the prefabricated sol for 5 minutes under ultrasonic environment. Then the glass fibers were pulled out smoothly, dried at 120℃, and heated at 500℃ for 1h. The coating and drying process has been repeated for several times, and finally the sample was placed in a box resistance furnace and heated at different temperatures to obtain barium ferrite film.

2.3. *Performance testing*

The coating state and the thickness of the barium ferrite film were observed by scanning electron microscope (SEM). According to the Chinese national military standard GJB2038A-2011, the attenuation efficiency of the samples was tested by the arch method, the test frequency was 2GHz-18 GHz, the incident angle was 8°, and the polarization direction was vertical polarization (VV). The mechanical properties of single fiber were examined using YG001A type electronic strength tester with tensile rate of 2mm/min.

3. Results and discussion

3.1. *Effect of heat treatment temperature on tensile strength and film quality*

As shown in Table 2, the film of samples exhibit a red-brown color when the heat treatment temperature was 600℃, with the increase of temperature, the sample color deepened, finally turned brown. In addition, SEM test results in Figure 1 showed that at 600℃ of the heat treatment temperature, the barium ferrite grains on the sample surface developed imperfect, and the particle size was small. When the heat treatment temperature was 640℃, the development of barium ferrite crystal tend to be complete, showing a hexagonal lamellar morphology, and the particle size increased significantly. With further increasing the heat treatment temperature, the barium ferrite crystals in the film developed more complete, and the particle size continued to increase. When the heat treatment temperature was raised up to 720℃, the complete barium ferrite particles of hexagonal lamellar morphology, which is 1-2μm, can be observed. Therefore, with the increase of heat treatment temperature, the crystal particles of the sample gradually grew up, showing hexagonal lamellar morphology. This is due to the lower the heat treatment temperature, the less the solid phase reaction occurred (see formula (1)-(4)), and there were Fe_2O_3, $BaFe_2O_4$ and other intermediate phase in the product. With the rising heat treatment temperature, the solid phase reaction is complete, and the product becomes only barium ferrite.

$$\text{dry gel} \rightarrow \gamma\text{-}Fe_2O_3 + BaCO_3 \quad (1)$$

$$\gamma\text{-}Fe_2O_3 + BaCO_3 \rightarrow BaFe_2O_4 + CO_2\uparrow \quad (2)$$

$$\gamma\text{-}Fe_2O_3 \rightarrow \alpha\text{-}Fe_2O_3 \quad (3)$$

$$BaFe_2O_4 + 5\gamma\text{-}Fe_2O_3 \rightarrow BaFe_{12}O_{19} \quad (4)$$

Table2. Effect of heat treatment temperature on sample's properties

Heat treatment temperature(℃)	Color	Film condition	Tensile strength(Mpa)
600	red-brown	well-distributed	1540
640	brown	well-distributed	1078
680	brown	well-distributed	484
720	brown	well-distributed	——

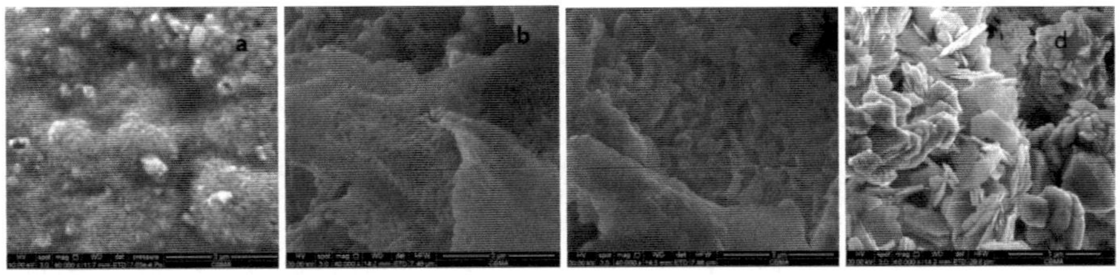

Figure 1. SEM images of samples sintered at different temperatures (a.600℃, b.640℃, c.680℃, d.720℃)

In addition, the test data from Table 2 showed that the heat treatment temperature had a significant influence on the tensile strength of $BaFe_{12}O_{19}$/ glass fiber. When the temperature was 600℃, the tensile strength of the sample was 1540 Mpa. With the rising heat treatment temperature, the tensile strength of the sample decreased rapidly, when the temperature was raised up to 680℃, the strength was only 484 Mpa, which could not meet the requirement of Chinese national military standard GJB 1872-94.When the temperature rose up to 720℃, embrittlement of the sample ocurred.

3.2. Effect of heat treatment time on tensile strength and film quality

Figure 2 and Table 3 is the test result of the samples sintered for different time at 640℃. SEM test results showed that when the heat treatment time was 5min, particles in the film were mostly acicular morphology. With the extension of heat treatment time, the crystallinity of barium ferrite crystals increased, crystal growth tended to be complete and the particle size increased. When the heat treatment time was 30min, particles in the film showed significant lamellar morphology, and thereafter, with the extension of the heat treatment time, the lamellar morphology developed more perfect and the particle size was larger. In addition, it is showed in Table 3 that with the extension of heat treatment time, sample color gradually deepened, from red-brown to brown and finally converted into dark brown. This is due to the extension of solid phase reaction between intermediate phases, which is beneficial to the formation of the barium ferrite crystals and the growth of the grains when the heat treatment time is prolonged.

Table3. Effect of heat treatment time on sample's properties

Heat treatment time(min)	Color	Film condition	Tensile strength(Mpa)
5	red-brown	well-distributed	1782
15	brown	well-distributed	1298
30	brown	well-distributed	1078
45	brown	well-distributed	671
60	dark brown	peel off	456

The data in Table 3 also showed that with the extension of the heat treatment time, the strength loss of the sample increased. When the heat treatment time was 5min, the tensile strength of the sample was 1782 Mpa, while the heat treatment time was extended to 45min, the tensile strength dropped to

671 Mpa, still meet the GJB1872-94 requirements. When the sample was sintered for 1h at 640℃, the tensile strength was only 456 Mpa, below requirement of GJB 1872-94 provisions, and the film peeled off.

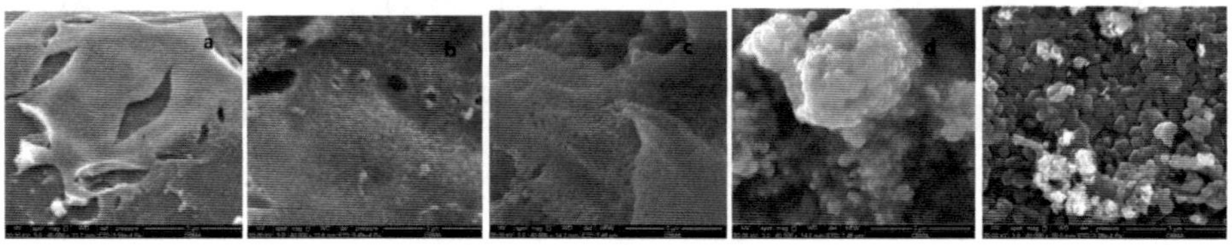

Figure 2. SEM images of samples sintered for different time (a.5min, b.15min, c.30min, d.45min, e.60min)

The reason of the strength loss of the sample after heat treatment can be explained from two aspects. On one hand, with the increase of heat treatment temperature, the thermal motion of particles increases, the molecular bond is weakened, and the strength of the glass fiber decreases. On the other hand, the surface of glass fiber is not smooth, but there are varying amounts of different sizes of micro-cracks. Especially after HF roughening treatment, micro-cracks and defects in the glass fiber surface are greatly increased. According to the Griffith's theory[10, 11], there is stress concentration near the micro-cracks and defects in the surface of the glass fiber because of the thermal stress during heat treatment. When the stress exceeds a certain critical value, cracks begin to expand and result in the strength decrease of the glass fiber. The higher the heat treatment temperature, the longer the heat treatment time, and the greater the stress of micro-cracks and defects, the more serious loss of the strength the glass fiber would suffer.

3.3. Effect of coating times on attenuation efficiency

Figure 3(a) was the reflectivity curve of the prepared samples. Data showed that as the frequency increases, the attenuation efficiency of the sample increased. When the frequency was 17GHz, the attenuation efficiency of the sample was best. But in the 2GHz-18GHz range, the overall attenuation efficiency of the sample was poor. Especially in the low frequency region, the attenuation efficiency was only about 6%.

There are many factors influence the attenuation efficiency of absorbing materials, in addition to the electromagnetic parameters and dimensions of the material itself, equalization of electromagnetic parameters, surface roughness, cracks, and parallelism of the surface also have some effects. The low attenuation efficiency of $BaFe_{12}O_{19}$-coated glass fiber prepared can be explained from the following three aspects:

A. Due to the large surface tension of glass fiber, big difference between interfacial thermal expansion coefficients and other factors, the film on the surface of the prepared material was rough with inhomogeneous thickness and cracks, resulting in deterioration of the absorbing properties of the material.

B. The distribution of absorbing agent in the sample was inhomogeneous, and the volume fraction of electromagnetic wave transparent material was larger than that of the absorbing agent, so that the power transmission coefficient of the material is larger, resulting in deterioration of the absorbing properties of the material.

C. The process of preparing thin films by sol-gel method is a process of non-equilibrium state and greatly influenced by the matrix, which is prone to abnormal crystal structure and microstructure, resulting in deterioration of the absorbing properties of the material. The main reasons are as follows: first, the lattice constant of the raw material does not match with the matrix material, in order to achieve the effective combination, crystal lattice distortion or crystal dislocation will occur near the interface between the film and the substrate. Second, because of the large curvature of the surface of glass fiber and its inconsistent thermal expansion coefficient with the film, the film is subjected to a large surface stress, resulting in the abnormal structure of the crystal.

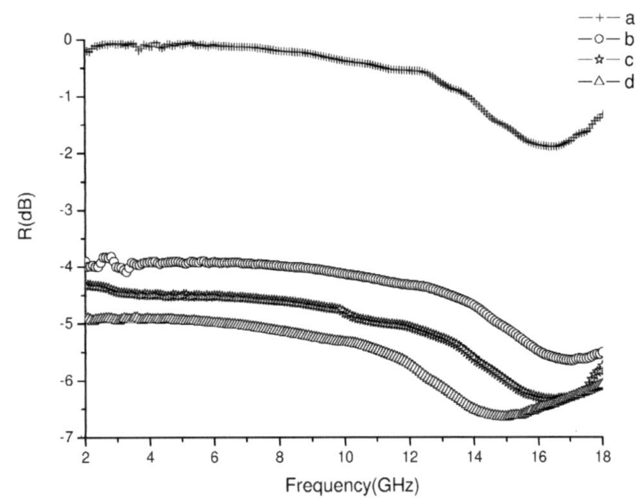

Figure 3. R-f curves of samples with different coating times (a.1, b.2, c.5, d.10)

In summary, under single coating conditions, due to the rough, thin film of $BaFe_{12}O_{19}$-coated glass fiber and barium ferrite lattice distortion, the effective absorbing agent content of the sample was low, resulting in the limited attenuation efficiency. In order to improve the film quality and the effective barium ferrite content, the film quality and absorbing properties of multi-coating $BaFe_{12}O_{19}$-coated glass fiber samples were studied.

Data showed that, with the increase of coating times, the attenuation efficiency of the sample had been improved, and the absorption peak shift towards lower frequency. The bandwidth of reflectivity lower than -5dB of 2 times coated samples was 3.2GHz (Figure 3(b)), while 6.8GHz of 5 times coated samples (Figure 3(c)), 11.8GHz of 10 times coated samples (Figure 3(d)). In addition, as the frequency increases, the reflectivity of the sample significantly reduced, the attenuation efficiency significantly improved and was up to 60%-80% within 2GHz-18GHz, having a high practical value.

In summary, increasing the coating times of the film can improve the attenuation efficiency and broaden the absorption band of the sample. And with the increase of the coating times, the absorption peak shift towards lower frequency. SEM test results in Figure 4 showed that with the coating times increased from 2 times to 10 times, the thickness of the barium ferrite film increased from 500nm to about 1.5μm. Therefore, increasing the coating times can increase the thickness of barium ferrite film and improve the surface state of the sample, which can improve the effective barium ferrite content and attenuation efficiency of the sample.

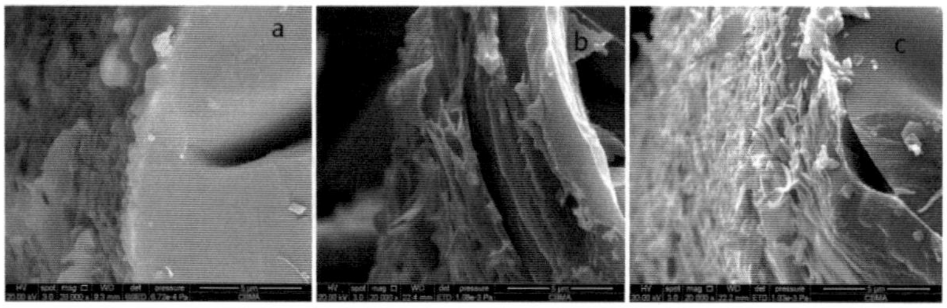

Figure 4. SEM images of multiple coating samples (a.2, b.5, c.10)

4. Conclusion

$BaFe_{12}O_{19}$-coated glass fiber composites prepared by sol-gel method have certain attenuation efficiency. Data show that the multi-coating samples have a best reflectivity of (15GHz, -6.65dB) and the bandwidth of reflectivity lower than -5dB is11.8 GHz, having a high practical value. The increase of heat treatment temperature and time was in favor of barium ferrite grain growth, and will result in the loss of strength of the sample. Increasing the coating times can improve the film quality and attenuation efficiency of the sample.

References

[1] Hu S J 2008 Trends of Progress in the Passive Interference Technology of Radar *Ship Electronic Engineering* **28**(6) 121-123,153

[2] William G R 1992 Millimeter wave screening cloud and method: America, US5148173. Sep.15, 1992

[3] Xu Y 2009 *Preparation and millimeter wave attenuation properties of carbon fiber/polyaniline composite*. Nanjing: Nanjing University of Science and Technology, 2009

[4] Wang B 2009 *Preparation and millimeter wave attenuation properties of retinyl Schiff base salts* Nanjing: Nanjing University of Science and Technology, 2009

[5] Chen X, Pan G P, Zhao J, Cao C X 2010 Millimeter wave attenuation performance of nickel coated bamboo fibre. *Journal of PLA University of Science and Technology(Natural Science Edition)* **11**(6) 664-667

[6] Han C, Zhao G Z and Huang X Z 2006 Millimeter wave absorption chaff study. *Journal of Projectiles, Rockets, Missiles and Guidance* **26**(2) 484-485

[7] Mali A, Ataie A 2005 Structural characterization of nano-crystalline BaFe12O19 powder synthesized by sol-gel combustion route *Scripta Materialia* **53** 1065-1070

[8] Zhong W, Ding W, Zhang N, Hong J, Yan Q and Du Y 1997 Key step in synthesis of ultrafine BaFe12O19 by sol-gel technique. *Magn. Magn. Mater* **168** 196-202

[9] Surig C, Hempel K A and Bonnenberg D *IEEE Trans Magn* **30** 4092

[10] Griffith A A 1920 Philosophical Transactions of the Royal Society. London, 1920, A221, 163

[11] Griffith A A ICAM. Delft, 1924, 55

IOP Conf. Series: Materials Science and Engineering **137** (2016) 012057 doi:10.1088/1757-899X/137/1/012057

Study on catalysis effect of TEPB on the curing reaction of HTPB binder system

S J Chang[1], J Tang, X Liu and W Yan

School of Chemical and Environmental Engineering, North University of China, Taiyuan 030051, China

E-mail: junchsh@163.com

Abstract: The catalysis effect of tri (exhoxyphenyl) bismuthine (TEPB) on the curing reaction of HTPB binder system was studied by using DSC method. The curing peak temperatures of the catalyst systems were measured to calculate kinetic parameters by using Kissinger and Crane methods, respectively. Two curing reaction kinetic equations were established. The results show that TEPB has high catalytic activity and can decrease the curing temperature of HTPB binder system, down to 35℃, in which the optimum volume of TEPB is 0.5% of HTPB binder system.

1. Introduction

Hydroxyl-terminated polybutadiene (HTPB) is a kind of telechelic polymer. It has low viscosity, oil resistant, aging resistant, good low temperature performance, and machining performance. Many kinds of polyurethane materials have been made by the cross-linking reaction of HTPB with toluene diisocyanate (TDI). Due to the convenience for casting and manufacturing, it has been applied to cast PBX binder [1-2].

The binder of HTPB imparts dimensional stability and structural integrity to the grain for appropriate mechanical properties. However, the reaction rate of hydroxyl functional groups of HTPB with isocyanates is usually low, leading to a high curing temperature and long manufacturing period [3]. When the casting PBX is cured, its internal residual stress could reduce the mechanical property of PBX after it is cooled to ambient temperature. High manufacturing temperature increase the danger of cast PBX productive process and increases the production cost, too.

Therefore, more and more attention has been paid to reduce the curing temperature of HTPB. The catalyst is the best choice to decrease curing temperature and shorten curing time[4]. Dibutyltin dilaurate (DBTDL), ferric tris (acetyl acetonate) ($Fe(AA)_3$) and triphenyl bismuth (TPB) are usually used as catalysts. The catalytic activity and mechanism have been investigated by using various techniques. However, the high curing temperature still was not solved after by using these catalysts [5-7]. In order to decrease the curing temperature and increase the mechanical properties, in this work, the catalysis effect of TEPB on HTPB binder system curing reaction was investigated.

2. Experimental

2.1. Formulation composition

HTPB binder system is made up of HTPB and DOA by the proportion of 1:1, with a certain amount of curing agent TDI. TEPB is put in additionally. HTPB is made by Shandong zibo qilu ethylene

IOP Conf. Series: Materials Science and Engineering **137** (2016) 012057 doi:10.1088/1757-899X/137/1/012057

chemical co., LTD, with a hydroxyl value of 0.77-0.82 mmol/mg. DOA is made by Liaoning yingkou tianyuan chemical co., LTD. TEPB was supplied by Shanghai Institute of Organic Chemistry, Chinese Academy of Sciences, with a purity of 99.54%.

2.2. Methond

By adding different amount of TEPB to HTPB binder system, the catalysis action may be assessed through testing its curing hardness. Catalyst influence of TEPB content on HTPB binder system was investigated at 0.3%, 0.5%, 0.7%. The curing reaction of HTPB binder system was studied at various heating rates, viz., 5, 10, 15 and 20℃·min^{-1}. Heating rates were varied for determining the kinetics by Kissinger and Crane methods, respectively. A certain amount of catalyst was put in the HTPB binder system. The mixture of about 20 mg was taken in alumina pans and subjected to DSC analysis in a Setaram DSC131 in nitrogen atmosphere at a flow rate of 50 mL·min^{-1}. The heating temperature was in the range of 20 ~ 300℃.

3. Results and discussion

3.1. The catalysis Effect of TEPB conten on HTPB binder system

HTPB binder system is one of the carbamates by the reaction of the hydroxyl group with isocyanate. Catalyst can reduce the reaction activation energy of HTPB binder system and accelerate reaction rate, leading to a shortened curing time. Catalyst and isocyanate may generate intermediate complex, which can react with hydroxyl compounds to generate polyurethane. The mechanism of reaction [8] is as follows:

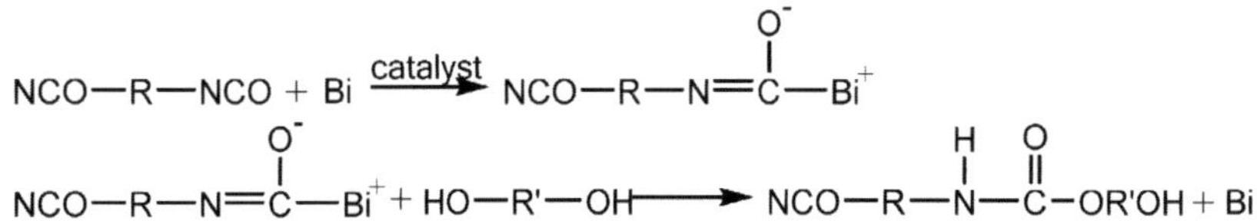

The curing property of HTPB binder system could be improved by the reaction of carbamate catalyst. Figure 1 shows the effect of curing hardness about different content TEPB on HTPB binder system. From Figure 1, when TEPB contents are 0.3%, 0.5%, and 0.7%, the curing hardness is higher than that without catalyst. The addition of catalyst accelerated reaction rate of HTPB binder system. As TEPB addition 0.5%, the hardness may reach to 35 HA after 72 h.

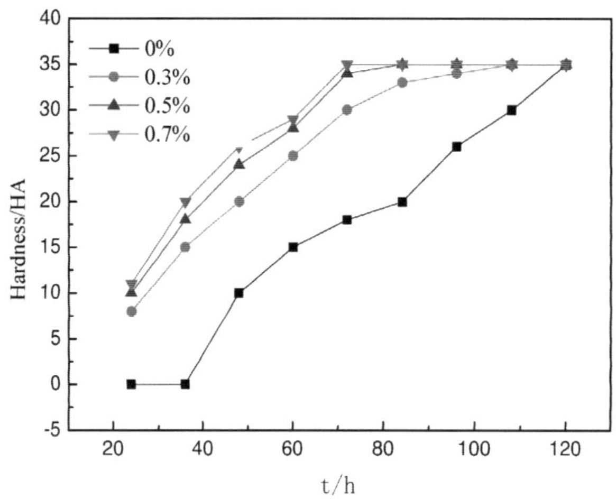

Figure 1. Hardness curve with the time change.

3.2. TEPB Catalyst to HTPB binder system

The curing reaction rates of HTPB binder system with TEPB and without catalyst were measured by the DSC method. Test results were shown in Figure 2. TEPB catalyst decreases the reaction temperature and makes exothermic peak of DSC curve translate forward. Two apparent exothermic peaks were found in the HTPB-TDI curing system with the catalyst TEPB.

According to the research on catalytic mechanism, TEPB could be combined with hydroxyl. The possible mechanism is through the activation on the isocyanate-hydroxyl complex, which speeds up the reaction proceeds. The first exothermic peak is due to that the combination of TEPB with hydroxyl hydrogen increases the nucleophilicity ability of the adjacent oxygen in the hydroxyl group. Therefore, the catalyzed reactions of hydroxyl with isocyanate may be suggested to involve the initial formation of a TEPB-hydroxyl intermediate, which reacts with isocyanate as a subsequent reaction. Thus, the urethane is formed. The catalyst is regenerated, leading to the formation of the second exothermic peak.

Table 1 and Table 2 give the initial curing temperature(T_i), the peak curing temperature (T_p) and the end curing temperature(T_f) from the DSC curves of the uncatalyzed and TEPB catalyzed HTPB-TDI curing system.

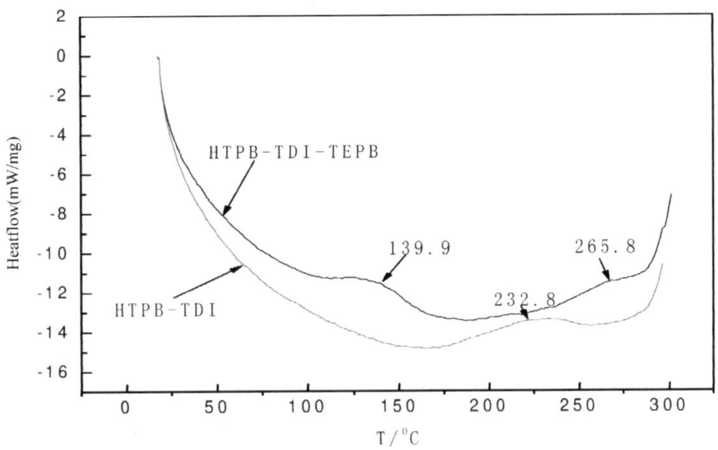

Figure 2. DSC curves of different curing system at 15℃·min^{-1}.

Table 1. DSC data of HTPB-TDI curing reaction.

β/℃·min^{-1}	T_i/℃	T_p/℃	T_f/℃
5	153.6	195.6	209.8
10	187.6	212.9	238.1
15	200.9	232.8	245.2
20	212.5	239.9	261.2

Note: β: heating rate; T_i: the initial curing temperature; T_p: the peak curing temperature; T_f: the end curing temperature.

There're apparent early and later stages in the DSC curve of TEPB catalyzed HTPB-TDI curing system. The peak temperature of the early stage was reduced 90℃ compared to uncatalyzed curing system. And the curing reaction peak temperature gradient (T_f -T_i) was decreased from 50℃ to 30℃. Above data shows that TEPB can decrease the curing reaction temperature and shorten reaction time.

The high peak temperature of later stage indicates that TEPB-isocyanate-hydroxyl associate or complex is quite stable. Although there're significant differences in catalytic mechanism of different catalysts, the reaction rate gets faster and catalytic activity gets higher when transient state is more stable. It shows the high catalytic activity of TEPB from the reaction mechanism.

Table 2. DSC characteristic temperature of early and later stage of TEPB catalysed HTPB-TDI reaction.

β/℃·min^{-1}	early stage reaction			later stage reaction		
	T_i/℃	T_p/℃	T_f/℃	T_i/℃	T_p/℃	T_f/℃
5	91.1	109.5	121.3	234.2	243.2	259.8
10	109.1	125.7	138.9	240.7	254.1	265.3
15	121.9	139.9	153.6	243.6	265.8	270.9
20	131.2	148.7	162.4	247.2	274.6	288.2

3.3. Kinetics of TEPB catalyzed HTPB binder system curing

Kinetic parameters of curing reaction are calculated by Kissinger and Crane method.

Kissinger formula:

$$d[\ \ln(\beta / T_P^2)\ ] / d(T_P^{-1}) = -Ea / R \qquad (1)$$

Crane formula:

$$d(\ln\beta) / d(T_P^{-1}) = -(Ea / nR + 2T_P) \qquad (2)$$

$$Ea / nR >> 2T_P,\ d(\ln\beta) / d(T_P^{-1}) = -Ea / nR$$

Where β is the heating rate, and Ea is the activation energy of curing reaction, and R is the universal gas constant, and T is the absolute temperature, and n is the order of reaction.

HTPB-TDI curing system abide n order reaction model according to the research on the curing reaction mechanism of HTPB [9,10], it's reaction kinetic model equation is:

$$d\alpha / dt = k(1 - \alpha)^n \qquad (3)$$

Where α is the degree of reaction, and t is the reaction time, and dα/dt is the reaction rate, and k is the reaction rate constant which following the Arrhenius formula $k = A\exp(-Ea/RT)$. For all the values of n except n=1.

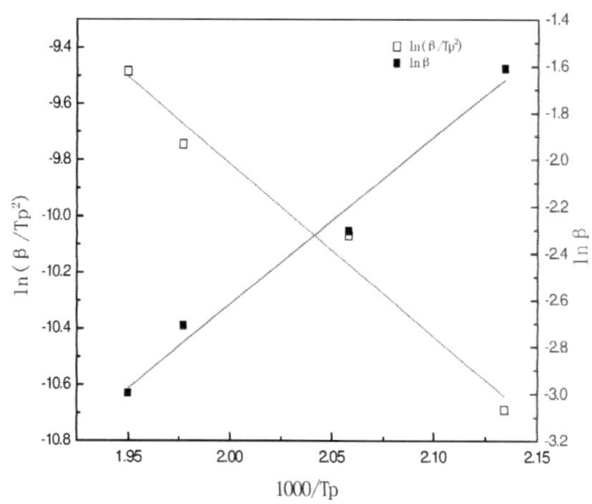

Figure 3 Linear fitting of -lnβ and ln(β/T_p^2) with 1000/T_p for uncatalysted curing system

Figure 3 and 4 show the fit linear plots of ln(β/T_P^2) and -lnβ *vs.*1000/T_P for uncatalyzed curing system and the early stage of TEPB catalyzed system, respectively. *Ea*, pre-exponential factor (*A*) and *n* can be calculated by Kissinger and Crane methods.

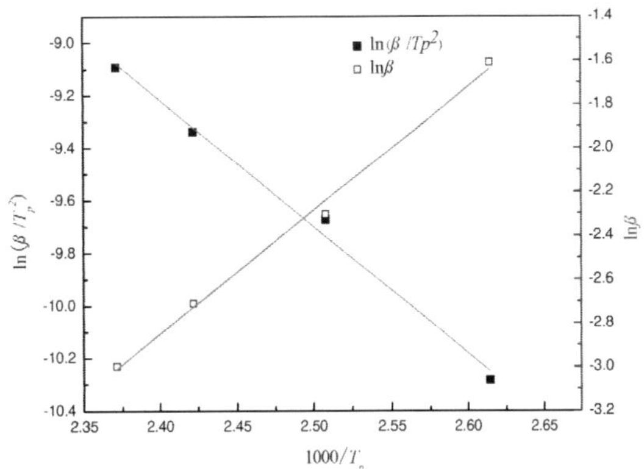

Figure 4 Linear fitting of -lnβ and ln(β/T_p^2) with 1000/T_p for TEPB catalyst system

The kinetic parameters of uncatalyzed HTPB-TDI curing system are Ea=51.29 kJ·mol^{-1}, A=7.7×10^4 s^{-1}, n=0.86; those in the early stage of TEPB catalyzed reaction are Ea=40.14 kJ·mol^{-1}, A=5.2×10^4 s^{-1}, n=0.85; those in the later stage are Ea=93.12 kJ·mol^{-1}, A=6.0×10^8 s^{-1}, n=0.91. Above parameters are substituted into the reaction kinetic model equation(3).

The kinetic equation of HTPB-TDI curing system is:

$$d\alpha / dt = 7.7\times10^4 \exp(-6169 / T)(1 - \alpha)^{0.86}$$

The early stage kinetic equation of HTPB-TDI-TEPB curing system is:

$$d\alpha / dt = 5.2\times10^4 \exp(-4828 / T)(1 - \alpha)^{0.85}$$

The later stage kinetic equation of HTPB-TDI-TEPB curing system is:

$$d\alpha / dt = 6.0\times10^8 \exp(-11199 / T)(1 - \alpha)^{0.91}$$

TEPB decreases the activation energy of curing reaction, which is the reason that TEPB can lower the reaction temperature. The reaction orders of two curing systems is similar and approximate 1, which shows that catalyst do not change the elementary reaction of HTPB-TDI and curing reaction abide *n* order reaction model correspondingly.

Table 3. Rate constants under different temperature.

catalyst system	k/s^{-1}			
	30℃	60℃	100℃	105℃
Null	1.11E-04	6.93E-04	5.06E-03	6.29E-03
TEPB$_{early}$	6.25E-03	2.63E-02	1.24E-01	1.48E-01
TEPB$_{later}$	5.33E-08	1.48E-06	5.48E-05	8.15E-05

Note: Null: uncatalyzed HTPB-TDI curing system.

The kinetic equations show that there is a linear relationship between reaction rate constants and reaction temperatures. Reaction rate constants decide the reaction rate under the certain temperature. Some reaction rates of uncatalyzed and TEPB catalyzed curing reactions under certain temperature are shown in table 3. The curing temperature of HTPB is 60℃ in actual production, and the reaction rate constant under this temperature is barely 6.93×10^{-4} s^{-1}. It causes the long curing period and low production efficiency of HTPB. And the reaction rate constant of TEPB catalyzed curing system under

35℃ is 6.25×10^{-3} s^{-1}, which is equal to that of uncatalyzed system under 105℃. The curing temperature of polyurethane is usually below 100℃. High temperature promotes isocyanate groups reacting with carbamic acid ester or urea linkages or generating cross bond, which decreases the properties of PU. High manufacture temperature reduces the safety of HTPB binder system curing process. This is just the reason that catalysts are added to decrease the manufacture temperature. TEPB can make the curing reaction rate reach the value at ambient temperature. Above data proves that the TEPB is available for ambient catalyst based on the reaction kinetic.

4. Conclusions

- The catalytic activity of TEPB increases with the concentration, and reaches the maximum when concentration is 0.5%. The appropriate concentration of TEPB is 0.5%.
- TEPB may decrease the curing reaction temperature and shorten reaction time by its high catalytic activity. The peak temperature using TEPB was reduced 90℃ compared to uncatalyzed curing system.
- HTPB-TDI-TEPB system curing reaction kinetic equation is obtained. The reaction rate constant at 35℃ reaches the value of uncatalyzed system at 105℃. TEPB can catalyze HTPB-TDI curing reaction at ambient temperature.

Acknowledgments

We are grateful to the Natural Science Foundation of Shanxi Province (2015011013).

References

[1] Wang J Y, An C W and Li G 2011 Preparation and performances of castable HTPB/CL-20 booster explosives *Propellants. Explos. Pyrotech.* **36** 34-41
[2] Adapaka S K, Vepakomma B R and Rabindra K S 2010 Evaluation of plastic bonded explosive (PBX) formulations based on RDX, aluminum, and HTPB for underwater applications *Propellants. Explos. Pyrotech.* **35** 359-64
[3] Chen C Y, Wang X F, Gao L L and Zheng Y F 2013 Effect of HTPB with different molecular weights on curing kinetics of HTPB/TDI system *Chin. J. Ener. Mater* **21** 771-6
[4] Korah B C, Kannan K G and Ninan K N 2004 DSC study on the effect of isocyanates and catalysts on the HTPB cure reaction *J. Therm. Anal. Calorim.* **78** 753-60
[5] Volkova E R, Tereshatov V V and Karmanov V 2011 I Influence of the concentration of iron(III) tris(acetylacetonate) used as urethane formation catalyst on the structure and properties of cold-cure polyurethanes *Russ. J. Appl. Chem.* **84** 1414-17
[6] Lu G L, Xia Q and Du J 1999 Catalysis effect of TPB on the cure of HTPB-based high burning rate propellant *Chin. J. Ener. Mater.* **7** 60-1
[7] Luo S G, Tan H M and Zhang J G 1997 Catalytic mechanisms of triphenyl bismuth, dibutyltin dilaurate, and their combination in polyurethane-forming *J. Appl. Polym. Sci.* **65** 1217-25
[8] He Y D, Sun X Y 2004 Study on the adjustment method of HTPB propellant binder matrix structure *J. Solid. Rocket. Techno.* **27** 294-7
[9] Singh M, Kanungo B K and Bansal T K 2002 Kintic studies on curing of hydroxy-terminated polybutadiene prepolymer-based polyurethane networks *J. Appl. Polym. Sci.* **85** 842-6
[10] Han J L, Yu C H and Lin Y H 2008 Kinetic study of the urethane and urea reactions of isophorone diisocyanate *J. Appl. Polym. Sci.* **107** 3891-902

IOP Conf. Series: Materials Science and Engineering **137** (2016) 012058 doi:10.1088/1757-899X/137/1/012058

Functional initiators for both ATRP and ROP catalyzed by iron(III) catalyst

J Li, C Yang and C J Cheng[1]

School of Chemistry and Chemical Engineering, Jiangxi Science & Technology Normal University, Nanchang, Jiangxi, 330013, P.R. China

E-mail: chengcj530@163.com

Abstract. α-Trichloromethyl benzyl alcohol was successfully used as initiators for both AGET ATRP and ROP reactions, which was catalyzed by only one non-toxic and very cheap catalyst iron(III) chloride. The corresponding polymers PMMA and PCL were characterized by ^{1}H NMR, and their molecular mass were calculated as 7.53 kDa and 10.08 kDa, respectively.

1. Introduction

Living polymerization methods have been extensively developed and applied in many areas such as polymer preparation, materials designing, life science, drug manufacturing and delivery, etc [1-3]. For different living polymerization approaches, living radical polymerizations, such as ATRP, reversible addition-fragmentation chain transfer polymerization (RAFT), and nitroxide mediated radical polymerization (NMP), are the mostly studied and used to prepare functional macromolecules [4-6]. Ring opening polymerization (ROP) is another powerful tool to produce many kinds of polymers, e.g., polycaprolactone (PLC), polyethylene oxide (PEO), etc [7-10].

To incorporate polymer segments which are incompatible into one macromolecule, strategies for combination of living polymerization and ROP are utilized, thus various functional polymers which are previously inaccessible are synthesized [11-12]. However, these reaction systems are always complex due to the presence of different components. To simplify the reaction system, multi-functional molecules are designed and used in literature [13].

In this study, a synthesized molecule is used as an initiator for both AGET ATRP and ROP reactions, which make it possible to prepare block copolymers simultaneously or in one pot. Furthermore, both activator-generated by electron transfer atom transfer radical polymerization (AGET ATRP) and ROP reactions can be catalyzed by only one catalyst.

2. Experimental

2.1. Instruments and reagents

Nuclear magnetic resonance (NMR) was measured by a BrukerAV 400MHz nuclear magnetic resonance meter ($CDCl_3$ as solvent, TMS as internal standard).

Chloroform, benzenaldehyde, tetrabutylammonium bromide (TBAB), Iron(III) chloride hexahydrate, ε-caprolactone (ε-CL), dichloromethane, petroleum ether (PE), ethyl acetate (EA), *N,N,N'N'*-tetramethyl ethylene diamine (TMEDA), ascorbic acid (AA), N,N-dimethyl formamide

(DMF) and dimethyl sulfoxide (DMSO) are all AR grade and used as received. Methyl methacrylate (MMA) was treated with basic ion-exchangeable resin to remove inhibitor before use.

2.2. Synthesis of α-trichloromethyl benzyl alcohol (1)
Synthesis of this initiator was easily finished according to our previous work [14], i.e., the mixture of benzaldehyde and chloroform under strong basic conditions furnished compound **1** in moderate yield.

2.3. Preparation of poly(methyl methacrylate) (PMMA) (2) using 1 as an initiator and $FeCl_3$ as a catalyst
The material MMA (10.0 g, 100 mmol), iron(III) chloride hexahydrate (0.54 g, 2 mmol), the initiator α-trichloromethyl benzyl alcohol (**1**) (0.23 g, 1 mmol), ascorbic acid (0.18 g, 1 mmol), and TMEDA (0.17 g, 1 mmol) were added to the solvent DMSO (80 mL). The resulting mixture was stirred at 80 °C for 12 h under the protection of N_2 atmosphere. After the reaction was completed, excess cold methanol was poured, and the resulting precipitate was filtered to afford crude polymer. The crude product was purified on neutral Al_2O_3 column to provide a white like solid (7.33 g), conversion 71%.

2.4. Preparation of poly(ε-caprolactone) (PCL) (3) using 1 as an initiator and $FeCl_3$ as a catalyst
Into a 250 mL of three-necked flask were added α-trichloromethyl benzyl alcohol (**1**) (0.23 g, 1 mmol), the monomer ε-caprolactone (1.14 g, 10 mmol), iron(III) chloride hexahydrate (0.54 g, 2 mmol) and the solvent DMSO (30 mL). The mixture was stirred at ambient temperature for 10 h under nitrogen protection. Then, the mixture was poured into cold methanol to form brown-red solid, after filtration and drying, the solid product (0.94 g) was obtained with conversion of 62%.

3. Results and discussion

3.1. Preparation of PMMA under AGET ATRP conditions catalyzed by $FeCl_3$
The compound α-trichloromethyl benzyl alcohol (**1**) proved to be a good initiator under SET-LRP conditions, and preparation of the corresponding polymer was discussed in our previous work [14]. However, in conventional ATRP or SET-LRP, copper catalysts (e.g., CuCl, CuBr, $CuCl_2$, $CuBr_2$, Cu(0)) are always involved. It is known that copper is toxic and excess copper content in bodies will lead to diseases such as copperiedus. Thus, copper-catalyzed polymerizations are limited especially in preparation of special polymers such as biomedical materials. Therefore, strategies for replacing copper catalysts in atom transfer radical polymerization (ATRP) or related polymerization reactions are developed in the past decade [15]. For example, Zhu's research group reported many work on iron-mediated living radical polymerizations in recent years. Their latest work is use of ppm level organometallic catalyst iron(III) acetylacetonate in toluene to prepare PMMA in homogenous phase [16]. Herein, we investigated polymerization of methyl methacrylate using iron(III) chloride as the catalyst and *N,N,N'N'*-tetramethyl ethylene diamine (TMEDA) as the ligand. When the reaction mixture was stirred at 80 °C for 12 h, the corresponding polymer PMMA was readily obtained in moderate conversion (Scheme 1). After purification on alumina column, the PMMA sample was characterized by ^{1}H NMR, and the spectrum was shown in Figure 1b. As comparison, Figure 1a indicated ^{1}H NMR spectrum of the initiator α-trichloromethyl benzyl alcohol (**1**). The number of proton on aryl ring is five, so the value of integration of peaks at 7.3-7.7 ppm is designated as five. Thus, integration value of the single peak at 3.60 ppm is 219, which is the signal of protons of methoxy groups in PMMA unit. Therefore, the number of MMA unit can be calculated as 219/3 = 73. The molecular mass of PMMA is $73 \times 100 + 226 = 7526 \approx 7.53$ kDa.

3.2. Preparation of PCL catalyzed by $FeCl_3$
Poly(ε-caprolactone) (PCL) is a useful polymer due to its bio-compatibility, bio-degradability, shape-memory, etc. and has been applied in tissue engineering. Toxic or expensive metal catalysts are usually used in conventional preparation of PCL, which limits its wide use in bio- or medical-related

areas. So, production of PCL with non-toxic and cheap catalysts is desirable. In this work, non-toxic and cheap iron(III) chloride as a catalyst is successfully utilized to prepare PCL. The method(Scheme 2) features simplicity and very mild reaction conditions (at ambient temperature). ^{1}H NMR spectrum of PCL is demonstrated in Figure 2. Similar to characterization of PMMA, the value of integration of peaks at 7.3-7.7 ppm is five, and that at 4.06 ppm is 154, indicating that there are totally 154/2 = 77 of methylene groups close to oxygen (i.e., OCH_2). So, the number of CL unit is about 77. Molecular mass of PCL can be achieved as: $128 \times 77 + 226 = 10082 \approx 10.08$ kDa.

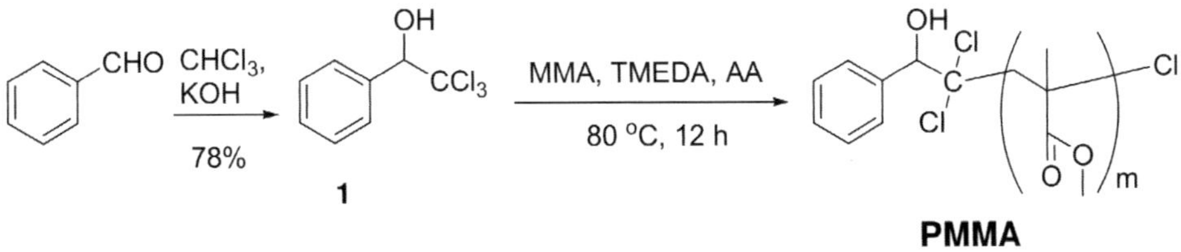

Scheme 1. Synthesis of PMMA under AGET ATRP conditions.

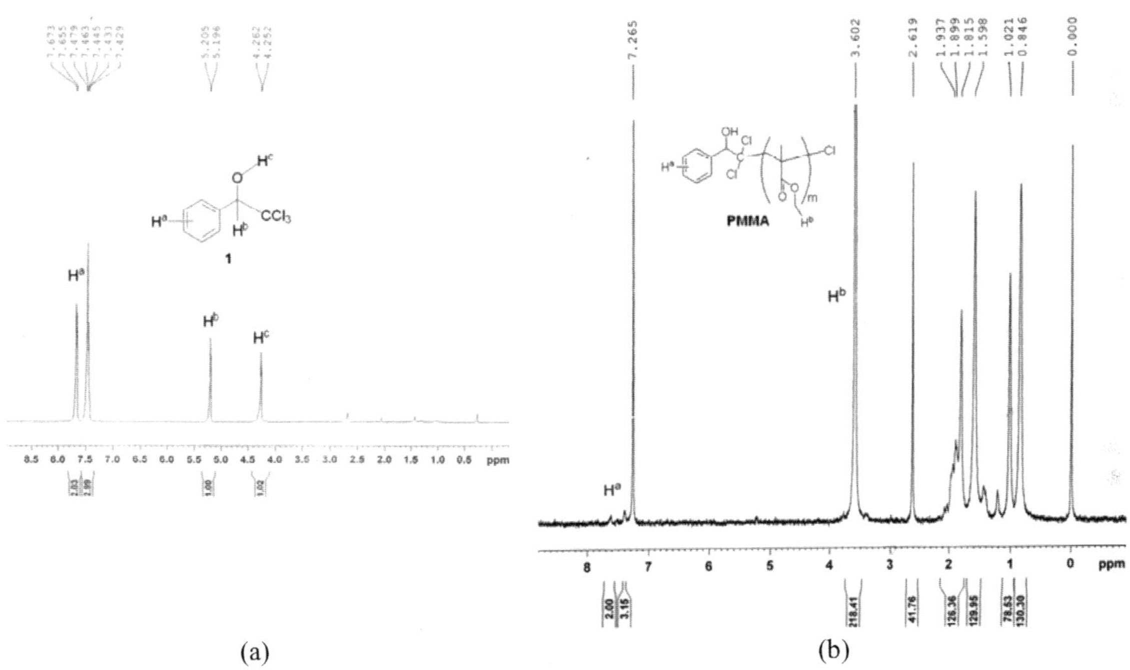

Figure 1. (a) ^{1}H NMR spectrum of the initiator α-trichloromethyl benzyl alcohol (**1**); **(b)** ^{1}H NMR spectrum of PMMA.

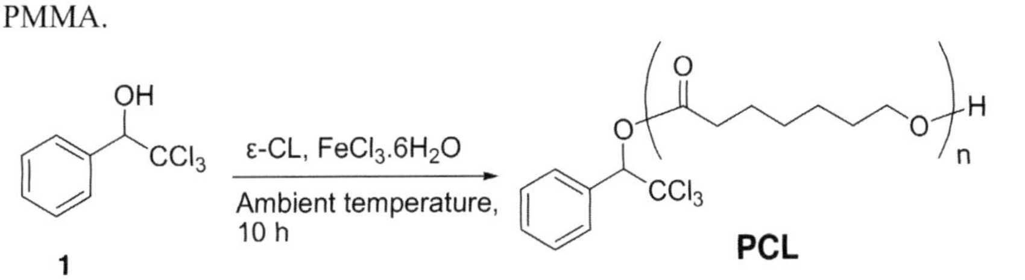

Scheme 2. Synthesis of PCL catalyzed by iron(III) chloride.

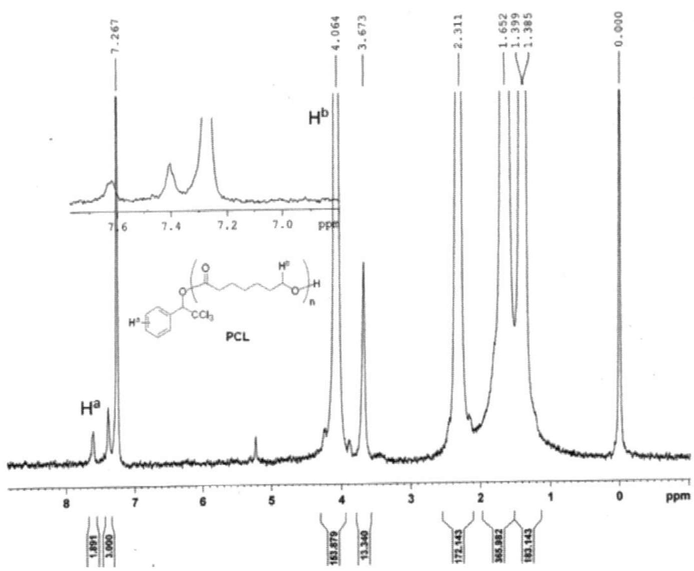

Figure 2. ^{1}H NMR spectrum of Poly(ε-caprolactone) (PCL).

4. Conclusions

Using the functional molecule α-trichloromethyl benzyl alcohol as initiators for both AGET ATRP and ROP, the corresponding polymers PMMA and PCL were readily prepared in moderate conversions catalyzed by only one catalyst $FeCl_3.6H_2O$. The method features mild reaction conditions, simple reaction system, green and cheap catalyst. The prepared polymers PMMA and PCL were characterized by ^{1}H NMR, and molecular weights were 7.53 kDa and 10.08 kDa, respectively. This approach is expected to prepare PMMA-*b*-PCL block copolymer simultaneously or in one pot, and this work is still under investigation.

Acknowledgments

The authors acknowledge financial support from the Natural Science Foundation of China (No. 21264008, 21564004) and the 9th Foundation of Creativity and Study for College Students in Jiangxi Science Technology Normal University.

References

[1] Aoshima S and Kanaoka S 2009 A renaissance in living cationic polymerization *Chem Rev.* **109** 5245-5287

[2] Perkowski A J, You W and Nicewicz D A 2015 Visible light photoinitiated metal-free living cationic polymerization of 4−methoxystyrene. *J. Am. Chem. Soc.* **137** 7580-7583

[3] Tonhauser C, Alkan A, Schömer M, Dingels C, Ritz S, Mailänder V, Frey H and Wurm F R 2013 Ferrocenyl glycidyl ether: a versatile ferrocene monomer for copolymerization with ethylene oxide to water-soluble, thermoresponsive copolymers. *Macromolecules* **46** 647-655

[4] Matyjaszewski K and Tsarevsky N V 2014 Macromolecular Engineering by Atom Transfer Radical Polymerization. *J. Am. Chem. Soc.* **136** 6513-6533

[5] Yamago S and Nakamura Y 2013 Recent progress in the use of photoirradiation in living radical polymerization. *Polymer* **54** 981-994

[6] Tasdelen M A, Kahveci M U and Yagci Y 2011 Telechelic polymers by living and controlled/living polymerization methods. *Prog. Polym. Sci.* **36** 455-567

[7] Blake T R and Waymouth R M 2014 Organocatalytic Ring-Opening Polymerization of Morpholinones: New Strategies to Functionalized Polyesters. *J. Am. Chem. Soc.* **136** 9252-9255

[8] Dove A P 2012 Organic Catalysis for Ring-Opening Polymerization. *ACS Macro Lett.* **1** 1409-1412

[9] Nuykenl O and Pask S D 2013 Ring-Opening Polymerization—An Introductory Review. *Polymers* **5** 361-403

[10] Kemo V M, Schmidt C, Zhang Y F and Beuermann S 2016 Low Temperature Ring-Opening Polymerization of Diglycolide Using Organocatalysts with PEG as Macroinitiator. *Macromol. Chem. Phys.* **217** 842-849

[11] Kryuchkova M A, Detrembleurb C and Bazuin C G 2014 Linear amphiphilic diblock copolymers of lactide and 2-dimethylaminoethyl methacrylate using bifunctional-initiator and one-pot approaches. *Polymer* **55** 2316-2324

[12] Kryuchkov M A, etrembleur C D, Jérôme R, Prud'homme R E and Bazuin C G 2011 Synthesis and Thermal Properties of Linear Amphiphilic Diblock Copolymers of L-Lactide and 2-Dimethylaminoethyl Methacrylate. *Macromolecules* **44** 5209-5217

[13] Cheng C J, Bai X X, Zhang X, Chen M, Huang Q H, Hu Z Y and Tu Y M 2014 Facile Synthesis of Block Copolymers from a Cinnamate Derivative by Combination of AGET ATRP and Click Chemistry. *Macromol. Res.* **22** 1306-1311

[14] Bai X X, Hu Y, Zhang X, Ai L L and Cheng C J 2014 Single electron transfer living radical polymerization via a new initiator. *IOP Conf. Series: Materials Science and Engineering* **62**

[15] Xue Z G, He D, Noh S K and Lyoo W S 2009 Iron(III)-Mediated Atom Transfer Radical Polymerization in the Absence of Any Additives. *Macromolecules* **42** 2949-2957

[16] Wu J, Jian X W, Zhang L F, Cheng Z P and Zhu X L 2016 Iron-Mediated Homogeneous ICAR ATRP of Methyl Methacrylate under ppm Level Organometallic Catalyst Iron(III) Acetylacetonate. *Polymers* **8** 29

Study on the behavior and mechanism of polycarbonate with hot-water aging

L P Kong[1], Y X Zhao[1,3], C H Zhou[1], Y H Huang[1], M Tang[1] and J G Gao[2]

[1]Key Laboratory of Rubber-Plastics, Ministry of Education/Shandong Provincial Key Laboratory of Rubber-Plastics, Qingdao University of Science and Technology, Qingdao 266042, China
[2]Qingdao Exit_&_Entry Inspection and Quarantine Bureau, Qingdao 266001, China

E-mail: zhaoyongxian@qust.edu.cn

Abstract. The present work was concerned with hot-water aging behavior and mechanism of Bisphenol A polycarbonate (PC) used as food and packaging materials. It indicated that with the aging time prolonged, PC sample had internal defects and the mechanical properties of PC materials changed not too much, molecular weight decreased, thermal stability declined. Phenolic hydroxyl absorption intensity enhanced in IR spectra and the maximum absorption wavelength red shift of benzene in UV-Vis spectra, the level of BPA increased. The color change of PC sample was not apparent.

1. Introduction

PC has outstanding impact resistance, good optical transparency, excellent thermal stability, cold resistance, creep resistance. It is widely used in food packaging field, such as tableware, cups, pure water bucket, etc. [1-2]. The ester group of PC molecule is easy to hydrolyze when exposed to warm, humid environment, or hot water for a long time, which quite reduced the transparency, toughness, etc. So that it is limited in the application [3-5].

To know the effect of hot-water aging on properties and structure of PC, the aging degree and the water resistance of the aging samples were evaluated by periodically examining the changes of the appearance and the performance. Study on aging mechanism and the relationship between aging precipitation BPA content and the degree of aging, providing a convenient to judge the aging properties of PC and the content of BPA, further providing a theoretical basis for the research of PC aging and aging protection.

2. Experimental

2.1. Sample preparation and aging process

The PC was produced by Taiwan's Chi Mei Industrial Co., Ltd. Dry the PC pellets at 120°C in a vacuum drying oven for 24h before injection molding, then made several PC standard samples with F2v130 type (Donghua Machinery company) injection molding machine. An annealing process lasting 4h at 135°C was performed to relieve the residual stress before aging tests. The PC specimens were exposed in glassware and added in de-ionized water. Put the glassware in aging oven GT-7017(GOTECH, Taiwan), aging instrument was under 95°C for 0h, 48h, 96h, 144h, 192h ,and 240h, respectively.

2.2. Characterization
The surface morphologies of PC samples were recorded on a 1500 stereomicroscope (Olympus Corporation). Infrared spectra of PC samples were carried out with a Vertex70 spectrometer (Bruker, Germany). The absorption peak at the 2968cm^{-1} was used as a reference which was asymmetric stretching vibration of the methyl, and that at the 3515cm^{-1} was characteristic peak of BPA. The ratio of their absorbance A(3515^{-1})/A(2696^{-1}) reflected the relative amount of precipitated BPA after PC degradation. Mechanical properties were measured on the universal testing machine GT-TCS-2000(GOTECH, Taiwan). A U-4100 UV–Vis spectrophotometer (Hitachi, Japan) was used to test the absorbance and the light transmittance of PC samples, and used light transmittance at wavelength 445 nm, 555 nm, 600 nm to calculate the yellow index YI, chromatism ΔC, transparency T% according to the formula in the literature [6]. An 1100 type GPC (Agilent, America) was used to measure molecular weight and molecular weight distribution of PC samples. DSC analysis of all samples was carried out with a DSC-Q20 (TA, America)

3. Results and Discussion

3.1. Surface morphology of PC after hot-water aging
Figure 1(a-f) shows the pictures of the surface morphology of the PC sample after hot-water aging time. There were lots of defects which became more and bigger in the surface of PC sample with the aging time.

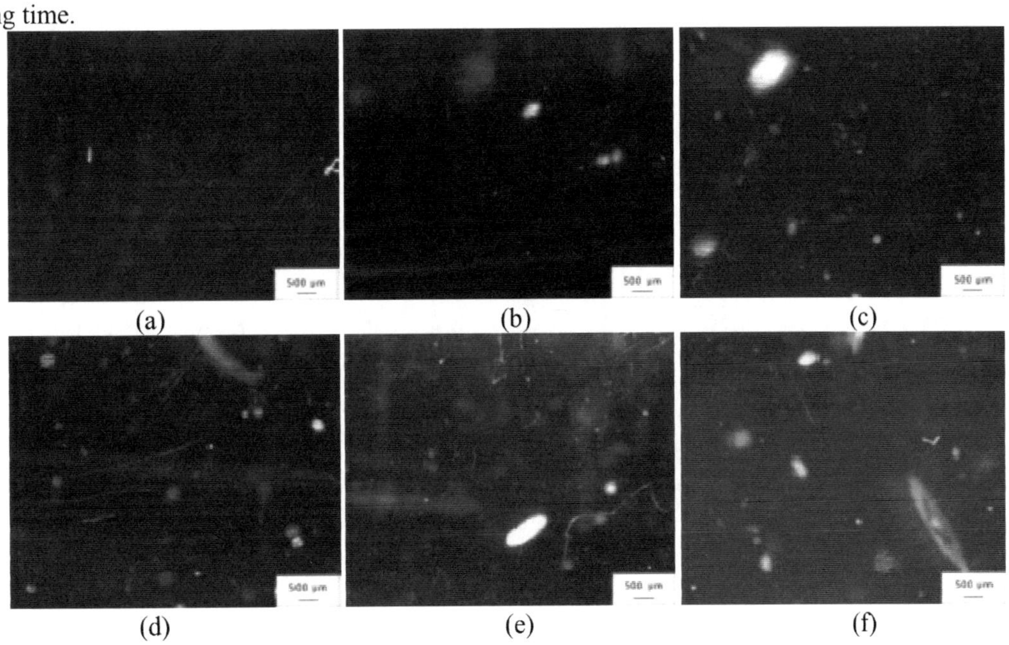

Figure 1. Surface morphology of PC after hot-water aging.
(a)0h, (b)48h, (c)96h, (d)144h, (e)192h, (f)240h

3.2. BPA relative content analysis of aged PC

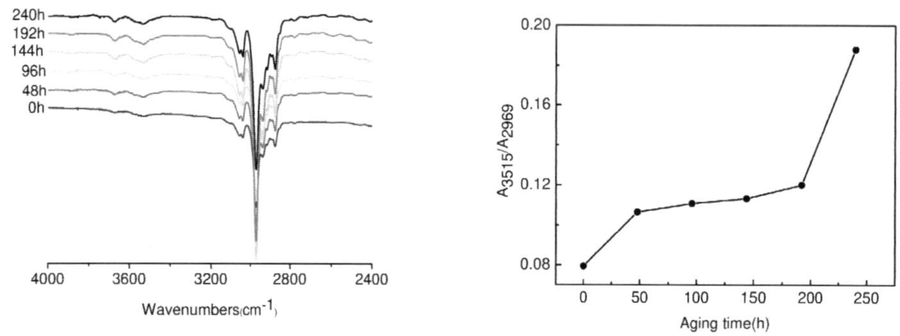

(a) (b)

Figure 2. FTIR spectra and BPA relative content of PC at the different aging time.

(a) FTIR spectra (b) BPA relative content

Figure 2(a, b) reports the effect of the aging time on the FTIR spectra and the BPA relative content of PC samples. As the extension of the aging time, the 3515cm^{-1} characteristic absorption peak which belonged to the -OH of phenolic groups became sharp, indicating the generation of BPA in the process of aging degradation [3]. The relative amount of BPA migration from PC samples enhanced with the aging time prolonging.

3.3. Mechanical properties of aged PC

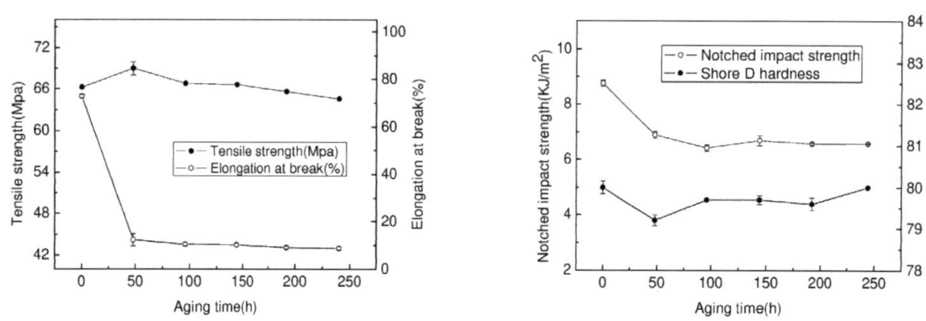

(a) (b)

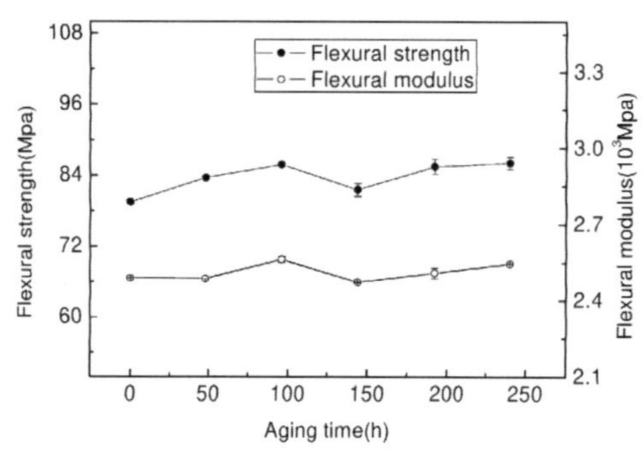

(c)

Figure 3. Mechanical properties of PC at different hot-water aging time.

(a) tensile properties (b) notched impact strength and hardness (c) flexural properties

Figure 3(a) reported that tensile strength of PC decreased slightly, and elongation at break of PC reduced with hot-water aging time. After aging for 48h, the elongation at break of PC descended mildly. Figure 3(b) reveals the notched impact strength and hardness of PC at the various aging time. The notched impact strength of PC decreased lightly after aging. The hardness of PC before and after aging showed almost no change. Figure 3(c) represents the effect of hot-water aging on PC flexural properties. The effect of aging time on flexural modulus and flexural strength of PC was basically consistent and had a small impact on the change of the whole rigidity.

3.4. UV-vis spectrum analysis of aged PC

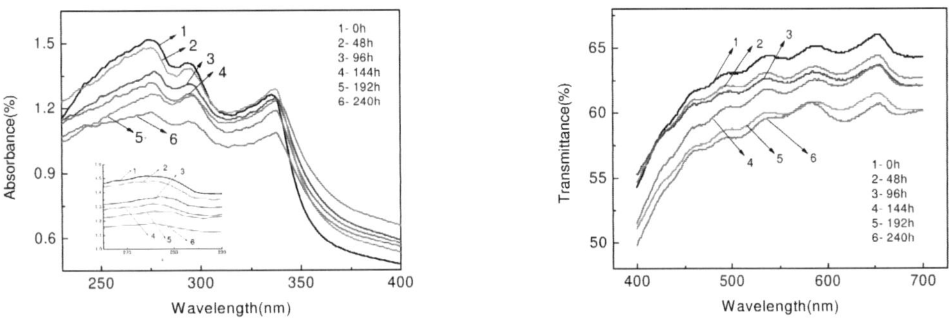

Figure 4. UV-vis spectra and light transmittance of PC at different hot-water aging time.

(a) UV-vis spectra (b) Light transmittance

Figure 4(a) reports the UV-Vis absorption spectrum of PC. It can be seen that the characteristic absorption peak of benzene ring (270nm) showed red shift. The red shift was arisen from n→π* conjugation between benzene ring (chromophores) and –OH (autochrome), that is the aging degradation migrated BPA. Red shift lessened with the extension of aging time, because water into the PC sample produced a blue shift solvent effect, blue shift increased with the prolonging of aging time. In a word, the red shift was larger than the blue shift, and the overall performance was a red shift.

From Figure 4(b) it can be seen that as the aging time prolonged, the UV transmittance of PC samples decreased.

Chromatism, yellow index and transparency of PC at the different aging time are represented in Figure 5.

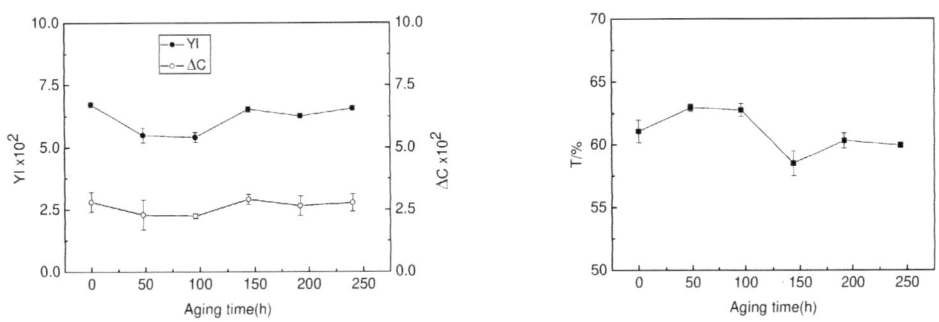

(a) (b)

Figure 5. Chromatism, yellow index and transparency of PC at different hot-water aging time.

a- chromatism, yellow index; b- transparency

With the hot-water aging time, chromatism of PC increased slightly, the yellow index of PC changed a trifle, and transparency of PC did not change obviously, the reason for this due to the increasing number of defects within the samples. And the hot-water aging made the PC specimen surface edge in fine lines. Visibly hot-water aging had a slight effect on the color change of PC.

3.5. Relative molecular weight and its distribution of aged PC

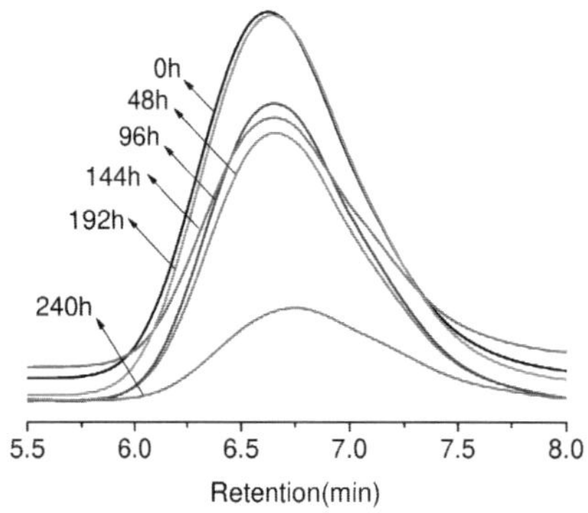

Figure 6. Chromatograms of PC at the different hot-water aging time.

Table 1. Molecular weight and its distribution of PC at the different hot-water aging time.

Ageing time	Mw×10^4	Mn×10^4	Mw/Mn
0h	6.972	6.114	1.140
48h	5.881	4.211	1.397
96h	5.804	4.293	1.352
144h	5.772	4.557	1.267
192h	5.417	3.585	1.14
240h	4.367	3.813	1.145

Figure 6 and Table 1 report the GPC analysis of PC sample with hot-water aging time. The weight average molecular weight (Mw) and the number average molecular weight (Mn) of aged PC decreased compared with that of the unaged PC when PC was aged for 240h, Mw reduced by 37.4%. The molecular weight distribution width index of hot-water aging PC increased than unaged PC, the relative molecular weight fell, which suggested that the degradation of PC occurred.

3.6. Glass transition temperature of aged PC

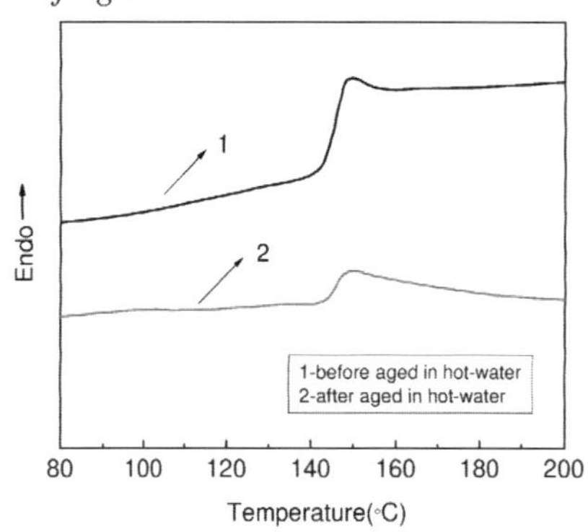

Figure 7. DSC curves of PC before/after aging in hot-water.

In Figure 7, Tg of PC had dropped after aging. When PC sample was immersed in hot water, hydrolysis made PC chain scission which generated the shorter molecular chain. Molecular weight abatement of the PC sample can increase the proportion of end segments and hence heightened the mobility of the whole molecule so that the Tg of PC decreased.

4. Conclusions

The pictures of the surface morphology of hot-water aging PC show that micro defects occur and its number raise with aging time. FTIR analysis reports that BPA migrates out of aged PC and BPA relative content increase with aging time. Hot-water aging has not obvious effects on the mechanical properties of PC. UV-Vis test shows the red shift of the characteristic absorption peak of the benzene ring. Hot-water aging has a small impact on the color change of PC, and chromatism, yellow index of PC have no obvious changes, transparency of PC decrease slightly. The DSC analysis reports that the Tg of PC drop after aging. GPC analysis indicates that the degradation of PC occurs during hot-water aging.

Acknowledgements

This work was financially supported by the National Natural Science Foundation of China (51243002,51373084), the Natural Science Foundation of Shandong Province (ZR2010EM025), the General Administration of Quality Supervision, Inspection, and Quarantine of the People's Republic of China (201410083).

References

[1] Kong L P 2015 Qingdao University of Science and Technology
[2] Li C Q, Li W and Liu G Y 2009 *J. Eng. Plast. Appl.* **37**(1) 60-63
[3] Zhao Y X, Zhao L, Zhao Y Y, Gao J G, Xiao Z L and Wang H 2013 *J. Adv. Mater. Res.* **690-693** 1590-93
[4] Gao W, Han S, Yang M, Long J and Dan Y 2009 *J. Polym. Degrad. Stab.* **94**(1) 13-17
[5] Kahlen S, Wallner G M and Lang R W 2010 *J. Sol. Energy* **84**(5) 755-62
[6] Chen W F. 1983 *China J. Hangzhou Chem. Ind.* **3** 17-24

A novel procedure for fabricating flexible screen-printed electrodes with improved electrochemical performance

C X Du[1], L Han[1, 3], S L Dong[1], L H Li[1] and Y Wei[1, 2]

[1] Beijing Engineering Research Center of Printed Electronics, Beijing Institute of Graphic Communication, Beijing 102600, China

[2] Department of Chemistry and Key Laboratory of Bioorganic Phosphorus Chemistry and Chemical Biology (Ministry of Education), Tsinghua University, Beijing 100084, China

E-mail: hanlu@bigc.edu.cn

Abstract. Screen-printed electrodes (SPEs) with improved electrochemical performance were fabricated in this study. The SPEs on hydrophilic surface of polyethylene ethylene terephthalate (PET) film showed better electrochemical behaviour than that on hydrophobic surface. The optimal condition of pretreating fresh SPEs was that alternately dealt with chemical treatment (soaked in 3M NaOH solutions for 1h) and high temperature curing (heated at 120 °C for 15 min) for two times. After chemical treatment, the electrochemical performance of self-made SPEs was better than the commercial three electrodes system. By analyzing cyclic voltammetry (CV) curves, we found that the oxidation peak currents and peak to peak separation reached 407.65 μA and 111.16 mV, which mean the sensitivity and electron transfer rate improved 1.9 times and 3.8 times compared with fresh SPEs, and 2 times and 3 times compared with commercial DropSens (DS) electrodes. The obtained SPEs were stable, convenient and inexpensive, which could be extensively applied for developing novel electrochemical sensors.

1. Introduction

Electrochemical sensors behave as a crucial analytical tool to detect a variety of analytes including antigens [1], enzymes [2, 3] and heavy metal ions [4], have received great interests due to their high efficiency and sensitivity [5]. Sensitive membranes and substrate electrodes are the core components of electrochemical sensors. Two categories of substrate electrodes are usually employed, namely, solid electrodes and screen-printed electrodes (SPEs). In contrast to solid electrodes, SPEs demonstrated more advantages in developing disposable electrochemical sensors due to its low cost, easy to manufacture and mass production [6-8].

However, in China, SPEs with excellent performance mainly relied on import from Spain and other countries, which leads to the higher prices of electrochemical sensors. Moreover, most of commercialized SPEs in the Chinese market are printed onto hard substrates, such as ceramics [9], glasses and so on, which limited its application in portable or relible electronic devices.

Considering the above-mentioned problems, in this work, a screen printing three-electrodes system containing working electrode, auxiliary electrode (counter electrode) and reference electrode was printed on a flexible PET film as substrates. We thoroughly evaluated the effects of surface hydrophilicity or hydrophobicity of PET substrates on electrochemical performance of SPEs.

Addtionally, the pre-treating methods including chemical treatment and anodizing processing for working electrodes of SPEs were studied by assessing the electrochemical behaviors of the screen printing three-electrodes system. Thus, the obtained SPEs could be extensively applied for developing novel electrochemical sensors.

2. Experimental section

2.1. Materials and reagents

Carbon ink (No.CH-8) and insulating ink (No.AC-3G) was obtained from Jujo Chemical Co., Ltd., Tokyo, Japan. Silver conductive ink (No.BY-2100) was purchased from Shanghai Baoyin Electronic Materials Ltd., Shanghai, China. PET films (100μm) were gained from Baoding Lucky Innovative Materials Co., Ltd., Hebei, China. Potassium hexacyanoferrate (III) and potassium hexacyanoferrate (II) trihydrate were purchased from Sigma–Aldrich (St. Louis, MO, USA). Sodium hydroxide was analytical reagents and purchased from Beijing Chemical Works. Ultrapure water obtained from a Millipore water purification system was used in all assays.

2.2. Fabrication and pretreatment of SPEs

High precision manual screen printing machine (Shanghai HOTING screen printing equipment Co. Ltd. Model: PHM Series) was used to fabricate SPEs. The process was shown in Figure 1. First, silver ink (a) was printed onto the PET films and supported as conductor and reference electrode, and then carbon ink (b) was printed on the silver layer and supported as working and auxiliary electrode. After each step, the film was heated at 120 °C for 1 h to cure the patterned paste. Finally, insulating ink (c) was printed on the carbon layer and then heated the film at 120 °C for 15 min to cure the insulating layer.

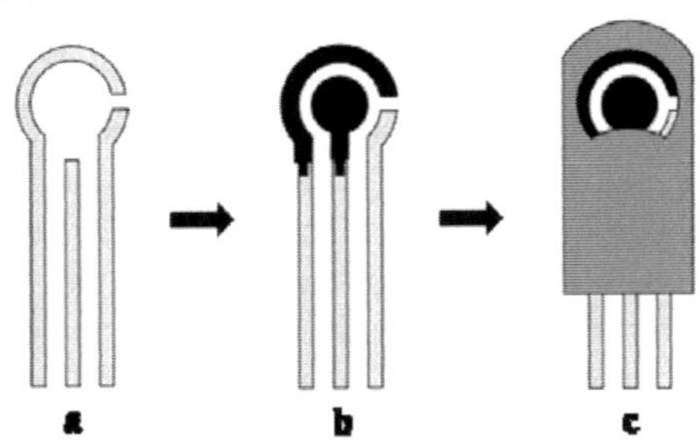

Figure 1. The schematic diagram of the process of printing SPEs.

In the procedure of pretreating fresh SPEs, we adopt two different approaches. In the first approach, the fresh SPEs were soaked into 3M NaOH solutions for 1 h as chemical treatment and then anodized in 0.5M NaOH solutions by applying 1.2V anodic potentials for 20s as electrochemical treatment. In the second approach, the fresh SPEs were alternately treated with chemical treatment and curing at 120 °C for 15 min for several times. After each alkaline solution treatment step, SPEs were rinsed using deionized water and then electrochemical testing was conducted.

2.3. Characterization

The contact angle of water on PET substrate was investigated by a video contact angle meter (Kruss, DSA100, Germany). Four-probe tester (Four probes tech, RTS-8, China) was used to detect the square resistance of SPEs. Electrochemical measurement was carried out using electrochemical workstation (Voltalab, PGZ402, France). The CV measurement of all SPEs samples were performed in a solution containing 1 mM $K_3[Fe(CN)_6]$ and $K_4[Fe(CN)_6]$ with the scan rate of 100 mV/s, and the performance of our SPEs were compared with import three electrodes systems (DropSens, DRP110, Spain). The surface morphology of SPEs was observed by D laser confocal topography measurement microscope (Keyence, VK-S105, Japan) and Scanning Electron Microscopy (SEM, Quanta 400, FEI) for different purposes.

3. Results and Discussion

3.1. Effect of hydrophilicity of PET substrate on electrochemical performance of SPEs

In order to ensure printing precision and quality, thermal stability and surface affinity of PET substrates are two crucial factors that we should consider seriously. After the cleaning of PET film, we heated it at 120 °C for 15 min. Considering ink adhesion can be seriously influenced by the hydrophilicity or hydrophobicity of substrates, the contact angle (θ) of water on both sides of PET film (denoting as PET-A and PET-B) was detected. In order to further improve the hydrophilicity of PET, O_2 plasma technique was employed to treat PET for 1 min. As shown in Figure 2, θ values of water on PET-A and PET-B are 52.5° and 75.7°, respectively, indicating PET-A surface displayed better hydrophilicity than PET-B. After O_2 plasma treatment, θ value of water decreased to 10.2°.

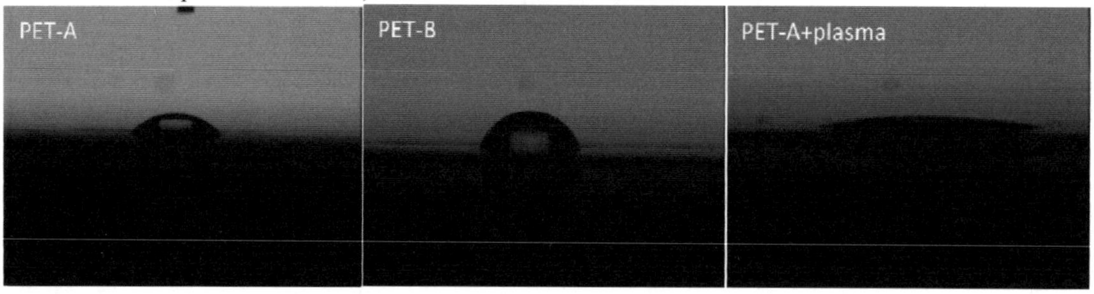

Figure 2. The contact angle of water on both sides of PET film and O_2 plasma treated PET-A surface.

The electrochemical behaviors of SPEs on PET-A, PET-B and plasma treated PET-A were thoroughly evaluated using commercial SPEs (DropSens, DRP110, Spain) as a control. As illustrated in Figure 3A, the electrochemical properties of self-made SPEs were inferior to DS electrodes. However, compared with SPEs on PET-B, SPEs screen-printed on PET-A and plasma treated PET-A possessed better performance, as shown in Figure 3B-C. Additionally, in terms of the oxidation peak current density and peak to peak separation, there was no significant difference between SPEs printed on PET-A and PET-A+plasma, elucidating that plasma treatment could not obviously improve the electrochemical performance of SPEs on PET-A. In order to further investigate the attachment results of carbon pastes on PET films, the morphology of the edges and surface planeness of carbon electrodes and were observed. As displayed in Figure 3D, an enlargement at the edge of SPEs on PET-A occurred more obviously than that of SPEs on PET-B. Moreover, there was a better surface planeness for SPEs on PET-A compared to that on PET-B, further demonstrating that the suitable hydrophilicity of substrates was propitious to the adhesion of carbon pastes, leading to their excellent electrochemical performance. Therefore, it was concluded that PET-A with contact angle of 52.5° was chosen as printing substrate and SPEs printed on PET-A were provided with better electrochemical performance (I_o=216.97μA, $\triangle E_p$=423.32mV).

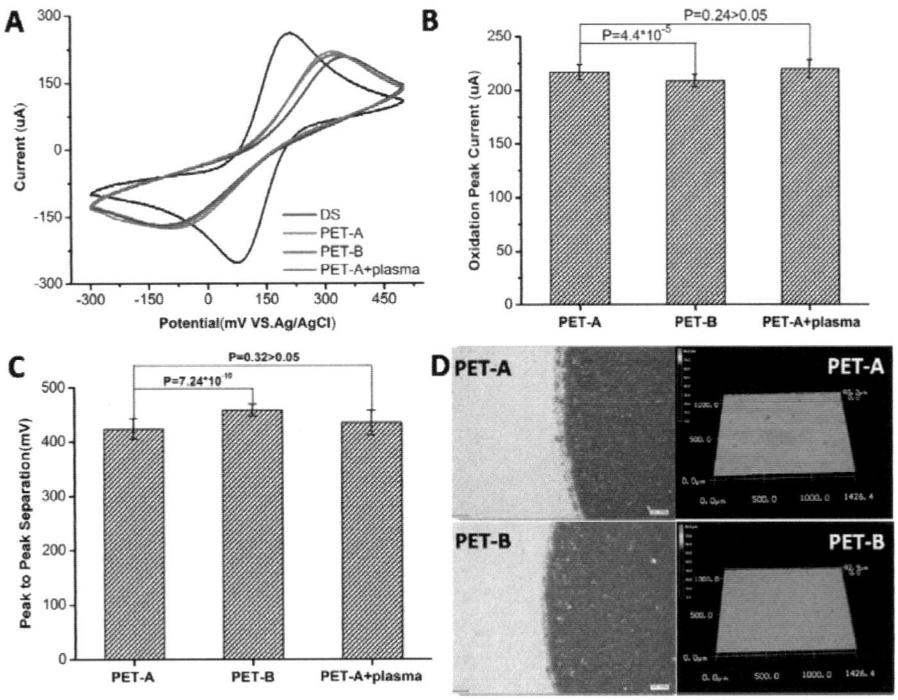

Figure 3. (A) Representative CVs of SPEs on PET-A, PET-B and plasma treated PET-A with comparison of DS electrodes; (B) oxidation peak current and (C) peak to peak separation of self-made SPEs (P is the probability value of F test statistics when significance level is a=0.05); (D) surface flatness of carbon working electrode on PET-A and PET-B.

3.2. Effect of pretreating procedures to SPEs

In order to improve the adhesion ability of carbon inks on substrates, insulating polymers were usually added, which might increase the electron transfer resistance [10, 11]. The main purpose of pretreatment on SPEs was to remove insulating polymers and to increase surface roughness. It can be seen from Figure 4A, before chemical treatment, the electrochemical performance of self-made SPE was inferior to DS electrode (Io=199.01μA, △Ep=341.40mV). However, it was significantly improved after chemical treatment and superior to DS electrodes. Fresh SPEs were then treated with NaOH solution, denoting as SPE-A. Similarly, fresh SPEs were treated not only with NaOH solution but also with anodizing treatment, denoting as SPE-B. The CV curves of SPE-A and SPE-B were plotted in Figure 4B. Compared with non-treated SPE, the electrochemical response of SPE-A and SPE-B was dramatically enhanced. As shown in Figure 4C and 4D, the oxidation peak current of SPE-A and SPE-B was remarkably higher than that of fresh SPE. Additionally, the peak to peak separation of SPE-A and SPE-B was evidently lower than that of fresh SPE. The superior electrochemical behaviors of SPE-A and SPE-B may be ascribed to the removal of insulating organic binders on surfaces of working electrodes. However, after further anodization treatment, the performance of SPE-B was inferior to SPE-A with largest oxidation peak currents (Io=324.79μA) and smallest peak to peak separation (△Ep=140.68mV), which was contrary to the reported results [12]. It may be owing to the difference in carbon ink composition and in printing and curing conditions. Finally, the surfaces of working electrodes on SPE, SPE-A and SPE-B were observed by SEM (Figure 5), illustrating that the rough surface exposed more carbon particulates which improved the electrochemical performance of SPEs. It was concluded that alkaline treatment on working electrodes of SPEs could significantly increase the electron transfer and then enhance the sensitivity of SPE, without the need of anodization treatment.

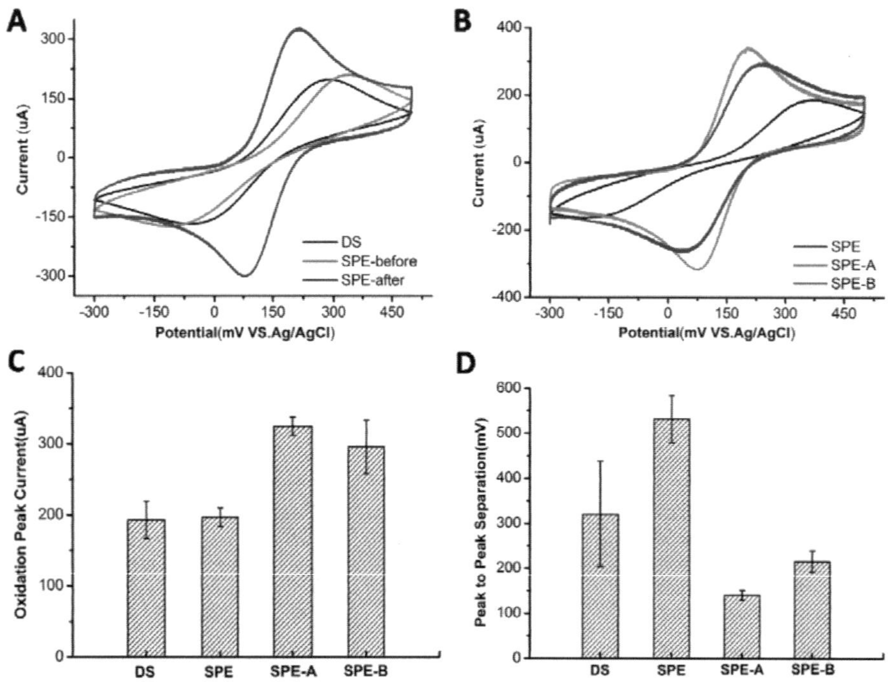

Figure 4. (A) Representative CVs of SPEs before and after chemical treatment with comparison of DS electrodes; (B) Representative CVs of fresh SPEs and SPEs suffering from different pretreating methods; (C) Oxidation peak current and (D) peak to peak separation of DS electrode, fresh SPEs and SPEs after pretreating with NaOH solution and/or anodizing treatment.

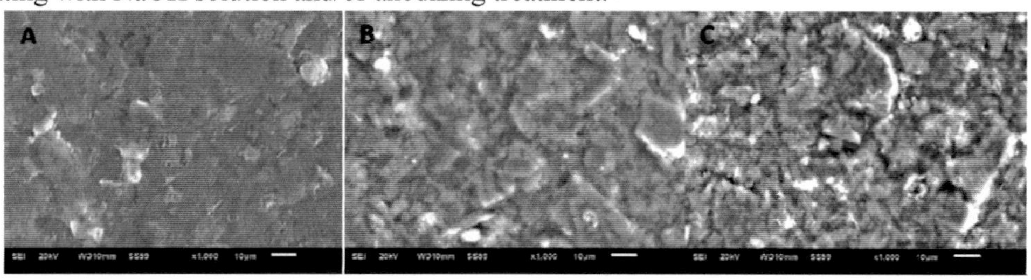

Figure 5. SEM images of working electrode surfaces of SPE (A), SPE-A (B) and SPE-B (C).

Whether increasing the times of alkaline treatment could further enhance the electrochemical performance of SPEs, the CVs of fresh SPEs and SPEs treated with different times were measured and illustrated in Figure 6A. It was demonstrated that, the peak current of treated SPEs increased obviously and the peak potential difference decreased slightly with increasing the corrosion times by 3M NaOH. Moreover, the square resistance of working electrodes exposed to alkaline solution for two times and three times decreased to 123.17 Ω/□ and 95.67 Ω/□, respectively, as shown in Figure 6B, which further confirm the results of CV curves in Figure 6A. However, we can observed from Figure 6C that there was a decline for carbon ink adhesion when the working electrode was treated by NaOH solution at the third time. Therefore, the optimal condition of pretreating procedure is that, fresh SPEs were alternately treated with chemical treatment and high temperature curing for two times, the oxidation peak currents of which is 407.65μA and peak potential difference of which is 111.16mV.

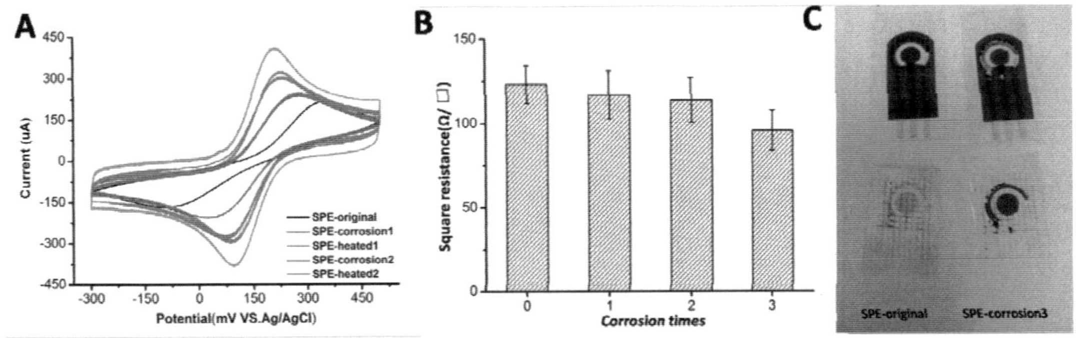

Figure 6. (A) CVs, (B) square resistance and (C) the result of 3M tape adhesion test of fresh SPEs and SPEs treated by alkaline solution for different times.

4. Conclusion

A new procedure for fabricating SPEs with excellent electrochemical behaviors was presented in this study. The SPEs printed on PET-A showed better performance than that on PET-B owing to the better hydrophilicty of PET-A than PET-B. When the SPEs was treated via the best condition, that is alternately treated with chemical treatment and high temperature curing for two times, the oxidation peak currents and peak potential difference of treated SPEs reached 407.65μA and 111.16mV, implying that the sensitivity and electron transfer rate improved 1.9 times and 3.8 times compared with fresh SPEs, and 2 times and 3 times compared with DS electrodes. Thus the obtained SPEs were stable, convenient and inexpensive, which could be extensively applied for developing novel electrochemical sensors.

Acknowledgements

This work was financially supported in part by grants from Chinese Natural Science Foundation (No.31300820), Beijing Municipal Commission of Education Project (No.KM20140015001), a project founded by Beijing Institute of Graphic Communication (No.27170115004/025), a foundation for raising scientific research level of materials and technologies for printed electronics (No.Eb201530) and a grant from Beijing collaborative innovation for green printing and publication.

References

[1] Yean C Y, Kamarudin B, Ozkan D A et al 2008 Enzyme-linked amperometric electrochemical genosensor assay for the detection of PCR amplicons on a streptavidin-treated screen-printed carbon electrode *ANAL CHEM* **80** 2774-2779

[2] Pereira S V, Bertolino F A, Fernández-Baldo M A et al 2011 A microfluidic device based on a screen-printed carbon electrode with electrodeposited gold nanoparticles for the detection of IgG anti-Trypanosoma cruzi antibodies *ANALYST* **136** 4745-4751

[3] Serafín V, Agüí L, Yáñez-Sedeño P and Pingarrón J M 2011 A novel hybrid platform for the preparation of disposable enzyme biosensors based on poly(3,4-ethylenedioxythiophene) electrodeposition in an ionic liquid medium onto gold nanoparticles-modified screen-printed electrodes *J ELECTROANALl CHEM* **656** 152-158

[4] Carapuça H M, Monterroso S C C, Rocha L S et al 2004 Simultaneous determination of copper and lead in seawater using optimised thin-mercury film electrodes in situ plated in thiocyanate media *TALANTA* **64** 566-569

[5] Taleat Z, Khoshroo A and Mazloum-Ardakani M 2014 Screen-printed electrodes for biosensing: a review (2008–2013) *MICROCHIM ACTA* **181** 865-891

[6] David H S, María D G, María BegoA G G and Agustín C G 2004 Enzymatic genosensor on streptavidin-modified screen-printed carbon electrodes *ANAL CHEM* **76** 6887-6893

[7] Dequaire M and Heller A 2002 Screen printing of nucleic acid detecting carbon electrodes *ANAL*

CHEM **74** 4370-4377

[8] Malhotra B D and Chaubey A 2003 Biosensors for clinical diagnostics industry *SENSOR ACTUAT B-CHEM* **91** 117-127

[9] Miserere S, Ledru S, Ruillé N, Griveau S, Boujtita M and Bedioui F 2006 Biocompatible carbon-based screen-printed electrodes for the electrochemical detection of nitric oxide *ELECTROCHEM COMMUN* **8** 238-244

[10] Del Carlo M, Di Marcello M, Perugini M et al 2008 Electrochemical DNA biosensor for polycyclic aromatic hydrocarbon detection *MICROCHIM ACTA* **163** 163-169

[11] Lucarelli F, Authier L, Bagni G et al 2003 Graphene–PEDOT:PSS on screen printed carbon electrode for enzymatic biosensing *ANAL LETT* **36** 1887-1901

[12] Wei H and Sun J J 2007 Enhanced electrochemical performance at screen-printed carbon electrodes by a new pretreating procedure *ANAL CHIM ACTA* **588** 297-303

IOP Conf. Series: Materials Science and Engineering 137 (2016) 012061 doi:10.1088/1757-899X/137/1/012061

Consistency analysis of plastic samples based on similarity calculation from limited range of the Raman spectra

B W Lai[1], Z X Wu[2], X P Dong[1,3], D Lu[1] and S C Tao[1]

[1]Lightwave Technology Institute, Information Science and Engineering College, Xiamen University, 361005, Xiamen, China
[2]Technology Center, Xiamen Entry-Exit Inspection and Quarantine Bureau, 361012, Xiamen, China

E-mail: xpdong163@163.com

Abstract. We proposed a novel method to calculate the similarity between samples with only small differences at unknown and specific positions in their Raman spectra, using a moving interval window scanning across the whole Raman spectra. Two ABS plastic samples, one with and the other without flame retardant, were tested in the experiment. Unlike the traditional method in which the similarity is calculated based on the whole spectrum, we do the calculation by using a window to cut out a certain segment from Raman spectra, each at a time as the window moves across the entire spectrum range. By our method, a curve of similarity versus wave number is obtained. And the curve shows a large change where the partial spectra of the two samples is different. Thus, the new similarity calculation method identifies samples with tiny difference in their Raman spectra better.

1. Introduction

The plastic safety parts are the critical parts of electrical products and have significant influence on the safety of electrical products. According to some regulations, China Compulsory Certification (CCC) for example, the plastic parts (non-metallic parts) of household electrical appliances or electrical products with similar applications are the critical part of safety. Moreover, the regulations required that all the materials and components of plastic parts for the productions be certified to be consistent with those of the test samples for type tests and declaration. Any changes to the materials or components of the plastic safety parts may affect the safety of electrical products. So the test of the consistency of the plastic safety parts is of great significance.

Chemical method is one of the most common methods for the consistency test of two plastic material samples. However, chemical method is of high cost and damages test samples. Raman spectrum detection method which uses Raman spectroscopy to detect the surface of samples and distinguish samples via spectrum analysis [1-2], has been widely study for its advantages of simplicity, convenience and high accuracy, and causing no damage to the samples [3].

To analyse the Raman spectrum data, one common approach is to extract the characteristic peaks and compare them with those of the standard samples in the standard database. It requires lots of work in building and improving the database. The other approach is to make similarity analysis of spectra between different samples [4]. It can distinguish samples with different materials more quickly and efficiently. However, this approach will be ineffective in distinguishing samples with same materials with different additives, such as additive type -- flame retardant in plastics. This is because the additive caused differences at only specific positions in Raman spectrum, but the similarity is calculated throughout the whole Raman spectrum range [5]. A simple but effective way to solve the problem is to calculate the similarity at a specific segment where the difference of the Raman spectrum of different samples appears. Nevertheless, it requires that we know the position of the calculation segment in the range [6].

Problems mentioned above are summarized as: 1) positions where differences appear in Raman spectra of different samples are unknown, and 2) the similarity calculated based on the whole range of Raman spectrum cannot distinguish samples with same materials but different additives. This paper proposed a novel spectrum analysis method named "moving window similarity calculation method." The method uses a window with certain width scanning across the entire Raman spectra to calculate the similarity between the spectra of different samples in the window. Then a curve of similarity versus wave number is obtained. Therefore, the curve will have a large fluctuation where the Raman spectra of the two samples are quite different. To prove the validity of our methods, the Raman spectra of two ABS samples, one not added with flame retardant (Sample A), and the other added with flame retardant (Sample B) are tested and analysed. The experimental results show that the moving window similarity analysis method is an effective method for distinguishing samples with same materials but different additives.

2. Experimental

2.1. Experimental equipment

Xplora ONETM Raman microscope (France HORIBA Scientific) has the advantages of non-destructive analysis, high spatial resolution and high sensitivity [7]. Software such asLabSpecAppllication and Matlab were used to process and analyse the spectra.

2.2. Experimental samples

All the starting materials were commercially available and were used without further publication. ABS samples were supplied by Defenghang Company (Xiamen, China).

2.3. Experimental procedure

The samples were put on the platform of the Raman microscope. Then the focal length was adjusted to set the focus onto the surface of the samples. The excitation wavelength is 785nm, the spectral sampling interval is 1.02cm^{-1}, the spectral scanning range is 200-2000cm^{-1}, the integral time is 10s, the aperture of the optical path is 300mm and the slit of the spectrometer is 100mm. In this experiment, we randomly chose three positions of the both samples for testing and obtained six groups of Raman spectrum data.

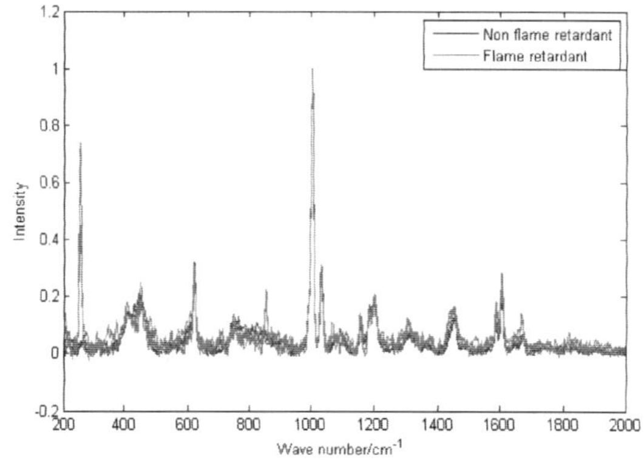

Figure 1. Spectrogram after preprocessing.

After testing, the software LabSpecAppllication was used to preprocess the data. The preprocessing procedure was divided into three steps: smoothing filtering, fluorescent background subtraction, and normalization. The spectra after preprocessing are shown in Figure 1. It can be seen from Figure 1 that the spectra of two samples show a good reproducibility, respectively. The spectra of the two samples are similar in most positions except on the wave length near 250cm^{-1} where the flame retardant ABS sample has an obvious peak and the non-flame retardant sample has not.

3. Results and analysis

3.1. Similarity calculation

Each group of collected data consists of two vectors. The first vector is the discrete intervals of wave numbers (200-2000cm^{-1}). The second vector $\boldsymbol{X}$ is the normalized intensities corresponding to each wave number. Since the first vector of each group is the same, $\boldsymbol{X}$ is only needed to be used in calculating the similarity.

The method of calculating similarity is to use similarity function. In this paper we use the following generalized Jaccard coefficient function [8]:

$$sim(X_1, X_2) = \frac{X_1^T \cdot X_2}{\| X_1 \|^2 + \| X_2 \|^2 - X_1^T \cdot X_2} \tag{1}$$

This function can better amplify the difference between vectors as compared to other similarity functions [9]. The value of this function is from 0 to 1. The higher the value is, the greater the similarity of the two sample show. The similarity computed by this function is 0.9268, which cannot explain the inconsistency of the two samples.

3.2. Moving window method

The moving window method uses a certain width of window to scan across the entire spectrum and calculate the similarity of the spectra in the window while scanning. From equation (1) we can see that when both $\boldsymbol{X_1}$ and $\boldsymbol{X_2}$ are close to 0, minor fluctuations will result in a significant difference in the degree of similarity. So before using this method, the data were transformed as follows:

$$x_i' = 1.2 - x_i \tag{2}$$

In the above equation, as x_i is the element of $\boldsymbol{X}$, the value of x_i' is between 0.2 and 1.2, which can avoid the situation mentioned above. Since the test result gives a good reproducibility, we use the first groups of data of each sample for calculation. The window width is set to be 5 cm^{-1}, the central wave length of the window range between 205cm^{-1} and 1995cm^{-1}. The spectra and calculation results are shown is Figure 2.

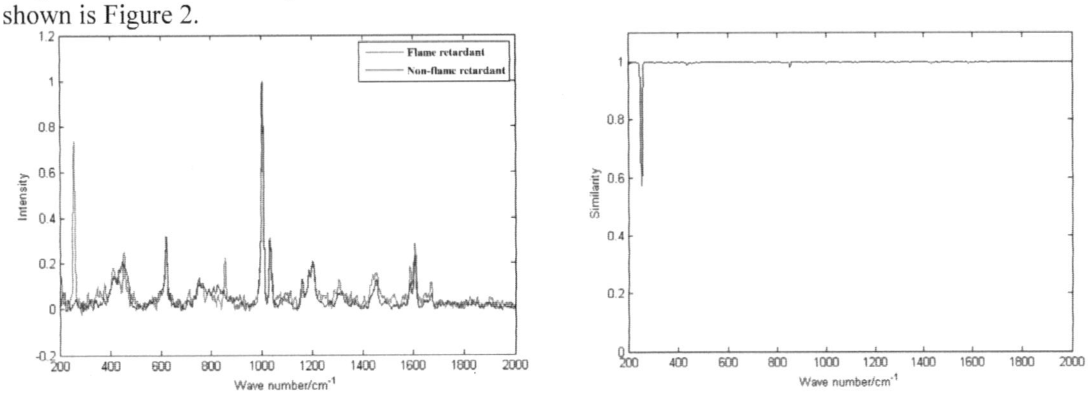

Figure 2. (a)Comparison of ABS samples' spectra with and without addition of flame retardant. (b) The results obtained by moving window similarity calculation method.

Figure 2 shows that, when the window moves to the peak which represents the flame retardant, the similarities decrease significantly, and that in other positions the similarities remain close to 1. In that case we can conclude that the two spectra belong to different samples. For comparison, we apply this method to two different test data of the same sample (without retardant), the results are shown in Figure 3.

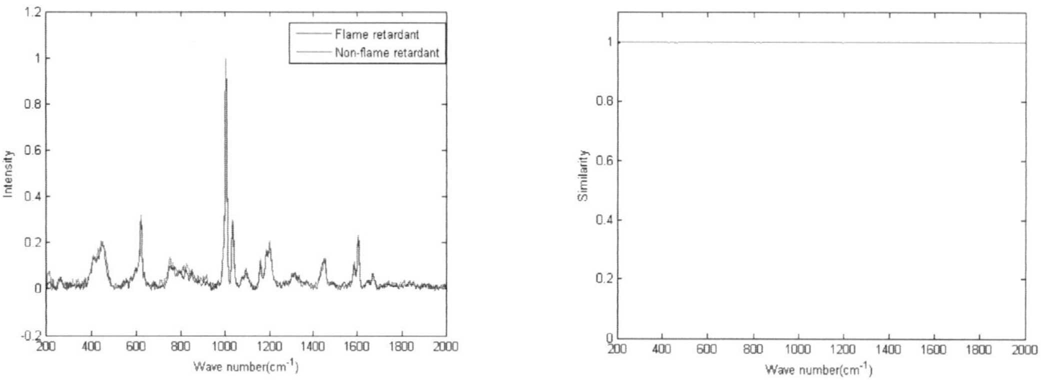

Figure 3. (a) Comparison of two tests on ABS sample without flame retardant. (b) The results obtained by moving window similarity method.

Figure 3 shows that, in all positions of the entire wave number, the similarities are close to 1. In that case we can conclude that the two spectra belong to the same sample.

4. Conclusion

This paper combines the Raman technique with similarity calculation in analysing consistency of the plastic safety parts of electrical products. As similarity calculated throughout the whole spectra from the samples of same material with different additives cannot be distinguished well, a method to use a window to scan the spectra and calculate the similarity is proposed. The experimental results show that this method can distinguish the ABS samples with and without flame retardant well, and its feasibility of this method is verified.

Acknowledgements

The project was financially supported by Chinese National Bureau of Quality Inspection (No.2014IK199).

References

[1] Peeters J R, Vanegas P and Tange L 2014 *Resources, Conservation and Recycling* **84** 35-43
[2] Florestan J, Lachambre A and Mermilliod N 1994 *Resources, Conservation and Recycling* **10** 67-74
[3] Wu L, Ouyang Z H, Cao S C, Yi D L, Qin X R, Sun S X and Liu X 2015 *Chinese Journal of Light Scattering.* **17** 180-186
[4] Zhuang Y, Liu Y D and Ji Z 2009 *Technical Acoustics* **28** 532-536
[5] Tsuchida A, Kawazumi H and Kazuyoshi A 2009 *Sensors* (Christchurch:New Zealand/IEEE) 1473-1476
[6] Liu C, Zhang Z H, Liu Y and Lu F 2011 *Pharm Care Res.* **11** 196-199
[7] Xu Y J, Luo R H, Guo M T and Wu J F 2007 *Laser Journal.* **28** 13-14
[8] Cheng Y Y, Chen M J and Wu Y J 2002 *Acta Chimica Sinica.* **60** 2017-2021
[9] Hamers L, Hemeryck Y and Herweyers G 1989 *Information Processing & Management.* **25** 315-318

Impact and dynamic mechanical thermal properties of textile silk reinforced epoxy resin composites

K Yang and J Guan[1]

School of Materials Science and Engineering, Beihang University, Beijing, China

E-miail: juan.guan@buaa.edu.cn

Abstract. Silk fabric reinforced epoxy resin composites (SFRPs) were prepared using simple techniques of hand lay-up, hot-press and vacuum treatment, and a series of volume fractions of silk reinforcements were achieved. The impact properties and dynamic mechanical properties of SFRPs were investigated using a pendulum impact testing method and dynamic mechanical thermal analysis (DMTA). The results suggest that silk reinforcement could greatly enhance the mechanical performances of SFRPs. The impact strength reached a maximum of 71 kJ/m^2 for 60%-silk SFRP, which demonstrated a potential of silk composites for defence and impact-resistant materials.

1. Introduction

Recently, natural fibre reinforced composites have been investigated with considerable interest owing to their advantages such as high specific strength, low density, promising biodegradability and environmental friendliness [1-6]. Generally, plant fibres (e.g. flax, hemp and jute) and basalt fibre reinforced composites have been the focus of natural fibres, whereas animal-sourced fibres (such as silk and wool) have seldom found to be utilised as composite reinforcements. Silk from domestic silkworm *Bombyx mori* (*B. mori*) is highly crystalline in the molecular structure and stands as the only continuous filament of natural fibrous fibre[7]. Mei-po Ho *et al.* [8] prepared short silk fibres reinforced poly (lactic acid) (PLA) composites via twin-screw micro-extruder, and they found the Young's modulus and flexural modulus were enhanced by 27% and 2% compared to the pure PLA. Zhao[9] studied the dynamic thermal mechanical properties of silk-PLA composites, and the storage modulus was shown to be improved at higher temperatures (from 70 to 160°C). Another biocompatible polymer, poly(ε-caprolactone) (PCL) was also widely used as the matrix of composites for biomedical applications. Li et al. [10] prepared silk fibre reinforced PCL and the 35 wt.%-silk showed the highest tensile strength while the 45 wt.%-silk showed the highest flexural strength. It is until very recently that Shah *et al.* [11] point out the unique properties of silk fibres could provide new opportunities for high-performance composites [12,13]. Despite the above mentioned specific studies on silk reinforced composites, the general effect of silk as reinforcements to improve the mechanical properties of plastics and composites has not been studied.

Epoxy resin is widely used with credentials such as easy processibility, low cost, balanced mechanical properties, good adhesive properties, good corrosion resistance and great high-temperature tolerance [14-19]. However, due to the high degree of crosslinking, many materials of epoxy resin are inherently brittle, and the impact properties are poor, often shown as catastrophic failure. These disadvantages limit the engineering applications of epoxy resin, which also drive the research on modifications of epoxy resin materials to improve their toughness [20-24].

This study intended to prepare silk fibre reinforced epoxy resin plastics/composites (SFRPs) via easily accessible hot-press technique, aiming to investigate the effect of silk reinforcements on improving the mechanical properties of the pure epoxy resin, especially toughness. The prepared SFRPs were tested for impact performances and dynamic mechanical thermal properties. The aim of the work was also to validate a hypothesis that silk fabric can greatly improve the toughness of the brittle epoxy resin, which may widen the engineering applications of these thermosets materials.

2. Materials and methods

2.1. Materials

Epoxy resin E-51 with curing agent DS-300G was supplied by Dasen Material Science & Technology. Inc. (Tianjin, China) which would be used as the matrix material. According to the product company, the cured density of this E-51 resin product was 1200 kg·m^{-3}. Tensile strength the impact strength was tested to be about 72-78 MPa and 12.5-13.1 kJ m^{-2}.

A plain woven fabric with an areal density of 90±5 g·m^{-2} was provided by Huzhou Yongrui Textile Co. Ltd. (Zhejiang Province, China). The fibre yarn includes nearly 80 threads of silk fibre of each strand. The density of the neat silk fibre was assumed to be 1300 kg·m^{-3}[25].

2.2. Fabrication of composites

SFRPs were fabricated by hand lay-up followed by a hot-press process. The specimens with size of 20 cm × 10 cm were cured at 120°C for 2 hrs with a 300 kPa pressure. The silk fabric reinforced composites (SFRPs) with silk fabrics of different volume fractions (v_f) from about 30% to 70% were prepared, and specimens were noted as 30%, 40%, 50%, 60% and 70%-silk for property analysis. v_f is calculated from the density and the weight of the silk, epoxy resin and the final prepared composites. The density of all specimens were tested by drainage method with Electronic Density Balance (FA1104J). Besides, a vacuum treatment (100 Pa for 1 hr at 40°C) was added after hand lay-up and prior to the hot-press process to facilitate the infiltration of epoxy resin and the removal of air bubbles.

2.3. Impact testing

The impact properties testing was conducted according to ISO 179:1997 with a Pendulum Impact Testing Machine (MTS ZBC 1000). The specimens (without notch) were loaded with a 2 J hammer.

2.4. Dynamic mechanical thermal analysis

Dynamic Mechanical Thermal Analysis (DMTA) for SFRPs of different v_f was conducted on a dynamic mechanical analyser (TA Instruments, Waters Ltd., DMA Q800) under 3-point bending mode at a dynamic strain of 0.2% and a fixed frequency of 1 Hz and a heating rate of 3 °C min^{-1} from 25 to 170 °C. The testing specimens were cut with rectangular shape: 20mm ×10 mm ×2 mm.

3. Results and discussion

3.1. Impact properties

A major limitation of the applications of epoxy resin composites is the low toughness and weak impact properties. The impact strength (IS) have therefore become a most important measure of our work that aimed to convert brittle epoxy resin into a tough material.

From Figure 1(a), we found that when v_f of silk reached 60%, the impact properties of the SFRPs surged to 70.70 kJ m^{-2} from about 12.80 kJ m^{-2}, whereas the trend of increasing was mild when v_f was below 60%. The results suggest there may be a critical v_f effect on the impact properties of silk reinforced composites. Shah's study[11] showed when v_f of woven silk fabric reached 58% a great improvement in quasi-static mechanical properties could be achieved. The hot-press technique allowed us to apply a higher pressure to achieve even higher v_f of silk. Thus we successfully prepared SFRPs with v_f as high as 70%. According to classical fibre reinforced composites fracture mechanics, when the reinforcement fibre's v_f reaches a critical value, e.g. 50%, the reinforcements could carry more than 10 times of the load carried by the matrix resin. This is because upon a critical v_f, the effect of crack inhabitation could be maximised and the fracture behaviour of the composites could switch to a totally different mode [26,27]. The large number of fibres could firstly inhibit the initiation of micro-cracks and secondly act as crack stoppers or deflect the crack path during crack growth. In this work, the critical v_f of silk reinforcement is between 50% and 60%, where the fracture mode of SFRPs transformed from brittle to tough.

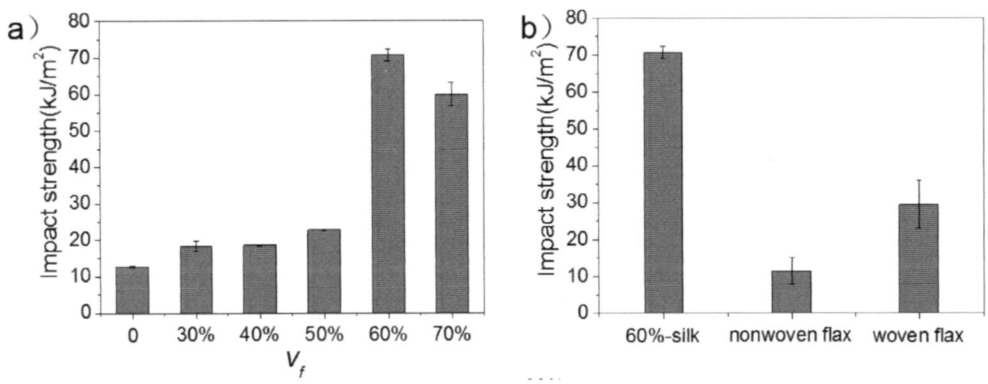

Figure 1. Impact properties of different composites (a) Comparing of different fibre v_f of SFRPs and (b) Comparing of 60%-silk and plant reinforced composites.

3.2. Dynamic mechanical thermal analysis (DMTA)

Dynamic mechanical thermal analysis (DMTA) is widely used to research the thermomechanical properties and viscoelastic behaviour of polymers and polymer composites such as epoxy-based composites[28-30]. Thermosetting resin used as matrix of composites would undergo a softening process which is called glass transition from glassy phase to rubbery phase, so the thermomechanical properties need to be estimated for reliable engineering applications.

Storage modulus E' and loss tangent $tan\delta$ are the two most important properties measured in DMTA. Figure 2 (a) showed E' of pure epoxy resin and SFRPs with different v_f in a temperature range from 25 to 170 °C; and Figure 2 (b) compared E' of each specimen at room temperature 25 °C before the glass transition and at 140 °C after the glass transition. E' was greatly enhanced after introducing silk fibres and it showed a linear increasing trend with increasing v_f. For 70%-silk SFRP at room temperature 25 °C, E' (8800 MPa) was three times of that for pure epoxy resin (2970 MPa). More importantly, E' could remain a high value of 1483 MPa for 70%-silk and 941 MPa for 60%-silk at 140 °C, while as it dropped to only 11 MPa for pure epoxy resin. This enhancement in E' for SFRPs especially at high temperatures would certainly widen the application of epoxy resin especially at $T>T_g$.

Figure 2 (c) showed the $tan\delta$ profiles of the pure epoxy resin and the SFRPs with different v_f. It was seen that the incorporation of silk fabric suppressed the $tan\delta$ peak height with increasing v_f of the fibre, as shown in figure 2 (d). The pure epoxy resin displayed the highest $tan\delta$ peak temperature at about 128 °C, which is defined as the T_g of this epoxy product. Interestingly, T_g shifted lower for all the SFRPs. The low $tan\delta$ values indicated better damping-resistant properties for SFRPs. Moreover, the good correlations between silk's v_f and the integrated $tan\delta$ area could be used to back calculate v_f of the silk reinforcement from property measurements for silk composite products. The enhanced post-T_g modulus and the low damping factor through T_g for the high v_f SFRPs could be attractive to particular applications.

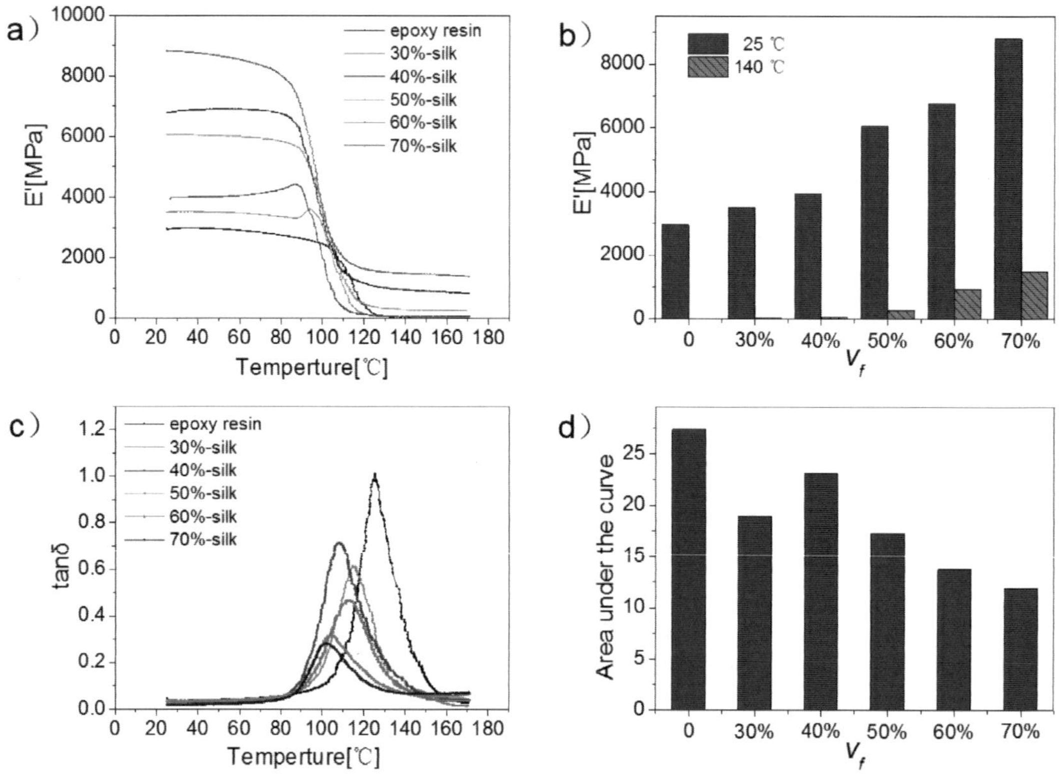

Figure 2. Dynamic mechanical property profiles of SFRPs on 3-point bending mode: (a), (b) storage modulus E' and (c), (d) loss tangent Tanδ as a function of temperature.

3.3. Comparing PFRPs and SFRPs

Compared to plant fibre reinforced plastics (PFRPs), SFRPs showed obvious advantages in impact performance. It can be found in the literature that the flax reinforced composites have an impact strength of 8-15 kJ m^{-2} for nonwoven flax and 23-36 kJ m^{-2}/[kg m^{-3}] for woven flax, which were lower than that of SFRPs reported here. One important reason was that we managed to achieve high v_f for silk reinforcements. According to Shah's study, the easy compressibility of silk fibres compared to plant fibres was owing to the irregular (almost-triangular) shape of the silk fibres and their concave and convex cross-sections. Moreover, it should also be noticed that the hexagonal shape of plant fibres could result in uneven packing and increased porous defects, which may become the controlling factor for impact resistance.

Considering the improved impact properties of SFRPs, crashworthy and impact-critical materials would be good choices for the use of SFRPs. Our new test record of IS for SFRPs indicates a good potential in the application of light-weight, defence and impact-resistance applications, such as wind turbine blades.

4. Conclusion

In our study, we have successfully prepared a series of silk fabric reinforced epoxy composites (SFRPs) by hand lay-up followed by hot-press techniques. The v_f of the SFRPs were achieved from 30% to a highest value of 70%. The impact properties were found to greatly enhance when v_f reached 60%. Moreover, dynamic mechanical thermal analysis showed the SFRPs with high v_f possessed high dynamic storage modulus (nearly 1.5 GPa at 140 °C) and a low loss factor (*tanδ* <0.3) through the glass transition of the matrix epoxy. Our findings would shed light on improving the brittle and weak epoxy resin through incorporating tough and strong silk fibres.

References

[1] Kakroodi A R, Kazemi Y, Rodrigue D 2014 Mechanical, rheological, morphological and water absorption properties of maleated polyethylene/hemp composites: Effect of ground tire rubber addition. *Composites Part B Engineering* **51** 337-44.

[2] Xue D, Hu H 2013Mechanical properties of biaxial weft-knitted flax composites. *Mater Design* **46** 264-9.

[3] Faruk O, Bledzki A K, Fink H and Sain M 2012 Biocomposites reinforced with natural fibers: 2000-2010. *Prog Polym Sci.* **37** 1552-96.

[4] Kuang Y, Yang A, Luan H, Mao H, Zhang W and Yao L, et al. 2016 Effects of structural parameters of textile substrates on the novel textile conformal half-wave dipole antenna. *J Ind Text.* (2016)10.1177

[5] Yao L, Jiang M, Zhou D, Xu F, Zhao D and Zhang W, et al. 2011 Fabrication and characterization of microstrip array antennas integrated in the three dimensional orthogonal woven composite. *Composites Part B: Engineering* **42** 885-90.

[6] Zhou D C, Yao L, Liang F, et al. 2010 Tensile and Shear Properties of Three Dimensional Orthogonal Woven Basalt/Kevlar Hybrid Composites **27** 38-42.

[7] Sashina E S, Bochek A M, Novoselov N P and Kirichenko D A 2006 Structure and solubility of natural silk fibroin. *Russ J Appl Chem* **79** 869-76.

[8] Ho M, Lau K, Wang H and Bhattacharyya D 2011 Characteristics of a silk fibre reinforced biodegradable plastic *Composites Part B: Engineering* **42** 117-22.

[9] Zhao Y, Cheung H, Lau K, Xu C, Zhao D and Li H 2010 Silkworm silk/poly(lactic acid) biocomposites: Dynamic mechanical, thermal and biodegradable properties. *Polym Degrad Stabil.* **95** 1978-87.

[10] Li W, Qiao X, Sun K and Chen X 2008 Mechanical and viscoelastic properties of novel silk fibroin fiber/poly(epsilon-caprolactone) biocomposites. *J Appl Polym* Sci **110** 134-9.

[11] Shah D U, Porter D and Vollrath F 2014 Can silk become an effective reinforcing fibre? A property comparison with flax and glass reinforced composites. *Compos Sci Technol.* **101** 173-83.

[12] Verpoest I, Van Vuure A W, El Asmar N and Vanderbeke 2010 *J Silk Fibre Cconposites.* EP2004733

[13] Vepari C and Kaplan D L 2007 Silk as a biomaterial. *Prog Polym SCI.* **32** 991-1007.

[14] Thomas R, Yumei D, Yuelong H, Le Y, Moldenaers P and Weimin Y, et al. 2008 Miscibility, morphology, thermal, and mechanical properties of a Dgeba based epoxy resin toughened with a liquid rubber. Polymer **49** 278-94.

[15] Varley R J 2004 Toughening of epoxy resin systems using low-viscosity additives. *Polym Int* **53** 78-84.

[16] Jain P, Choudhary V and Varma I K 2003 Effect of structure on thermal behaviour of epoxy resins. *Eur Polym J.* **39** 181-7.

[17] Mimura K, Ito H 2002 Characteristics of epoxy resin cured with in situ polymerized curing agent *Polymer* **43** 7559-66.

[18] Ferrari M, Vichi A and Garcíagodoy F 2000 Clinical evaluation of fiber-reinforced epoxy resin posts and cast post and cores. *Am J Dent* **13** 15B-18B.

[19] Gupta V B, Drzal LT , Lee Y C and Rich M J 1958 The temperature-dependence of some mechanical properties of a cured epoxy resin system. Polymer Engineering & Science. **25** 812-23.

[20] Francucci G and Rodriguez E 2016 Processing of plant fiber composites by liquid molding techniques: An overview. *Polym Composite* **37** 718-33.

[21] Niedermann P, Szebényi G and Toldy A 2015 Characterization of high glass transition

temperature sugar-based epoxy resin composites with jute and carbon fibre reinforcement. Composites Science & Technology **117** 62-8.

[22] Kongkaew P and Choawanklang W 2015 *Fiber orientation effecting the mechanical properties at coconut fiber reinforce epoxy resin composite*. International Conference on Science and Technology

[23] Niedermann P, Szebényi G and Toldy A 2015 Effect of epoxidized soybean oil on mechanical properties of woven jute fabric reinforced aromatic and aliphatic epoxy resin composites. *Polym Composite*. (2015) 10.1002

[24] Gassan J and Bledzki A K 1999 Possibilities for improving the mechanical properties of jute/epoxy composites by alkali treatment of fibres. *Composites Science & Technology* **59** 1303-9.

[25] Bunko K and Kennedy J F 2007 Handbook of fiber chemistry, third ed. CRC Press LLC. **69** 612.

IOP Conf. Series: Materials Science and Engineering **137** (2016) 012063 doi:10.1088/1757-899X/137/1/012063

Study on preparation and mechanical performance of TPU/nonwoven composites

X C Sun[1] and B J Xi

Shaoxing University, Zhejiang, China

E-mail: xichaosun@126.com

Abstract. In order to study the influence of resin content and layer sequence parameters on the mechanical properties of TPU/non-woven composite materials synthesized by moulding pressing technology. The effects of the resin content and layer sequence on composites were discussed. Through experiments and theoretical analysis, it was revealed how resin content, layer sequence impact on mechanical properties of composite. The mechanics properties of TPU/non-woven composite materials are improved. The process is pressure 0.5 MPa, temperature 110 ℃ and time 120s min. The melting of the TPU infiltrated into the fabric and filled the space between the fibers.

1. Introduction

Resin matrix composite materials have high strength ratio, excellent designability and corrosion resistance properties, etc [1, 2]. They have a wide range of applications in aerospace, automotive and marine industries. The design of the layer sequence can not only meet the thickness requirement of composite materials in practical applications, but also compensate the performance limitations of singularity element composites [3]. At present, domestic and foreign scholars have been studying the structure and performance of thermoplastic fiber reinforced composite materials, polymer matrix reinforced composites and the three-dimensional textile composite material, in which significant breakthroughs were achieved [4, 5]. However, the research on reinforced non-wovens for composite materials is relatively small. Based on previous literature, this paper uses thermoplastic polyurethanes (TPU) as matrix and hot pressing molding process development of resin matrix composites to observe and analyze the mechanical properties of TPU/non-woven composite.

2. Materials and instrumentations

2.1. Raw materials

TPU, with a surface area of 83.8 g/m^{2}and thickness 0.06 mm, was provided by Shanghai Anlao industrial co., LTD., China. It has a good hot melt, a strong stick relay and a high mechanical strength. Needle-punched non-wovens were purchased from Zhe Jiang Fusen spunlace non-woven co., LTD., China. The main specifications are shown in table 1.

2.2. Main equipments

The property of mechanic was studied by Electronic fabrics strength machine (YG065H, Shaoxing Yuanmao mechanical and electrical equipment co., LTD., china). The morphology of materials was characterized by Scanning electron microscope (SNG-3000, Korean SEC CO., LTD., Korea).

Table 1. Material specifications.

Material	Compo sition	Surface area (g/m^2)	Thickne ss (mm)	Breaking stress (MPa)		Elongation at break (%)	
				Crosswise	Lengthways	Crosswise	Lengthways
Non-woven	PET	448.00	3.30	4.06	4.90	90.64	50.92

3. Layer lay-up design

In order to study the relationship between layer sequence and mechanical property of composite materials, the authors used non-wovens and TPU placed in layers of pure non-wovens in order to make resin matrix composite materials with 4 layers, according to 4 kinds of layer scheme shown in figure 1. 0^0fiber as number B in longitudinal direction is defined as a composite material, and 90^0fiber as number A latitudinal direction is defined as a composite material.

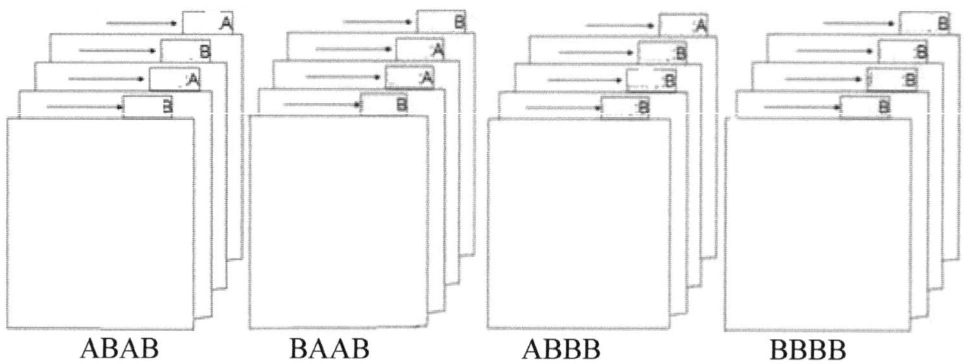

Figure 1. Non-woven layer sequence

4. Process of resin matrix composites molding

TPU is a kind of thermoplastic resin, so the fabrication of the composite material is made by a hot-pressing process. Reading reference literature of hot-pressing molding process parameters [5-7], the authors determined the optimum hot-pressing parameters as being: pressure 0.5 MPa, hot-pressing time 120s and hot-pressing temperature 110 ℃. The non-woven was decorated in processing mould plate vulcanizing machine, when the temperature rose to 110 ℃. Then the authors set the hot-pressing to 0.5MPa, normally keeping the composite materials for 120s in constant temperature hot pressing. The process is shown in figure 2 and figure 3.

Mold release → Warm-up → Loading mode → Prepressing → Hot-pressing

Figure 2. Composite extrusion molding process.

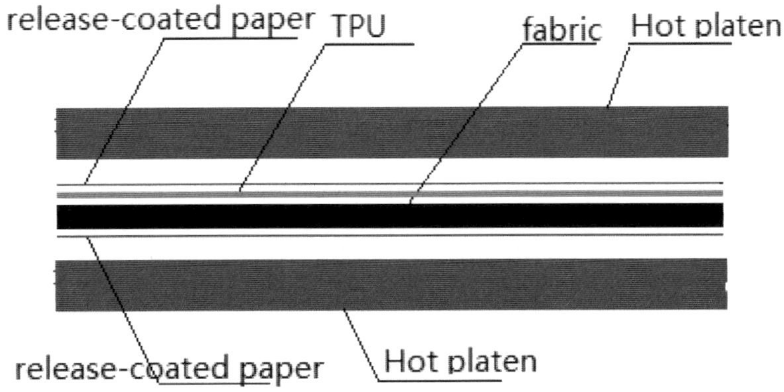

Figure 3. Preparation of composite material.

5. Analysis

5.1. Impacts of resin on the mechanical properties of the composites

The mechanical properties of the composites were tested by the testing standard GB/T 1447-1447, which is used to characterize the tensile performance of fiber reinforced plastics [8-10]. The non-woven material was tested by the electronic fabrics strength machine, and mechanical properties of composite materials were determined for layer BBBB. Test results are shown in table 2.

Table 2. Tensile properties of the material data

Name of fabric	Number of plies	Thickness(mm)	Breaking stress(MPa)		Elongation at break(%)	
			Crosswise	Lengthways	Crosswise	Lengthways
Non-woven	1	3.30	4.06	4.90	90.64	50.92
Composite	1	2.70	5.90	7.80	16.70	19.10
Composite	2	5.30	5.90	7.80	28.40	31.10
Composite	4	10.80	5.70	6.80	105.1	56.80

According to table 2, it can be concluded that the fracture stress is higher than the pure resin reinforced composites non-wovens, and the longitudinal stress is greater than the transverse fracture, but with the increase of composite material layers, the fracture stress does not increase accordingly. This is because the molten resin infiltrates the internal structure, and the space between the fiber and the fiber form become tightly coated, causing the fiber bundle to form a continuous phase [11, 12]. If connected at the same time, the TPU on each layer, causes interlayer structure formation and consolidation, improving its mechanical properties, and with the increase of composite material layer, the fracture stress of composite material change is not large, showing that the resin can evenly permeate into the non-woven, indirect composite extrusion molding process parameters being reasonable [13, 14].

5.2. Impacts layer sequence on the mechanical properties of the composites

The longitudinal tensile properties of the tests performed on the composite materials are shown in table 3.

From table 3, it can be concluded that the mechanical properties of samples 1 and 2 are similar .The fracture stress of sample 3 is the worst, but the elongation at break is the best. The fracture stress of sample 4 is the best, but elongation at break is the worst. The main cause of this phenomenon is the positive correlation between the material fracture stress and the number of samples participating in longitudinal tensile, the larger the number, the greater the fracture stress. The elongation at break of composite materials has an opposite trend, namely the mechanical properties of composite materials is not related with the layer order, but is related to the number of longitudinal tensile.

Table 3. The tensile properties of the composite material test results

Marking sample	Lay-up	Breaking stress (MPa)	Elongation at break (%)
1	ABAB	5.00	70.60
2	BAAB	5.10	72.80
3	ABBB	4.70	77.90
4	BBBB	6.80	56.80

5.3. Microstructure analysis of the composite materials

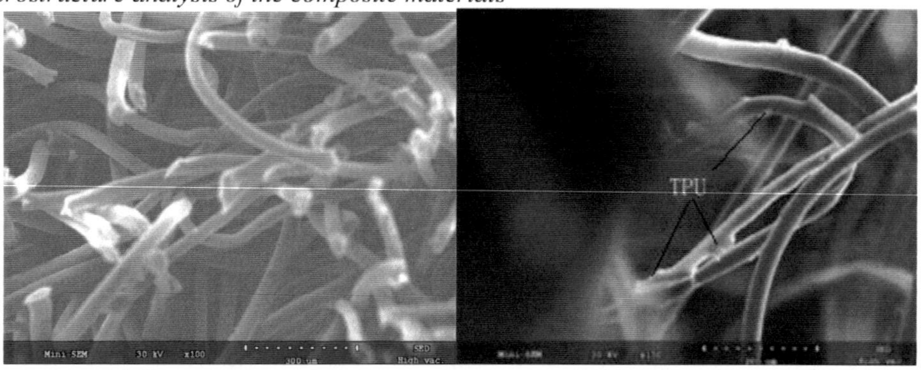

(a) pure non-wovens (b) composite materials

Figure 4. SEM images of different materials.

Figure 4(a) and (b) show SEM images of pure non-wovens and composite materials, respectively. It can be seen from the diagram, that the pressure is 0.5 MPa, 120 s hot-pressing time, hot-pressing temperature is 110 ℃, the melting of the TPU takes place under pressure, infiltrating into the fabric, filling the space between the fibers. Material deformation with pressure takes place at the same time, discharging gas inside the fabric, reducing the gap and improving the shape and structure of the composite materials [15, 16]. Thus, the resin reinforced composites determines the mechanical properties of fiber in a close extent, and contribute to the optimization of the important reference indicator of composite extrusion molding process.

6. Conclusions

We studied the preparation of resin matrix composites. The influence of resin content and layer order on the materials was further explored. Results show that the fracture stress of composite materials is proportional to the resin content, but has very little to do with the layer order. Further research shows that part of the resin infiltration of composite materials causes a major improvement on its mechanical properties.

Acknowledgement

The authors gratefully acknowledge the financial supports by the important subject fund of Zhe Jiang Province Science and Technology Hall (2014C31067).

References

[1] Yan L, Shi Y Y and Duan J H 2011 Review of Manufacturing Technology for Advanced Resin-Matrix Composites *Aeronautical Manufacturing Technology* **3** 55-58.

[2] Xu Y H, Yuan X L and Hu H 2011 Tensile Properties of co-Woven-Knitted fabric reinforced composites *Journal of Donghua University (Natural Science)* **37**(1) 42-45.
[3] Zhu C Y, Gao Z Y and Zhu J P 2005 Tensile strength property of combinational 3D woven composites *Journal of Textile Research* **26**(5) 13-15.
[4] Zhu C Y, Shen X H, Li Y Q and Zhu J P 2005 Resin permeability of combined 3D woven reinforced composites *Journal of Textile Research* **26**(4) 21-23.
[5] Rudov C S, Mouritz A P and Lee L, et al. 2003 Fiber damage in the manufacture of advanced three-dimensional woven composites *Compos. Part A-Appl. S.* **34** 963-970.
[6] Lee L, Rudov C S and Mouritz A P, et al. 2002 Effect of weaving damage on the tensile properties of three-dimensional woven composites *Compos. Struct.* **57** 405-413.
[7] Palanivelu K, Balakrishnan S and Rengasamy P 2000 Thermoplastic polyure-thane toughened lyacetalblends *Polym. Test.* **19** 75-83.
[8] Wen B Y, Zhang X D and Li Y C. Study of POM /TPU high toughened alloy 2000 *China Plastics Industry* **28**(1) 7-9
[9] Chiang W Y and Huang C Y 1989 The effect of the soft segment of polyure-thane on copolymer-type polyacetal/polyurethane blends *J. Appl. Polym. Sci.* **38** 951-968.
[10] Wu S H 1985 Phase structure and adhesion in polymer blends: A criterion for rubber toughening *Polymer* **26** 1855-1963.
[11] Varma D S, Varma M and Varma I K 1985 Coir fibers, Part II: Eva-luation as a reinforcement in unsaturated polyester resin composites *J. Reinf. Plast. Compos.* **4**(4) 419-4301.
[12] Bledzki A K and Gassan J 1999 Composites Reinforced with Cellu-lose Based Fiber *Prog. Polym. Sci.* **24**(2) 221-226.
[13] Ishida H and Koenig J 1980 Effect of hydrolysis and drying on the silox-ane bonds of a silane coupling agent deposited on E-glass fibers *J. Polym. Sci. Pol. Phys.* **18**(2) 233-237.
[14] Chien C, Dresselhans O and Endo M 1982 Effect of uniaxial stress on the Raman spectra of graphite fibers *J. Appl. Phys.* **26**(12) 5867- 5870.
[15] He F X, Bian J and Lin H L, et al. 2015 Preparation and characterization of functionalized nano-graphene sheet/PP-PP-g-MAH composites *Acta Materiae Composite Sinica* **32**(1) 47-53.
[16] Lei Y, Wu Q and Zhang Q 2009 Morphology and properties of microfibrillar composites based on recycled poly (ethylene terephthalate) and high density polyethylene *Compos. Part A-Appl. S.* **40**(6) 904-912.

Studies on green and efficient catalytic oxidation of a triazole compound

J Luo[1], **Y C Liu**[1,4], **K H Huang**[2], **T Chai**[1], **J H Wang**[1], Y W **Yu**[1], **J M Yuan**[1], **S J Chang**[1], J H **Guo**[1] **and J Zhang**[1,3]

[1]School of Chemical Engineering and Environment, North University of China, TaiYuan 030051, China
[2]Longhui Vocational and Technical Institute, ShaoYang 422200, China
[3]Hubei Space Sanjiang Honglin Detection and Control Co.Ltd, XiaoGan, 432000, China

E-mail: liuyucun204@sina.com

Abstract. 1-Methyl-3,5-dinitro-1,2,4-triazole is an insensitive energetic compound that can be prepared by oxidizing the nitrate salt of 1-methylguanazole. The influence of the reaction time, reaction temperature, reactant ratio, feeding method and catalytic oxidation method on the yield were discussed. The results show that the optimum reaction conditions are as follows: mass ratio of sodium tungstate to nitrate salt to 1-methylguanazole, 4:4.4; time, 5.5h; and temperature, 65-75°C. The yield of this oxidation reaction reached 51.36%.

1. Introduction

Oxidations are important and widely used reactions. H_2O_2 is a popular oxidizer because it is inexpensive, non-toxic, safe, and environmentally friendly. Many reports have been published that investigate molybdate and tungstate for catalytic wet peroxide oxidation of alcohols and alkenes[1-13]. 1-Methyl-3, 5-dinitro-1,2,4-triazole (DNMT) is an insensitive energetic compound with a low melting point. Although information on the synthesis of DNMT has been reported, no information has been made available on the optimization of the process. Price and DiStasio [14] synthesized DNMT in two steps, but the yield was unsatisfactory, and heavy metal wastewater was obtained, which, if not effectively discarded, could cause environmental contamination. The cause of this low yield has not yet been determined. In this study, we investigate the effect of the oxidation stage on the yield in order to determine this cause. Several determining factors of the oxidation stage were studied. The highlights of this study are as follows: (a) optimizing the parameters of the process; (b) determining the optimal catalytic oxidation method and feed ratio; (c) exploring the reuse of the wastewater generated by the process.

2. Results and discussion

2.1. The catalytic oxidation system

The presence of oxalate or hydrochloric acid improved the yield to different degrees. The oxidizing ability is stronger in an acid solution. For example, sodium tungstate reacts with H_2O_2 to form peroxotungstate. In the presence of oxalate, $[WO(O_2)_2C_2O_4]^{2-}$, a coordination complex was formed [15]. Because of its lipophilicity, this kind of catalyst more easily passes into the oil phase, which allows the reaction to be fully carried out. As shown in table 1, the catalytic oxidation was more efficient when *p*-toluene sulphonic acid (TsOH) was used as an additive. The results shown in table 2 also indicate that the highest yield was obtained when TsOH was 1.2 mmol (of sodium tungstate).

Table 1. Effect of acidic additives to the yield

Additives	H_2SO_4	HCl	oxalate	TsOH	acetic acid	H_3PO_4
Yield (%)	36.82	42.27	38.64	50.22	34.09	30.17

Table 2. Effect of TsOH dosage to the yield

Dosage of TsOH(mmol)	0.6	1.2	1.8	2.4	3.0
Yield (%)	50.22	51.36	45.45	42.50	23.18

2.2. *Properties of the catalyst for cycling studies*

At the end of reaction, most of the catalyst was dissolved in mother liquors. In order to study the usable life of the catalyst, the recovered mother liquors were enriched to 10ml. Then, 25ml (35%) H_2O_2 was added, and the reaction was left to run for 5.5 h at 65°C. The results indicated that when the catalyst was used repeatedly for 5th, the yield still reached 22.7%, as shown in table.3.This illustrates the catalytic oxidation system has better catalytic activity and more effective recycling performance.

Table 3. Effect of repeated use of the catalyst to the yield

Time	1st	2nd	3rd	4th	5th
Yield (%)	35.6	33.4	31.5	28.3	22.7

2.3. *Process parameters studies*

2.3.1. Reaction temperature. In order to investigate the effect of the temperature on the oxidation process, oxidation was conducted at 60, 65, 70, 75 and 80°C. The process was found to be safe below 70°C, and the yield was also larger in the temperature range of 65-70°C, as shown in figure 1.

2.3.2. Reaction time. The reaction time is critical to any organic reaction, especially when the reaction is performed in an environment with an excessive amount of acid under an elevated temperature. If the reaction time is too long, the product may be degraded, causing unnecessary waste. However, if the time is too short, full synthesis may not be accomplished. According to the data shown in figure 2, 5.5 h is the optimal amount of time.

2.3.3. Ratio of the reactant and the consumption of materials. As a mild oxidizer with weak acidity, H_2O_2 has significant advantages and does not cause impurities. However, it decomposes at high temperatures, which directly affects its utilization ratio. The results shown in figure 3 indicate that H_2O_2 values of greater or less than 25 ml were not beneficial to the yield.

The present research indicated that the mass ratio of sodium tungstate and DAMT had a significant impact on the reaction. If too little sodium tungstate was added, a low oxidizing degree resulted, reducing the yield and leading to unwanted by-products. However, too much sodium tungstate addition accelerated the decomposition of the hydrogen peroxide, which also resulted in a low oxidizing degree. Figure 4 shows the percentage of 4:4.4 is the most beneficial to the yield.

IOP Conf. Series: Materials Science and Engineering **137** (2016) 012064 doi:10.1088/1757-899X/137/1/012064

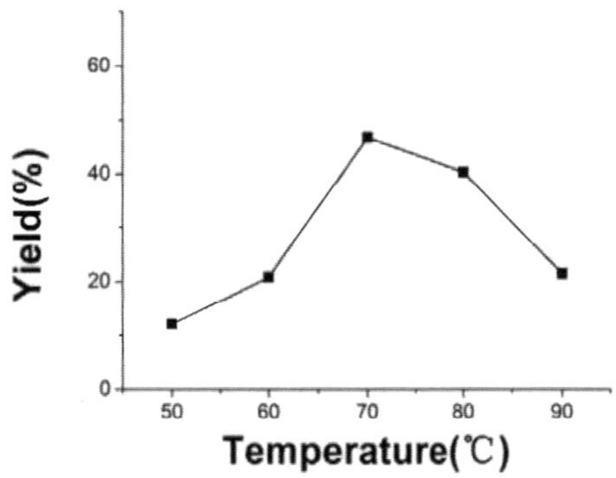

Figure 1. Temperature of the reaction.

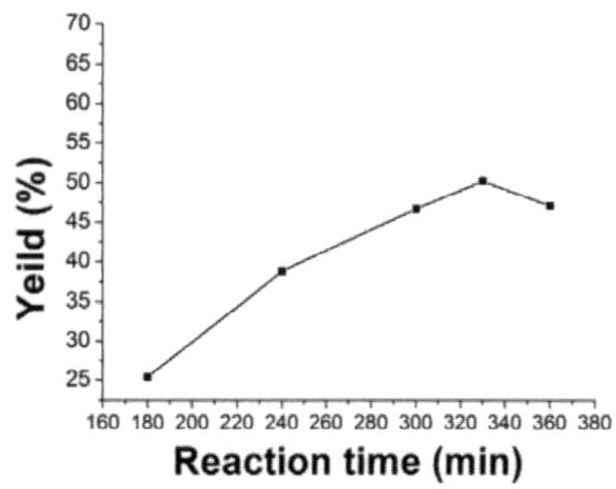

Figure 2. Time of the reaction.

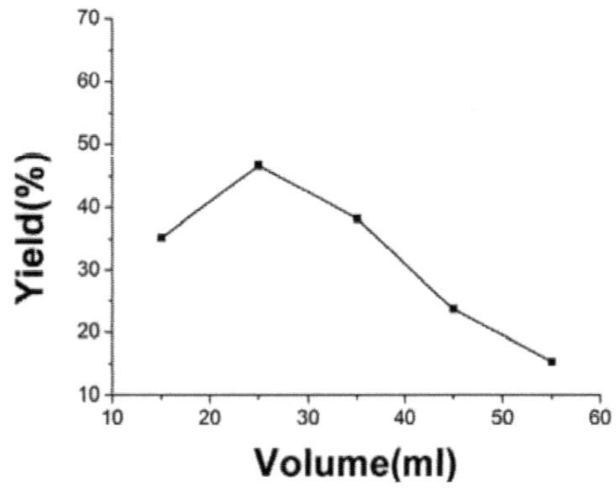

Figure 3. Dosage of H_2O_2.

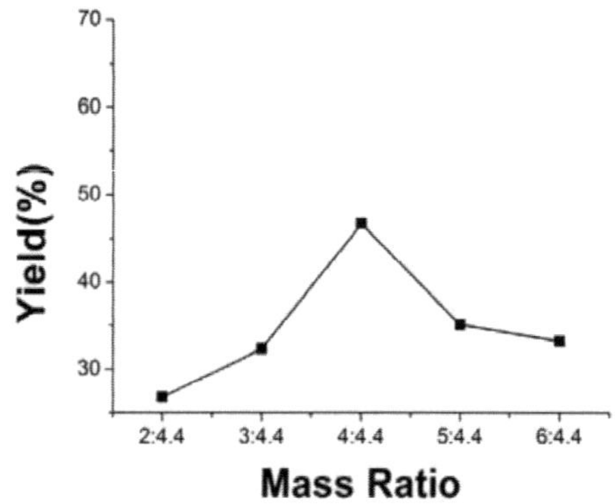

Figure 4. Sodium tungstate : DAMT.

3. Experiment

3.1. General

The ^{1}H and ^{13}C NMR spectra were recorded on a Bruker-Avance DRX 500 MHz NMR instrument with DMSO-d_6 solvent. The IR spectra were tested by a Perkin Elmer Spectrum 100 instrument. Elemental analyses were performed on a HERAEUS 1106 elemental analyzer (Germany). All the reagents were of analytical grade, purchased from commercial sources, and used as received. The dicyandiamide, aqueous solution of methyl hydrazine (40%), ethyl acetate, acetone, sodium bicarbonate and sodium nitrite were purchased from Nan Tong Dong Li Corporation (China).

3.2. Synthesis

1-Methyl-3, 5-dinitro-1,2,4-triazole. Nitrate salt of 1-methylguanazole, Na_2WO_4, TsOH and (35%) H_2O_2 was mixed in the mole ratio of 0.023:0.012:0.0012:0.439. The resulting mixture was heated to 65°C and stirred for 5.5 h. After repeated extraction by ethyl acetate, the extract was separated, dried with anhydrous Na_2SO_4, filtered, concentrated under reduced pressure, and dried in a vacuum to yield a light yellow solid product (2.21 g, 50.22%). mp 93-95°C. ^{1}H NMR (500MHz, DMSO-d_6) δ: 4.4755(s, 3H, CH_3). ^{13}C NMR(500MHz, DMSO-d_6) δ: 155.772, 153.423, 39.024. IR (KBr): 2925, 1373 (-CH_3), 1514, 1319(-NO_2).

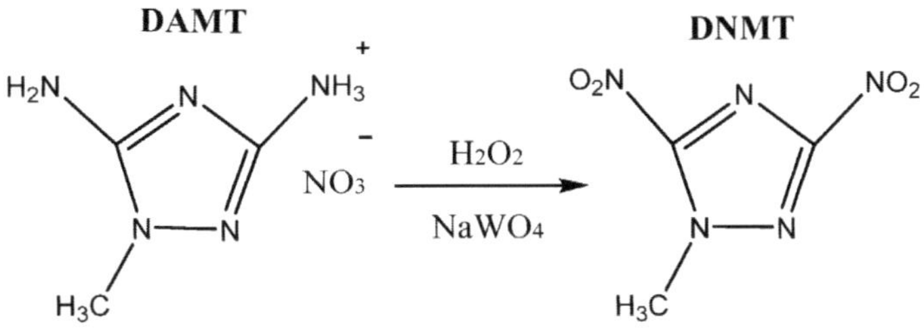

Figure 5. Preparation of DNMT.

4. Conclusions

In this paper, the oxidation of DNMT was studied. The process parameters were optimized through single factor experiments. The results showed that the best conditions for the synthesis are to mix Nitrate salt of 1-methylguanazole, Na_2WO_4, TsOH and (35%) H_2O_2 with a molar ratio of 0.023:0.012:0.0012:0.439 for 5.5 h at a temperature between 65~70°C. Under these conditions, the yield reached up to 51.36%. In addition, the catalytic system of Na_2WO_4 and TsOH reduces the discharge of waste water and shows high cyclability.

Acknowledgments
This project is financially supported by NSAF (No. U1330135).

References

[1] Jacobson S E, Muccigrosso, D A and Mares F 1979 Oxidation of alcohols by molybdenum and tungsten peroxo complexes. *Journal of Organic Chemistry* **44** 921-924.

[2] Trost M B and Masuyama Y 1984 Chemo selectivity in molybdenum catalyzed alcohol and aldehyde oxidation. *Tetrahedron Letter* **25** 173-176.

[3] Dengel A C, Griffith W P and Parkin B C 1993 Studies on polyoxo and peroxo- metalates. Part1. Tetrameric heteropoly peroxo-tungstates and heteropoly peroxo molybdates. *Journal of Chemical Society Dalton Transaction* **18** 2683-2688.

[4] Neumann R and Gara M 1995 The manganese-containing polyoxo metalate [WZnMn II (ZnW9O34)2]12-,as a remarkably effective catalyst for hydrogen peroxide oxidation. *Journal of American Chemical Society* **117** 5066-5074.

[5] Zhang H X, Gao J J and Meng H 2012 Removal of thiophenic sulfurs using an extractive oxidative desulfurization process with three new phosphotungstate catalysts. *Industrial & Engineering Chemistry Research* **51**(19) 6658-65.

[6] Ding Y X, Zhu W S and Li H M 2011 Catalytic Oxidative Desulfurization with a Hexatungstate/Aqueous H_2O_2/Ionic Liquid Emulsion System. *Green Chem* **13**(5) 1210-16.

[7] Wen S Z, Guo P Z and Hua M L, 2011 Oxidative desulfurization of fuel catalyzed by metal-based surfactant-type ionic liquids. *J Mol Catal A: Chem* **347** 8-14.

[8] Tang L, Luo G Q and Kang L H, 2013 A novel [Bmim] PW/HMS catalyst with high catalytic performance for the oxidative desulfurization process. *Korean J Chem Eng* **30**(2) 314-320.

[9] Huang W L, Zhu W S and Li H M 2010 Heteropolyanion-based ionic liquid for deep desulfurization of fuels in ionic liquids. *Ind Eng Chem Res* **49**(19) 8998 -9003.

[10] Leng Y, Zhao P P and Zhang M J 2012 A dicationic ionic liquid-modifi ed phosphotungstate hybrid catalyst for the heterogeneous oxidation of alcohols with H_2O_2. *Sci China Chem* **55**(9) 1796-1801.

[11] Isbii Y, Yamawaki K and Ura T 1988 Hydrogen peroxide oxidation catalyzed by heteropoly acids combined with cetylpy ridinium chloride : epoxidation of olefins and allylic alcohols, ketonization of alcohols and diols, and oxidative cleavage of 1 , 2-diols and olefins. *Journal of Organic Chemistry* **53** 3587-3593.

[12] Bortolini, O, Conte V and Furia F D 1986 Metal catalysis in oxidation by peroxides. Molybdenum and tungsten catalyzed oxidations of alcohols by diluted hydrogen peroxide under phase-transfer conditions. Journal of Organic Chemistry **51** 2661-63.

[13] Bortolini O, Campestrini S and Furia F D 1987 Matel catalysis in oxidation by peroxides anionic molybdenumpicolinate N-oxido-peroxo complex: an effective oxidation of primary and secondary alcohols in nonpolar solvent. *Journal of Organic Chemistry* **52** 5467-69.

[14] Price W and DiStasio A R 2010 Synthesis of DNMT: A new energetic melt-pour ingredient, *in 2010 insensitive munitions and energetic materials technology symposium.*

[15] Dickman M H and Pope, M T 1994 Peroxo and superoxo complexes of chromium, molybdenum and tungsten. *Chemical Review* **94** 569-584.

IOP Conf. Series: Materials Science and Engineering **137** (2016) 012065 doi:10.1088/1757-899X/137/1/012065

The breakage of nanopore in AAO template

X R Jia, H Wang[1] and Y Zhen

Department of Physics, Capital Normal University, Beijing Key Laboratory of Metamaterials and Devices, Key Laboratory of Terahertz Optoelectronics, Ministry of Education, and Beijing Advanced Innovation Center for Imaging Technology, Beijing, 100048, P.R. China

E-mail: wanghai@cnu.edu.cn

Abstract. In the present work, AAO template is fabricated in oxalic acid solution under a constant voltage by several steps. By the Bernoulli principle, the pressure on the wall of hole increases which lead to the breakage of nanopore as a result of the reducing effective migration rate of Al^{3+}. The quantity of the breakage of nanopore rises with the increase of the concentration of Al^{3+}. Further, nanopore is closed by oxide due to the decrease of effective migration rate of Al^{3+}. Finally, a "nanoflower-like" shape can be observed in experiments.

1. Introduction

AAO films with a hexagonal array structure of mono-disperse nanopore is a template which can be used for the fabrication of nano-materials, such as nanodots [1], nanowires [2] and nanotubes [3], because of the simple preparation technology, the cheap cost, highly ordered pores and with a large aspect ratio.

During the fabrication of AAO template, ions pass through the nanopore array participating electro-chemical reaction. Under the same applied voltage, the density of the ions and its Debye screening length are constant in nanopore. Thus the flow of the ions is incompressible approximately in nanopore. The nanopore array modifies the velocity of ions and the change of velocity may exert a traverse pressure on the wall of nanopore, which leads to the breakage of nanopore array.

In present work, we fabricate AAO template in oxalic acid solution with different Al^{3+} concentration. The concentration of Al^{3+} in solution exerts an influence on the concentration diffusion direction. When the direction of concentration diffusion of Al^{3+} opposites with the directional migration of Al^{3+} under applied electric field, the breakage of nanopores can be easily observed.

2. Experimental

Firstly, high-purity (99.999%) aluminum foil of 0.1mm thickness are degreased in acetone, and then the aluminum foil are washed with deionized water. Secondly, the aluminum foil are electrochemically polished in a 1:4 volume mixture solution of perchloric acid and ethanol to the surface roughness at 10

V for 3 min at 6℃. Thirdly, the AAO films are fabricated by several-step anodization. All the samples are anodized in a 0.3 M oxalic acid solution at 10℃ at 40 V, afterward the formed porous oxide layer are chemically removed in a mixture solution of H_3PO_4 (6wt%) and H_2CrO_4 (1.8wt%) at room temperature. The anodizing time and the time of removing the oxide layer vary with the anodizing step. The anodizing time of from the first step to the sixth step anodic oxidation are 6h, 15min, 5min, 2min, 44s and 16s respectively, and the time of removing the oxide layer formed from the first step to the fifth step are 12h, 2h, 44min, 16min and 6min respectively [4]. Finally, the morphology of the films are observed by scanning electron microscope (SEM) from the Figure 2 to the Figure 6.

The electropolished aluminum are anodized in 0.3M oxalic acid solution at 0℃at 45 V for 6.2h, afterward the formed porous oxide layer are chemically removed in a mixture solution of H_3PO_4 (6wt%) and H_2CrO_4 (1.8wt%) at room temperature for 13h, then the second anodizing step in a 0.3M oxalic acid solution at 0℃ at 45V for 15min, the oxide layer formed in the second steps are removed with a mixture solution of H_3PO_4 (6wt%) and H_2CrO_4 (1.8wt%) at room temperature for 2.3h. The aluminum with hexagonal ordered concave of the second step anodization are anodized in a 0.3M oxalic acid containing saturated Al ion concentration solution at 0℃at 45V for 5min, 8min and 10min respectively. The morphology of the films are observed by scanning electron microscope (SEM) from the Figure 7 to the Figure 9.

3. Discussion

During the process of electrochemical reaction, the velocity of ions are determined by applied voltage and ions concentration gradient. During the preparation of AAO template, the electric field accelerates oxide dissolution to producing Al^{3+}, and Al^{3+} under the drive anodizing electric current and ion concentration gradient current migration from nanopore to the solution [5, 6]. At the interface of nanpores and the solution, in Figure 1, the effective diameter is transforming from pore diameter D_p to interpore distance D_{int}, so the effective cross-sectional area transforms from $S_{pore} = \pi(D_p/2)^2$ to $S_{solution} = \pi(D_{int}/2)^2$, because the flow of the Al^{3+} is incompressible approximately in nanopore. For incompressible fluid, when the velocity of fluid increases, its pressure decreases. On the contrary, the velocity of fluid decreases, the pressure increase. It is the principle of Bernoulli equation. For the flow of the Al^{3+}, the mass of the Al^{3+} is equal, $S_{pore} * v_{pore} = S_{solution} * v_{solution}$, because of $S_{pore} < S_{solution}$, according to $D_P < D_{int}$, $v_{pore} > v_{solution}$, that is the velocity of Al^{3+} at the interface of nanopore and solution slow down, in addition the density of the Al^{3+} ρ is constant in nanopore, and the central axis of the Al^{3+} is the center of the pore diameter D_p and the interpore distance D_{int} respectively, so the height of center of gravity h is constant, according to Bernoulli equation $\frac{1}{2}\rho v^2 + \rho gh + P = const$, thus $P_{pore} < P_{solution}$ on account of $v_{pore} > v_{solution}$, that is the pressure enhances at the interface from nanopore to solution.

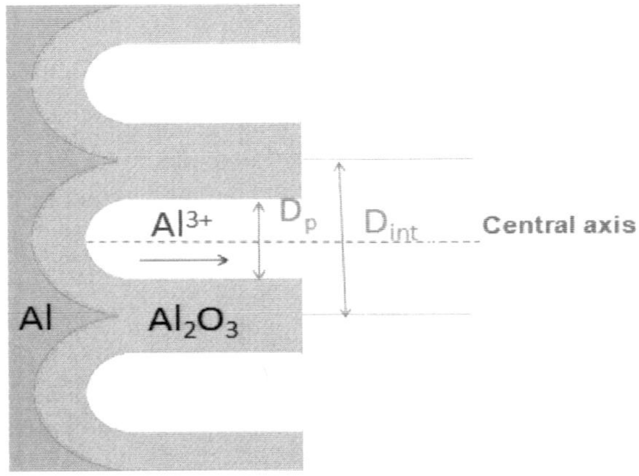

Figure 1. The cross-sectional schematic of the porous alumina film.

During the fabrication of AAO template, the effective velocity of the Al^{3+} are under the influence of the electric field and the ion concentration gradient in the electrolyte. The direction of Al^{3+} migration is from nanopore to solution under the action of the electric field; the electric field accelerates oxide dissolution to producing a number of Al^{3+} spreading into the solution, then the concentration of Al^{3+} in oxalic acid solution increases with time. when the concentration of the Al^{3+} in the pore is larger than in solution, the direction of Al^{3+} migration is from nanopore to solution under the action of ion concentration gradient, the effective velocity of the Al^{3+} is augment on account of the same direction of concentration diffusion of Al^{3+} and the directional migration of Al^{3+} under applied electric field. However when the concentration of the Al^{3+} in the pore is smaller than in solution, the direction of Al^{3+} movement is from solution to nanopore under the action of ion concentration gradient, so the effective velocity of Al^{3+} is decrease on account of the direction of concentration diffusion of Al^{3+} opposite with the directional migration of Al^{3+}.

From the Figure 2 to the Figure 4 show typical SEM images of AAO films after two, three and four-step anodic oxidation. A common feature is a hexagonal array structure of mono-disperse nanopores, while the quantity of the breakage of nanopores is more and more with the increase of oxidation step. We consider the electric field accelerates oxide dissolution to producing a number of Al^{3+} spreading to the solution during the fabrication of AAO template, that result in the concentration of Al^{3+} in oxalic acid solution increase with the increase of oxidation step, and then the effective velocity of Al^{3+} is lower and lower with the increase of oxidation step under the influence of electric field and ion concentration gradient, the effective velocity of the Al^{3+} reducing result in the enhancement of pressure on the wall of nanopore by the Bernoulli principle, so the quantity of the breakage of nanopores is more and more with the increase of oxidation step. Meanwhile at the interface between nanopore and solution, the effective cross-sectional area from $S_{pore} = \pi(D_p/2)^2$ to $S_{solution} = \pi(D_{\text{int}}/2)^2$, the mass of the Al^{3+} is equal, $S_{pore} * v_{pore} = S_{solution} * v_{solution}$. $S_{pore} < S_{solution}$ according to $D_P < D_{\text{int}}$, so $v_{pore} > v_{solution}$. From two-step to four-step anodic oxidation, the concentration of Al^{3+} in oxalic acid solution is not saturated, so the effective velocity of Al^{3+} in the nanopore is not smallest, therefore the oxide at the surface of nanopores that are fabricated by Al^{3+} from nanopore to solution and the inverse migration oxygen ions is difficult, but Al^{3+} spreading into solution.

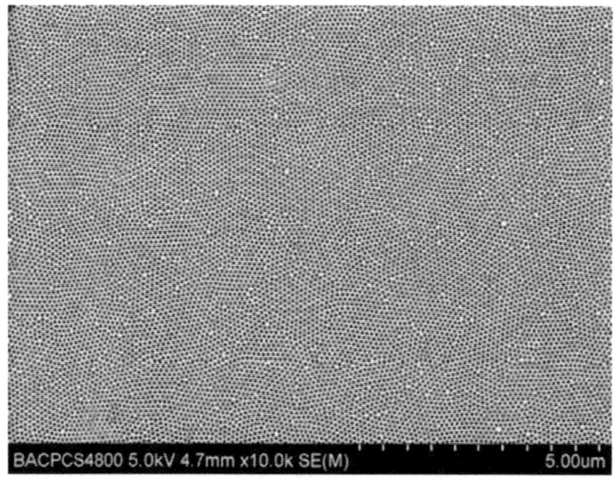

Figure 2.SEM images of AAO film made for 15min by two-step.

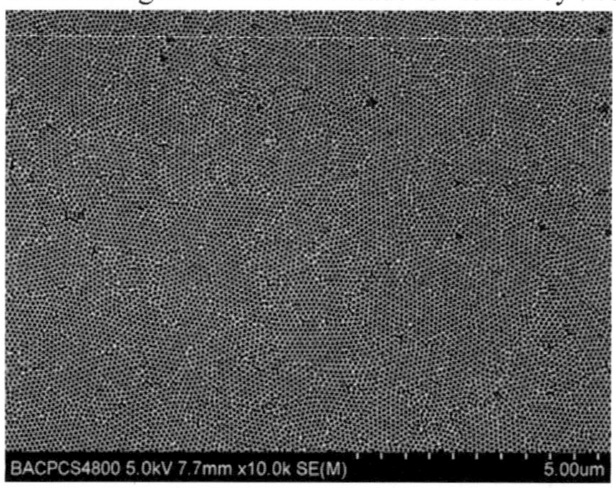

Figure 3.SEM images of AAO film made for 5min by three-step.

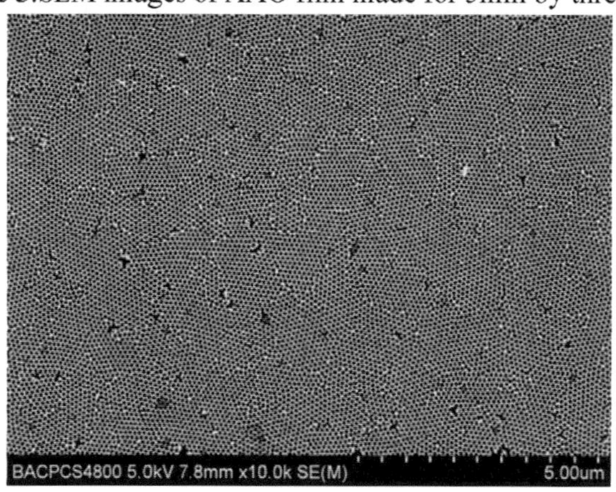

Figure 4. SEM images of AAO film made for 2min by four-step.

The Figure 5 and the Figure 6 show typical SEM images of AAO films by five-step and six-step anodic oxidation, the oxide at the surface of nanopores are fabricated. The Figure 5 show a number of the "line-like" oxide at the surface of nanopores, although the "line-like" oxide is disorder distribution, its distribution density is almost the same, in addition the "line-like" oxide distribute on the boundary of the domain, and below the "line-like" oxide is mono-disperse nanopore with a hexagonal array structure. We consider the increase of Al^{3+} concentration in oxalic acid solution with the increase of oxidation step, the effective velocity of Al^{3+} is smaller on account of the direction of concentration diffusion of Al^{3+} opposite with the directional migration of Al^{3+} under applied electric field. At the interface of nanopore and solution, the effective cross-sectional area transform from $S_{pore} = \pi(D_p/2)^2$ to $S_{solution} = \pi(D_{\text{int}}/2)^2$, the mass of the Al^{3+} is equal, $S_{pore} * v_{pore} = S_{solution} * v_{solution}$. $S_{pore} < S_{solution}$ according to $D_P < D_{\text{int}}$, so $v_{pore} > v_{solution}$. So the effective velocity of the Al^{3+} in the nanopore decrease, therefore the oxide at the surface of nanopores that are fabricated by Al^{3+} from the nanopore to the solution and the inverse migration oxygen ions is easy, especially on the boundary of the domain, it lead to a part of Al^{3+} stacking due to disorder nanopores, so the stacking point of the oxide form "line-like" oxide. The Figure 6 show typical SEM images of AAO films by six-step anodic oxidation, we can observe a large number of the "nanoflower-like" oxide at the surface of nanopores, and the "nanoflower-like" oxide are composed of the middle of "trunk-link" and around "beard-link", also its distribution density is almost the same, and below the "nanoflower-like" oxide is mono-disperse nanopore with a hexagonal array structure. The shape of the oxide in the Figure 6 is larger than in the Figure 5, because the effective velocity of Al^{3+} is lower due to the Al^{3+} higher concentration by six-step anodic oxidation.

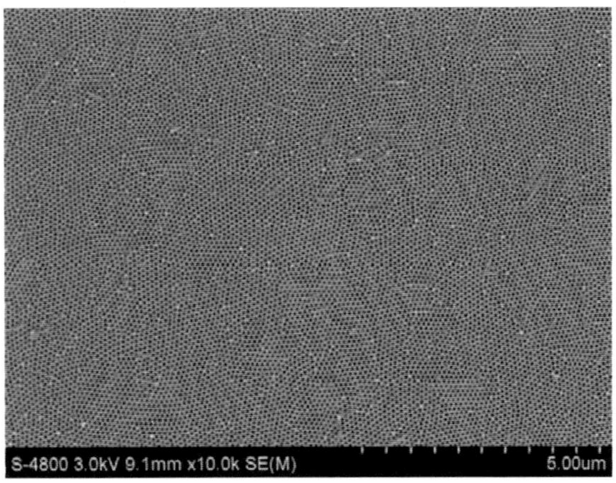

Figure 5. SEM images of AAO film made for 44s by five-step.

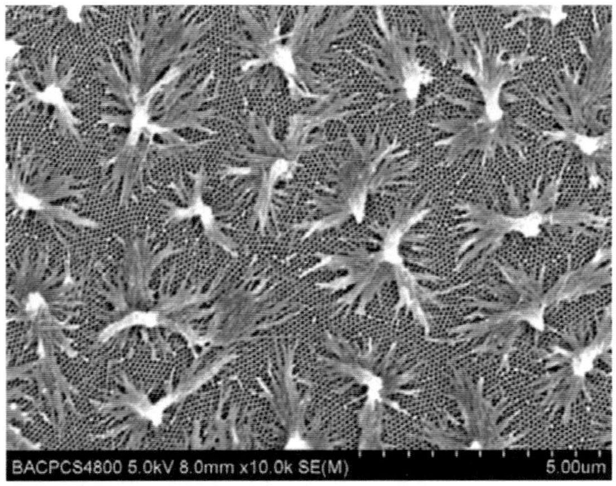

Figure 6. SEM images of AAO film made for16s by six-step.

To prove the conclusion of the oxide at the surface of nanopores that are fabricated by Al^{3+} from the nanopore to the solution and the inverse migration oxygen ions but that are not affected by the oxidation step and time is correct, we change the oxidation parameter. The aluminum with hexagonal ordered concave of the second step anodization are anodized in a 0.3 M oxalic acid solution containing saturated concentration of Al^{3+} at 0℃ at 45 V for 5min by three-step anodic oxidation.

The Figure 7 show typical SEM images of AAO films in oxalic acid solution containing saturated concentrcation of Al^{3+} for 5min by three-step anodic oxidation. We can observe the some of nanopores is closed by oxide at the surface of nanopores, in addition some of the breakage of nanopores in AAO template. When the concentration of Al^{3+} is saturated in the oxalic acid solution, the direction of concentration diffusion of Al^{3+} opposites with the directional migration of Al^{3+} under applied electric field, so the effective velocity of Al^{3+} is the smallest. At the interface of nanopores and solution, the effective cross-sectional area transform from $S_{pore} = \pi(D_p / 2)^2$ to $S_{solution} = \pi(D_{\text{int}} / 2)^2$, the mass of the Al^{3+} is equal, $S_{pore} * v_{pore} = S_{solution} * v_{solution}$, $S_{pore} < S_{solution}$ according to $D_P < D_{\text{int}}$, so $v_{pore} > v_{solution}$. Therefore the effective velocity of Al^{3+} in the nanopore is the lowest, the result in the oxide in surface nanopores are formed by Al^{3+} from the nanopores to the solution and the inverse migration oxygen ions.

Figure 7. SEM images of AAO film made in saturated Al^{3+} concentration for 5min by three-step.

Figure 8 show typical SEM images of AAO films in oxalic acid solution containing saturated concentration of Al^{3+} for 8min by three-step anodic oxidation.We can observe the breakage of oxide at the surface of nanopores and below the breakage of nanopores is mono-disperse nanopore with a hexagonal array structure. We thought the effective velocity of Al^{3+} reduce due to the Al^{3+} limited in closed nanopores, so the pressure on the wall of nnopore enhance by Bernoulli principle, and then the nanopore break and below the breakage of nanopore is order nanopores. Although the literature are also reported the breakage of nanopore and the reason of the breakage of nanopores is the wall of naopore corroded in acid and alkali solution [7, 8], the theory that the breakage of nanopore the reason why the effective velocity of Al^{3+} reducing lead to the enhancement of pressure on the wall of nanopore is not reported.

Figure 8. SEM images of AAO film made in saturated Al^{3+} concentration for 8min by three-step.

The Figure 9 show typical SEM images of AAO films in oxalic acid solution containing saturated concentration of Al^{3+} for 10min by three-step anodic oxidation, we can observe a large number of the "nanoflower-like" oxide at the surface of nanopores, and it is composed of the middle of "trunk-like" and around "beard-link", and its distribution density is almost the same, in addition below the "nanoflower-like" oxide is mono-disperse nanopore with a hexagonal array structure. We consider the breakage of nanopore largen when the Al^{3+} in the nanopores pass through the nanopore to solution, finally the alumina of the surface will shrink to form different patterns "nanoflower-like". The middle of "trunk-link" are originate from the closed oxide at the surface of nanopores shrinking, and the around "beard-link" are originate from the breakage of nanopores.

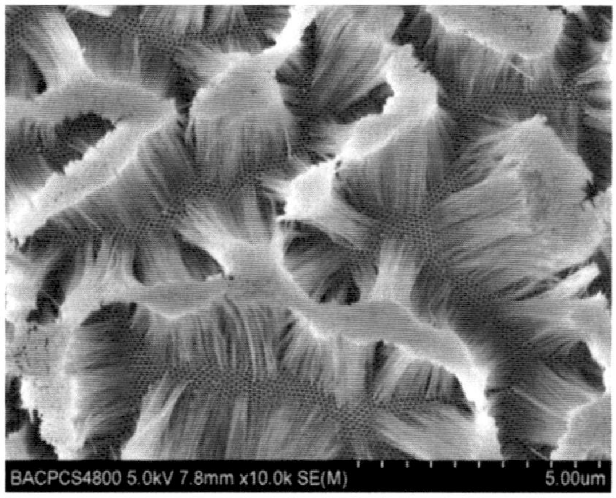

Figure 9. SEM images of AAO film made in saturated Al^{3+} concentration for 10min by three-step.

We can observe the "nanoflower-like" shape are similar from the Figure 6 and the Figure 9. The Figure 9 show typical SEM images of AAO films in oxalic acid solution containing saturated concentration of Al^{3+} under 45V for 10min by three-step anodic oxidation, while the Figure 6 show typical SEM images of AAO films under 40V for 2s by six-step anodic oxidation. Because the oxidation step and time are different, the "nanoflower-like" oxide are not affected of the oxidation step and time. The same parameter is the high concentration of Al^{3+} in oxalic acid solution, so the direction of concentration diffusion of Al^{3+} opposite with the directional migration of Al^{3+} under applied electric field, the result in the effective velocity of Al^{3+} is decrease in the nanopores. Therefore the oxide at the surface of nanopores that are fabricated by Al^{3+} from the nanopore to the solution and the inverse migration oxygen ions is easy. The effective velocity of Al^{3+} reducing lead to the enhancement of pressure on the wall of nanopore by the Bernoulli equation $\frac{1}{2}\rho v^2 + \rho gh + P = const$. Therefore the breakage of nanopores and the "nanoflower-like" oxide are formed.

4. Conclusion

We study the different AAO template fabricated by changing the concentration of Al^{3+} in oxalic acid solution. When the concentration of Al^{3+} is low, the effective velocity of Al^{3+} is large due to the direction of the direction of concentration diffusion of Al^{3+} same with the directional migration of Al^{3+} under applied electric field in the electrolyte. Al^{3+} spreads into the solution easily, which benefits for the order AAO template formation. However, in the case of high Al^{3+} concentration in acid, the direction of the electric field opposites with the ion concentration gradient in the electrolyte. Therefore, the effective velocity of Al^{3+} becomes smaller, which results in the pressure enhancement on the wall of nanopore and leads to the breakage of nanopore finally.

Acknowledgments

This work was financially supported by the National Natural Science Foundation of China (NSFC) with grant No.11274233 and Beijing education committee under grant KM201610028004.

References

[1] Kikuchi N, Okamoto S and Kitakami O 2009 Generation of nanosecond magnetic pulse field for switching experiments on a single Co/Ptnanodot *J Appl Phys.* **105**(07) 506
[2] Yin A, Li J, Jian W, Bennett A and Xu M 2001 Fabrication of highly ordered metallic nanowire arrays by electrodeposition *Appl Phys Lett.* **79**(7) 1039
[3] Kouklin N 2005 Self-assembled network of carbon nanotubes synthesized by chemical vapor deposition in alumina porous template *Appl Phys Lett.* **87**(20) 3105
[4] Zhen Y, Wang H, Jia X, Jiang H and Liu J 2011 The Effect of Multi-Step Anodic Oxidation *Materials Science Forum.* **694** 585
[5] Osullivan J and Wood G 1970 The morphology and mechanism of formation of porous anodic films on aluminum *Proc R Soc London A.* **317** 511
[6] Thompson G 1997 Porous anodic alumina: Fabrication, characterization and applications *Thin Solid Films.* **297**(1-2) 192
[7] Zhang L,Yao S, Zhang W and Wang H 2005 Preparation and Formation Mechanism of Alumina Nanowires *Acta Phys Chim Sin.* **21**(11) 1254
[8] Sun X, Liang J, Zhao J, Ma Q and Xu B 2010 Preparation of alumina nanowires, nanorods, and nanowalls by chemical etching *Appl Phys A.***98**(10) 263

IOP Conf. Series: Materials Science and Engineering **137** (2016) 012066 doi:10.1088/1757-899X/137/1/012066

Preparation research of Nano-SiC/Ni-P composite coating under a compound field

H Z Zhou[1, 2, 3], W H Wang[2], Y Q Gu[2], R Liu [2] and M L Zhao[2]

[1] Key Laboratory of Advanced Structural Materials and Application Technology of Jiangsu Province, China

[2] Department of Material Engineering, Nanjing Institute of Technology, Nanjing 211167, Jiangsu, China

E-mail: zhzmsc@njit.edu.cn

Abstract. In this paper, the preparation process of Ni-P-SiC composite coatings on 45 steel surfaces with the assistance of magnetic and ultrasound fields was researched. The influence of external field on the surface morphology and performance of the composite layer is also discussed. Experimental results showed that when prepared under magnetic and ultrasonic fields, composite layers are significantly more dense and uniform than coatings made without external fields. Nano-SiC particles, dispersed uniformly in the layer, significantly improve the hardness of the composite layer, and the composite layer under the external field had the highest hardness at 680 HV. The external fields can also accelerate deposition and increase the thickness of the layer. Compared to layers processed without the assistance of external fields, the thickness of the layers increased by nearly ten μm.

1. Introduction

With the addition of nanoparticles [1-5], the wear resistance, hardness, corrosion resistance and other properties of electroless composite coatings (ECCs) have been improved [6-9]. The preparation process of nano ECCs has therefore become one of the research hot spots in this field [10-14]. An amorphous Co-Ni-B-Ce coating with fine grains was prepared by Xuan et al. [10] under the condition of a magnetic field. This coating showed an increased ability to resist deformation as well as improved toughness and wear resistance. Fan et al. [11] improved the deposition rate, grain refinement, and inhibition of nanoparticle aggregation of layers by using an ultrasonic assisted preparation method. As a result, the hardness and corrosion resistance of the coating were significantly improved. A smooth and flat Ni-P/SiC coating surface of low porosity and a fine cellular structure was prepared by Wang et al. [12] by utilizing ultrasonic mechanical stirring composite preparation technology. Even so, there have been few studies to date on fabrication technology under an extra field for ECCs.

Because of its excellent corrosion and erosion resistance properties, SiC is also a useful kind of electronic material. Therefore, nano-SiC particulates are commonly used as reinforcement phases for electroless deposited Ni-P alloys, and the fabrication and performance of electroless Ni-P-SiC composite coatings have been widely investigated [2, 15-16]. In the present work, nano-SiC/Ni-P coatings on the carbon steel surface prepared with the assistance of an ultrasonic and magnetic field were studied. Based on micro-morphology observation, structure analysis, and thickness and micro-hardness determination, the role of the compound field in the preparation process is summarized.

2. Experimental process and characterization methods

2.1. Experimental

45 steel matrix samples were polished and placed in 15% HCl solution for 1 min. The plating solutions were made according to the formula in Table 1 with a pH value of approximately 10. Then, the solutions were stirred for 5 min by an ultrasonic generator. Finally, the samples were processed at 25 ℃ under the conditions of Table 2.

Table 1. Solution composition of electroless nano-composite plating.

Reagent name	Content(g/L)
$NiSO_4 \cdot 7H_2O$	25
$NaH_2PO_2 \cdot H_2O$	25
$NH_3 \cdot H_2O$(ml/L)	22
$Na_4P_2O_7 \cdot 10H_2O$	50
20 nm-SiC	5

Table 2. Fabrication processes of different samples.

Sample number	SiC	Magnetic field	Ultrasonic field	Process time
1#	-	-	-	1 h
2#	5g/L	-	-	1 h
3#	5g/L	7.5T/m	-	1 h
4#	5g/L	7.5T/m	-	1.5 h
5#	5g/L	7.5T/m	40 Hz, 100 W	1 h
6#	5g/L	7.5T/m	40 Hz, 100 W	1.5 h

2.2. Characterization methods

The coating samples' micro-morphology was observed by SEM (JSM-6460lv). The ECCs' composition was analysed by EDS (GENESIS 2000XMS60). Also, the phase was detected by XRD (UltimaIV) with the scanning angle from 10 DEG to 90 DEG, scanning step 0.02 DEG and scanning speed 1.2 DEG/min. The thickness of the samples was also measured by TT206 coating thickness tester. The hardness was measured with an hvs-1000 digital micro hardness meter under a holding pressure of 100g for 10s.

3. Results and analysis

3.1. Microstructure of the coating samples

Figure 1 shows SEM photos of the coating samples. Figure 1 (a) and (b) indicate that the surfaces of sample coatings 1# and 2# are incompact microstructures. From the high magnification SEM photos in Figure 1 (a) and (b), it also can be seen that the surface coating of the 2# sample has a nonuniform and loose particle aggregation state with many gaps, which results in the coating surface being very rough. On the other hand, from the high magnification SEM photos in Figure 1 (c), (d), (e), and (f), we can see that with the addition of the different extra fields, the surfaces of the coating samples became more and more continuous and compact. The coating surfaces of samples 3# and 4# were denser than sample 2#, and the surface morphologies of 5# and 6# were more compact and smooth as well. This also shows that the extra ultrasonic and magnetic field played a very important role in the coating deposition and was helpful in forming more compact and flat nano-composite coatings.

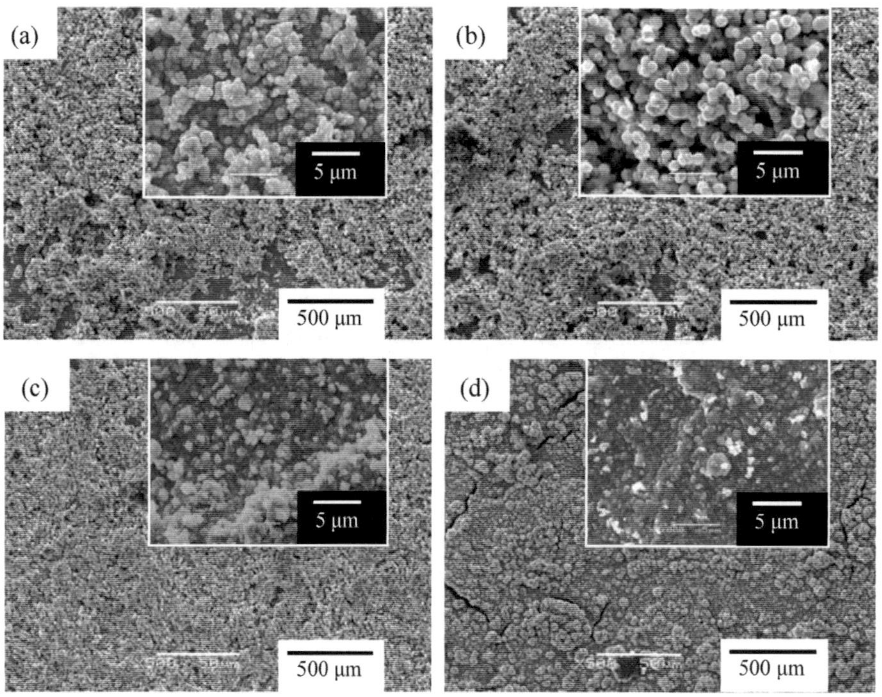

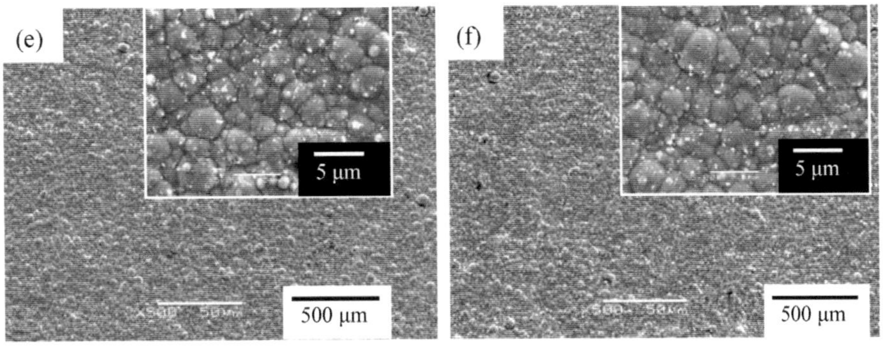

Figure 1. SEM morphology of coatings
(a) 1#, (b) 2#, (c) 3#, (d) 4#, (e) 5#, (f) 6#.

3.2. Composition and phase of the coating samples

Because samples 3# and 4# and samples 5# and 6# differed only in deposition time, only samples 1#, 2#, 3# and 6# were selected as targets for EDS and phase analysis.

Figure 2 shows the EDS analysis results of the samples made under different preparation conditions. From Figure 2 (a), it is clear that no nano-SiC particles existed in the 1# sample, and the coating composition was mainly dominated by Ni and P. From Figure 2 (b), (c), it is clear that the nano-SiC particles added into the solution deposited in the coating with the Ni, P elements, and the deposition of Ni, P element remained mostly unchanged. Thus, it may be concluded that the added SiC particles, just as an ideal nucleation point for Ni-P do not affect the deposited ratio of the Ni, P elements. Figure 2 (d) shows the EDS analysis results of the high magnification SEM morphology on Figure 1 (d). The Fe element signal is from the matrix.

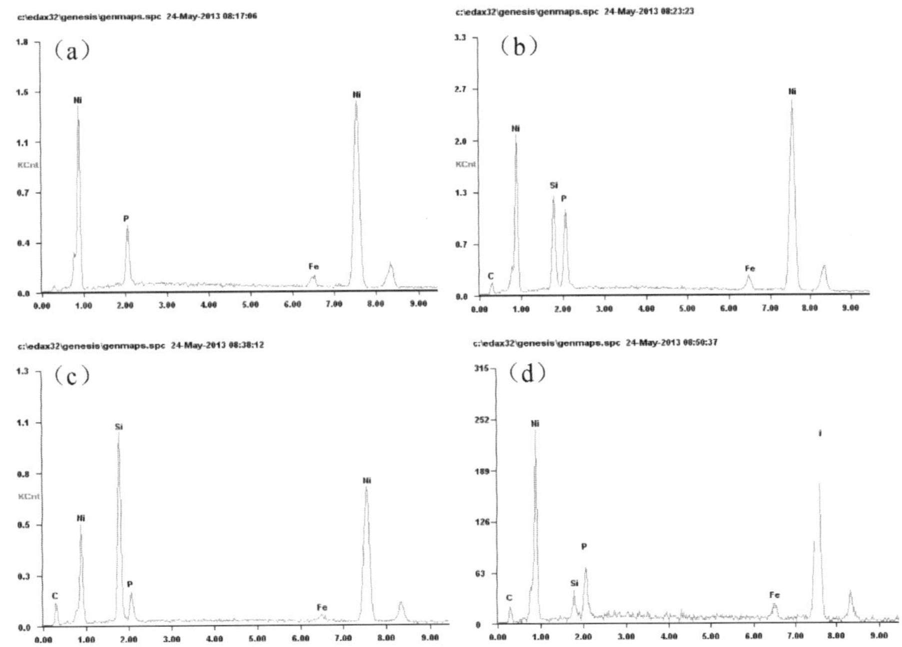

Figure 2. EDS analysis results of the coatings (a)1#, (b)2#, (c)3#, (d)6#.

Figure 3 displays the XRD analysis results of the coatings prepared under four different conditions. In the same way, it can be seen from the XRD result of the 1# sample that there were no SiC particles. On the other hand, the results of the 2#, 3#, 6# sample show diffraction peaks of the SiC phase (JCPDS-ICDD: 00-001-1118), which complies with the experimental conditions. According to JCPDS-ICDD: 00-003-0953, the peaks at 35.351°, 44.613°, 57.392° and 72.721° can be ascribed to the Ni_2P phase. Compared with the 1# sample, the position of the Ni-P's peaks of 2#, 3#, 6# samples do not change much, which indicates that the addition of SiC particles did not change the crystallization behavior of the Ni-P alloy, and that the intensities of the Ni_2P phase diffraction peaks were higher than those of the 1# sample. This clearly indicates that the addition of the external fields has contributed to the crystallization of Ni-P. The diffraction peaks of Fe (JCPDS-ICDD: 00-001-1252) in the samples derive from the matrix.

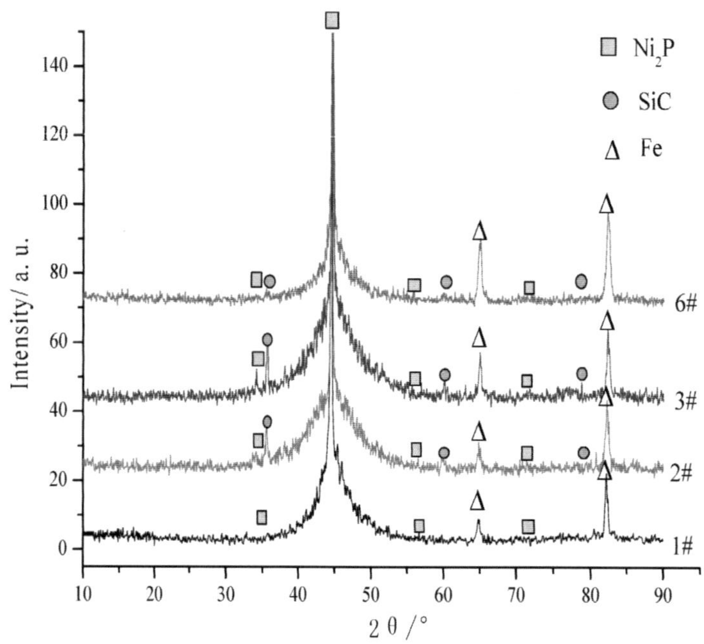

Figure 3. XRD spectra of the coatings.

3.3. Thickness of the coating samples

Figure 4(a) shows the average thickness values of the coating samples. It is clear from Figure 4(a) that the thickness values of the coatings underwent an increasing trend from 17.2 μm to 29.6 μm. Compared with the thickness value of 2# coating, the thickness values of the 3#, 4#, 5#, 6# samples all increased. This result shows that the addition of the external fields not only accelerated the deposition of the nano-composite coatings, but also thickened the composite coatings. It can also be found from the thickness values of 3# and 4# samples that the coatings' thicknesses increased as the process time was extended.

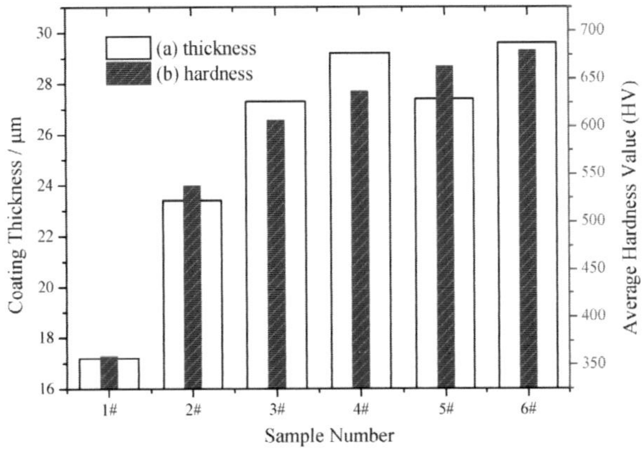

Figure 4. Average thickness and hardness values of coating samples (a) average thickness values, (b) average hardness values.

3.4. Hardness of the coating samples

Figure 4(b) shows the mean micro-hardness values curve of the different coating samples. All data were averaged over the selected five dispersed points on the each coating's surface. In Figure 4(b), it can be seen that the Victorinox hardness values of the 2#, 3#, 4#, 5# and 6# samples were significantly higher than the value of the 1# sample. In particular, the hardness difference between the 1# and 2# samples was near 200 HV, which shows that the addition of nano-SiC can significantly improve the hardness of the coatings. By this method, the wear resistance of the nano-composite coatings can also be greatly improved because the SiC particles themselves are hard particles. Also, the exotic SiC particulates improve the ability of the plastic deformation of the Ni-P matrix. As the main point of contact in the friction process, the protruding SiC particles bear the loading and shear force. Thereby, the wear abrasion of the composite coatings will be reduced. At the same time, the hardness values of samples 5# and 6# were higher than those of samples 3# and 4#. These results not only illustrate that the thicker the coatings are, the greater the hardness value and wear resistance are, but they also indicate that the hardness values correspond to the uniformity and compactness of the composite coatings.

4. Conclusions

In this work, nano-SiC/Ni-P coatings were prepared with the help of a magnetic and ultrasonic compound field. Then, the surface micro-morphology and microstructure of the composite coatings were observed and analyzed. The thickness and micro-hardness of the coatings were also measured. Therefore, the following conclusions can be drawn.

1) The surface density and flatness of the composite coatings that were prepared under the conditions of an external magnetic and ultrasonic field were better than that of the composite coatings prepared without any external field.

2) The composite coatings were mainly composed of Ni_2P and SiC phases.

3) The compound field can effectively promote electroless deposition and increase the thickness of the coatings. The thickness of the composite layer made under the external field showed an increase of nearly ten μm.

4) The addition of nano-SiC significantly increased the micro-hardness of the composite coatings. The maximum micro-hardness value of nano-SiC/Ni-P coating reached nearly 680 HV.

Acknowledgments
The work is supported by National Natural Science Foundation of China (Grant No. 51301088) and the innovation practice training projects for the college students of Nanjing Institute of Technology (Grant No. TB20160227).

References

[1] Chen W, Gao W and He Y 2010 A novel electroless plating of Ni-P-TiO_2 nano-composite coatings *Surface and Coatings Technology* **204**(15) 2493-2498
[2] Gao J Q, Liu L and Wu Y et al. 2006 Electroless Ni-P-SiC composite coatings with superfine particles *Surface and Coatings Technology* **200**(20-21) 5836-5842
[3] Khalifa O R M, El-Wahab E A, Tilp A H 2010 The corrosion behavior of electroless Ni-P-SiC and Ni-Sn-P-SiC nano-composite coating *Journal of Applied Sciences Research* **6**(12) 2280-2289
[4] Balaraju J N, Narayanan T S N S and Seshadri S K 2003 Electroless Ni-P composite coatings *Journal of Applied Electrochemistry* **33** 807-816
[5] Chen S Y, Fan H and Liang J et al. 2011 Preparation of Electroless Composite Ni-P-nano-SiC Plating on 45# Steel *Journal of Northeastern University* **32**(6) 810-805
[6] Gou Y, Huang W and Xue Y 2009 Recent Progress in Electroless Ni-P-SiC Composite Plating *Materials Review* **23**(23) 65-68
[7] Wang H 2011 The Effects of the Process Parameters on Micro-Hardness for Electroless Ni-P-Nano SiC Composite Plating *Advanced Materials Research* **366** 187-191
[8] Yan Y T, Huo Y Q and Sun Z L et al. 2011 Preparation of Electroless Ni-P Composite Coating and Its Tribological Properties *Advanced Materials Research* **199** 696-699
[9] Yu A Z, Li M D and Lin J M et al. 2012 Mechanical Milling Prepared Nano-Sized Al2O3 Particles Utilized in Electroless Plating Ni-P-Al_2O_3 Composite Coatings *Advanced Materials Research* **602** 1700-1705
[10] Xuan T P, Zhang L and Huang Q H 2005 Micromorphology and Wear Resistance of Chemical Plating Co-Ni-B-Ce Alloy under Magnetic Field Chinese *Journal of Applied Chemistry* **22**(3) 265-271
[11] Fan Y Z, Ma L and Cao X M 2011 Effect of Ultrasonic Waves on Ni-P-Al_2O_3 Electroless Composite Coating on Magnesium Alloy *Advanced Materials Research* **383** 953-957
[12] Wang Y, Ma C Y and Ma C H et al. 2012 Surface topographies of Ni-P-SiC coatings prepared by ultrasonic-mechanical stirring-electroless plating *Ordnance Material Science & Engineering* **35**(5) 80-82
[13] Shang B, Zhang L X and Dong Y H 2008 Simulating and analysing temperature field of the concrete base slab during construction period based on finite element *Science Technology and Engineering* **8**(16) 4742-4744
[14] Zhao Y H, Zhang Z G and Zhao Y Q 2011 Ultrasonic scattering technology for chemical composite plating with Ni-P nanoSiC *Journal of Lanzhou University of Technology* **37**(2) 26-29
[15] Soleimania R, Mahboubia F and Armana S Y et al. 2015 Development of mathematical model to evaluate microstructure and corrosion behavior of electroless Ni–P/nano-SiC coating deposited on 6061 aluminum alloy *Journal of Industrial and Engineering Chemistry* **23** 328-337
[16] Wang Q Z, Callistia M and Greera J et al. 2016 Effect of annealing temperature on microstructure, mechanical and tribological properties of nano-SiC reinforced Ni-P coatings *Wear* **356-357** 86-93

IOP Conf. Series: Materials Science and Engineering **137** (2016) 012067 doi:10.1088/1757-899X/137/1/012067

Dissolving cellulose with twin-screw extruder in a NaOH complex aqueous solution

Y P Yang, Y Zhang, A Dawelbeit and M H Yu[1]

State Key Laboratory for Modification of Chemical Fibers and Polymer Materials, College of Materials Science and Engineering, Donghua University, Shanghai 201620, People's Republic of China

E-mail: yumuhuo@dhu.edu.cn

Abstract. Novel cellulose dissolution method with twin-screw extruder was developed in order to improve the dissolution property, to simplify production procedure and to produce cellulose spinning dope which is stable and which has a higher concentration of cellulose. Therefore, the extrusion conditions on the cellulose dissolution in NaOH/thiourea/urea were extensively studied in this work. The resulted extrudates of twin-screw extruder dissolution method were characterized by polarized optical microscope image, the solubility experiment and the apparent viscosity. The results revealed that the screw revolution speed of such process could improve the solubility value (S_a) of cellulose, and the solubility of cellulose reached a maximum value of 7.5 wt% at higher revolutions 450 rpm. On the other hand, the cellulose solutions were more transparent and balanced with its apparent viscosity values lower and more stable compare to stirring method, which indicated dissolving cellulose with twin-screw extruder was reliable. Moreover, the whole dissolving time is quite short, and the process is simple. The soluble effect of twin screw extrusion was far superior to traditional stirring, and the most suitable temperature was -2℃.

1. Introduction

Cellulose is one of the plentiful natural and linear macromolecule polymers. The chemical structure of cellulose consists of β-glucopyranose units with 1-4-glycosidic bond linked. It has been classified as environmentally friendly, inexpensive, biodegradable, may and biocompatible be chemically modified to produce functionally products [1-3]. Cellulose, however, cannot melt, dissolution is a key issue. There are a limited number of solvent systems to directly dissolve cellulose, such as inorganic metal complexes (cuam, cadoxen, cuen etc.) [4], lithium choride/N,N-dimethylactamide (LiCl/DMAc) [3], N-methylmorpholine-N-oxide (NMMO) [5], dimethylsulfoxide/tetrabutylammonium fluoride (DMSO/TBAF) [6], ionic liquids [7, 8] and aqueous NaOH systems [9-11]. Thus it is a growing urgency to build a new, green, inexpensive solvent for cellulose. Recently, Zhang and co-workers had developed some "green" solvent systems like NaOH/urea [12], LiOH/urea [13], and NaOH /thiourea aqueous solutions [14]. The three solvent systems can dissolve cellulose quickly. Furthermore, Lixia Gu and Jianyong Yu [15, 16] developed a NaOH/urea/thiourea solvent system that is more effective in dissolving cellulose. Moreover, higher concentration cellulose spinning solution is more stable than NaOH /thiourea or NaOH/urea aqueous solution systems. However, cellulose dissolve in the above solvents should be precooled to -12 ℃ or -10 ℃ with vigorously stirring, which is not favored for scale up of the cellulose dissolution technology, and is not yet applied for industrial use. Herein, there is a large space to improve the dissolution property, decrease dissolution time, simplify production procedure, and produce cellulose spinning dope-which is stable and has a higher concentration of cellulose.

Twin screw extrusion is a short-time, high-shear process with the benefit of high can and versatility overcome the problems of dissolving cellulose by conventional stirring method [17].

In this work, we present a new method to dissolve cellulose with twin-screw extruder in an optimal NaOH (9 wt%)/urea (3 wt%)/thiourea (5 wt%) aqueous solution at -2 ℃. The extrudates' quality obtained was stable and more productive. Nevertheless, extrusion conditions' effects on cellulose dissolve in NaOH complex aqueous solution have not been discussed before. Moreover, dissolution reliability and dissolving capacity of co-rotating twin-screw extruder are discussed here in detail.

2. Materials and methods

2.1. Materials
Cellulose pulp [α-cellulose > 95%, DP=550] was provided by Longma Corporation (Jiangsu Province, China). As received, all chemicals were of analytic grade and were used.

2.2. Cellulose solution preparation
Dried cellulose pulps were dispersed into NaOH (9 wt%)/urea (3 wt%)/thiourea (5 wt%) aqueous solution at room temperature to prepare cellulose suspension. Then the cellulose suspension was extruded by a twin-screw extruder (D = 16 mm & L/D = 40) precooled to -2 ℃ to obtain transparent cellulose extrudate very quickly.

2.3. Solubility test
The transparent cellulose solution was obtained through centrifuging the prepared extrudate at 1100 rpm for 10 min at 5℃. Undissolved fractions were divided and then washed using acetone and water, respectively. Finally, undissolved fractions were dried for 24 h at 60 ℃ in a vacuum oven. Therefore, the solubility (S_a) value of cellulose in NaOH complex aqueous solution was expressed as

$$S_a = [(W_1 - W_2)/ (W_1 - W_2+W_0)] *100 \quad (1)$$

Where W_1 was the original cellulose pulp's weight, W_2 was the undissolved fraction's weight and W_0 was the NaOH complex aqueous solvent's weight.

2.4. Characterization
Polarized optical microscope (DM2500P, Leica, Germany) was utilized to observe cellulose's morphological change in NaOH/thiourea/urea aqueous solution at ambient temperature. The cellulose solutions were observed and photographed by pressed between two superimpose glasses.

The apparent viscosity of each prepared solution was carried out on a rotational viscometer (SNB-1, Shanghai Jingtian Electronic Instrument Co., Ltd, China) at 20 ℃. The cellulose solution was degassed before testing.

3. Results and discussion

3.1. Effect of screw revolution speed on solubility
Figure 1 showed the dependence of cellulose's S_a values and the cellulose extrudates' polarized optical microscopic pictures on the revolutions of the extruder. Cellulose pulp dissolved in aqueous NaOH/thiourea/urea solution by using a twin screw extruder at -2 ℃. The S_a curves sharply increased and reached a maximum value of 7.5 wt% at more than 450 rpm. This stated that the rise in the rates and the shearing force of extruder could reinforce the solubility and dispersion [18]. Also it could be seen from the polarized optical images that cellulose's solubility in NaOH complex aqueous solution could be enhanced considerably by increasing the extruder revolutions.

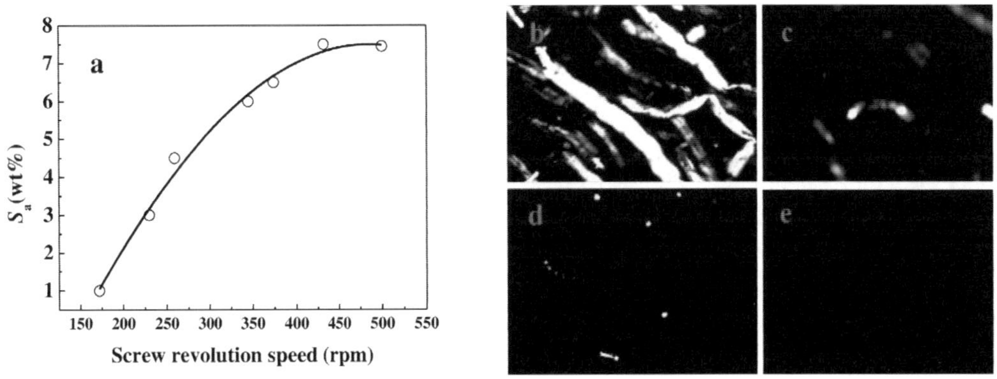

Figure 1. The dependence of cellulose's S_a values and the cellulose extrudates' polarized optical microscopic pictures at 0 rpm (b), 200 rpm (c) , 350 rpm (d) and 450 rpm (e) with twin screw extruder at -2 ℃.

3.2. The dissolution reliability of twin-screw extruder

Twin screw extrusion was a high efficiency and high-shear process, while the related research of the resulted extrudate's uniformity was relatively less. According to this consideration, the polarized optical microscopic image and the apparent viscosity of each prepared cellulose solution were studied to prove the dissolution reliability of twin-screw extruder, which can be seen in figure 2 and 3.

In NaOH/thiourea/urea solvent system, two 7.0 wt% cellulose solutions were prepared. Solution 1 was dissolved by twin screw extruder as mentioned previously and solution 2 was dissolved through stirring method [16]. Figure 2 showed that solution 1 was entirely dissolved, however, solution 2 was not completely dissolved remaining some undissolved cellulose pulp. It is obvious that twin-screw extruder dissolution method is more efficient than the stirring method. Moreover, twin-screw extruder dissolution method can overcome the problems of dissolving cellulose by stirring method and solution 1 was more transparent than solution 2. The undissolved cellulose pulp in solution 2 appeared to be surrounded by the viscous solution, so it was difficult for solvent to enter into the pulp[16]. The solvent seemed to lose its dissolution ability even with prolonged stirring. Therefore, twin-screw extruder dissolution method is conducive to scaling up of the cellulose dissolution technology.

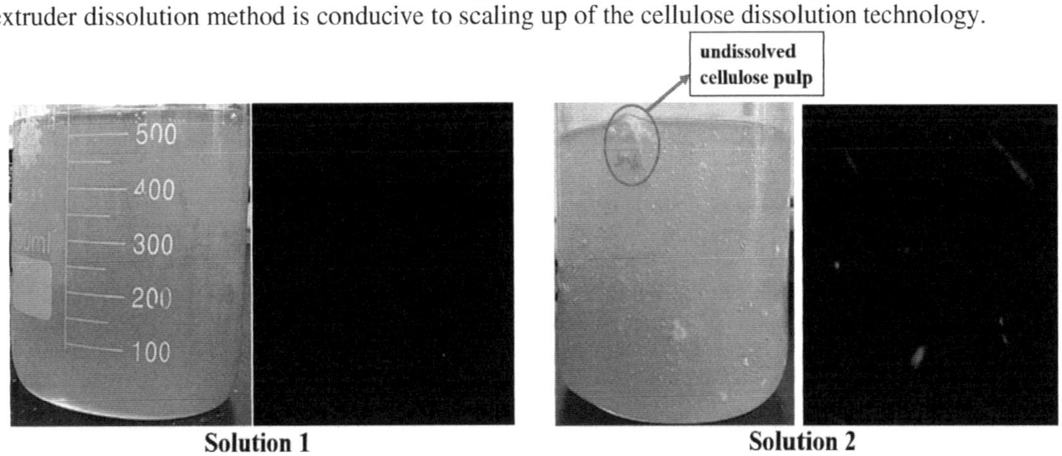

Figure 2. Digital and polarized optical microscopic images of two cellulose solutions.

In figure 3, curve *a* and *b* indicated that the viscosity results of cellulose extrudate were lower than the solution dissolved by conventional stirring. It is necessary for wet-spun process and industrial applications. Curve *c* and *d* represented the random S_a values of the cellulose solution dissolved by twin screw extruder and conventional stirring, respectively. It could be observed that the S_a values of the cellulose extrudates were relatively more stable than the solution dissolved by conventional stirring. The twin screw extruder has more powerful shearing force, and makes the macromolecular chain less of mutual entanglement and then makes the chain easy to stretch. Meanwhile, the reverse thread element of twin screw extruder can provide enormous pressure to make the small solvent molecules easy to penetrate into the cellulose pulp and promote the solvation effects. However, the mixing dissolution is limited by motor power and the blade structure, lead to the cellulose pulp local diffusion uneven. In order to improve the dissolve ability, conventional stirring need to increase motor power, design the blade structure and avoid the phenomenon of climbing pole. Obviously, twin screw extruder takes more advantages in the preparation of high concentration and high quality spinning cellulose solution.

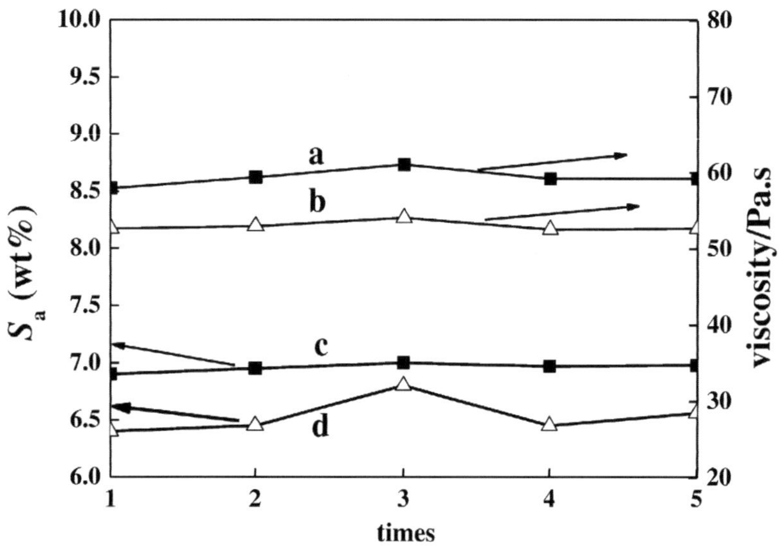

Figure 3. Differences of random Sa values and the apparent viscosity of two cellulose solutions (the results were from 5 times random sampling among cellulose dissolution process).
a, d – stir dissolution method; b, c – twin-screw extruder dissolution method

4. Conclusion

Cellulose pulp (nonactivated or untreated) is able to dissolve in NaOH complex solution directly. The low toxicity's solvent system has more excellent solubility ability for cellulose in comparison with NaOH/urea and NaOH/thiourea. The novel dissolution approach with twin-screw extruder has higher efficiency, and it can obtain an excellent cellulose extrudate, further to obtain a high quality cellulose spinning dope. Furthermore, the whole dissolving time is very short, and the process is simple. The soluble effect of twin-screw extrusion was far superior to traditional stirring, and the most suitable temperature was from -12℃ to -2℃. Regenerated cellulose fibers and films with better mechanical properties, which is prepared by this simple, environmentally friendly and inexpensive technology, will be discussed in the further research. Therefore, it is hopeful to replace the viscose production technology that involves serious pollution problems.

References

[1] Kim J and Sungryul Y A 2006 *Macromolecules* **39** 5583-5583
[2] Cuissinat C and Navard P 2008 *Cellulose* **15** 67-74
[3] Schult T, Hjerde T and Optun O I 2002 *Cellulose* **9** 149-158
[4] Heinze T and Liebert T 2001 *Prog. Polym. Sci.* **26** 1689-1762
[5] Fink H P, Weigel P and Purz H J 2001 *Prog. Polym. Sci.* **26** 1473-1524
[6] Koehler S and Heinze T 2007 *Macromol. Biosci.* **7** 307-314
[7] Swatloski R P, Spear S K, Holbrey J D and Rogers R D 2002 *J. Am.Chem. Soc.* **124** 4974-4975
[8] Gutowski K E, Broker G A and Willauer H D 2003 *J. Am. Chem. Soc.* **125** 6632-6633
[9] Isogai A and Atalla R H 1998 *Cellulose* **5** 309-319
[10] Isogai A, Chiba J, Atalla R H and Verona W 1995 United State Patent No.5410034
[11] Sobue H, Kiessig H and Hess K 1939 *Zeitschrift für Physikalishe Chemie* **B43** 309
[12] Cai J, Zhang L N and Zhou J P 2007 *Adv. Mater.* **19** 821-825
[13] Cai J, Liu Y T and Zhang L N 2006 *J. Polym. Sci. Pol. Phys.* **44** 3093-3101
[14] Ruan D, Zhang L N and Zhou J P 2004 *Macromol. Biosci.* **4** 1105-1112
[15] Jin H J, Zha C X and Gu L X 2007 *Carbohyd. Res.* **342** 851-858
[16] Zhang S, Li F X, Yu J Y and Hsieh Y L 2010 *Carbohyd. Polym.* **81** 668-674
[17] Dai Y Y, Dong H Z, Hou H X, Qi X Y and Zhang H 2012 *Starch-Starke* **64** 374-381
[18] Qin X Z, Lu A and Zhang L N 2012 *J. Appl. Polym. Sci.* **126** 470-477

Crack prediction in EB-PVD thermal barrier coatings based on the simulation of residual stresses

J W Chen[1], Y Zhao[1, 2, 3], S Liu[1], Z Z Zhang[1] and J Ma[1]

[1] NDT Lab, Laser Institute of Shandong Academy of Sciences, LISADA, Jinan, China
[2] Centre of NDT Engineer and Technique in Shandong, Jinan, China

E-mail: yang.aisin.zhao@gmail.com

Abstract. Thermal barrier coatings systems (TBCs) are widely used in the field of aerospace. The durability and insulating ability of TBCs are highly dependent on the residual stresses of top coatings, thus the investigation of the residual stresses is helpful to understand the failure mechanisms of TBCs. The simulation of residual stresses evolution in electron beam physical vapor deposition (EB-PVD) TBCs is described in this work. The interface morphology of TBCs subjected to cyclic heating and cooling is observed using scanning electron microscope (SEM). An interface model of TBCs is established based on thermal elastic-plastic finite method. Residual stress distributions in TBCs are obtained to reflect the influence of interfacial roughness. Both experimental and simulation results show that it is feasible to predict the crack location by stress analysis, which is crucial to failure prediction.

1. Introduction

Thermal barrier coatings (TBC) are deposited on the surface of super alloy to protect the substrate, which can improve the thermal efficiency of engine more than 60% at 1600°C [1-4]. TBCs are widely used in the field of aerospace. The major failure form of TBCs is the ceramic layer fall off from the bond coat during service. When the ceramic layers fall off, the high temperature gas will corrode the substrate which may have catastrophic consequences [5]. Hence, the study on the TBCs' failure mechanism has an important significance.

TBCs usually consist of a super alloy substrate (S), a metal bond coating (BC), a thermal grown oxide (TGO) and a top ceramic coating (TC). At present, preparation methods of TBC are commonly electron beam-physical vapor deposition (EB-PVD) and air-plasma spray (APS) in industry production [6]. Compared to APS, the EB-PVD method has the characteristics of high cost and good spraying quality [7]. In addition, TBCs prepared by the two methods have different interfacial morphology between BC and TC.

Because of the mismatch of fast contraction of sprayed coating layer, thermal expansion and possible phase transition, the residual stresses of thermal barrier coatings will increase during the preparation [8-10]. Furthermore, the shape of TGO and parameters of each layer have great influences on the stresses of TBCs, due to the complex interaction among thermal, elastic, plastic, oxidation, creep, sintering and phase transition. By studying the stress evolution, it is possible to predict the crack initiation and growth. Thus, the study on the stress field is very meaningful to understand the failure mechanism of TBCs.

The residual stresses in TBCs prepared by APS have been investigated using theory and experiment methods [11-14]. The main objectives of the present work are to obtain the stress filed of TBCs prepared by EB-PVD using finite element method and to predict the crack initiation and propagation based on the simulation.

2. Experiment

2.1. Specimens and heat treatment

TBCs were composed of a metal bond coating and a top ceramic coating deposited on a GH3030 super alloy substrate. The size of the substrate was $\Phi 30\times 4$mm. The main component of the metal bond coating was a NiCoCrAlY alloy, and the thickness of BC layer was about 120μm. On top of the bond coating, a ZrO_2-7wt%Y_2O_3 top coating with the thickness of 120μm was deposited. During the coating deposition using EB-PVD method, the TBCs were cooled from the preparation temperature of 850°C (the initial stress was zero) to room temperature (20°C). After the coating deposition, the specimens were subjected to thermal cycle treatments in a furnace. A single thermal cycle included three stages: the specimens were heated up to 1000°C in 10mins, kept at 1000°C for 1h, and cooled down the specimens to room temperature in 10mins at last. There were 5 groups of samples, and the number of thermal cycles was 0, 1, 10, 50 and 100h, respectively.

2.2. Microstructure

After the isothermal cycling treatments, the TBCs samples are cut into four equal parts along two diameters. The cross-sections are polished and then corroded using HCl(20ml) + $CuSO_4$(5g) + H_2O(100ml). Figure 1 indicates the scanning electron microscopy (SEM) results of TBCs on cross sections. It is shown that the TBCs consist of three layers. It is known that there is almost no TGO without oxidation treatment. After isothermal cycling treatments for 100h, a TGO layer related to the oxygen ingress through the porous TC layer and outward diffusion of Al in the BC layer is formed between the TC and BC layer.

More than 20 equidistant points distributed along the BC/TC interface were selected to measure the TGO thickness perpendicular to the interface. The thickness of TGO as a function of thermal cycle time is expressed by the power law equation below [15-16]

$$h^2=k_p\times t \tag{1}$$

where h represents the thickness of TGO, t represents the holding time at high temperature, and k_p represents a regression constant, which is relevant to the porosity and thickness of TBC. Based on the experimental results, k_p equals to about 0.053μm²/h. By comparison with the results (~0.17μm²/h) of Zhu [17], the growth rate of TGO thickness prepared by EB-PVD is slower than that by APS. TC layer begins to fall off when the thickness of TGO is greater than 8-10μm [18]. Therefore, TBCs prepared by EB-PVD have longer service life.

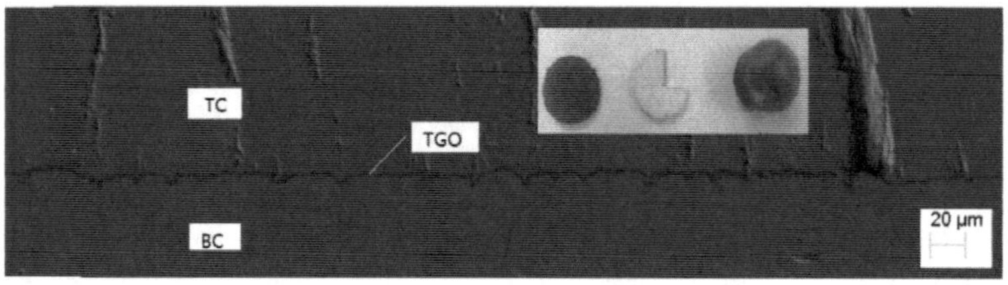

Figure 1. SEM images of TBCs after isothermal cycling treatments for 100h.

3. Finite element model

As illustrated in Figure 2, the simulation model corresponding to the scanning electron microscopy patterns in Figure 1 is shown. The model is composed of a TC, a TGO and a BC. The thickness of BC and TC are both 120μm. The thickness of TGO ranges from 0 to 10μm. The model is simulated using the finite element software ANSYS, and an axisymmetric analysis based on the coupled thermal and structure element PLANE 13 is used to reduce the data processing time.

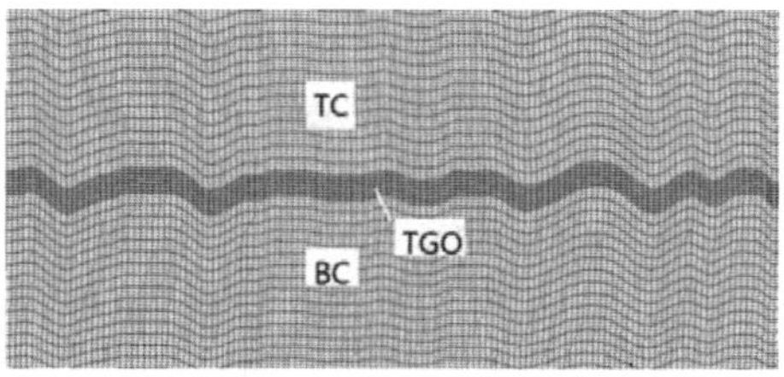

Figure 2. The finite element model corresponds to the scanning electron microscopy patterns.

Based on the experimental process, the temperature of isothermal cycling is loaded on the TBCs. The temperature load of one thermal cycle is composed of a heating process from 20 to 1000°C for 10mins, a holding process at 1000°C for 1h and a cooling process from 1000 to 20°C for 10 mins.

Table 1. Material property parameters [11, 19].

	T(°C)	E(GPa)	v	alpx(10^{-6}/°C)	ρ (kgm^{-3})	k(Wm^{-1}K^{-1})	C(Jkg^{-1}K^{-1})	σ_Y(MPa)
	20	220		14.8				
S	600	170	0.33	16.9	7900	16.3	437	
	1100	120		18				
	20	200		13.2				400
BC	600	160	0.31	15.2	7320	16.1	501	300
	1100	110		17.6				50
TG		360	0.27	8	3980	30	800	
	20	48		9				
TC	600	40	0.11	10.1	5600	1.8	500	
	1100	22		12.2				

In this work, it is assumed that the TC and TGO are elastic materials and the BC is an elastic-plastic material with temperature dependence of yield strength. The property parameters are shown in Table 1. The thermal load is agreement with the experimental situation. And just one cycle process is simulated because the creep characteristic is not taken into account.

4. Results and discussion

Compared with the mode II fracture resulted from shear stress component σ_{xy}, the normal stress components σ_x and σ_y more likely lead to the mode I fracture [20]. The component σ_x is a cause of a vertical surface crack, while σ_y can be used to predict the crack initiation and propagation along the BC/TC interfaces.

4.1. Effect of interfacial roughness

The stress field of the BC and TC layers in the x direction with different interfacial roughness is shown in Figures 3 and 4. The results are parts of the interface in Figure 1. As shown in Figure 3, the maximum value of tensile stress occurs at the peak region and the maximum compressive stress lies in the valley region. By comparing Figures 3(a), (b) and (c), it is shown that the stress levels are dependent on the curvature of the interface. The maximum compressive stress at the valley increases with the increase of curvature. The stress in the valley ranges from -116Mpa to -341Mpa and the stress at the peak ranges from -71.4Mpa to 63.5Mpa. Therefore, the tension stress at the peak may cause the generation of vertical surface cracks, and the probability of crack generation increases with the increase of curvature.

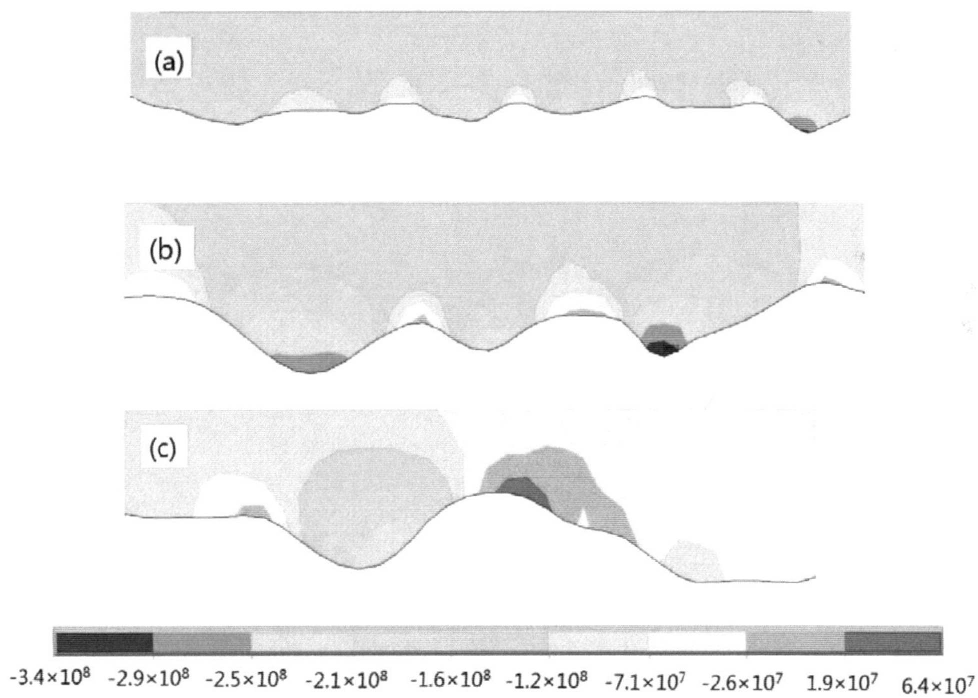

Figure 3. The contour plot of σ_x at the TC layer after cooling.

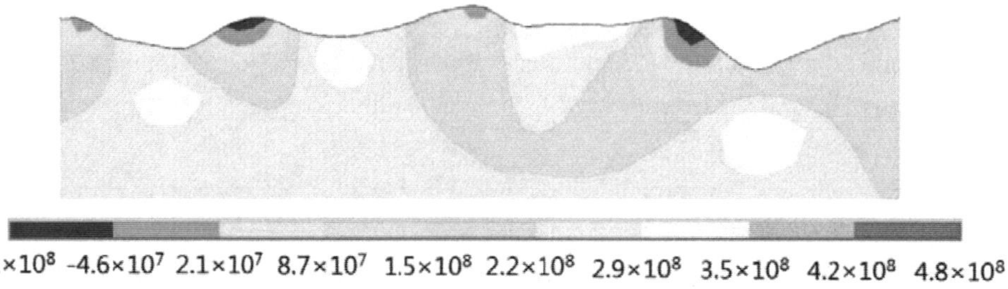

Figure 4. The contour plot of σ_x at the BC layer after cooling.

As shown in Figure 4, contrary to the stress distribution in TC layer, the maximum value of tension stress occurs at the valley region and the maximum compressive stress lies in the peak region. The stress level also increases with the increase of curvature. The maximum tension stress in the valley ranges from 153Mpa to 484Mpa. Therefore, the vertical surface cracks are most likely to initiate at the valley.

The stress contours of the BC and TC layers in the y direction with different interfacial roughness are shown in Figure 5. It is shown that the maximum tensile stress occurs at the valley region and the maximum compressive stress lies in the peak region at the TC layer, and the regulation is contrary at the BC layer. The maximum tension stresses at the TC and BC layers are 123Mpa and 484Mpa, respectively. Consequently, the cracks in the TBCs start at the valley of TC layer and at the peak of BC layer.

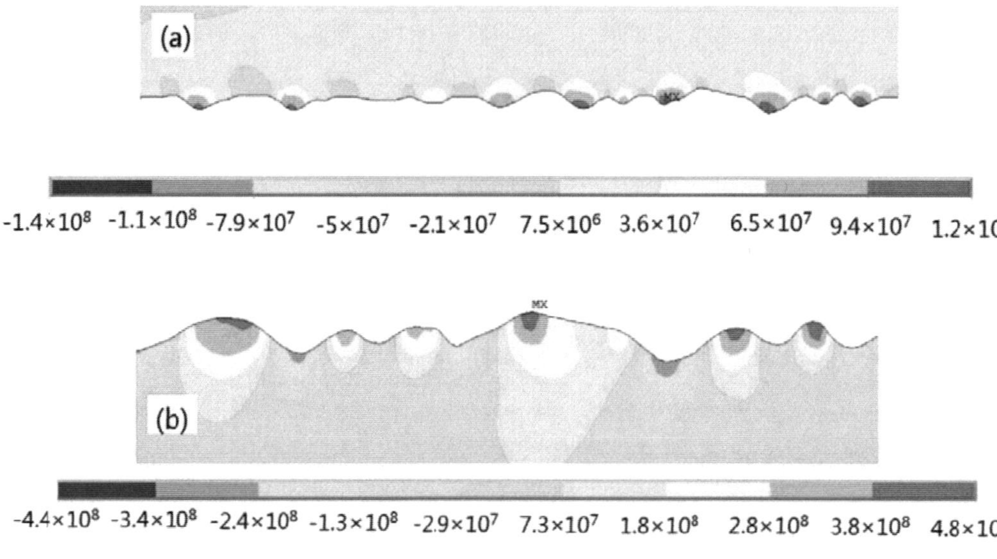

Figure 5. The contour plot of σ_y after cooling (a) at the TC layer and (b) at the BC layer.

4.2. Crack prediction

In view of the stress analyses on the TC/BC interface, the vertical surface cracks at the TC and BC layer occur at the peak and valley, respectively. This is consistent with the experimental results shown in Figure 6. Figure 6(a) contains two cracks: one located at the peak (in the TC layer) and the other located at the valley (in the BC layer). All the cracks are located at the valley in the BC layer as shown in Figure 6(b).

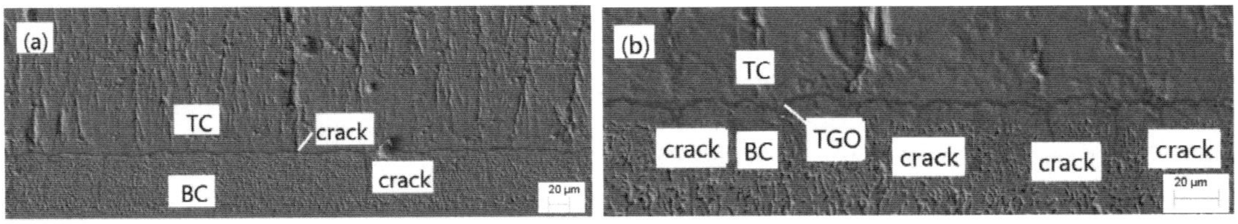

Figure 6. SEM of an EB-PVD TBCs after isothermal cycling treatments for (a)50h and (b)100h.

The crack in Figure 7(a) originates from the interface between TC and TGO, and extends into the TC layer along the lateral direction. The crack in Figure 7(b) originates from the interface between BC and TGO, and extends along the interface. It is hard to tell the initial position of both the cracks. Based on the analyses above, it can be predicted that the cracks in Figure 7(a) and (b) occurs at the valley (in the TC layer) and peak (in the BC layer), respectively.

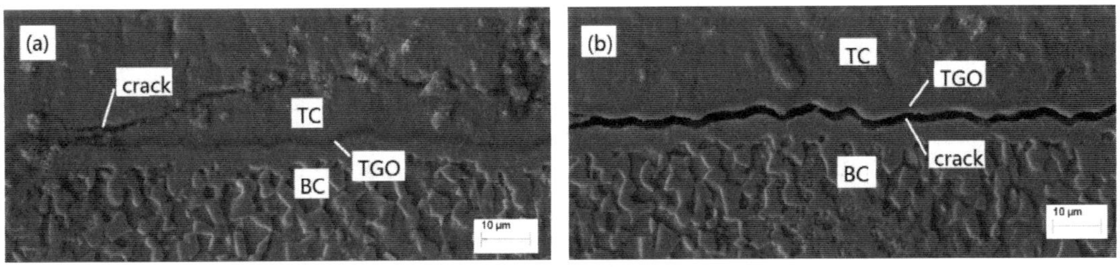

Figure 7. SEM of an EB-PVD TBCs after isothermal cycling treatments for 100h.

5. Conclusions

In this work, the finite element method based on thermal elastic-plastic has been used to establish interface model of TBC. Then, we have calculated the residual stress field of EB-PVD TBC to analysis the influence of interface morphology on residual stress. The influence of interface morphology on residual stress is analyzed via the calculation of the residual stress field of EB-PVD TBC enduring cycle heating and cooling. The main conclusions are summarized as follows:

(1) In the TC layer, it is more likely to be in the state of tension stress at the peak, which may result in the generation of vertical surface cracks. And the probability of crack generation increases with the increase of curvature. In the BC layer, the initiation of vertical surface cracks is most likely to occur at the valley. In addition, the crack path at TC/BC interfaces and separation of the TBC layers are more likely to start at the valley of TC layer and at the peak of BC layer.

(2) By comparing the experimental and simulation results, it is feasible to predict the crack location by stress analysis. Therefore, the simulation of residual stresses evolution is valuable for studying the failure mechanism of TBCs.

Acknowledgments

This work has been supported by the National Natural Science Foundation of China (No. 51205240), the Outstanding Young Scientists Program of Shandong Academy of Sciences (No. 201505) and Independent Innovation Projects of Jinan (No. 201407).

References

[1] Zhao Y, Chen J and Zhang Z 2015 *Opt. Eng* **54** 094104
[2] Miller R 1987 *Surf. Coat. Technol* **30** 1
[3] Meier S and Gupta D 1994 *J. Eng. Gas Turb. Power* **116** 250
[4] Miller R 1997 *J. Therm. Spray Technol* **6** 35
[5] Padture N, Gell M and Jordan E 2002 *Science* **296** 280
[6] Chen D and Gell M 2007 *J. Amer. Chem. Soc* **90** 3160
[7] Schulz U and Fritstcher K 1997 *Mater. Sci. Forum* **957** 251
[8] Wenzelburger M, Escribano M and Gadow R 2004 *Surf. Coat. Technol* **180-181** 429
[9] Zhang X, Xu B and Wang H 2005 *Thin Solid Films* **488** 274
[10] Yang L, Liu Q, Zhou Y, Mao W and Lu C 2013 *J. Mater. Sci. Technol* **30** 371
[11] Mao W, Chen Q, Dai C, Yang L, Zhou Y C and Lu C 2010 *Surf. Coat. Technol* **204** 3573
[12] Song M, Ma Y and Gong S 2011 *Prog. Nat. Sci* **21** 262
[13] Mao W, Jiang J, Zhou Y and Lu C 2011 *Surf. Coat. Technol* **205** 3093
[14] Ranjbar-Far M, Absi J, Mariaux G and Dubois F 2010 *Mater. Des* **31** 772
[15] Rosler J, Baker M and Aufzug K 2004 *Acta Mater* **52** 4809
[16] Ranjbar-Far M, Absi J, Mariaux G and Dubois F 2010 *Mater. Des* **31** 772
[17] Zhu J, Chen W and Xie H 2015 *Sci. China Phys. Mech. Astr* **58** 1
[18] Hernandez M, Karlsson A and Bartsch M 2009 *Surf. Coat. Technol* **203** 3549
[19] Wang Q, Jiang J and Mao W 2010 *Mater. Rev* **24** 259
[20] Ranjbar-Far M, Absi J and Mariaux G 2012 *J. Therm. Spray Technol* **21** 1234

Rheokinetic analysis on the curing process of HTPB-DOA-MDI binder system

T Chai[1,3], Y C Liu[1], H Ma[1], Y W Yu[1], J M Yuan[1], J H Wang[1], J H Guo[1,2,3]

[1] Chemical Engineering and Environment College, North University of China, Taiyuan 030051, China

[2] Department of ESH, Science and Technology University of Sichuan Staff, Chengdu 610101, China

E-mail: lyc2ct@sina.com; guojiahuphd@163.com

Abstract. Polymer bonded explosives (PBXs) are a kind of rubbery explosives, in which the explosive components are bound together by a polymeric binder. The hydroxyl terminated polybutadiene (HTPB)-based polyurethane binder systems possess a unique combination of properties, including superior adhesion, excellent mechanical properties and have been widely used in rocket propellants, PBXs and other weapons. The cure reaction between HTPB and diisocyanates in the castable PBXs plays an important role in the adjustment of process parameters as well as properties of explosive formulations. In this paper, viscosity measurements were carried out at different temperatures (15°C, 25°C, 35°C, 45°C, 55°C) to a series of systemic studies on the viscosity build-up and rheological reaction rate constant during the curing process of the of HTPB- dioctyl adipate (DOA)-diphenyl methane diisocyanate (MDI) binder system, discovered and explained the phenomenon that the pot-life of the binder system was prolonged when the curing temperature raised and the curing reaction rate was accelerated. And the rheology kinetic model of the HTPB-DOA-MDI binder system was established.

1. Introduction

Polymer bonded explosives (PBXs) are kinds of particulate composites, which usually consist of explosive crystals cured together by a macromolecular binder. For the PBXs decreased the risk of accidental detonations during storage, utilizing and transportation, PBXs have been drawn high degree of attention by the governments and scientists around the world in the fields of insensitive munitions (IM) [1-3]. The hydroxyl terminated polybutadiene (HTPB)-isocyanates binder systems possess a unique combination of properties, including high solid loading, long pot life, good flow-ability, superior adhesion, which impart dimensional stability, structural integrity and excellent mechanical properties to the grain, and have been widely used in PBXs, ballistic missiles, launch vehicle, composite solid propellant, nuclear warhead detonating device and underwater weapons [4-6]. There is no doubt that, the formulation of polymeric binder and the properties of the cured polymer matrix have significant impact on the properties of the PBXs product.

Currently, castable PBXs based on HTPB typically require curing under high temperature conditions (60°C~80°C), and the curing time usually takes 5 to 7 days [1]. Because the cross-linking reactions between HTPB and isocyanates curing agents are exothermic reactions, therefore, it is easily lead uneven distribution of the cross-linking density under high temperature curing. Moreover, during the cooling of the casted grain from high temperature to ambient temperature, it usually generates a thermal stress (i.e. shrinkage stress), which leads to a substantial decline in the performance of the explosive product. The reaction mechanism and rheokinetic of the cure process determine the morphology of the polyurethane network, which in turn affects the physical, mechanical and explosive properties of the cured product [7-14]. Therefore, an understanding of the rheokinetics of polyurethane network formation is of vital importance to improve the safety of the curing process, improve the mechanical properties of the grain and the explosion performance of the cured product.

In this paper, viscosity measurements were carried out to a series of systemic studies on the viscosity build-up and rheological reaction rate constant during the curing process of HTPB- dioctyl adipate (DOA)-diphenyl methane diisocyanate (MDI) binder system. And the rheology kinetic model (i.e. Arrhenius model and Eyring model) of the HTPB-DOA-MDI binder system was established.

2. Experimental

2.1. Materials

Hydroxyl terminated polybutadiene (HTPB) was supplied by Liming Research Institute of Chemical Industry (China). The curing agent, diphenyl methane diisocyanate (MDI) was purchased from Bayer (Germany). Plasticizer, dioctyl adipate (DOA) was obtained from Sinopharm Chemical Reagent (China).

HTPB was dried under vacuum (1-2mm Hg) at 100°C for two hour, and then put under a nitrogen atmosphere prior to use. MDI and DOA were dehydrated with 4-Å molecular sieves for 3 days before use.

2.2. Sample preparation

For the preparation of the samples studied in the experiment, HTPB and MDI were precisely weighed with a stoichiometric ratio (r = n[NCO] : n[OH] = 1.0). The weight ratio of HTPB:DOA is 5:1 in all the tests. Firstly, HTPB and DOA were thoroughly mixed for 30min, then, mixed and homogenized for 5 minutes by a mechanical stirrer at high-speed stirring (1,000 rpm). The materials were mixed at room temperature. Having completely mixed all materials, samples were utilized for immediate rheokinetic test.

2.3. Viscosity build-up measurement

The rheokinetic analysis of the HTPB-DOA-MDI binder system was investigated by monitoring the viscosity build-up of the curing reaction in the SNB-1A Rotational Viscometer (Shanghai FangRui Instrument, China). The samples measured in the viscosity build-up test were kept in a water bath in the range of 15°C~55°C with an accuracy of ± 0.5°C. And the viscosity data of the binder system were recorded every 2 min by the software automatically.

3. Results and discussion

3.1. Determination of the rate constants and the pot-life of the binder

Figure 1 shows the viscosity build-up of the HTPB-DOA-MDI binder system under different curing temperatures. From the figure, we can find that curing temperature has significant impact on the reaction rate of the curing process. As the curing reaction temperature rise, the viscosity build-up of the binder system grew significantly faster. The reasons may be that, with the increase of the curing temperature, the rate of formation of urethane groups from –NCO groups and –OH groups was speed up, which accelerated the viscosity build-up of the binder system.

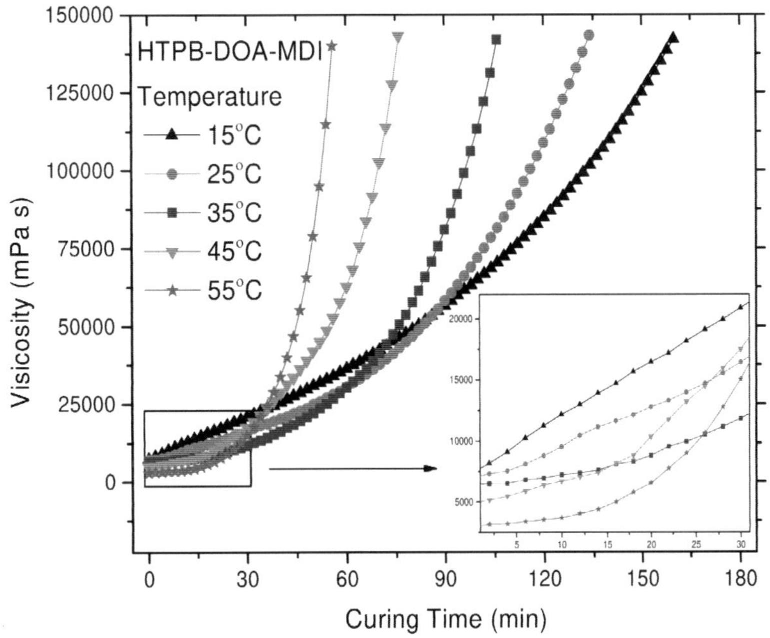

Figure 1. Viscosity build-up of the HTPB-DOA-MDI binder system under different curing temperatures.

The pot-life of the binder system is usually defined as the time interval that the viscosity was increased to a specific value (for example, 20,000mPa·s). The pot-life of the HTPB-DOA-MDI binder system under different temperature was analyzed and summarized in Table 1. From the table, when the curing temperature was in the range of 15°C~35°C, the pot-life of the HTPB-based binder system was increased from 30min to 48min; and in the range of 35°C~55°C, the pot-life of the HTPB-based binder system was decreased from 48min to 34min. From the enlarged partial view of Figure 1, we can find that the initial viscosity of the binder system was decreased from about 6,000mPa·s to 1,000mPa·s with the temperature increased from 15°C to 55°C. And the reason for this phenomenon is mainly due to the inherent nature of HTPB, i.e. as the temperature increases, the viscosity of HTPB decrease.

The rheokinetic parameters of the HTPB-DOA-MDI binder system were calculated by applying different single linear regressions, as shown in Figure 2. The corresponding data, such as reaction rate constant (k_η), correlation coefficient, and the ratio of $k_\eta(T_c)$ /$k_\eta(T_{15})$ are shown in Table 1. From the data in Table 1, one can find that the reaction rate constant (k_η) increase from 0.01362 min^{-1} to 0.07594 min^{-1} with the increase of curing temperature from 15°C to 55°C. What's more, $k_\eta(T_c)$ /$k_\eta(T_{15})$ increase from 1.34 to 4.65, this result indicates that with the increase of the curing temperature, the formation of urethane network was significantly accelerated. And the similar phenomena were reported in the literature of Lucio [14], Sekkar [15], Jiahu [16], et al.

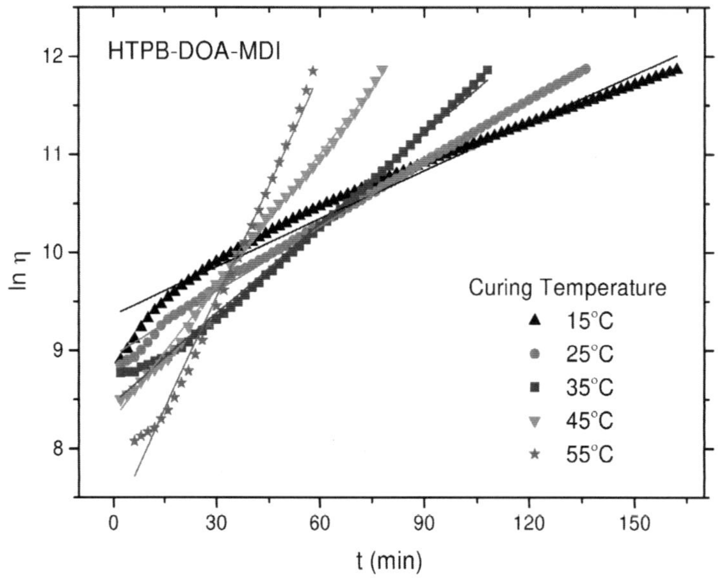

Figure 2. ln η versus curing time at different curing temperatures for the HTPB-DOA-MDI binder system.

Table 1. Rate constants and pot-life for the curing process of HTPB-DOA-MDI binder system at different isothermal curing temperatures.

Curing temperature T (°C)	Curing temperature T (K)	Rate constant k_η (min^{-1})	Correlation coefficient	$k_\eta(T_c)/k_\eta(T_{15})$	Pot-life (min)
15	288.15	0.01632	0.9742	—	30
25	298.15	0.02187	0.9972	1.34	40
35	308.15	0.03047	0.9951	1.87	48
45	318.15	0.04194	0.9972	2.57	38
55	328.15	0.07594	0.9889	4.65	34

3.2. Kinetic and thermodynamic studies

To calculate the thermodynamic parameters, e.g., pre-exponential factor (A) and activation energy (ΔE_η) are computed from the Arrhenius formula (1). Correspondingly, activation entropy (ΔS_η) and activation enthalpy (ΔH_η), are computed from the Eyring equation (2):

$$\ln k_\eta = -\frac{\Delta E_\eta}{RT} + \ln A \tag{1}$$

$$\ln \frac{k_\eta}{T} = -\frac{\Delta H_\eta}{RT} + \frac{\Delta S_\eta}{R} + \ln \frac{R}{Nh} \tag{2}$$

where k_η is the rheo-reaction constant and T is the curing temperature (K).

From the Arrhenius diagram (Figure 3) for the HTPB-DOA-MDI cure reaction, the pre-exponential factor (A) and activation energy (ΔE_η) are calculated with a first-order liner regression. An analogous evaluation of the rheokinetic data in the Eyring diagram (Figure 4) is carried out to calculate the value of activation enthalpy (ΔH_η) and activation entropy (ΔS_η). The values of ΔE_η, A, ΔH_η and ΔS_η are calculated and shown in Table 2.

From the Arrhenius diagram (Figure 3) and Table 2, we can find that activation energy of the curing reaction is rather low (ΔE_η = 28.96 kJ·mol^{-1}), which is related to a fast urethane reaction rate. And the pre-exponential factor A (also is called frequency factor) is equal to 2687.37.

It is well known that activation enthalpy (ΔH_η) and activation entropy (ΔS_η) are two thermodynamic parameters which are very important to reveal the reaction mechanism. From the Eyring diagram (Figure 4) and Table 2, the activation enthalpy (ΔH_η) is equal to 26.41kJ·mol^{-1}. And the activation entropy (ΔS_η) is equal to -187.83J·K^{-1}·mol^{-1}. According to transition state theory, activation entropy is usually regarded as the degree of disorder in the reaction system [17]. In this work, large negative values of ΔS_η are found, which demonstrate of an associative mechanism in the transition states.

From the above data, the Arrhenius model and Eyring model for the curing reaction of the HTPB-DOA-MDI binder system can be written as:

$$\ln k_\eta = -\frac{28.96}{RT} + 7.8963 \tag{3}$$

$$\ln \frac{k_\eta}{T} = -\frac{26.41}{RT} - \frac{187.83}{R} + \ln \frac{R}{Nh} \tag{4}$$

where k_η is the reaction constant and T is the curing temperature (K).

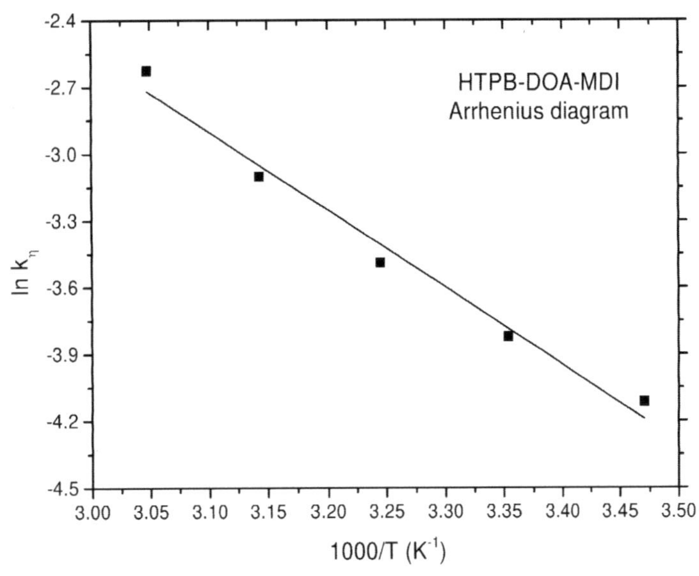

Figure 3. Arrhenius diagram for the curing reaction of the HTPB-DOA-MDI binder system.

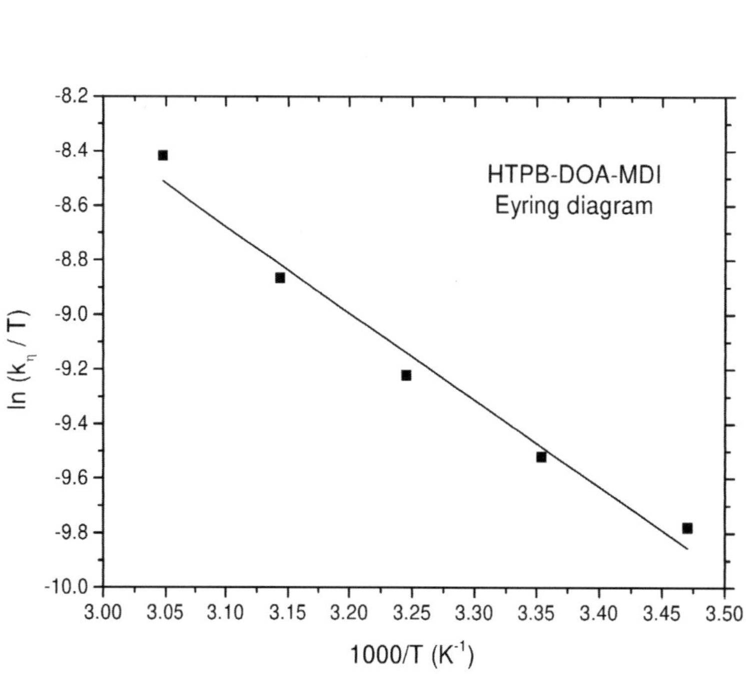

Figure 4. Eyring diagram for the curing reaction of the HTPB-DOA-MDI binder system.

Table 2. Rheokinetic parameters for the curing reaction of the HTPB-DOA-MDI binder system.

ΔE_η (kJ·mol^{-1})	A (L·mol^{-1}·min^{-1})	ΔH_η (kJ·mol^{-1})	$\Delta S\eta$ (J·K^{-1}·mol^{-1})
28.96	2687.37	26.41	-187.83

4. Conclusions

In this study, viscosity measurement was utilized to monitor the curing reaction of the HTPB-DOA-MDI binder system. Curing temperature showed significantly influence on the reaction rate constant and the pot-life of the binder system. Kinetic and thermodynamic parameters, e.g., activation energy (ΔE_η), activation enthalpy (ΔH_η) and activation entropy (ΔS_η) are computed from the Arrhenius formula and Eyring equation. The results of which were ΔE_η = 28.96 kJ·mol^{-1}, ΔH_η = 26.41 kJ·mol^{-1} and ΔS_η = −187.83 J·K^{-1}·mol^{-1}, respectively.

Acknowledgments
This work was supported by NSAF (Grant No. U1330131).

References

[1] Daniel M A 2006 Polyurethane binder systems for polymer bonded explosives DTIC Document
[2] Fuente J L 2009 An analysis of the thermal aging behaviour in high-performance energetic composites through the glass transition temperature *Polym. Degrad. Stab.* **94**(4) 664-669
[3] Ye R, Yu Y G and Cao Y J 2013 Analysis of Micro-scale Flame Structure of AP/HTPB Base Bleed Propellant Combustion *Def. Technol.* **9**(4) 217-223
[4] Krishnan P S G, Ayyaswamy K and Nayak S K 2013 Hydroxy terminated polybutadiene: Chemical modifications and applications *J. Macromol. Sci., Part A: Pure Appl.Chem.* **50**(1) 128-138
[5] Badgujar D M, Talawar M B, Asthana S N and Mahulikar P P 2008 Advances in science and

technology of modern energetic materials: an overview *J. Hazard. Mater.* **151**(2-3) 289-305
[6] Jawalkar S N, Mehilal, Ramesh K, Radhakrishnan K K and Bhattacharya B 2009 Studies on the effect of plasticiser and addition of toluene diisocyanate at different temperatures in composite propellant formulations *J. Hazard. Mater.* **164**(2-3) 549-554
[7] Amrollahi M, Sadeghi G M M and Kashcooli Y 2011 Investigation of novel polyurethane elastomeric networks based on polybutadiene-ol/polypropyleneoxide mixture and their structure-properties relationship *Mater. Des.* **32**(7) 3933-3941
[8] Sheikhy H, Shahidzadeh M, Ramezanzadeh B and Noroozi F 2013 Studying the effects of chain extenders chemical structures on the adhesion and mechanical properties of a polyurethane adhesive *J. Ind. Eng. Chem.* **19**(6) 1949-1955
[9] Heijden A E D M and Leeuwenburgh A B 2009 HNF/HTPB propellants: Influence of HNF particle size on ballistic properties *Combust. Flame* **156**(7) 1359-1364
[10] Sikder A K and Sikder N 2004 A review of advanced high performance, insensitive and thermally stable energetic materials emerging for military and space applications *J. Hazard. Mater.* **112**(1-2) 1-15
[11] Toosi F S, Shahidzadeh M and Ramezanzadeh B 2015 An investigation of the effects of pre-polymer functionality on the curing behavior and mechanical properties of HTPB-based polyurethane *J. Ind. Eng. Chem.* **24** 166-173
[12] DeLuca L T, Galfetti L, Maggi F, Colombo G, Merotto L and Boiocchi M 2013 Characterization of HTPB-based solid fuel formulations: Performance, mechanical properties, and pollution *Acta Astronaut.* **92**(2) 150-162
[13] Gui J H, Liu Y C, Chai T, Jing S M, Ma H and Qin N 2015 Synthesis and properties of a nano-silica modified environmentally friendly polyurethane adhesive *RSC Adv.* **5**(56) 44990-44997
[14] Lucio B and Fuente J L 2014 Rheokinetic analysis on the formation of metallo-polyurethanes based on hydroxyl-terminated polybutadiene *Eur. Polym. J.* **50**(0)117-126
[15] Sekkar V, Krishnamurthy V and Jain S 1997 Kinetics of copolyurethane network formation *J. Appl. Polym. Sci.* **66**(9) 1795-1801
[16] Guo J H, Liu Y C, Chai T, Ma H, Jing S M and Liu D C 2015 Kinetic and Thermodynamic Studies on Curing Reaction of HTPB-IPDI Binder System by FT-IR Measurements *Proceedings of the 2015 International Autumn Seminar on Propellants, Explosives and Pyrotechnics* 539-544
[17] Yang P F, Yu Y H, Wang S P and Li T D 2011 Kinetic studies of isophorone diisocyanate-polyether polymerization with in situ FT-IR *Int. J. Polym. Anal. Charact.* **16**(8) 584-590

IOP Conf. Series: Materials Science and Engineering **137** (2016) 012070 doi:10.1088/1757-899X/137/1/012070

Research on interfacial polymerization of pyrrole assist with Span80 system

Q H Yang[1], Z Y Tu and N Y Zhao

College of Materials Science and Engineering, Xi'an University of Science and Technology, Xi'an, 710054, China

E-mail: yangxjtu@hotmail.com

Abstract. With assistance of surfactants, self-assembled Polypyrrole (PPy) film was prepared via oil / water interfacial polymerization. The chemical structure and morphologies of the obtained samples have been characterized by Fourier transform infrared (FT-IR) and Scanning Electron Microscope (SEM), respectively. The electrochemical performance recorded on an electrochemical workstation, mainly includes cyclic voltammetry (CV) tests. The prepared PPy film has its own extremely vesicular structures from results and indicates by using different concentration surfactant Span80. The PPy film prepared 25 ℃ with 3.32 g/L Span80 (surpass its critical micelle concentration) as a surfactant possess a supernal specific capacitance of 368.18 F/g at a scan rate 50 mV/s in 1 M $NaNO_3$ aqueous solution at.

1. Introduction

Recent advances in the field of electrically conducting polymers have aroused substantial scientific and technological interest to research the conducting polymers which characterized with increasing interests in the control of their morphology at the different level. Conductive polymers are prospective substitution for extensive applications in electrochemical energy storage devices, like sensors, energy storage, electrocatalysis, electrorheology, electromagnetic interference shielding, and biomedicine, or in the conversion to nitrogen-containing carbon materials [1-7] and so forth. Among them, polypyrrole (PPy) received much concern lied in its environmentally friendly, preferable electrical conductivity and relative ease of synthesis [8-10]. The polymerization method of polypyrrole involves the electrochemical process of oxidation and reduction [11]. Sahoo et al. [12] choose a facile route to synthesize a nanocomposite which based on graphene and Polypyrrole nanofiber. The maximum capacitance of this composite was 466 F/g and also presented the highest energy density of 165.7 Wh/Kg at 10 mV/s scan rate in 1 M KCl solution. Pierce et al. [13] deposited pyrrole on the surface of graphene oxide (GO) platelets with ammonium persulfate (APS) as an oxidant, generated PPy/GO hybrid assemblies, offered a way to generate new solid hybrids and change the physical properties systematically. Our laboratory [14] used $FeCl_3$ as oxidants, Tween80 as a surfactant to polymerize polypyrrole by interfacial polymerization; an optimized specific capacitance of 261 F/g can be obtained at sans rate 50 mV/s and retains 75% of the initial specific capacitance value after 1000 cycles.

In consideration of the pre-existing achievements, we know surfactants can form micelles that act as a template for synthesizing poriferous materials. So we choose long-chain surfactants Span80 to assist the polymerization of polypyrrole with particular morphology. The optimal PPy film presented improved specific capacitance in 1M $NaNO_3$ aqueous solution. The impacts of surfactant concentration on the microstructure and surface morphology and faradaic pseudocapacitance behavior of as-prepared PPy films were explored.

2. Experimental

2.1. Reagents
Pyrrole was bought from Shanghai Sinopharm Chemical Reagent, purified by distillation under reduced pressure and stored in a refrigerator not exceeding 4 ℃ prior utilization. $FeCl_3$ was pure analysis and purchased from Tianjin Fuchen chemical reagents factory; Span80 was analytically pure and obtained from Tianjin Fuchen chemical reagent company; Hydrochloric acid (HCl), which was bought from Xian city west new fine chemical industry.

2.2. Preparation of PPy
Polypyrrole (PPy) was prepared in chloroform ($CHCl_3$) / water interface and used $FeCl_3$ as oxidant via interfacial polymerization, and 0.52 ml pyrrole was dissolved in 50 ml chloroform ($CHCl_3$) solution which includes Span80 to develop the oil phase, and oxidant $FeCl_3$ is dissolved in deionized water to compose aqueous phase followed Table 1. With a glass rod making oxidant solution to drainage into the organic phase, form the interface. Polymerization conditions choose static interface react 5h at 25 ℃, a black self-assembled polypyrrole freestanding film was displayed in interface. The synthetic samples was disposed with deionized water and ethyl alcohol for many times and dried in a vacuum oven for 24h at 60 ℃.

Table1. Experimental formula of Span80 system.

Number	1	2	3	4	5
Span80（g/L）	0	0.003	0.065	0.323	3.32

2.3. Characterization of PPy
We used SEM (JSM-6460LV, JEOL, and Japan) to observe the products micro morphologies. The FT-IR spectra were list on a Bruker FT-IR Equinox 55 spectrophotometer with KBr pellets. A total of scans arranges from 400 to 5000 cm^{-1}.

Cyclic Voltammetry (CV) Tests and Electrochemical Impedance Spectroscopy (EIS) detection were obtained by employing Model CS350 Electrochemical workstation (Wuhan, China) PPy was served as a working electrode. The counter electrode was platinum and the reference electrode was calomel electrode, respectively. The electrolyte is 1 M $NaNO_3$, CV tests were done between -0.4 and 0.6 V at a scan rate of 50 mV/s. EIS tests were carried out with an AC perturbation of 10 mV in the frequency range from 10^5 to 0.01HZ at open circuit.

3. Results and discussions

3.1. Microstructure of PPy
The microstructure of PPy films is presented in Figure 1. They are corresponding to the samples of 1, 2, 4 and 5 in Table 1. PPy will express a micro globular together or spread and aggregation morphology clutter without surfactants added in Figure 1 a. The morphology of PPy will be mostly changed when Span80 added. A morphological structure with different vesicles of varying size existing, and they are merging to a certain content when the concentration of surfactants is high. What can explain the phenomena is that the concentration of Span80 addition surpassing the critical micelles concentration of Span80. The increased volume and numbers of micelles induce to the simple merging mutually. It follows that polypyrrole can be induced to grow in this model, like Figure 1 b and c. When the concentration of Span80 increasing sequentially, the micelle shape can be changed and result in granular PPy, and a relatively compact skeletal structure presented, as shown in Figure 1 d.

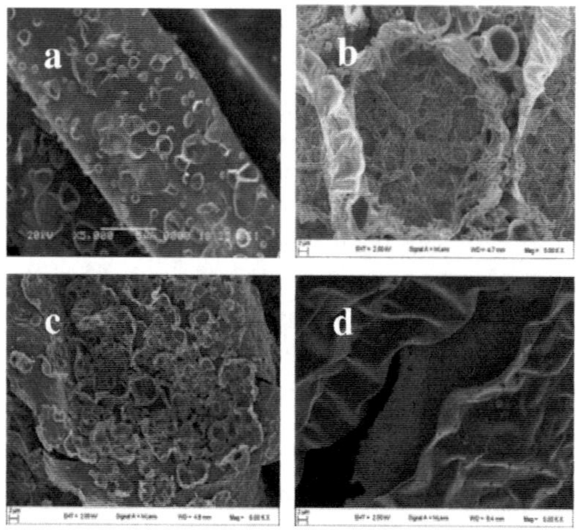

Figure 1. SEM images of PPy prepared with Span80 (a, b and c are corresponding to 1, 2, 4 and 5.).

As shown in Figure 2, the absorption peaks at 1400 cm^{-1} and 1540 cm^{-1} belong to the asymmetric and symmetric pyrrole ring-stretching modes, respectively. Peaks at 1300 cm^{-1} is associated with the C-N stretching vibration, 1041 cm^{-1} and 1138 cm^{-1} are ascribed to the =C-H in-plane vibration. Peaks at 905 cm^{-1} and 960 cm^{-1} are observed as PPy doping state respectively and the =C-H out-of-plane vibration, The N-H bond in-plane vibration showed at 781 cm^{-1}. Near infrared area, 3128 cm^{-1} is caused by the residual surfactant and hydroxyl of water. The shifting of the peaks indicates the change in the chemical environment during the formation of the samples. Thus, the FTIR results reveal the successful preparation of PPy.

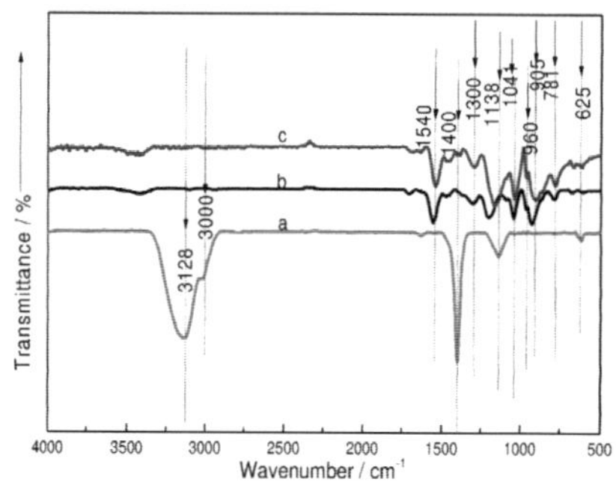

Figure 2. FTIR spectra of PPy (a, b and c are 1, 3 and 5).

3.2. *Electrochemical Performance of PPy*

Figure 3 reveals the CV curves of the specimen; line 1 displays the cyclic voltammetry curve of sample 1. All symmetric curves demonstrate the excellent reversibility of polypyrrole in a reasonable scanning stable potential. A rectangular shape is obviously different from the conventional conductive polymer, mainly depends on the amounts of chain segments participated in whole electrochemical reaction during the circular scanning process. The more chain segments take part in, the better rectangle shape present. According to the equation (1) [15], the specific capacitance of sample 1 obtained is 90.26 F/g. Because each of the polypyrrole chains can't for charge and discharge completely, a low active material utilization ratio during the cyclic voltammetry (CV) scanning process inducing a small specific capacitance. With the increasing of Span80 concentration, the cyclic voltammetry curve shape of the active materials changed observably and the trend of the curves show symmetrical distribution approximately, without any apparent oxidation and deoxidization peaks, corresponding to the flats of charge and discharge. It all comes to describe a stable exchange of electrolyte and electrode in the process of a given charging and discharging process, a higher electrochemical reversibility, a fine supercapacitors electrode material properties emerging. The area results of oxidation and deoxidization curves confirm the value of the specific capacitance. The calculation of 2, 3, 4 and 5 are 132.75 F/g, 154.87 F/g, 220.13 F/g and 368.18 F/g. The calculations are respectively consistent with the galvanostatic charge-discharge test results we used later.

$$C_S = \frac{\int_{V_a}^{V_c} I(V)dV}{mv(V_c - V_a)} \tag{1}$$

Where Cs, I, m, and v are the specific capacitance (F/g), sweep current (A/cm^2), the mass of the active electrode materials (electrode area of PPy is about 1 cm^2, g), and voltage sweep rate (mV/s), respectively.

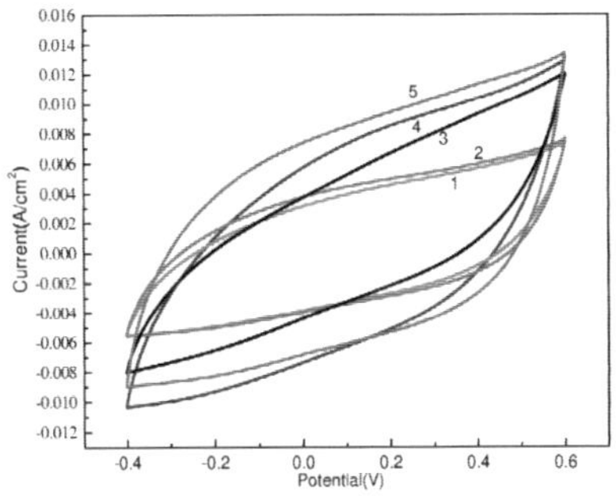

Figure 3. CV curves of PPy.

3.3. Conclusions

A non-ionic surfactant system was used, PPy electrode materials were prepared by interfacial polymerization, and the influence of synthesis conditions on PPy films performance was studied, then we come to the conclusion that the chemical composition of PPy can't be totally changed by adding Span80, but changed significantly in morphology. With increasing Span80 concentration, the morphology of PPy changed from disorderly deposits microspheres to vesicle accumulation formed, until a relatively compact skeletal structure presented at last. The electrochemical tests manifest the rising specific capacitance of PPy with adding Span80 increasing. When Span80 concentration is 3.32 g/L, we received optimized specific capacitance is 368.18 F/g.

Acknowledgments

This work was financially supported by National Natural Science Foundation of China (21204072). The Key Scientific and Technological innovation team Program of Shaanxi Province, China (2014KCT-04).

References

[1] Ates M 2013 A review study of (bio)sensor systems based on conducting polymers *Mater. Sci. Eng. C* **33** 1853-59

[2] Ramya R, Sivasubramanian R and Sankaranarayanan M V 2013 Conducting polymers-based electrochemical supercapacitors-Progress and prospects *Electrochim. Acta* **101** 109-29

[3] Yuan X X, Ding X L and Wang C Y et al. 2013 Use of polypyrrole in catalysts for low temperature fuel cells *Energy Environ. Sci.* **6** 1105-24

[4] Lee B M, Kim J E and Fang F F et al. 2011 Rectangular-Shaped Polyaniline Tubes Covered with Nanorods and their Electrorheology *Macromol. Chem. Phys.* **212** 2300-07

[5] Tantawy H R, Aston D E, Smith J R and Young J L 2013 Comparison of Electromagnetic

Shielding with Polyaniline Nanopowders Produced in Solvent-Limited Conditions *ACS Appl. Mater. Interf.* **5** 4648-58

[6] Tran H D, Li D and Kaner R B 2009 One-Dimensional Conducting Polymer Nanostructures: Bulk Synthesis and Applications *Adv. Mater.* **21** 1487-99

[7] Lu X F, Zhang W J and Wang C 2011 One-dimensional conducting polymer nanocomposites: Synthesis, properties and applications *Prog. Polym. Sci.* **36** 671-12

[8] Vernitskaya T V 1997 Polypyrrole: A conducting polymer; its synthesis, properties and applications *Russ. Chem. Rev.* **66** 443-57

[9] McCullough R D 1998 The Chemistry of Conducting Polythiophenes *Adv. Mater.* **10** 93-16

[10] Parakhonskiy B and Shchukin D 2015 Polypyrrole Microcontainers: electrochemical synthesis and characterization *Langmuir* **31** 9214-18

[11] Agha H, Fleury J B and Galerie Y 2014 Polypyrrole coating films for nanoparticles *Colloids and Surf., A: Physicochem. Eng. Aspects* **462** 217-24

[12] Sahoo S, Dhibar S and Hatui G et al. 2013 Graphene/polypyrrole nanofiber nanocomposite as electrode material for electrochemical supercapacitor *Polymer* **54** 1033-42

[13] Pierce S, Howe L and Ferguson K et al. 2015 Hybrid assemblies by pyrrole polymerization on nano graphene oxide platelets *Polymer* **70** 139-48

[14] Yang Q H, Hou Z Z and Huang T Z 2015 Self-assembled polypyrrole film by interfacial polymerization for supercapacitor applications *J. Appl. Polym. Sci.* **132** 41615

[15] Shinde S S, Gund G S and Kumbhar V S et al. 2013 Novel chemical synthesis of polypyrrole thin film electrodes for supercapacitor application *Eur. Polym. J.* **49** 373-3739

IOP Conf. Series: Materials Science and Engineering **137** (2016) 012071 doi:10.1088/1757-899X/137/1/012071

The rheological properties of bamboo cellulose pulp/ionic liquid system

Y F Zhang[1], P R Zhang, J Wu, Q X Jia and X Y Liu

School of Material Science and Engineering, Beijing Institute of Fashion Technology, Beijing 100029, China

E-mail: zhang-y-f@163.com

Abstract. In this study, two kinds of spinning solutions were prepared by dissolving bamboo cellulose pulp into 1-ethyl-3-methyl imidazole chloride salt ([EMIM] Cl) and 1-butyl-3-methyl imidazole diethyl phosphate salt ([BMIM]DEP) ionic liquids, respectively. Furthermore, the rotational rheometer was used to test the steady-state rheological properties of above as-prepared spinning solutions. The research results show that both of these two ionic liquids exhibit better solubility to the bamboo cellulose pulp. The apparent viscosities(η_a) decrease with the increased temperature(T) and shear rate(γ) and increase with the increased concentration. The non-Newtonian index(n) declined with the increase of both shear rate and concentration, as well as increased with the build-up temperature. The structural viscosity index($\Delta\eta$) increased with the increased concentration and tended to decrease with temperature rise. Meanwhile, viscous flow activation energy($E\eta$) decreases with the increased share rate as well as the concentration. According to the results, it can be seen that the bamboo cellulose pulp/[EMIM]Cl with the concentration of 6% at 70℃ exhibits better spinnability.

1. Introduction

Cellulose is a kind of renewable and widely applied resources with environmental protection and biocompatibility. It can be used for energy, materials, chemicals, textiles, paper and other fields[1]. In cellulose's family, plenty of produces are made from the so-called "cellulose generation". The recycled fiber is one of the most important products. There are many kinds of methods to obtain the regenerated cellulose fibers, such as direct melt spinning, enzyme cellulose spinning, Copper-ammonia cellulose solution spinning technology, etc. Among various kinds of such fabrication methods, the application research of new type ionic liquid in this area is more active. For example, Ingildeev[2] reported that the cellulose -ionic liquid was spun by the wet and the dry-wet spinning methods, and found that the regenerated cellulose fibers by the dry-wet spinning were preparing exhibited better strength, crystal orientation degree as well as easier prone to fibrillation. Electrospinning was also used to fabricate cellulose- ionic liquids by Quan S L and co-workers [3]. Furthermore, in order to decrease the viscosity of electrospun solutions, dimethyl formamide(DMF) and dimethyl sulfoxide(DMSO) were used as the cosolvent during the electrospinning process by Härdelin, etc [4]. It is well known that the performance of spinning solutions provide influences not only on the machining performances but also affect the choices of spinning process as well as the fibrous produces' qualities. Accordingly, deep research of the rheological properties of the spinning solutions is particularly important in the process of development of new type of cellulose fibers. China is rich in bamboo resources, relatively high levels in development and utilization of the world [5]. In this paper, by the solubility of bamboo cellulose pulps in different ionic liquids, the influential factors of the spinning solutions rheological properties were studied.

2. Experiment

2.1. Preparation of bamboo cellulose pulp/ionic liquid

The bamboo cellulose pulp/ionic spinning solutions with different concentrations were obtained by mixing and stirring different amounts of bamboo cellulose pulps (Tangshan sanyou group Dongguang pulp Co., Ltd. DP=500， α-cellular content~90%) with 10 g ionic liquid in the three-flake container at 80℃and the certain degree of vacuum until dissolved. Over here, bamboo cellulose pulps were activated by 18% NaOH solution.

2.2. Instruments and characterization

Rotational Rheometer (Bohlin Gemini HR nano-Peltier, clamp: ECT-PP20. The distance: 1000 μm, steady shear rate 0.1-250 S^{-1}) was used to measure the rheology properties of as-prepared spinning solutions (concentration~8%, temperature 80℃). In this work, the rheology properties of bamboo cellulose pulps dissolved in different ionic liquids ([EMIM]Cl and [BMIM]DEP (Ke-neng Material Technology Co., Ltd., Linzhou, China), with different concentration and at different temperatures were studied.

3. Results and discussion

3.1. Apparent viscosity

η_a is an important indicator which reflects the liquidity. It is closely related to temperature, concentration, molecular structures of the liquids.

3.1.1. The influence of temperature. Figure 1 is apparent viscosities of the bamboo cellulose pulp/[EMIM]Cl and bamboo cellulose pulp/[BMIM]DEP with the concentration of 8% between 30℃ to 80℃, respectively. It can be seen obviously that with the increased temperature, the rheological curve moved down for both of above two solution systems, which indicate the behavior of shear-thinning as well as the typical pseudoplastic non-Newtonian fluid. Furthermore, the value η_a decreases at the beginning and then gradually converge by increasing γ. For polymer molecule, when the molecular weight over more than a certain critical value, the flexible long-chain molecules between each other twisted into a knot, forming entanglement between molecular chains. According to the polymer network structure theory, there are two kinds of entanglement between large molecular chains: the tangles of geometry and the bonding of intermolecular. The bound nodes are all with instantaneous degeneration in the polymer fluid that can divide, reconstruct as well as issue in a particular condition to reach the dynamic balance [6]. When the temperature increased, the bound nodes solve quickly by random thermal motion of molecules, and make the gravity center of these molecular chains flowing along the direction of displacement and mutual slippage between the chains, which decrease the entanglement effect between molecules and friction, and enhance the liquidity of the solution and η_a decreased. When the shear rate increases, the disentanglement rate of the bound node is greater than the reconstruction rate, which leads to both of the numbers of bound nodes and apparent viscosity decrease. But when the shear rate increases to threshold, the bound node to reconstruction was restricted, resulting in the viscosity dropping to the minimum value without any further changes.

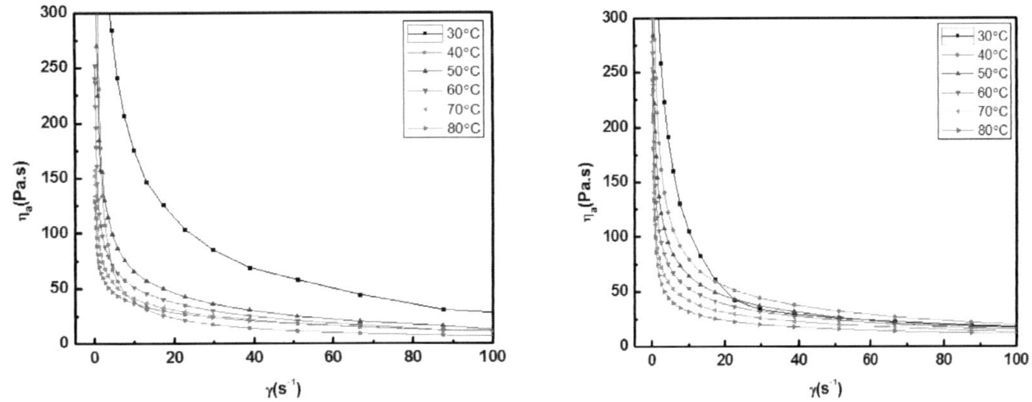

a-[EMIM]Cl b-[BMIM]DEP

Figure 1. η_a-γ curves of bamboo cellulose pulp/ionic liquid solutions (concentration~8%)

3.1.2. The influence of Concentration. The relationship between η_a and γ of bamboo cellulose pulp/[EMIM]Cl and bamboo cellulose pulp/[BMIM]DEP with different concentrations at 80℃ are shown in Figure 2. It can be seen that the rheological curves move up when concentrations increase. Meanwhile, with the increase of γ, the value η_a down first and then stabilize. However, the value η_a of bamboo cellulose pulp/[BMIM] DEP system is significantly higher than bamboo cellulose pulp/[EMIM] Cl system at the same concentration and shear rate. As we were known, random coil state macromolecules are isolated in dilute solution. When reaching to a certain concentration, polymers contact and through overlapping each other that make the chain sections roughly uniform and form tangles network in such solution. In this kind of uniform distribution of polymer solution, the total number of molecules per unit volume chain should be proportional to the square of the concentration [7]. The greater the concentration, the more molecules bound nodes and friction forces between each node, which increase the apparent viscosity value of the spinning solution. Because the different molecular structure of [BMIM]DEP and [EMIM]Cl, the solvation structural level of cellulose molecule chain in both the liquid is different, which lead to differences in phase geometry bound node and the bonding force of intermolecular, and impact on the flexibility and kinematic of molecular chain in the fluid. Because of the branched chain length of [BMIM]DEP molecules are longer, the dissolved cellulose chain segment motion resistance is larger, so the apparent viscosity of bamboo cellulose pulp / [BMIM] DEP system is relatively larger as well.

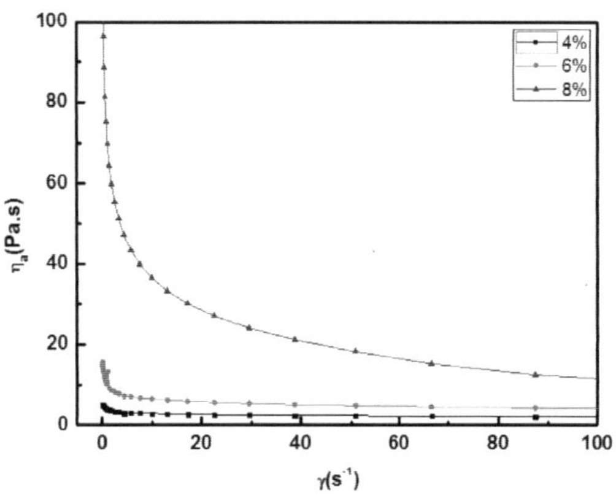

a-[EMIM]Cl

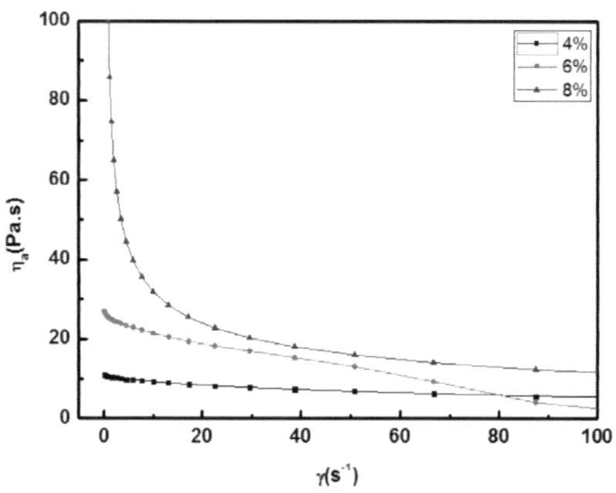

b-[BMIM]DEP

Figure 2. η_a-γ curves of bamboo cellulose pulp/ ionic liquid solutions at 80℃

3.2. Non-Newtonian index

The non-Newtonian index is used to represent the degree of polymer fluid deviates from the Newtonian fluid, it can reflect the viscoelastic of spinning solution, and it is usually less than 1. When the non-Newtonian index is much more close to 1, it indicates that spinnability of the spinning solution is better. The non-Newtonian index can be expressed by power law function formula (τ= K γ n), in which the value n can be obtained according to lgτ-lgγcurve graph [8].

Table 1 shows the non-Newtonian index values of two kinds of bamboo cellulose pulp/ionic liquids with different concentrations and temperatures. With the increase of γ, the non-Newtonian index of solution exhibits a trend of decrease. When the temperatures increase, the non-Newtonian index increased as well. However, under the condition of lower shear rate, the value n changes with some volatility. It is mainly due to that macromolecular chain segment motion are accelerated when the temperature increase, which is advantageous to the molecular chains disentanglement, and makes a decline in sensitivity of the shear rate to the liquidity of the solution. According to Table 1, it can be found that when the concentration is less than 6%, both of above two solutions become much more like the Newtonian fluid. However, with higher concentrations, the non-Newtonian index is still smaller even under low shear rate, indicating a far deviation from the Newtonian fluid. Such above are fit well with the steady-state rheological curve with different spinning fluid concentration. By comparing the solubility and rheological properties, [BMIM]DEP and [EMIM]Cl can be used to preparing cellulose spinning solutions.

Table 1. Non-Newtonian Index of bamboo cellulose pulp/ionic liquid solutions

Ionic liquid	Concentration %	Shear rate s-1	Temperature ℃					
			30	40	50	60	70	80
[EMIM]Cl	4	0.1～1	0.98	0.99	0.98	0.99	0.99	0.98
		1～10	0.74	0.86	0.91	0.93	0.92	0.86
		10～100	0.49	0. 76	0.79	0.83	0.86	0.86
	6	0.1～1	0.95	0.96	0.98	0.98	0.99	0.96
		1～10	0.69	0.76	0. 78	0.79	0.91	0.82
		10～100	0.42	0.57	0.67	0.74	0.82	0.82
	8	0.1～1	0.47	0.48	0.56	0.69	0.76	0.73
		1～10	0.46	0.47	0.52	0.61	0.65	0.70
		10～100	0.20	0.26	0.34	0.34	0.42	0.49
[BMIM]DEP	4	0.1～1	0.94	0.97	0.98	0.99	0.99	0.97
		1～10	0.65	0.80	0.84	0.89	0.93	0.94
		10～100	0.37	0.57	0. 65	0.73	0.79	0.84
	6	0.1～1	0.93	0.95	0.97	0.99	0.99	0.97
		1～10	0.61	0.73	0.75	0.86	0.89	0.93
		10～100	0.35	0.45	0.51	0.63	0.73	0.80
	8	0.1～1	0.38	0.53	0.55	0.58	0.61	0.65
		1～10	0.35	0.48	0.54	0.55	0.60	0. 54
		10～100	0.26	0.41	0.40	0.46	0.51	0.56

3.3. Structural viscosity index

Δη can be used to characterize the spinning fluid structured degree which is an important parameter of the spinnability of spinning solution. It is expressed as $\Delta\eta=(-\ d\lg\eta/d\gamma^{1/2})\times 10^{2}$. According to the linear fitting curve slope (from lgη- $\gamma^{1/2}$ curve), Δη can be calculated under different temperatures and concentrations. In the non-Newtonian area, Δη＞0. The smaller of Δη and the smaller of the spinning fluid structured degree and the better spinnability.

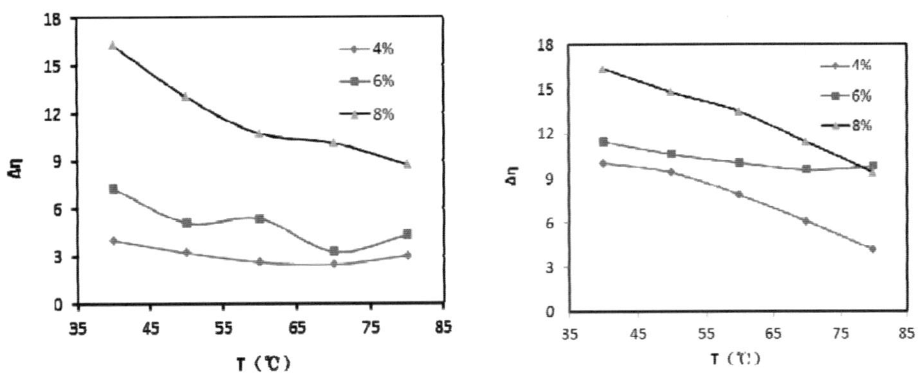

a-[EMIM]Cl b-[BMIM]DEP

Figure 3. Δη-T curves of bamboo cellulose pulp/ionic liquid solutions

Figure 3 exhibit the relationships between Δη and T of two kinds of bamboo cellulose pulp/ionic liquid systems with different concentrations. With the increases concentration, Δη increase. And Δη decreases when the temperature increases. When the temperature reaches at 70℃, Δη is relatively lower. Thus, in this work, it can be seen that 70℃ is the better spinning temperature. Furthermore, the value Δη of bamboo cellulose pulp/[EMIM]Cl is smaller than that of bamboo cellulose pulp/[BMIM]DEP at the same temperature and concentration, indicating a relatively good spinnability.

3.4. Viscous flow activation energy

Eη is a physical quantity to describe the fluid viscosity on the temperature sensitivity, which represents the polymer chain segment that be used to overcome the barrier (the minimum energy of chain segment transition from original location to near "holes"), and can be affected by polymer molecular structures, intermolecular forces, molecular weight, shear rate, polymer concentration, temperature, solvent and so on. If Eη is larger, the viscosity will be more sensitively to temperature changes, which is not conducive to the fiber forming. Within a small scope of temperature interval, the spinning solution viscosity can be expressed by Arrhenius formula, $\eta = Ae^{E\eta/RT}$. Based on logarithm on the equation, figure lnη～1/T, the Eη can get from the straight slope [9].

Figure 4 are Eη of bamboo cellulose pulp/ionic liquid solutions. It can be seen that with the increase of spinning solution concentration as well as the shear rate, Eη decrease. especially in the low shear rate range that the change of Eη is more.. Accordingly, the value Eη of the bamboo cellulose pulp/[EMIM]Cl is larger than that of bamboo cellulose pulp/[BMIM]DEP at the same concentration, indicating that the viscosity of bamboo cellulose pulp/[EMIM]Cl is more sensitive to temperature. To the bamboo cellulose pulp/[BMIM]DEP system, there is only a bit influence of the concentration to Eη when the solution concentration exceeds 6%. Such phenomena might be due to that the temperature plays an important role in macromolecular chain disentanglement at low shear rate. However, when the shear rate is higher, γ plays the main role in disentanglement, so that leads to the Eη lower, and the sensitivity to the temperature is decreased also. When solution concentration was increasing, the number of tangle molecule point also increased accordingly, due to the limited effect of temperature on the disentanglement, the sensitivity of viscosity to temperature is reduced. Since the anion of [EMIM]Cl exhibit strong ability to accept the hydrogen bonds, and the alkyl chain of its cationic groups is shorter, so the bonding force between [EMIM]Cl and cellulose molecules is relatively larger, when the cellulose chain segment transit from the state of in situ to the near "holes", a larger barrier must to be overcome, which needs more energy, thus the Eη value is relatively higher. Therefore, in the process of spinning, both temperature and uniformity of the solutions should be kept stabilization to prevent the excessive volatility of the system viscosity.

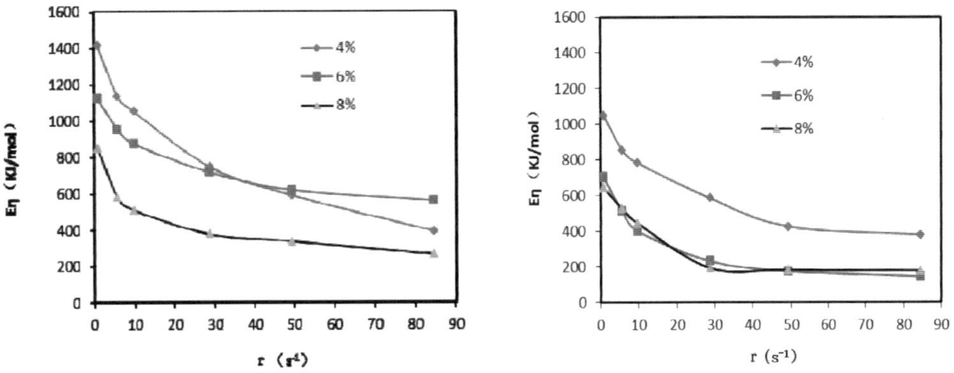

a-[EMIM]Cl b-[BMIM]DEP

Figure 4. Eη-γ curves of bamboo cellulose pulp/ionic liquid solutions at 80℃

4. Conclusions

- In the range of 30℃～80℃, η_a of bamboo cellulose pulp/[EMIM]Cl and bamboo cellulose pulp/[BMIM]DEP solutions decrease with the increase of temperature as well as shear rate, and increase along with the increased concentration, which shows as shear thinned, and can be classified as a typical non-Newtonian fluid.
- In the range of concentration 4%～8%, the non-Newtonian index decrease with the increased shear rate and concentration, and also exhibits an increasing trend when the temperature increased. Meanwhile, Eη declines when the shear rate and concentration declined. Δη reduce with the temperature up and increase with the increased concentration.
- At the same temperature and concentration, Δη of the bamboo cellulose pulp/[EMIM]Cl system is lower than bamboo cellulose pulp/[BMIM]DEP, and Eη is larger. According to analysis results mentioned above, bamboo cellulose pulp/[EMIM]Cl solution exhibits good spinnability with the concentration of 6% at 70℃.

Acknowledgments
In this paper, by the Beijing Natural Science Foundation Project, Beijing municipal education commission of science and technology plan to support key projects, serial number: KZ201510012013.

References

[1] Edgar K J, Buchanan C M, Debenham J S, etc. 2001 Advances in cellulose ester performance andapplicatio *Prog Polym Sci* **26**(9) 1605-1688.

[2] Ingildeev D, Effenberger F, Bredereck K and Hermanutz F 2013 Comparison of direct solvents for regenerated cellulosic fibers via the lyocell process and by means of ionic liquids *J. Appl. Polym. Sci.* **128**(6) 4141-4150.

[3] Quan S L, Kang S G and Chin I J 2009 Characterization of cellulose fibers electrospun using ionic liquid *Cellulose* **17**(2) 223-230.

[4] Härdelin L, Thunberg J, Perzon E, Westman G, Walkenström P, Gatenholm P. 2012 Electrospinning of cellulose nanofibers from ionic liquids: the effect of different cosolvents J. *Appl. Polym. Sci.* **125**(3) 1901-1909.

[5] Dou Y, Yu X J, Fumiyo I 2010 *The Current Situation of Bamboo Resource Development and Utilization of China and Countermeasure to Upgrade*. Nanchang: The sixth China Bamboo Industry Academic Conference 469-474.

[6] X Y Shen, X D Wu, Y L Li 2000 *Polymer Processing principle* Beijing: China textile & apparel press 137.

[7] Y Z Xu 1988 *Polymer structural rheology*. Chengdu: Sichuan education press 190-234.

[8] Malkin A, Ilyin S, Roumyantseva T and Kulichikhin V 2012 Rheological Evidence of Gel Formation in Dilute Poly(acrylonitrile) Solutions *Macromolecules* **46**(1) 257-266.

[9] Wang D, Sun D P, Shen X Y 2008 Rheological behavior of bacterial cellulose/LiCl/DMAc solution *Cellulose science and technology* **16**(3) 60-63.